modern welding

COMPLETE COVERAGE OF THE WELDING FIELD IN ONE EASY-TO-USE VOLUME

by

ANDREW D. ALTHOUSE

Technical-Vocational Education Consultant
Member of American Welding Society

and

CARL H. TURNQUIST

Career Education Consultant
Member of American Welding Society

and

WILLIAM A. BOWDITCH

Vocational Supervisor
Warren Consolidated Schools
Warren, Michigan
Member of American Welding Society

South Holland, Illinois

THE GOODHEART-WILLCOX CO., INC.
Publishers

Library of Congress Cataloging in Publication Data

Althouse, Andrew Daniel.
   Modern welding.

   Includes index.
   1. Welding.   2. Metals.   I. Turnquist, Carl
Harold, joint author.   II. Bowditch,
William A., joint author.   III. Title.
TS227.A369 1979   671.5'2   79—15573
ISBN 0—87006—279—4

INTRODUCTION

In the manufacturing, construction, and service industries, the forming and joining of metals by welding is the number one metal joining process.

There is a need in modern industry for craftsmen with varying degrees of technical training and skill, and for welding engineers who are able to design weldments, and write specifications.

This text, MODERN WELDING, which is written in nontechnical language, is intended for those who need a substantial background in welding fundamentals. It will help the student obtain a working knowledge of the properties and characteristics of metals, and common testing procedures. It is also intended for those now engaged in welding who want to increase their skills, and for those in industrial plants who are responsible for training welding and cutting operators. Diagrams showing the various welding and cutting processes and stations are printed in full color to add greatly to the teaching value of the text.

MODERN WELDING, which was prepared with the needs of both the apprentice and the journeyman in mind, provides up-to-date coverage of the welding field in one easy-to-use volume. It covers a multitude of welding procedures; not only the more common gas, arc and resistance processes, but those used in aerospace and other modern industries as well.

A. D. Althouse

C. H. Turnquist

W. A. Bowditch

CONTENTS

ACKNOWLEDGMENTS

The publishing of a book of this nature would not be possible without the assistance of the many segments of the Welding Industry.

The authors gratefully acknowledge the cooperation of the following:

A—P Controls Corp., Ace-Sycamore, Inc., ITT Holub Industries, Acorn Iron and Supply Co., Acro Welder Mfg. Co., Adjustable Clamp Co., Aeroquip Corp., Airco Welding Products, Div. of Airco, Inc., Airmatic/Beckett-Harcum, Ajax Magnethermic Corp., Allegheny Ludlum Steel Corp., Allison-Campbell Division, ACCO, Aluminum Co. of America (ALCOA), American Pullmax Co., Inc., American Society of Mechanical Engineers, American Welding Society, Ames Precision Machines, Ampower Products, Inc., Arcair Co., Arcos Corp., Argopen, Aronson Machine Co., Atlas, Auto Arc-Weld, AVA/Welding Industry Council, Baldwin-Lima-Hamilton Corp., Bastian-Blessing Co., Bausch and Lomb, Inc., Bernard Co., Div. of Dover Corp., The Black and Decker Mfg. Co., S. Blickman, Inc., Boeing Co., Buffalo Forge Co., Bundy Tubing Div., Bundy Corp., Cam-Lok (Div. of Empire Products, Inc.), Champion Blower and Forge Co., Inc., Chemetron Corp., Cincinnati Electrical Tool Co., Cleanweld Products, Inc., Clements Mfg. Co., Columbia Electric Mfg. Co., Contour Sales Corp., Craftsweld Equipment Corp., De-Sta-Co Div., Dover Corp., Detroit Board of Education, Detroit Public Schools, Detroit Testing Machine Co., Diamonite Products Mfg. Co., Diano Corp., W. C. Dillon and Co., Inc., Dispatch Oven Co., Dockson Corp., Dow Metal Products, Duffers Associates, Inc., Duro Engineering Co., Dynabrade, Inc., Eisler Engineering Co., Electric Controller and Mfg. Co., Erico Products, Inc., Exomet, Inc., Falstrom Co., Fenway Machine Co., Inc., Fibre-Metal Products Co., Ford Motor Co., Fusion, Inc., The Gasflux Co., General Electric Co., General Electric X-Ray Corp., General Motors Corp., Goodyear Tire and Rubber Co., Greene Mfg., Inc., Gregory Industries, Inc., H and M Pipe Beveling Machine Co., Inc., Hamilton Standard Div., United Aircraft Corp., Handy and Harman, Harnischfeger Corp., Harris Calorific Div., Emerson Electric Co., Haynes Stellite Co., Heath Engineering Co., Henes Mfg. Co., Hercules Welding Products Co., Hobart Brothers Co., James Hoerner, Pioneer H. S., San Jose, Calif., ITT Holub Industries, Instrument Control Co., Invincible Vacuum Corp., Jackson Products, Jewel Mfg. Co., Jimmie Jones Co., KGM Equipment Co., Kamweld Products Co., Inc., Kedman Co., Koldweld Div., Kelsey-Hayes Co., Kolene Corp., Laramy Products Co., Inc., Lenco, Inc., Lincoln Electric Co., Linde Div., Union Carbide Corp., Magnaflux Corp., R. C. Mahon Co., Maitlen and Benson, Inc., Mallory Metallurgical Co., P. R. Mallory and Co., Inc., Marquette Mfg. Co., Div. of Applied Power Industries, Inc., McDonnell and Miller, Inc., Metallizing Company of America, Inc., Metco, Inc., Michigan Seamless Tube Co., Miller Electric Mfg. Co., Miller Falls Co., Modern Engineering Co., Inc., National Carbon, National Electronics, Inc., National Tube Div., U. S. Steel, National Welding Equipment Co., Nelson Stud Welding, Niagara Machine and Tool Works, Norton Co., J. B. Nottingham and Co., Omark Industries, Inc., Cecil C. Peck Co., Peterson Mfg. Co., Inc., Philips Electronic Instruments, Phillips Petroleum Co., Physmet Corp., Div. of Manlabs, Inc., Precision Welder, Flexopress Corp., Pressed Steel Tank Co., Inc., Production Technology, Inc., a Caterpillar Subsidiary, Progressive Machinery Corp., Red-D-Arc, Ltd., Resistance Welder Corp., Rexarc, Inc., Riehle Testing Machines Div., Ametek, Inc., Robotron Corp., Royco Instruments, Inc., Ruemelin Mfg. Co., Sciaky Bros., Inc., Shore Instrument and Mfg. Co., Singer Safety Products, Inc., A. O. Smith Corp., Smith Welding Equipment, Div. of Tescom Corp., Sonobond Corp., Sperry Div., Automation Industries, Inc., Stoody Co., Sylvania Electric Products, Taylor-Winfield Corp., Tec Torch Co., Inc., Tempil Div., Big Three Industries, Inc., Thermacote Co., Thermal Dynamics Corp., Tinius Olsen Testing Machine Co., Tube Turns, Div. of Chemetron Corp., Tuffaloy Products, Inc., Tweco Products, Inc., Tweezer-Weld Corp., United Clamp Mfg. Co., United States Steel Corp., Vega Enterprises, Inc., Vickers, Inc., Victor Equipment Co., Vogel Tool and Die Corp., Wales Strippet, Inc., Wall Colmonoy Corp., Weldex Div., Metal Craft Co., Welding Engineer, Welding Equipment and Supply Co., Weldit/Winona, Weldma Co., Weltronic Co., Edwin L. Wiegand Div., Emerson Electric Co., Wilson Div., American Chain & Cable, Wyzenbeek and Staff.

INTRODUCTION TO
WELDING AND CUTTING PROCESSES

This introductory section illustrates the many welding and cutting processes used in modern industry. Full color drawings of every type welding station are color keyed to the various materials used. To follow the process more easily, the student should refer to the drawings frequently while reading the text.

American Welding Society (AWS) standard terminology is used as much as possible in these descriptions.

Each welding process is explained as follows:

1. Its application and purpose.
2. Colored illustration(s) of the process.
3. Heat (energy) source used.
4. Controls.
5. Operation of the process.
6. Safety.
7. References to text chapters and paragraphs.

The color illustrations, preceded by the letter "P" in this introductory section, are intended to show only the general process. In each illustration the complete welding station is diagrammed. Some details have been enlarged to help explain the process.

P-1. OXYGEN-ACETYLENE WELDING (OAW)

The oxygen-acetylene welding process combines an oxygen-acetylene mixture to provide a high-temperature flame for welding. This flame provides enough heat for welding most metals and for all types of brazing.

In this process, an oxyacetylene station is used, as shown in Fig. P-1A. Acetylene is supplied from one cylinder; compressed oxygen is supplied from another cylinder. Both cylinders must be equipped with a pressure reducing regulator. Each is fitted with two gauges.

One pressure gauge (HIGH) indicates the pressure in the cylinder. The other pressure gauge (LOW) indicates the pressure of the gas being fed to the torch.

Separate flexible hoses carry the gases to the torch. The torch has two needle valves. One valve controls the rate of flow of the oxygen; the other controls the rate of flow of the acetylene to the torch tip. The mixed gases burn at the torch tip orifice. See Fig. P-1B.

Acetylene burns in the atmosphere with a yellow-red flame. A carburizing flame is blue with an orange and red end. It may release black smoke. A neutral flame has a quiet, blue-white inner cone. This is the flame used in most welding processes. An oxidizing flame results in a short, noisy, hissing inner cone. It tends to burn the metal being welded.

Other fuel gases can be used in place of acetylene. These include propane, LP and MAPP.

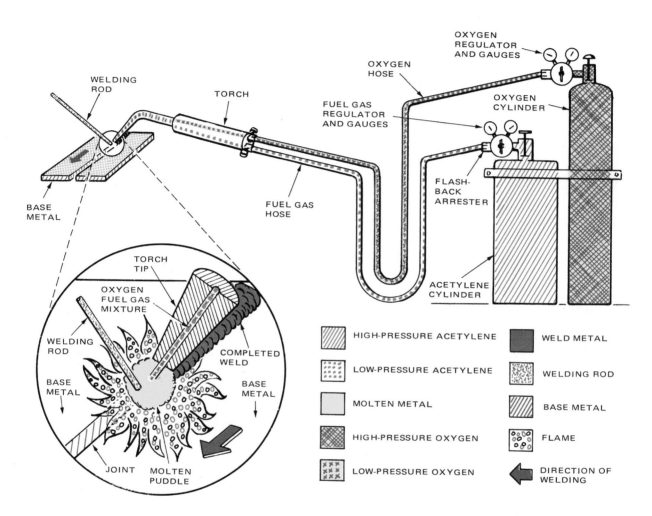

WELDING ROD

TORCH

OXYGEN HOSE

OXYGEN REGULATOR AND GAUGES

FUEL GAS REGULATOR AND GAUGES

OXYGEN CYLINDER

BASE METAL

FUEL GAS HOSE

FLASH-BACK ARRESTER

ACETYLENE CYLINDER

TORCH TIP

OXYGEN FUEL GAS MIXTURE

WELDING ROD

COMPLETED WELD

BASE METAL

BASE METAL

JOINT

MOLTEN PUDDLE

HIGH-PRESSURE ACETYLENE

LOW-PRESSURE ACETYLENE

MOLTEN METAL

HIGH-PRESSURE OXYGEN

LOW-PRESSURE OXYGEN

WELD METAL

WELDING ROD

BASE METAL

FLAME

DIRECTION OF WELDING

Fig. P-1A. Oxy-acetylene welding (OAW). The oxygen and fuel gas are mixed in a torch. The mixture burns at the torch tip. The heat from this flame is used to melt the base metal and welding rod. This melted material forms a welded joint.

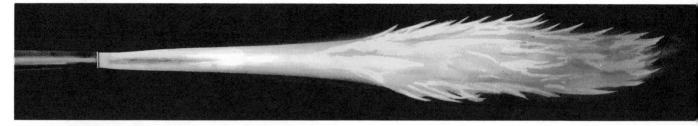

Acetylene Burning in Atmosphere
Open fuel gas valve until smoke clears from flame.

Carburizing Flame
(Excess acetylene with oxygen.) Used for hard-facing and welding white metal.

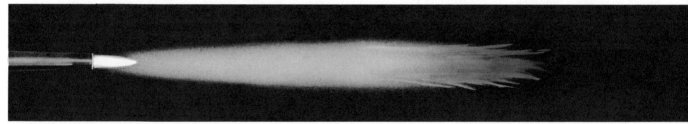

Neutral Flame
(Acetylene and oxygen.) Temp. 5600 F. For fusion welding of steel and cast iron.

Oxidizing Flame
(Acetylene and excess oxygen.) For braze welding with bronze rod.

Fig. P-1B. Color appearance of oxygen-acetylene flames.
(Smith Welding Equip. Div., Tescom Corp.)

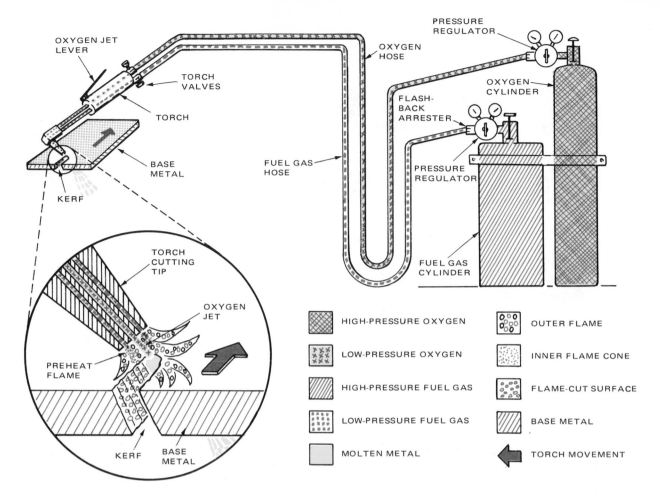

Fig. P-2A. Oxygen-fuel gas cutting (OFC). Oxygen and fuel gas are mixed in the torch. The mixture burns at several orifices in the torch tip. When the flame has heated the base metal to a dull cherry red, a lever on the torch is pressed. This allows a jet of oxygen to rush out of an orifice. This oxygen jet quickly oxidizes the heated base metal and blows it away. This removal of material leaves a cut (kerf) in the base metal.

Welding goggles should be worn for eye protection. Gloves, nonflammable clothing and all other required safety clothing should be worn to protect against burns. Good fire safety and prevention techniques should be employed. Provide good ventilation.

See Chapters 1 and 2 for additional oxyacetylene welding information.

P-2. OXYGEN-FUEL GAS CUTTING (OFC)

Oxygen-fuel gas cutting process uses an oxygen-fuel gas flame to heat metal. Then an oxygen jet cuts the heated metal. It is one of the most popular methods of cutting and shaping steel.

It is possible to burn (rapidly oxidize) iron or steel. The oxy-fuel gas flame raises the temperature of the metal to a cherry red color (1472 to 1832 F.). Then, a high-pressure jet of oxygen is directed at the metal, causing metal to burn (oxidize) and blow away very rapidly. This is why the term "burning" is sometimes used in connection with oxygen-fuel gas cutting process.

The process requires cylinders of oxygen and fuel gas, as shown in Fig. P-2A. Each cylinder has a regulator and two pressure gauges. One pressure

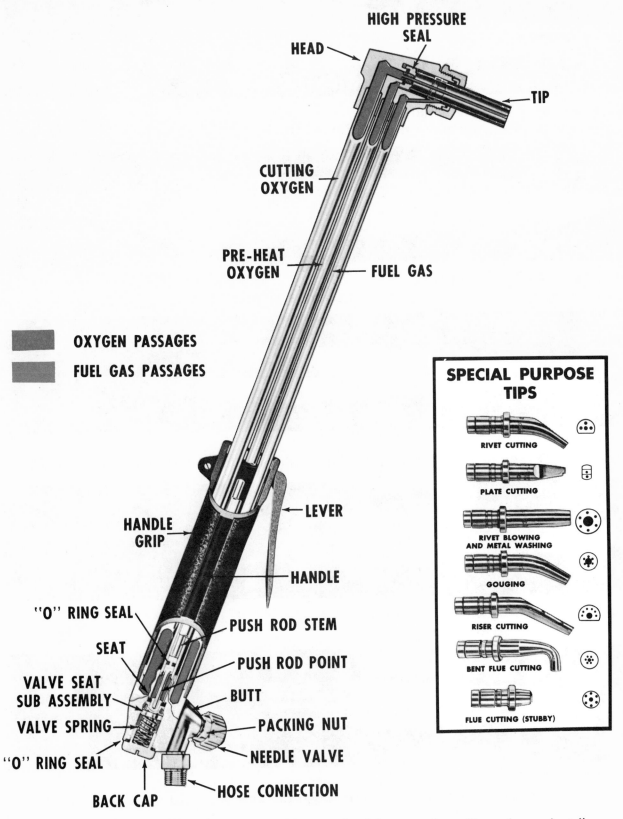

HIGH PRESSURE SEAL

HEAD

TIP

CUTTING OXYGEN

PRE-HEAT OXYGEN

FUEL GAS

OXYGEN PASSAGES

FUEL GAS PASSAGES

LEVER

HANDLE GRIP

HANDLE

"O" RING SEAL

PUSH ROD STEM

SEAT

PUSH ROD POINT

VALVE SEAT SUB ASSEMBLY

BUTT

VALVE SPRING

PACKING NUT

"O" RING SEAL

NEEDLE VALVE

BACK CAP

HOSE CONNECTION

SPECIAL PURPOSE TIPS

RIVET CUTTING

PLATE CUTTING

RIVET BLOWING AND METAL WASHING

GOUGING

RISER CUTTING

BENT FLUE CUTTING

FLUE CUTTING (STUBBY)

Fig. P-2B. Cutting torch. Oxygen-acetylene gases are mixed and then carried to orifice to form preheat flames. Oxygen, carried directly to tip, oxidizes metal and blows it away to form cut. (Smith Welding Equip. Div., Tescom Corp.)

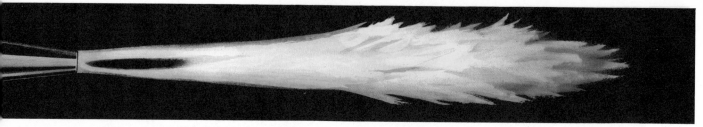

Acetylene Burning in Atmosphere
Open fuel gas valve until smoke clears from flame.

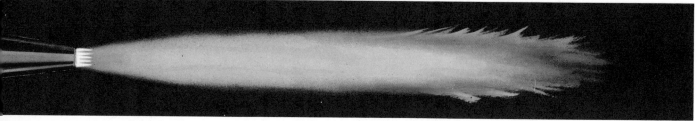

Carburizing Flame
(Excess acetylene with oxygen.) Preheat flames require more oxygen.

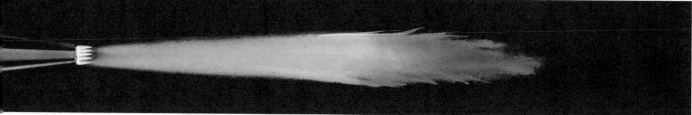

Neutral Flame
(Acetylene with oxygen.) Temp. 5600 F. Proper preheat adjustment for cutting.

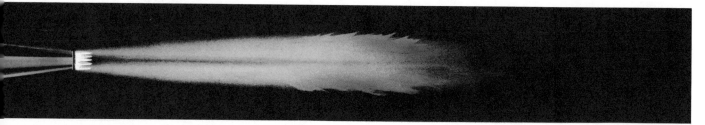

Neutral Flame with Cutting Jet Open
Cutting jet must be straight and clear.

Oxidizing Flame
(Acetylene with excess oxygen.) Not recommended for average cutting.

Fig. P-2C. Conditions of oxyacetylene cutting flame when adjusting the torch.
(Smith Welding Equip. Div., Tescom Corp.)

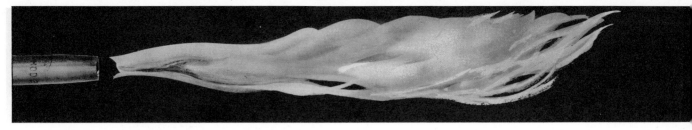

LP Gas Burning in Atmosphere
Open fuel gas valve until flame begins to leave tip end.

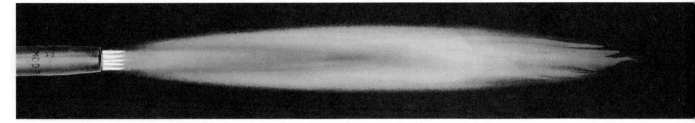

Reducing Flame
(Excess LP-Gas with oxygen.) Not hot enough for cutting.

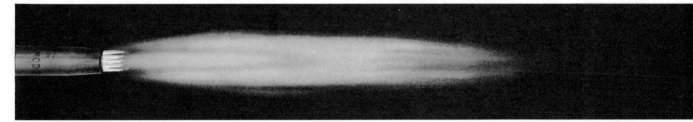

Neutral Flame
(LP-Gas with oxygen.) For preheating 1/8th in. and under prior to cutting.

Oxidizing Flame with Cutting Jet Open
Cutting jet stream must be straight and clean.

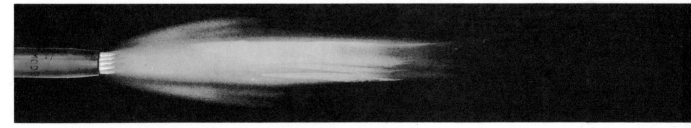

Oxidizing Flame Without Cutting Jet Open
(LP-Gas with excess oxygen.) The highest temperature
flame for fast starts and high cutting speeds.

Fig. P-2D. Conditions of oxygen-LP Gas cutting flame when adjusting the torch.
(Smith Welding Equip. Div., Tescom Corp.)

gauge indicates cylinder pressure. The other gauge indicates the pressure of the gas being fed to the torch. Flexible hoses carry the gases to the torch. Construction details of a typical oxy-fuel gas cutting torch are shown in Fig. P-2B.

Several different fuel gases may be used in this process. The flame adjustments used when acetylene is the fuel gas are shown in Fig. P-2C. The flame adjustments for liquefied petroleum (LP) are shown in Fig. P-2D.

Welding goggles should be worn for eye protection. Approved gloves and

proper clothing must be worn. It is important that the area of work be cleared of combustible material. It is also suggested that a fire watch be posted. Good ventilation should be provided.

See Chapter 3 for more detailed instructions.

P-3. SHIELDED METAL ARC WELDING (SMAW)

The shielded metal arc welding process uses an electric arc between a flux covered electrode and the metal

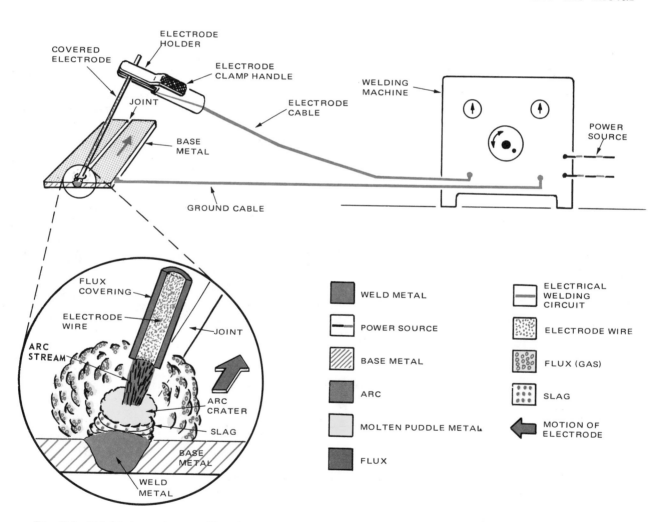

Fig. P-3. Shielded metal arc welding (SMAW). An electric arc is drawn between the covered electrode and the base metal. The heat of the arc melts the end of the electrode and the base metal where the arc contacts it. The metal from the electrode provides the filler metal for the weld.

being welded (base metal). Heat from the electric arc melts both the end of the electrode and the base metal to be joined. This process is most often used for maintenance and small production welding.

The equipment used in this welding process provides a welding current which may be either AC or DC. The amount of current is adjustable. The operator controls the movement of the hand-held electrode holder. The electrode is a flux covered metal wire. A cable connects the electrode holder to the power source.

Fig. P-3 illustrates a typical station for shielded metal arc (stick) welding.

The heat of the electric arc may be controlled by the current intensity and by the arc length. The diameter and material of the electrode will determine the kind and amount of welding current required.

The arc between the welding electrode and the base metal is struck by the operator. The correct arc length must also be controlled by the operator.

Some of the covering on the electrode turns into a gas shield which surrounds the arc as the electrode melts. Some of the covering melts and covers the completed weld with a protective layer (slag) while it cools.

The operator must wear an approved helmet, gloves and protective clothing. The work station must be well ventilated.

Chapters 5 and 6 give more detailed information on shielded metal arc welding.

P-4. METAL ARC CUTTING (MAC)

The metal arc cutting process uses an arc between a metal electrode and the base metal. The electrode is heavily covered with flux. This heats the base metal. The molten metal flows from the base metal to form a cut (kerf). This process is used mainly for small maintenance jobs.

The arc cutting process requires a current source, either AC or DC. A manually operated electrode holder provides a grip for controlling the electrode. Electric current flows through the electrode holder lead and arcs between the electrode and the workpiece. A ground cable (work lead) between the workpiece and the power source completes the circuit.

Fig. P-4 illustrates an arc cutting station. Heat from the arc is controlled by the arc length, current and electrode material.

In operation, the equipment, both electrical and mechanical, is adjusted to provide the desired arc. The operator strikes the arc between the electrode and the base metal and the cutting is started. The operator moves the electrode as the cut (kerf) progresses.

Protection is needed from intense heat, the light of the arc and some sparks. This requires wearing approved helmets, gloves and welders' clothing. The equipment must include the necessary shielding and safety devices. Excellent ventilation is needed. A fire watch is recommended.

See Chapter 9 for more detailed instructions.

P-5. AIR CARBON ARC CUTTING (AAC)

The air carbon arc cutting process uses an electric arc to heat and melt the base metal. A jet of air then blows the melted metal away. Air carbon arc cutting may be used on many metals.

Fig. P-5 shows a typical station for air carbon arc cutting.

The recommended electrical supply is direct current, reverse polarity

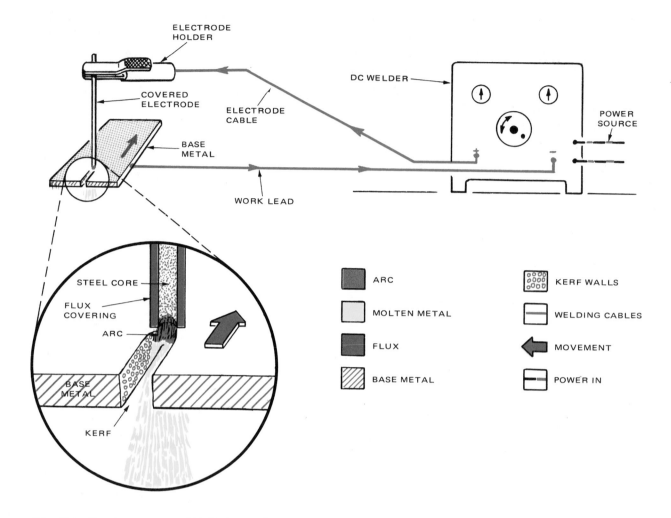

Fig. P-4. Metal arc cutting (MAC). An arc between a very heavily covered metal electrode and the base metal melts the end of electrode and the base metal. The metal electrodes melt far back into the covering, and this produces a jet action of gases which blows the molten base metal away.

(DCRP) or AC. A flexible cable (electrode lead) connects the electrode holder to the welding machine. A ground cable (work lead) connects the base metal to the welding machine.

The air jet may be supplied from either a compressed air cylinder or an air compressor. The air line is attached to the electrode lead. A lever-operated valve in the electrode holder controls the air flow. The welder operates the electrode holder manually. This process can be used for either cutting or gouging metal.

The length of the carbon electrode between the air jet nozzle and the arc must be maintained at such a distance that the air jet will be effective in blowing away the molten metal.

The current and voltage supplied to the arc may be regulated by the adjustments on the welding machine.

This cutting process produces considerable sparking. The operator must be protected by gloves, helmet and clothing. Excellent ventilation is needed. A fire watch is recommended.

See Chapter 9 for more details.

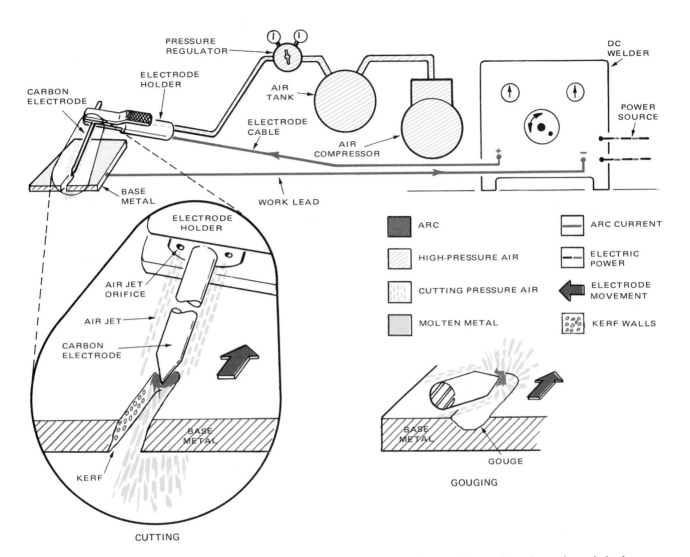

Fig. P-5. Air carbon arc cutting (AAC). The electric arc between the metal or carbon electrode and the base metal melts the base metal. Manually operated air jets attached to the electrode holder blow the molten metal away. This process is used for gouging base metal as well as for cutting.

P-6. OXYGEN ARC CUTTING (OAC)

The oxygen arc cutting process uses an electric arc to heat the base metal. Then a jet of oxygen cuts the heated metal. It is used for cutting cast iron, steel and many other metals. It is a rapid metal-cutting process.

Fig. P-6 illustrates a typical station for oxygen arc cutting.

The equipment includes a special electrode holder and a hollow metal electrode. An oxygen cylinder and regulator provide a controlled flow of pressurized oxygen through the electrode. A flexible hose carries the oxygen to the electrode holder. The operator controls the oxygen through a hand valve on the electrode holder. A welding machine supplies an arc current to the electrode. Current is conducted to the electrode holder through an electrode holder cable. A work lead connects the metal workpiece to the welding machine.

Welding and Cutting Processes 17

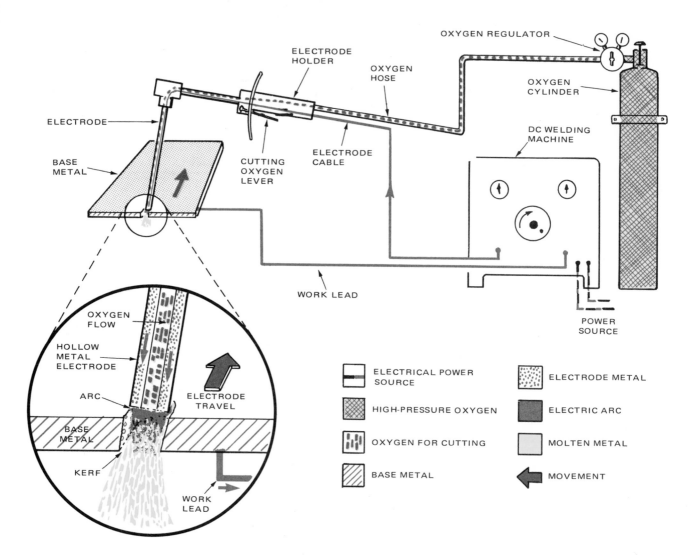

Fig. P-6. Oxygen arc cutting (OAC). An electric arc is drawn between a hollow electrode and the base metal. A lever on the holder allows a jet of oxygen to flow through the electrode. This jet cuts the base metal by rapid oxidization (combining the metal with oxygen).

To make the cut, the operator strikes an arc and, as soon as the metal surfaces to be cut have a heated spot, the operator opens the oxygen valve. The flow of oxygen into the arc and against the metal very rapidly heats and burns away the metal. This process may be used either in air or under water.

The heat of the arc is adjusted by a current adjustment control on the welding machine.

This cutting process produces a great shower of sparks. The operator must wear a helmet, gloves and protective clothing. There is danger from sparks flying around the legs and feet. Good ventilation and good fire protection are important.

See Chapters 9 and 19 for more details on this process.

P-7. GAS TUNGSTEN ARC WELDING (GTAW)

Gas tungsten arc welding uses the heat of an electric arc between a tungsten electrode and the base metal. A separate welding filler rod is fed into the molten base metal if needed. A shielding gas flows around the arc to keep away air and dirt.

This process is sometimes called TIG (Tungsten Inert Gas) welding.

This process is particularly desirable when welding stainless steel, aluminum, titanium and many other nonferrous metals.

Fig. P-7 illustrates a typical station for gas tungsten arc welding.

An AC-DC welding machine may be used with a regulated flow of a shielding gas, such as argon or helium. The shielding gas flows from a cylinder through a regulator, flow meter and a hose to the workpiece.

The welder normally operates the torch (electrode holder) and the filler metal. This torch has a gripping device to hold the tungsten electrode and a heat-resistant gas flow cup or nozzle surrounding the electrode.

Some small capacity torches are

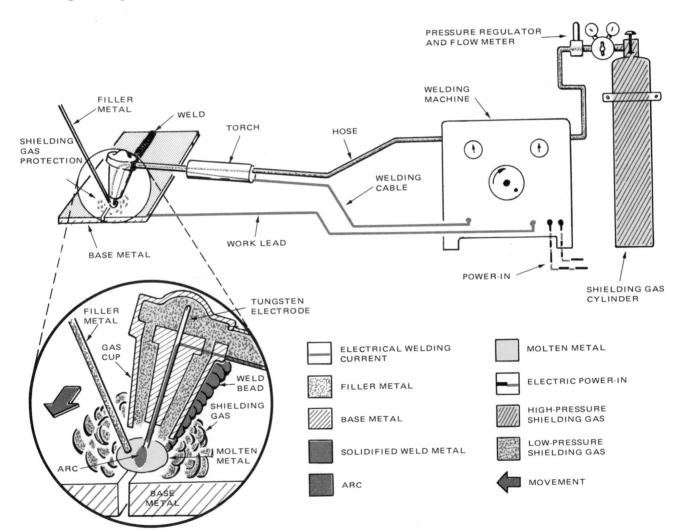

Fig. P-7. Gas tungsten arc welding (GTAW). An electric arc is drawn between the end of the tungsten electrode and a spot on the base metal. Only the base metal melts. If filler metal is needed in the joint, a separate filler metal is used. The base metal puddle melts the filler metal as needed. Shielding gas flows out a nozzle around the tungsten electrode.

Welding and Cutting Processes 19

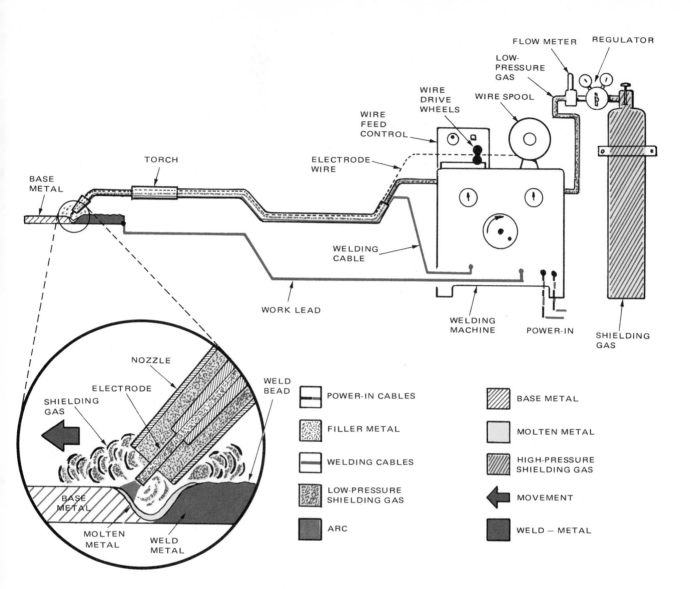

FLOW METER   REGULATOR

LOW-
PRESSURE
GAS

WIRE
DRIVE
WHEELS   WIRE SPOOL

WIRE
FEED
CONTROL

TORCH   ELECTRODE
WIRE

BASE
METAL

WELDING
CABLE

WORK LEAD

WELDING
MACHINE   POWER-IN

SHIELDING
GAS

NOZZLE

ELECTRODE   WELD
BEAD

SHIELDING
GAS

BASE
METAL

MOLTEN   WELD
METAL   METAL

	POWER-IN CABLES		BASE METAL
	FILLER METAL		MOLTEN METAL
	WELDING CABLES		HIGH-PRESSURE SHIELDING GAS
	LOW-PRESSURE SHIELDING GAS		MOVEMENT
	ARC		WELD — METAL

Fig. P-8. Gas metal arc welding (GMAW). An electric arc is drawn between a metal electrode and the base metal. Heat melts the end of the electrode wire and a spot on the base metal. A shielding gas flows out of the torch nozzle. This gas keeps the oxygen and impurities in the air from contacting the weld.

air-cooled. Large torches are water-cooled.

Heating properties of the arc may be controlled by changing current and arc length. Diameter of tungsten electrode, thickness, and kind of base metal, will determine welding amperage.

This process is easier on a flat (downhand) surface. However, such welding is possible in other positions.

Gas tungsten arc welding generates intense heat and light as well as metal spatter. Operator must wear welding helmet, gloves and welder's clothing. See Chapters 11 and 12.

P-8. GAS METAL ARC WELDING (GMAW)

In gas metal arc welding an electric arc between a continuously fed metal electrode and the base metal produces heat. The arc is shielded by a gas.

This process is popular in production and repair shops. It is often called MIG (Metal Inert Gas) welding. See Fig. P-8.

Power source is a DC welding current. A shielding gas cylinder, a regulator and a hose provide a flow of shielding gas to the arc. Shielding gases such as carbon dioxide, argon, or helium may be used. An electrode feeding device supplies metal electrode continuously. A torch and cable carry the electrode wire, current and the shielding gas to the arc. The torch usually has a trigger switch for controlling electrode feed and gas flow.

Speed controls for wire feed are usually mounted in the wire feed mechanism normally located on top of the welding machine. Current controls are on the standard arc welding machine. Voltage control is on the constant voltage (cv), also known as constant potential (cp). Current is changed by adjusting the wire feed speed. Shielding gas volume adjustments are made at gas flow meter on regulator. The kind of shielding gas used usually depends on the metals being welded.

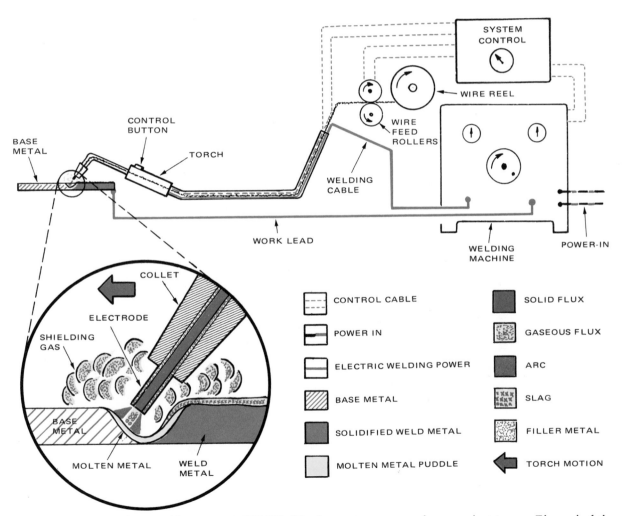

Fig. P-9. Flux-cored arc welding station (FCAW). The heat energy comes from an electric arc. The end of the electrode and a spot on the base metal are melted to form the weld. The flux core provides a gaseous shield around the arc and also provides a slag covering to keep the air away from the weld until it cools.

The operator:
1. Selects the electrode size.
2. Adjusts the current.
3. Adjusts the shielding gas flow.
4. Adjusts the rate of electrode feed.
5. Controls the arc length and torch movement.

The operator must wear an approved helmet, gloves and welder's clothing. The welding area must have good ventilation. See Chapters 11 and 12.

P-9. FLUX-CORED ARC WELDING (FCAW)

In flux-cored arc welding, Fig. P-9, heat comes from an arc between a flux-cored electrode and the base metal. This process is particularly desirable for welding structural steel and in other low-carbon applications.

Some flux-cored wires are used without CO_2. Others use CO_2.

A constant voltage DC arc is usually used. A welding machine furnishes the electric power.

The electrode is a hollow metal tube with the center (core) filled with a flux material. An electrode feeding machine feeds the electrode from a large spool to the electrode holder.

The heat of the arc depends upon the arc length, voltage and the amperage setting of the welding machine. The higher the current, the greater the heat of the arc. The speed of the electrode feeding machine may be adjusted.

The operator is exposed to heat and

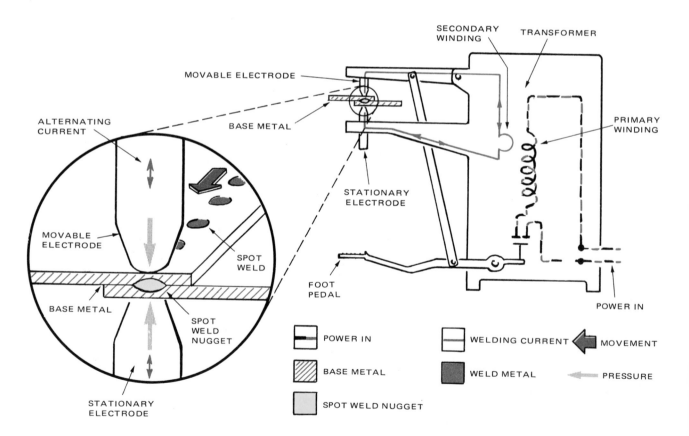

Fig. P-10. Resistance spot welding (RSW). A step-down transformer provides a low voltage, high current electrical flow. Through the electrodes, it heats a small area on two sheets of metal as these sheets are pressed together between the spot welding electrodes. The metals become hot enough to fuse together.

light from the arc. An arc welder's helmet, leather gloves and protective clothing should be worn. Excellent ventilation should also be provided.

Chapters 11 and 12 give more detailed information concerning the operation of this process.

P-10. RESISTANCE SPOT WELDING (RSW)

Resistance spot welding, Fig. P-10, passes an electric current through the metal. Resistance to the electrical flow heats the metal to welding temperature. The process is used to weld together two or more overlapping pieces. It is well suited to automatic welding. Spot welding is commonly used to join auto

body sections, cabinets and other sheet metal assemblies.

A step-down transformer converts fairly high voltage-low amperage current to a low voltage-high amperage current. The weld is made between two electrodes which press the metals together. A heavy electrical current flows from one electrode through the metals to be welded together, to the second electrode.

These electrodes are special metal alloys which can carry the high current and still have physical strength to operate under high pressures. The electrodes on small spot welders, used to weld thin materials, may be air-cooled. Electrodes for welding thicker metals are water-cooled.

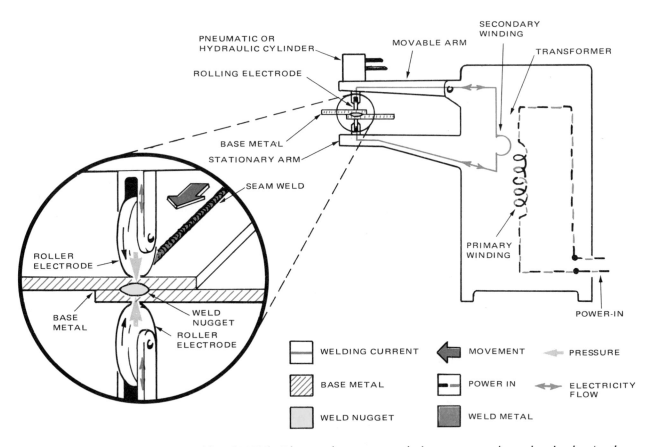

Fig. P-11. Resistance seam welding (RSEW). Electrical energy travels between two electrode wheels. As the wheels travel, they clamp two sheets of base metal together. As the electricity travels through the base metal, the metal becomes hot enough to fuse the two sheets together and form a seam weld.

Resistance spot welding is controlled by the amperage, the electrode pressure and the length of time the current flows.

In an automatic spot welder, an electronic controller controls the amperage, pressures and timing of the current.

The operator must wear flash goggles. If the metals must be manually handled, special gloves are needed to prevent injury to the hands.

See Chapters 13 and 14 for detailed information concerning spot welding.

P-11. RESISTANCE SEAM WELDING (RSEW)

Resistance seam welding, Fig. P-11, is a special application of spot welding.

It is often used to weld joints in containers and other products which require an airtight or vapor-tight seam.

The electrodes are wheels. The work to be welded is passed between the revolving wheel electrodes. A timing device turns on the welding current at controlled but rapidly repeating intervals. The rapidly repeating current flow makes a series of overlapping spot welds which appear to be a continuous line of welding. These machines are usually automatic. The timing of the current flow, the amperage, and the pressure on the electrodes is regulated by electronic controllers.

The operator must wear flash goggles and all other required safety clothing. If the metal must be handled, he or she should wear special, approved gloves.

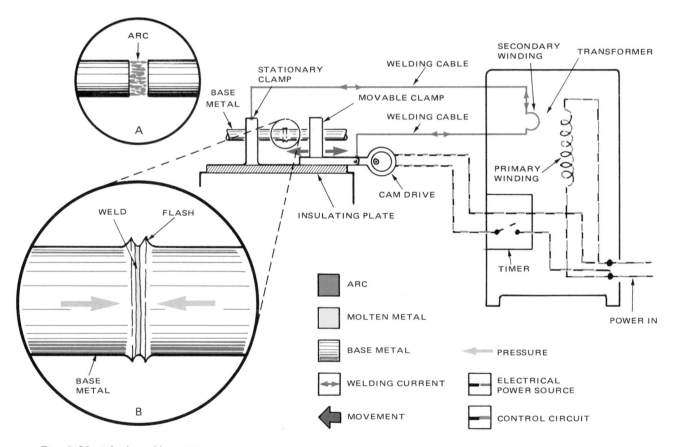

Fig. P-12. Flash welding (FW). A—Electrical energy creates an arc which melts the ends of two pieces of base metal. B—When the ends are molten, they are pushed together to fuse into one piece.

See Chapters 13 and 14 for additional information.

P-12. FLASH WELDING (FW)

The flash welding process uses an electric arc to heat the base metals. It provides a strong, clean weld joint. Its chief use is in production welding. It combines resistance welding, arc welding and pressure welding.

As shown in Fig. P-12, a step-down transformer provides the welding current. The two pieces to be welded are held in current-conducting movable clamps.

To make a flash weld, the workpieces are brought together under light pressure. A heavy low-voltage current travels between the two base metals.

As soon as the current is established, the two pieces of metal are drawn apart very slightly. At this point an electric arc passes between them. The arc heats the surfaces of the two metals. When the surfaces are sufficiently heated, they are forced together under very high pressure. This pressure causes a slight flow of heated and somewhat dirty metal from the joining surfaces. The clean, heated subsurface metal brought into contact produces a good weld. The finished weld will have a flash or enlargement at the joint.

Welding heat is controlled by the rate of current flow. The quality of the weld is controlled by the current flow, the length of time of the arc and, finally, the pressure at the time the two surfaces are brought together.

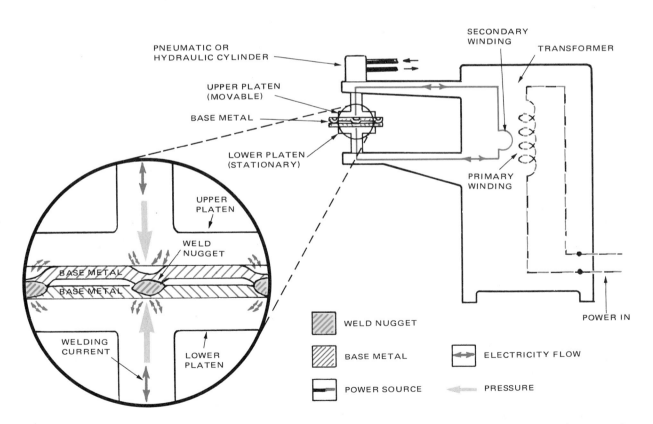

Fig. P-13. Resistance projection welding (RPW). Electrical energy heats projection in one sheet of base metal as these projections touch another sheet of base metal. Both pieces become hot enough at the contact spots to fuse together as pressure is applied.

When flash welding manually, the operator must be able to judge the metal temperatures, time that the welding surfaces must be brought together and the proper welding pressure.

It is necessary that protective clothing, face shields and gloves be worn. Flying sparks (metal expelled at joint) are produced during this process.

See Chapters 13 and 14 for more detailed information.

P-13. RESISTANCE PROJECTION WELDING (RPW)

The projection welding process uses resistance to the flow of electricity to create heat for welding. See Fig. P-13. It is similar to spot welding. It is commonly used in production welding.

One of the two pieces of metal is run through a machine which makes bumps or projections in the metal of a designed shape and size.

The welding machine electrodes are flat plates called platens. The two pieces of bare metal are placed together between the platens. They touch only at the projections.

Welding current is supplied by a resistance welder transformer. The welding current flows through the pieces to be welded while they are clamped between the platen plates. The current only flows through the pieces where they touch each other. Due to the projections, the current is concentrated at the points of contact. These points heat up and fuse.

The welding current flows for a short time and, at the same time, pressure is applied between the two platens. This completes the weld. The timing of the current flow and the application of welding pressure is an important part of this welding process.

Some flash and sparking may take place during the welding. The operator should wear flash goggles, safe clothing, safety shoes and leather gloves.

A more complete explanation of this process is given in Chapters 13 and 14.

P-14. TORCH SOLDERING (TS)

The torch soldering process, Fig. P-14, uses an air-fuel gas flame which heats the base metal enough to bond molten solder to it. This is a popular method of joining metals in manufacturing and service operations. Torch soldering is used to fill a seam or to make an airtight joint of some strength.

The air-fuel gas torch can be used for soft soldering. It can also be used to braze small metal parts.

In this soldering process, the amount of heat is controlled by the amount of gas flowing through the torch. Larger torch tip orifices are used when more heat is needed. The rate of gas flow is usually controlled by a needle valve on the torch. Atmospheric air is drawn into the torch through holes in the torch tip, as shown in Fig. P-14.

An acetylene regulator is mounted on the acetylene cylinder. The torch flame is controlled by the acetylene pressure at the torch tip. The final flame adjustment is made with the torch valve.

The acetylene cylinder may be replaced by a cylinder containing MAPP gas or liquefied petroleum (LP). Some torches may burn any of these gases. Some torches, however, can only be used with one type of fuel.

Safety goggles or flash goggles are required to protect the eyes. Use pliers to handle the hot metal. Keep moisture away from molten solder. Moisture in contact with molten solder instantly changes to steam. This may cause molten solder to fly in all directions.

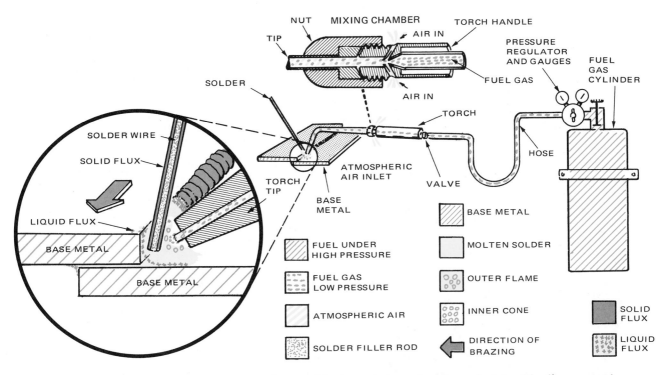

Fig. P-14. Torch soldering (TS). A mixture of air and fuel gas is burned at the end of tip. The flame provides the heat to warm the base metal and to melt the solder. A flux is needed to keep the base metal and solder clean enough to allow the solder to adhere (stick).

See Chapter 15 for more detailed instructions about soldering.

P-15. TORCH BRAZING (TB)

Brass and bronze metals have a lower melting temperature than steel. In torch brazing an oxy-fuel flame heats the base metal, and the heated base metal melts the brazing rod. There is less warping of the base metal with brazing. Fig. P-15A illustrates a typical torch brazing station.

The oxy-fuel gas flame is adjusted to provide a neutral, oxidizing or reducing flame, depending on the behavior of the metal in the joint. This means that the flame adjustment is very important. See Fig. P-15B.

Metal parts to be brazed are first carefully cleaned, then heated with the torch. The brazing filler metal and flux is then brought into contact with the metals at the point where the braze is to be made. Flux is added in powder or paste form or as a coating on the brazing filler metal.

Flux keeps the metal clean. It does not preclean the metal.

The brazing filler metal is touched to the base metal and it is heated by the base metal. It then melts and flows into the joint. If the mating surfaces were properly cleaned, a good brazed joint should result.

The operator should wear goggles, gloves and fire-resistant clothing. There may be a severe health hazard if brazing filler metal contains zinc, cadmium, phosphorous or beryllium. Overheating the brazing alloy can create similar health hazards. Because of fumes from the flux, the brazing operation must be well ventilated.

See Chapters 16 and 21 for more detailed instructions about brazing.

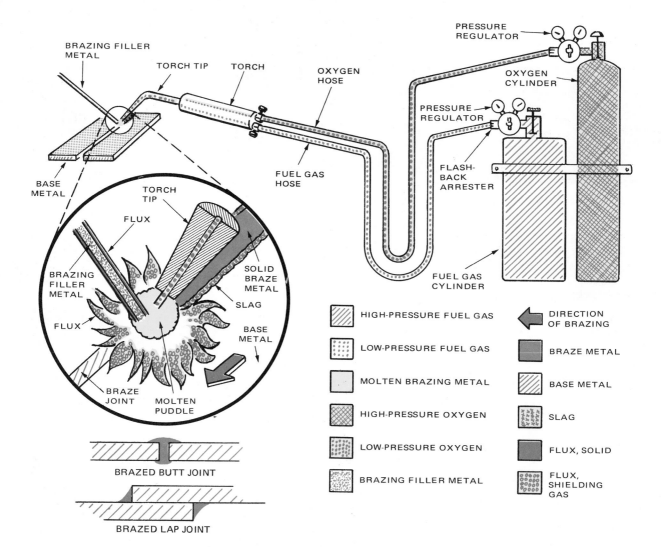

Fig. P-15A. Torch brazing (TB). *Oxygen and a fuel gas are mixed in a torch. The mixture is burned at the torch tip. Heat raises the temperature of a spot on the base metal until the melted brass or bronze filler metal adheres to the base metal. A flux is used to keep the base metal and filler metal clean during the operation.*

P-16. SUBMERGED ARC WELDING (SAW)

In submerged arc welding an electric arc between an electrode and the base metal produces welding heat. The arc is submerged in a granular flux. This process is more popular when welding thick plate joints. The equipment is usually automatically or semiautomatically operated. The electrode also feeds automatically into the arc. See Fig. P-16.

A hopper and feeding mechanism are used to provide a flow of flux over the joint being welded. The arc, generated by AC or DC current, is submerged in the flux.

The chemical composition of the flux will affect the composition of the completed weld. Alloy elements can be added to the weld by controlling the chemical composition of the flux.

Welding heat is regulated by changing the voltage and current provided by the welding machine.

LP Gas Burning in Atmosphere
Open fuel gas valve until flame begins to leave tip end.

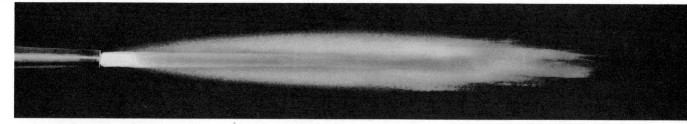

Reducing Flame
(Excess LP-Gas with oxygen.) For heating and soft soldering or silver brazing.

Neutral Flame
(LP-Gas with oxygen.) For brazing light material.

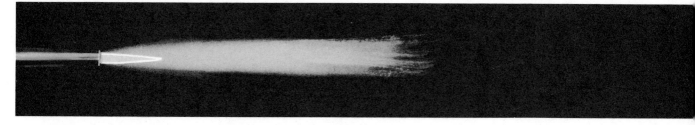

Oxidizing Flame
(LP-Gas with excess oxygen.) Hottest flame (about 5300 F.).
For fusion welding and heavy braze welding.

Fig. P-15B. Color appearance of oxygen-LP gas flames.
(Smith Welding Equip. Div., Tescom Corp.)

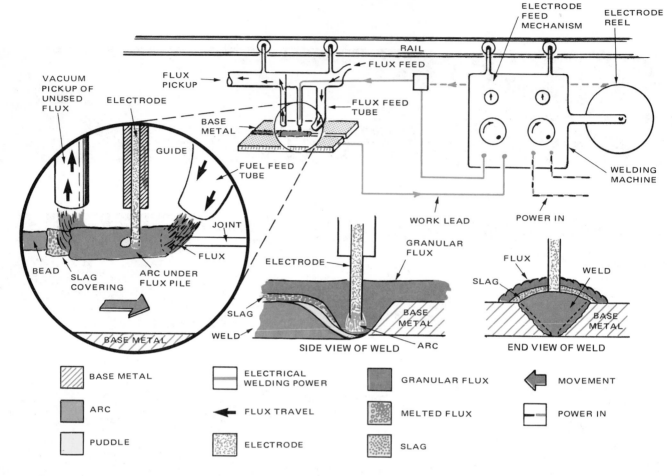

Fig. P-16. Submerged arc welding (SAW). Electrical current creates an arc to heat and melt the end of the electrode wire and the base metal. The arc is covered with grains of flux. Some of this flux turns into a gas shield surrounding the arc. Some flux melts and forms a slag over the weld.

The electrode, usually power fed and made of mild steel, extends through the shielding material to a point just above the base metal. The correct arc length is automatically maintained underneath the shielding material.

As the weld progresses, some of the flux melts and forms a slag over the weld. A vacuum machine picks up what flux is unused.

Since the arc flame is submerged in the shielding material, it cannot be seen during welding. This reduces, to some extent, the hazard of burns and flying sparks. Still, the operator must wear approved goggles, gloves, clothing and provide good ventilation.

See Par. 17-2 for more details on submerged arc welding.

P-17. ELECTROSLAG WELDING (ESW)

Electroslag welding uses one or more electric arcs between continuously fed metal electrodes and the base metal. The arcs provide heat to weld vertical butt joints.

Movable, water-cooled molds keep molten metal in place as weld moves upward. A molten flux (slag) covers molten weld metal. This process is

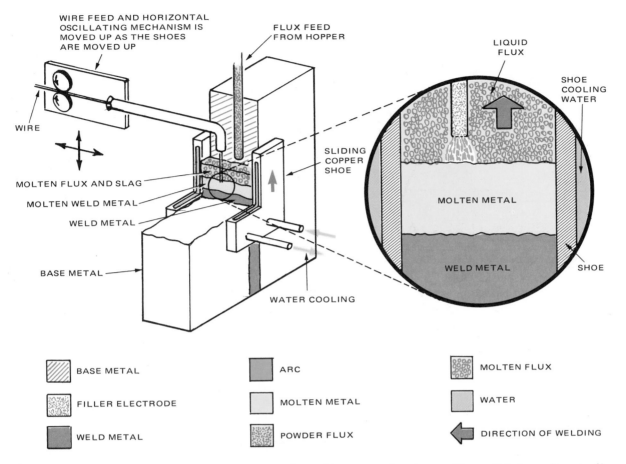

WIRE FEED AND HORIZONTAL OSCILLATING MECHANISM IS MOVED UP AS THE SHOES ARE MOVED UP

FLUX FEED FROM HOPPER

LIQUID FLUX

SHOE COOLING WATER

WIRE

SLIDING COPPER SHOE

MOLTEN FLUX AND SLAG

MOLTEN WELD METAL

WELD METAL

MOLTEN METAL

BASE METAL

WELD METAL

SHOE

WATER COOLING

BASE METAL

FILLER ELECTRODE

WELD METAL

ARC

MOLTEN METAL

POWDER FLUX

MOLTEN FLUX

WATER

DIRECTION OF WELDING

Fig. P-17. Electroslag welding (ESW). The energy for welding is an electric arc. The weld is formed vertically. Water-cooled shoes contain the molten metal until it solidifies.

used when making welds on thick, heavy structures. Basically it is a form of arc welding.

Fig. P-17 illustrates a typical setup for electroslag welding.

Flux completely covers the welding surface and a metal electrode is passed through the flux to a point just above the molten metal. In large welds, more than one metal electrode is used. The electrode is often automatically oscillated (moved back and forth) to weld a larger area. The mold is water-cooled. Otherwise, it would melt and fuse with the weld.

The arc, electrode and flux material are raised as the weld moves up. If the weld is very large, it may be necessary

to drain off the molten slag and replace it with new flux material occasionally.

The heat source is controlled by the current intensity and the physical characteristics of the flux. Operators must wear protective clothing and helmets. Heat from this type of weld is quite intense.

See Par. 17-3 for more details.

P-18. STUD WELDING (SW)

Stud welding is a semiautomatic arc welding process. It is used to attach metal fastening devices to metal plates or beams without drilling and tapping. Bolts, screws, rivets and spikes may be attached in this way.

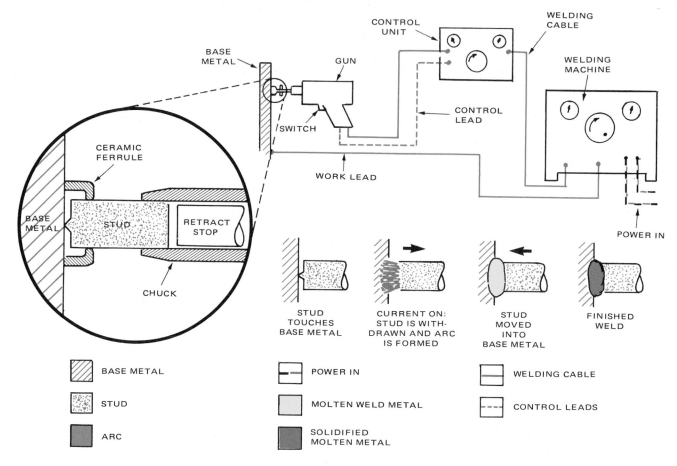

BASE METAL

CERAMIC FERRULE

BASE METAL

STUD

RETRACT STOP

CHUCK

CONTROL UNIT

GUN

WELDING CABLE

WELDING MACHINE

SWITCH

CONTROL LEAD

WORK LEAD

POWER IN

STUD TOUCHES BASE METAL

CURRENT ON: STUD IS WITH-DRAWN AND ARC IS FORMED

STUD MOVED INTO BASE METAL

FINISHED WELD

BASE METAL

STUD

ARC

POWER IN

MOLTEN WELD METAL

SOLIDIFIED MOLTEN METAL

WELDING CABLE

CONTROL LEADS

Fig. P-18. Stud welding (SW). Electrical energy creates an arc which melts the end of a special electrode (a bolt, a pin, a clip or a stud) and the base metal. The special electrode is then moved into the base metal puddle. The assembly is held until the weld metal cools.

Fig. P-18 shows a stud welding station. The heat source is an electric arc. The enlargement shows the stud, stud chuck and ceramic ferrule. Another enlargement shows the four steps in making the welded joint.

The energy source (welding machine) is an electric welding transformer. The control on the welding machine governs the current intensity in the electric arc. Current varies with the size of the stud and the kind of metal. The control unit has a timer which controls the duration of the arc.

The operator installs the stud in the gun. The gun is then positioned on the base metal. A switch on the gun starts the stud welding operation.

The operator must wear gloves, a face shield with flash goggles and fire-resistant clothing. Hard hat and safety shoes are also recommended.

A more complete explanation of stud welding is given in Par. 17-4 and 17-5.

P-19. FORGE WELDING (FOW)

In forge welding a furnace heats two pieces of metal to a plastic temperature. Pounding (hammer blows) fuses the two pieces together. This process is used to join wrought iron, low carbon steel and medium carbon steel pieces. It can be used in places where there is

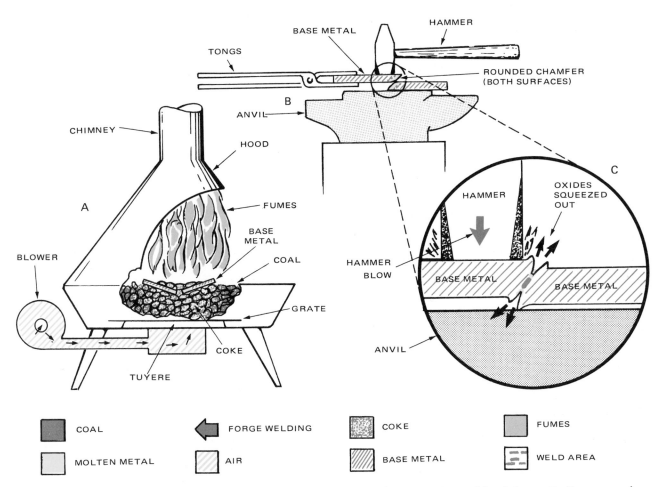

Fig. P-19. A forge welding station. A—Cross section through blacksmith's coal-fired forge. B—Hammer and anvil being used to make forged weld. C—Enlarged view of welding action.

no electricity or fuel gas available. Fig. P-19 illustrates a typical blacksmith's forge welding station.

The forge has a cast iron pan or tub. A blower supplies air at the bottom of the forge through an opening called a tuyere. The fuel is usually a good grade of soft coal. Heat changes the burning coal to coke near the center of the fire. The fire is carefully tended by feeding coal from the outside toward the center. Coking means driving out the combustible gases from the coal. As this coal is being coked, the gas that is released will burn in an open flame over the fire. A thick bed of coked coal should be made before any welding is

attempted by this process.

Heat is controlled by the airflow through the tuyere (air inlet). Increasing airflow increases the rate of combustion.

Metal parts to be welded are placed down in the coke at the point of combustion. When the parts reach the forging temperature - indicated by the color of the metal, a bright red - they are withdrawn.

Heating may also be done in a furnace burning a fuel gas. When heated, the parts are placed together on an anvil and pounded together to make a weld. The lap joint is usually used when forge welding.

The blacksmith welder should wear eye protection, safety shoes and flame-resistant clothing and gloves. This process produces a great deal of heat. Sparks are often created when the hot metal is pounded.

See Par. 17-14 and 26-11 for more details.

P-20. COLD WELDING (CW)

Cold welding uses very high pressures to force two metal parts together. No outside heat source is used. High pressure forces the very clean metals together and the surface molecules only are fused. The method is used mainly on softer metals such as aluminum-to-aluminum, copper-to-copper, and aluminum-to-copper. The joint is formed by pressure. Good fusion occurs resulting in a strong weld. Butt welds and lap welds may be made with this process.

The source of energy to produce the weld is a tremendous pressure usually produced by using hydraulic cylinders. See Fig. P-20. The weld is controlled by the size of the die surfaces in contact with the metal and the amount of hydraulic pressure.

Metal surfaces being joined must be very, very clean. The process adjustments include the size of the die surfaces and the amount of the hydraulic or leverage pressure.

The metal must be cleaned immediately before welding. High speed

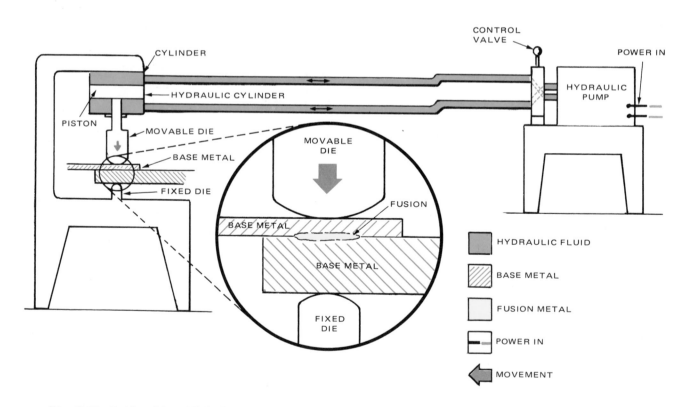

Fig. P-20. Cold welding (CW). Pressure creates the welding energy. In this illustration a thin sheet of base metal is lap welded to a thicker sheet of base metal. Note that the upper die is larger than the bottom die. This is because the top base metal is softer than the bottom base metal. A hydraulic piston is used to force the upper die against the metal. The hydraulic pressure used is only enough to fuse the surfaces of the base metals being welded.

wire wheels are often used. The operator should wear gloves, a face shield or safety goggles and safety shoes.

See Par. 17-15 for more details on the cold welding process.

P-21. ULTRASONIC WELDING (USW)

In ultrasonic welding very high sound frequencies excite metal surface molecules. This movement among the molecules produces fusion. This process is most often used to join very light materials, for example, attaching fine wires to foil or wires to wires.

Fig. P-21 illustrates the ultrasonic welding station. Since no outside heat is applied, this process is particularly desirable where the control of the heat zone is important. A transducer transforms the high frequency vibration into heat.

Small rods contact the two metals to be welded. One of these rods vibrates at a very high frequency, much above the normal sound range (ultrasonic). This causes the material being joined to vibrate at a corresponding rate. During the vibrations some molecules of the two surfaces become intermixed and form a strong joint.

This type of welding is controlled by the rate of vibration and the pressure exerted by the vibrating elements on the parts being welded. Surfaces to be joined by ultrasonic welding must be very clean and free of oxidation.

Operators should wear gloves and

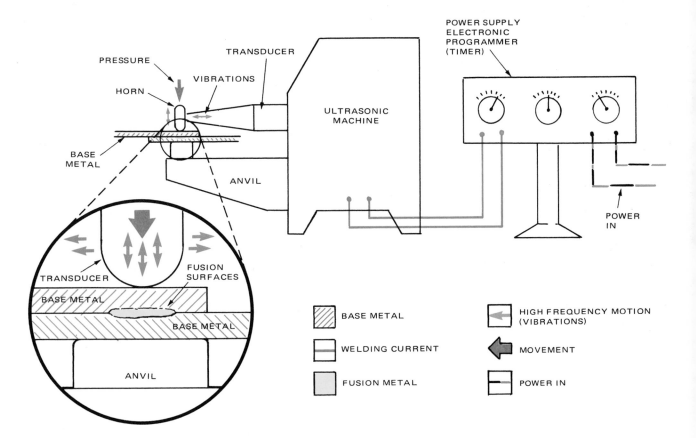

Fig. P-21. Ultrasonic welding (USW). Welding energy is created by very, very high frequency vibrations of the machine horn. The molecules at the surfaces of the two pieces of base metal are made to move so rapidly that the surfaces fuse.

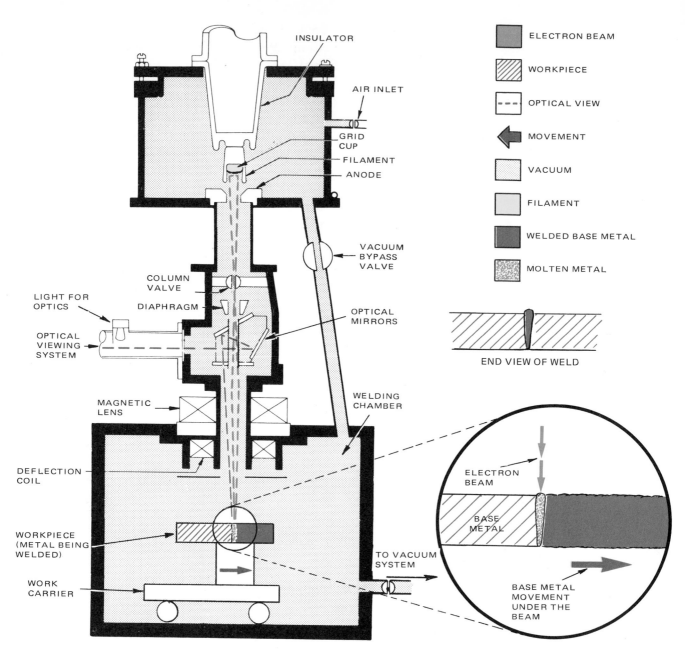

Fig. P-22. Electron beam welding (EBW). Welding energy is created by hitting the base metal with a beam of electrons. The heat energy is very intense. Welds formed are very narrow.

goggles. Fine particles could be thrown off. See Par. 17-6.

P-22. ELECTRON BEAM WELDING (EBW)

Electron beam welding uses energy from a focused stream of electrons to heat and fuse metals.

It is a good process for welding heavy parts when distance between them is small. Weld can be made in deep, narrow space and narrow weld zone. Fig. P-22 shows a setup used to join metals difficult to weld by other processes or for welding metals at very high speed.

Because equipment is large and expensive, it is used only where other processes cannot do the job.

The machine uses an electronic tube which gives off (emits) electrons. These streams of electrons are controlled (focused and concentrated) by electromagnets called a magnetic lens. The electron beam is generated in much the same way as the light beam in a television receiver. It can be bent by a magnetic field as in the television set. Heat intensity is regulated by the amount of power generated in the electron gun filament.

The electron beam weld is usually made in a vacuum, as the air molecules tend to interfere with the beam. The vacuum chambers are also a shield against radiation.

The operator watches the weld through a safe optical system and directs the beam with remote controls. Some modern electron beam welding equipment may be used under atmospheric pressure conditions.

The surfaces to be joined should be cleaned prior to welding.

The operator and other persons near the machine must be protected from the radiation given off by the beam. Most machines use lead as shielding material.

See Par. 17-17 for more details.

P-23. FRICTION WELDING (FRW)

Friction welding uses friction to create enough heat to fuse two pieces of metal together. This process is used

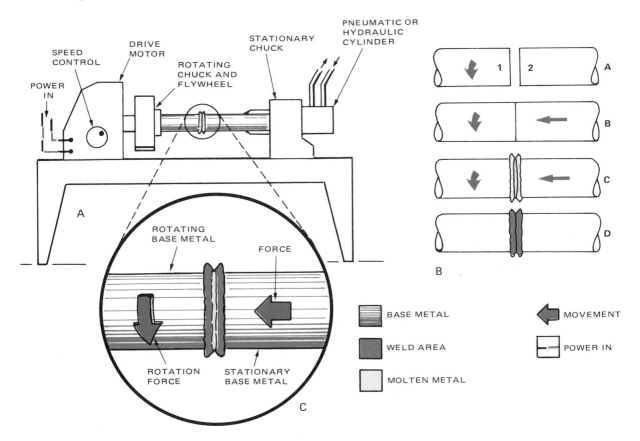

Fig. P-23. Friction welding (FRW). A—Schematic of friction welding station. B—Four steps in producing friction weld. C—Section through completed friction weld.

chiefly in butt welding rather large, round rods or cylinders. Fig. P-23 illustrates a friction welding station.

No outside heat is supplied. The ends of the parts to be joined are brought together under pressure. Heat is generated at the welding surface as one of the parts is made to revolve. The resulting friction between the stationary and revolving parts develops the heat needed to form the weld. As the metal surfaces reach the plastic state, they are forced together under great pressure. The process creates a clean metal-to-metal welded surface.

Equipment includes the necessary clamping devices, a mechanism for revolving one part and a method for subjecting the friction surfaces to a high pressure.

The control of this type of weld is based on the pressure and the speed at which the surfaces rotate against one another.

Friction welding produces considerable sparking. The operator needs to wear goggles or a face shield to avoid injury.

See Par. 17-18 for more details.

P-24. EXPLOSION WELDING (EXW)

Explosion welding, Fig. P-24, joins metals together by using a powerful

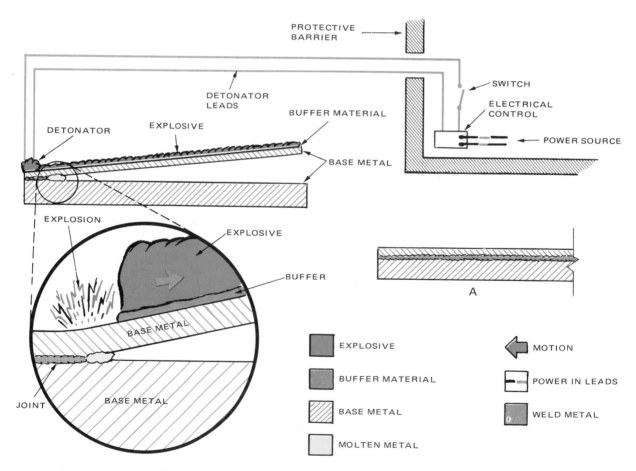

Fig. P-24. Explosion welding (EXW). A buffer material is applied to the top surface of one of the metal sheets. Welding energy comes from explosive material placed on top of buffer material. An igniter (detonator) is operated from behind a barrier. The explosion proceeds from left to right and welds the top plate to the bottom plate almost at once, without deforming either piece of metal. A—Completed weld.

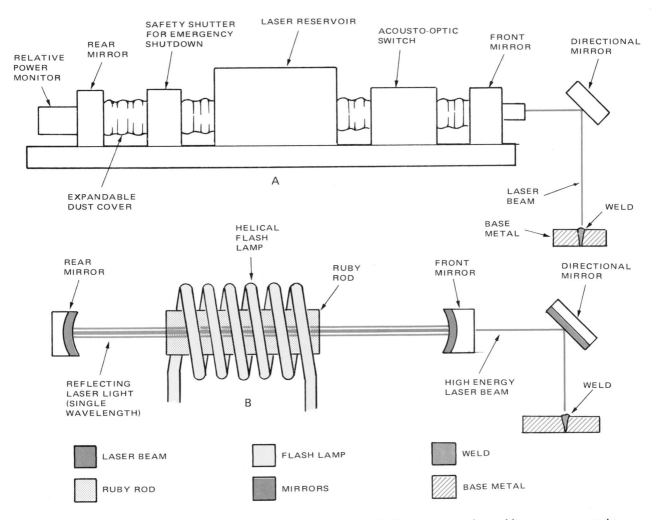

Fig. P-25. *Laser beam welding (LBW). A single wavelength light beam creates the welding energy as it hits the base metal. The laser beam can also be used for cutting and piercing base metals. A—Laser beam station. B—Laser beam light source.*

shock wave. This creates enough pressure between two metals to cause surface flow and cohesion. It is often used to weld large sheets together. In one common application, it welds thin stainless steel sheet to mild steel sheet. It is also used for welding aluminum and molybdenum together.

The energy source is the tremendous shock wave caused by igniting an explosive material. The operation requires very careful setup. Bonding takes place in an instant. Such welding is done either in a safety chamber or, less frequently, under water.

Safety is a very important matter.

Both explosive and fixture must conform to approved written specifications. Welder must be protected from sound of explosion by wearing industrial "ear muffs" and/or ear plugs. Face shields, safety helmets, and safety shoes should be used. Special permits are required from government authorities because of explosives used.

Par. 17-20 explains this special type of welding in more detail.

P-25. LASER BEAM WELDING (LBW)

Laser beam welding uses a single frequency light beam. This beam puts

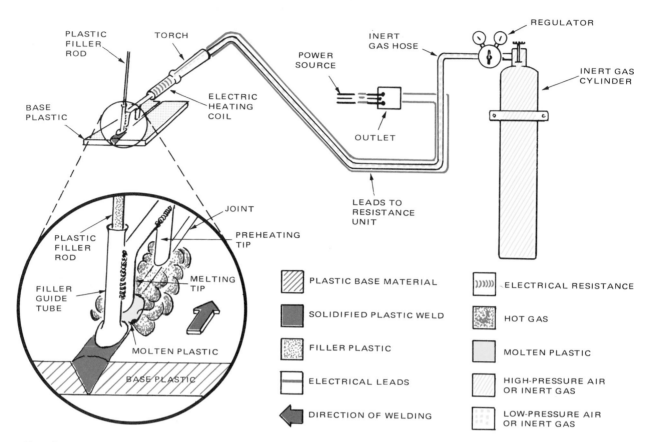

Fig. P-26. Torch plastic welding (TPW). Heat source is heated air or a heated inert gas. Heated gas first preheats the joint, then a second orifice melts plastic as plastic filler is fed through tube to molten plastic.

energy into a metal causing it to heat up to its melting temperature. Laser beam is useful in welding small, light materials, particularly in locations where it is very difficult to weld with any other process. Fig. P-25 shows typical laser beam welding setup.

The beam is created by putting light or heat energy into a single molecule of a substance (ruby or carbon dioxide). The single frequency energy of the single molecule substance increases in intensity by traveling between two mirrors until it passes through the weaker or poorer of the mirrors.

Heat of laser beam is very intense and is easily directed to spot where needed. Since laser beam can be reflected, it can be directed to a weld by any combination of mirrors. It can

be either continuous heat source or pulsed beam. There is an instantaneous heat release when beam contacts base metals. Control of heat is by means of control of input to laser beam source.

Laser beams must be very carefully guarded from being directed toward any part of body or anything of heat-sensitive nature. Operators must wear special goggles designed for laser operators. See Par. 17-21 for more details.

P-26. TORCH PLASTIC WELDING (TPW)

In torch plastic welding heated air or heated shielding gas melts and fuses plastic base materials and plastic filler materials together. The use of plastics in modern industry makes considerable amounts of plastic welding necessary.

Fig. P-26 shows a typical torch plastic welding setup.

The temperature at which plastics are joined is relatively low - between 400 and 800 deg. F.

The gas is heated by an electrical heating coil. The heated gas flows through the heating nozzles and heats the parts to be welded. The welding tip is designed to press the heated plastic filler material into the weld area.

The heat source is controlled by adjusting the resistance unit and/or gas flow. The operator must manipulate both the torch and the filler material.

The temperature of the filler material during the welding is hot enough to cause severe burns. It is advisable to wear gloves and safety goggles. Good ventilation is recommended.

See Chapter 18 for more details.

P-27. OXYGEN LANCE CUTTING (LOC)

In the oxygen lance cutting process, an oxygen-fuel flame heats the base metal while a jet of oxygen is directed at the heated metal to cut (burn) it. This method has been used for many years to cut heavy steel sections.

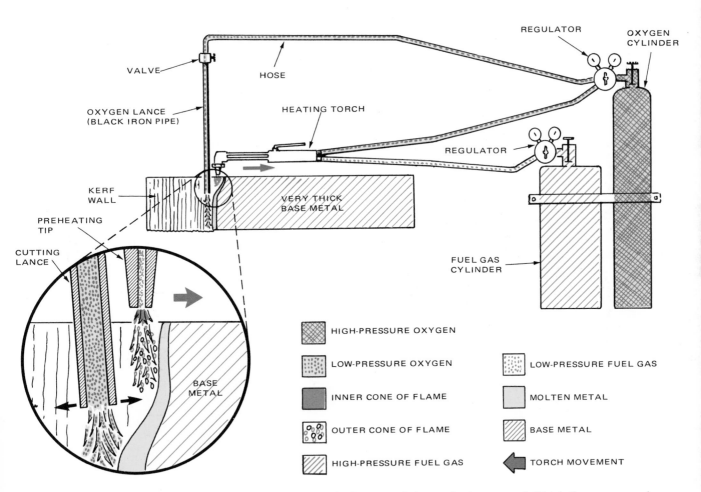

Fig. P-27. Oxygen lance cutting (LOC). An oxygen-fuel gas torch heats the base metal. The valve on a metal pipe is then opened and a jet of oxygen rapidly oxidizes the heated base metal and blows it away. The pipe is slowly consumed.

Fig. P-27 illustrates a typical oxygen lance cutting station.

The oxygen lance is a straight piece of iron pipe with a hand valve. It is attached by a hose to one or more oxygen cylinders equipped with regulators and gauges.

The oxygen lance is used along with an oxy-fuel gas cutting torch (see Par. P-2). After the cut is started with the cutting torch, the lance oxygen valve is opened. The stream of oxygen from the iron pipe flows into the kerf started by the cutting torch. It can cut sections of great thicknesses. The lance is slowly melted away during the cutting

and must be replaced frequently.

As with all cutting operations, wear proper eye protection, gloves and clothing to prevent personal injury. It is also important to cover or clear the work area of combustible materials. Post a person to watch for possible fires. Good ventilation is required.

See Chapter 19 for more information.

P-28. OXYGEN-FUEL GAS UNDERWATER CUTTING (OFGUC)

To heat the base metal, the underwater oxy-fuel gas cutting process uses

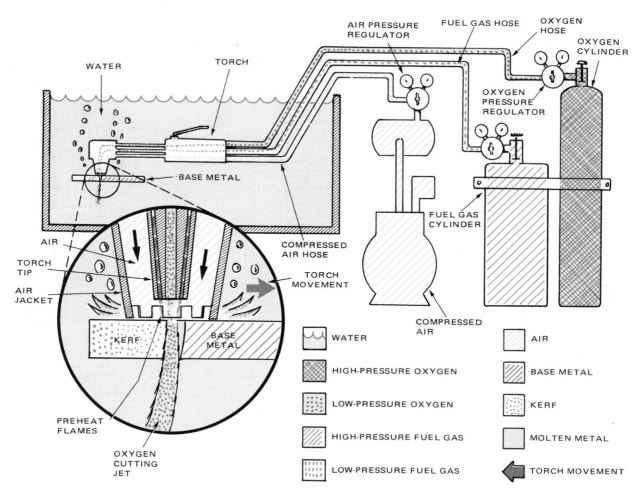

Fig. P-28. Oxygen-fuel gas underwater cutting (OFGUC). The oxygen-fuel gas flames burn in a "bubble" of compressed air which keeps the water away from the flame. The heated metal is rapidly oxidized and is blown away by the oxygen jet.

an oxy-fuel gas flame, surrounded by compressed air. An oxygen jet oxidizes the heated metal and blows it away.

The oxy-hydrogen flame plus an oxygen jet has long been used to cut steel under water.

The process is used for salvage operations and underwater construction.

Fig. P-28 shows an underwater cutting outfit.

Acetylene gas cannot be used as a fuel gas because acetylene is unstable at pressures above 14 psi (pounds per square inch). This means that, at most, acetylene could not be used safely in water deeper than 10 or 12 ft. Oxyhydrogen can be used at any depth up to 200 ft.

The proprietary gas called MAPP, manufactured by Dow Chemical Co., may also be used for this purpose. It is stable under high pressure.

In addition to oxygen and fuel, compressed air is required for underwater cutting. This air forms an air pocket in which the flame burns under water.

The equipment for using either oxyhydrogen or oxygen-MAPP is much the same. However, different cutting tips are used.

The oxygen and the fuel gas cylinders are connected to the torch with regulators and suitable lengths of hose. The valves on the underwater cutting torch control the torch flame.

The fuel gas is usually lighted under

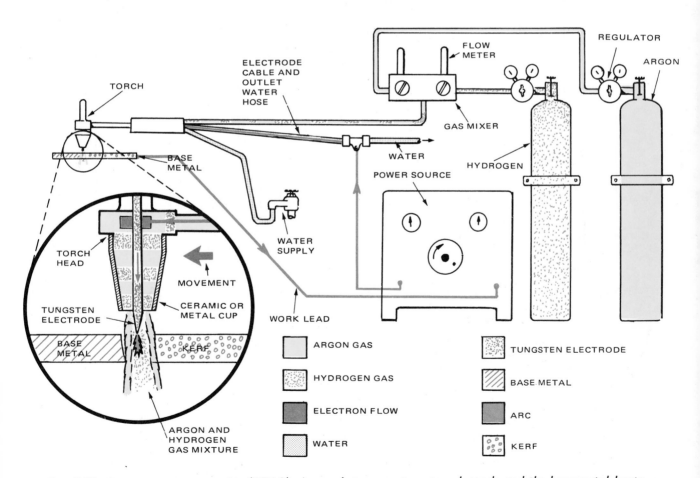

Fig. P-29. Gas tungsten arc cutting (GTAC). An arc between a tungsten electrode and the base metal heats and melts the base metal. A torch lever then turns on a shielding gas mixture which blows the melted metal away to form a cut (kerf) in the base metal.

water by an electric ignitor built into the air jacket. Gas cutting under water requires suitable diving gear and extensive training as a diver.

See Chapter 19 for more details.

P-29. GAS TUNGSTEN ARC CUTTING (GTAC)

In gas tungsten arc cutting, an arc between a tungsten electrode and the base metal heats the metal. The shielding gas then blows the melted metal away. This process is used on aluminum, stainless steel, nickel and many other metals. Since the electrode being used to melt the metal is tungsten, the electrode does not burn away rapidly.

Fig. P-29 illustrates a typical station for gas tungsten arc cutting.

Equipment includes a water-cooled electrode holder, a current supply and regulated supply of shielding gas, usually argon, helium or carbon dioxide.

The current supply is usually direct current straight polarity. The DCSP is furnished by an arc welder generator or rectifier. Current and voltage may

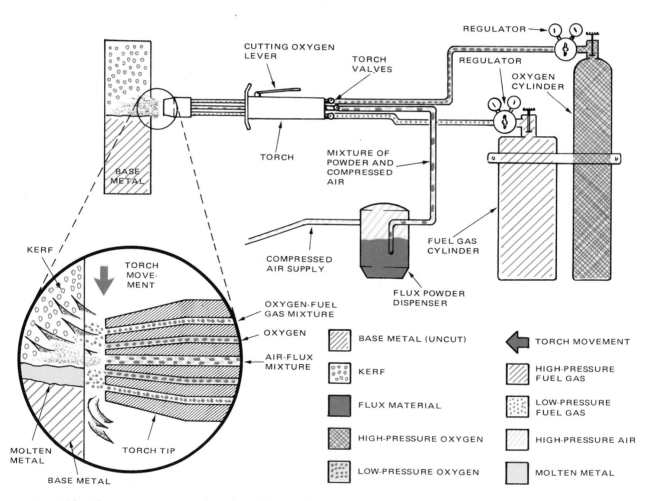

Fig. P-30. Flux oxygen cutting (FOC). A flame fed by heating orifices in the tip heats the metal. A torch lever is then pressed and an air jet mixed with a powdered chemical rapidly melts the base metal. The flux makes this melted metal very fluid and the air jet, plus the oxygen jet, blows the metal and flux away to form a cut (kerf) in the base metal. This process is used mainly for alloy steels.

be adjusted to job requirements.

The operator adjusts the flow of current, water and shielding gas. The cutting torch can be either manually or automatically operated.

Shielding gas serves two purposes. It blows the molten metal away from the cutting area and keeps the surfaces of the cut from oxidizing.

Gloves, helmet and protective clothing must be worn. There is considerable sparking with this type of cutting.

See Chapter 19 for more details.

P-30. FLUX OXYGEN CUTTING (FOC)

Powdered flux cutting uses an oxygen-fuel gas torch to heat the base metal. This process is used for cutting alloy steels, cast iron and nonferrous metals which form oxides with a very high melting temperature. Fig. P-30 shows a basic flux cutting station.

Powdered flux cutting requires an oxygen cylinder and a fuel gas cylinder, both fitted with regulators and gauges. Compressed air carries the flux powder to the torch tip. An oxygen jet flows against the heated metal to cut it.

In powdered flux cutting, flux powder is introduced into an oxyacetylene cutting torch flame. The flux powder decreases the formation of refractory oxides (solids). The molten metal is then easily removed to form a kerf.

A very similar process uses an iron powder. The iron powder increases the total heat of the flame. The iron in the powder also absorbs the oxygen in the area of the cut, reducing the alloy oxides. Almost the same equipment is used as in powdered flux cutting.

Because of the hazard of sparks, the operator must wear goggles, gloves, high shoes and protective clothing. It is advisable to post a fire watch when

cutting. The area must be cleared of combustible material.

See Chapter 19 for more details.

P-31. PLASMA ARC CUTTING (PAC)

Plasma arc cutting uses an electric arc and fast-flowing ionized gases to melt and cut metals. This process cuts aluminum, stainless steel and most other metals rapidly. Fig. P-31 illustrates a station for plasma arc cutting.

The metal is melted by the heat of the plasma arc. Then the molten metal is blown away by the high velocity of the shielding gas.

Plasma arc cutting requires a special water-cooled cutting nozzle. It makes use of a tungsten electrode connected to a source of DC power, compressed gas and suitable controls.

Plasma arc cutting is usually used along with automatic cutting devices. The current is controlled by devices on the power source. Water flow to cool the torch is usually manually adjusted by the operator.

Plasma arc cutting is a very noisy process. The operators must be protected from the noise by the use of ear plugs or industrial "ear muffs." It is sometimes necessary to use a "walkie-talkie" type of communication system where plasma arc cutting is being done. The operator must also be protected with an approved helmet, gloves, protective clothing and other required safety equipment.

See Chapter 19 for more details.

P-32. STEEL TEMPERATURE - COLOR RELATIONSHIPS

It is important to understand what happens when metals are heated, melt-

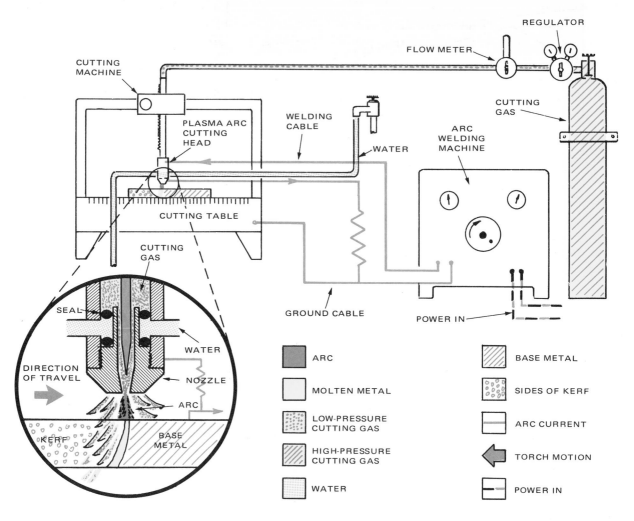

Fig. P-31. Plasma arc cutting (PAC). An electric arc between a tungsten electrode and the torch body ionizes some of the cutting gas. This ionized gas (plasma) leaves the torch and hits the base metal. The very high plasma temperatures superheat the base metal and very rapidly form a cut (kerf) in the base metal.

ed and then cooled. The physical properties of the steel are determined by these temperatures and actions.

There are thermometers which can accurately measure surface temperatures of steels (melting crayons, pyrometers and thermocouples). However, the color of the steel is a very popular and quite accurate way to judge the temperature. Fig. P-32 illustrates the color range for steel.

At low temperatures, steel looks gray to the naked eye. When heated above room temperature, the steel surface starts to oxidize. First the oxide coating is black (0 to 100 F.). Then the oxide surface turns blue (300 to 700 F.), purple, and then red. At 1000 to 1900 F., the red becomes brighter. The red blends into orange at 1700 to 2200 F. When it is bright yellow the steel starts to melt and become a liquid (welding temperatures).

Note that the temperature affects steel properties differently as the steel's carbon content varies.

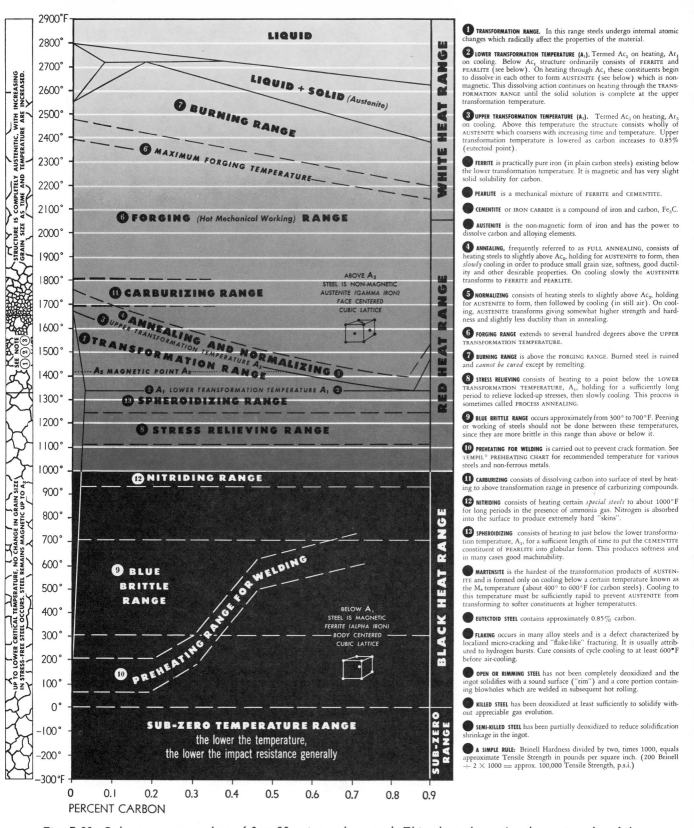

Fig. P-32. Color-temperature chart of 0 to 90 point carbon steel. This chart shows the change in color of the steel surface as the temperature changes from sub-zero to molten metal.
(Tempil° Div., Big Three Industries, Inc.)

Because metals are good conductors of heat, always use extreme caution when handling heated pieces of steel. Temperatures above 120 F. will cause severe burns. Protect the body by wearing approved goggles, leather gloves, and protective clothing. Handle hot metals with a pliers or tongs. See Chapter 25 for more details.

P-33. WELDING AND ALLIED PROCESSES

The American Welding Society recognizes 12 major heading for welding and allied processes. See Fig. P-33. Each process is explained in detail elsewhere. Refer to Chapter 29 for information on welding symbols.

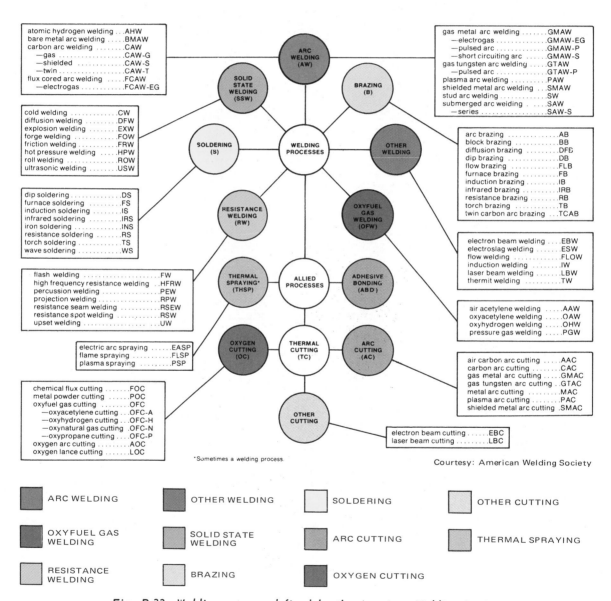

Fig. P-33. Welding process defined by the American Welding Society.

Chapter 1
OXYACETYLENE WELDING THEORY AND PRACTICE

1-1. DEFINITION OF WELDING

Welding may be described as a metal-working process in which metals are joined by heating them to the melting point, and allowing the molten portions to fuse or flow together.

1-2. SOLDERING AND BRAZING

Two other metal joining processes which are often confused with welding are Soldering and Brazing.

When two metals, which are not melted, are joined by a third metal which has a melting point below 800 deg. F., this process is called soldering. An example of soldering is the joining of copper to steel using a lead-tin alloy.

Brazing is done when two metals, which are not melted, are joined with a third metal which melts at temperatures above 800 deg. F. An example of brazing is the joining of two pieces of steel with a silver alloy.

Manual welding may be defined as an art. The skill to weld and solder metals together can only be obtained after a diligent study of the methods and after careful and correct practice. In classifying welding as an art, it is meant that some persons can do welding better than others because of a seemingly natural gift, although it has been found that any normal person can, under good instruction and by following the correct pro-

cedure, in time become a successful welder. By saying that manual welding is an art, it is further meant that continuous practice is necessary in order to maintain a high standard of skill in this kind of work. It is, therefore, recommended that only proper equipment be used when learning welding, that only the proper metals be used for practice, that a thorough, fundamental procedure be followed, and that very careful supervision be given the trainee during the first practice sessions in order to correct early mistakes.

1-3. DIFFERENT TYPES OF WELDING AND CUTTING

The most common types of welding are: Gas Welding, Arc Welding, Gas Arc Welding, and Resistance Welding. Other types include: Atomic-hydrogen Welding, Thermit Welding, Cold Welding, Ultrasonic Welding, Electron Beam Welding, Friction Welding, Laser Welding, and Plasma Welding.

Two popular types of thermal cutting are Gas Cutting and Arc Cutting. All of these processes will be explained in detail in the chapters of this book. The oxygen-acetylene process will be studied first because:

(1) The fundamentals of gas welding include fundamentals important to most other forms of welding.

(2) The oxygen-acetylene process is

a popular manual welding process. It is slower and easier to control than some other processes.

Figs. 1-1 and 1-2 illustrate three methods of adding oxygen to a fuel gas to support combustion.

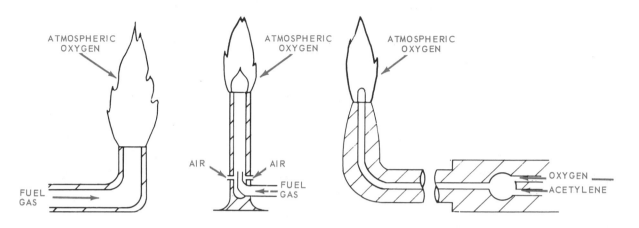

Fig. 1-1. Two methods of combining oxygen and fuel gases to produce fuel gas flames.

Fig. 1-2. In this application, pure oxygen under pressure is supplied to the fuel gases.

1-4. GAS WELDING

One of the most popular welding methods is to use a gas flame as the source of heat. This flame is produced by burning a fuel gas in the presence of the oxygen from the air, or from a pure oxygen source, or both. Fuel gas flames may receive their oxygen in three ways:

1. From the surrounding atmosphere which:

    A. Gives lowest temperature.

    B. Is the least clean.

    C. Produces the least heat.

2. Air, containing oxygen, is drawn in through holes in the torch. This process:

    A. Gives a higher temperature.

    B. Is cleaner.

    C. Gives more heat.

3. By supplying pure oxygen under pressure to the fuel gases before they burn which:

    A. Gives the highest temperature.

    B. Is cleanest.

    C. Gives the most heat.

1-5. WELDING FLAMES

In gas welding, various metals are joined by melting and fusing. A very intense, concentrated flame is applied to the metal until a spot under the flame becomes molten and forms a liquid puddle. When two metals melt or puddle, and the molten pools run together, the edges of the two pieces

Fig. 1-3. Three oxygen-fuel gas flame adjustments. A—Carburizing flame has three distinct flame sections. B—Inner cone on oxidizing flame is sharp and pointed. C—Inner cone on neutral flame is smooth and rounded. Neutral flame is correct.

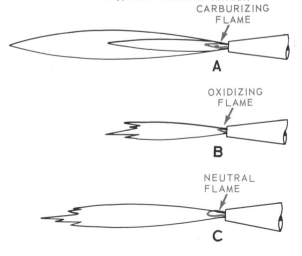

become one. This process must be performed carefully to minimize damage to the metals.

Certain conditions are necessary for a good weld:

A. Temperature of flame must be high enough to melt the metals.

B. Enough heat must be supplied to overcome the heat losses.

C. The flame must not burn (oxidize) the metal.

D. The flame must not add dirt or foreign material to the metal.

E. The flame must not add carbon to the metal.

F. The products of combustion should not be toxic (poisonous).

The quantity of heat is determined by the amount (cu. ft. per hr.) of gases burned. To obtain more heat, the tip orifice is made larger and more pressure is supplied to feed gas to the tip. Whether a large torch tip or a small torch tip is used, the temperature of the flame is the same.

It should be remembered that the amount of heat generated, and therefore the thickness of the metal which may be welded, will depend on the amount of fuel gas burned per unit of time. Therefore the amount of heat depends on the size of the torch orifice.

There are several commercially used gas welding and cutting flames:

1. Oxygen-acetylene (oxyacetylene), used for welding and cutting.

2. Oxygen-hydrogen, used for welding and cutting.

3. Oxygen-natural gas or artificial gas, used for cutting.

4. Oxygen-liquefied petroleum gas, used for cutting.

5. MAPP, used for welding and cutting.

6. Linde FG-2, used for cutting. For flame temperatures, see Par. 28-10.

1-6. THE OXYACETYLENE FLAME

Harmful oxidizing or carburizing flames result when the wrong proportions of the two gases, oxygen and acetylene, are used. If too much oxygen is used, an oxidizing flame results. If too much acetylene is used, a carburizing flame results. Fig. 1-3 shows the various flames. The two harmful flames are easily recognized. See Par. 1-13. The correct flame heats the metal and does not carburize or oxidize it. It is called a NEUTRAL FLAME. A neutral flame is the result of a perfect proportion and mixture of acetylene and oxygen. In a neutral flame, these two gases unite so that the oxygen burns up the carbon and the hydrogen in the acetylene and releases only heat and harmless gases. The colors of the FLAMES are shown in P-2C. (Refer to CHAPTER 28 for more on the chemistry of the welding flame.)

In chemical terms, acetylene + oxygen = carbon dioxide + water (vapor) + heat. The two gases formed, CO_2 (carbon dioxide) and H_2O (water in the vapor form), are considered harmless.

Oxygen in the air surrounding the flame is also used to complete the burning process. In crevices and corners, where the air has difficulty getting to the flame, additional cylinder oxygen must be fed to the flame.

The effect of an improper mixture of gases in the welding flame is easily recognized, and the final test for a neutral flame is determined by the manner in which the melting metal reacts to the flame.

Dirt in a welding flame may come from two sources:

A. Dirty gases.

B. Dirty equipment.

Good quality gases should always be used. The purity of the gases made by the manufacturers should be noted and taken into consideration. A neutral oxyacetylene welding flame will produce a temperature of 5,600 to 5,900 deg. F. An oxidizing flame will produce a slightly higher temperature.

Temperatures required to melt various metals are listed in the table, Fig. 1-4. An oxyacetylene welding flame, at temperatures of 5,600 deg. F. and up is high enough to melt the common metals.

METAL	MELTING TEMP. DEG. F.
Aluminum	1215
Brass (yellow)	1640
Bronze (cast)	1650
Copper	1920
Iron, Gray cast	2200
Lead	620
Steel (.20%) SAE 1020	2800
Solder (50-50)	420
Tin	450
Zinc	785

Fig. 1-4. Melting temperatures of some of the commonly used metals.

1-7. OXYGEN-ACETYLENE WELDING EQUIPMENT

Before discussing welding procedures, it is advisable to find out about welding equipment so its limitations and its possibilities may be kept in mind when learning the welding processes.

Basically, oxyacetylene welding equipment consists of a source of supply of two gases, oxygen and acetylene, and a mechanism where the gases are safely mixed and supplied to a torch tip, at which point they are ignited and a high temperature flame is produced. Fig. 1-5 shows a complete gas welding station. In the order of flow of the gases through the equipment, the following apparatus will be found in common use:

A. Gas cylinders; oxygen cylinder, acetylene cylinder.

B. Pressure regulators and gauges;

oxygen regulator, acetylene regulator.

C. Hose; oxygen hose, acetylene hose.

D. Welding torch.

There are two types of oxyacetylene welding torches is common use:

1. Equal-pressure type torch.

2. Injector type torch.

As the name implies, the equal-pressure type torch operates on approximately equal, or the same, pressure for both oxygen and acetylene. This type equipment is the most common.

The injector type torch is usually used in connection with an acetylene generator. The torch operates on a relatively low acetylene pressure and a much higher oxygen pressure.

See CHAPTER 2 for further details concerning the construction and operation of each of these torches.

1-8. ASSEMBLING OXYACETYLENE WELDING EQUIPMENT

Proper handling of welding equipment is essential to safety, and to obtaining good welds and a reasonable amount of service from the equipment.

Oxygen and acetylene cylinders are usually owned by the companies which furnish the gases contained in the cylinders. A small rental, or demurrage, is charged for the use of the cylinders after a reasonable rent-free period. The remainder of the equipment, however, is usually the property of the operator.

Because of high pressure in the oxygen cylinder, and the ease of combustion of acetylene, it is necessary that great care be used when handling cylinders. See PAR. 1-9 for further information on handling cylinders.

Always wear approved welding goggles when welding. See PAR. 2-25, for specifications of welding goggles.

Starting up and shutting down the equipment is the same regardless of the

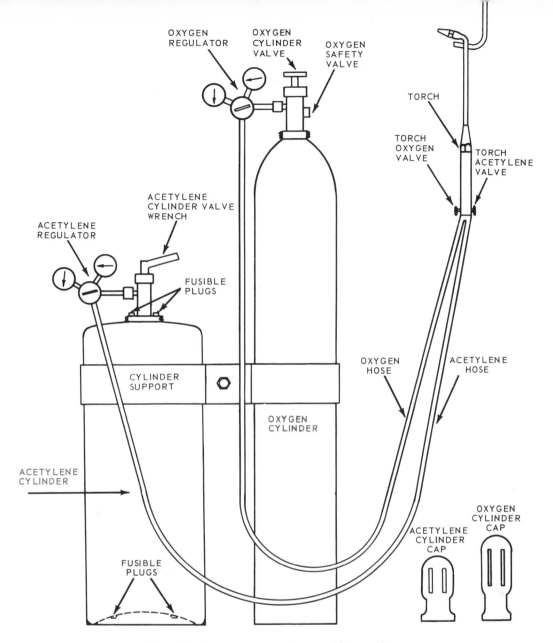

Fig. 1-5. An oxygen-acetylene welding outfit.

different operations. The recommended safety practices must be followed at all times.

Before welding equipment is used, it is very important to make sure the equipment is properly assembled. Check to see if the gas cylinders are in good condition. The cylinders should be fastened securely so there is no chance of upsetting or dropping them.

If the apparatus is portable, the cylinders should be fastened to the truck by

means of steel straps or chains. The truck should be so designed that it is almost impossible to upset it accidentally.

For stationary welding stations, the cylinders should be securely fastened to the wall, floor, or posts, by means of chains or steel straps. These safety devices should be of a design which permits quick change of cylinders.

Before attaching a regulator to a cylinder, the cylinder opening should be blown out, by opening the cylinder valve slightly (cracking) and allowing some of the high pressure gas to blow through the valve opening to remove dirt particles. Inspect the sealing surfaces and the fittings. Avoid using damaged or worn parts. The regulators may then be attached to the cylinder.

Only fixed-end wrenches, having wide jaws, provided for the purpose, should be used on the fittings. Be sure that the regulator nut fits the cylinder valve fitting properly. Fuel cylinder valves are usually fitted with left-hand threads. Oxygen cylinder valves are fitted with right-hand threads. The thread diameters on the two cylinder valves are also different. These are safety precautions as they make it impossible to change the regulators from one type cylinder to another, and thus mix the gases. Many different varieties of cylinder and regulator fittings are in common use.

The hose connecting the regulators to the torch should be fastened firmly to the fitting. It should be installed in a manner that does not twist the hose when the torch is held in the welding position. The torch when held in a welding position should not strain the operator's hand, or make it necessary for the operator to twist the torch to get it in place.

Before attaching the hose to the torch, the hose should be blown out. With the regulators attached, open the cylinder valves, then gently open and close the regulator valves, first the acetylene regulator, then the oxygen regulator. This brief purging will clear the hose lines. Where pipe threads are used, they should be sealed with pipe thread-sealing compound (such as glycerine and litharge paste) when assembling.

Following the hose purging, the torch should be attached to the hose. Note that on oxyacetylene welding equipment, the acetylene hose nipple nuts have left-hand threads, and the oxygen hose fittings have right-hand threads. Use only wide jaw, correct fitting wrenches. After the welding equipment is assembled, test for leaks.

Leak testing is a very essential checking procedure. This testing must be done when installing any new cylinder or any new part of apparatus.

To test for leaks, the recommended procedure is to put soapsuds on the outside of the joints suspected of leaking. OIL OR FLAMES OF ANY KIND SHOULD NEVER BE USED. To test for leaks (read PAR. 1-11 first), turn the regulator screws out all the way, open the cylinder valve, and build up from 5 to 15 lbs. of pressure in the regulator and hose by turning the regulator screw in (clockwise) slowly. Then, apply the soapsuds solution to the joints. Leaks, if any, will be indicated by bubbles.

When first using welding equipment, remember that the procedure is:

1. Learn the proper ways to prepare the station for use.

2. Learn the proper methods of igniting the torch.

3. Adjust gas flows for proper flame.

4. Learn proper method of shutting down the equipment.

The proper procedure for handling the torch and size tip to use are dependent on the type weld desired, kind of metal used, thickness of the metal, and structure (shape and position) of the metal in general. Instructions are included in PAR. 1-10 and 1-11.

1-9. HANDLING FUEL AND OXYGEN CYLINDERS

Welding cylinders, when properly handled, are quite safe. Improperly handled, they may be very dangerous.

Cylinders should never be dropped or allowed to tip over. The cylinder cap, enclosing and protecting the cylinder valve, should always be screwed on the cylinder when the cylinder is not in use, or when it is being moved.

A cylinder in use should be firmly anchored in an upright position, and in such a way it cannot be tipped. Cylinders in storage should be kept at near room temperature. Avoid the storing or use of cylinders in extremely hot locations. Check on local community building and fire codes and see to it the cylinders are used and stored according to those codes.

numbers corresponding to number drill orifice size. When an operator becomes familiar with the operation of a certain manufacturer's torches and numbering system, it is seldom necessary to refer to orifice number drill sizes.

Since the orifice size determines the amount of acetylene and oxygen fed to the flame, the orifice therefore determines the amount of heat produced by the torch. The larger the orifice the greater the amount of heat generated.

For practice purposes with a balanced pressure type torch, the tip sizes shown in Fig. 1-6 should give satisfactory results.

METAL THICKNESS	DIAMETER WELDING ROD	TIP DRILL SIZE	PRESSURES OXYGEN-ACETYLENE	
1/16	1/16 - 3/32	60 - 69	4	4
1/8	3/32 - 1/8	54 - 57	5	5
1/4	5/32 - 3/16	44 - 52	8	8
3/8	3/16 - 1/4	40 - 50	9	9

Fig. 1-6. A table of welding rod sizes and tip sizes used to weld various thicknesses of metal. Sizes listed in this table are approximate and will give satisfactory results. The size of the piece welded will govern the choice. When welding small pieces use the smaller tip and welding rod. When welding larger pieces use the larger size tip and welding rod.

1-10. SELECTION OF WELDING TORCH TIP SIZE

Welding torch tip size is designated by a number stamped on the tip. The tip size is determined by the size of the orifice. There is no standard system of numbering welding torch tip sizes. Manufacturers have their own numbering system. For this reason, in this text, tip size instructions are given in orifice "number drill" size. Number drills consist of a series of eighty drills number one through 80. The diameter of a number one drill is .2280 in. The diameter of a number 80 drill is .0135 in. Note that the larger the number, the smaller the drill. See Chapter 28 for table of number drill sizes. See Fig. 28-1 for manufacturers' welding tip

If the torch tip orifice is too small, not enough heat will be available to bring the metal to its melting and flowing temperature. If the torch tip is too large, poor welds will result because the weld will have to be made too fast, the welding rod melting will be hard to control, and the appearance and quality of the weld will be poor.

1-11. LIGHTING EQUAL-PRESSURE TYPE TORCHES

To light the torch, turn on the gases and adjust them to the proper pressures for the tip size:

1. Check equipment for condition.
2. Inspect regulators. Turn adjusting screws all the way out (counterclockwise) before opening the cylinder valves.

This prevents damage to the regulator diaphragm. Stand to one side of the regulator when opening the cylinder valves. A burst regulator could cause severe injury.

3. Open the oxygen cylinder valve very slowly (counterclockwise) to prevent damage to the regulator diaphragm from the pressure and heat of 2000 psi. When the regulator high-pressure gauge reaches its highest reading, turn the cylinder valve all the way open. This is necessary because the oxygen cylinder valve has a double seat or back seat. In the all-out position, this seat closes any possible opening along the valve stem through which the high-pressure oxygen could escape. (The operator should stand to one side of the gauges while opening the cylinder valve to avoid injury from a burst gauge.)

4. Slowly open the acetylene cylinder valve 1/4 to 1/2 turn counterclockwise. Use the proper size wrench and leave it on the valve in case an emergency shutoff is needed.

5. Adjust the oxygen torch pressure. Open the torch oxygen valve one turn. Turn the oxygen regulator adjusting screw in (clockwise) until the low-pressure oxygen gauge indicates a pressure corresponding to the size of the tip orifice. The flow pressure is always lower than the pressure when the torch valve is closed. Then turn off the oxygen valve on the torch. To avoid needle damage, use only finger tip force to close the torch valves. Too much force could destroy valve seats also.

6. Open the acetylene torch valve one turn. Turn the acetylene regulator adjusting screw in slowly (clockwise) until the low-pressure acetylene gauge indicates a pressure corresponding to the tip size. An approximate setting may be arrived at as shown in the table, Fig.

1-6. Turn off the acetylene torch valve using finger tip force only. The regulator pressures have now been adjusted to approximately the proper levels.

7. PURGE THE SYSTEM BEFORE LIGHTING THE TORCH. To ensure that the proper gases are in the respective hoses (no air or oxygen in the acetylene hose and no acetylene in the oxygen hose), the system must be purged. This is done by allowing the acetylene to flow through the acetylene hose and oxygen to flow through the oxygen hose for a short time before lighting. Note: If Steps 1 through 6 are followed, the system will have been purged and the torch is ready for lighting.

8. After purging, crack the acetylene torch valve no more than 1/16 turn. Using a flint lighter, ignite the acetylene gas coming out of the tip.

9. Next, turn on the acetylene torch valve slowly until the acetylene flame jumps away from the end of the tip slightly. This indicates that the proper amount of acetylene is being fed to the tip. A quick flip of the torch should make the flame leap away from the tip 1/16 in. and come back again. If the flame will not move back to the tip, too much acetylene has been turned on. (If the tip is worn, it may be difficult to make the flame jump away from the tip.)

Another method for determining the correct amount of acetylene is to increase the flow until the flame becomes turbulent (rough) a distance of 3/4 to 1 in. from the torch tip. With the right amount of acetylene, the flame will no longer smoke, or release soot. Look at Fig. 1-7 and compare the flames.

10. After the acetylene is regulated, slowly open the oxygen valve on the torch. As the oxygen is fed into the

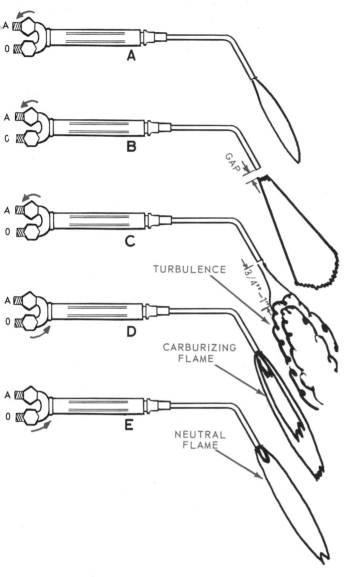

TURBULENCE

GAP

3/4"–1"

CARBURIZING
FLAME

NEUTRAL
FLAME

Fig. 1-7. Recommended steps for lighting oxyacetylene welding torch. A. Open the acetylene torch valve slightly and light the acetylene with a spark lighter. B. The correct amount of acetylene is flowing if the flame jumps away from the tip when the torch is shaken. Or, C. As shown here, a turbulence is created in the acetylene flame and the sooty smoke is eliminated. D. Begin turning on the oxygen by opening the torch oxygen valve. E. Continue to turn on the oxygen torch valve until the middle flame is eliminated and a rounded inner cone is seen.

flame, the brilliant acetylene flame turns purple and a small inner cone starts to form. This inner cone is light green in color. When first formed, the extremity of this inner cone will have a blurred and irregular contour.

As the oxygen is turned on slowly, the inner cone loses its blurred edge and becomes a round, smooth cone. Stop the adjustment at this point. Any increase in oxygen will result in an OXIDIZING FLAME. (Too much oxygen will burn or oxidize the metal being heated.) The tip of this inner cone is the hottest part of the flame.

The correct quantities of gases for the smaller tip sizes may also be detected by listening to the torch flame. It should emit a soft purr, not a sharp irritating hiss, when correctly adjusted.

11. If the torch burns with an irregular contour (feather) to the cone, the flame is called a CARBURIZING FLAME. There is an excess of acetylene. But if the inner cone has a very sharp point and if it hisses excessively, it usually means that too much oxygen is being used. If the flame has a smooth inner cone, the flame is called NEUTRAL.

There is another method of adjusting the welding torch:

1. Turn on the cylinder gases as described previously.

2. Open the acetylene torch valve one turn. Slowly turn in the adjusting screw on the acetylene regulator. When the acetylene starts flowing, light the acetylene. Resume turning the acetylene regulator screw in until the acetylene is made to jump away from the torch, or the turbulence is correct as in the first method.

3. Open the oxygen torch valve one turn. Turn the oxygen regulator adjusting screw in slowly until enough oxygen is being fed to the torch to completely consume all the acetylene and a neutral flame is obtained. (The neutral flame is described earlier. Also see the color illustration on Page 29.)

METAL THICKNESS	TORCH TIP NUMBER DRILL SIZE	OXYGEN REGULATOR PRESSURE PSI	ACETYLENE REGULATOR PRESSURE PSI
1/16	56	8 - 20	5
1/8	53	11 - 25	5
1/4	48	12 - 23	5

Fig. 1-8. Table shows correct O_2 and acetylene pressures for welding different thicknesses of metal using an injector type torch.

4. This method may be used in place of the first method; but in cases where the operator uses a long hose, and where the hose is bent in many different directions, the pressure drop in the hoses may vary. Because of this the pressures at the tip may vary. The first method assumes the torch valves are in excellent condition and will not go out of adjustment. The operator may choose either of the two methods. Results of either methods are generally satisfactory.

1-12. LIGHTING INJECTOR-TYPE WELDING TORCH

The steps for lighting an injector-type torch, in general are:

1. Inspect the equipment to make sure all parts are in good operating condition.

2. Inspect the regulators. The adjusting screws of the regulators should be turned all the way out (counterclockwise), before the cylinder valves are opened. This prevents damage to the regulator diaphragm when the cylinder valve is opened. Never stand in front of the regulator when opening the cylinder valves.

3. Open the oxygen cylinder valve very slowly until the regulator high-pressure gauge reaches its maximum reading, then turn the valve all the way open. The oxygen cylinder valve is turned all the way out because this valve has a double seat or back seat. In the all-out position, this seat closes any possible opening through which the high-pressure oxygen might escape along the valve stem. (The operator should stand to one side of the gauges while opening the cylinder valve to prevent possibility of injury in case the gauge should burst.)

4. With an acetylene cylinder wrench, slowly open the acetylene cylinder valve a quarter to a half turn. Leave the wrench on the acetylene cylinder valve stem so the cylinder may be shut off quickly in an emergency.

5. To adjust the oxygen torch pressure, open the torch oxygen valve wide open (one to one and one half turns). Turn the pressure adjusting screw in on the oxygen regulator until the low-

Fig. 1-9. Oxyacetylene welding a butt joint. Steel plates are 5/16 in. thick. Note position and angle of torch flame and filler rod. (Dockson Corp.)

pressure (delivery) gauge registers the approximate pressure shown in Fig. 1-8. Then close the oxygen valve. Use only finger tip force to close the torch valves. Too much force will damage the needle valves.

6. Open the torch acetylene valve 1/2 turn. Turn in the pressure adjusting screw on the acetylene regulator until the low-pressure (delivery) gauge shows a pressure of 5 psi. Close the torch acetylene valve.

By following this procedure, the system will have been purged.

7. To light after purging, open the torch oxygen valve 1/4 turn. Open the torch acetylene valve 1/2 turn. Light the tip with a lighter. Open the torch oxygen valve wide (1-1/2 turns) and adjust the torch acetylene valve to the desired flame.

8. Should acetylene delivery pressure become so low that a delivery pressure of 5 psi can no longer be obtained, open the torch acetylene valve all the way, and turn the oxygen regulator pressure adjusting screw to the left until the flame shows an excess acetylene feather about four times as long as the inner cone. Then, adjust the torch acetylene valve to give the desired flame. Fig. 1-9 shows a butt joint being welded on a steel plate.

1-13. TORCH ADJUSTMENTS

The torch may be adjusted to produce the following flame characteristics:
1. Neutral flame.
2. Carburizing flame.
3. Oxidizing flame.
In general, the neutral flame is the one desired. However, in welding aluminum, in brazing, and in some other operations where oxidizing of the metals would interfere with welding, a slightly carburizing flame is often used.

Fig. 1-3 illustrates the appearance of each of these flames. While a slightly carburizing flame may be recommended for certain work, usually a neutral flame will do just as well. However, because of the slight fluctuation in gas pressures, it is difficult to maintain a perfectly neutral flame and it may vary from neutral to slightly oxidizing or carburizing. Therefore, in order to avoid the possibility of running into an oxidizing flame, a slightly carburizing flame is safer.

The torch may occasionally "pop" (backfire). This small explosion at the flame may be the result of several avoidable conditions. The most frequent cause is preignition of the gases. Some causes of backfiring are:

1. The gas is flowing out too slowly and the pressures are too low for the size tip (orifice) used. The gases are therefore burning faster (flame propagation) than they can flow out of the tip. This trouble may be corrected by adjusting to a slightly higher pressure for both the oxygen and the acetylene.

2. The tip may become overheated from overuse, from operating in a hot corner, or from being too close to the weld. Cool the tip.

3. The inside of the tip may have carbon deposits or a hot metal particle may be lodged inside the orifice. These particles become overheated and act as ignitors. Correct this by cleaning the tip. See PAR. 2-22, and 2-24.

Torch popping happens rarely but could occur when the inner cone of the flame is submerged in the puddle.

A FLASHBACK occurs when the gas burns back to the regulator. If this happens, the hose, the torch, and the regulators are usually damaged and should be replaced or overhauled.

Flashbacks occur from one of four conditions following:

1. Flame backs into the acetylene hose. When oxygen feeds back into the acetylene hose, a combustible mixture forms and a violent explosion may result. A clogged barrel or mixture passage, along with an excessive oxygen pressure, may cause this trouble.

2. In the oxygen hose, organic oxides form inside the hose. If the hose is heated to the ignition point, an explosion may result.

3. An excessive flow from the cylinder of acetylene may also cause a flashback. If the rate of flow is greater than the volume of gas released from the acetone in the cylinder, the gas, instead of burning outside the tip, will tend to burn inside the torch and hose.

4. Torch tip overheating will cause a type of flashback. The torch will be heard to pop. This may happen if the torch is operated in a hole where its heat cannot be removed by radiation.

1-14. SHUTTING OFF TORCH

If the operator wishes to leave the station for just a few minutes, it is only necessary to close the torch valves and lay the torch aside. However, if the equipment is not to be used soon the outfit should be completely shut down. To do this:

1. Close the hand valves on the torch, preferably the acetylene valve first (this is to eliminate the soot).

2. Close the cylinder valves (tightly).

3. Open hand valves on the torch.

4. Wait until both the high and low-pressure gauges on both the acetylene and oxygen regulators read zero.

5. Turn the adjusting screws on both the acetylene and oxygen regulators all the way out.

6. Close both hand valves on the torch (lightly) and hang up the torch.

These instructions may be followed

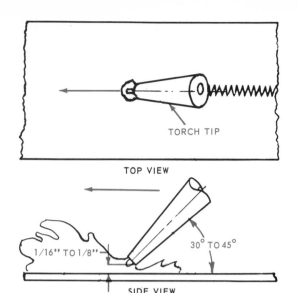

Fig. 1-10. Recommended torch angle, direction and flame distance from the metal when puddling in the flat position.

for both the balanced-pressure and the injector-type torch.

1-15. TORCH POSITIONS AND MOVEMENTS

Forehand welding, Fig. 1-10, provides that the torch be held at an angle of 15 to 75 deg. to the work. This will depend on the tip size used, metal thickness, and other welding conditions. A 30 to 45 deg. angle is typical. The flame spreads over the work in the direction of the weld preheating the metal before it comes under the high

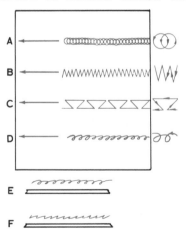

Fig. 1-11. Common types of flame movement patterns used when gas welding.

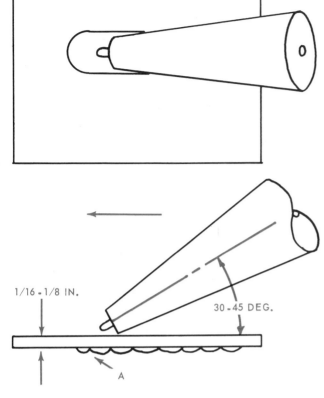

1/16 - 1/8 IN.

30 - 45 DEG.

A

Fig. 1-12. Puddling procedure. This illustration shows the correct position of the torch in relation to the base metal during a puddling exercise. Detail A shows the penetration (sag) below the under surface of the base metal.

temperature flame.

Use either the oscillating or the circular torch motion. In either case, the cone should never go outside of the puddle. Fig. 1-11 shows different torch motions used by welders.

1-16. PUDDLING

Before attempting a weld of any kind, it is recommended that the beginner practice "puddling." Puddling is an important and fundamental part of welding, because in most welding operations a molten puddle of metal is carried along the seam of the parts to be welded together. This melting is true in most forms of welding, both gas and electric arc. The characteristics of the puddle of molten metal indicate the penetra-

tion, torch adjustment, torch handling and torch movement. These puddle characteristics, which are judged by watching the condition of the puddle, guide the experienced welder when producing a good weld.

The size (diameter) of the puddle will be in proportion to its depth; therefore, the operator may judge the depth, or penetration, of a weld by watching and controlling the size of the puddle of molten metal. On very thin metal, the penetration or depth of the puddle will be greater in proportion to the width, than will be the case with thicker metal.

The appearance of the surface of the puddle will indicate the condition of adjustment of the torch. The neutral flame, when melting a good grade of metal will give a smooth, glossy appearance to the puddle. The edge of the puddle away from the torch will have a small bright incandescent spot, which will move actively around the edge of the puddle. If this spot is oversize, the flame is NOT NEUTRAL. Also, if the weld puddle bubbles and sparks excessively, there is either a poorly adjusted flame; or, a poor quality and/or dirty metal is being welded. The puddle will have a dull and dirty (soot) appearance, if the flame is carburizing to any great extent.

The tip of the inner cone of the torch flame must be held within the boundary of the puddle at all times. The correctly adjusted flame over the puddle prevents the oxygen in the atmosphere from coming in contact with the surface of the puddle and causing an oxidizing condition.

Hold the torch far enough away from the puddle surface to keep the tip of the inner cone from touching the puddle. The tip of the inner cone should be held 1/16 - 1/8 in. from the puddle surface as shown in Fig. 1-12. If the puddle sinks or sags too far, indicating too much penetration, lower the angle of the torch rather than draw the torch away from the surface of the puddle. Fig. 1-13

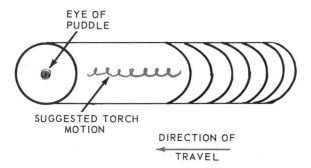

Fig. 1-13. *Puddling procedure. This illustration shows the puddle in progress.*

shows a puddle in progress. Note the width of the puddle and the torch movement which controls the width of the puddle. Note also the eye of the puddle which is a bright flake of oxide which indicates the movement of the molten metal. A partially completed puddling practice piece is shown in Fig. 1-14. Note the side view showing the penetration.

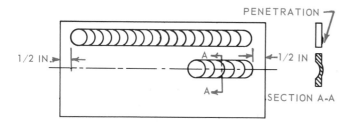

Fig. 1-14. *Puddling procedure. This illustrates a partly completed puddling exercise piece. Note the penetration. This operator has obtained uniform width and has carried the puddle in a straight line, and the penetration is uniform.*

Before starting to practice with welding rod, the learner should be able to produce five consecutive beads by puddling, each at least 5 in. long, without melting any holes through the metal,

and at the same time secure good penetration. The beads must also be straight (in line) and even in width. Beginners who can do this have become familiar with torch operation. They now know about the theory and practice of the weld puddle sufficiently to proceed with learning the manipulation of the welding rod.

1-17. TYPE OF WELDS MADE WITHOUT THE USE OF WELDING ROD

An interesting exercise in welding sheet metal, which is somewhat different because no welding rod metal is needed, is called the OUTSIDE CORNER JOINT WELD. This exercise teaches how to weld by using some of the parent metal as the filler metal. The pieces are placed one against the other, at right angles so the vertical piece extends beyond the surface of the horizontal sheet approximately 1/32 to 1/16 in. as shown in Fig. 1-15. The two pieces are then tacked together at their ends. The extended metal serves as the filler metal.

The weld must have good penetration, but the penetration should not show on the inside corner. The operator will find that very little torch motion is needed for this exercise. Further, the torch tip should be slightly tilted, making the flame point inward toward the flat or horizontal surface. The weld should be all on the horizontal surface. None of it should run over on the vertical edge. This is necessary, because in many cases of metal finishing, the weld is made into a right angle corner by grinding the excess metal from the one side. After checking the weld for appearance, the penetration may be tested by bending the two pieces of metal open like the pages of a book. Any cracking or breaking of the metals at the joint will indicate a lack of penetration or fusion.

Another good exercise to help in learning the use of a welding torch when welding without welding rod, is the FLANGE WELD.

To prepare the metal for this weld,

be classified according to the type of joint and position of the weld.

The basic joints are:
1. Butt.
2. Lap (fillet weld).
3. Outside corner.
4. Inside corner (fillet weld).

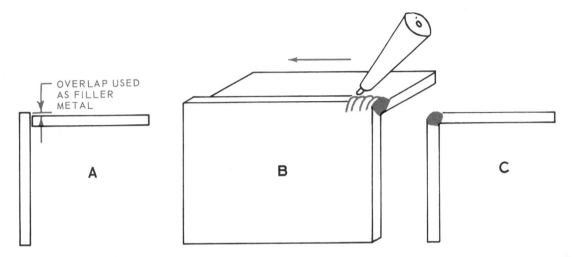

Fig. 1-15. Steps in performing the outside corner joint without a welding rod. A. Metal in position to weld. B. Weld in progress. C. Appearance of the finished weld.

bend a 90 deg. flange about 1/4 -1/2 in. long on two pieces of sheet steel about 1/32 to 1/16 in. thick. Be sure the lengths of the two flanges are equal. Place the flanges together along their lengths, and fuse the two flanges with the welding torch using the flanges as the filler metal in the same manner as was done when welding the outside corner joint. Fig. 1-16 shows a flange joint weld in progress.

1-18. TYPES OF OXYACETYLENE WELDS MADE WITH WELDING ROD

To become proficient in the art of gas welding, certain fundamental exercises must be planned and practiced until satisfactory welds can be performed consistently. The different fundamental gas welding operations may

The basic welding positions are:
1. Horizontal on a horizontal surface (flat or downhand position).
2. Horizontal on a vertical surface.
3. Vertical on a vertical surface.
4. Overhead.

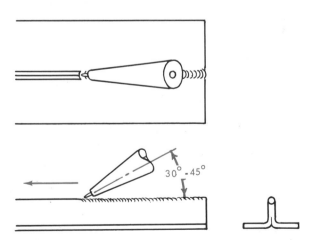

Fig. 1-16. A flange joint weld in progress. Note that the base metal is used as the filler metal, therefore a welding rod is not needed.

The welding of each of the above joints should be practiced in each of the above positions. Fig. 1-17 illustrates

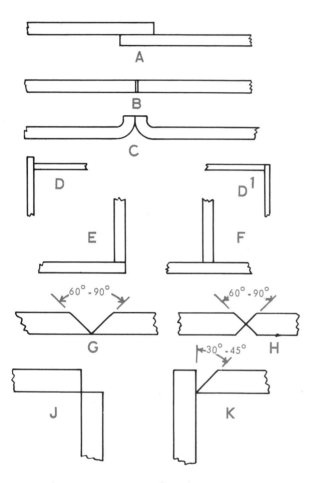

Fig. 1-17. Some typical welding joint designs; A. Sheet steel lap joint in the flat position. B. Sheet steel butt joint in the flat position. C. Flange joint in the flat position. D and D¹ Outside corner joints. E and F. Inside corner joints (F is sometimes called a T-Joint.) G, H, J and K Joint designs for metal plate. Note that when welding joints A, B, and D¹, welding rod is used as the filler metal. When welding the joints at C and D no welding rod is required as a filler metal, because the metal pieces themselves are melted to form the bead and to join the pieces together.

some typical types of joints. Welding of these joints should be performed on both thin sheet steel and finally on steel plate of at least 3/8 in. in thickness. After obtaining the necessary skill on steel plate with these exercises, the welder may then proceed to study special welding applications, i. e., pipe welding, aluminum welding, cast iron welding, etc.

1-19. USE OF WELDING ROD IN RUNNING A BEAD

Puddling without the use of welding rod is used mainly on outside corner or flange joints. WELDING ROD (FILLER METAL) IS ADDED WHEN EXTRA METAL IS NEEDED TO CREATE THE CORRECT WELD SHAPE AND STRENGTH. Welding these joints with the puddle method only causes a thinning of the metal in the weld area. In order to obtain a strong weld, metal from a welding rod is fed into the puddle to increase its depth and to increase the thickness of the metal at the weld. See PAR. 2-29 for specifications of welding rods. The bead in a correctly made weld should be slightly crowned (convex) for extra thickness and strength of the weld. Fig. 1-18 illustrates a method by which welding rod material is added to a weld to make a slight crown to the bead (face of weld).

To weld using welding rod, bring the torch to that part of the joint where the weld is to start. Melt a small puddle on the surfaces of the two pieces and allow the puddles to flow together. At the same time, with the other hand, bring the welding rod to within about 3/8 in. of the torch flame and 1/8 in. from the surface of the puddle. In this position, the welding rod will become preheated, and it will melt sooner when it is dipped into the puddle. When the operator judges that the puddle needs additional metal, the end of the welding rod is inserted into the puddle, and some of the welding rod will melt and mix with the molten parent metal. Enough welding rod metal should be added to raise the puddle to a slight crown. At the same time, continue the torch motion without interruption. Torch control at this time is very important. Control over puddle condition and welding rod melting can be done by small changes of torch position. As soon as enough welding rod metal has been added, withdraw the welding rod to the slightly withdrawn position described previously which should maintain the end of the welding rod in a preheated condition.

If the welding rod is withdrawn too far from the torch, it will become too cold, and will cool and chill the puddle

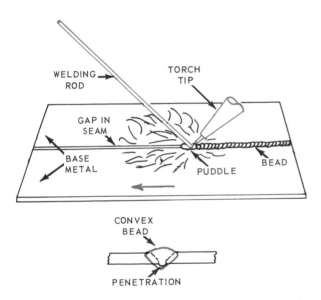

Fig. 1-18. Recommended torch and welding rod positions for welding a butt joint in the flat position.

when it is again inserted into it. On the other hand, if the welding rod is held too close to the flame of the torch, it will become too hot. If it should become molten, drops of molten welding rod will be blown by the flame upon the cooler parts of the metal being welded. Such a condition will result in a very uneven bead, poor fusion, and probably poor penetration.

The beginner is sometimes tempted to use different size welding rods for the same weld. This change should be avoided because:

1. If 3/32-in. welding rod is correct, and if a change is made to 1/16-in. welding rod it will be extremely difficult to add enough welding rod to obtain a good weld.

2. The smaller rods will make it more difficult to control the puddle.

3. There will be a tendency to burn (oxidize) the smaller size welding rod.

If the operator changes to 1/8-in. welding rod when 3/32-in. welding rod is correct, the following troubles might result:

1. The larger welding rod will cool the puddle too much while it is being added and prevent consistent welding.

2. There will be a tendency to add too much welding rod, destroying penetration and building the weld higher than it should be on the top surface.

Whether to use a 1/16, 3/32 or 1/8-in. welding rod is not so important as it is to adhere to one size after becoming used to it for a certain thickness of metal and a certain size tip. See Fig. 1-6 for a table of recommended welding rod sizes and CHAPTER 2 for more detailed information on welding rod choices.

It should be pointed out that a good weld, meaning good fusion, good bead, and good penetration, is obtainable only by attaining skill in the handling of the welding torch and the welding rod simultaneously and in harmony one with the other. This means that the torch motion should be constant in forward speed, in the width of the motion, and with the proper distance between the cone and the metal. The slant of the torch in respect to the surface should always be the same. The filler metal additions should be made at consistent regular intervals.

1-20. BUTT JOINT WELDING

Butt joint welding is one of the most common welds made with the oxyacetylene torch. The instructions which follow will aid the beginner in making this weld on thin steel.

Procure two pieces of mild steel approximately 1 in. wide and 5 in. long. The pieces should be clean, flat, and the edges should be straight. Place the two pieces of metal across two firebricks, permitting the bricks to support the

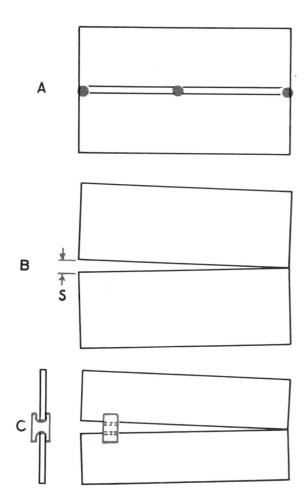

Fig. 1-19. Some methods used to maintain correct position of welded pieces, since the weld metal shrinks as it solidifies and cools: A. "Tacking" pieces together before welding. B. Allowing for shrinkage (S). C. Use of special wedges.

ends of the metal. Place the edges of the two pieces of metal together at the end where the weld is to start. As the weld proceeds along the joint, the metal shrinks in cooling from a molten state and tends to pull the two pieces of metal together. This shrinkage may cause the edges to lap one over the other, or warp the metal. The operator may prepare the metal for this expansion and contraction by:

1. Fusing (or tacking) the ends of the two pieces of the metal together before proceeding with the welding as in Fig. 1-19, Part A. This method will produce some internal strain, but will keep the ends sufficiently in line to enable the operator to make a good weld. Tack every inch or so along the joint.

2. Tapering the gap between the two pieces of metal to allow for contraction as in Fig. 1-19, Part B. The approximate contraction is from 1/8 to 1/4 in. per foot of length (the wider the puddle, the greater the contraction).

3. Using especially prepared wedges which may be placed between the two pieces of the joint to prevent the contraction of the metals as the weld cools as in Fig. 1-19, Part C. This method is more generally used with long joints.

4. Clamping the metal in a heavy fixture to minimize movement.

Refer to Fig. 1-18 for information concerning the correct position of the torch and the welding rod.

Light the torch, adjust it to a neutral flame, and proceed as follows: Bring the torch to the point where the weld is to start, holding the torch at a 75 deg. angle with the tip pointing along the direction of the joint. Hold the inner cone approximately 1/16 in. away from the metal. With the other hand, bring the end of the welding rod approximately 3/8 in. away from the welding torch

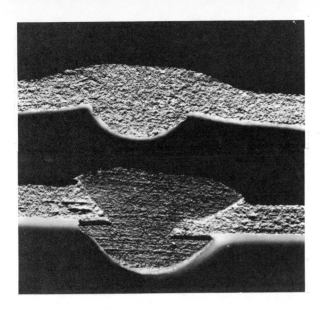

Fig. 1-20. Macrographs (4X) of two butt joint welds in mild steel. The metals have been etched to show the grain and fusion. The top weld used filler metal of the same composition as the base metal; while the bottom weld used a commercial welding rod as the filler metal. Note that the penetration in both cases is excessive in relation to the thickness of the base metals.

and just above the metal (about 1/8 in.). The torch flame will melt a puddle on the edges of each of the pieces of metal which should extend equally over the two pieces of metal. Apply the welding rod metal as directed in PAR. 1-19.

Advance the torch a very short distance, as the motion continues, until the puddle again reaches the size of the previous puddle. The welding rod is again dipped into the puddle and the puddle is built up to a crown. Continue this procedure throughout the length of the weld joint, using a continuous torch motion, Fig. 1-11.

The tip should be kept at a uniform distance from the weld, and the torch angle with the metal should remain unchanged. Welding rod metal should be added in uniform amounts, and at regular intervals. After the weld has been finished, allow it to cool, and then inspect it. Fig. 1-20 is a magnified photograph of a good butt weld in mild steel.

There is a type of butt joint welding in which the heat source penetrates completely through the workpiece. Called keyhole welding, it forms a hole at the leading edge of the weld metal. As the heat source progresses, the molten metal fills in behind the hole to form the weld bead.

1-21. LAP JOINT WELDING

A rather common welding procedure, used extensively in industrial work, is the WELDED LAP JOINT. The joint consists of lapping one piece of sheet metal over the one to which it is to be welded, as shown in Fig. 1-21. This

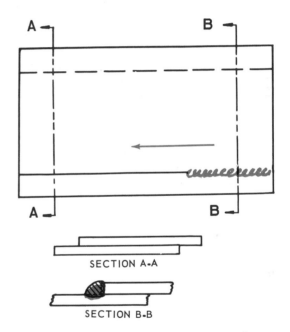

Fig. 1-21. Lap joint welding Section A-A shows position of the two pieces in relation to each other. The two pieces should be placed as close together as possible (no gaps). Section B-B shows the finished weld.

weld should be first performed in the flat position. Although the welding technique is typical, several things must be kept in mind in order to obtain a satisfactory lap weld:

1. It will be found difficult to heat the

bottom piece of metal to a molten state before the top metal edge melts too much, making the weld very ragged. The way to prevent this is to concentrate the torch flame on the lower surface. The bottom piece of metal requires two-thirds of the total heat, as shown in Fig. 1-22.

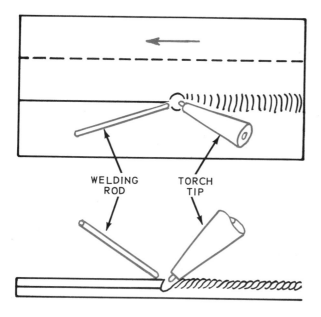

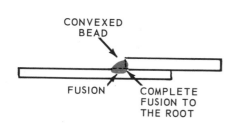

Fig. 1-22. *Recommended position of the torch when fillet welding a lap joint. Note the position of the torch and welding rod in relation to the base metal.*

2. The welded portion (the weld nugget) must be at least as thick as the original metal. To provide this thickness, add enough welding rod metal to crown the weld slightly.

The beginner usually has a tendency to perform this weld without heating

the bottom metal sufficiently to obtain fusion. The destructive separation test (CHAPTER 22) will quickly show the lack of fusion at this point. Before testing the specimen, the appearance of the weld should be closely inspected for such things as an even bead, consistency of width, cleanliness, and good general appearance. The metal should not sag on the reverse side of the bottom piece (too much penetration), and the bead should be straight.

A special form of lap welding is known as PLUG WELDING. The American Welding Society defines a plug weld as follows:"Plug welding is a method of attaching a lining by welding it to the base plate through holes punched or cut through the lining. The holes which vary in size according to the thickness of the liner are filled with weld metal which bonds to the base plate and to the liner around the edges of the hole." See Fig. 1-23 for an example. The torch flame

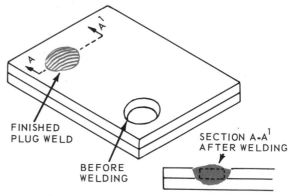

Fig. 1-23. *A plug weld. This is a special application of lap welding.*

should be concentrated on the base plate in order to bring it to its melting temperature at the same time that the edges of the hole in the liner melts.

In aircraft tubing repairs, the plug weld is known as a ROSETTE WELD.

A SLOT WELD is the same in all respects as a plug weld, except that instead of a hole being drilled in the liner metal, an elongated hole or slot is cut.

Fig. 1-24. In this method, the two pieces of metal do not overlap. In the welding operation, the edges of both

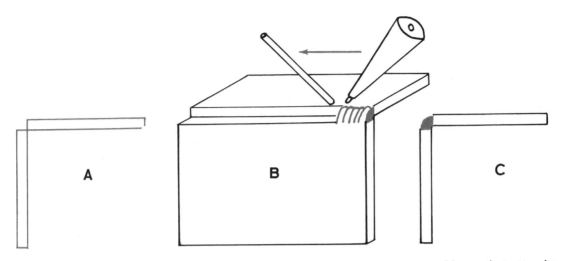

Fig. 1-24. *Recommended steps when welding an outside corner joint using a welding rod: A. Metal is in position. B. Weld is in progress. C. Weld is completed.*

1-22. OUTSIDE CORNER JOINT WELDING

Outside corner joints may be made without using welding rod (see PAR. 1-17). An alternate method of performing an outside corner weld is shown in

Fig. 1-25. *Cross sections of outside corner welds (a 4X macrograph) show a properly fitted weld at the top and a joint with a poor fit-up at the bottom. Metal has been etched to emphasize grain lines and fusion.*

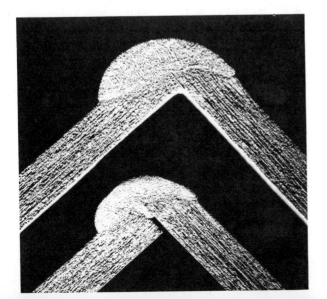

pieces of metal are melted and the joint is strengthened by the addition of filler metal. Fig. 1-25 shows a macrograph of a completed outside corner joint made with welding rod.

1-23. INSIDE CORNER AND T-JOINT WELDING

This is a fairly easy exercise to perform, but one in which the operator may find it difficult to obtain sufficient penetration. Two pieces of 1/16-in. stock are placed with their surfaces at right angles to each other, either in an L-formation (corner joint) or in an inverted T-formation (T-joint), and one inside corner is welded as shown in Fig. 1-26. When welding in this position, the torch flame is placed inside of the L. As it is difficult to obtain additional oxygen from the air to complete the combustion of the acetylene, when welding in this position, it is sometimes necessary to open the oxygen torch valve slightly to provide enough oxygen

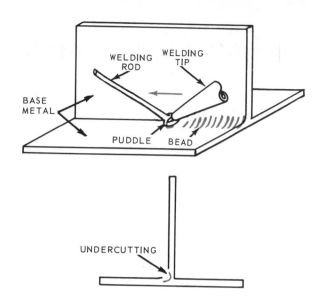

Fig. 1-26. Fillet welding of a T-joint in horizontal position.

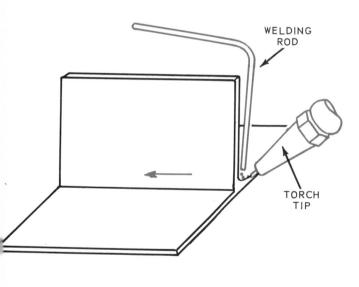

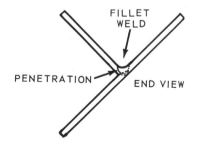

Fig. 1-27. Fillet welding of a T-joint in flat (down hand) position.

for complete combustion. This adjustment would result in an oxidizing flame under normal circumstances, but in this special case, it produces a neutral flame. The exercise may be set up in two different ways:

1. The preferred method is to have the two pieces set at an angle of 45 deg. from the horizontal, forming a trough as shown in Fig. 1-27. The operator will find that the exercise is easier with the specimen set up in this manner. This position means that the joint is in a 45 deg. position, and that the weld is in the flat position. The two pieces should be tacked at both ends of the joint before making the weld.

2. A second method is to have one piece horizontal and the other vertical, with the edge of the vertical piece touching the middle of the surface of the horizontal piece. Before welding the pieces in this position, the operator should tack the ends to keep them in line while welding. It will not be necessary to provide for expansion and contraction, in this case, inasmuch as the metals are pulled one against the other as the weld cools and solidifies.

The technique of handling the torch and welding rod will be almost the same in this exercise as in the previous one. The operator will find that very little torch motion is necessary; also, it is very important to produce a puddle before attempting to add the welding rod; otherwise, insufficient penetration will result. To secure fusion in the vertex of the weld, the torch should be held as close as possible without allowing the inner cone of the flame to touch the metal, then draw it back slightly when adding the welding rod. With the joint in this position the operator is actually making a horizontal weld on a vertical surface (the vertical piece). After the weld is completed, inspection should

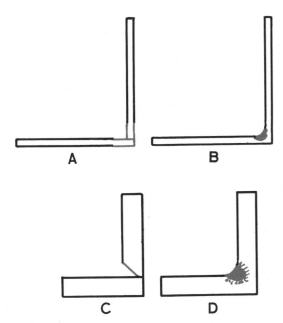

Fig. 1-28. End views of some popular corner joint preparations for fillet welds. A. Thin metal corner joint in position for horizontal fillet welding. B. Completed fillet weld or a thin metal corner joint. C. Thick metal single bevel groove corner joint. D. Completed fillet weld on a thick metal single bevel groove corner joint.

show a good bead, good fusion, consistent width of the weld, a clean appearance, and an equal distribution meaning that half of the weld is on one piece, and half on the other. Also, the vertical piece of metal especially should not show any indications of having some of its metal melted away leaving it thinner (undercutting).

To test the weld for penetration, the two pieces of metal are closed together like the pages of a book. Lack of penetration, or fusion, is indicated by the added metal peeling away from the parent metal. Fig. 1-28 shows some alternate setups for inside corner joint designs while Fig. 1-29 illustrates some typical T-joint designs.

1-24. WELDING POSITIONS

Welds made in the flat position (downhand) are the easiest to make.

However, it is often necessary to make welds in various other positions.

The recognized positions are:

FLAT POSITION WELDING--The position of welding where welding is

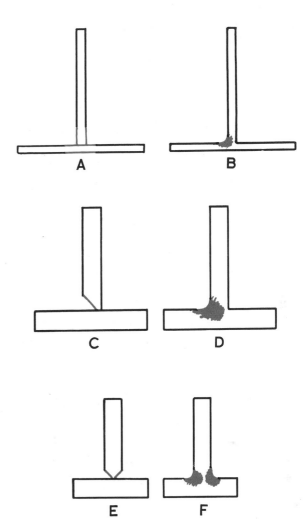

Fig. 1-29. End views of some popular T-joint preparations for fillet welds. A. Thin metal T-joint. B. Completed thin metal T-joint fillet weld. C. Single bevel groove T-joint on thick metal. D. Completed fillet weld on a single bevel groove T-joint. E. A double bevel groove T-joint in thick metal. F. Completed fillet welds in a double bevel groove T-joint.

performed from the upper side of the joint, and the face of the weld is approximately horizontal.

HORIZONTAL POSITION WELDING--Fillet Weld--The position of

welding where welding is performed on the upper side of an approximately horizontal surface, and against a surface which is approximately vertical.

GROOVE WELD--The position of welding where the axis of the weld lies in a plane that is approximately horizontal and the face of the weld lies in a plane that is approximately vertical.

VERTICAL POSITION WELDING-- The position of welding were the axis of the weld is approximately vertical.

OVERHEAD POSITION WELDING-- The position of welding where welding is performed from the underside of the joint, where the axis of the joint and base metal are both approximately horizontal.

1-25. OXYACETYLENE WELDING IN HORIZONTAL POSITION

This position consists of welding a horizontal joint on a vertical surface as shown in Fig. 1-30. Welding in this

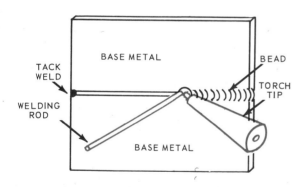

Fig. 1-30. Horizontal butt joint weld on a vertical surface. Note that the torch flame is directed upward to help hold the molten metal in place.

position should be practiced on butt joints, inside corner joints, and lap joints. A precaution to be noted in order to produce an excellent weld is that the torch tip, instead of pointing directly along the weld joint, should point at a slightly upward angle. This tip po-

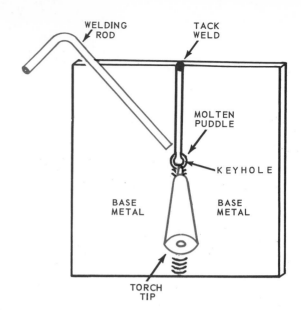

Fig. 1-31. A butt joint being welded while in a vertical position. Note the position and angle of the welding rod and torch tip with respect to the weld.

sition will enable the force from the velocity of the gases to keep the molten metal from sagging. It will be found that it is easier to obtain more consistent penetration, in both the horizontal position and the vertical position welding, than in flat position welding. The operator will have little difficulty in obtaining a good looking bead, but he will find it more difficult to make a vertical or a horizontal weld with as good an appearance as a horizontal weld on a horizontal surface (downhand).

1-26. OXYACETYLENE WELDING IN THE VERTICAL POSITION

Vertical welding consists of welding in a vertical direction on a vertical surface. The exercises should be performed in a vertical position, including the butt joint, Fig. 1-31. The welding will not be found difficult after the first few attempts if the following precautions are observed:

1. The weld should proceed upward. The torch is inclined from the surface of the metal at an angle of approximately 15 to 30 deg. The torch tip is pointed

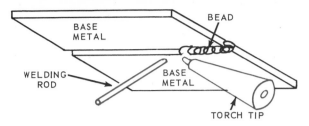

Fig. 1-32. A butt joint being welded while in the overhead position. Caution: Wear gloves, cap, and long sleeves. Be sure all clothing is fire resistant.

up. This tip position enables the force from the gas velocities to keep the molten metal from falling, or sagging (due to gravitational pull).

2. A very small torch motion should be maintained to enable the gas velocities to keep the molten metal continually in position, and not allow it to sag.

3. Bead, fusion, and penetration should be checked according to the methods previously recommended.

1-27. OXYACETYLENE WELDING IN THE OVERHEAD POSITION

Overhead welding is the most difficult welding position. The operator should be skilled in welding in all other positions before starting overhead welding as shown in Fig. 1-32. This requires consistent, diligent practice to reach a satisfactory degree of skill. The operator should protect the body with fire resistant work clothes and gauntlet gloves. He or she should wear high-top shoes or welding spats when performing this exercise. Trousers should have covered pockets and no cuffs. Overhead welding should be practiced on all the standard joints. There should be at least two excellent samples of each type of joint before practice welding is concluded. The exercise should be mounted approximately from 6 to 12 in. above the operator's head to be most comfortable, and the welder should stand to one side of the seam, welding paral-

lel to the shoulders.

The position of the welding rod and the torch is almost the same in reference to the puddle as in welding in other positions. The force of the torch flame gas velocities, plus the surface tension (the attraction of the molecules for each other), overcomes to a great extent the pull of gravity on the molten metal. However, be careful to keep the metal as close to its lowest flow temperature as possible. Any superheating of the metal produces a more fluid condition and may cause the molten metal to fall.

A quick removal and return of the flame to the puddle (flip action) and good timing when inserting the filler rod in the puddle will result in welds of comparable quality to flat position welds.

1-28. JOINT PREPARATION FOR PLATE WELDING

Welders should be able to successfully weld both thin and thick metal (plate). The American Welding Society has standardized the names of the joints and the names of those parts of the joint which are in the weld zone. Fig. 1-33 shows standard terms for various parts of a groove weld.

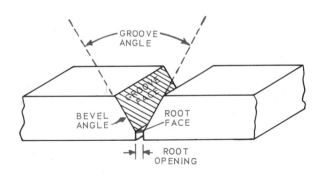

Fig. 1-33. Standard terms for the various parts of a plate joint.

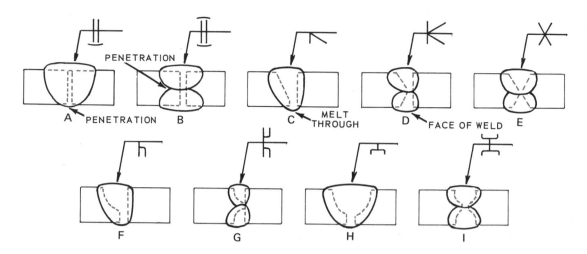

Fig. 1-34. Methods used to prepare metal edges before butt welding. Method depends on metal thickness and whether it can be welded from both sides. A—Uncut edges welded from one side. B—Uncut edges welded from both sides. C—Single bevel groove. D—Double bevel groove welded from both sides. E—Double V groove welded from two sides. F—Single J groove welded from one side. G—Double J groove welded from both sides. H—U groove welded from one side. I—Double U groove welded from both sides. See Chapter 29 for explanation of weld symbols shown.

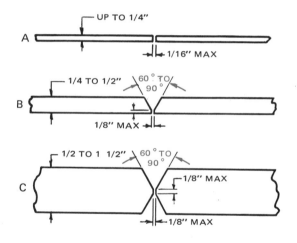

Fig. 1-35. Specific angles and dimensions for butt joints.

Several methods have been developed to prepare the edges of metal for welding as the metal increases in thickness. Fig. 1-34 shows edges prepared for various types of butt joints. The straight bevel and/or the V-joint preparation is most common. It can be done with ordinary shop tools. J and/or U groove preparation requires special equipment. The latter grooves are more economical in respect to quantity of welding rod metal used, gas or electricity cost, and in time saved.

Fig. 1-35 shows some specific angles and measurements used when preparing metal edges for butt joint groove welds.

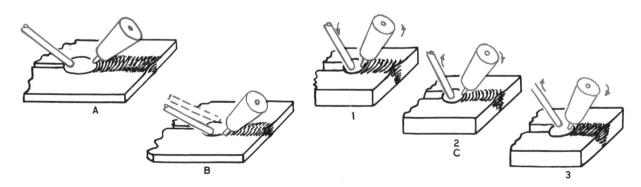

Fig. 1-36. Torch and welding rod positions and motions used when welding thicker metals, where wide puddles are required.

When oxyacetylene welding the plate joints, the welding rod motion and torch motion must be used in a related fashion. Fig. 1-36 shows how the welding rod is moved to one side of the puddle as the torch is brought to the other side. The oscillating motion is then reversed. This technique enables excellent control of the puddle as to fusion and build up.

1-29. BACKHAND WELDING

Backhand welding, as shown in Figs. 1-37 and 1-38, provides that the torch be held at an angle of 30 to 45 deg. with the work, and the flame directed back

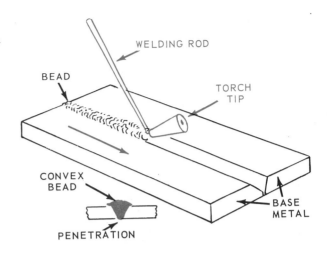

Fig. 1-37. Welding a single V-groove joint using backhand method.

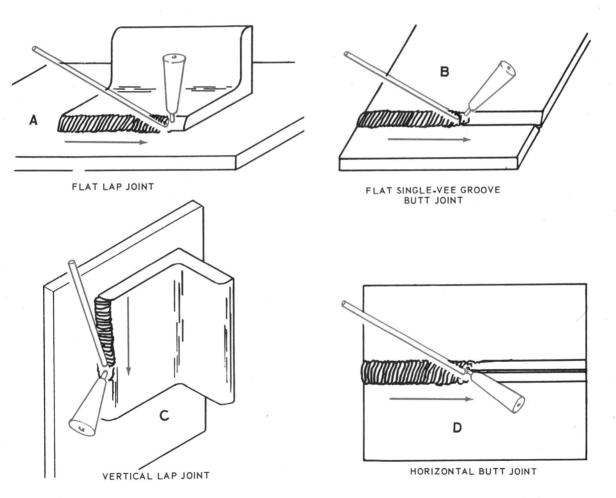

Fig. 1-38. Positions of welding rod and flame when welding, using backhand method.

over the portion of the work that has been welded. The directing of the flame in this manner tends to anneal the weld

continuously for the full length, or the welding may be done by the "step" method illustrated in Fig. 1-39.

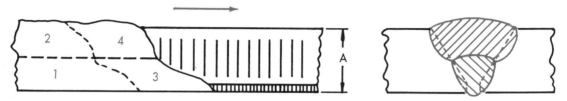

Fig. 1-39. The "step" method of multi-layer welding showing the sequence of the "step" welds by number.

and relieve the welding stresses to a great extent. In addition, the direction of the flame tends to help the welder form a good bead, and attain good penetration of the weld. Backhand welding is commonly used in welding cast iron or drain pipe, and in welding thick, heavy sections in which the stresses created by welding are relieved. In backhand welding, the continued flare of the torch flame on the hot portion just welded tends to maintain a rather large puddle of molten metal. In order that the edge of the puddle may solidify into a bead, it is necessary to move the flame upward, at frequent regular intervals. This action allows the edge of the puddle to cool slightly and solidify.

1-30. MULTI-LAYER WELDING

Multiple-layer welding may be used in place of single-layer welding for all positions of welding and thicknesses of metal, but is especially well-suited for welds on thick materials. The usual criterion is the size of puddle that can most effectively be handled or controlled by the welder for the particular position of welding. If the completion of a weld in one-layer requires a puddle larger than indicated by this criterion, two or more layers should be made.

The layers of the weld may be made

1-31. APPEARANCE OF A GOOD WELD

The weld is usually inspected by careful optical examination. Some inspectors use a magnifying glass of 2 to 10 diameters magnification.

The weld should be of consistent width throughout its length.

It should be straight so the two edges form two straight lines, one parallel to the other.

The weld should be slightly crowned or convex (built up above the surface of the parent metal). This crown should be consistent (even).

The weld should have the appearance of being fused into the base metal, not having any distinct line of demarcation. That is, it should have a blended appearance, and not have a distinct edge between it and the parent metal.

The surface of the weld should have a ripple throughout its length. The ripples should be evenly spaced.

The weld should have a clean appearance. There should be no color spots, no scale on the weld, and no rough pitty appearance to the weld. Lap welds and corner welds should normally show no visible penetration on the side opposite the bead. On a butt weld turn the specimen over and check the penetration. The degree of penetration will be indicated by the sag of the lower surface of the weld. The sag should be slight, and yet penetration should be obtained along the complete length of the weld. The amount of sag should be approximately 1/64 to 1/32 in. The penetration is hard to determine; the easiest way to test for it is to place the specimen in the vise, or jig, with the upper surfaces of the weld up to the correct height. The weld will be of sufficient strength, but a weld of this nature will cost more than the weld described previously. It is only necessary to produce a weld as strong, or a little stronger, than the original metal. It is not possible to produce sufficiently

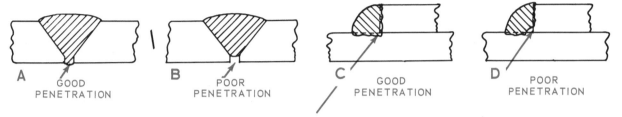

Fig. 1-40. Examples of butt (A and B) and lap (C and D) joint welds in cross section. Note in B and D that the fusion has not been completed to the root (bottom) of the joint.

weld held at the edges of the jaws. The upper half of the specimen is then bent, closing the two edges upon the welded part like a book. If the weld has not penetrated satisfactorily, it will crack open at the joint as it is being bent.

The welder should obtain consistent and efficient penetration. It is possible to secure a very good weld with abnormal penetration, and still build the

strong butt joints if penetration is not obtained.

The quality of the resultant welds can be inspected by noting the size, shape and condition of the joint.

Fig. 1-40 shows the cross section of both butt joint and lap joint welds. Note that B and D show poor penetration.

Fig. 1-41 shows the cross section of both good and poor lap and butt joints.

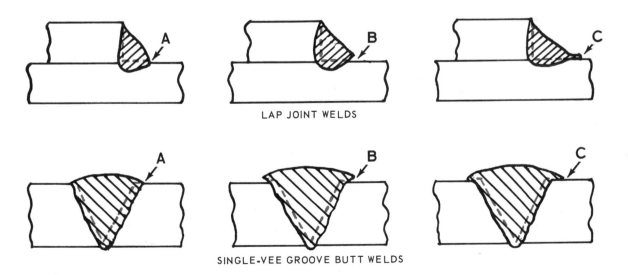

LAP JOINT WELDS

SINGLE-VEE GROOVE BUTT WELDS

Fig. 1-41. Cross sections of good and poor butt and lap joints. A. Properly made joint with proper fusion at edges of weld. B. Poor fusion at edges of weld. C. Poor fusion and overlapping of filler metal at edges of weld.

In a lap joint, the main fault is lack of fusion along the toe of the bead, as shown at B. To correct, direct more heat toward the bottom metal. The overlap at C is caused when too much welding rod is added to the puddle.

Fig. 1-42, in cross section, shows severe undercutting. In an inside corner weld, this condition should be improved by directing more heat on the bottom

1-32. METAL FUME HAZARD

When heated, many metals release irritating and toxic fumes. The metals which release fumes that are most dangerous are:

Cadmium.

Zinc.

Lead.

Cadmium is widely used as a rust

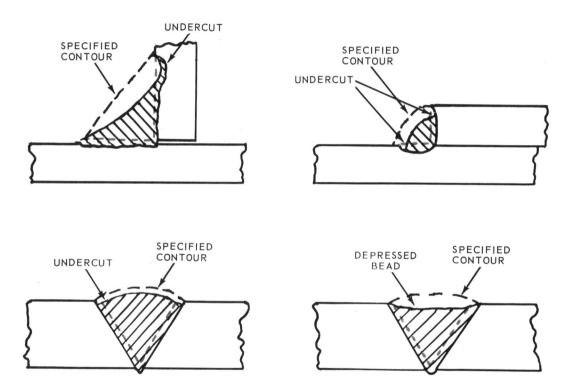

Fig. 1-42. Cross sections of some typical faults in finished welds. Shape of bead contour is usually specified. When undercutting occurs, or, if face of weld is not built up to specifications, welded joint is weakened.

piece, and less on the vertical piece. In the lap weld, undercutting is caused by overheating the weld metal either by poor alignment of the torch, improper torch movement, too small a tip using too high gas velocities or using undersized welding rod.

A specially designed welding bench may be used to practice welding under good conditions. See Figs. 1-43 and 1-44.

protective plating over steel. It is white in color. The fumes generated by welding, brazing or cutting cadmium coated metal are very dangerous. If it is necessary to perform welding operations on cadmium-coated metals, very thorough ventilation must be provided. Cadmium coatings may usually be identified by the fact that the metal, when gently heated with a torch, will turn to a yellow-gold color.

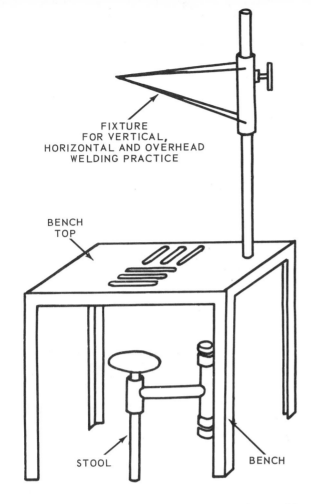

Fig. 1-43. A well-planned welding bench or table. Note the facilities for holding work when welding in the overhead position.

Fig. 1-44. Well-designed bench enables a person to practice welding in any position and at various heights. A—Clamp is used for holding welding exercises flat, horizontal or vertical. B—Height adjustment enables one to position welding exercise at various heights above the bench. C—This fixture may be removed from the bench. (Greene Mfg., Inc.)

Zinc is widely used in die-cast metal parts. Galvanized sheet metal is an example of zinc-coated metal. It is white in color and melts at a rather low temperature, about 787 deg. F., although when alloyed with other metals this melting temperature may be either raised or lowered. When heated with either a welding torch or an electric arc, it gives off a white vapor which is very irritating to the respiratory system. Very thorough ventilation must be used when welding or heating zinc parts or zinc-coated materials.

Lead is not as widely used as cadmium or zinc. However, some plating tanks are lead lined. Lead is also used for tanks and pipes for handling certain liquids and gases. Most storage batteries (electric) have lead plates, posts and cell connectors. Many paints contain lead oxide pigments. Most solders contain lead in varying amounts.

When heating any substances containing lead, thorough ventilation is required. Lead poisoning may occur without the operator being warned in any way of the danger since its fumes are not irritating. The human body does not appear to be able to throw off lead taken into the body either through the lungs or the digestive system. Therefore a person exposed to lead fumes or lead in any form may slowly build up an accumulation of lead until acute lead poisoning develops.

1-33. REVIEW OF SAFETY IN OXYACETYLENE WELDING

Oxyacetylene welding equipment is safe to use if it is properly used, but it potentially possesses great destructive power if carelessly used. Therefore, it is important that the operator be familiar with all of the potential dangers in the welding processes. Most of the safety hazards of oxyacetylene welding were pointed out as the operation of the equipment was explained in this Chapter. Welders should know all these precautions, and follow them for their own safety, for the safety of fellow workers, and to protect equipment. The following precautions are reviewed and elaborated on, in the interest of safety:

1. Protective clothing and shields.
   A. Wear goggles. Various shades are required for various welding applications. In general, the heavier the metal being welded or cut, the darker, high numbered, shade protection lens will be required.
   B. Wear a pair of heavy leather or asbestos gloves.
   C. Wear leather or asbestos apron or jacket for overhead welding, and for other applications where clothing is in danger of being exposed to sparks.
   D. Remove combustibles from pockets, especially matches.
   E. Avoid wearing trousers with cuffs and open pockets into which sparks may fall. Avoid oily articles.
   F. Use only fireproof materials for the support of articles being welded or cut.

2. Welding or Cutting Tanks or Containers.
   A. Do not weld or cut a tank or container, unless the device has been processed to be safe for such operations. See American Welding Society Recommendations describing procedures to be followed in preparing for welding and/or cutting certain types of containers which have held combustibles.
   B. Remember many substances which are not usually considered flammable or explosive, become vaporized and therefore explosive when heated to a high temperature.

3. Handling Oxygen Cylinders and Equipment.
   A. All equipment used for oxygen such as cylinders, regulators, valves, and torches, must be kept free of oil.
   B. Check for leaks, using soap and water. Never use gas welding equipment with leaks.
   C. Open oxygen cylinder valves slowly. Open fully when in use, to eliminate possible leakage around the cylinder valve stem.
   D. Purge oxygen valves, regulators, lines, and torches before use.
   E. Most fabrics, in an atmosphere of oxygen, will burn with explosive force.
   F. Support gas cylinders so they cannot tip over. A valve broken from an oxygen cylinder will cause it to become a rocket with tremendous force. An acetylene cylinder will behave like a flame thrower, if the cylinder valve is broken off, and if the gas is ignited.
   G. Stand to one side of oxygen regulators when opening the oxygen cylinder valve.
   H. Always call oxygen "oxygen." It should never be called air. Call acetylene "acetylene," not "gas." Identify the content of cylinders by the name marked on the cylinder. If a cylinder is unnamed, do not use it.

I. Keep cylinders away from exposure to high temperatures. Remember the pressure in an oxygen cylinder increases with the temperature.

J. Store "full" and "empty" cylinders separately in a well-ventilated space.

K. Mark empty cylinder "MT" with chalk as soon as they are taken out of service.

L. If a cylinder leaks around a valve or a fuse plug, tag it to indicate the fault, move it to a safe area and immediately notify the supplier to pick it up.

M. Keep cylinder caps screwed on all cylinders that are not in use and particularly while they are being moved.

N. Keep the regulators in good repair. Do not use a leaking (creeping) regulator.

4. The Flame.

A. Always turn off the torch flame unless the torch is held in your hand.

B. Be sure that no combustible material is in the area in which an oxyacetylene torch is to be lighted.

5. Handling the Oxyacetylene Welding Station.

A. Blow out the cylinder valves before attaching the regulators. Caution: When blowing out acetylene valves, regulator and hoses, be sure no open flames are near because acetylene is very flammable.

B. Use a well-fitting wrench to attach regulators.

C. Blow out the hose before attaching a torch.

D. Acetylene hose fittings use left-hand threads. A groove on the periphery (around the sides) of the nut indicates that it has a left-hand thread. Oxygen hose fittings have right-hand threads. Do not interchange fittings on hoses.

E. To avoid burns, use a spark lighter for lighting a torch.

F. Keep all hose away from oil and grease.

Examine the hoses regularly for leaks. Leaking acetylene may cause a severe explosion or fire.

If a hose is burned or injured by a flashback, replace it. A flashback usually burns the inner wall of the hose and makes it unsafe to use. Never repair a hose by binding it with tape.

The adjusting screw on the regulator must be turned out (counterclockwise, regulator closed) before opening the cylinder valve. Open the cylinder valve slowly.

Open the acetylene cylinder only 1/4 to 1/2 turn and leave the wrench on the valve stem.

Adjust oxygen and acetylene regulator pressures as recommended by the manufacturer of the torch.

Never use acetylene pressure in excess of 14 psig (measurement of pressure - pounds per square inch gauge).

A backfire is caused by an instantaneous extinguishment and reignition of the flame at the torch tip. Usually the trouble will clear itself immediately. If it does not, carefully inspect the equipment, purge the lines and relight. A backfire is usually caused by overheating the tip.

A flashback is a burning back of the flame into the tip, torch, or even into the hose. It is characterized by a squealing or sharp hissing sound and by a smoky or sharp pointed flame. In case of a flashback, immediately extinguish the flame by first closing the torch oxygen valve and then the torch acetylene valve. Wait a few minutes for the torch to cool before relighting it. Flashbacks indicate something radically wrong with the equipment. Before relighting, purge each line separately, and readjust the regulator pressures to the recommended pressures.

The Occupational Safety and Health Administration (OSHA) has established many compulsory safety requirements for the welding industry. Most of the safety precautions mentioned in this text will help one conform to OSHA requirements.

1-34. TEST YOUR KNOWLEDGE

Write answer to questions on a separate sheet of paper. Do not write in this book.

1. Give a definition of the term, welding.

2. What size torch tip and welding rod is suggested for welding 1/8-in. steel?

3. What is meant by undercutting a welded joint?

4. What are the gases produced by the combustion of the oxygen-acetylene flame?

5. What is the approximate temperature of the oxyacetylene flame?

6. Does oxygen burn?

7. Name the four main parts of an oxygen-acetylene welding station.

8. What should be used when checking an oxyacetylene station for leaks?

9. What is a number drill?

10. How are the sizes of welding tip orifices designated?

11. What are two popular types of welding torches?

12. List the steps required for lighting an oxyacetylene torch.

13. Where is the hottest part of the flame?

14. List, in the proper order, the six steps to be followed when shutting off a torch.

15. How does the "puddle" indicate the penetration being obtained?

16. What is the approximate angle between the welding tip and the weld when welding flat stock on a horizontal surface?

17. How is welding rod metal added to a weld?

18. Name three things which may cause a torch to "pop."

19. What is the position of the torch in forehand welding? Backhand welding?

20. Name four basic types of welded joints.

21. Name or illustrate four methods used to compensate for shrinkage when welding a butt joint.

22. When welding an inside corner joint, a flame adjustment of a slightly oxidizing nature is used. Why?

23. What are the four recognized welding positions?

24. What are some of the personal safety precautions which should be observed when doing overhead welding?

25. Why must the regulator adjusting screws be turned out before opening the cylinder valves?

26. Why should cylinder valves always be opened slowly?

27. How does the welding position affect torch position?

28. Why should a spark lighter always be used for lighting a welding torch?

Chapter 2

OXYACETYLENE WELDING
EQUIPMENT AND SUPPLIES

The operator should have a thorough knowledge of the purpose, design, construction, and operation of welding equipment, and supplies. This knowledge is necessary to secure safe, conscientious use of the equipment, and to obtain the best results from the materials on hand.

2-1. COMPLETE OXYACETYLENE WELDING OUTFIT

The term oxyacetylene outfit refers to the basic equipment needed to weld. An oxyacetylene station includes an oxyacetylene outfit, plus a welding table, ventilation, lighting, and other necessary room equipment.

The equipment necessary for a complete oxyacetylene outfit may vary depending upon the welding operations which are to be performed.

In general, a welding outfit will consist of the following:

1. Gas supply.
    A. Oxygen, cylinder (compressed gas) or tank (bulk liquid).
    B. Acetylene, cylinder or generator.
2. Regulators; complete with high-pressure and low-pressure indicators or gauges, and cylinder and hose fittings. Two regulators with gauges are required; one for the oxygen cylinder, and one for the acetylene cylinder.
3. Hoses.
4. Torch; complete with welding tip, mixing chamber, needle valves, and hose fittings.
5. Goggles; welding, with either a No. 4, 5 or 6 tinted filter shade.
6. Lighter spark.

The equipment for a complete outfit is shown connected and ready for use in Fig. 1-5.

2-2. OXYGEN SUPPLY

Oxygen used for welding is stored in cylinders of various sizes, which are usually painted. Since there is no standard color code, the cylinders are painted a color selected by the manufacturer.

Oxygen is stored in cylinders at a pressure of from 2,000 to 2,640 psig (pounds per square inch gauge), depending on the cylinder material. The pressure will vary also according to room temperature.

Oxygen is obtained by three different processes. One process consists of the liquefying of atmospheric air by compression and cooling. The atmospheric air consists of approximately 21% oxygen, 78% nitrogen, 1% other gases (by volume). Liquid air rapidly evaporates if the pressure, approximately 1,000 psig, is reduced. Oxygen and nitrogen have different boiling temperatures so it is easy to separate the oxygen from the nitrogen. The nitrogen will boil off first because oxygen boils at a higher temperature

(-295 deg. F.) than nitrogen (-317 deg. F.) at atmospheric pressure.

The 2 percent by volume of other gases consists mainly of water vapor, carbon dioxide, argon, hydrogen, neon, and helium. Water vapor and carbon dioxide are removed during the compression and liquefying process. The other gases are removed at the time the oxygen and nitrogen are separated, by evaporation. Nitrogen and the other gases have a lower boiling temperature than oxygen. As these gases boil away, the oxygen remains and is stored either as a gas or as a liquid, depending on how it is to be used.

Another method of producing oxygen is by electrolysis of water. In this process, an electric current is passed through water causing the water to separate into its elements, which are oxygen and hydrogen. In the electrolytic process, oxygen will collect at the positive electrode, and hydrogen will collect at the negative electrode.

Fig. 2-1 illustrates an experiment

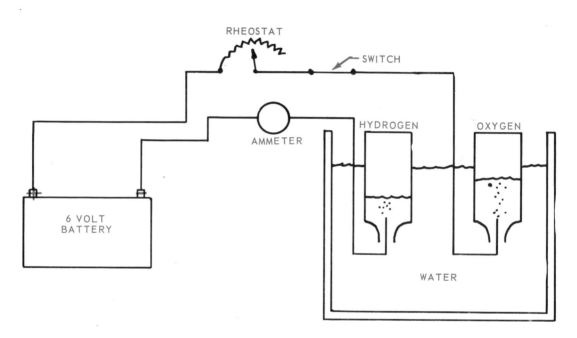

Fig. 2-1. An experiment illustrating the electrolysis of water. To operate the experiment:
1. Fill the bottles (small) with the water and invert them over the copper wires (electrodes) which extend upward from the bottom of the dish. Be sure that the bottles are completely filled with water.
2. Turn on the switch. Soon small bubbles will be seen rising from the electrodes and displacing the water in the bottles. Do not remove the bottles from the electrodes until each is completely filled with gas. The bottle over the positive (+) electrode should be about one half the size of the other bottle. This is because approximately twice as much hydrogen is produced compared to the oxygen produced.
NOTE: 1. Only direct current can be used.
        2. The process can be speeded up by using a higher voltage (12 volts).
TO TEST OXYGEN:
    Remove the bottle from the positive (+) electrode and quickly insert a burning match in the neck of the bottle. The match will burn very rapidly showing that oxygen is present.
TO TEST FOR HYDROGEN:
    Remove the bottle from the negative (−) electrode and quickly bring a burning match to the neck of the bottle. The gas (hydrogen) in the bottle will burn.
    Caution: Perform this immediately on removing the bottle from the electrolyte. If there is a delay and air is allowed to mix with the hydrogen in the bottle the combustion will be very rapid. EXPLOSION.

which may be performed to show how water is separated into its elements, by electrolysis.

A third method used to produce oxygen for welding requires the heat of an oxygen-bearing pellet.

Commercial oxygen, sold on the American market, is close to being 100 percent pure. A popular method of distribution consists of shipping oxygen as a liquid. Such installations are chiefly used in steel mills and by steel fabricators that operate large automatic cutting and welding equipment. Normally, the oxyacetylene welding operator will use the gaseous cylinder. Oxygen sold in liquid form is in large thermos bottle-like tanks; however, it is not held under very high pressure. The evaporation of some of the liquid keeps the temperature of the liquid very low, approximately -295.4 deg. F. At this low temperature, the oxygen remains a liquid under normal atmospheric pressure.

Some portable liquid oxygen cylinders have been developed, and experience indicates that their use will increase. The advantage of liquid oxygen storage is chiefly with the saving in size and weight of the container. Fig. 2-2 illustrates a liquid oxygen storage cylinder.

2-3. OXYGEN CYLINDERS

Since the oxyacetylene welding operator will be using oxygen from cylinders, he must know how to use these cylinders and the various safety precautions which must be taken.

Because of the high pressure of the oxygen stored in cylinders, the pressure being equivalent to one ton of pressure upon one square inch of area, the cylinders must be of very sturdy construction. The Interstate Commerce

Commission (I.C.C.) has prepared specifications for the construction of these cylinders. They are forged in one

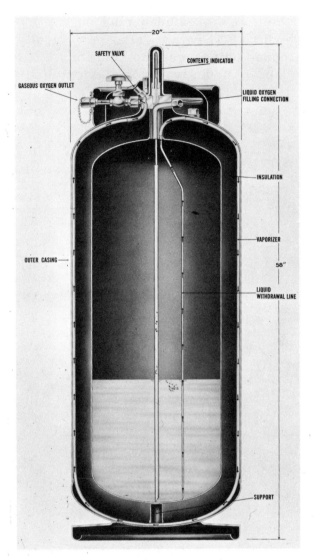

Fig. 2-2. Liquid oxygen cylinder, section view. Note that liquid oxygen drawn from the tank passes through insulating space surrounding the cylinder. Heat required to vaporize the liquid is drawn from this space, thus helping to keep the liquid oxygen cold. (Linde Div., Union Carbide Corp.)

piece, no part of which is less than 1/4 in. thick. The steel used is armor plate type, high-carbon steel. Fig. 2-3 illustrates an oxygen cylinder which has a section cut away to show the construction of the cylinder and the valve. Oxygen cylinders are tested regularly and must withstand hydrostatic pressures of over 3,300 psig. These cylinders are periodically annealed to relieve stresses created during on-the-job handling. They are also periodically cleaned using a caustic solution.

The cylinder valve is of special de-way open and never used in a partly open position. Fig. 2-4 shows the construction of an oxygen cylinder valve. The valve is fastened to the oxygen cylinder by 1/2 or 3/4 in. pipe threads. The 3/4 in. size is the most popular. The valve, located in the upper end of the cylinder, incorporates a pressure

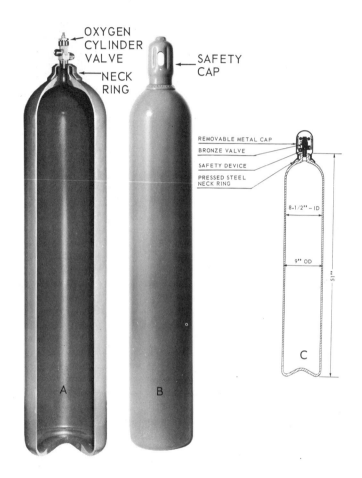

Fig. 2-3. A typical oxygen cylinder with a 244 cubic foot capacity. A. Internal construction of the cylinder. Note the one-piece forged construction. B. Exterior of the cylinder. Note the cap over the valve. C. Dimensions of a 244 cu. ft. cylinder. (Pressed Steel Tank Co.)

sign to withstand high pressure and is made of forged brass. It is a "back seating valve," meaning that when the valve is turned all the way open, the stem is sealed to prevent the leakage of oxygen around the stem. When in use, this valve should be turned all the safety device. The safety device consists of a pressure disc which will burst before the pressure becomes great enough to rupture the cylinder as shown in Fig. 2-5. The valve outlet fitting is a standard male thread to which all standard U. S. made pressure regu-

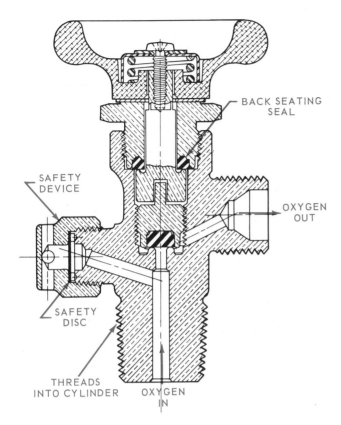

Fig. 2-4. An oxygen cylinder valve, internal construction. (Bastion-Blessing Co.)

BACK SEATING SEAL

SAFETY DEVICE

OXYGEN OUT

SAFETY DISC

THREADS INTO CYLINDER

OXYGEN IN

dropped. The cylinders, when full or partially full, should never be allowed to stand by themselves without adequate support. The pressure in the oxygen cylinder will indicate the amount of oxygen remaining in the cylinder (Boyle's Law).

CYLINDERS SHOULD ALWAYS BE KEPT VALVE END UP, AND THE VALVE SHOULD BE CLOSED WHEN THE CYLINDER IS NOT IN USE, WHETHER THE CYLINDER IS FULL OR EMPTY.

Oxygen is purchased by the cubic foot measured at atmospheric pressure. The cylinders usually remain the property of the oxygen manufacturer and are loaned to the consumer. The price of an oxygen cylinder alone is quite high and it is generally to the advantage of the consumer to use the

lators may be attached. A handwheel for operating the valve is permanently attached to the valve stem.

Threads on the upper body of the cylinder provide a means whereby a heavy steel cap is screwed over the valve to protect it from injury during shipment. The thread size is 3-1/8 in. diameter, 7 or 11 threads per inch. If the cylinder valve should ever be broken off, the very high pressure of the gas in the cylinder, upon escaping, would burn any material it touched, and it would also tend to give the cylinder rocket propulsion. Because of this danger, it is always recommended that the cylinders be handled by two persons per cylinder. It is also recommended that when being shipped, the cylinder should be clamped to a structure to eliminate the danger of its being tipped over or

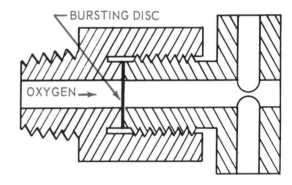

BURSTING DISC

OXYGEN →

Fig. 2-5. A schematic of an oxygen cylinder safety plug. Note the safety disc.

rental system. Most companies will not charge any rental for a cylinder if it is returned within thirty days from date of delivery. After thirty days, a small fee is charged per day (demurrage charge).

Two sizes of oxygen cylinders are in common use in the welding shop. These are:

122 cu. ft.
244 cu. ft.

Other sizes available are 55, 80, 125,

150, 250, and 300 cu. ft. capacities as shown in Fig. 2-6. Under full pressure, approximately 1 cu. ft. of gas is stored in each 10 cu. in. of space in the cylinder.

Many newer cylinders permit a pressure increase of 10 percent with an equal increase in the amount of oxygen in the cylinder.

There is no national color code governing the color of oxygen cylinders. The user should become familiar with the color code used by his supplier.

2-4. THE OXYGEN MANIFOLD

In many large industries and in many school welding shops, an oxygen cylinder is not made a part of each welding station; rather the oxygen is piped to the welding stations. In such installations one or more oxygen cylinders are attached to a manifold from which the oxygen is piped to the welding stations. The oxygen leaving the manifold is at a reduced pressure usually between 30 and 100 psig depending on the length and size of the pipe and the amount being used.

The oxygen manifold is usually located in the oxygen cylinder storage room, so it is not necessary to move cylinders outside of this room. The shop is safer and space is saved by not having the cylinders in the shop.

With a manifold system, a line oxygen regulator is used at each welding station in order to control the oxygen pressure to each torch. A line regulator has only one pressure gauge and this gauge indicates the pressure on the delivery (torch) side of the system. See Fig. 2-7, which shows a typical oxygen manifold installation. Note that the manifold has pressure up to the manual control valve which is kept closed when the oxygen system is not in use. A master

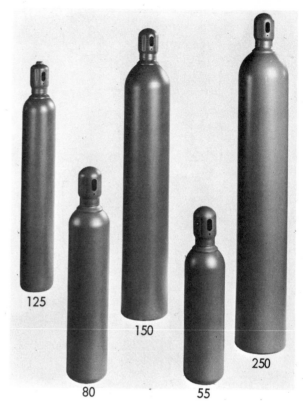

Fig. 2-6. Five different capacity oxygen cylinders. They range from 55 cu. ft., to 250 cu. ft. capacity, and under certain conditions they may be charged with 10 percent more oxygen (i.e. 220 cu. ft. cylinder may be charged with 242 cu. ft. of oxygen).
(Pressed Steel Tank Co.)

regulator controls the pressure of the oxygen in the piping system after it leaves the manifold. This regulator has two gauges. One gauge shows the pressure in the manifold, and the other gauge shows the pressure in the piping.

Fig. 2-7. An oxygen manifold for five cylinders. Note that a master shutoff valve is located between the regulator and the cylinders.

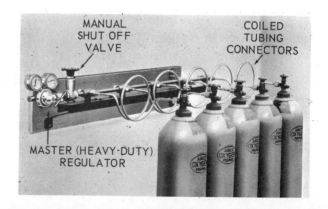

See PAR. 2-13, for information concerning the construction and operation of these regulators.

In manifold installations, the copper

further technical information concerning the chemical structure and nature of acetylene.

Acetylene is made available for oxy-

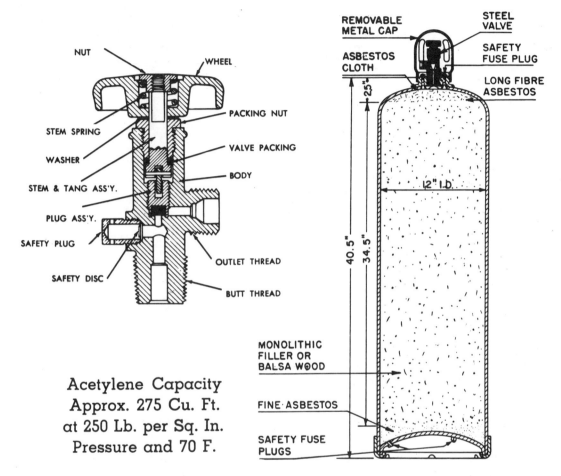

NUT

WHEEL

STEM SPRING

PACKING NUT

WASHER

VALVE PACKING

STEM & TANG ASS'Y.

BODY

PLUG ASS'Y.

SAFETY PLUG

OUTLET THREAD

SAFETY DISC

BUTT THREAD

Acetylene Capacity
Approx. 275 Cu. Ft.
at 250 Lb. per Sq. In.
Pressure and 70 F.

REMOVABLE
METAL CAP

STEEL
VALVE

ASBESTOS
CLOTH

SAFETY
FUSE PLUG

25"

LONG FIBRE
ASBESTOS

12" I.D.

40.5"

34.5"

MONOLITHIC
FILLER OR
BALSA WOOD

FINE ASBESTOS

SAFETY FUSE
PLUGS

Fig. 2-8. Cut-away view of acetylene cylinder showing the porous filler. Note the fuse plugs at the top and bottom of cylinder.

tubing (pigtail), connecting the cylinders to the manifold, should be frequently annealed. This is because the tubing is subjected to high cylinder pressure, and it therefore may become brittle and more subject to breakage.

2-5. ACETYLENE SUPPLY

Acetylene is produced by the chemical combination of calcium carbide with water. The chemical formula for acetylene is C_2H_2. See CHAPTER 28 for

acetylene welding using two different methods:

1. Acetylene storage cylinder.
2. Acetylene generator.

2-6. ACETYLENE CYLINDERS

Acetylene gas may be stored in cylinders specially designed for this purpose. The gas is first passed through filters and purifiers. The storage of acetylene in its gaseous form under pressure is not safe at pressures above 15 psig. The method used to safely store acetylene in cylinders is as follows:

1. The cylinders are filled with a monolithic filler (cement) which cures to a porosity of 85 percent as required by federal safety regulations. This prevents large acetylene gas accumulations.

2. The cylinders are then charged with acetone. Acetone absorbs acetylene.

The theory is that the acetylene molecules fit in between the acetone molecules. Using both of these techniques prevents the accumulation of a pocket of high pressure acetylene.

These cylinders, like oxygen cylinders, are fabricated according to I.C.C. specifications.

The base of this cylinder is concave and it usually has two plugs threaded into it (pipe threads) as shown in Fig. 2-8. These threaded plugs (fuse plugs) have a center made of a special metal alloy which will melt at a temperature of approximately 212 deg. F. If the cylinder should be subjected to a high temperature, the plugs will melt and allow the gas to escape before the pressure builds up enough to burst the cylinder. Fig. 2-9 shows the construction of an acetylene cylinder fuse plug. These pre-

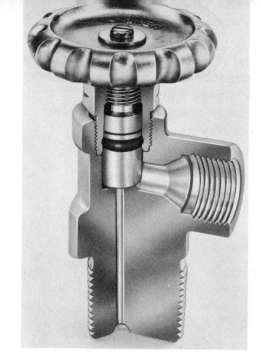

Fig. 2-10. An acetylene cylinder valve. (Chemetron Corp.)

cautions are necessary as the pressure in an acetylene cylinder builds up rapidly with an increase of temperature.

Acetylene cylinder valves come in two types. A common type is provided with a 3/8-in. square shank. It is turned by means of a 3/8-in. square box socket wrench.

It is recommended that this cylinder valve be opened only 1/4 to 1/2 turns. The wrench should be left on the valve stem at all times that the valve is open, in order that the valve may be closed quickly in case a hose or some other part catches on fire.

Another type of acetylene cylinder valve is fitted with a handwheel. The regulator fitting is a female fitting. Fig. 2-10 shows the construction of this second type of acetylene cylinder valve.

The amount of acetylene in a cylinder cannot be estimated by the pressure in the cylinder because the pressure of the acetylene gas coming out of the acetone solution will remain fairly constant (depending on the temperature) until most of the gas is consumed. The

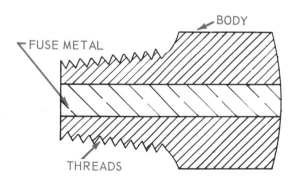

Fig. 2-9. Acetylene cylinder fuse plug. The body of the plug is usually made from brass. The fuse metal has a low melting temperature, (212 deg. F.).

amount of acetylene in a cylinder can be determined accurately by weighing the cylinder. The tare weight (weight without the acetylene gas) of the cylinder is always stamped on the cylinder. The gas weighs 1 lb. per 14-1/2 cu. ft.

Acetylene cylinders should always be used in an upright position; otherwise some of the acetone is likely to escape with the acetylene and contaminate the equipment and the flame. Each time that an acetylene cylinder is refilled, the tare weight is checked and acetone is added if necessary.

Acetylene cylinders are available in a variety of sizes. The most common for welding use are sizes 130, 290 and 330 cu. ft. Other available sizes are: 75, 190 and 360 cu. ft. capacity.

Two small acetylene cylinders are designed for portable welding and cutting equipment. They are: size B, 40 cu. ft., and size MC, 10 cu. ft. These cylinders (sometimes called tanks) use Prest-O-Lite (P.O.L.) fittings.

All acetylene cylinders are designed and constructed according to I.C.C. Specifications No. 8. All designs must be tested by the Bureau of Explosives and must pass these tests before they can be used commercially.

It is important to remember that an acetylene cylinder must absorb heat as acetylene is released from it. Therefore, the rate of acetylene flow from a cylinder is somewhat limited. Cylinders exposed to freezing temperatures will release acetylene gas very slowly. However, cylinders cannot be heated to high temperatures. The fuse plugs melt at 212 deg. F.

In describing the construction of the acetylene cylinder earlier, it was pointed out that the acetylene is dissolved in acetone. This means that acetylene cannot be drawn from the cylinder any faster than it can be released from the

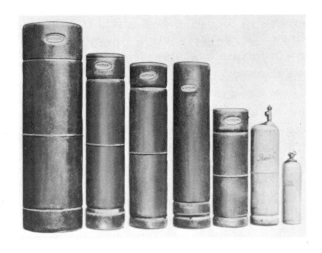

Fig. 2-11. Various types and sizes of acetylene cylinders. (Linde Div., Union Carbide Corp.)

acetone. (This release is a kind of boiling action.) If one attempts to draw off acetylene more rapidly than it is being released from the acetone, a considerable amount of acetone will be drawn from the cylinder along with the acetylene.

The maximum safe rate for drawing acetylene from a cylinder is one-seventh of the cylinder's per-hour capacity. Thus, a single 250 cu. ft. cylinder can supply acetylene at about 35 cu. ft./hr. If a greater flow rate is needed, cylinders must be connected to a manifold to supply the required flow.

AN OXYACETYLENE FLAME CONSUMING SOME ACETONE WILL BURN WITH A PURPLE COLOR.

Acetone in the flame is not desirable. It lowers the flame temperature and increases gas consumption. Moreover, the quality of the weld is affected.

As was previously learned, low or freezing temperatures can cause flow of acetylene to decrease as heat is needed to boil off the acetylene. Another factor causing a slowdown in acetylene flow is a nearly exhausted cylinder. As the cylinder approaches the empty or discharged condition, the acetylene is released more slowly. Additional flow must be added from another source. This is achieved by combining the flow from several cylinders that feed into a manifold. This kind of installation makes it possible to use up more of the acetylene from each cylinder. This, of course, also cuts down the cost of the acetylene used.

The rate at which acetylene may be drawn off depends on three things:

1. The temperature of the cylinder.
2. The amount of charge remaining in the cylinder.
3. The number of cylinders providing the flow.

Too rapid removal of acetylene can be dangerous. Such pressure drop could cause a flashback.

Fig. 2-11 illustrates several sizes of acetylene cylinders. Some have a recessed top. This recess protects the cylinder valve which has a female regulator connection or fitting. The cylinder on the extreme right is the MC size, while the one next to it is the B size. The valves in the larger cylinders shown in Fig. 2-11 are known as P.O.L. Commercial.

2-7. ACETYLENE MANIFOLD

As explained in PAR. 2-4, there are several advantages in a manifold system in which the fuel and oxygen cylinders are not located at the welding stations in the room. In the case of acetylene, however, there is one additional reason for having a manifold installation. For safety reasons, the flammable code in some cities does not allow the storage and use of acetylene cylinders in rooms located below other rooms which are used for classes, meetings, or assemblies. In such instances, a manifold system is used to carry the acetylene fuel from the cyl-

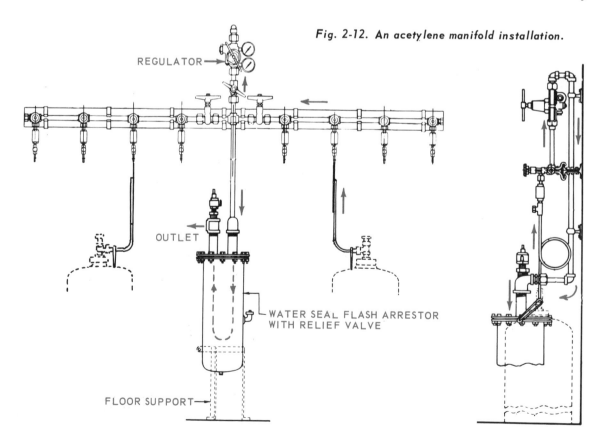

Fig. 2-12. An acetylene manifold installation.

REGULATOR

OUTLET

WATER SEAL FLASH ARRESTOR WITH RELIEF VALVE

FLOOR SUPPORT

inders to the point of use.

In a manifold installation, the manifold room should maintain an approximate temperature of 70 deg. F.

Fig. 2-12 illustrates a modern acetylene cylinder manifold installation. It should be noted that the acetylene manifold system incorporates some features not included in an oxygen manifold system. In addition to regulator and control valves, the acetylene manifold also has a waterseal type flash arrester.

BRASS PIPING MAY BE USED WITH ACETYLENE. COPPER PIPING IN THE PRESENCE OF ACETYLENE FORMS COPPER ACETYLIDE, AN UNSTABLE COMPOUND THAT DISASSOCIATES VIOLENTLY AT THE SLIGHTEST SHOCK, AND MUST NOT BE USED.

2-8. ACETYLENE GENERATOR

As was mentioned in PAR. 2-5 acetylene is produced by the chemical action of water with calcium carbide. A generator brings the water and calcium carbide together in the amounts and at the rate required to safely generate the amount of acetylene needed.

Acetylene generators are available in two types: a low-pressure type, and a medium-pressure type. The low-pressure type generates acetylene at approximately 1/4 psig (4 to 6 in. of water column). With this type of generator it is necessary to use an injector-type torch. See PAR. 2-20.

The medium-pressure type acetylene generator generates acetylene up to 15 psig which is the maximum safe pressure at which free acetylene gas may be stored. With this type of generator, equal pressure type torches may be used. See PAR. 2-19 for a description of the equal pressure type torch. Fig. 2-13 illustrates an acetylene generator. The use and installation of an acetylene generator is controlled by safety codes which should be carefully studied and rigidly followed. A generator installation will require regular care which includes checking and replenishing the water and calcium carbide supply, and removal of the calcium hydroxide sludge. The removal and disposing of the sludge is perhaps the greatest problem in the use of the acetylene generator.

Referring to Fig. 2-13, it will be noted that the generator is partially filled with water. The calcium carbide is stored in the hopper at the top of the generator. A feed mechanism feeds the calcium carbide into the water. Acetylene is generated by the action of the water on the calcium carbide. As soon as the predetermined acetylene pressure has been generated, the feed of calcium carbide is stopped and no more can be fed into the water until the acetylene has been drawn off and the acetylene pressure in the generator decreases. A water level supply device stops the feeding of calcium carbide if insufficient water is in the base of the generator.

2-9. PRESSURE REGULATOR PRINCIPLES

All gases are commonly stored in cylinders at pressures considerably above the working or flame pressures. Most welding torches operate at pressures of between 0 and 30 psig. Therefore, it is necessary to provide a pressure regulating mechanism to reduce and otherwise regulate the pressure as needed. This mechanism is called a pressure regulator. Every mechanism which requires the control of liquid and/or gas pressures uses this type regulator. It performs two services:

1. It reduces the high storage cylin-

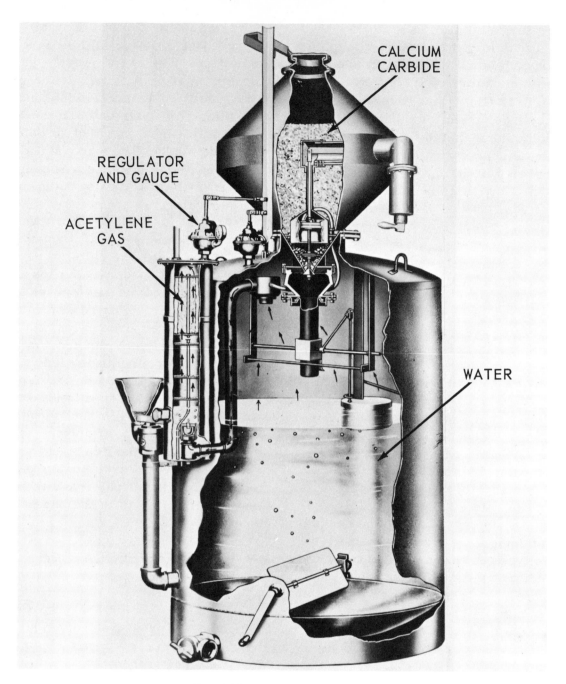

CALCIUM
CARBIDE

REGULATOR
AND GAUGE

ACETYLENE
GAS

WATER

Fig. 2-13. An acetylene generator.
(Rexarc, Inc.)

der pressure to suitable working pressure.

2. It maintains a constant gas pressure at the torch (even though the cylinder pressure may vary).

Fig. 2-14 shows a pressure regulator complete with gauges and fittings.

There are two basic types of regulator mechanisms:

1. Nozzle-type.
2. Stem-type.

Fig. 2-15 is a diagrammatical cross section drawing showing constructon of the nozzle-type of regulator. Fig. 2-16

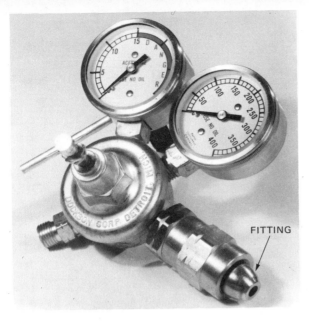

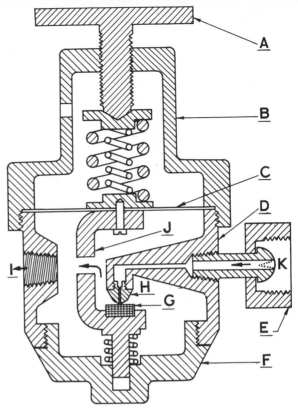

FITTING

Fig. 2-14. Single stage acetylene cylinder regulator. The high-pressure gauge shows cylinder pressure; low-pressure gauge shows pressure delivered to torch. Note that low-pressure gauge indicates that it is dangerous to use acetylene at pressures above 15 psi.

·illustrates a cross section of a stem-type regulator.

Most regulators have two gauges. One gauge (high-pressure) shows the pressure in the cylinder and the other gauge

Fig. 2-15. A schematic drawing, a nozzle-type pressure regulator in cross section. A. Adjusting screw. B. Bonnet. C. Diaphragm. D. Body. E. Cylinder nut. F. Body cap. G. Seat. H. Nozzle. I. Hose fitting opening. J. Cage. K. Gas opening from cylinder.

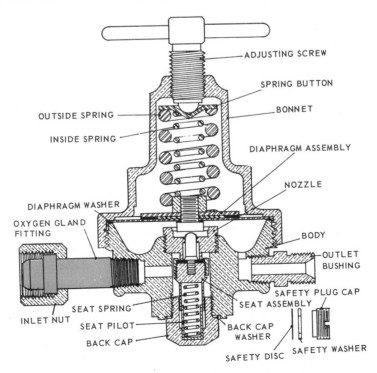

Fig. 2-16. Cross section drawing of stem-type pressure regulator. Note shape of oxygen gland fitting which, with the inlet nut, connects regulator to cylinder.

(low-pressure) shows the pressure of gas being delivered to the torch.

Regulators are also obtainable as:

1. Single-stage regulators which reduce supply pressure to working pressure in one step. Example - 3000 psi to 5 psi.

2. Two-stage regulators which reduce supply pressure to working pressure in two steps. Example - 3000 psi to 200 psi to 5 psi.

2-10. NOZZLE-TYPE PRESSURE REGULATOR

See Fig. 2-15 for an illustration of a nozzle-type pressure regulator.

This regulator consists of a forged, die cast, brass or aluminum body (D), having fixed into the body a means whereby the regulator may be attached to the cylinder valve (K). It also has openings for both a high-pressure gauge and a low-pressure gauge. A third opening (I) connects the regulated low-pressure gas to the hose that carries the gas to the torch.

The front on the regulator has a flexible wall called a diaphragm (C), which is sealed firmly to the regulator body. A spring is mounted between the outside of the diaphragm and the bonnet (B). The force on the diaphragm is adjusted by means of an adjusting screw (A) threaded into the bonnet and pressing against the spring. An arm (J) is attached to (or touches) the inside of the diaphragm, and this arm curves down into the regulator body chamber and around to form a seat (G), which presses up against a nozzle (H) of the regulator. The opening is automatically controlled by the diaphragm to allow the gases to come from the cylinder when needed. The line that leads to the nozzle comes from the cylinder; and to this line is attached the high-pressure gauge. A fine mesh screen or ceramic

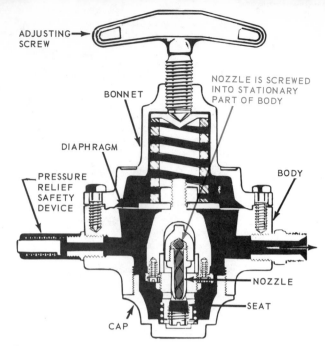

Fig. 2-17. A cross section drawing of a single-stage, nozzle-type pressure regulator. (Chemetron Corp.)

filter is commonly located in this line to keep dirt from entering, and injuring the regulator. The screen also serves as a flame arrester and should be always left in place. The arm that comes from the diaphragm is backed by a spring (or is attached to the diaphragm). The spring continually tends to push the seat of the valve firmly against the nozzle to stop the gas flow.

If the adjusting screw in the body is turned "in" (clockwise), the heavy spring on the outside of the diaphragm will overcome the body spring, move the seat away from the nozzle, and allow some gas to pass from the cylinder into the regulator chamber. As this gas enters the regulator body, it tends to build up a pressure in the body. The force created by this pressure tends to push the diaphragm out against the diaphragm spring and close the nozzle. The pressure tends to fall as the gas is released from the regulator and flows through the hose to the torch. This action will allow the diaphragm spring to move the diaphragm slightly, open the valve and allow more gas to come

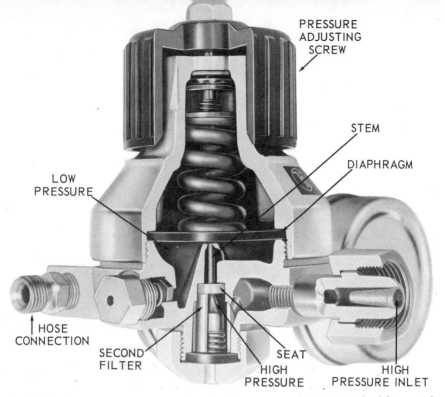

Fig. 2-18. A cross section view of a single-stage regulator. Note the filters at the high-pressure inlet fitting, and in the high-pressure section below the diaphragms. (KGM Equipment Co.)

into the regulator body. The balance of the diaphragm spring pushes the diaphragm (and nozzle) open and the pressure under the diaphragm pushes the nozzle closed which tends to keep the pressure flowing through the regulator constant at the preset pressure.

The pressure in the body cannot increase or decrease from a certain setting because of this compensating action. A near constant pressure, therefore, is maintained in the regulator body, independent of what the pressure is in the cylinder. At a particular setting of the adjusting screw (A), a specific and constant pressure will be maintained in the body as long as the cylinder valve is open. If the adjusting screw (A) is turned in, this pressure will increase somewhat and then stay constant. If the adjusting screw (A) is turned all the way out, it will stop the flow of gas from the cylinder completely.

These regulators come in various gas-flow capacities and nozzle orifice sizes. The diaphragm size and the spring size are changed according to the volume of gas desired. Master regulators are designed to allow a large amount of gas to flow through. Line regulators allow the flow of a relatively small amount of gas.

The springs are made of a good grade of spring steel, while the diaphragm may be made of brass (phosphor bronze), sheet spring steel, stainless steel or rubber.

In modern practice the diaphragm is sealed at the joint between the diaphragm and the regulator body, by means of suitable gaskets, and the clamping action between the body and the bonnet.

The nozzle is usually made of bronze, while the seat may be made of various materials such as casein, hard rubber, plastic, or fiber.

A cross section of a nozzle-type regulator is shown in Fig. 2-17.

2-11. STEM-TYPE PRESSURE REGULATOR

The stem-type regulator works on the same principle as the nozzle-type, but instead of using a nozzle and seat, it uses a poppet valve and seat (stem and seat). Fig. 2-18 illustrates a typical stem-type regulator. The operation of this type regulator is as follows:

The high-pressure gas enters the bottom chamber when the cylinder valve is opened. The diaphragm movement is controlled by the pressure in the top, or low-pressure chamber and the spring forces. Gas from this low-pressure chamber is fed to the torch. The construction is such that the high-pressure tends to force the valve against its seat. The valve stem, therefore, does not need to be attached to the diaphragm. This type regulator lends itself to installations requiring a rather high rate of flow, and is commonly used on manifolds and flame cutting machines.

The materials of construction are similar to the nozzle-type. The seat is constructed of the same material, while the stem (pin) is usually made of stainless steel. The stem and seat are designed to enable the complete assembly to be removed as a cage, permitting easy servicing. Fig. 2-19 shows an exploded view of a stem-type regulator.

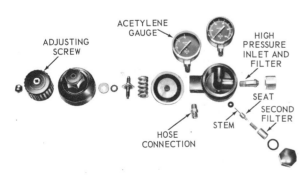

Fig. 2-19. The same regulator shown in Fig. 2-18 showing the parts in an exploded view.

2-12. TWO-STAGE PRESSURE REGULATOR

The two-stage pressure regulator may be considered to be two regulators in one. In the first stage, the high pres-

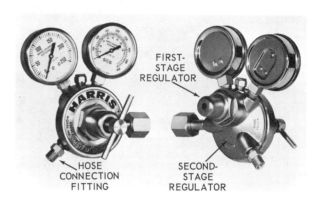

Fig. 2-20. Two views of a two-stage oxygen pressure regulator. (Harris Colorific Div., Emerson Electric Co.)

sure is reduced and regulated to an intermediate pressure by a valve and diaphragm mechanism, which has a fixed pressure adjustment. Figs. 2-20 and 2-21 illustrate the external appear-

Fig. 2-21. Oxygen regulator. Adjustment is on second stage. (Linde Div., Union Carbide Corp.)

ance of typical two-stage oxygen regulators, while Fig. 2-22 illustrates a gaugeless two-stage regulator. The second stage, which is adjustable, regulates the pressure and flow of the gas to the torch. A two-stage regulator is operated in the same manner as a single-stage regulator.

In a two-stage oxygen regulator, the pressure is reduced through the first stage to about 200 psig in the intermediate stage. This means that in the second stage the pressure need only be regulated from 200 psig to the torch pressure.

It is claimed that the two-stage regulator will provide a more constant torch pressure, especially when large volumes of gas are being consumed. Fig. 2-23, illustrates in color, the mechanical operation and the pressure conditions in this type regulator. A high performance single stage regulator is shown in Fig. 2-24. This illustration also shows the details of construction of this regulator.

In a two-stage acetylene regulator, the first stage reduces the pressure to approximately 50 psig.

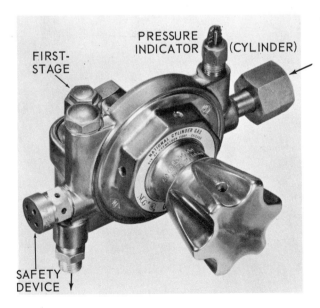

Fig. 2-22. An illustration of a gaugeless two-stage regulator. The pressure is indicated by the height of a pin pushed up by oxygen pressure.

The two-stage regulator may be constructed to use either the nozzle-type or stem-type mechanism. Fig. 2-25 illustrates a two-stage stem-type regulator. Some two-stage regulators have been made which are a combination of the two styles as shown in Fig. 2-26.

2-13. MASTER SERVICE REGULATOR

As stated in PAR. 2-4 and 2-7, manifold systems for both oxygen and acetylene require the use of large master regulators which control the flow of the gases from the manifold to the welding station line. These regulators are basically of the same construction as described in PAR. 2-10, 2-11, and 2-12, however, some modifications are made to adapt the regulator to its particular job. The regulator must be capable of controlling a large quantity of gas through the regulator even when the difference between the cylinder pressure and the line pressure is small. These regulators always have two gauges, a high-pressure gauge which indicates the manifold (cylinder) pressure, and a low-pressure gauge which indicates the line pressure. The typical cylinder regulator usually has insufficient gas flow capacity to be used as a master regulator on a manifold installation.

2-14. LINE STATION REGULATOR

Line station regulators are used in connection with manifold systems.

These regulators have the usual adjustment in order to control the gas pressure to the torch and in every way they are handled the same as other regulators. They are usually equipped with only one gauge. This gauge is connected to the discharge side of the regulator and indicates the pressure of

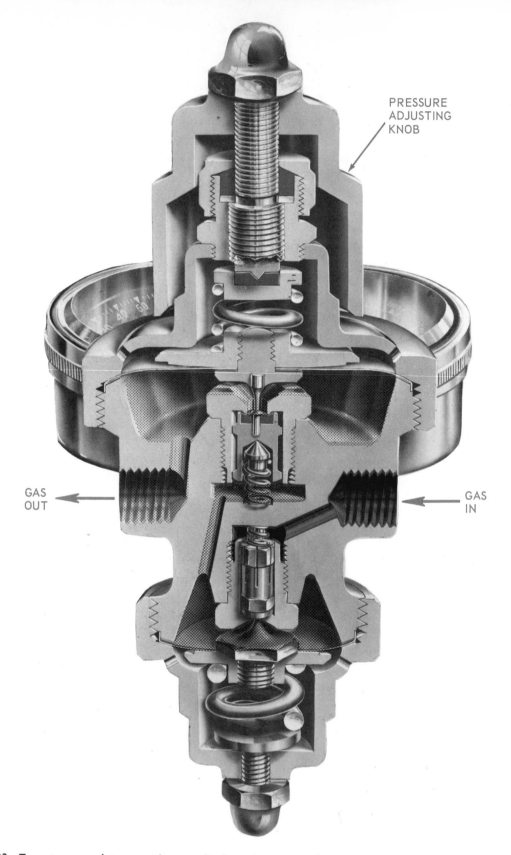

PRESSURE
ADJUSTING
KNOB

GAS
OUT

GAS
IN

Fig. 2-23. Two-stage regulator provides two diaphragms, two needles, and two seats. The first-stage reduces the high gas pressure (solid red) as it comes from the cylinder to some intermediate pressure (dark tint). The second-stage is the low-pressure stage which reduces the intermediate pressure to some constant pressure needed by the torch (light tint). (National Welding Equipment Co.)

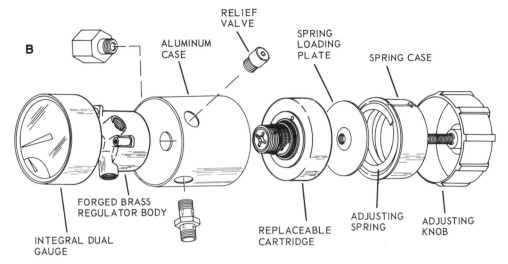

Fig. 2-24. A high performance single stage regulator. A shows the exterior of the regulator. B shows the internal construction. Note the replaceable cartridge which makes replacing the needle and seat quick and easy.

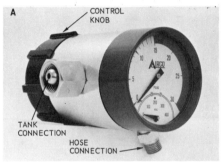

the gas being delivered to the torch.

Since the inlet pressure is lower than with regulators attached to cylinders, the regulator will have larger orifice nozzles and a more flexible, sensitive diaphragm. Since these regulators usually connect to the gas distribution pipe, the inlet connection may be a standard pipe fitting, and not the usual tank fitting. The discharge fitting will be the same as the usual regulator, and will be a right-hand thread fitting for oxygen, and a left-hand thread fitting for acetylene or fuel gases.

2-15. WELDING REGULATOR SAFETY

The regulator is perhaps the single most important device contributing to safety in oxyacetylene welding. If handled properly, its safety qualities will be preserved, but if it is abused, it may fail in its safe operation. To get

Fig. 2-25. Cross section view of a two-stage regulator which uses stem-type valves in both stages. (Victor Equip. Co.)

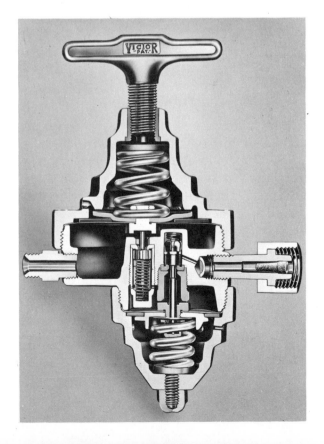

both life and safety from a regulator, be sure to observe the following rules:

A. Always crack (open slightly) the cylinder valves and blow out the cylinder fitting passages before attaching the regulator fitting to the cylinder. Be sure that the wrench fits the regulator nut properly. NEVER USE COMMON PLIERS OR PIPE WRENCHES.

D. Be sure that the regulator adjust-

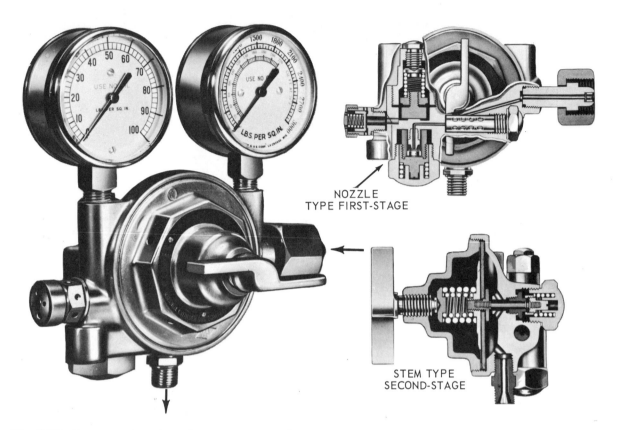

NOZZLE TYPE FIRST-STAGE

STEM TYPE SECOND-STAGE

Fig. 2-26. Cross section view of a two-stage regulator which has a nozzle-type first stage and a stem-type second stage. (Chemetron Corp.)

a regulator. Face fitting away from the operator during this operation.

B. Examine the condition of the threads on both the regulator and the cylinder fittings. The regulator fitting should screw on the cylinder valve easily. Have the fittings repaired rather than use great force to assemble. NOTE: Acetylene fuel gas cylinders have left-hand threads. All oxygen cylinders have right-hand threads. Regulators, hoses, torches, etc., have left-hand threads for fuel gas and right-hand threads for oxygen.

C. Use the proper wrench to tighten

ment is turned all the way out before opening the cylinder valve. In this position no gas will flow through the regulator into the low-pressure side.

E. Open the cylinder valve slowly. If the cylinder valve were to be opened suddenly and the cylinder had maximum pressure, the sudden buildup of pressure in the regulator may cause considerable heat of compression within the regulator. The temperature at the regulator seat may approach 1000 deg. F. The seat could fail.

All approved type pressure regulators incorporate a safety pressure disc on the low-pressure side. This disc is designed to burst at a pressure between 100 and 200 psig. This is below the pressure at which the diaphragm would burst. If the nozzle or seat should leak severely, such a safety device will keep the regulator diaphragm from bursting.

If the seat is in poor condition or leaking, the heat of friction along with the heat of compression may cause the regulator seat to melt or decompose. This action can cause the regulator to explode. Do not stand in front of a regulator as the cylinder valve is opened even when opening it very slowly.

F. Never allow oil or grease to come in contact with any part of oxygen welding equipment. Where pure oxygen is present, it may burn violently (explode). Never wear oily or greasy gloves or clothing when working with any welding equipment.

G. Test for leaks ONLY with a soap and water solution.

H. Never interchange oxygen and acetylene regulators or gauges.

I. Around electrical equipment, insulate the tanks with wool or rubber to prevent grounding. A grounded cylinder may allow an electrical arc to form and the resulting heating or burning of the cylinder may cause an explosion.

2-16. PRESSURE GAUGES

As mentioned in a previous paragraph, the gauges are mounted on the regulators. The high-pressure gauge is connected into the regulator between the regulator nozzle and the cylinder valve. It registers the cylinder pressure when the cylinder valve is opened. The low-pressure gauge is connected into the diaphragm chamber of the regulator, and registers the pressure of the gas flowing to the torch. The gauges are built with gears and springs similar to a pocket watch. Being of delicate construction, they must be handled accordingly. Refer to Fig. 2-27, for details of the construction of pressure gauges.

The basic principle of operation of the pressure gauge depends on the Bourdon tube. This tube which is made

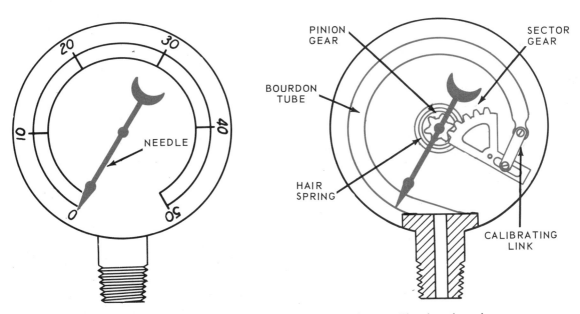

Fig. 2-27. Pressure gauge. Left. Exterior. Right. Internal view. The thread on the gauge is national pipe thread, 1/4 in.

of phosphor bronze is flat in cross section, and is bent to fit inside the circular case. It is closed at one end and is connected with the pressure to be measured at the other. As the pressure in the tube increases, it tends to straighten. As it straightens, it operates the gear and pointer mechanisms, and the dial is calibrated to indicate corresponding pressures. The gauge is usually fastened to the regulator body using 1/4-in. dia. pipe threads.

A heavy glass is used to cover the dial face and needle and is attached to the body by means of a large threaded clamp ring (called a bezel). The gauges come in various sizes; the 2-1/2, 3, and 3-1/2-in. diameter dials are the most popular. The calibration of the gauges depends entirely upon the pressure to be used, and the usual recommendation is that a gauge be obtained with a dial indication of at least 50 percent more than the highest pressure to be used with the gauge.

The oxygen high-pressure gauge usually has a 3-1/2-in. diameter dial, and is calibrated from 0 psig to 3000 psig. There are two scales on the high-pressure oxygen gauge dial. One is the pressure scale. The other scale is calibrated in cubic feet, and indicates the amount of gas left in the cylinder under various pressures which are indicated on the dial. This type of scale is suitable for the oxygen, because oxygen is stored under direct pressure, and the amount of gas remaining in the cylinder is proportional to this pressure (Boyle's Law).

The high-pressure acetylene gauge is usually of 3 in. diameter size, and is calibrated up to 400 or 500 psig.

The low-pressure acetylene gauge is usually of 2-1/2-in. dial size and is calibrated from 0 to 30 psig or 50 psig although many gauges are only cali-

brated up to 15 lbs. leaving the space from 15 to 50 psig blank.

REMEMBER, ALWAYS KEEP THE TORCH PRESSURE OF ACETYLENE BELOW 15 psig.

The oxygen low-pressure gauge has a variety of dial calibrations. For light welding, the dial (2-1/2 in.) is calibrated up to 50 psig, but for heavy welding and for cutting, the gauge may read as high as 200, 400, or even 1000 psig. The diameter of the gauge used in cutting is usually 3 in. in diameter. Fig. 2-28 illustrates some typical pressure gauge calibrations.

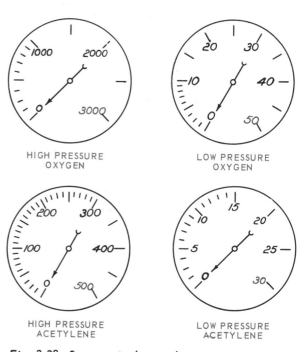

Fig. 2-28. *Some typical regulator pressure gauge calibrations.*

Some rules to be followed when handling gauges:

1. Always have the regulator adjusting screw turned out when opening the cylinders, otherwise the excessive pressure may rupture the Bourdon tube and permanently damage the low-pressure gauges; it might also injure the regulator.

Fig. 2-29. A single-stage regulator with a calibrated diaphragm spring. Note the large readings on the bonnet and the fewer calibrations on the dial that turns with the adjusting screw. (Dockson Corp.)

2. When opening the cylinder valve, or when turning the adjusting screw in on the regulator, turn these stems slowly because, if the pressures are allowed to enter the gauges too suddenly, even though the pressure is not excessive, it will strain the mechanism of the gauge and eventually destroy its accuracy.

3. The pressure to be used in a gauge should always be one-half to two-thirds of the maximum calibration of the dial, meaning that if the gauge is calibrated up to 300 psig, a 200 psig reading should be the maximum used.

4. Teflon tape or, preferably, a paste made of glycerine and litharge, should be used for sealing the threads that connect the gauge to the regulator. Do not use pipe compounds containing oil.

Some regulators use a calibrated spring-loaded diaphragm in place of a gauge, Fig. 2-29. These devices use less space than gauges, are more rugged, need less service, but are usually not as sensitive or accurate as a Bourdon Tube type gauge.

2-17. WELDING HOSE

For most oxyacetylene welding, a flexible device must be used to carry the gases from the regulators to the torch. The popular means used is reinforced rubber hose. This hose must be flexible and strong, and the gases must have no deteriorating effect on the materials of construction. The hose is built of three principal parts: the inner lining which is composed of a very good grade of gum rubber; this in turn is surrounded by layers of rubber-impregnated fabric, while the outside cover, or wearing cover, is made of a colored vulcanized rubber, plain, or ribbed, to furnish a long-wearing surface, see Fig. 2-30. Hose is manu-

Fig. 2-30. A cutaway of a single welding hose. Note the inner rubber, the two layers of fabric and the outer rubber cover. (Goodyear Tire and Rubber Co.)

factured in three common colors; black, green, and red. The use of these colors is not standardized, but the red is usually used for carrying acetylene or other fuel gases, while either the green or the black is used to carry the oxygen. The hose is specified according to its inside diameter and it comes in several sizes. The most common are 3/16, 1/4, and 5/16 in. ID (inside diameter). The size to be used depends on the size of the torch, and the length of the hose to be used. The 3/16 in. ID hose is very flexible and light, and is used extensively for light duty welding such as aircraft tubing. The hose can also be obtained as double hose to minimize entanglement as shown in Fig. 2-31. Suppliers usually furnish the hose in 25 ft. lengths.

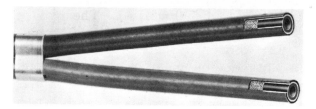

Fig. 2-31. A cutaway of a double welding hose. These hoses are fastened together to eliminate entanglement. Double hoses have one layer of fabric reinforcement.

The hose should never be used alternately, carrying first one gas and then another gas. If oxygen were to pass down a used acetylene hose, a combustible mixture might form. To prevent this, special precautions are used when attaching the hose to the regulators and torch. The hose is clamped to a nipple by means of a hose clamp, and this nipple is fastened to a regulator or torch by means of a nut. The nut and fitting have right-hand threads, if they are to be used with oxygen, the nut is marked OXY. If the fitting and nut are for acetylene, the threads are left-hand and the nut is marked ACE.

The nut also has a groove machined around its six sides. Both the oxygen nipple and the acetylene nipple use a rounded face as a sealing surface. Fig. 2-32 illustrates some typical hose fittings.

The hose must be carefully handled to prevent accidents. It should not be allowed to come in contact with any flame or hot metal. Care should be taken that the hose is not kinked sharply, as this might crack the fabric and permit the pressure to burst the hose. A kink will also hinder the gas flow. When the equipment is not being used, the hose should be hung away from the floor, and away from other things that might injure it. When welding, the hose should be protected from falling articles, from vehicles running over it, and from being stepped on, as these actions might injure the hose.

Hose reels are available. They are usually spring loaded and roll up the hose into the container when the station is not being used.

2-18. TYPES OF OXYACETYLENE TORCHES

Two types of torches (sometimes called blowpipes) are in use at the present time. These two types come in a variety of sizes and designs:

A. Welding torch.

B. Cutting torch (sometimes incorrectly called a burning torch).

Both of these torches are somewhat similar in construction, but the cutting

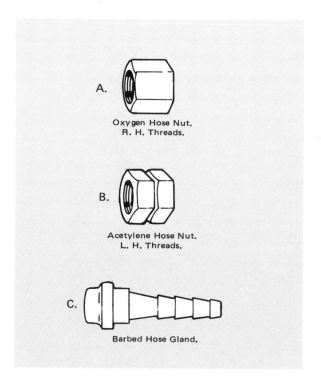

Fig. 2-32. Hose fittings. Fuel gas (acetylene) hose nut has groove cut into it. It also has left-hand thread (L.H.).
(Airco Welding Products, Div. of Airco, Inc.)

torch is provided with a separate control valve which controls an oxygen jet which does the cutting. See CHAPTERS 3 and 4.

Fig. 2-33. An oxyacetylene welding torch.
(Linde Div., Union Carbide Corp.)

and/or stainless steel. The various parts are threaded and silver brazed together.

The hand valves are located at the end of the torch where the hose is attached to the handle or are at the tip end of the handle. They are of the

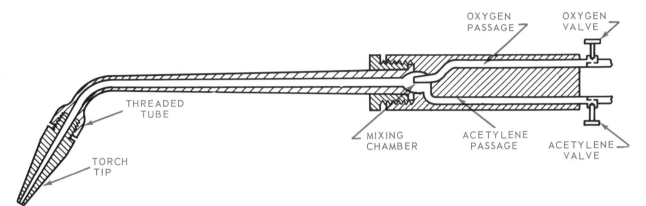

Fig. 2-34. A schematic drawing of an oxyacetylene welding torch.

There are two types of welding and cutting torches in common use:

A. Equal-pressure type (medium-pressure type). This is sometimes called balanced-pressure type.

B. Injector-type.

2-19. EQUAL-PRESSURE TYPE WELDING TORCH

The gases are mixed in the welding torch, and the gases are burned at the end of the torch tip. The welding torch consists of four main parts as shown in Fig. 2-33.

1. Body.
2. Hand valve.
3. Mixing chamber.
4. Tip.

The equal-pressure torch is used with cylinder gases. Its construction necessitates that each gas be supplied under enough pressure to force it into the mixing chamber, as illustrated in Fig. 2-34. Torches are made of several materials including brass, aluminum,

needle-type design. They are usually made of yellow brass with a packing of asbestos twine, or impregnated leather, as shown in Fig. 2-35. These hand shut-off valves are used chiefly for shutting

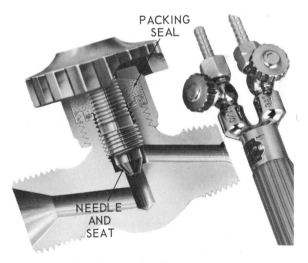

Fig. 2-35. A torch valve cross section.
(National Welding Equip. Co.)

the gas off and turning it on; however, many operators use these valves to throttle (make the final adjustment) the

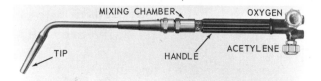

Fig. 2-36. A medium-duty equal-pressure type welding torch. (Chemetron Corp.)

gases being fed to the torch. Figs. 2-36 and 2-37 illustrate the construction of a medium-duty equal-pressure type welding torch, while Fig. 2-38 shows a light-duty torch with two different sets of tips.

The mixing chamber is usually located inside the torch body, although some torches incorporate the mixing chamber in the torch head. Gases are fed to this chamber through two brass tubes leading from hand valves. The size and design of the mixing chamber depends on the size of the torch. Some torch designs change the size and shape of the mixing chamber at the same time the tip is changed. The size and shape of

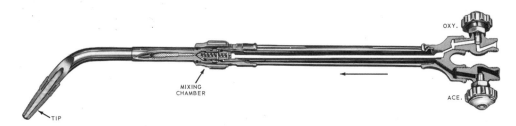

Fig. 2-37. A cutaway of a medium-duty equal-pressure oxyacetylene welding torch.

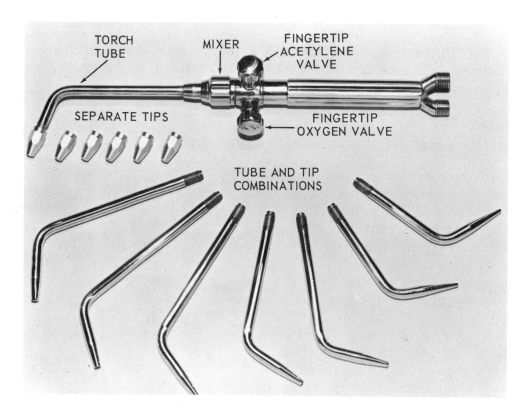

Fig. 2-38. A light-duty equal-pressure torch. Note the two different sets of tips and also note the location of the torch valves. One mixer is used for all tips. (KGM Equip. Co.)

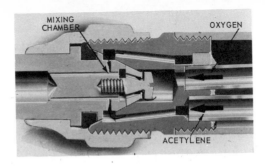

Fig. 2-39. A typical mixing chamber as used on equal-pressure type welding torches.

the holes and the chambers should never be altered, nor should the parts be abused. Fig. 2-39 illustrates a well designed mixing chamber. The gases, after being mixed, are fed through the barrel of the body to the tip where combustion takes place. The barrel is made separate from the tip. Sometimes the barrel and the tip are one piece.

The orifice, or hole drilled in the tip, must be of accurate size. Tip size or number is stamped on the tip and indicates its size in terms of its ability to allow welding gases to flow through it. The tips are usually made of copper, but some are nickel-plated as a means of reflecting the heat and staying cooler. PAR. 1-10 explains various systems used for indicating torch tip size. In this text torch tip size is indicated in number drill sizes. Refer to Fig. 28-1 for the corresponding tip size for a particular manufacturers torch.

Fig. 2-40 gives useful information concerning metal thickness, tip orifice sizes, welding rod diameter, and other facts concerning gas welding, using the equal-pressure type torch.

METAL THICKNESS	SIZE * WELDING TIP ORIFICE	WELDING ROD DIAMETER	OXYGEN		ACETYLENE		WELDING SPEED FT/HR.
			PRESSURE	CU.FT./HR.	PRESSURE	CU.FT./HR.	
1/32	74	1/16 in.	1	1.1	1	1	
1/16	69	1/16 in.	1	2.2	1	2	
3/32	64	1/16 in. or 3/32 in.	2	5.5	2	5	20
1/8	57	3/32 in. or 1/8 in.	3	9.9	3	9	16
3/16	55	1/8 in.	4	17.6	4	16	14
1/4	52	1/8 in or 3/16	5	27.5	5	25	12
5/16	49	1/8 in. or 3/16 in.	6	33.	6	30	10
3/8	45	3/16 in.	7	44.	7	40	9
1/2	42	3/16 in.	7	66.	7	60	8

* Note the tip orifice size as shown is the number drill size. These recommendations are approximate. The torch manufacturers' recommendations should be carefully followed.

Fig. 2-40. A table showing the relationships between welding gas pressures, welding rod diameters and metal thicknesses.

2-20. INJECTOR-TYPE WELDING TORCH

The injector (low-pressure) type welding torch looks much like the equal-pressure type torch.

However, the internal construction of the injector type torch is somewhat different. The chief characteristic of the injector type torch is its ability to operate using very low acetylene pressure, and in general the acetylene pressure remains practically constant regardless of the size tip or thickness of the metal being welded.

The ability of this torch to operate on low acetylene pressure has certain advantages. It is particularly desirable for use in connection with acetylene generators which by their nature can only supply acetylene at a low pressure. Also they have the advantage of being able to draw, more completely, the charge from acetylene cylinders.

Fig. 2-41 shows the internal con-

lene along with it. Handling the valves and the other operation of the torch is much the same as with the equal-pressure type torch. It should be noted that the oxygen pressure used in these torches is considerably higher than with the equal-pressure type torches. Torch adjustments should be as recommended by the torch manufacturer. The materials of construction are usually the same as those materials used in the equal-pressure torch.

2-21. WELDING TIPS

The solid copper welding tip supplies the flame for gas welding. Since copper conducts heat rapidly, there is little danger of overheating and backfire. Tip size and condition are most important. There are two types of tips:

1. Detachable type (tip and tip tube are separate).

2. Tip and tip tube are one piece (integral).

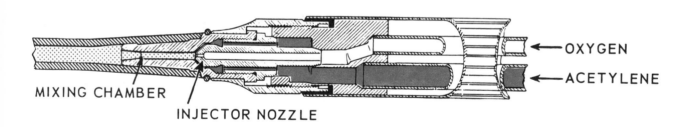

MIXING CHAMBER

INJECTOR NOZZLE

OXYGEN

ACETYLENE

Fig. 2-41. Cross section drawing of mixing chamber area of an injector-type welding torch. The acetylene is induced (drawn) into the mixing chambers by the pulling action (suction) of the oxygen jet. Injector torches are particularly adaptable for use with acetylene generators which operate under low-pressure.

struction of the mixing chamber portion of an injector type welding torch. It should be noted that the oxygen line enters the mixing chamber through a jet which is surrounded by the acetylene passage. As the oxygen flows from the jet it draws (or injects) the acety-

Fig. 2-42 shows the basic differences in the two types.

The tip is subjected to both mechanical wear and flame erosion. As tips are removed and installed in a tip tube, the attaching threads may be subjected to considerable wear and abuse. Wrenches

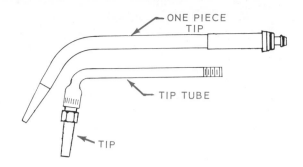

Fig. 2-42. *Two commonly used welding tip designs.*

Fig. 2-44. *Welding torch tube and tip brazed to form a one piece unit. (National Welding Equip. Co.)*

used on these tips should be of the box-end type. Pliers should never be used. Do not try to remove a hot tip from a tip tube. Allow the tip and tip tube to cool first. Also, do not install a cold tip in a hot tip tube.

While welding, the molten metal may "pop" and throw molten droplets into the tip orifice where they may remain until removed with a tip cleaner. Also, the heat from the flame will cause some tip erosion. Always use tips which are made for a particular torch as shown in Fig. 2-43. Some makes of torches use a

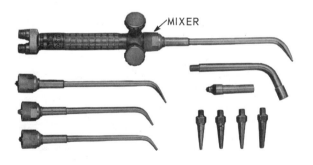

Fig. 2-43. *Light-duty equal-pressure torch. Attachments differ from those in Fig. 2-37. A separate mixer is provided with each one-piece tip.*
(Airco Welding Products, Div. of Airco, Inc.)

pliable, heat resistant synthetic gasket to seal the joint between the tip tube and the torch. With this type of torch, a wrench is not needed to install or service a tip. Hand tightening is sufficient. Separate tip tube and tip instructions:

Avoid dropping a tip as the seat which seals the joint may be damaged. The

flame end of the tip may receive mechanical damage by being allowed to come in contact with the welding work, the bench or firebricks. This damage may roughen the end of the tip and cause the flame to burn with a "fish-tail."

Fig. 2-44 illustrates a popular type welding tip.

2-22. WELDING TIP CLEANERS

As mentioned in the previous PAR. (2-21) welding tips are sometimes subject to considerable abuse. However, the orifice must be kept smooth and clean if the tip is to perform satisfactorily. When cleaning a welding tip, the orifice must not be enlarged nor scarred. Carbon deposits and slag must be removed regularly, if good performance is expected.

Special welding tip cleaners have been developed which perform this service operation satisfactorily. The cleaner consists of a series of broach-like wires which correspond in diameter to the diameter of the tip orifices, Fig. 2-45. These wires are packaged in

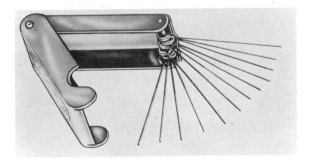

Fig. 2-45. *Welding tip orifice cleaner. (Thermacote)*

Fig. 2-46. A welding tip orifice cleaner in use. (Maitlen & Benson)

Fig. 2-47. Reconditioning the orifice end of a torch tip.

a holder which makes their use safe and convenient. Fig. 2-46 illustrates a tip cleaner in use.

Some welders prefer to use a number drill, the size of the tip orifice, to clean welding tip orifices. If a number drill is used it must be used very carefully so that the orifice diameter is not enlarged, bellmouthed, reamed out-of-round, etc.

The flame end of the tip must be clean and smooth. Its surface must be at right angles to the center line of the tip orifice, if a correctly shaped flame is desired. A 4-in. mill file is commonly used to recondition this surface, as shown in Fig. 2-47.

2-23. AIR-ACETYLENE TORCH

The air-acetylene torch is often used where a light portable flame of medium temperature (2500 deg. F.) is required. This torch is used extensively in copper plumbing (soft soldering), refrigeration lines (silver brazing), and to solder or braze small parts. If large parts are to be silver brazed, the oxy-acetylene torch is recommended.

The air-acetylene torch receives its acetylene from a cylinder, through a regulator and hose. As the acetylene flows through the torch it draws air from the atmosphere into it in order to supply the oxygen necessary for combustion. The torch operates on the same principle as a Bunsen Burner used in chemistry laboratories.

Fig. 2-48, illustrates an air-acetylene torch outfit.

The same precautions should be observed when using the acetylene cylinders, regulators and torches as required when handling the oxyacetylene torch.

Fig. 2-48. An air-acetylene torch. (Chemetron Corp.)

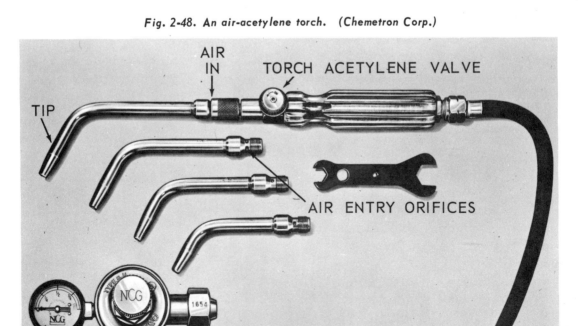

AIR IN

TORCH ACETYLENE VALVE

TIP

AIR ENTRY ORIFICES

ACETYLENE REGULATOR

Cylinders for small portable air-acetylene torches come in various capacities as illustrated in Fig. 2-11. Regular sizes are 10 cu. ft., and 40 cu. ft. The 10 cu. ft. cylinder is called the MC size. The valve fittings on both of these cylinders are Prest-O-Lite fittings.

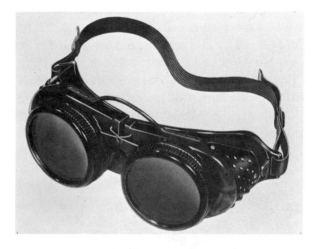

Fig. 2-49. Welding goggles of the 50 mm diameter lens type. These goggles fit over prescription glasses. (Jackson Products)

2-24. REVIEW OF SAFETY WHEN HANDLING TORCHES

The following are some pointers concerning safety when handling torches. These recommendations concern safety both to the equipment and to the operator:

1. Do not put a cold tip in a hot tip tube since the hot tip tube will produce shrinking action on the tip as it cools.

2. Use only a clean wood surface or a leather surface to clean the end of a tip. Keep the oxygen flowing during this operation to prevent plugging the orifice.

3. Be careful when cleaning a tip with a tip cleaner not to increase the size of the orifice, or to cause it to become out-of-round or tapered.

4. Always extinguish a torch whenever it is not in your hands.

5. The torch hand valve should only be turned with the fingertips.

6. If a torch backfires find the trouble and remedy it before continuing to use the torch.

7. Each welding station should be provided with a hook upon which to hang the torch.

8. Be careful that a torch is not directed toward another person while it is being lighted.

9. Be sure no flammable material is near the welding station.

2-25. WELDING GOGGLES AND PROTECTIVE CLOTHING

The operator must wear suitable goggles when doing oxyacetylene welding. The flame and puddle of molten metal emits both ultraviolet and infrared rays both of which may cause eye injury if viewed at a close distance. Goggles also protect eyes from flying sparks. The glare is reduced too, and the operator will be able to see the weld puddle more clearly.

The common welding goggle has a sparkproof frame for the lens, and uses an elastic to hold the goggles securely on the operator's head. The welding lens is 50 mm (millimeter) in diameter, and each goggle holds two pairs of lenses. The outer pair is clear glass, of optical quality, and 1/16 to 3/64 in. thick. This is called the cover lens and is to protect the inner or filter lens from metal spatter. These cover lenses need to be replaced frequently otherwise the operator's view of the weld will be dimmed.

The filter lenses are tinted either green or brown and are made in a variety of shade intensities. The shades intensities are indicated by shade number. These range from No. 1 to No. 14.

APPLICATION	BASE METAL THICKNESS	SUGGESTED SHADE NO.
Shielded metal-arc welding 1/16, 3/32, 1/8, 5/32 in. dia. electrodes		10
Gas-shielded arc welding (nonferrous) 1/16, 3/32, 1/8, 5/32 in. dia. electrodes		11
Gas-shielded arc welding (ferrous) 1/16, 3/32, 1/8, 5/32 in. dia. electrodes		12
Shielded metal-arc welding 3/16, 7/32, 1/4 in. diameter electrodes		12
5/16, 3/8 in. dia. electrodes		14
Atomic hydrogen welding		10-14
Carbon-arc welding		14
Soldering		2
Torch brazing		3 or 4
Light cutting	Up to 1 in.	3 or 4
Medium cutting	1 in. to 6 in.	4 or 5
Heavy cutting	Over 6 in.	5 or 6
Gas welding (light)	Up to 1/8 in.	4 or 5
Gas welding (medium)	1/8 in. to 1/2 in.	5 or 6
Gas welding (heavy)	Over 1/2 in.	6 or 8

Fig. 2-49A. Recommended lens shade numbers for various arc welding and oxy-gas cutting and welding applications. (ANSI Z87.1 – 1968)

Filter lenses must conform to the American National Standards Institute (ANSI) requirements for eye protection for welding (ANSI Z87.1 - 1968). Approved lenses carry a shade number and manufacturer's mark.

Standard welding goggles, Fig. 2-49, are often designed to fit over glasses. Fig. 2-49A lists recommended shades for various welding applications. If operations are short duration, lighter shade indicated may be used. For longer or continuous operation, the darker shade should be used. In general, shade number should increase with tip size.

Cover lenses are rather inexpensive. Filter lenses are quite expensive. It is therefore necessary that the filter lenses ALWAYS be protected by cover lenses. Many cover lenses are protected by a thin layer of transparent plastic which keeps metal spatter from pitting and adhering to them. The life of these plastic coated lenses is greater than that of clear glass.

Some oxyacetylene welding operators prefer to use the eye shield type of eye protection. These shields also use both the cover glass (plates) and filter glass. The lenses are the same size as those used in arc welding helmets. These shields not only fit over spectacles well but also give a good range of vision. Fig. 2-50 illustrates the eye shield type of eye protection.

Fig. 2-50. Eye shield type eye protection. The size of the cover glass and filter lens is 2 x 4-1/4 in. (Dockson Corp.)

The filter plates are 2 x 4-1/4 in. in size and are marked with both the shade number and the manufacturer's mark the same as with the 50 mm welding lens.

The rectangular type eye shield may be fitted with a headband, as shown in Fig. 2-51, for added comfort. Always wear eye protection. Eyes can never be replaced.

The welding operator must wear protective clothing. The hands should be protected with leather or asbestos gloves. The cuffs should be either the gauntlet type, or an elastic band which makes a tight seal between the glove and the coat sleeve. Jackets should be either leather or of a fabric treated to be nonflammable or slow burning. Trousers should be without cuffs and the fabric should be treated to resist burning.

2-26. TORCH LIGHTERS

Matches or burning paper should not be used for lighting a welding torch. Carrying matches or other combustible material such as combs or pens in pockets while welding is dangerous. If a spark should enter a pocket a serious burn might result before the fire can be extinguished.

A flint-and-steel lighter is perhaps the most popular type. Fig. 2-52, shows this type lighter. Pistol grip spark lighters are also available.

Many establishments which use a number of gas welding stations provide pilot lights which use either city gas or acetylene. The city gas is piped to an outlet near the welding station, and a small flame is kept burning continuously. The flame outlet should be located overhead where it will not have any chance of igniting anything on the living

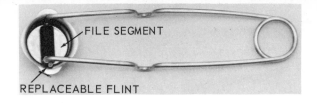

Fig. 2-52. Flint-and-steel lighter. The steel cup tends to trap the gas. When the flint is rubbed on the file segment, the spark quickly and safely ignites the fuel gas.

level of the room. There are two types of acetylene pilot lights; one leaves a very small acetylene flame burning at the torch tip when the torch valves are turned off (not the cylinder valves). When the operator turns the acetylene on, the flame grows to the desired size immediately. The other type provides a torch holder which incorporates a pilot light. Fig. 2-53 shows a combination economizer and lighter. It consists of a mechanism through which the oxygen and acetylene are first fed before going to the torch. This mechanism is also used as the torch holder, when the torch is not being used. Before the torch is placed in this holder, it is lighted and adjusted. When put in the holder, it presses a lever which turns off both

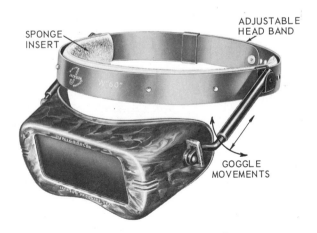

Fig. 2-51. A 2 x 4-1/4-in. eyepiece welding goggle with a special type headband holding device.

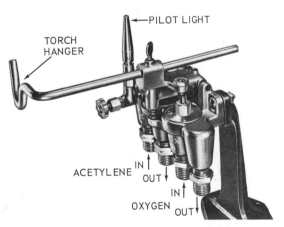

Fig. 2-53. Combination economizer and torch lighter. (Chemetron Corp.)

the oxygen and acetylene, leaving a very small acetylene flame burning at a special outlet, or pilot light, as shown

Oxyacetylene Equipment, Supplies 115

in Fig. 2-54. When the torch is lifted from this holder, gas flow starts, the gas is ignited by the pilot light and the torch is ready to be used for welding. The device saves considerable gas and time. Safety is also improved, as the chance of having the torch laid aside and continue to burn is minimized.

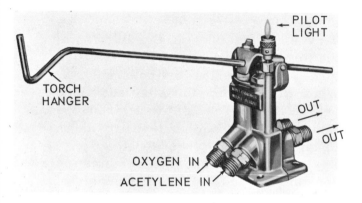

Fig. 2-54. Combination economizer and torch lighter. (Weldit/Winona)

2-27. OXYACETYLENE WELDING SUPPLIES

Many supply items are needed in order to perform the usual oxyacetylene welding operations. The more common supplies needed are:
1. Welding Gases.
   A. Oxygen.
   B. Acetylene.
2. Welding Rod (Filler metal) for,
   A. Steel.
   B. Stainless steel (CHAPTER 18).
   C. Cast iron (CHAPTER 18).
   D. Aluminum (CHAPTER 18).
   E. Hard surfacing (CHAPTER 20).
3. Fluxes.
   A. Cast iron welding (CHAPTER 18).
   B. Aluminum welding (CHAPTER 18).
   C. Stainless steel welding (CHAPTER 18).

Refer to CHAPTERS 15 and 16 for lists of supplies needed for soldering and brazing.
4. Firebrick.
5. Carbon paste and forms.
6. Asbestos.
   A. Sheet.
   B. Powder.
7. Glycerine.
8. Litharge.
9. White lead.

2-28. OTHER FUEL GASES

Handling oxygen and acetylene is explained in PAR. 2-3 and 2-5. Other fuel gases in common use are:
1. Hydrogen.
2. LP (liquefied petroleum) propane and butane.
3. Natural gas.
4. Methylacetylene-propradiene (MAPP).
5. Polypropalene based fuel gas (FG-2).

Hydrogen may be used with oxygen instead of acetylene. The resultant flame does not produce as high a temperature as the oxyacetylene flame; however, the oxyhydrogen flame is very clean and is recommended for welding aluminum and magnesium. Because it can be used at a higher pressure than acetylene it is also recommended for underwater welding and cutting. Since hydrogen is itself a reducing agent, this flame if properly adjusted, minimizes oxidation. A regular oxyacetylene torch may be used with hydrogen as the fuel gas. Hydrogen is supplied in cylinders similar to oxygen and the pressures in the cylinders are about the same. The standard sizes of hydrogen cylinders are:

200 cu. ft.
100 cu. ft.

Hydrogen cylinders are fitted with special fittings and the regulators used on these cylinders must be provided with proper mating attachments. Hydrogen has no odor and with either air or oxygen it forms a possible powerful explosive mixture. Hydrogen connections should be regularly checked for leaks using a soap-and-water solution.

Liquefied petroleum is sold under a variety of names. It has some variations in chemical analysis. For welding use, the general title of liquefied petroleum (LP) gas is used. This fuel is supplied in liquid form and is under a positive pressure which varies with the temperature. LP gas is used mostly for cutting, soldering, and brazing. Most oxygen cutting torches can use LP fuel by the use of the correct tip. An atmosphere (air feed) type torch is commonly used for general heating purposes using this fuel.

LP gas is sold by the pound. Common sizes of tanks are:

20 lb.

35 lb.

60 lb.

100 lb.

The customer usually purchases the 20 and 35 lb. size and returns them to the dealer for refilling. The 60 and 100 lb. sizes are usually leased. Industries which use large quantities of LP gas provide their own bulk storage. The fuel is delivered to them from bulk tank cars or trucks.

When using LP gas, the pressure must be regulated using an LP gas regulator. These regulators are usually supplied with two gauges; one gauge to show the pressure in the storage tank, and the other to indicate the pressure in the torch or burner line. These regulators are usually attached to the tank using a standard POL Commercial fitting for this purpose.

Natural gas, as now piped to most communities, is an excellent fuel for certain uses. It is particularly adaptable for cutting, soldering, brazing and preheating.

Because natural gas is delivered at a rather low pressure, injector type torches are used both for cutting and for general heating. Some small torches have been developed which use compressed air and natural gas particularly for soldering and brazing. Natural gas serves very nicely for many preheating operations. Natural gas piping installations which use either compressed air or oxygen should be protected by a water seal or a blow back valve to keep air and oxygen from backfiring into the gas supply line. Always consult with local safety authorities on this matter.

Stabilized methylacetylene-propadiene is a fuel gas sold under the trade name of MAPP. This a trademark of the Dow Chemical Co. It should be noted that this fuel gas is not a mixture of Calcium Carbide generated acetylene with other substances but rather it is a compound of methylacetylene. The fuel has the safety and ease of handling of liquefied petroleum gas (LPG) with a heating value approaching that of acetylene.

This fuel may be stored and shipped in the liquefied state. The cylinders are available in various sizes. The usual acetylene regulator may be used with this gas and the cylinders have the same thread as acetylene cylinders.

This fuel has some advantages over acetylene when used for cutting. It has a narrower explosive range, it being 3.4 percent to 10.8 percent in air compared to 2.5 percent to 80 percent for acetylene.

It is used for underwater cutting since it may be used at pressures of over 15 psig and acetylene cannot.

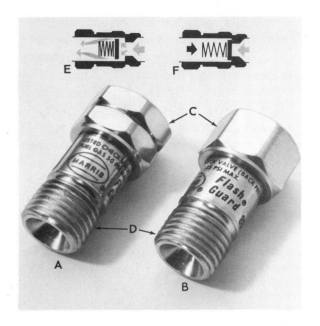

Fig. 2-54A. Flashback arrestors are check valves that prevent mixture of fuel gas and oxygen from flashing back (burning) in welding hose and regulator. A. Fuel gas check valve. B. Oxygen check valve. C. Torch or regulator connection. D. Welding hose connection. E. Normal flow-valve open. F. Reverse flow-valve closes.
(Harris Calorific Div., Emerson Electric Co.)

All usual cutting torches may be used with the fuel. However, special tips made for MAPP gas should be used. These tips are available for all common cutting torches.

OXY-MAPP cutting is rapid and a very liquefied slag is formed which flows away leaving a clean cut.

The customer usually owns the storage cylinders so there are no demurrage costs. Since the fuel is in the liquid form, a cylinder of MAPP fuel contains many more cubic feet of gas than an acetylene cylinder of equal size.

Cutting four duplicate parts from steel plate. The cutting torches are guided by electric motors which are controlled by an electronic device which follows the templet.
(Airco Welding Products, Div. of Airco, Inc.)

2-28A. FLASHBACK ARRESTORS

If a torch tip becomes clogged, burning fuel gas and oxygen may flow back into the hose and regulator. Use flashback arrestors to reduce this danger. These check valves close to prevent gases from reversing, Fig. 2-54A. They usually are installed between torch and welding hose, or at outlet of fuel and oxygen regulators. Different arrestors are used for oxygen and for fuel gas. Some cutting attachments have a check valve built in.

2-29. GAS WELDING ROD (Filler rod metal)

The American Welding Society defines welding rod as follows: "Filler metal, in wire or rod form, used in gas welding and braze welding, and those arc welding processes wherein the electrode does not furnish any or all of the filler metal." In this paragraph only the gas welding rods will be studied. The electric arc processes are explained in CHAPTERS 5 to 8 incl.

Some common gas welding rods are:
1. Mild steel.
2. Cast iron.
3. Stainless steel.
4. Braze welding alloys.
5. Aluminum.
   A. Drawn.
   B. Extruded.
   C. Cast.

The mild steel, braze welding alloys, stainless steel, and some aluminum rods are usually made in 36 in. lengths and are available in the following diameters:

1/16 in.	5/32 in.	5/16 in.
3/32 in.	3/16 in.	3/8 in.
1/8 in.	1/4 in.	

They are packaged in 50 lb. bundles. Mild steel welding rods are copper coated to keep them from rusting. Aluminum rods or wires are packed in 36 in. lengths or in coils. Some aluminum rods are flux coated. The coated rods are sold in 28 in. lengths.

Iron and steel gas welding rod specifications are sometimes confusing, although the problem is somewhat simplified by the fact that the old Army Air Corps, Navy, and Federal specifications are now combined into one name and one series of numbers under military (MIL) specifications. However, both the MIL and the American Welding Society (AWS) numbers are still used. In this text specification numbers refer to AWS numbers.

The following are the AWS numbers for oxyacetylene steel welding rods:

Number	Use	Tensile Strength
RG 45	mild steel rod	45,000 psi
RG 60	low alloy steel rod	65,000 psi
RG 65	low alloy high strength steel rod	105,000 psi

In these numbers, the letter "R" stands for a welding rod. The letter "G" stands for gas welding. The numbers 45, 60, and 65 indicate the approximate tensile strength of the weld in thousands of pounds per square inch (45 = 45,000 psi).

Fig. 2-55 lists the characteristics and recommended use of the RG-45 gas welding rod.

2-30. WELDING FLUXES

The American Welding Society defines flux as follows: "Material used to prevent, dissolve or facilitate removal of oxides and other undesirable substances."

Fluxes are required when gas heating certain metals with the intent to join them with solder, silver alloys, copper

TENSILE STRENGTH LBS./SQ. IN.	CARBON MAX. PERCENT	MANGANESE MAX. PERCENT	SULPHUR MAX. PERCENT	PHOSPHORUS MAX. PERCENT	SILICON MAX. PERCENT
45,000	0.06	0.25	0.035	0.025	0.03

Application: A general purpose rod recommended for welding low-carbon steel sheet, plate, and pipe. It is widely used for automotive repair work. Because of its non-flaking copper coating, it is ideal for ornamental art purposes.

Fig. 2-55. Characteristics of the AWS RG 45 Gas Welding Rod.

alloys or to weld them. The composition of a flux is determined by the specific application for which it is made. The general classification for fluxes is:
Soldering:
    Aluminum.   Sheet steel.
    Copper alloys.   Stainless steel.
    Galvanized sheet.
Brazing:
    Aluminum.   Cast iron.
    Steel.   Copper alloy.
    Steel alloy (such as stainless steel.)
Welding:
    Aluminum welding. Cast iron welding.
    Braze welding.   Stainless steel welding.

Note from the flux classifications that no flux is required for mild steel. Protection of the weld from oxidation is not so critical. The iron oxides melt at a much lower temperature than the mild steel and float to the surface.

2-31. FIREBRICKS

Firebricks are used to form welding table tops, to build forms around articles to aid with preheating, and/or stress relieving. They are useful for building up supports for articles to be welded or brazed. They can be safely used even in places which become heated, or, are exposed to the welding or cutting flame.

Firebricks are made of refractory (difficult to burn) materials.

The size of firebricks in common use is 8-3/4 x 4-1/2 x 2-1/2 in.

2-32. CARBON PASTE AND FORMS

There are many places where the welder will find this material useful. A few typical uses are:

1. For building dams to contain the molten metal when building up a broken section.

2. For protecting drilled or threaded holes which are in or adjacent to a weld or braze.

3. For building up a support for an uneven surface which is to be welded.

4. As a protecting cover over metal which is adjacent to a weld but which might be injured by spatter or heat.

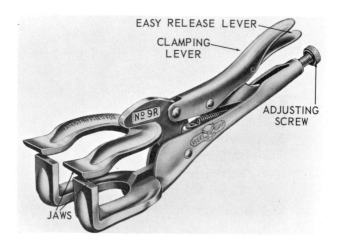

Fig. 2-56. Clamping pliers for holding stock being welded or brazed. (Peterson Mfg. Co.)

The paste form is available in various size cans. The formed carbon is available in round or square rods and in plate form. See CHAPTER 18 for more information.

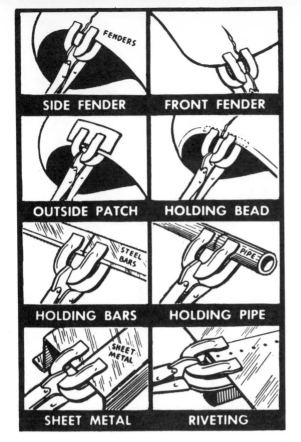

Fig. 2-57. *Application of the clamping pliers shown in Fig. 2-56.*

2-33. ASBESTOS

Caution: There may be danger in handling asbestos in any form. Inhaled powdered asbestos may cause cancer of the lung. This danger may be avoided by wearing a respirator. A temporary furnace can be constructed using sheet asbestos and firebricks. Sheet asbestos is supplied in 1/16 and 1/8 in. thickness and in 36 in. rolls.

Powdered asbestos is sometimes used in a metal box where small malleable iron castings are placed after welding or braze welding, in order that they may cool slowly. Powdered asbestos may be mixed with water and molded over surfaces to be protected during welding or brazing. This asbestos paste serves the same purpose as carbon paste.

Asbestos sheet or molded transite board is often used to support welding

exercises as they are being welded. The asbestos acts as an insulator and permits better heat control of the metal.

2-34. SEALING COMPOUNDS

To make an airtight seal on threaded pipe joints, certain proprietary compounds are recommended. Apply these sealing compounds on the male threads only. In this way, there is little danger of compound entering the pipe.

Sealing compounds are available in either paste form or tape form. One should avoid using sealing compounds that have a lead content.

SEALING COMPOUNDS WITH AN OIL CONTENT SHOULD NEVER BE USED ON OXYGEN LINES.

2-35. CLAMPS AND CLAMPING FIXTURES

The success or failure of many welding operations depends on how the metals are held in place during the welding operation. Many varieties of clamps and clamping devices have been developed for this purpose.

Pliers with clamping jaws of special design for holding and aligning parts are shown in Fig. 2-56. The pliers have deep jaws to enable clamping around obstructions. Fig. 2-57 shows various applications for this type of clamping pliers. Various shapes may be clamped for welding with the chain-clamp pliers, shown in Fig. 2-58. C-clamp pliers have

Fig. 2-58. *Quick release chain-clamp pliers. This clamp is capable of clamping material of almost any shape.*

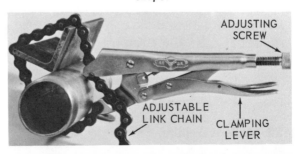

also been developed to hold parts firmly and in alignment as shown in Fig. 2-59.

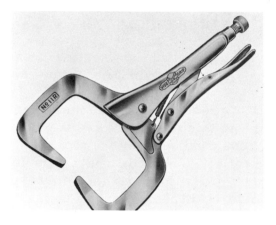

Fig. 2-59. C-clamp pliers.

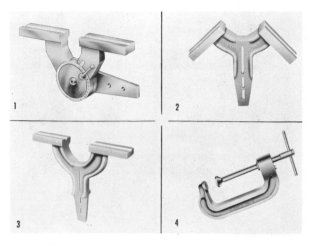

Fig. 2-60. Special alignment fixtures for holding stock which is being welded. (Wales Strippet, Inc.)

Fixtures of various designs are very convenient for aligning, holding, and positioning various shaped metals, as shown in Fig. 2-60. No. 1, in the illustration, shows a special double fixture with a protractor scale which makes quick and accurate aligning of the parts possible.

2-36. TEST YOUR KNOWLEDGE

1. What type safety device is used on an oxygen cylinder?

2. What type safety plug is used on an acetylene cylinder?

3. What type threads are used on acetylene fittings?

4. What type threads are used on oxygen fittings?

5. Of what materials are pressure-regulator diaphragms made?

6. Why must oil not be used on welding station fittings?

7. How is the hose fastened to the torch?

8. Why are the welding hoses colored?

9. What calibration gauge is used for the low-pressure oxygen when ordinary welding is being done?

10. What is the maximum pressure which should be used with a gauge whose highest calibration is 400 psig?

11. What is a two-stage regulator?

12. How may one determine how much oxygen and acetylene pressure to use with a particular tip?

13. What protection do welding goggles provide?

14. What are the two types of torches?

15. Why is a flux necessary for certain types of welding?

16. Is it necessary to use a flux when welding mild steel? Why?

17. Describe an application for carbon paste.

18. What is the size of the regulation firebrick?

19. Name some uses for asbestos sheet and packing.

20. Why are matches not recommended for lighting welding torches?

Chapter 3

OXYACETYLENE CUTTING

In oxyacetylene cutting of metal, an oxyacetylene flame is used to heat the metal, and an oxygen jet to perform the cutting. The terms "flame cutting" and "flame machining," may also be used to correctly identify the process.

The art of oxyacetylene cutting has progressed rapidly, and it is now possible to accurately flame cut both very thin, and very thick steel sections. For production work many layers of metal may be cut at the same time, which reduces both time and costs.

Oxyacetylene cutting is particularly useful when shaping metal parts prior to the fabrication of machine frames and building structures. In such fabrication, standard rolled plates or sections may be cut to accurate size, and then welded together to form a solid steel structure. Such structures are strong, economical to build, and present an attractive appearance.

3-1. HEAT OF COMBUSTION, STEEL

Steel may be thought of as a combustible material if sufficiently heated. During the process of burning, it releases a considerable amount of heat measured in British thermal units. This is called heat of combustion. The burning metal helps maintain the high temperature required in the oxyacetylene cutting process.

3-2. OXYACETYLENE CUTTING PROCESS

The oxyacetylene cutting process consists of using one or more oxy-

Fig. 3-1. A portable oxyacetylene welding and cutting outfit. (Modern Engineering Co.)

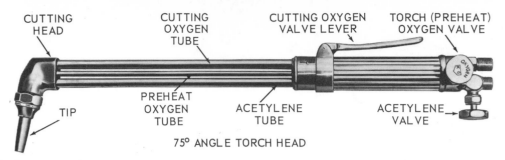

CUTTING HEAD | CUTTING OXYGEN TUBE | CUTTING OXYGEN VALVE LEVER | TORCH (PREHEAT) OXYGEN VALVE

TIP

PREHEAT OXYGEN TUBE | ACETYLENE TUBE | ACETYLENE VALVE

75° ANGLE TORCH HEAD

Fig. 3-2. Cutting torch with three tubes, which has the preheat flame mixing chamber in the torch head just above the cutting tip. (KGM Air Products)

acetylene flames to heat a spot on a piece of steel to a "cherry red" temperature (approximately 1800 deg. F.). The oxyacetylene flames are adjusted and used in the same manner as when these flames are used for welding. When the spot in the metal reaches the "cherry red" temperature, the oxygen jet is turned on, and as rapidly as the jet action cuts the metal, the torch is moved in the direction the operator wishes the cut to travel. The preheat flames are kept operating during the cutting action, as the cutting action may cease unless the heat from the preheat flames provide extra heat (the oxidation action alone will not usually supply enough heat to permit cutting to continue). As the cutting torch is moved along the line to be cut, a kerf (slot) is created behind the tip.

3-3. CUTTING OUTFIT

An outfit used for manual oxyacetylene cutting, except the torch, is similar to the oxyacetylene welding outfit shown in Fig. 3-1. The term cutting outfit is used to include all equipment required to perform a cut. A cutting

station would include the outfit, lighting, ventilation, a cutting table, and possibly a booth.

Since the torch must provide an oxygen cutting jet, it is quite different than a welding torch.

Because the oxygen pressures for cutting are usually higher than the pressures used when welding, an oxygen regulator with a higher working pressure should be used. Also, use a heavier duty oxygen hose.

Carefully review CHAPTER 1, before connecting and operating the oxyacetylene cutting outfit. Information concerning cylinders, manifolds, regulators, hoses, torches, and tips applies to the oxyacetylene cutting outfit.

3-4. CUTTING TORCH

A cutting torch is similar to a welding torch but, in addition, has a separate passageway for the oxygen jet. See OXYGEN CUTTING CHEMISTRY, CHAPTER 28.

This chapter will deal with only the procedure for oxyacetylene cutting. Also see CHAPTER 19 for other methods of cutting.

Fig. 3-3. Schematic cross sectional drawing of cutting torch with preheat mixing chamber located between the torch tubes and handle. (Linde Div., Union Carbide Corp.)

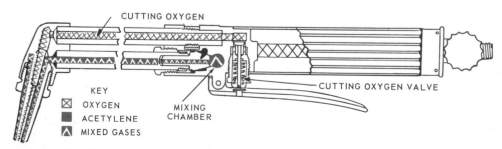

CUTTING OXYGEN

CUTTING OXYGEN VALVE

KEY
⊠ OXYGEN
■ ACETYLENE
◭ MIXED GASES

MIXING CHAMBER

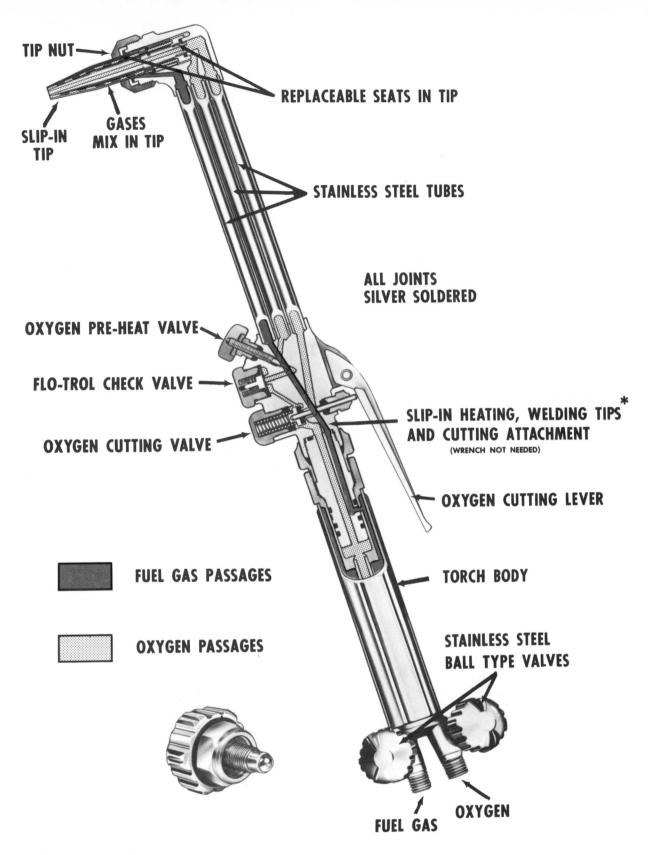

TIP NUT

REPLACEABLE SEATS IN TIP

SLIP-IN TIP

GASES MIX IN TIP

STAINLESS STEEL TUBES

ALL JOINTS SILVER SOLDERED

OXYGEN PRE-HEAT VALVE

FLO-TROL CHECK VALVE

OXYGEN CUTTING VALVE

SLIP-IN HEATING, WELDING TIPS* AND CUTTING ATTACHMENT

(WRENCH NOT NEEDED)

OXYGEN CUTTING LEVER

FUEL GAS PASSAGES

OXYGEN PASSAGES

TORCH BODY

STAINLESS STEEL BALL TYPE VALVES

FUEL GAS

OXYGEN

Fig. 3-4. A sectioned view of cutting torch attachment mounted on a welding torch handle. The preheat gases are mixed in the head of the torch. (Smith Welding Equipment Div., Tescom Corp.)

Oxyacetylene Cutting 125

METAL THICKNESS	PREHEAT ORIFICE DRILL SIZE	CUTTING ORIFICE DRILL SIZE	OXYGEN PRESSURE PSIG	ACETYLENE PRESSURE PSIG	SPEED INCHES PER MIN.
1/8 - 3/8 in.	70	67	20 - 30	3	14 - 18
3/8 - 3/4 in.	58	62	30 - 40	5	12 - 15
3/4 - 1 in.	57	54	40 - 45	5	10 - 12
1-1/2 - 2 in.	68	51	45 - 50	5	9 - 10

Fig. 3-5. Table showing approximate oxygen and acetylene pressures used when cutting sheet steel with the equal-pressure type cutting torch. NOTE: Most cutting tip manufacturers will recommend a range of at least three sizes for any cutting condition. One end of range gives most economy, other end maximum speed. For the learner the middle size is usually the best. Table is for torch using tip which provides four preheat orifices. For heavier or faster cutting a tip which provides 6-8-10-12 preheat orifices may be used.

In an oxyacetylene cutting torch the heating flame comes from several orifices arranged in a circle around a center oxygen orifice, Fig. 3-2. Fig. 3-3 shows a different design. The operator controls the cutting operation through the use of a cutting oxygen lever. In operation, a preheating flame is maintained at the tip through small orifices arranged around the center orifice. Cuts are made by depressing the cutting oxygen lever on the torch which controls the flow of oxygen from the center orifice as shown in Fig. 3-4. As in welding, cutting torch is connected to oxygen and acetylene cylinders. See PAR. 1-7 and 1-8.

Two different types of flame cutting torches are in use:

1. Equal-Pressure Torch.
2. Injector Type Torch.

3-5. LIGHTING EQUAL-PRESSURE TYPE OXYACETYLENE CUTTING TORCH

To light the equal-pressure type hand cutting torch:

1. Check the equipment for condition.
2. Inspect the regulators. Adjusting screws should be turned all the way out.
3. Open the oxygen cylinder valve very slowly until the regulator high-pressure gauge reaches its maximum reading. Then, turn the cylinder valve all the way open to close the double-seating valve. While doing this, the operator should stand to one side of the gauges.

4. Open the acetylene cylinder valve slowly 1/4 to 1/2 turn. Leave the acetylene cylinder valve wrench in place so the cylinder valve may be shut off quickly if necessary.

5. Open the torch oxygen valve one

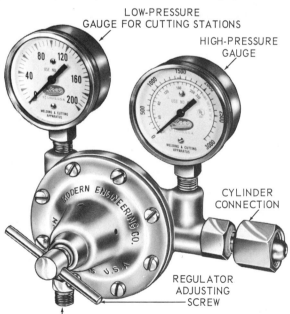

Fig. 3-6. Single-stage nozzle type oxygen regulator. The low-pressure gauge is calibrated up to 200 psig indicating this regulator is used in cutting operations.

LOW-PRESSURE GAUGE FOR CUTTING STATIONS

HIGH-PRESSURE GAUGE

CYLINDER CONNECTION

REGULATOR ADJUSTING SCREW

HOSE CONNECTION

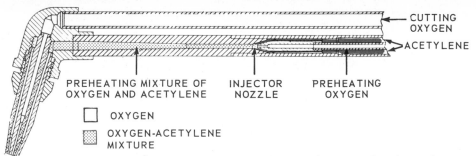

CUTTING OXYGEN

ACETYLENE

PREHEATING MIXTURE OF OXYGEN AND ACETYLENE

INJECTOR NOZZLE

PREHEATING OXYGEN

☐ OXYGEN

▨ OXYGEN-ACETYLENE MIXTURE

Fig. 3-7. Cross section of an injection type cutting torch. The acetylene is induced into the mixing chambers by the pulling action (suction) of the oxygen jet. This injector type cutting torch is particularly adaptable for use with acetylene generators which operate under low pressure.

turn. With this valve open, next open the oxygen cutting valve and adjust the oxygen regulator to give the desired operating cutting pressure. If the oxygen cutting valve is not opened while adjusting the oxygen pressure, a drop in oxygen pressure will occur when this valve is opened during the cutting operation. Result is a reduced preheat flame, and, possibly, a poor cut. See table Fig. 3-5 for oxygen and acetylene cutting pressures. Close the torch oxygen valves.

6. Open acetylene torch valve one turn. Slowly turn in acetylene regulator adjusting screw until low-pressure acetylene gauge indicates pressure corresponding to tip size, Fig. 3-5. Close torch acetylene valve. Regulator pressures are now adjusted. See single-stage regulator in Fig. 3-6.

By following this procedure, the system will have been purged.

7. To light the torch, open the torch acetylene valve approximately 1/16 turn. Then, use a flint lighter to ignite the acetylene.

8. Next increase the flow of acetylene (open the torch acetylene valve) until the acetylene flames jump away from the end of the tip slightly and back again when the torch is given a shake or whipping action. An alternate method of adjusting the acetylene after the torch is lighted, is to turn on the acetylene until most of the smoke clears from the flame. See Fig. P-2C for illustrations of

various conditions of the flame adjustment.

9. Now open the torch oxygen valve, and adjust it to a neutral flame. Open the cutting oxygen valve, and readjust the preheat flame to compensate for any oxygen pressure drop through the torch. The torch is now adjusted and is ready to be used as a cutting torch.

3-6. LIGHTING INJECTOR TYPE OXYACETYLENE CUTTING TORCH

Fig. 3-7 illustrates the typical gas flow through an injector type oxyacetylene cutting torch. The following procedure is for lighting such a torch:

1. Check the equipment to make sure all parts are in good operating condition.

2. Inspect the regulators. The adjusting screws of the regulators should be turned all the way out.

3. Open the oxygen cylinder valve very slowly until the regulator high-pressure gauge reaches its maximum reading, then turn the valve all the way open.

4. Using an acetylene cylinder wrench, slowly open the acetylene cylinder valve 1/4 to 1/2 turn. Leave the wrench in place on the acetylene cylinder valve.

5. Open the torch oxygen valve 1/4 turn. Open the torch oxygen cutting

orifice lever wide open. Adjust the oxygen regulator screw to give the correct delivery pressure, as shown in the table in Fig. 3-8. Close the torch oxygen valves.

To cut, bring the tip of inner cone of the preheating flames to the edge of the metal to be cut. The cutting torch should be held so that the inner cone on the preheat flames are about 1/16 to

METAL THICKNESS	PREHEAT ORIFICE DRILL SIZE	CUTTING ORIFICE DRILL SIZE	OXYGEN REGULATOR PRESSURE PSIG	ACETYLENE REGULATOR PRESSURE PSIG*
1/8 - 1/4 in.	75	67	15 - 20	1
1/4 - 3/8 in.	74	62	20 - 25	1
3/8 - 1/2 in.	72	59	25 - 30	1
1/2 - 3/4 in.	71	55	30 - 35	1
3/4 - 1 in.	70	54	35 - 40	1
1-1/2 - 2 in.	68	51	45 - 50	1

* The pressure as shown may be necessary in order to light the torch. After the torch has been lighted, it may be necessary to readjust the acetylene regulator so the low-pressure gauge shows no pressure.

Fig. 3-8. Table of the approximate oxygen and acetylene pressures used when cutting steel with injector type cutting torch.

6. Open the torch acetylene valve fully. Adjust the acetylene regulator delivery pressure to give the pressure required in the above table. Close the torch acetylene valve.

7. The pressures are now adjusted, and the cutting torch is ready to be lighted.

The foregoing procedure will have purged the system.

8. To light, open the torch oxygen valve 1/4 turn. Open the torch acetylene valve fully. Use a spark lighter to ignite the fuel gas.

9. Press down the oxygen cutting orifice lever and adjust the torch acetylene valve until the preheating flames are neutral.

10. The torch is now ready for oxyacetylene cutting operations.

3-7. USING A CUTTING TORCH

The cutting torch must be carefully used if cuts are to be clean and accurate. The tip must be in excellent condition, the preheat flames must be correctly adjusted, and the cutting oxygen pressure must be correct.

1/8 in. from the surface of the metal being cut, as shown in Fig. 3-9.

As soon as the surface of the metal has been heated to a cherry red or white heat color, open the oxygen cutting

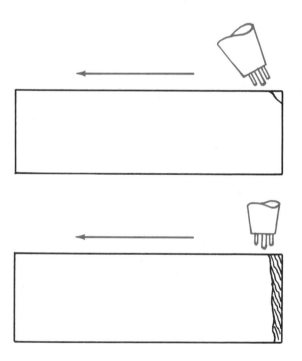

Fig. 3-9. Drawing which shows correct oxyacetylene cutting torch position for starting cut, also the required change in the torch position as the cut progresses.

valve all the way. The jet of oxygen coming through the center of the tip (oxygen jet) will cause the heated metal to burn (oxidize) away, forming the cut (kerf).

One of the best indications of a good cutting operation is the appearance of the slag stream at the bottom of the cut. The ideal slag stream slants a little in the direction of the torch motion. If the slag stream lags behind the torch tip travel, the flame adjustment, and/or the cutting pressure adjustment is too little (not enough heat and/or not enough oxygen) or, the tip travel is too fast.

3-8. CUTTING ATTACHMENTS

Most manufacturers of oxyacetylene welding and cutting torches market a cutting attachment, which may be attached to the welding torch to change it into a cutting torch. The cost of such an attachment and the welding torch on which to attach it, is usually less than the cost of a separate welding torch and a cutting torch. For portable kits, such an attachment saves space. To attach a cutting attachment, it is only necessary to remove the welding tip tube, and screw on the cutting attachment.

The operation of the torch with the cutting attachment is the same as the operation of a regular cutting torch. Fig. 3-10 illustrates a cutting attachment. Remember to open the torch handle oxygen valve wide-open and adjust the oxygen regulator to the proper cutting jet pressure.

3-9. CUTTING TIPS

The cutting tip will normally have at least two orifices. One orifice, which is usually in the center of the tip, is for the cutting oxygen and one or more smaller orifices are for preheating the metal to be cut, as shown in Fig. 3-11 (upper left corner).

Cutting tips may be of one or two piece construction. One piece tips are used for oxyacetylene cutting only. Two piece tips are used for all other gas cutting: natural gas, propane, MAPP, etc.

For satisfactory service, tips must be kept in good condition. The orifice end must be clean; the surface must be at right angles to the orifices so the preheat flame is shaped and aimed properly. The sealing faces of that part of the tip, (which fastens into the torch body) must be clean and free from scratches, burrs, nicks, etc., or these joints may leak.

Clean the tips as described in CHAPTER 4.

Always store the extra tips in soft holders (such as holes drilled in a wood block).

See CHAPTER 4 for more detailed information.

3-10. CUTTING STEEL WITH OXY-ACETYLENE CUTTING TORCH

Metals which may be cut with the oxyacetylene cutting torch may be divided into two classes:

1. Metals whose oxides have a lower

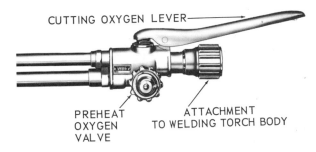

CUTTING OXYGEN LEVER

PREHEAT OXYGEN VALVE

ATTACHMENT TO WELDING TORCH BODY

Fig. 3-10. Cutting attachment which is used to convert an oxyacetylene welding torch into a cutting torch. (Victor Equipment Co.)

melting temperature than the metal.

2. Metals whose oxides have a higher melting temperature than the metal.

Practically all steels fall under the first classification and, therefore, cutting presents little difficulty. When the cutting jet is turned on, the iron oxides which form melt at a lower temperature than the base metal and are blown away, leaving a clean and straight cut. Fig. 3-11 shows a cut in progress. With an expert handling a cutting torch, or in

impossible to cut an even kerf. It is very important that these refractory oxides be reduced by chemical action or be prevented from forming. See CHAPTER 19.

Items of importance to be watched in cutting are:

1. Pressure of the oxygen fed to the cut.

2. Size of the oxygen jet orifice.

3. Speed of the cutting torch across the metal.

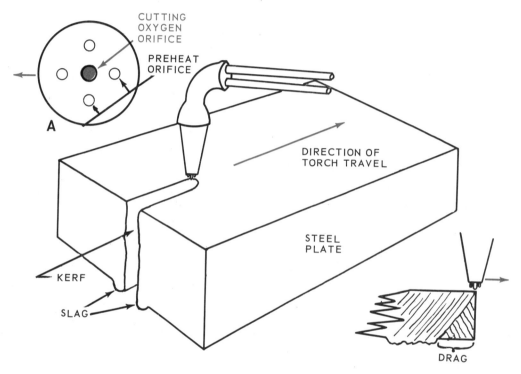

Fig. 3-11. An oxyacetylene cutting torch being used to cut a steel plate. Detail A shows end view of a typical cutting tip with a center oxygen orifice, and four preheat orifices.

automatic machine cutting, the kerf or slot formed during the cut has a smoothness of almost machine like quality.

The second group, which includes cast iron, some alloy steels, such as stainless steel, and nonferrous metals presents a complication because the oxide has a higher melting temperature than the metal. It is almost

4. Distance of the preheat flame from the metal.

5. Size (amount of heat) of the preheat flames.

6. Torch tip position (angle) relative to the metal.

It should be noted that the oxygen pressure will determine the velocity of the oxygen jet. The orifice size will determine the amount of oxygen CFM (cubic feet per minute) at any particular pressure.

The cut should proceed just fast enough to provide a slight amount of drag at the line of cutting. If the drag is too small, the oxygen consumption is too great. If the drag is large, the cutting tip orifices may be too small for the job.

Fig. 3-12 shows the result if too much oxygen is fed to the steel being cut. The cut widens out as the jet penetrates the thickness of the metal, leaving a bellmouth on the side of the metal away from the torch.

Fig. 3-13 shows the result if the torch is moved too rapidly across the work. The metal at the bottom, or far side of the cut, will not be burned away, since it did not receive enough heating and oxygen to complete the cut. Since the metal is not cut away completely, the resulting turbulent action of the torch gases will leave a very rough and irregular shaped kerf.

If the torch is moved too slowly across the work, the preheated metal will be completely burned away, and the preheating flame wasted. In such a case, the oxygen to the cutting orifice should be closed off, and the metal heated to the proper temperature by the preheating flames, before turning the cutting oxygen on again. These starting and stopping periods may also cause an irregular cut.

Metal which is very dirty and rusty should be cleaned before starting the cutting operation, as the impurities will slow the cutting speed, and may cause a rough and irregular kerf.

The torch motion to be used in cutting is a matter of the operator's own experience. Usually no motion is used. In some cases the thickness of the metal necessitates an oscillating motion in order to obtain the necessary width of cut.

The operator, when cutting, should stand in a comfortable position, and

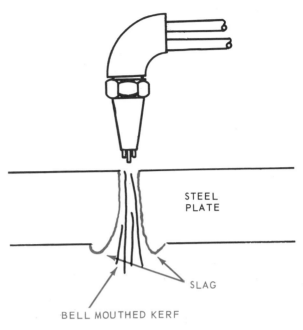

Fig. 3-12. *The effect of using too much oxygen when cutting steel. Note how the kerf widens at the bottom of the plate.*

stand where he can look into the cut as it is being formed. This positioning also means that the torch movement should

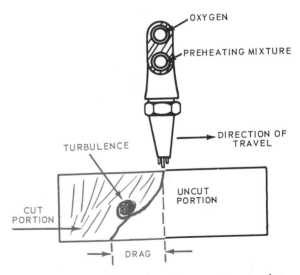

Fig. 3-13. *The effect of moving a cutting torch too rapidly across the work.*

be away from the operator, rather than toward him, in order that he may see into the kerf. The torch is usually held with both hands.

drops from the cut. A container should be placed under the cut, to catch the very hot liquid slag. It is preferred that this container be refractory lined.

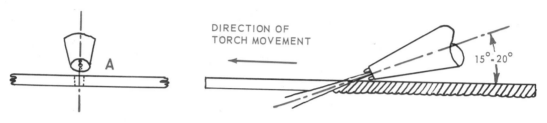

Fig. 3-14. A recommended procedure for cutting thin steel. Notice that the two preheat flames are in line with the cut (kerf).

Normally the tip is perpendicular to the surface being cut. The end of the preheating flame, inner cone, should just avoid touching the metal.

The operator should wear asbestos leggings, and the cuffs of the trousers should be covered to keep them from catching the white hot metal slag as it

3-11. CUTTING THIN STEEL

Cutting steel of 1/8 in. thickness or less requires the use of the smallest cutting tip available. In addition, the tip is usually pointed in the direction the torch is traveling. By angling the tip, the preheating flames have a chance

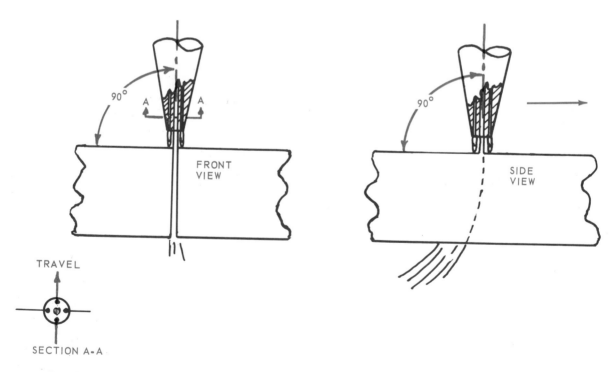

Fig. 3-15. A recommended technique for cutting thick steel. Note the position of the torch tip preheat orifices in relation to the line of the cut (kerf). Two preheat flames are in the line of the torch progress. This position enables one preheat flame to be ahead of the cut, two flames to heat the sides of the cut, and one flame to heat down in the kerf.

to heat the metal ahead of the oxygen jet, as shown in Fig. 3-14. If the tip is held vertical to the surface, the small amount of metal preheated may be cooled by the adjacent metal, and prevent a smooth cutting action. Many welders actually rest the edge of the tip on the metal during this process. Be careful to keep the end of the preheating inner cone just above the metal.

cutting progress. A torch movement from right to left enables the operator to follow the guide lines on the metal. Fig. 3-16 shows the progress of a cut in thick steel.

After the edge has been heated to a dull cherry red, the oxygen jet should be opened all the way, by pressing on the cutting lever. As soon as the cutting action starts, move the torch tip at a

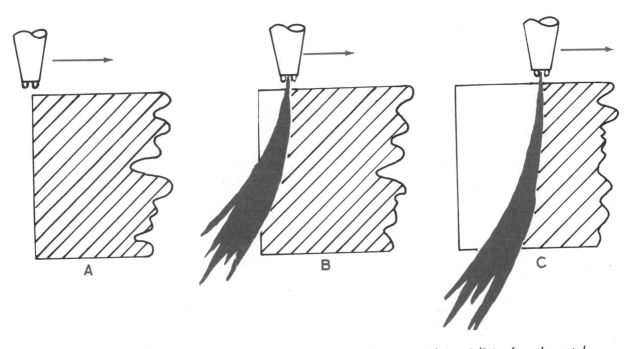

Fig. 3-16. Progress of a cut in thick steel. A. Preheat flames are 1/16 to 1/8 in. from the metal surface. The torch is held in this spot until the metal becomes cherry red. B. Note that the torch is moved slowly to maintain the rapid oxidation, even though the cut is only part way through the metal. C. Note that as the cut is made through the entire thickness, the bottom of the kerf lags behind the top edge slightly.

3-12. CUTTING THICK STEEL

Steel over 1/8 in. thickness may be cut by holding the torch so the tip is almost vertical to the surface of the base metal being cut. Fig. 3-15 shows the position of the cutting torch tip orifice when cutting thick steel.

One way is to start at the edge of the stock, and move from left to right, if right-handed, as this permits the operator to look into the kerf and check

steady rate. Avoid an unsteady movement of the torch, or the cut will be irregular and the cutting action may stop. Fig. 3-17 shows how to cut thick steel.

To start a cut faster in thick plate, the operator may start at the corner of the metal by slanting the torch in a direction opposite the direction of travel, as shown in Fig. 3-9. As the corner is cut, the operator moves the torch to a vertical position, until finally the total

Fig. 3-17. A cut in progress on thick steel.

thickness is cut, also shown in Fig. 3-9, and the cut may proceed.

Two other methods used to start cuts are as follows:

One is to nick the edge of the metal where the cut is to start, with a cold chisel. The sharp edges of the metal upset by the chisel will preheat and oxi-dize rapidly under the cutting torch, and the cut may be started without preheating the entire edge of a thick plate.

The second method is to place an iron filler rod under the preheating flames at the edge of a thick plate. The filler rod will reach the cherry red temperature quickly and when the cutting oxygen is turned on the rod will oxidize and cause the thicker plate to start oxidizing.

3-13. CUTTING CHAMFERS (BEVELS)

Another important torch cutting operation is the cutting of chamfers or bevels for preparing the edges of steel plate being made ready for joining by welding, as shown in Fig. 3-18. The thicker pieces of steel must have the chamfered edge preparation, so weld will penetrate through the thickness of the metal.

The bevel angles may be cut at the same time the metal is being cut to size and shape, or, they may be cut as a separate operation.

Instructions relative to cutting thick metal, given previously, are usable when cutting bevel angles.

It is important to obtain a high quality cut when making the bevel, to minimize any further plate preparation, and to insure good fit-up of the joint to be welded.

3-14. CUTTING PIPE OR TUBING

One of the most popular uses of the oxyacetylene cutting torch is the preparation of pipe for joining by welding. See CHAPTER 18, for more pipe welding instructions. The cutting torch is especially useful for preparing odd-shaped joints on the job. Also, it may be used extensively to chamfer the edges of thick pipe to provide a V-joint for the welder.

The exact procedure to follow when cutting pipe depends on the diameter. For small diameter pipe, it is best to keep the torch tip almost tangent to the circumference of the pipe (same distance from surface when moving around

the pipe) during the complete cutting operation, instead of attempting to cut through the two thicknesses of the pipe simultaneously. With a pipe diameter of approximately four inches and up, it is possible to keep the torch tip perpendicular to the pipe surfaces while cutting, without the torch having a

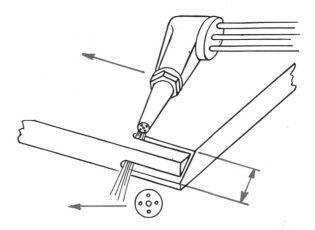

Fig. 3-18. A method of obtaining a beveled edge on thick metal plate using a cutting torch.

tendency to burn through the other side. Of the two methods, the perpendicular position permits a cleaner and straighter cut. If the welder's helper rotates the pipe as it is being cut, a very clean cut can be obtained. Most welders start the cut at the extreme edge of the pipe, and then cut back to the marked chamfer ring.

When chamfering pipe by hand, it is best to point the torch toward the end of the pipe. This procedure provides a clean chamfer, permits more accurate cutting, and should not leave any excessive oxide clinging to the pipe when the cut is completed. Cutting machines which use a mechanism to revolve the torch around the pipe produce excellent chamfered edges.

Fig. 3-19 shows how to use a cutting torch to bevel or chamfer the end of a pipe which is to be welded. It must be remembered that when cutting pipe, it is not the size of the pipe that determines the size of the cutting torch tip, but rather the thickness of the pipe wall which is the controlling factor. When steel or alloy steel pipe is to be used for high-pressure purposes, it is very important that the approved procedures be carefully followed.

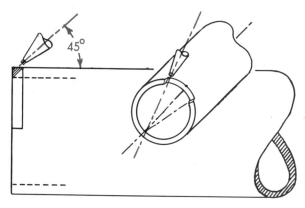

Fig. 3-19. A cutting torch being used to cut a bevel on a steel pipe.

3-15. PIERCING AND CUTTING HOLES

Holes may be pierced in steel plates rapidly and with accurate results. To pierce is to produce a small hole (in comparison to the size of the metal surface) through a steel plate. The process consists of holding the cutting torch with the nozzle perpendicular to the surface of the metal, and preheating the spot to be cut until it is a bright, cherry red. After the metal is brought up to the proper temperature, the oxygen jet may be turned on very slowly. At the same time, the nozzle should be raised enough to eliminate the slag being blown back into the nozzle orifices. Because of the great amount of heat required in an operation of this kind, it is recommended that the operator use at least the next larger tip in

relation to the thickness of the metal, than recommended in Figs. 3-5 and 3-8. See PAR. 19-3 for use of oxygen lance for cutting holes in thick plate.

To cut larger holes in steel plate, the typical steel-cutting method described in this Chapter is recommended. It is good practice, however, to outline the hole first, using special chalk so this line may be used as a guide to permit the operator to cut as accurate a hole as possible. If the size of the hole warrants it, it is best to do the cutting with an automatic machine, or with a radius bar attachment clamped to the torch head.

3-16. CUTTING AND REMOVING RIVET HEADS

The cutting torch, in salvage operations, is frequently used for removing rivet heads when dismantling large fabricated structures which have been assembled with rivets.

Two typical rivet shapes are:
1. Button head rivet.
2. Countersunk rivet.

The procedure for removing these heads by cutting is fundamentally the same as for any cutting operation, but one additional precaution should be observed. If possible, the operator should do the cutting without damaging the steel plate. To do this, it is very important that the size of the tip is carefully selected. If too large a tip is used, the steel plate will be injured at the same time the rivet head is being removed. If too small a tip is used, the method becomes too slow. Practically all welding equipment companies recommend special shaped cutting tips for rivet cutting.

The procedure for cutting the button head rivet is to preheat the head of the rivet to a bright cherry red, the body

An automatic flame cutting machine equipped with two cutting torches.
(Airco Welding Products, Div. of Airco, Inc.)

of the steel plate usually being adequately protected from the preheating by a coating of scale. The rivet head is then cut straight through its thickness. If the rivet is in the horizontal position, one-half the head is removed, preferably the lower half, and finally the upper half. The appearance of an accurately performed job shows a clean removal of the head without any score marks from the cutting jet on the steel plate, Fig. 3-20.

rivet heads may be removed with very little, if any, damage to the steel plate.

3-17. GOUGING WITH CUTTING TORCH

Cutting a curved groove on the edge or surface of a plate, and removing bad spots in welds to prepare for rewelding are additional uses for the cutting torch. The principle of groove cutting or

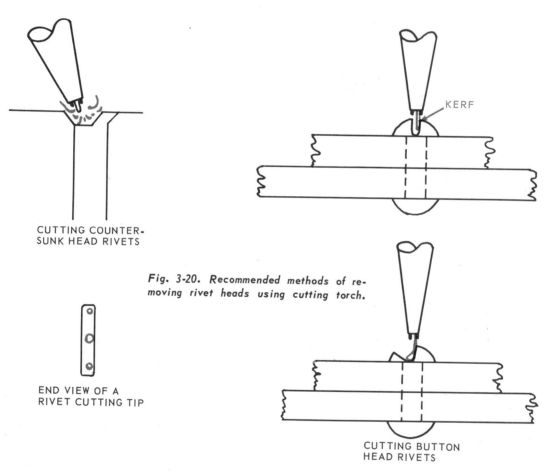

CUTTING COUNTER-SUNK HEAD RIVETS

END VIEW OF A RIVET CUTTING TIP

KERF

CUTTING BUTTON HEAD RIVETS

Fig. 3-20. Recommended methods of removing rivet heads using cutting torch.

Removal of a countersunk rivet head is more difficult than the button head rivet. The rivet is usually tightly embedded in the plate. However, by carefully selecting the tip size, and by proceeding with the cutting from the bottom of the rivet head upward, after the head has been preheated, countersunk

gouging is that instead of using a high-velocity jet as in a cutting tip, a large orifice low-velocity jet is used. The slower moving oxygen stream oxidizes the surface metal only and penetrates very slowly enabling the operator to gouge or groove with considerable accuracy.

In a gouging tip, there are usually five or six preheat orifices to provide for even preheat distribution. In an automatic machine, a gouging tip will do very accurate gouging to exact depths to remove bad spots, and to quickly and conveniently prepare the edges of metal

Fig. 3-21. A typical oxygen gouging operation. A low-velocity cutting jet is used to enable better control of the gouge width and depth.

for welding. Fig. 3-21 illustrates a typical oxygen gouging operation.

If the gouging cut is not started properly it is possible to cut too deep and actually cut through the entire thickness. It is also possible to cut too shallow and cause the operation to stop. The speed at which the torch is moved along the gouge line is important. Too rapid a movement will create a narrow, shallow gouge, and too slow torch movement will create a gouge which is too deep and wide.

3-18. CUTTING FERROUS ALLOY METALS

The introduction of many alloy steels into industry has made it necessary for new cutting techniques to be developed in order that these metals may be successfully and economically cut. Of the alloy steels, stainless steel is perhaps the most widely used. Stainless steel consists of chromium, nickel, and mild steel, see CHAPTERS 18, 23, and 24. Many of these metals have melting temperatures below that of steel. The oxides formed have a melting temperature higher than that of the original metal, therefore, the oxides must be reduced and/or removed from the cut immediately, or the cutting action will stop. It has been found that for the same relative thickness of metal, stainless steels need approximately 20 percent more preheating flame and 20 percent more oxygen for the cutting. It has also been found that it is a good practice to use a slightly carburizing flame when preheating.

The metal to be cut should be so placed that the cutting tip and flame are in a horizontal position. The cut should start at the top of the metal and proceed downward in a vertical line. A slight, but quick, up-and-down motion of the torch facilitates the removal of the slag. Fig. 3-22 shows how

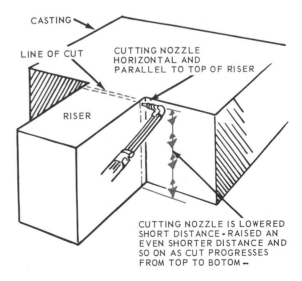

CASTING

LINE OF CUT

CUTTING NOZZLE HORIZONTAL AND PARALLEL TO TOP OF RISER

RISER

CUTTING NOZZLE IS LOWERED SHORT DISTANCE - RAISED AN EVEN SHORTER DISTANCE AND SO ON AS CUT PROGRESSES FROM TOP TO BOTOM —

Fig. 3-22. A technique used for the oxyacetylene cutting of chromium steel. Note how the torch is raised and lowered to assist in removing the slag from the cut.

the torch should be moved up and down to facilitate the slag removal. It will be found difficult to obtain as clean and narrow a kerf when cutting alloy metals as when cutting straight carbon steels, using a regular cutting torch.

As in the case of steel, the alloy metals must be preheated before the cutting operation is started. The stainless steels especially, must be preheated to a white heat before the cutting is successful. The cutting action is much more violent with stainless steels than with straight carbon steels. Cutting takes place with considerable sparking and blowing of the slag.

In situations where the progress of the cutting is frequently interrupted by the presence of unmeltable slag, the operator may find it advisable to hold a mild steel welding rod in the kerf of the metal. This mild steel rod, upon being melted by the cutting torch mixes with the alloy steel, and thus dilutes or reduces the percentage of the alloys in the area of the cut. The cutting properties of the alloy metal in the area of the cut thus becomes more like those of mild steel and the cut proceeds more smoothly.

The adding of welding rod to the cut is also applicable to poor grade steels, cast irons, and to old, oxidized, steel castings.

Powder cutting, plasma arc, and inert gas arc cutting have proven much more practical for cutting steel alloys and nonferrous metals. For full details refer to CHAPTER 19.

3-19. CUTTING CAST IRON

As mentioned previously, it is more difficult to cut cast iron than steel, because iron oxides of cast iron melt at a higher temperature than the cast iron itself. However, successful cut-

ting has been performed on cast iron in many salvage shops and foundries. It is important, when cutting cast iron, to preheat the whole casting before the cutting is started. The metal should not be heated to a temperature that is too high, as this will oxidize the surface and make cutting difficult. A preheat temperature of about 500 deg. F. is usually satisfactory.

When cutting cast iron, the preheating flame of the torch should be adjusted to a carburizing flame to prevent oxidation from forming on the surface before the cutting starts. The cast iron kerf is always wider than a steel kerf because of the oxidation difficulties. If it is desirable to have gray cast iron (machinable) the casting should be cooled very slowly.

Since it is difficult to cut cast iron with the usual oxyacetylene cutting torch, other more satisfactory methods of cutting have been developed, such as the oxygen arc, carbon arc powder cutting and inert gas cutting. See CHAPTER 19, for a description of these cutting processes.

3-20. AUTOMATIC CUTTING

Automatic cutting machines are being constantly improved. In addition to the powder and flux methods of cutting stainless steels, low alloy steels, cast iron, etc., cutting machines have been improved by the use of electric solenoid valves to control gas flow and electronic and magnetic devices for controlling the torch movement. The electronic tracers enable extremely accurate following of a template, and consequently the torch will produce almost perfect duplicate shapes. A light beam is focused downward from the apparatus onto a line drawing of the part to be cut. A photoelectric cell picks up the

TEMPLATE
AND
TRACER

Fig. 3-23. A three-torch automatic oxy-fuel flame cutting machine in operation. Note the template and the electronic template tracer on the left. (Airco Welding Products, Div. of Airco, Inc.)

light reflected upward from the line. Adjustments are made so that once the machine is set to trace a line the machine automatically follows this line with great accuracy. Electronic devices have also been developed to maintain a constant tip height over the metal being cut. This device also helps insure good quality cuts at all times.

There are many automatic mechanisms available, which perform automatic cutting operations, using the oxyacetylene cutting torch. Special mechanisms are used for cutting definite numbers of certain shaped objects; for cutting a number of the same size objects at the same time; for cutting irregular shaped articles, using an inexpensive template as a guide; for cutting straight kerfs and for chamfering metal. CHAPTER 19 describes semiautomatic and automatic cutting processes.

Practically all of the automatic cutting torches are driven by variable-speed electric motors. Those used in production of duplicated articles by the thousands are usually cam-actuated, while those used to reproduce a certain number of the same size articles simultaneously, generally use multiple torches.

A popular type of automatic cutting machine is one which follows the contour of a steel, wood, or aluminum pattern, and cuts a duplicate shape in the metal. Fig. 3-23 shows an automatic cutting machine in operation. When steel patterns are used, a magnetic tracer can be used as the guide device. The variable-speed, electric motor permits various speeds in feet per minute. This variable speed device enables the machine to adapt to different thicknesses of metal. The guide for the torch is a small rail which has been shaped to conform to the wood or metal pattern. The machine has a specially calibrated tachometer (speed counter) which registers in feet per minute.

The operation of these automatic machines necessitates three important adjustments:

1. Adjustment for the rapidity of the cut, which is controlled by the speed of the motor through a gear train.

2. Gas pressure must be carefully adjusted to insure a clean cut through the thickness of the metal without wasting gas.

3. Distance the torch tip is from the metal being cut must be carefully adjusted to obtain the best results. This adjusting is done by means of a graduated scale on the torch body, and a gear vernier mechanism (measuring device which permits fine adjustment), to raise and/or lower the torch.

Once set up, the apparatus is self-operating. However, the initial adjustments must be very carefully made.

3-21. REVIEW OF SAFETY IN OXYACETYLENE CUTTING

Oxyacetylene cutting may be safely performed. However, as in all oxyacetylene welding and cutting, proper procedures must be followed in order to eliminate dangers which might be caused by wrong handling or carelessness. Certain potential hazards exist and the operator should always observe approved procedures to eliminate these dangers. In addition to observing the correct handling procedures for oxygen and acetylene equipment, the following points should be carefully observed:

Considerable sparking and flying of sparks, which are globules (round-shaped particles) of molten metal, accompanies cutting operations. In order to avoid accidents from this hazard:

A. Floors on which cutting is done should be concrete or other fireproof material.

B. Workbenches and other necessary furniture should be of metal or other fire-resistant material.

C. Oil, paper, wood shavings, gasoline, lint, flammable materials should not be in the room in which flame cutting is performed.

D. Certain clothing features should be carefully checked. Leather or other slow-burning fabrics should be worn. Trousers should be without cuffs. In many situations, the operator and his clothing may be protected by a sheet metal shield which will deflect the sparks from the operator. Pockets and clothing should be inspected for possible flammable materials such as buttons, combs, celluloid rules, matches, pencils and the like.

E. Objects being flame cut may present inherent hazards. Tanks and containers may be welded or cut only by an experienced welder who has the equipment required to steam the tank or pass an inert gas through the tank as it is being worked on. The tank may also be filled with water except in the area of the work.

In all cases the tank must be vented to prevent the entrapment (holding) of potentially explosive gases.

This work should never be done except under the supervision of a qualified safety engineer. When flame cut, a small amount of flammable material in such an object may cause a powerful explosion. Certain metals such as magnesium may burn with an explosive force, if flame cut. Be certain of what is being cut.

F. Face and hands must be protected from metal splatter.

G. A fire extinguisher should always be at hand while flame cutting, for use in possible emergencies.

H. In CHAPTER 1, the correct procedures for setting up and handling cylinders, regulators, hoses and torches were explained. Be sure to know these procedures. It is a good idea to review this Chapter.

I. Review PAR. 1-33, in CHAPTER 1, on metal fume hazards, and be sure to know what metal is being worked on before performing any oxyacetylene cutting operation.

J. The normal atmosphere contains about 21 percent oxygen by volume. As the oxygen content in an enclosed space is increased above this percentage, there is an increasing danger of a spark or flame causing a fire or explosion.

3-22. TEST YOUR KNOWLEDGE

Carefully review these questions. Answering them will check your knowledge of flame cutting as covered in this Chapter.

1. Which is the correct word to describe this process "Flame cutting" or "burning"?

2. In what chapter are other methods of cutting explained?

3. How does a Cutting Torch differ from a Welding Torch?

4. What are the two basic types of flame cutting torches in common use?

5. In adjusting the pressures before lighting a cutting torch, why is it necessary to check the pressures with the torch cutting lever in the fully open position?

6. Why is it necessary to provide a higher oxygen pressure with the cutting torch than with a welding torch?

7. How may some welding torches be adapted to do oxyacetylene cutting?

8. In what way does a cutting tip differ from a welding tip?

9. What is the purpose of the preheating jets or orifices in a cutting torch tip?

10. How should the cutting torch be held, in relation to the work, when cutting thin sheet steel?

11. What controls the size tip to be used when cutting pipe?

12. Why is cast iron difficult to cut with the oxyacetylene flame?

13. What procedure is recommended when it is necessary to cut stainless steel with the oxyacetylene cutting torch?

14. What are the advantages in using automatic flame-cutting machines?

15. What safety precautions must be considered before starting a flame-cutting operation?

16. What is the name of the slot created by a cutting jet?

17. Why must the oxygen jet pressure be accurately adjusted?

18. What is drag?

19. How do you determine when to open the oxygen cutting jet?

20. Why must the end of the cutting tip be square or perpendicular to the orifices?

21. What happens to the material cut away by the oxygen jet?

22. Why is it possible to cut off rivet heads without injuring metal plates riveted together?

23. What happens to the gouge shape when the torch movement is made slower?

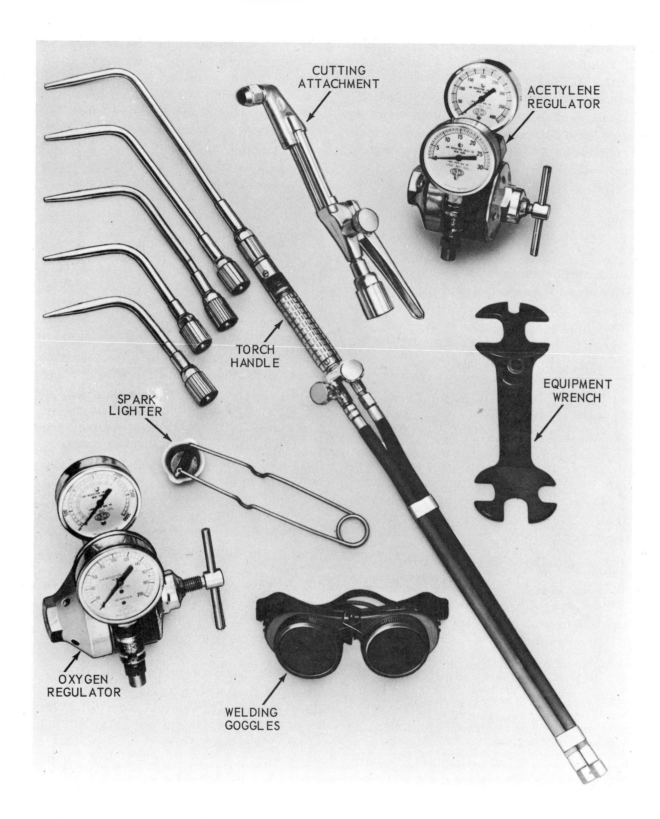

CUTTING
ATTACHMENT

ACETYLENE
REGULATOR

TORCH
HANDLE

EQUIPMENT
WRENCH

SPARK
LIGHTER

OXYGEN
REGULATOR

WELDING
GOGGLES

Fig. 4-1. Oxyacetylene welding and cutting outfit less cylinders.
(Airco Welding Products, Div. of Airco, Inc.)

Chapter 4

OXYACETYLENE CUTTING EQUIPMENT AND SUPPLIES

Oxyacetylene cutting deals with high pressures, flammable gases, flying sparks, rough handling, and other severe services. Therefore, the equipment must be reliable and rugged, and yet quick to respond to changes in demand.

The design and construction of an oxyacetylene cutting outfit is similar to an oxyacetylene welding outfit. There is a major difference, however, in the design of the torch and in the oxygen regulator for heavy-duty cutting.

It is very important, for the sake of safety, for quality of results and for economical operation, that the operator become thoroughly familiar with the design and construction of this equipment, and its correct use.

4-1. COMPLETE GAS CUTTING OUTFIT

The complete gas cutting outfit consists of:
Cylinder truck.
Oxygen cylinder.
Oxygen regulator.
Oxygen hose and fittings.
Acetylene cylinder.
Acetylene regulator.
Acetylene hose and fittings.
Cutting torch.
This equipment less the truck and cylinders is shown in Fig. 4-1.

To describe the equipment, it may be best to first describe each part of the complete apparatus in the order in which the complete outfit will be assembled.

4-2. CYLINDER RACK OR CYLINDER TRUCK

The cylinders must be securely fastened in an upright position with a chain or clamping device. If the cutting outfit is portable, the cylinder truck must be designed in such a way that it is resistant to tipping. The cylinders must be securely fastened to the cylinder truck as shown in Fig. 4-2. The possi-

Fig. 4-2. Oxygen and acetylene cylinders mounted on a portable truck and secured by a safety chain. (Victor Equipment Co.)

bility of injury to the cylinders must be reduced to a minimum.

4-3. THE CUTTING TORCH

The torch must mix the acetylene-oxygen gases and carry the gases to the orifices, where they combine chemically to become the preheat flames. It must also carry the oxygen to an orifice where, as it emerges, it oxidizes the metal and blows it away to form a clean cut (kerf).

The torch is constructed mainly of a yellow brass body, and a copper tip. It is equipped with three valves:

1. An acetylene valve similar to the welding torch.

2. An oxygen valve similar to the welding torch.

3. A cutting oxygen valve, button or lever operated with an automatic spring closing device.

cutting operations only, as shown in Fig. P-2B.

The combination welding and cutting torch is most popular in small shops where welding and cutting are auxiliary operations to the work in the shop.

4-4. CUTTING TORCH TIPS

Just as in welding, the proper size cutting tip is very important, if quality work is to be done. The preheat flames must furnish just the right amount of heat, and the oxygen jet orifice must deliver the correct amount of oxygen at just the right pressure and velocity to produce a clean cut (kerf). All of this must be done with a minimum consumption of oxygen and fuel gases. All too often, careless workers or ones not acquainted with the correct procedures, waste both oxygen and acetylene.

Each manufacturer has cutting tips

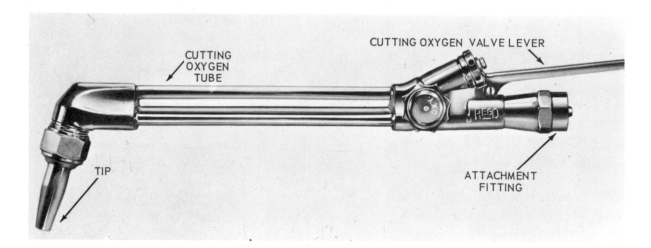

Fig. 4-3. *Cutting torch attachment. This unit may be attached to a welding torch handle after the tip and tube have been removed from the welding torch. Note: This torch is designed for cutting only.*
(Chemetron Corp.)

A cutting torch may consist of an attachment which may be fastened to a welding torch body as shown in Fig. 4-3, or it may be a torch designed for

of different designs. The orifice arrangements and the tip material, copper alloy, are much the same among various manufacturers, however that

NUMBER OF PREHEAT ORIFICES	DEGREE OF PREHEAT	APPLICATION
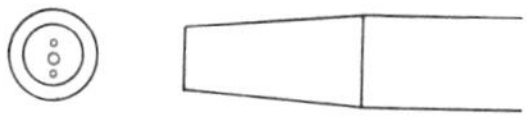 2	Medium	For straight line or circular cutting of clean plate.
 2	Light	For splitting angle iron, trimming plate and sheet metal cutting.
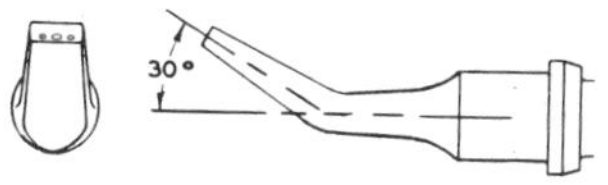 2	Light	For hand cutting rivet heads and machine cutting 30 deg. bevels.
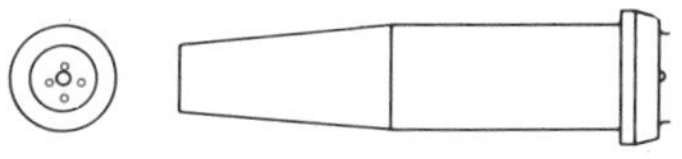 4	Light	For straight line and shape cutting clean plate.
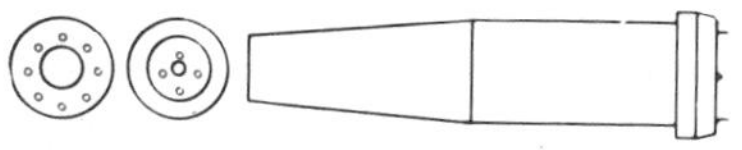 4, 6, 8	Medium	For rusty or painted surfaces.
 6	Heavy	For cast iron cutting and preparing welding V's.
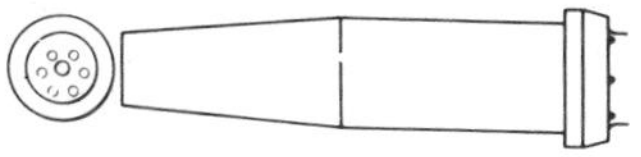 6	Very Heavy	For general cutting also for cutting cast iron and stainless steel.
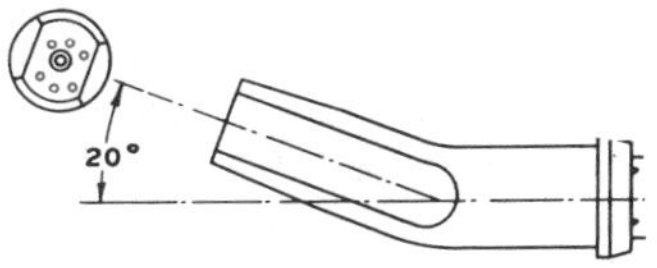 6	Medium	For grooving, flame machining, gouging and removing imperfect welds.
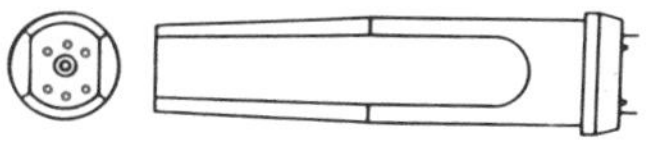 6	Medium	For grooving, gouging or removing imperfect welds.
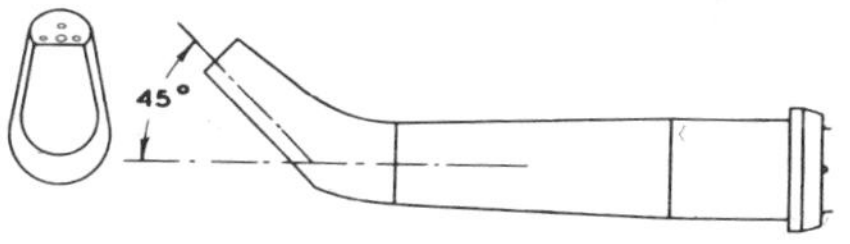 3	Medium	For machine cutting 45 deg. bevel or hand cutting rivet heads.
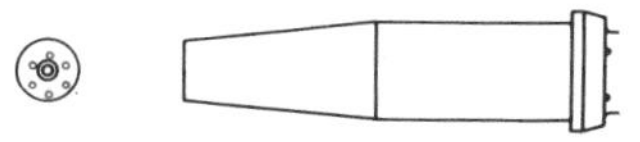 6	Heavy	Flared cutting orifices provides large oxygen stream of low velocity for rivet head removal (washing).

Most of the above cutting tips are available in two or more sizes and should be selected on the basis of the thickness of the metal and the job to be performed.

Fig. 4-4. Table of some common cutting torch tips and their uses.

part of the tip which fits into the torch head often differs in design. Fig. 4-4 is a table showing several different orifice arrangements and their uses, while Fig. 4-5 illustrates several cutting tip designs.

The internal construction of one type cutting tip and head which is designed to mix the preheat gases at the tip is shown in Fig. 4-6. Some cutting torches are designed to mix the preheat flame gases in the handle. A typical tip design

Oxyacetylene Cutting Equipment 147

Fig. 4-5. Several types of cutting tips.
(National Welding Equip. Co.)

for this type cutting torch is shown in
Fig. 4-7. The tips and seats are designed and constructed to produce a
good flow of gases, to keep the tips as

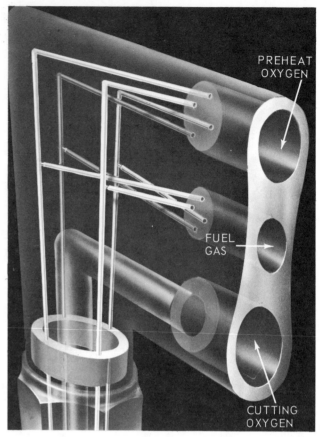

Fig. 4-6. Phantom view of cutting torch head showing gas passageways.

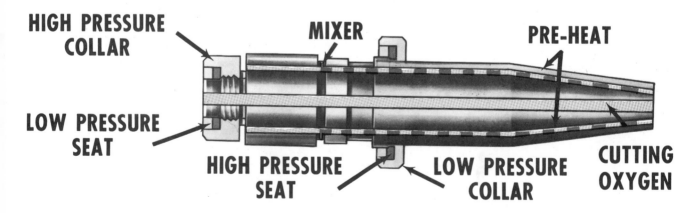

Fig. 4-7. A cutting tip cross section.

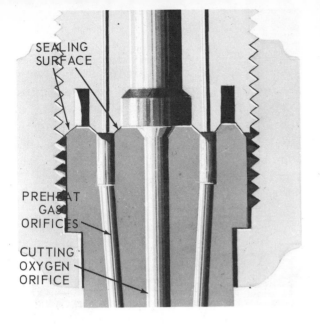

Fig. 4-8. Cutting tip cross section which shows details of the cutting tip seating design.

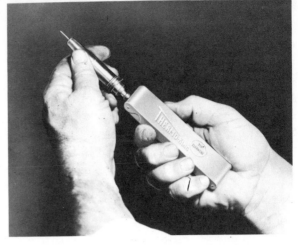

Fig. 4-9. Cutting tip orifice being cleaned. (Thermacote Co.)

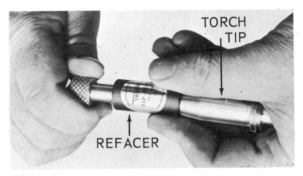

Fig. 4-10. Special tool used to recondition flame end of cutting tip. (Thermacote Co.)

cool as possible, and to produce leak-proof joints. If these joints leak, the preheat gases may mix with the cutting oxygen or they may escape to the atmosphere. Fig. 4-8 shows the details of a cutting tip seating design.

It is very important that the orifices and passages be kept clean, free of burrs to permit free gas flow and a well-shaped flame. Fig. 4-9 shows a cutting tip orifice being cleaned. Fig. 4-10 shows a tool used to recondition the flame end of a cutting tip. Fig. 4-11 shows several tips; one in need of repair, two beyond repair, and one in good

Fig. 4-11. Four cutting tips. One is in good condition, one is repairable, and two must be discarded. (Airco Welding Products, Div. of Airco, Inc.)

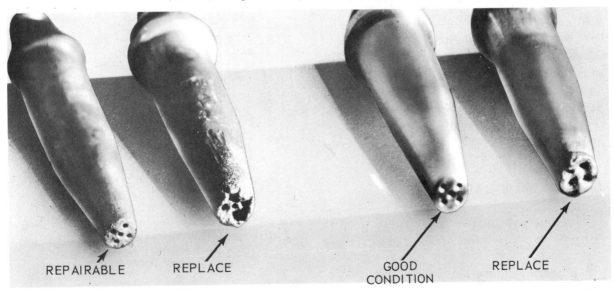

condition. Since it is extremely important that the sealing surfaces be kept clean and free of scratches or burrs, the tips should be stored in a container that cannot scratch the seats, preferably an aluminum or wood rack.

4-5. REGULATORS

The standard acetylene regulator is usable for cutting stations. However, the oxygen regulator is quite often a heavy-duty type designed for cutting applications. Since the oxygen cutting pressures sometimes reach as much as 100 to 150 psig and because the gas flow is quite high, the oxygen regulators usually have heavy-duty springs and a high capacity regulator orifice. The two-stage regulator is often preferred for cutting applications. The regulator fittings used to attach the regulator to the cylinders are standard and so are the hose fittings. If heavy cutting is to be done it is advisable to check the oxygen regulator by model number and manufacturer's specifications to be sure the correct regulator is being used.

4-6. TORCH GUIDES

The operator should always try to cut a smooth kerf and to cut accurately to a dimension. Freehand cutting (holding the torch in your hands) makes both of these objectives very difficult. Many mechanical, electrical and electronic devices have been developed to help produce clean cuts, accurate size cuts, and exact duplicate pieces. Fig. 4-12 shows a simple torch guide.

4-7. MECHANICAL GUIDES

Mechanical guides are used to help control the position of the torch. They do not control the speed of the cut, and therefore, the operator must be very skilled, otherwise rough, ragged cuts may result. To cut straight edges, a piece of angle iron may be clamped to the work and the tip moved along the edge of the angle iron to insure a straight cut as shown in Fig. 4-13.

Fig. 4-12. Aid to cutting straight lines with torch. Put a 3/4-in. band-type clamp on cutting torch tip, as shown. Fasten clamp so end of band will ride on top of piece of angle iron clamped onto stock to be cut. Locate so tip of torch is right distance above metal. To cut in straight line, hold torch tip against angle iron as you cut.
(James Hoerner, Pioneer H. S., San Jose, Calif.)

Another popular type guide makes use of a small steel wheel mounted on the torch tip to reduce the friction. The

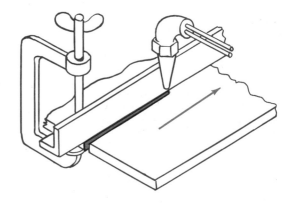

Fig. 4-13. Straightedge used as a cutting torch guide. Use two or more clamps to hold angle iron.

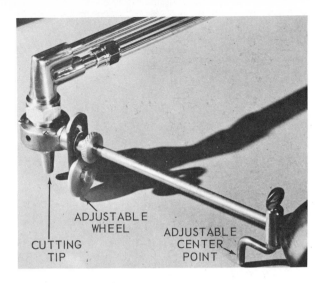

Fig. 4-14. Circle-cutting guide. The center of the circle to be cut should be marked with a center punch. (Victor Equip. Co.)

wheel can be used to control the height of the tip above the metal, or it can be used to allow the tip to move more smoothly along the guide.

A metal, plaster, or wood template is often used when the cut is to be irregular.

To cut arcs or circles, a circle guide may be used. This device consists of a rod, adjustable in length, with a center pivoting point and a means of holding the tip. Fig. 4-14 illustrates such a device. Fig. 4-15 shows the circle guide in use.

Fig. 4-15. Circle cutting guide in use. Note safety clothing worn by this operator. (Smith Welding Equip. Div., Tescom Corp.)

Guides are also obtainable which may be adjusted to produce a torch movement smaller, the same size, or larger than the template: In these, torches are mounted on a pantograph frame, and as the operator moves a tracer around a template or sample, the torch moves to produce a duplicate shape in some ratio to the size of the template. For example, the cut piece may be one half the size of the template or the machine can be adjusted to produce a cut piece twice the size of the template. Fig. 4-16 illustrates a pantograph device.

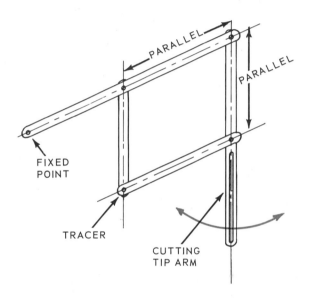

Fig. 4-16. Pantograph cutting torch holder used with a cutting torch.

4-8. ELECTRICAL GUIDES

An improvement over the mechanical guides is an electric motor driven guide. This unit has a variable speed motor that not only enables a welder to cut to a dimension, but also to cut at a certain set speed. This device usually has four wheels, one driven by a reduction gear, two on swivels (caster style) and one freewheeling. The torch is mounted on the side of the body and is adjusted up and down by a gear and

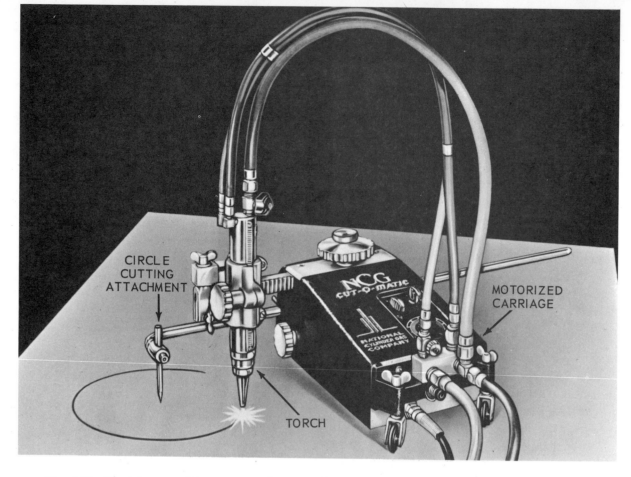

Fig. 4-17. Electric motor driven carriage being used to cut circle in steel plate.   (Chemetron Corp.)

rack. The rack is a part of the special torch. The torch can also be tilted to cut bevels. The apparatus is obtainable with a straight two-groove track, and it also has a radius arm for use when cutting circles and arcs. A motor driven cutting torch guide being used to cut arcs is shown in Fig. 4-17. There is an off-and-on switch, a reversing switch, clutch, and a speed adjusting dial calibrated in feet per minute.

Fig. 4-18 shows an electric drive carriage being used on a straight track. The operator must be sure the electric cord and the hose will not become caught on some obstruction during the cutting operation. The best way to check hose and electric cord clearance, and the clearance of the torch, is to free-wheel the carriage the full length of the track by hand with the clutch released

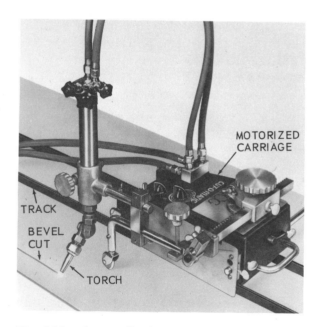

Fig. 4-18. Electrically driven cutting torch carriage being used on straight track to cut beveled edge on steel plate. (Chemetron Corp.)

to check the smoothness of operation.

A type carriage which is manually directed but which has a controlled power speed is shown in Fig. 4-19. Such a unit may be used to cut straight edges, circles, arcs and irregular shapes.

Fig. 4-19. *Electrically driven, manually guided cutting torch carriage. (Linde Div., Union Carbide Corp.)*

4-9. TRACER DEVICES

In an electronic guided carriage a light source is used to follow a trace line on a pattern or drawing. The trace line may be as thin as .040 of an inch.

A pinpoint of light is directed onto the trace line on a template. Some of this light is reflected into a photoelectric cell which will send signals through an amplifier to a motor which will cause the steerable wheel on the carriage to turn and follow the pattern, Fig. 4-20. The pattern used may be a drawing or it may be cut from black colored material and mounted on a white background.

The electronic circuit is so designed that the direction the steering motor turns is determined by the intensity of the light reflected from the template surface as shown in Fig. 4-21. When

Fig. 4-20. *Electronic device uses a light beam to sense changes in line direction on black and white line pattern. Unit electronically controls movements of cutting torch to cut metal to shape of pattern. (Chemetron Corp.)*

the light source reflects off a dark surface to the photoelectric cell the steering motor turns in one direction. And when the light is reflected from a white surface the steering motor turns in the opposite direction. The steering motor is not operated when the light source reflects from a black surface and a

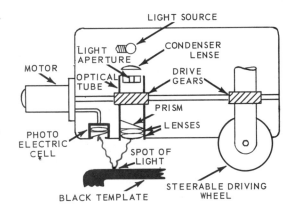

Fig. 4-21. *Diagrammatic sketch of an electronic directional control which uses an optical system in scanning bead for controlling torch movements.*

white surface equally, as it will when the light source is right on the edge of the black template.

The light spot which follows the template is beamed to lead the position of the carriage slightly. This gives time for the steering motor to correct the position of the steering wheel as changes in the pattern occur.

If the light was following a straight line, the reflected light would be equal from the black surface of the template and the white background surface, and the steering motor would be neutral. When the light source reaches a curve it leaves the pattern and the photo-electric cell picks up a more intense reflected light. This causes the steering motor to energize and change the position of the steerable wheel to bring the light source back to the edge of the pattern. As the curve continues or changes, the steerable wheel is constantly bringing the carriage and light source back to the neutral position, at the edge of the pattern. With this type of electronic carriage, a pattern can be followed to a very precise degree.

Two newer devices used to guide a cutting torch are the Magnetic Follower, and the Electronic Follower. These depend on very sensitive response to the movement of a stylus (tracer roller) or electric eye. Two servo motor systems are used, one for lateral movement, and one for longitudinal movement. Fig. 4-22 shows a machine using a magnetic tracer. Some tracer movements control the minute current through the screen grid of an electronic amplifier tube, others through a transistor in the central control unit. The larger current flow developed in the amplifiers is then used to operate the servo motors which move the cutting torch head. The cutting apparatus shown in Fig. 4-22 mounts the pattern and the magnetic tracer

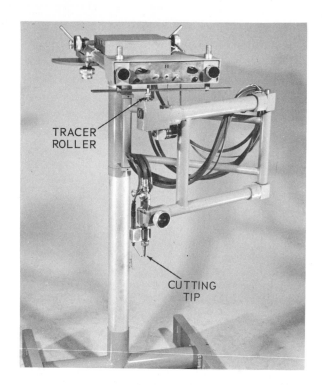

Fig. 4-22. Cutting machine which mounts tracer unit above cutting tip. As revolving magnetic tracer follows pattern, cutting torch travels duplicate path. (Heath Eng. Co.)

immediately above the cutting torch. The unit is either mounted on its own stand, or, can be set up on a cutting table. The unit is shown in operation in Fig. 4-23.

Some units have a tracer roller which is rotated by a small motor. The roller is magnetized to keep the roll in contact with the steel pattern. As the tracer is rotated, it follows the exact shape of the pattern and causes the cutting torch to follow this pattern on the metal being cut.

4-10. MULTIPLE TORCHES

There are many cases when it is necessary to make more than one piece of a specially shaped metal part. To do this, two or more torches may be mounted on the cutting machine. Fig. 4-24 shows such an arrangement. The

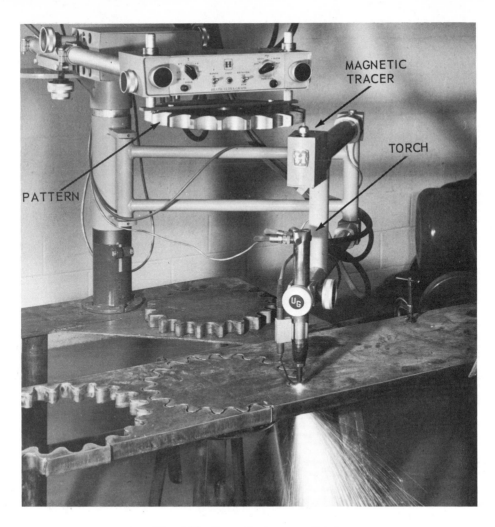

Fig. 4-23. Magnetic tracer unit in use.

Fig. 4-24. Shape cutting eight parts simultaneously from one pattern. (Chemetron Corp.)

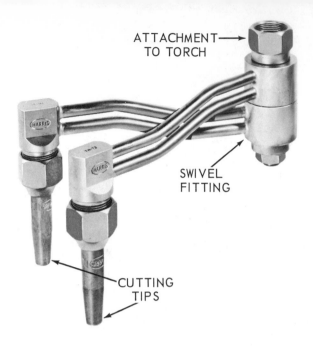

ATTACHMENT
TO TORCH

SWIVEL
FITTING

CUTTING
TIPS

Fig. 4-25. Attachment for manual flame cutting torch which enables two cuts to be made at one time. Swivel fitting permits placing two tips in various positions. (Harris Calorific Div., Emerson Electric Co.)

torches have to be carefully mounted to insure that the cuts are accurate. Fig. 4-25 illustrates a device which allows two cuts to be made at the same time using a hand or mechanically operated cutting torch.

4-11. SPECIAL CUTTING AND GOUGING TIPS

Many special cutting tips have been designed for special applications. Special tips for special preheating fuels such as liquefied petroleum (LP) and propane are some examples. Tips for cutting rivets or for gouging metal are other examples of special tips. See Fig. 4-4 for a table of special cutting and gouging tips and their suggested uses.

4-12. TEST YOUR KNOWLEDGE

1. If a cutting torch tip has three orifices, what are the names of the orifices?

2. Of what material is a cutting tip made?

3. Why must a cutting tip flame end be free of dirt, particles and scratches?

4. How should a tip orifice be cleaned?

5. Under what conditions are cutting torch attachments generally used?

6. How many hoses are connected to a cutting torch?

7. Does the oxygen regulator for cutting stations differ from one used on a welding station? How?

8. How may one accurately cut circles or arcs with a cutting torch?

9. What is the best way to store cutting tips when they are not in use?

10. How many valves does a cutting torch have?

11. Describe the two different type valves used on a cutting torch.

12. How does an electronic tracer follow a pattern?

13. How many duplicate parts may be cut at one time with a multiple torch machine?

14. Describe an alternate method of cutting duplicate parts with a cutting torch.

15. Why is it so important that the seating end of the cutting tip be free of dirt, scratches, or dents?

16. When using a multiple torch cutting machine, how many patterns must be used to operate all the cutting torches?

17. How may a welder cut a straighter kerf without using a cutting machine?

18. How are the acetylene and oxygen cylinders fastened to the cutting station or portable truck?

19. How many preheat holes and what degree of preheating are recommended for cutting cast iron and stainless steel?

20. How many preheat holes and what degree of preheating are recommended for cutting rivet heads?

Chapter 5

DIRECT CURRENT
ARC WELDING

Two basic types of welding use electricity as their energy source: RESISTANCE WELDING and ARC WELDING.

Electric arc welding is based on the fact that, as electricity passes through a gaseous gap from one electric conductor to another, a very intense and concentrated heat is produced. The temperature of the spark, or the arc, jumping between the two conductors, is approximately 6,500 to 7,000 deg. F. The action of the arc is the basis for the following arc welding methods:

A. Carbon-arc welding.

B. Metal-electrode arc welding.

C. Gas metal-arc welding.

D. Atomic-hydrogen welding.

E. Gas tungsten-arc welding.

This chapter will discuss direct current (DC) arc welding. Electric Resistance welding is explained in CHAPTERS 13 and 14.

5-1. DC ARC WELDING MACHINE CHARACTERISTICS

It is important to understand the voltage-current characteristics of the DC arc welding machine. Under a no-load (or open circuit) condition, the voltage of the machine will increase. As the load on the machine increases, the current will go up and the voltage will come down. It is important to understand this voltage-current relationship.

Ohm's Law for electricity states that voltage in a closed circuit has a constant relationship to the current and the resistance of the circuit. The Ohm's Law formula is $E = I \times R$. "E" is the electromotive force (voltage). "I" is the intensity of current in amperes. "R" is the resistance in the circuit.

If one assumes that the only adjustment on a welder is for amperage, various DC arc welder characteristics may be shown. In a DC arc welder, if the electrode gap or the resistance in the circuit is held constant, when the amperage is increased, the voltage must be increased. This may be seen in the formula below:

$$E = I \times R \text{ (constant)}$$

The amperage output is limited by the voltage from the main power source. Amperage is also limited by the wiring of the welder circuits.

If the voltage is held constant and the R, resistance, in the circuit (arc gap) increases, the amperage will decrease. If the resistance (arc gap) decreases, the amperage will increase with a constant voltage. This is illustrated by the formula below:

$$E \text{ (constant)} = I \times R \text{ (increases)}$$

Therefore, with a constant voltage arc welder, the most important control of the heat generated in the electric arc is the length of the arc gap. The arc gap must be held constant once the amperage is set on the machine. If the arc gap changes, the amperage across the gap will vary and so will the welding heat developed.

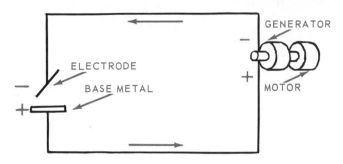

Fig. 5-1. Metal electrode arc welding circuit.

5-2. FUNDAMENTALS OF DC ARC WELDING

Electric arc welding is defined by the American Welding Society:

"A group of welding processes wherein coalescence is produced by electric heating with an electric arc or arcs, with or without the application of pressure, and with or without filler metal."

The metal electrode arc method of welding consists of producing an arc between the metal to be welded, and the metal to be added to the base metal (metal electrode). This is usually done by having the operator place a metal electrode in a holder and with it, strike an arc on the base metal. Fig. 5-1 shows the electrical circuit for metal arc welding.

To make a good weld the operator must be concerned with the following:

1. Welding machine and circuit.
2. Electrode.
3. Arc and its manipulation.

A generator or rectifier (device which converts AC to DC) is used to produce a direct current of sufficient strength to travel through the welding circuit, jump the arc gap, melt the electrode, and create a molten puddle in the base metal. The molten portion of the electrode will fuse into the puddle of the base metal, producing a weld. The current (amperes-See Par. 6-2) of the

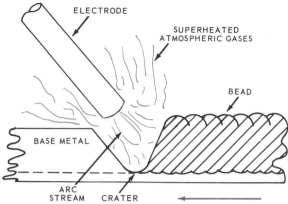

Fig. 5-2. A bare electrode arc weld in progress.

machine and the direction of current travel may be varied by the operator to suit the job being done.

The electrode wire may be ferrous, ferrous alloy, with some of the non-ferrous metals also being used.

Electrodes are manufactured as bare wire, wire with a light coating of flux material, or wire with a heavy covering of fluxing material.

Bare electrode arcs are the most difficult to maintain, and the finished weld is not normally as sound as a weld made with covered electrode as shown in Fig. 5-2.

A lightly coated (dusted) electrode will give the arc some stability. Fluxing materials in the heavily covered electrode will give the arc considerable stability, help to float out impurities from within the molten puddle, and will develop inert gases which will keep the outer surfaces of the molten weld from oxidizing. Fluxes in the heavily covered electrode will also harden on the surface of the weld to form a crust or slag. This protects the weld metal from oxidation as it cools as shown in Fig. 5-3. Electrodes are manufactured in a number of different diameters. As the metal thickness varies, current flow in the welding circuit varies and the electrode diameter is changed accordingly.

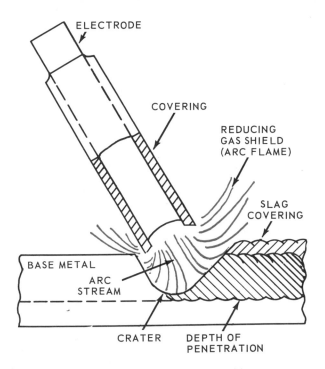

Fig. 5-3. A covered electrode arc weld in progress.

METAL THICKNESS	ELECTRODE SIZE	AMPERES (FLAT WELDING)	VOLTAGE (APPROXIMATE)
1/16 - 1/8	3/32	50 - 90	15 - 17
1/8 - 1/4	1/8	90 - 140	17 - 20
1/4 - 3/8	5/32	120 - 180	18 - 21
3/8 - 1/2	3/16	150 - 230	21 - 22
1/2 - 3/4	7/32	190 - 240	22
3/4 - 1	1/4	200 - 300	22

Fig. 5-4. A table of metal-arc electrode sizes and applications. For lap or T-joint welds, increase the current values slightly. These values are approximate and must be varied to apply to the particular case.

The arc, when viewed through the helmet lens, is seen to be divided into two separate parts: The Stream, and the Flame, as shown in Fig. 5-3. The arc flame consists of neutral gases which appear to be pale red. The vaporized metal in the stream appears yellow, while the liquid metal in the stream appears green. If the arc is longer than normal, the flame gases can no longer protect the stream.

The metal will form oxides and nitrides, resulting in a very weak and brittle weld. The voltage used, therefore, is very important as too much voltage (pressure) enables a longer arc, resulting in considerable spattering. If the weld progresses too rapidly, the weld will not penetrate; if too slowly, either too much penetration or too much buildup will result.

If the correct current flow and arc length are maintained, a good weld should result with direct current. The voltage and amperage required for any particular weld may be easily obtained from established tables. The correct arc length is entirely the operator's responsibility. See Fig. 5-4.

5-3. FUNDAMENTALS OF DIRECT CURRENT STRAIGHT POLARITY ARC WELDING (DCSP)

The welding circuit shown in Fig. 5-5 is known as a straight polarity

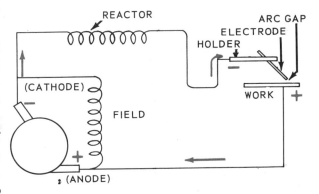

Fig. 5-5. Wiring diagram of a direct current, straight polarity (DCSP) arc welding circuit. The field winding is needed to create the current while the reactor is needed to produce an even, steady flow of current. The electrons flow from the electrode to the base metal.

circuit. It is understood that the electrons are flowing from the negative terminal (cathode) of the machine to the electrode. The electrons continue to travel across the arc into the base metal and to the positive terminal (anode) of the machine.

Approximately two-thirds of the total heat produced with DCSP is released at the base metal while one-third is released at the electrode.

The choice of direct current straight polarity depends on many variables such as material of the base metal, position of the weld, as well as the electrode material and covering. Additional information given in Chapter 6 will help in determining when DCSP should be used.

5-4. FUNDAMENTALS OF DIRECT CURRENT REVERSE POLARITY ARC WELDING (DCRP)

It is possible, and sometimes desirable, to reverse the direction of electron flow in the arc welding circuit. When electrons flow from the negative terminal (cathode) of the arc welder to the base metal, this circuit is known as direct current reverse polarity (DCRP). In this case, the electrons return to the positive terminal (anode) of the machine from the electrode side of the arc, as shown in Fig. 5-6.

When using DCRP, one-third of the heat generated in the arc is released at the base metal and two-thirds is liberated at the electrode.

With two-thirds of the heat released at the electrode in DCRP, the electrode metal and the shielding gas are superheated. This superheating causes the molten metal in the electrode to travel across the arc at a very high rate of speed. Deep penetration results from the force of the high velocity arc.

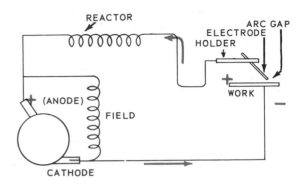

Fig. 5-6. A wiring diagram of a direct current, reverse polarity (DCRP) arc welding circuit. Note that the electrons flow from the base metal to the electrode.

There is a theory that, with a covered electrode, a jet action and/or expansion of gases in the metal at the electrode tip causes the molten metal to be propelled with great impact across the arc.

The choice of direct current reverse polarity depends on many variables such as material of the base metal, position of the weld, as well as the electrode material and covering. Information in Chapter 6 will help in determining when DCRP should be used.

5-5. SAFETY, PROTECTIVE CLOTHING AND SHIELDING

Before proceeding further with the study of arc welding operation this safety lesson should be studied and observed.

Arc welding performed with proper safety equipment presents no great safety hazards. However, the beginner should be made aware of the hazards, and should be taught the correct procedures with arc welding in order that the hazards which exist may be properly observed and injury avoided.

The chief hazards to be avoided in arc welding are:

1. Radiation from the arc, ultraviolet and infrared rays.

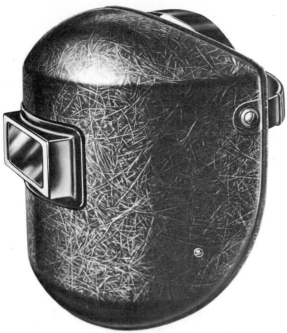

Fig. 5-7. An arc welding helmet. It is used to protect the eyes and face while arc welding. The lens reduces the amount of harmful rays which would reach the welder's eyes, but will still allow the welder to see the weld and arc crater.
(Fibre Metal Products)

2. Flying sparks, globules of molten metal.

3. Electric shock.

4. Fumes.

5. Burns.

Radiation from the arc presents some dangers. Eyes must be protected from radiation from the arc by the use of an arc welders helmet or face shields with approved lenses. Never look at an arc from any distance, unless eyes are protected by approved lenses, Fig. 5-7.

Face, hands, arms and other skin surfaces must be covered. Gloves should be worn and other parts of the body covered by clothing of sufficient weight to shut out the rays from the arc, otherwise burns (comparable to sunburn) will result.

The arc welding operation should be shielded so that no one may accidentally look directly at the arc or have it shine in the eyes. If someone is severely "flashed," special treatment should be administered at once by a physician.

Arc welding is usually accompanied by flying sparks. These present a hazard if they strike unprotected skin, lodge on flammable clothing, or hit other flammable material. It is advisable to wear suitable weight clothing and cuffless trousers.

Pockets should be covered so they will not collect sparks. Remove flammable materials, such as matches, plastic combs or fountain pens. High shoes with safety toes should be worn.

The possibility of dangerous electric shock can be avoided by working on a dry floor, using insulated electrode holders and wearing dry welding gloves. Avoid using arc welding equipment in wet or damp areas.

The health hazard of fumes, developed by the electrode covering and molten metal, may be avoided by the use of proper ventilating equipment. Certain special jobs require forced ventilation into the welder's helmet. The fumes generated in the welding arc may contain poisonous metal oxides. Arc welding should never be done in an area which is not well ventilated.

Hot metal will cause severe burns. Use asbestos or leather gloves with tightfitting cuffs which fit over the sleeves of the jacket. Many welders wear an apron of leather or other heavy material for protection. Hot metal should be handled with tongs or pliers. In a welding shop all metal should be first cooled in the quenching tank before it is handled with bare hands.

5-6. ASSEMBLING AND INSPECTING ARC WELDING STATION

The typical direct current arc welding station consists of:

HELMET
CAR
GLOVES
WOODEN
PLATFORM
HIGH
SHOES

Fig. 5-8. A complete DC arc welding station. Notice the protective clothing worn by the welder. The welder stands on a wooden platform to reduce the danger of an electric shock and to reduce fatigue.

A. Power supply.
    1. Welding generator.
    2. Rectifier welding machine.
B. Ground cable, and an electrode cable, which have flexible strands and are rubber covered.
C. Electrode holder.
D. Steel bench about 30 in. high.
E. Stool.
F. Special clamp device or C clamp.
G. Electrodes.
H. Electrode holder hanger (insulated).
  I. Booth.

The booth should be well ventilated and well lighted. Cable should be protected from abuse. The arc welding machine should be placed close to the booth to keep the arc cables as short as possible and to permit quick and easy adjusting of the machine. Fig. 5-8 shows a complete arc welding station.

Before starting the machine, check all items to see that the equipment is in usable condition.
  A. Welding operator:
    1. Gloves in good condition.
    2. Helmet or shield with proper lens.
    3. Apron or other additional protective clothing.
    4. Cuffless trousers.
    5. Empty, covered pockets.
    6. Cuffless sleeves.
    7. Dry oil-free garments.
    8. No jewelry.
  B. The station:
    1. Ground and electrode cables, for wear and tightness of connections.
    2. Booth curtains in good condition.
    3. Fuses of right capacity in main switch.
    4. Ampere adjustment set to minimum on the machine.
    5. Ventilation systems in good working order.
  C. Supplies:
    1. Correct size electrodes.
    2. Clean welding stock.

Carefully inspect, clean and oil the machines according to manufacturer's recommendations. Generally, a medium grade of automobile engine oil is used in plain motor bearings. Ball bearing units often require special grease. In some arc welding installations, an electrician or maintenance mechanic looks after these duties.

Inspect the ground electrode cables, and electrode holder. Then hang the electrode holder away from the work to be welded, preferably on a fiber or wood hanger. If the machine is a motor-generator unit, inspect the commutator after starting the machine; if the brushes are arcing badly the machine should not be used until the trouble is corrected. All arc welding should be performed in a booth, or portable curtains should be used to keep passers-by from observing the arc.

5-7. ADJUSTING AND STARTING ARC WELDING MACHINE

When the station is properly equipped and inspected, the welder is ready to start the arc welding machine. The steps usually followed are:

1. See that the electrode holder is hung on an insulated hanger. The machine should never be started with the electrode holder on a table or base metal, since this closes the arc welding circuit and causes the arc welder to start under a full load, which is hard on most machines.

2. Close the main power switch or circuit breaker to the machine.

3. Adjust the machine to the proper current setting depending on the base metal and selected electrode sizes. Refer to Fig. 5-4 for suggested arc welding machine settings.

4. Press the magnetic starter ON button on the machine. On some machines it is possible to adjust both amperage and voltage; on other machines it is possible to set only amperage or voltage. After running a few practice beads it may be necessary to readjust the fine (vernier) adjustment to suit the individual welders needs to perform a more perfect weld.

5-8. SELECTING THE ELECTRODE

Selection of the proper electrode for a given job is one of the most important decisions the welder must make.

Electrodes may differ in these ways:

1. Light coatings and heavy coverings, see Fig. 5-9.

2. Chemical composition of the covering may vary to give desired re-

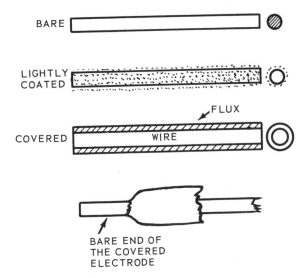

Fig. 5-9. A comparison of bare, lightly coated, and heavily covered electrodes.

sults when welding different metals and metal alloys, or to improve welding characteristics in different positions.

3. Electrode coating may be designed for use with DCSP, DCRP or AC.

4. Compositions of electrode metals.

5. Electrode diameters desired.

Heavily covered electrodes will normally produce welds of superior strength and appearance, but they are more expensive.

Electrodes are designed and designated by color codes and numbers, such as, E6010, E6011, etc. Each electrode has qualities which may make it more desirable than another for a particular job. See CHAPTER 6 for a more complete discussion of electrode number designations.

Electrode metal is made in a variety of alloys and sizes to best suit all welding applications.

See Fig. 5-4 for recommended electrode diameter to be used for different metal thicknesses. It may be desirable when making a weld on thick metal, to use a smaller electrode than recommended for the first bead of a multiple bead weld.

The welder must inspect each electrode to make certain it is clean, dry, and not cracked, since this may affect the quality of the finished weld.

To place an electrode in a holder, the jaws of the holder are opened, and the bare section of the covered electrode is placed between the jaws. See Fig. 5-10. If, after the jaws are closed, the

Fig. 5-10. A typical electrode holder used in metallic arc welding. By pressing down on a lever, the jaws are opened for insertion of the electrode. A powerful spring keeps the jaws closed tightly.
(Jackson Products)

electrodes can be easily moved, the holder is loose and needs repair.

5-9. STRIKING AN ARC

One of the first lessons to be mastered when learning to arc weld, is to produce an arc between the metal electrode and the base metal. To strike an arc, the electrode must first touch the base metal, and the end must then be withdrawn to the correct arc distance or length.

At the first attempt to strike the arc the electrode may tend either to stick (weld itself to the base metal), or when the electrode is withdrawn, the movement may be too great and the voltage cannot maintain the arc, and the arc will break (go out). Only experience will overcome these difficulties.

The arc should be started by drawing a rather long arc momentarily to preheat the base metal before deposition of

filler material (electrode) starts, to insure thorough fusion when beginning to weld. This may also be accomplished by moving a short arc rapidly back and forward over a limited area at the start.

There are two common methods of producing the arc. The welder may use a glancing or scratching motion with the end of the electrode, or, he may use a straight down-and-up motion. Fig. 5-11 illustrates both methods.

5-10. ELECTRODE MOTIONS

The manipulation of the electrode when arc welding is very important. Remember that any changes of motion, angle, and arc length may affect the quality of the weld.

On single layer welds, the arc may progress with either a straight forward motion or an oscillatory motion depending on the size of the joint to be filled to bring it up to proper contour, and insure proper fusion in the root zone (bottom of the weld).

As the thickness of the metal increases, the electrode must be moved in some definite pattern, to secure the proper fusion of the electrode metal with the base metal, and to obtain the proper weld face. Many operators use a small oscillating (back-and-forth) motion along the line of the weld to control and produce a nice ripple. Small circular motions may be used. Some prefer the back-and-forth motion along with the half-circle across the weld on a 45 deg. angle to it. Some arc welders use figure-eight motion; some use a triangular motion; while most of them use all of these motions, depending upon the type of seam, and the position in which it must be welded. See Fig. 5-25.

In multi-layer welding, the first layer should be deposited by moving the arc in a straight forward motion to secure

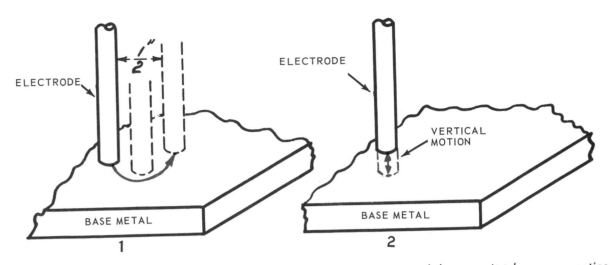

Fig. 5-11. Methods of striking an arc. The second method is most often used; however, it takes some practice to be skillful in its use.

proper fusion in the root zone, while subsequent layers should be deposited with an oscillating motion to build up the weld to the proper contour. The first layer should be made with an electrode small enough to get thorough penetration at the root zone.

Each layer should be thoroughly cleaned by brushing or chipping, to remove all excess scale and oxides, before depositing succeeding layers. The surface of each layer should be reasonably smooth and uniform in contour.

Various movements of the electrode may be made to form the seam or weld face. In general the movement of the electrode will be governed by the type of joint, the size of the weld, and the position of the weld.

5-11. RUNNING METAL ARC BEAD ON FLAT SURFACE

Before attempting to weld any type of seam in any position, practice laying arc beads on a similar piece of metal of the same thickness as that to be welded. You must be able to lay several excellent arc beads before pro-

ceeding to make the various types of welds.

An arc bead is produced by first creating the arc. The arc is made by touching the end of the electrode lightly to the metal, and withdrawing it to the proper distance (gap). This gap varies with the size and the type of electrode used. Bare electrodes should have a gap of 1/8 to 3/16 in. which necessitates a voltage of 20 to 24 volts across the arc. Covered electrodes use a longer gap, 3/16 to 1/4 in. necessitating 30 to 40 volts across the arc. An easy way to check on the proper length of an arc is to listen to the sound of the arc. The proper arc gap when using bare electrodes provides a distinct, crackling or frying sound. Covered electrodes also give distinct indications of proper arc length. The arc has a distinct hissing sound, usually without a crackling sound. Further, the color of the arc changes, due to the influence of the longer arc on the gas envelope and slag.

Balls of molten metal should not form on the end of the electrode as these globules drop into the puddle or crater. This condition indicates too long an arc. By carefully observing these conditions, the beginning arc welder will soon become adept in maintaining the correct arc length.

To practice, secure a piece of mild steel approximately 1/8 to 1/4 in. thick, 2 in. wide and 6 in. long. Ground this metal to the table, preferably at the end where the practice is to begin, as shown in Fig. 5-12. The grounding may be done by clamping it to the table with copper coated clamps.

The amount of current depends on the thickness of the metal, and size of electrode used. Sizes and current values that have been found suitable for welding are given in Fig. 5-4. In special cases other approved sizes and currents may be specified. The current and voltage used should be measured by suitable meters, either mounted on the arc welding machine panel, or portable ones for use at the point of welding. Allowance should be made for voltage drop where the meters are not at the point of welding.

A. Start and adjust the welding machine:

1. Inspect the station and repair any items which require maintenance.

2. Adjust the machine to the proper settings according to the base metal thickness and electrode size. See Fig. 5-4. Always have an experienced operator try the machine before attempting to weld with it, as many hours may be wasted by trying to learn arc welding with a machine that is out of adjustment. Having an experienced arc welder observe one's technique when starting will correct faults which, if they become habits, will be hard to change.

B. Insert the bare end of the electrode in the electrode holder.

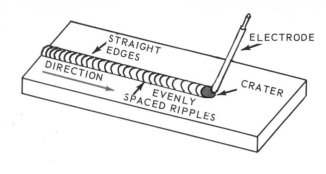

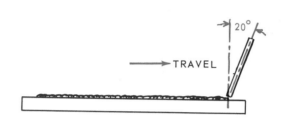

Fig. 5-12. An arc bead in progress. Note that the electrode is inclined 20 deg. in the direction of travel. The completed arc bead should have straight edges, evenly spaced ripples, and uniform height.

C. Assume a comfortable sitting or standing position.

D. Lower the helmet or place the hand shield in front of the face.

E. Grasp the electrode holder in one hand in a relaxed manner, and strike the arc. A slight semicircular vertical motion may be used (a glancing blow). Lower the electrode slightly for each motion until the electrode contacts the parent metal. As soon as the arc is produced, lower the electrode toward the work until the arc emits a hissing sound (like frying steak).

F. Hold the electrode in near vertical position with the electrode leaning about 20 deg. from vertical in the direction of travel.

G. Use no circular, oscillating, or saw-tooth motion for this thickness of metal.

H. Progress either to the right, if right-handed, or the left, if left-handed, gradually lower the hand as the electrode is used up.

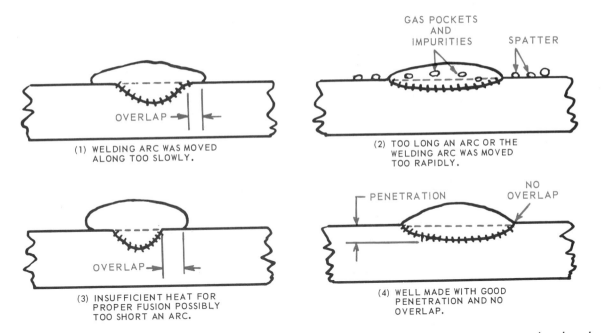

(1) WELDING ARC WAS MOVED ALONG TOO SLOWLY.

(2) TOO LONG AN ARC OR THE WELDING ARC WAS MOVED TOO RAPIDLY.

(3) INSUFFICIENT HEAT FOR PROPER FUSION POSSIBLY TOO SHORT AN ARC.

(4) WELL MADE WITH GOOD PENETRATION AND NO OVERLAP.

Fig. 5-13. This sketch shows the effects on the finished arc bead of welding speed, current setting, and arc length.

I. Keep watching the puddle to check its width, and also note the length of the arc, Fig. 5-12.

J. A chalk line put on the metal before the weld is started will help the beginner to maintain a straight line. To stop the arc, move the electrode to the back of the crater and raise the electrode.

K. Upon completing the bead, remove helmet. Put on chipping or grinding goggles and chip off the slag. Then use a wire brush to complete the cleaning job and inspect the bead for:

1. Alignment. 4. Gas holes.
2. Width of bead. 5. Spatter.
3. Penetration. 6. Even bead.

A good bead on 1/8-in. plate will look very similar to an oxyacetylene weld. It will have a clean, shiny surface, of even width, good penetration, and good fusion. Figs. 5-13 and 5-14 show the effects of arc length on the appearance and penetration of an arc bead.

A practical application of arc bead

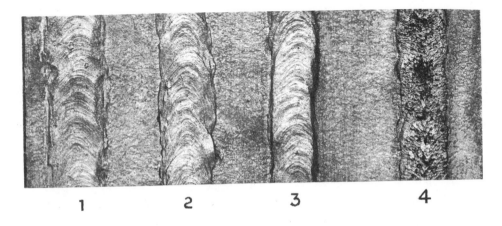

1   2   3   4

Fig. 5-14. The appearance of finished welds. 1, 2 and 3. Various covered electrode beads after slag removal. 4. Bare electrode bead. (General Electric Co.)

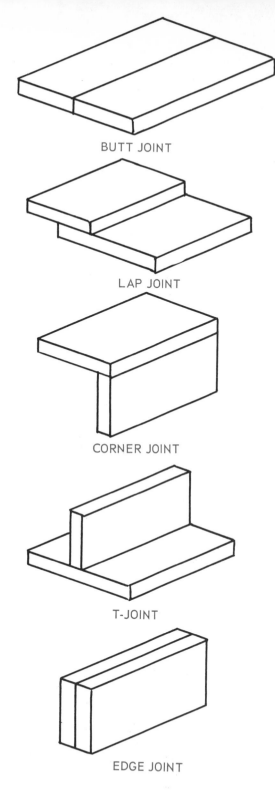

BUTT JOINT

LAP JOINT

CORNER JOINT

T-JOINT

EDGE JOINT

Fig. 5-15. Common arc welding joint designs.

exercises is the rebuilding of worn surfaces in welding maintenance work. Shafting, excavation implements, gear teeth, wheels of various kinds, journals, etc., frequently are worn to the extent that they must either be discarded or salvaged. Rebuilding these surfaces by laying arc beads side by side over the worn surface in one or more layers, then refinishing, has become an important arc welding maintenance operation.

Another application of bead work is hard surfacing or wear-resistant surfacing. Laying beads of special metallic alloys side by side on a soft steel surface provides a surface that is extremely hard and resistant to abrasion. See CHAPTER 20.

5-12. TYPES OF JOINTS

In arc welding, there are five basic types of joints, or ways of arranging the base metals to be joined in relation to one another:

1. Butt joint.
2. Lap joint.
3. Corner joint.
4. T-joint.
5. Edge joint.

See Fig. 5-15.

The edge preparation and method of welding may vary with each joint classification.

5-13. POSITION OF WELDS

Welds made in the flat position (downhand) are the easiest to make. However, it is often necessary to make welds in various other positions.

The recognized positions are:

FLAT POSITION WELDING.

The position of welding where welding is performed from the upper side of the joint, and the face of the weld is approximately horizontal, as shown in Fig. 5-16.

HORIZONTAL POSITION WELDING.

Fillet Weld. The position of weld-

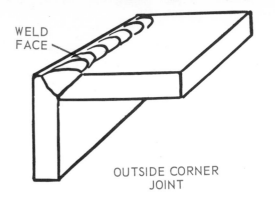

WELD FACE

OUTSIDE CORNER JOINT

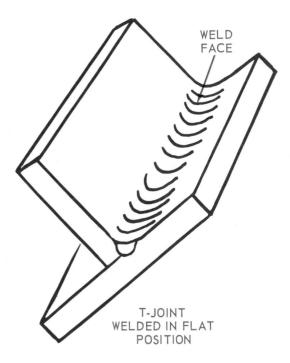

WELD FACE

T-JOINT WELDED IN FLAT POSITION

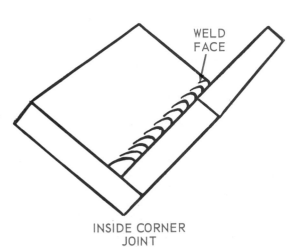

WELD FACE

INSIDE CORNER JOINT

Fig. 5-16. Examples of welds made in the flat (down-hand) position. Note that the face of each weld is in a horizontal position.

ing where welding is performed on the upper side of an approximately horizontal surface, and against an approximately vertical surface.

Groove Weld. The position of welding where the axis of the weld lies in an approximately horizontal plane, and the face of the weld lies in an approximately vertical plane.

VERTICAL POSITION WELDING.

The position of welding where the axis of the weld is approximately vertical.

OVERHEAD POSITION WELDING.

The position of welding where welding is performed from the underside of the joint, and the axis of the joint and base metal are both approximately horizontal.

5-14. ARC BLOW

One characteristic of DC arc welding, not found when using AC arc welders, is the occasional fluctuating arc, or arc instability. This action is usually due to the forces in the magnetic field built up around the arc. All conductors of electricity are surrounded by magnetic lines of flux when current is flowing. It is the effect of the magnetic flux surrounding the armature windings in an electric motor which produces the force which causes the motor to rotate. The welding arc is a very flexible conductor. Any magnetic effect which might cause the arc to move will be successful. Iron and steel are much better conductors of magnetic flux than air. Magnetism will travel in the metal rather than air. The lines of flux become crowded as they approach the ends of a weld and there will be a more dense or concentrated magnetic force at the end of the weld, tending to deflect or move the arc from its normal path. This movement is called arc blow.

Fig. 5-17 illustrates the magnetic flux path around a DC arc. Note that at the beginning and end of the weld, the

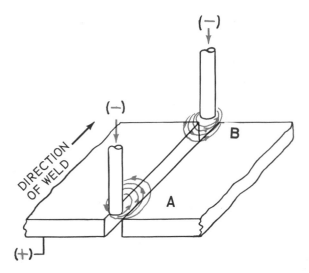

Fig. 5-17. How magnetic field around electrode is deflected at ends of joint, as it attempts to flow in the metal and not through the air. This concentration of magnetic flux forces arc toward center of base metal. Thus arc is seen to "blow" away from area directly under electrode.

trouble caused by arc blow is greatest.

When welding together two pieces of metal as shown in Fig. 5-18, the flow of welding current may set up a field around the work which may make the arc difficult to control. With the arc as shown in position C, and if the base metal grounds at A and B, due to an

uneven worktable, the current flow in the base metal will be from A to C and from B to C. As the arc path weaves back and forth, the strength of the magnetic field around the two pieces of base metal will alternately increase and decrease. This fluctuation in magnetic

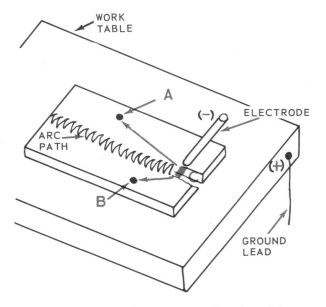

Fig. 5-18. How welding current may flow through base metal from arc to point where metal is touching grounded worktable as weld progresses. The magnetic field surrounding the current paths will fluctuate as arc weaves along arc path. This fluctuating magnetic field may cause "arc blow" or unwanted arc movements.

field strength around the two pieces of metal may cause the arc to fluctuate from the desired path. Arc blow may be

A. FLAT FACE FILLET B. CONVEX FACE FILLET C. CONCAVE FACE FILLET

Fig. 5-19. Cross sections of acceptable fillet welds (etched and magnified four times). (U. S. Steel)

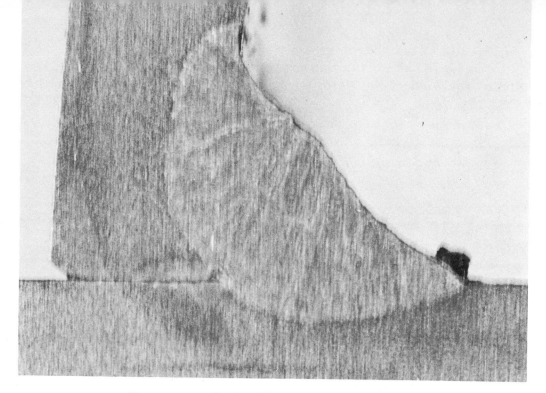

Photomacrograph of a fillet weld on a T-joint using DC. Note the undercut on the vertical piece of metal.

reduced by carefully clamping the work to the welding table in several places. In welding large pieces it may be desirable to move the ground clamp or cable as the weld progresses.

If the current flow changes direction rapidly, as in AC current arc welding, the magnetic field will also change directions very rapidly and this change will cancel the "arc blow" effect and stabilize the arc.

5-15. UNDERCUTTING

Undercutting is defined as A GROOVE MELTED INTO THE BASE METAL ADJACENT TO THE TOE (EDGE) OF A WELD AND LEFT UNFILLED BY WELD METAL.

Since the base metal is thinner in the area of the undercut, this area becomes an area of possible weld failure.

It is essential that the bead have no indications of undercutting, and that thorough fusion is evident along the

length of the bead, as shown in Fig. 5-19. Undercutting usually indicates that the welder has either been using the wrong electrode motion, or that the machine is not adjusted correctly. Fig. 5-20 illustrates beads which have poor

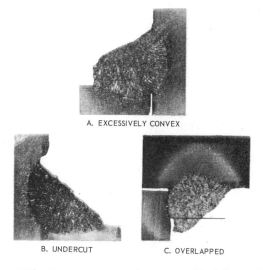

A. EXCESSIVELY CONVEX

B. UNDERCUT

C. OVERLAPPED

Fig. 5-20. Cross sections of unacceptable fillet welds (etched and magnified four times). These are unacceptable. A. Excessive contour. B. Undercutting. C. Overlapping (This weld was made in the overhead position.)

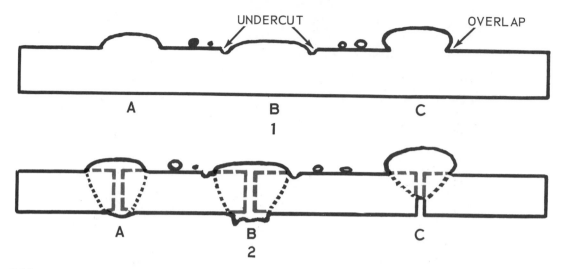

Fig. 5-21. A properly made bead is shown at A, it has good contour and penetration. At B, the bead is undercut and the base metal is thus weakened. At C, the bead was made with insufficient heat and the bead is overlapped with poor fusion and penetration.

contour, are undercut or are overlapped. Fig. 5-21 also illustrates undercutting and overlapping.

5-16. ARC WELDING AN EDGE JOINT

The edge weld is defined by AWS as A JOINT BETWEEN THE EDGES OF TWO OR MORE PARALLEL OR NEARLY PARALLEL MEMBERS. See Fig. 5-15.

The edge weld may be made in any position.

This type of joint is the easiest of the various joints to arc weld.

On thin metal no edge preparation is necessary. On thicker pieces of base metal, the edge should be ground or machined to provide a V, U or J-groove.

As the first experience in joining two pieces together by arc welding, the welder should obtain two pieces of metal approximately 1/4-in. thick. This thickness of metal should require no edge preparation. The pieces should be clamped together and grounded to the table. Firebricks may be used to prop the pieces up so that a flat weld posi-

tion is possible. The arc is struck and a bead laid in the same manner as when making a bead on a flat piece, as shown in Fig. 5-22.

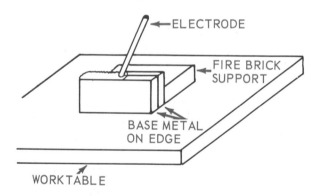

Fig. 5-22. A suggested setup for arc welding an edge joint in the flat position.

5-17. ARC WELDING BUTT JOINT IN FLAT POSITION

To arc weld two 1/8-in. thick, mild steel plates together, using 1/8-in. diameter covered electrodes, the operator should adjust the machine to approximately 80-100 amperes. To adjust for the correct amperage, the beginner should start the machine, insert a mild steel 1/8-in. diameter electrode in the holder, and strike an arc while

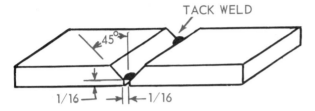

Fig. 5-23. Preparing a V-groove butt joint for welding. The edges have been beveled and the metal tack welded to hold it in proper alignment.

someone watches the ammeter. Do not look at the arc rays without correct eye protection. It is better to have another person watch the ammeter while striking the arc. The arc should be emitting a frying sound, which indicates the proper arc length for covered electrodes, when the ammeter is read. The

correct machine setting to be used for the different exercises may be obtained from Fig. 5-4. The voltage is read at the same time and should be between 17 and 20 volts.

After practicing enough to obtain consistent good welds with the 1/8 in. thick mild steel, continue to practice on thicker plates.

Procure two pieces of 3/8 or 1/2 in. thick cold-rolled steel and grind or cut the metal to a V (45 deg.), which should extend within 1/16 in. of the bottom. Tack the ends, leaving a 1/16 in. gap at the bottom of the V as shown in Fig. 5-23.

The base metal joint should be tapered, or tacked, to prevent or control distortion while welding. This welding may be done in one, two, or three passes, as shown in Fig. 5-24. If done in one operation, the welder must

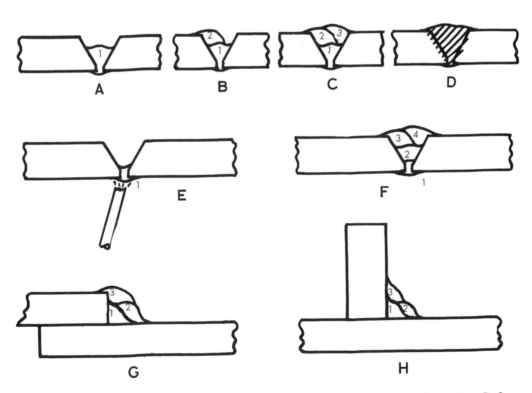

Fig. 5-24. Arc welded joints welded in multiple passes. A. First or root pass in a butt joint. B. Second pass. C. Third pass. D. Finished weld. E and F. Show butt joint where root pass is welded from the penetration side of the joint. G. Shows a lap joint welded in three passes. H. Shows a T-joint welded in three passes.

use an oscillating motion to distribute the filler metal in the V and to secure adequate fusion. If two or more beads are used to complete the weld, the operator should clean the first bead before attempting to make the next bead to prevent including slag in the next bead or pass. This weld should be penetrated slightly, and it should also be built above the original metal.

In case of a long weld where more than one electrode is used, the arc shall be restarted about 1/4 to 3/8 in. ahead of the crater, going back over the crater to properly fill it and keep a uniform contour, thus assuring proper preheating as in the start of the weld.

Rusty, dirty, or greasy metals are difficult to arc weld as the arc will be hard to maintain. Also the resulting weld will be poor. A steel brush is a handy tool with which to remove rust and dirt from the metal. Such a brush should be kept in a convenient place in the booth.

Some operators use steel wool for metal cleaning when preparing for high quality welds. When welding stainless steel, stainless steel wool is suggested for the cleaning. Always remember, the cleaning device must be clean too.

Place the two pieces of metal to be welded on the table with their longer edges touching (butt joint). Allow for metal contraction by "V-ing" the gap lengthwise, or, by tacking the two pieces at both ends of the joint. Put the face shield in place, strike the arc, and reduce the arc length until the frying sound is heard. Slowly move the electrode and puddle along the seam. Control the puddle size by the movement of the electrode along the seam. Be sure the puddle is equally divided between the two pieces. Use very little motion.

When the end of the seam is reached, break the arc by moving the arc to the back of the crater and lifting the electrode.

Cool the weld, remove the slag, if any, and wire brush the weld surface. Guard your eyes against flying particles. ALWAYS USE SAFETY GOGGLES WHEN CLEANING METAL.

Inspect the weld for straightness, constant width, smoothness, penetration, gas bubbles, fusion, spatter, and buildup. It should have a clean-looking bead with straight edges and constant width of bead. The height of the bead should be constant. The ripples should be evenly spaced. The penetration should just show through the under part of the metal joint. The weld should be without small cavities which would indicate too long an arc. It should have good fusion. Fusion means a good bond between the added metal and the original steel plates. It is indicated by a perfect blend, not by a distinct edge between the added metal and the parent metal. There should be little or no spatter. Spattering is the result of too long an arc, or, too high voltage (potential) and too high current. If the metal is loosely connected to the table, the resulting wandering arc may give poor fusion and weld appearance.

5-18. ARC WELDING LAP JOINT IN FLAT POSITION

This type of seam is common although it is not the best weld for strength. Fig. 5-25 illustrates various lap joint assemblies. To arc weld this joint successfully, remember that the piece which is not having its edge welded will require the greater portion of the heat. To distribute the heat, a weaving motion must be used with most of the motion taking place over the bottom piece. The electrode must be held in a position to point it at a slight angle into the joint. To keep the arc length constant, the welder must raise the electrode slightly as the arc travels over the edge of the upper piece. The finished bead must be slightly crowned (convex) and must be straight, even

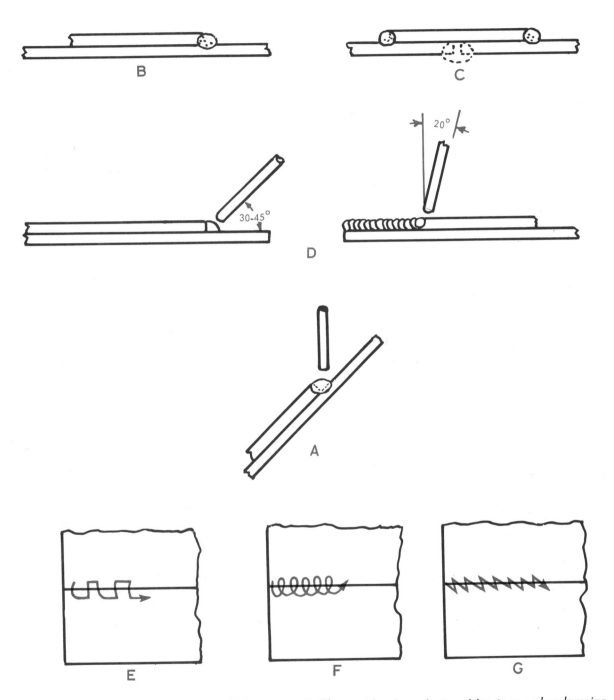

Fig. 5-25. Various ways used to arc weld lap joints: A. Flat position is easiest position to use when learning to weld lap joint. B. Lap weld in this position is also considered to be in flat position, but is a little more difficult than (A) above. C. This is a design which presents two lap joints and one butt joint. D. Proper position of electrode for lap welding is shown in front and end view of a lap weld in progress. E, F, G. Various electrode motions used when lap welding.

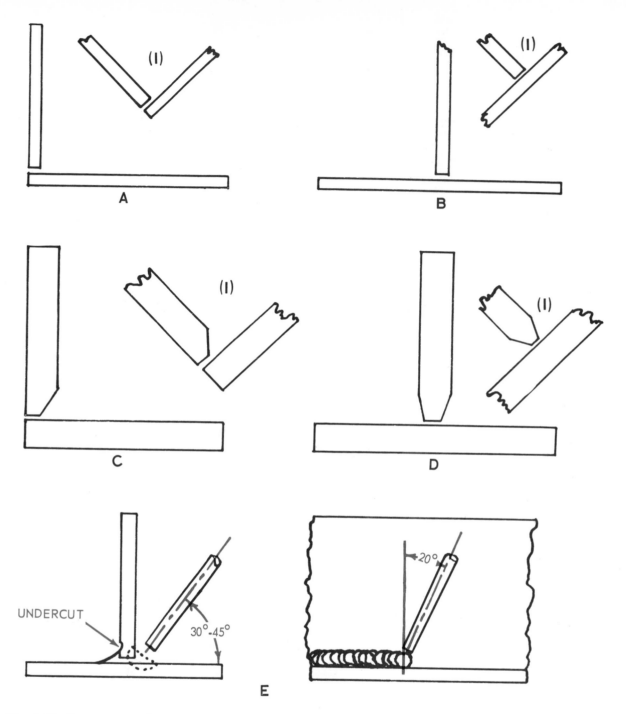

Fig. 5-26. Preparation and arc welding of inside corner and T-joints. If possible the metal should be placed in a position shown in (I) on each type of joint. A. Thin metal corner joint welded in inside. B. Thin metal T-joint. C. Thick metal corner joint welded on the inside. D. Thick metal T-joint. E. Electrode angle used when making fillet weld on corner and T-joint.

in width, smooth, and clean. It should show good fusion between the bead and the parent metal.

The beginner will be able to make a better weld if the joint is positioned as shown in Fig. 5-25 Part A. In welding exercises the best results are obtained if the electrode is as close to vertical as possible. Therefore, the joint should be positioned to make this electrode position possible.

5-19. ARC WELDING A CORNER JOINT IN FLAT POSITION

This joint may be welded from the inside with a fillet weld, or from the outside of the corner using a butt weld. When welding these joints in the flat position the base metals should be placed as shown in Figs. 5-16 and 5-26, for the easiest welding position. The electrode must be held at an equal angle between the pieces (45 deg.), and the holder end should be tilted slightly in the direction the electrode is traveling. The finished weld should be concave for the fillet weld buildup for the outside corner joint, and both welds should be straight, even in width, smooth, clean, and evenly placed, half on one piece and half on the other piece.

5-20. ARC WELDING T-JOINT IN FLAT POSITION

The T-joint is formed by placing one of the base metal pieces in the center, or near the center of the other piece of base metal and at a right angle to it, to form a T-shape. This joint may be welded either on one side only or on both sides. To weld this joint in the flat position, the base metals should be placed and tacked as shown in Figs. 5-16, and 5-26.

5-21. ARC WELDING IN HORIZONTAL POSITION

When welding a butt, lap, edge or outside corner joint in the horizontal position, the electrode should be pointed upward at about 20 deg. to counteract the sag of the molten metal in the

crater. The electrode is also inclined about 20 deg. in the direction of travel of the weld as shown in Fig. 5-27. The

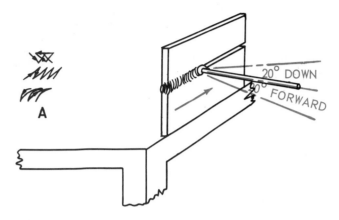

Fig. 5-27. Arc welding horizontally positioned butt joint on vertical surface. The electrode should be held 20 deg. down from joint and 20 deg. forward of puddle in direction of travel. Recommended electrode motions are shown at A.

arc gap (low voltage) should be shortened so the molten filler metal will travel across the horizontally positioned arc. Be sure to eliminate undercutting at the edge of the bead, which is the result of excess current for the size of the electrode used or poor electrode motion. When welding a T or inside corner joint with one piece of the base metal in a near vertical position, and one in a near horizontal position, the electrode is inclined 20 deg. in the direction of travel. The electrode is also positioned at about a 45 deg. angle to the horizontal piece of metal as shown in Fig. 5-28. The motion used is usually some type of weave motion with the forward motion taking place on the vertical piece.

5-22. ARC WELDING ON VERTICAL SURFACE

Whenever possible, welds should be made with the seams in the flat position. In factories, special turntables are used

Fig. 5-28. A welder about to tack weld T-joint prior to making a fillet weld in horizontal position. Note position of electrode. (Lincoln Electric Co.)

Fig. 5-30. A welder shown making fillet weld in vertical position on a large field welded structure. (Hobart Brothers)

to move the work so that this position may be obtained. However, many welds have to be done in a vertical, horizontal, or in an overhead position because of the size of the project or because turntables are not available. These welds must be of the same quality and strength as welds done in a flat position. Fig. 5-29 illustrates metal and electrode positions for vertical butt joint welding.

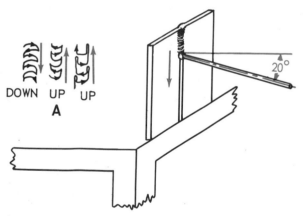

Fig. 5-29. Method of welding butt joint in vertical position. The electrode is held at 20 deg. from horizontal in the direction of travel. Recommended electrode motions are shown at A.

When welding a vertical seam, the molten metal tends to flow down the seam. This flow or sag can be minimized and even eliminated by pointing the electrode slightly (approximately 20 deg.) upward. Further, a short arc must be maintained, and the motion

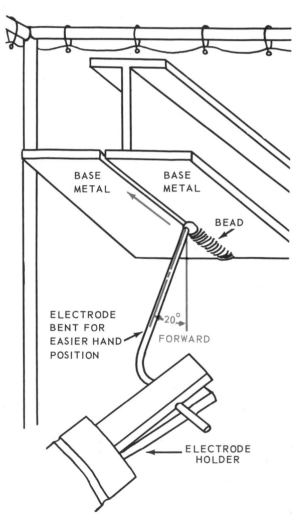

Fig. 5-31. Method of welding a butt joint in overhead position. Only covered electrodes should be used and the welder should use a short arc.

must be such that the force of the arc will oppose the sagging. Fig. 5-30 shows a welder making a vertical weld.

5-23. ARC WELDING OVERHEAD

This method of arc welding is the most difficult. It can also be dangerous for any operator who is not wearing the correct protective clothing. The beginner should practice making beads in the overhead position before attempting to weld seams. Fig. 5-31 shows a butt

joint overhead arc weld in progress. Fig. 5-32 shows an overhead weld being made on a T-joint.

5-24. CARBON ARC WELDING PRINCIPLES

The carbon arc method of welding produces an arc between the metal and a carbon electrode or between two carbon electrodes. The arc thus produced will melt the base metal, and the operator may add metal to this molten

Fig. 5-32. Making a practice overhead arc weld. Note position of electrode in electrode holder. This welder is wearing a cape to protect his arms and shoulders, apron, and gauntlet gloves.

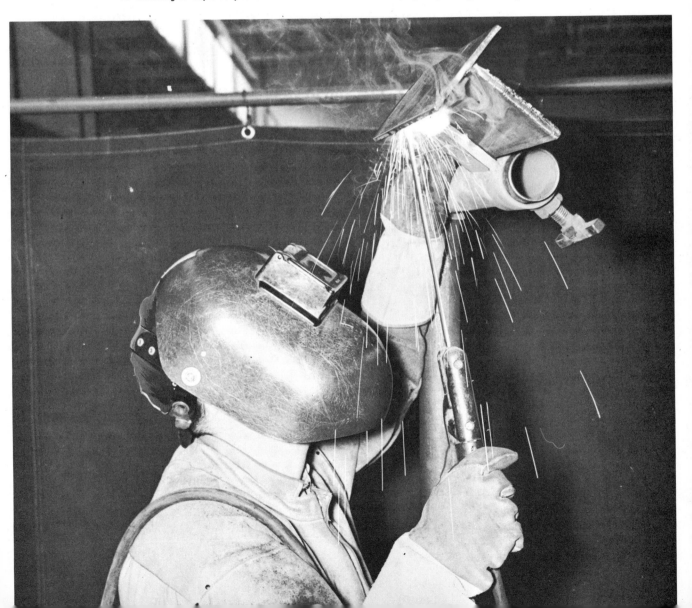

puddle by means of a welding rod (filler rod), as shown in Fig. 5-33.

In the type of carbon arc welding making use of two carbon electrodes,

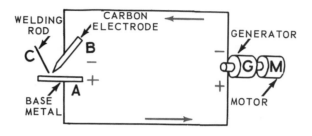

Fig. 5-33. Single carbon electrode arc circuit. An arc between the carbon electrode and the work furnishes the heat. The welder must add welding rod to the weld as in gas welding.

an arc is struck only between the two electrodes. The heat from this arc is high enough for heat treating, soldering, brazing, and light gauge metal welding. A double carbon arc welding electrode holder is shown in Fig. 5-34.

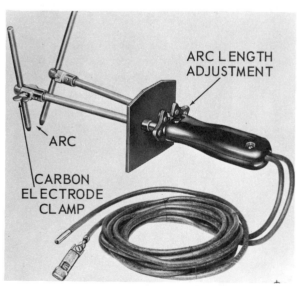

Fig. 5-34. A twin carbon electrode holder with carbon electrodes in place. Two leads are required because the arc is created between the two electrodes.

The carbon electrode is prepared for use by grinding the electrode to the shape of a cone. The taper of the cone

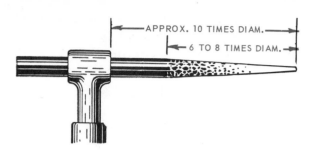

Fig. 5-35. Method of preparing carbon electrode for use in carbon arc welding. The carbon should taper a distance equal to 6-8 times its diameter, and should stick out from holder a distance equal to 10 times its diameter. (National Carbon)

should be 6-8 times the diameter of the electrode. When placing the carbon electrode in the holder the electrode should stick out from the holder a distance equal to 10 times the electrode diameter as shown in Fig. 5-35.

5-25. SHUTTING DOWN ARC WELDING MACHINE

Shutting down of an arc welding station is usually accomplished by:

A. Hanging the empty electrode holder on its insulated hook.

B. Pressing the OFF button on the machine.

C. Pulling the manual switch or circuit breaker to the OFF position.

D. Cleaning the station including:

    1. Electrode holder.

    2. Bench.

    3. Floor.

E. Storing electrodes of usable length.

F. Hanging helmet in its assigned place.

It is very important that the equipment be kept clean, since during arc welding oxide dusts are formed which, if allowed to accumulate, will collect on the equipment and will cause them to deteriorate rapidly. The machine should also be kept away from moist or corrosive locations. Practically all manufacturers have special features constructed in their machinery, for which reason it is recommended that the instructions accompanying each machine be read carefully and followed.

5-26. REVIEW OF SAFETY IN ELECTRIC ARC WELDING

Following are some safety rules that must be carefully observed, if accidents when welding are to be prevented:

1. The eyes and face must be protected from harmful rays and sparks by using an approved type of helmet. See Fig. 5-7.

2. Recommended clothing and shoes must always be worn.

3. Open pockets and cuffs not permitted, because they may catch hot sparks and the clothing may be ignited.

4. The floor on which the operator stands should be kept dry, to eliminate the chance of an electrical shock.

5. Only an experienced electrician should work on electrical power connections used in the electric arc welding machine.

6. The operator should wear heavy, gauntlet-type gloves.

7. When arc welding, all skin should be covered to prevent burns from the arc rays.

8. The operator should have adequate ventilation to protect the nose, throat, and lungs from harmful and irritating fumes generated in the electric arc.

5-27. TEST YOUR KNOWLEDGE

1. What is the recommended voltage when metallic arc welding?

2. What polarity does anode signify?

3. Should an arc weld penetrate through the thickness of the metal?

4. Does an arc weld usually have a ripple bead similar to acetylene welding?

5. Why must the operator wear gloves when electric arc welding?

6. Why are Nos. 10 and 12 safety lenses used in helmets and hand shields?

7. Is it necessary to vary the voltage for different thicknesses of metal? Why?

8. In what direction do the electrons travel when using straight polarity?

9. What is the distinct sound indication of the correct arc length when welding with covered electrodes?

10. At what angle should the electrode be held in relation to the metal for flat position butt joint welding?

11. How much of the heat used for arc welding is liberated at the electrode when using straight polarity?

12. What is the result of maintaining a welding arc that is too long?

13. Why is it necessary that all electric connections be tight and clean?

14. What is the function of the coating on a shielded-arc electrode?

15. Why doesn't the ammeter register unless an arc is being maintained?

16. What produces the heat in electric welding?

17. What is meant by the arc stream?

18. Name four different types of electric arc welding.

19. What are the two fundamental types of electric welding?

Chapter 6
DC ARC WELDING
EQUIPMENT AND SUPPLIES

In this Chapter we will discuss the construction, operation and maintenance of various kinds of direct current, electric arc welding apparatus.

6-1. ARC WELDING STATION

A complete arc welding station consists of a booth, DC power source, welding table, ventilating system, electrode cable, electrode holder, and a ground cable, as shown in Fig. 6-1.

There are several types of DC power sources:

1. Motor generator.
2. Engine driven generator.
3. Rectifier.
4. AC/DC arc welder combinations.

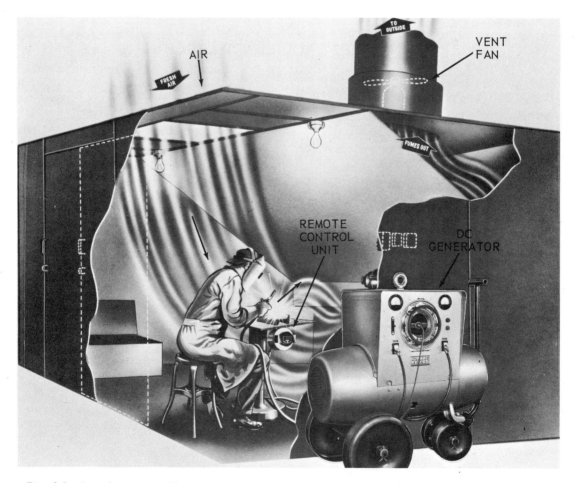

Fig. 6-1. Complete arc welding station. Note that the ventilating air enters above the welder and is directed down and away from the welder's face. (Hobart Brothers Co.)

6-2. DIRECT CURRENT, ARC WELDING GENERATOR

The generator used in the motor generator unit and in the engine driven unit is basically the same. Those who use the machine should become familiar with its construction and operation.

The task of the arc welding generator is extremely difficult. As the operator strikes an arc, the electrical pressure (potential, volt reading) drops and the current flow goes to a maximum. The generator is necessarily of special design, and incorporates features not found in any ordinary electric generator.

The welding generator must produce enough current (DC) at a constant rate to furnish heat required to melt whatever thickness of metal is to be welded. The apparatus must be safe to use. It must be of sturdy construction so it will stand years of use. It must furnish

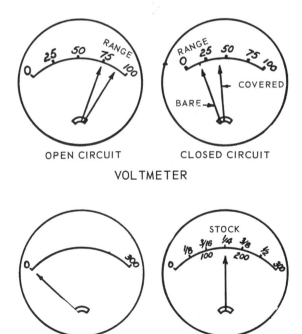

OPEN CIRCUIT   CLOSED CIRCUIT

VOLTMETER

OPEN CIRCUIT   CLOSED CIRCUIT

AMMETER

Fig. 6-2. Typical arc welding meter readings.

electricity for welding at a constant rate up to its rated capacity, regardless of how heavy the demand, or how irregular the conditions.

The generator may be equipped with either one or two regulating mediums. If one medium is used, this adjustment regulates both the ampere flow and the voltage (potential). If two adjustments are used, one is used to adjust the current, and the other to control the potential (volts).

To understand these adjustments and to realize what the various readings mean, it must be remembered that the ampere flow denotes the quantity of electricity. Upon it depends the amount of heat being produced at the weld. The voltmeter reading denotes the pressure of the electricity, and upon it depends the ability to strike and hold an arc. The higher the voltage reading or potential, the longer the arc which may be maintained.

When the machine is not being used, even though it is running, no current flows. The ammeter will read zero except when the machine is being used for producing an arc, or when current is flowing. The voltmeter will register at all times except when the machine is not running. The voltmeter, therefore, will indicate two potentials: (A) open-circuit, (B) closed-circuit potential as shown in Fig. 6-2.

The open-circuit voltmeter reading is the reading the meter will have when the machine is running, but not being used to produce an arc potential. The open-circuit and closed-circuit voltage of the generator is adjustable, however the open-circuit voltmeter reading rarely exceeds 60-80 volts. The closed-circuit voltage is much lower than the open-circuit reading, meaning that when the machine starts to produce an arc, the pressure which forces the amperes through the arc gap decreases considerably.

The amount of the closed-circuit voltage is important. If the closed-circuit potential is too high, the weld metal will be brittle and will contain pits and air holes. If it is too low, it

Welding generators are specially constructed to produce high current flow at low potential. A variety of generator sizes are available. Fig. 6-3 shows a motor generator on wheels. The current produced by the generator should be steady, and the potential

Fig. 6-3. DC motor-generator welding machine on wheels.

will be almost impossible to strike (create) and maintain an arc.

The generator produces electrical energy (watts). Watts are equal to the product of the potential (volts) and the current flow (amperes), both are normally indicated on dials mounted on the machine. The dials give the relative values of these two. The formula for finding watts is: Watts = Voltage x Amperage.

must not fluctuate during the welding procedure. A steady current is maintained by special devices incorporated in the design of the generator. Compensation poles are built into the generator so that the coils overlap the main field coils, producing a more stable arc as shown in Fig. 6-4. A reactor is also used. This is an electrical device which acts as a current shock absorber. It absorbs the current fluctuations and

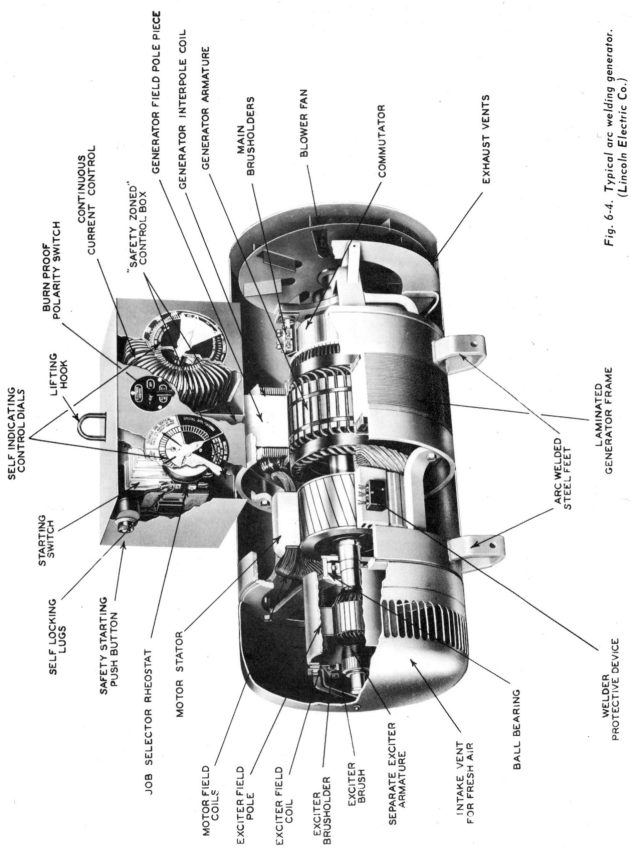

SELF INDICATING CONTROL DIALS

LIFTING HOOK

STARTING SWITCH

SELF LOCKING LUGS

SAFETY STARTING PUSH BUTTON

JOB SELECTOR RHEOSTAT

BURN PROOF POLARITY SWITCH

CONTINUOUS CURRENT CONTROL

"SAFETY ZONED" CONTROL BOX

GENERATOR FIELD POLE PIECE

GENERATOR INTERPOLE COIL

GENERATOR ARMATURE

MAIN BRUSHOLDERS

BLOWER FAN

COMMUTATOR

EXHAUST VENTS

Fig. 6-4. Typical arc welding generator. (Lincoln Electric Co.)

LAMINATED GENERATOR FRAME

ARC WELDED STEEL FEET

WELDER PROTECTIVE DEVICE

BALL BEARING

INTAKE VENT FOR FRESH AIR

SEPARATE EXCITER ARMATURE

EXCITER BRUSH

EXCITER BRUSHOLDER

EXCITER FIELD COIL

EXCITER FIELD POLE

MOTOR FIELD COILS

MOTOR STATOR

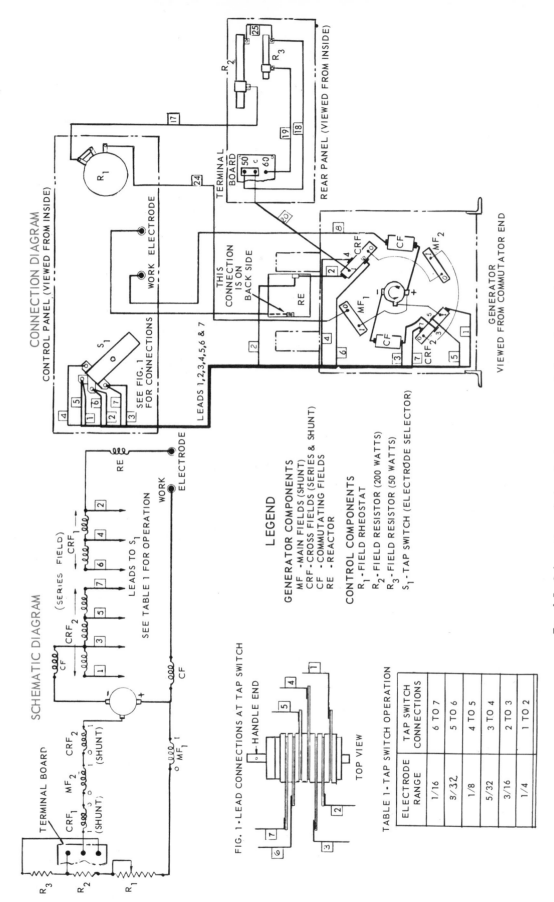

Fig. 6-5. Schematic and electrical connection diagram for a DC motor-generator arc welder.

smooths out the flow of current to the arc. It consists essentially of a huge coil of wire, wound on a laminated metal core and connected in series with the arc, as shown in Fig. 6-5.

Some machines use a separate exciter to maintain good potential and current flow characteristics. An exciter is a small generator electrically connected to the field windings of the large generator. The exciter keeps a constant potential on the main fields and also prevents them from reversing their polarity.

The electric arc length varies slightly, depending on the steadiness of the operator. When the arc length changes, the potential and the current tend to fluctuate. Most of the machines produce consistent arcs under varied conditions. Fig. 6-6 shows the electrical wiring of a typical DC motor generator

gasoline engine. Some models are built in a vertical position, some in a horizontal position. If a separate housing is used to house the generator and motor or engine, a flexible coupling is used to connect the two. The generator bearings are usually ball or roller bearings. The commutator bars are usually assembled into a cylindrical shape. The brushes must make good contact with the commutator, and the commutator must be in good condition to operate satisfactorily. See CHAPTER 26, for service information on arc welding equipment.

6-3. GENERATOR AND MOTOR AIR FILTERS

To reduce bearing and commutator maintenance costs to a minimum, and to increase the high efficiency life of

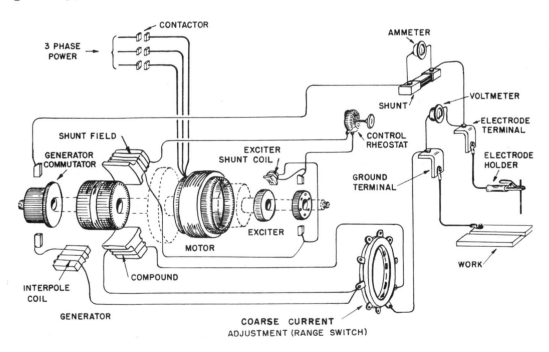

Fig. 6-6. Pictorial wiring diagram of DC motor-generator welding machine, with electric motor drive.

arc welder. The generator may be a separate part, or it may be built into the same housing as the motor or

a motor generator arc welder, an air filtering accessory has been developed. The air filters or cleaners are made

ON-OFF SWITCH   POLARITY SWITCH

AIR FILTER

Fig. 6-7. An air filter has been installed on a DC motor-generator arc welder to prevent air-borne dirt from damaging internal parts of the welder.

to fit practically all models of arc welders. Fig. 6-7 shows an air filter installation. The filters may be cleaned by blasting with high pressure air, with steam, or by rinsing in a commercial solvent.

6-4. DC ARC WELDING GENERATOR DRIVES

As mentioned before, the generator may be driven by a gasoline engine, or an electric motor. A gasoline engine is used in remote localities, or on construction work where electricity is not available. Where practical the electric motor drive is used. Its construction, housing, bearings, etc., are similar to the generator. Electrically, it is usually an induction motor using 60 cycle, 3 phase current of either 230 or 460 volts. The motor is equipped with a manual switch, magnetic starter, and an overload, cut-out safety switch. The motor size varies with the generator size. A 10 to 15 HP motor is usually used to drive a 200 amp; 20 to 40-volt generator.

Gasoline engines used are of typical construction. They come in a variety of sizes. Some are water-cooled, some

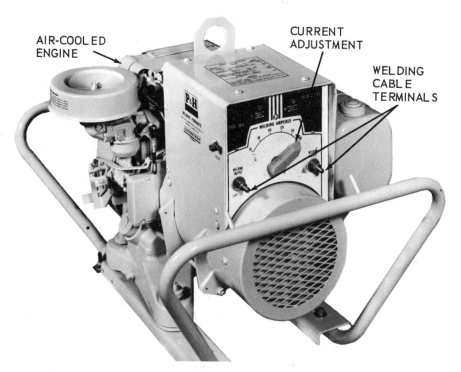

AIR-COOLED ENGINE   CURRENT ADJUSTMENT   WELDING CABLE TERMINALS

Fig. 6-8. A direct current arc welding generator driven by an air-cooled engine.
(Chemetron Corp.)

air-cooled. All have automatic throttle controls and governors to control the flow of power to the generator according to the welding demand. Fig. 6-8 illustrates a DC generator driven by an air-cooled engine. A water-cooled engine and DC generator combination is shown in Fig. 6-9.

Fig. 6-9. Water-cooled engine used to drive a DC generator.

6-5. ENGINE DRIVEN ARC WELDING EQUIPMENT

The welding generator is basically a source of electrical energy. Several of the welding companies now supply dual-purpose units that are usable both as a welding machine and as a source of regular electric power. Construction companies, farmers, remotely located activities, and concerns that find it wise to have stand-by electrical service use such machines.

The machine fundamentally is an internal combustion engine driven generator. It is powered by an automatically governed gasoline engine. Units of this type usually operate at 1800 RPM. The typical generator can furnish up to 200 or 300 amperes of direct current for welding, and from 5 to 8 KW at 120 volts single phase (AC). Units of both smaller and larger capacity are obtainable. This type is the more simple of the dual-purpose units as it uses only one generator.

Some of the larger units require up to 70 HP at 1800 RPM, and can provide

Fig. 6-10. A 300-amp. DC Welder with 6-12 KW of 115-230 V auxiliary power, driven by a water-cooled engine. This type unit may be mounted on a truck for portability.

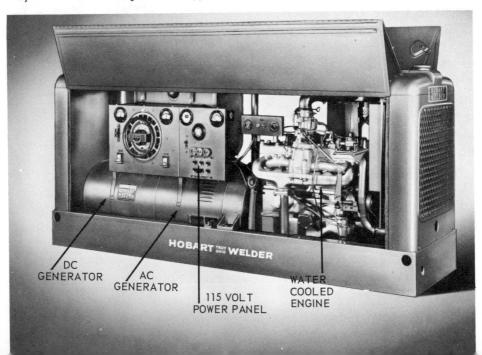

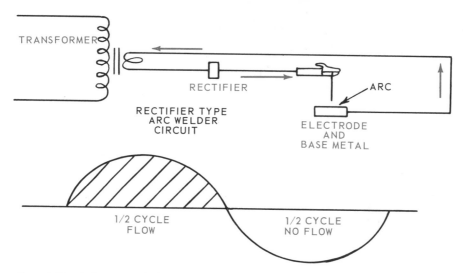

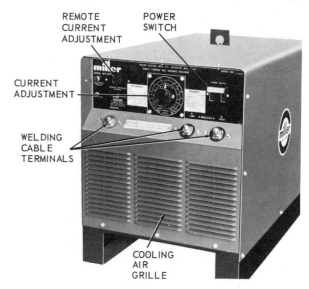

Fig. 6-11. Schematic of the current flow during half cycle of DC arc welder. The electrons can flow only in the direction of the arrows. When the flow attempts to reverse, the rectifier resists the flow.

either single or three phase 115 volt or 230 volt current up to 25 KW (kilowatts) (25000 watts), as shown in Fig. 6-10. These units furnish up to 600 amperes DC at 40 volts for welding purposes. To furnish both DC and AC, the generator unit consists of two separate generators, one to create welding current (DC), the other to produce auxiliary power current (AC). Such a machine can be used both as a welder, and as a power source at the same time.

6-6. RECTIFIER TYPE DC ARC WELDER

The rectifier type DC arc welder consists of a transformer, and a silicon or selenium rectifier to convert the alternating current to direct current. Silicon and selenium are semiconductors of electricity which permit easy flow of electrons in one direction but poor flow in the other. Fig. 6-11 shows a schematic wiring diagram.

The rectifier type arc welder is a DC welder without major moving parts. It can deliver either straight or reversed polarity direct current. It is

very quiet in operation, and maintenance is reduced to a minimum. Fig. 6-12 illustrates a rectifier type DC welder. In this welder, the only moving

Fig. 6-12. DC arc welder uses a rectifier to change AC input to DC output. Polarity is changed by reversing welding leads on machine. (Miller Electric Mfg. Co.)

part is an air circulation fan which serves to cool the mechanism. The machines are built in various ampere capacities. Fig. 6-13 shows the internal construction of a rectifier type

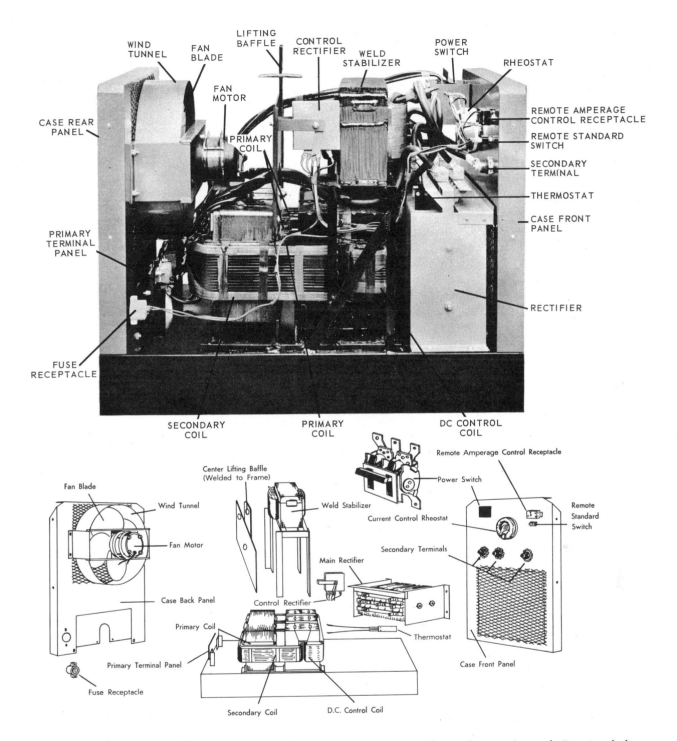

WIND TUNNEL

FAN BLADE

LIFTING BAFFLE

CONTROL RECTIFIER

WELD STABILIZER

POWER SWITCH

RHEOSTAT

FAN MOTOR

CASE REAR PANEL

PRIMARY COIL

REMOTE AMPERAGE CONTROL RECEPTACLE

REMOTE STANDARD SWITCH

SECONDARY TERMINAL

THERMOSTAT

CASE FRONT PANEL

PRIMARY TERMINAL PANEL

RECTIFIER

FUSE RECEPTACLE

SECONDARY COIL

PRIMARY COIL

DC CONTROL COIL

Fan Blade

Wind Tunnel

Fan Motor

Center Lifting Baffle (Welded to Frame)

Weld Stabilizer

Remote Amperage Control Receptacle

Power Switch

Current Control Rheostat

Remote Standard Switch

Secondary Terminals

Case Back Panel

Main Rectifier

Control Rectifier

Primary Coil

Thermostat

Primary Terminal Panel

Fuse Receptacle

Secondary Coil

D.C. Control Coil

Case Front Panel

Fig. 6-13. Rectifier type DC arc welder. Above photo shows the welder with case removed. Drawing below illustrates the major parts of the welder shown in photo above. (Chemetron Corp.)

DC arc welder. Fig. 6-14 illustrates the wiring diagram of a three-phase input rectifier type DC arc welder.

The primary circuit of this type arc welder, shown in Fig. 6-14 is three-phase. The fan motor is single-phase

DC Arc Welding Equipment 191

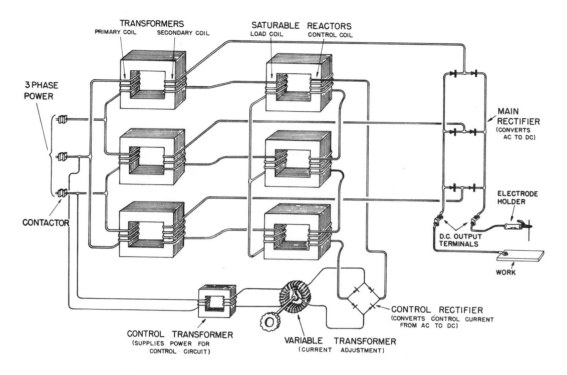

Fig. 6-14. A wiring diagram for rectifier type DC arc welder.

with a capacitor start device. Because of the three-phase input, three rectifiers are needed.

6-7. DUAL-PURPOSE ARC WELDERS

The use of rectifiers to change AC to DC has resulted in quiet operating machines. Because there are rectifiers of such efficiency that one square inch of rectifying surface will carry as much as 200 amperes, some companies are offering machines that can be used for either AC welding, or DC welding. The use of selenium and silicon products has resulted in more efficient and more durable machines. This type machine, because it is both DC and AC uses only single-phase primary current (power source). The appearance, operation, and maintenance of these welders is similar to the other transformer types, as shown in Fig. 6-15.

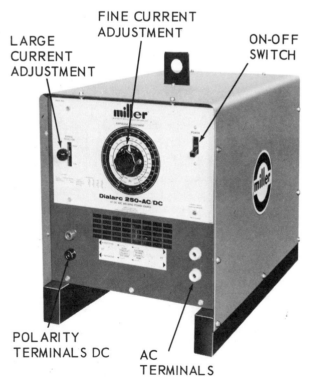

Fig. 6-15. A combination AC and DC arc welder transformer-rectifier type. (Miller Electric Mfg. Co.)

6-8. THREE-PURPOSE ARC WELDERS

Inert gas arc welding usually requires automatic devices to control the inert gas flow, water flow, control of the high frequency arc starting device when AC is used, and automatic starting and stopping of the welding current.

Some equipment manufacturers are offering a self-contained welder, that can be used for DC welding, AC welding, and for inert gas arc welding with all the controls contained in one unit, such as shown in Fig. 6-16. This welder is a versatile welding station usable for practically all the forms of arc welding.

6-9. OTHER TYPES OF DC ARC WELDERS

The rectifier tube type of welder uses the principle of the rectifier tube found in storage battery chargers and alternating current input radio receivers, to supply current for welding. The principle of operation is to pass the current through a gaseous gap from a sharp electrode to a flat-faced electrode. The current goes through the gas in one direction easily, but has difficulty in traversing from the flat face (carbon) to the tungsten wire point. One-half the wave of the AC is permitted to go on, but the other one-half wave is stopped almost completely. By using two tubes, both sides of the AC wave can be rectified into DC. The size of the tubes or the number of tubes determines the amount of current or capacity.

Storage batteries are sometimes used as a source of welding current, either for arc or resistance welding. Edison cells, or lead-acid cells, may be connected in series, or parallel, to pro-

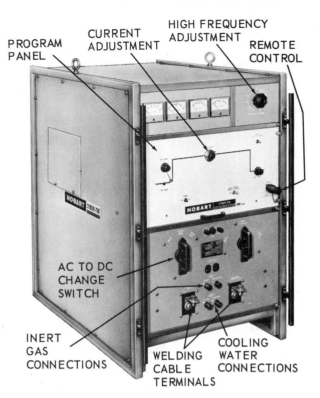

Fig. 6-16. Combination AC-DC arc welder with inert gas welding facilities built in.

duce high current flow at safe potentials. The size and weight of the complete apparatus, and the matter of maintaining and of recharging the batteries periodically are drawbacks to the extensive use of this type welding equipment.

6-10. LEADS, COPPER

Large diameter, superflexible leads (cables) are used to carry the current from the welding machine to the work and back. The lead from the machine to the work is known as the electrode lead, and the lead from the work to the machine is known as the work (ground) lead. These leads are well insulated with rubber and a woven, fabric-reinforcing layer, as shown in Fig. 6-17. The leads are usually subjected to considerable wear and should be checked periodically for breaks in

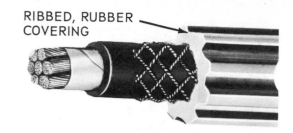

RIBBED, RUBBER COVERING

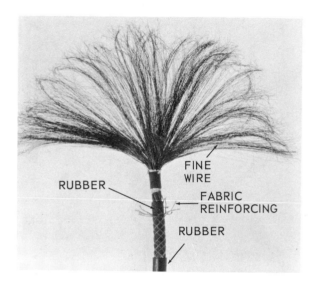

RUBBER

FINE WIRE

FABRIC REINFORCING

RUBBER

Fig. 6-17. Above. Welding lead. Note the heavy ribbed outer rubber covering. Below. The hundreds of individual wires used to form the complete lead (cable). This construction produces the flexibility needed in these leads. (Lincoln Electric Co.)

Leads are produced in several sizes. The smaller the number, the larger the diameter of the lead. See Fig. 6-18. The lead must be flexible in order to reduce the strain on the arc welder's hand as he welds, and also to permit easy installation of the cable. To produce this flexibility, as many as 800 to 2500 fine wires are used in each cable.

The length of the lead cables has considerable effect on the size to be used for certain capacity machines. Fig. 6-18 shows some arc welding lead recommendations.

The wiring to the motor, in case a motor-driven generator is used, must conform to local, state and national electrical codes.

6-11. LEADS, ALUMINUM

Electrode and work leads (cables) until recently were made only of flexible copper cable. Aluminum is now also available for this purpose. Even though the next size larger cable should be used, for equal current carrying

LEAD NO.	LEAD DIA.	LENGTH IN FEET		
		0-50 FT.	50-100 FT.	100-250 FT.
		CURRENT CAPACITY*AMPERES		
4/0	.959	600	600	400
3/0	.827	500	400	300
2/0	.754	400	350	300
1/0	.720	300	300	200
1	.644	250	200	175
2	.604	200	195	150
3	.568	150	150	100
4	.531	125	100	75

*Lengths given are for the total combined length of the electrode and work leads.

Fig. 6-18. Arc Welding Lead Recommendations: The voltage drop in these copper leads will be approximately 4 volts with all connections clean and tight.

the insulation. The potential carried by the leads is not excessive, varying between approximately 14 and 80 volts.

capacity if long runs are used, the aluminum lead is considerably lighter than the copper lead. Each individual wire in the standard aluminum lead is slightly larger than in the copper lead. For example, aluminum AWG 1 lead uses 840 wires, and each wire is #30 B&S gauge. Basically, the aluminum wire is only one-third the weight of the same size copper. Its current-carrying capacity is 61 percent as good as copper. With the proper insulation on the cable, the aluminum cable weighs about one-half that of the copper cable. Aluminum lead is made from pure electrolytic aluminum and is semiannealed.

6-12. CONNECTIONS FOR LEADS

To consistently carry the heavy currents used in welding, all parts of the welding circuit must be of heavy-duty construction and all connections of heavy-duty design.

The copper or aluminum leads are fastened to the various devices by means of lugs and are soldered or mechanically attached to the leads, as shown in Fig. 6-19. The lugs provide

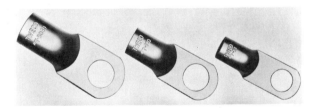

Fig. 6-19. Lugs for welding leads. The three sizes will fit cables from No. 6 to 4/0. They may be connected to the lead by soldering or mechanical crimping.

a firm means of attaching the leads to the welder terminals and to the ground (work lead) terminals. These connections must be durable and must have low resistance or the joint will overheat during welding and also result in unsatisfactory current flow. Connections are also available for connecting one lead to another, as shown in Fig. 6-20.

Some of the processes used are:
1. Mechanical.
2. Soldering.
3. Brazing.
4. Welding.

In the case of aluminum cables, it is claimed that it is best to clamp the aluminum to the electrode holder and to the other terminals. However, the lead (cable) can be successfully aluminum brazed to either aluminum connections or copper connections. It is recommended that twisting or torsion on the lead, especially at the connections, be avoided as the lead will tend to separate from its terminals, if this is done. Mechanical connections must be tight and clean.

Soldered connections must be done skillfully or only a portion of the electrical flow area will be connected.

Silver brazing must also be done by an experienced person to insure enough electrical flow area.

Another method of connecting copper welding cables is the copper welding of the cable ends to lugs or to each other.

The process used is a method of welding copper in which no outside source of heat is required. Powdered copper oxide and powdered aluminum are placed in a small graphite crucible

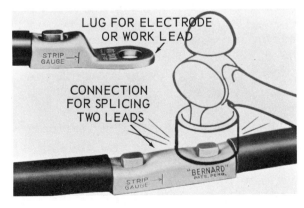

Fig. 6-20. Welding lead connections. The connections are made mechanically in this type connection by striking the malleable punch area with a hammer.

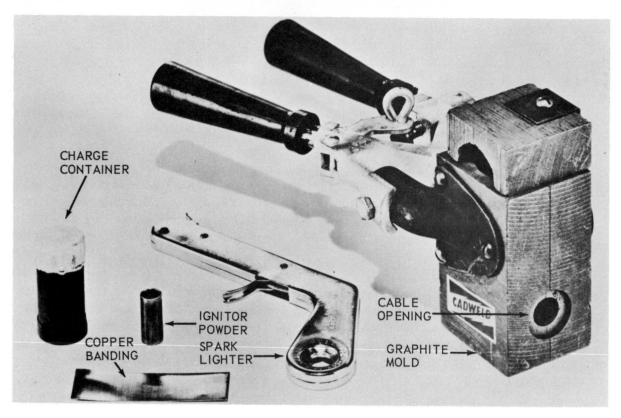

CHARGE
CONTAINER

COPPER
BANDING

IGNITOR
POWDER
SPARK
LIGHTER

CABLE
OPENING

GRAPHITE
MOLD

CADWELD

Fig. 6-21. Tools needed to make a fused copper cable splice or a cable to lug weld. The cable is clamped in the graphite mold. The charge and ignitor are placed in the top of the mold and the charge is ignited. Melting of the charge and the weld which results is similar to the reaction in thermit welding. (Erico Products, Inc.)

and ignited by means of a spark. The process is exothermic (gives off heat). The rapid oxidation of the powder creates enough heat to melt the copper, and to melt the ends of the leads and connectors, thus producing a sound weld. The molten copper flows over the cable ends in the graphite mold; the cable ends become molten, and in a few seconds are securely welded together in a solid copper nugget, as shown in Fig. 6-21.

The cable is prepared by stripping approximately one inch of insulation from each end to be joined. Both ends are then placed in the welder, butted together under the center of the tap hole, and locked by the clamp-type crucible. The flint spark gun ignites the mixture; and in about ten seconds,

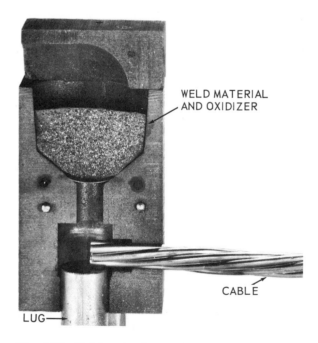

WELD MATERIAL
AND OXIDIZER

CABLE

LUG

Fig. 6-22. Welding lead parts being joined by means of thermit type welding process.

the weld is completed. The same procedure and equipment may be used to attach terminal lugs, as shown in Fig. 6-22.

The preparation of the cable prior to welding and two examples of finished cable welds are shown in Fig. 6-23.

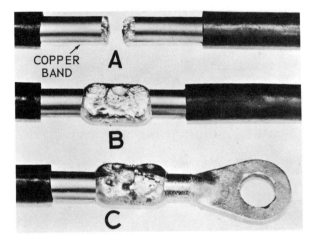

Fig. 6-23. *Fused copper lead splice (B) and a lead to lug joint (C). The lead is prepared by removing insulation, and wrapping a thin band of copper around the strands to keep them in place during the thermit welding operation (A).*

6-13. GROUND ATTACHMENTS

Good ground connections are important when welding with either AC or DC. Occasionally the work lead is permanently fastened to the welding bench or table by bolting or welding. This practice is practical when small pieces are positioned on a table.

Frequently, however, the ground cable must be fastened to the article being welded, due to its size or location. Clamping, bolting, or welding the work lead to the piece is still practiced on this type of job, see Fig. 6-24. It is sometimes difficult to use a clamping device on a metal fabrication. Use clamps carefully on finished surfaces to avoid marring the surface. A magnetic ground is available which

permits quick fastening of the work lead to the work to be welded. This makes it easy to change the position of the ground to obtain better arc characteristics, and does not injure or mar the article to be welded. The work lead is either soldered or mechanically fastened to this permanent magnet grounding device, and the operator easily positions it on any ferrous surface. The magnets are replaceable and are quite powerful.

6-14. ELECTRODE HOLDERS

The electrode holder is the part of the arc welding equipment held by the operator when welding. It is used to hold metallic or carbon electrodes. Many different styles and models have been produced, but they all have similar characteristics. The electrode lead is fastened to the electrode holder either inside of the handle or to a lug. The most common arrangement is to attach the lead to a metal connection inside the handle. The handle itself is made of an insulating material which has high heat and electrical resistance qualities. Electrode holders are built to produce a balanced feeling when held in the operator's hand, with the cable draped over the operator's arm, and with the average length of the metallic electrode in the holder.

Two means are commonly used to clamp the electrode in the holder. One is of pincher construction and has a

Fig. 6-24. *A spring-loaded ground clamp for the work lead. (Lenco, Inc.)*

coil spring to produce the necessary pressure to obtain a good contact between the holder and the electrode, see Fig. 6-25. The other uses a cantilever spring, as shown in Fig. 6-26. There

brush. The electrode clamps should also be kept clean by using a file, sandpaper, or other suitable means. When lightly coated or covered electrodes are used, it is necessary that the

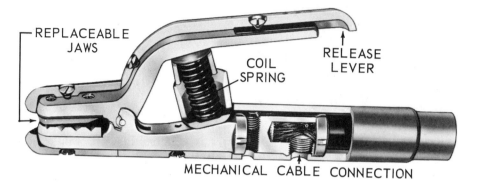

Fig. 6-25. Electrode holder which uses a coil spring for clamping pressure. (Fibre-Metal Products)

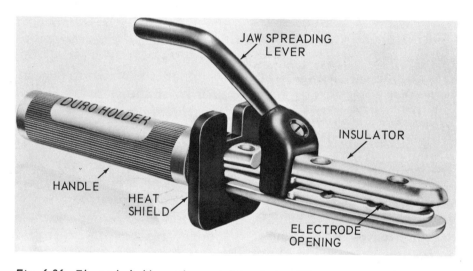

Fig. 6-26. Electrode holder with special insulation. The jaws are made of a special copper alloy. The electrode lead may be soldered or clamped to the holder. The holder uses a cantilever spring design to clamp the electrode in the jaws. (Duro Engineering Co.)

are two pieces of spring-like metal extending from the electrode handle and equipped with notches. The operator slides the electrode between the two pieces of metal and the electrode is firmly clamped in place.

It is advisable to clean the electrode at the point where it is to be connected. This may be done by using a clean wire

covering material be scraped off the electrode at its point of contact with the electrode holder.

When welding heavy work the electrode holders are sometimes equipped with shields, i.e., a small heat-resisting plate to prevent the radiation of heat from the work directly onto the operator's hand. Another way of in-

Fig. 6-27. *Storing electrodes. Each bin is marked with AWS number and electrode diameter.*

creasing the operator's comfort is to water-cool the handle of the electrode holder. This is done by circulating water through the handle by means of flexible rubber hose. Water-cooled holders are popular on carbon arc welders.

6-15. METALLIC ELECTRODES

Metallic electrodes of many types are available. One of the more common ways to classify electrodes is by the covering (coating) on the electrode. This includes:

1. Bare electrodes.
2. Dusted electrodes.
3. Flux-dipped electrodes.
4. Extruded electrodes and coverings.

Of these types, the electrode with an extruded or flux-dipped covering is most popular. The bare electrode is the least expensive. A slightly more expensive electrode is dust-coated with a flux to help reduce the oxidizing action of the arc. For such welding applications as high-temperature steels,

tool steels, molybdenum steels, and for especially strong, mild-steel welds, covered (heavily coated) electrodes should be used. Special electrodes have been developed for each kind of material and type of weld. Fig. 6-27 shows a bin used to store the variety of electrodes used in a typical job shop.

Some covered electrodes are coated by dipping them in a liquid flux solution to produce a coating of sufficient thickness. The flux may also be applied to a covered electrode while the wire is being extruded (forced through a forming die). When a thick covering of special material or fluxes is required on an electrode as in hard surfacing applications, the coating is usually applied by dipping or extruding.

Most of the mild steel electrodes are physically similar in their specifications with the exception of the flux used.

The common electrode wire sizes are: 1/8, 5/32, 3/16, 7/32, 1/4, 5/16, and 3/8 in. diameter. These rods come in lengths of 14 in. for all sizes and may also be obtained in 18-in. lengths in some sizes.

Electrodes are usually packed in 50 lb. packages.

The most commonly used electrodes are made of mild steel, but electrodes of many metal alloys may be purchased, including:

1. Mild steel.
2. Low alloy steel.
3. Nickel steel.
4. Chrome-moly steel.
5. Manganese-moly steel.
6. Nickel-manganese-moly steel.
7. Nickel-moly-vanadium steel.
8. Aluminum.
9. Copper-aluminum (aluminum bronze).
10. Lead-bronze.
11. Phosphor bronze.

6-16. BARE ELECTRODES

Steel wire electrodes without any coating or covering are still used for some welding operations. It is more difficult to produce a satisfactory weld using a bare electrode. However, they are still used especially in places where post-cleaning is difficult. A competent welder can produce satisfactory welds with them, but the strength and durability of the joint is less than that produced with covered electrodes.

In coil form bare electrodes are used with inert gases.

The welding process is known as Inert Gas Metal Arc Welding (MIG) (See CHAPTERS 11 and 12).

Bare electrodes usually have the following chemical composition:

Carbon	.08 to .13 %
Manganese	.30 to .40 %
Phosphorus	.012 to .018%
Silicon	maximum .03 to .30%
Sulphur	.026 to .028%

6-17. ELECTRODE COVERING FUNDAMENTALS

The basic function of the covering (coating) on electrodes is as follows: During the arc process the covering changes to neutral or reducing gases such as carbon monoxide, or hydrogen (CO or H_2). These gases, as they surround the arc proper, prevent air from coming in contact with the molten metal, and thus prevent oxygen from the air which may approach the molten metal, from combining with it. However, this action usually does not protect the hot metal after the arc leaves that point. The flux on covered electrodes, therefore, also, requires a chemical compound which, in addition to providing a neutral or reducing shield around the arc, also contains special ingredients which promote fusion and tend to remove impurities from the molten metal.

Flux which forms the covering on an electrode commonly consists of feldspar, mica steatite, titanium dioxide, calcium carbonate, magnesium carbonate and various aluminas. The neutral or reducing gas-producers are: Carbon hydrates such as paper, cotton, wood flour, cellulose, starch, and dextrin. Some special electrodes have metallic salts included in the covering to produce the correct alloy metal in the weld.

In addition to the action of the flux in the arc proper and with the molten metal, the residue forms a coating of material over the weld after solidification, which prevents the air from contacting the still hot metal. This coating over the bead is called slag. A good flux-covered electrode can produce a weld that is similar in appearance to the oxyacetylene weld, and

TEMPERATURE
CONTROL

*Fig. 6-28. Electrode drying oven.
(Dispatch Oven Co.)*

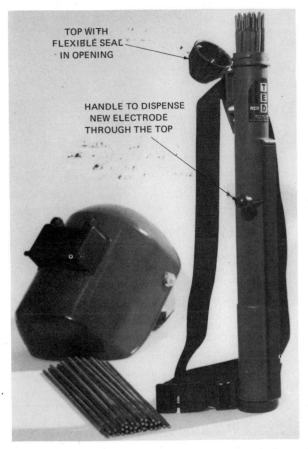
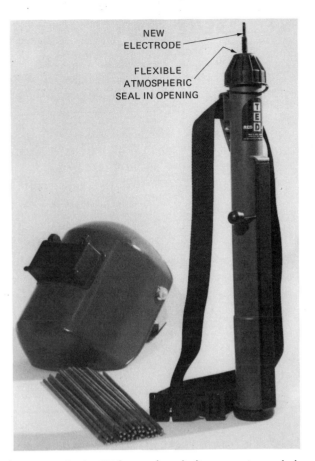

Fig. 6-29. Electrode dispenser. Left. Loaded electrode dispenser. Right. With top closed, dispenser is sealed. Moving lever feeds new electrode through top seal. (Red-D-Arc Ltd.)

which has excellent chemical and physical properties.

Since dampness destroys the effectiveness of most electrode coverings, electrodes are often stored in a specially built drying oven as shown in Fig. 6-28.

Covered electrodes must be handled carefully to prevent breaking or cracking the coverings. An electrode carrier may be used to minimize the chance of damage to the electrodes. Such a carrier is shown in Fig. 6-29.

6-18. ELECTRODE CLASSIFICATION

There are a large variety of welding electrodes on the market. They vary considerably in the composition of metal and the composition of the coating (shielding) of the electrode, according to the planned use of the electrode.

The American Welding Society has developed a series of identifying number classifications. See Figs. 6-30, 6-31, and 6-32.

The letter E preceding the four or five digit number (EXXXXX) indicates a welding electrode used in arc welding. This is contrasted with the letters RG, which indicate a welding rod used for gas welding.

The meaning of digit numbers in the AWS designation follows: The first two or three digits of the four or five digit number (E60XX) or (E100XX) represents the tensile strength. That is, 60

ELECTRODE AWS NUMBERS	POSITION	USE	TYPE CURRENT USED
E 4520	all		DCSP
E 6010	all	Penetration	DCRP
6011	all	Penetration	DCRP or AC
6012	all	Production	DCSP or AC
6013	all	Sheet Metal and Fillets	DCSP, DCRP or AC
6020 (iron oxide)	H (Fillets) / F		DCSP or AC / DCSP, DCRP or AC
6027 (iron powder)	H (Fillets) / F		DCSP or AC / DCSP, DCRP or AC
E 7010	all		DCRP
7011	all		DCRP or AC
7014 (iron powder)	all		DCSP, DCRP or AC
7015 (low hydrogen)	all		DCRP
7016 (low hydrogen)	all		DCRP or AC
7018 (iron powder, low hydrogen)	all		DCRP or AC
7020	H (Fillets) / F	Chrome-Moly Steel	DCSP or AC / DCSP, DCRP or AC
7024 (iron powder)	H (Fillets) / F		DCSP, DCRP or AC
7027 (iron powder)	H (Fillets) / F		DCSP or AC / DCSP, DCRP or AC
7028 (iron powder, low hydrogen)	H (Fillets) / F		DCRP or AC
E 7010-X	all		DCRP
7011-X	all		DCRP or AC
7014-X (iron powder)	all		DCSP, DCRP or AC
7015-X (low hydrogen)	all		DCRP
7016-X (low hydrogen)	all		DCRP or AC
7018-X (iron powder, low hydrogen)	all		DCRP or AC
7020-X	H (Fillets) / F	Chrome-Moly Steel	DCSP or AC / DCSP, DCRP or AC
7024-X (iron powder)	H (Fillets) / F		DCSP, DCRP or AC
7027-X (iron powder)	H (Fillets) / F		DCSP or AC / DCSP, DCRP or AC

ELECTRODE AWS NUMBERS	POSITION	USE	TYPE CURRENT USED
7028-X (iron powder, low hydrogen)	H (Fillets) / F		DCRP or AC
E 8010-X	all	Chrome-Moly Steel	DCRP
8011-X	all	Chrome-Moly Steel	DCRP or AC
8013-X	all		DCSP, DCRP or AC
8015-X (low hydrogen)	all		DCRP
8016-X (low hydrogen)	all	Nickel Alloy	DCRP or AC
8018-X (iron powder, low hydrogen)	all		DCRP or AC
E 9010-X	all	Chrome-Moly Steel	DCRP
9011-X	all	Chrome-Moly Steel	DCRP or AC
9013-X	all		DCSP, DCRP or AC
9015-X (low hydrogen)	all		DCRP
9016-X (low hydrogen)	all		DCRP or AC
9018-X (iron powder, low hydrogen)	all		DCRP or AC
E 10010-X	all	Chrome-Moly Steel	DCRP
10011-X	all		DCRP or AC
10013-X	all		DCSP, DCRP or AC
10015-X (low hydrogen)	all		DCRP
10016-X (low hydrogen)	all	Nickel Alloy	DCRP or AC
10018-X (iron powder, low hydrogen)	all		DCRP or AC
E 11015-X (low hydrogen)	all		DCRP
11016-X (low hydrogen)	all		DCRP or AC
11018-X (iron powder, low hydrogen)	all		DCRP or AC
E 12015-X (low hydrogen)	all		DCRP
12016-X (low hydrogen)	all	Nickel Alloy	DCRP or AC
12018-X (iron powder, low hydrogen)	all		DCRP or AC

NOTE: The suffix X stands for the weld metal chemical composition. See Fig. 6-31.

Fig. 6-30. Various AWS electrode classifications and recommended positions, applications, and polarity to use for each.

means 60,000 psi and 100 means 100,000 psi. The tensile strength may be given in the "as welded" or "stress relieved" condition. See the electrode manufacturer's specification to determine under what condition the indicated

—A1	1/2% Mo
—B1	1/2% Cr, 1/2% Mo
—B2	1 1/4% Cr, 1/2% Mo
—B3	2 1/4% Cr, 1% Mo
—C1	2 1/2% Ni
—C2	3 1/4% Ni
—C3	1% Ni, .35% Mo, .15%Cr
—D1 and D2	.25—.45% Mo, 1.25—2.00% Mn
—G	.50 min Ni, .30 min Cr, .20 min Mo, .10 min V (Only one of the listed elements is required)

Fig. 6-31. Approximate chemical composition of suffix numbers of the AWS Electrode Numbering System.

tensile strength occurs. "As welded" means without post-heating. "Stress relieved" means the welding is given a heat treatment after welding to relieve stress caused while welding. See Para. 28-7 for an explanation of stress caused by welding. The second digit from the right indicates the recommended position of the joint that the electrode is designed to weld. For example, EXX1X: This electrode will weld in all positions; EXX2X electrodes are used for welds in the flat or horizontal position; the EXX3X is recommended for flat position welds only.

The right-hand digit indicates the power supply (AC, DCSP, or DCRP), type of covering, and presence of iron powder or low hydrogen characteristics, or both.

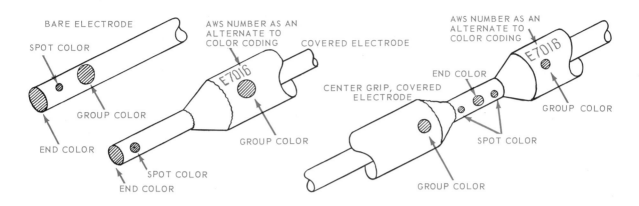

Fig. 6-32. The American Welding Society has standardized a numbering system for identifying welding electrodes. This electrode number is placed on the covering near the end of the electrode. Some companies still use the color code for identifying electrodes.

The last two digits should be looked at together to determine the proper application and the covering composition for an electrode. For example:

ELECTRODE NUMBER	COVERING COMPOSITION
EXX10	High cellulose, sodium
EXX11	High cellulose, potassium
EXX12	High titania or rutile, sodium
EXX13	High titania or rutile, posassium
EXX14	Iron powder, titania
EXX15	Low hydrogen, sodium
EXX16	Low hydrogen, potassium
EXX18	Iron powder, low hydrogen
EXX20	High iron oxide
EXX24	Iron powder, titania
EXX27	Iron powder, iron oxide
EXX28	Iron powder, low hydrogen

See Fig. 6-30 for a table which shows the recommended position, application, and polarity for various electrodes.

For a more complete description of the electrode covering chemical composition and electrode uses, see Para. 6-19.

Occasionally, an electrode number may have a letter and number after the normal four numbers, such as E7010-A1 or E8016-B2. This letter and number combination or suffix is used with low alloy steel electrodes. The suffix indicates the chemical composition of the deposited weld metal. See Fig. 6-31. The letter "A" indicates a carbon molybdenum steel electrode. The letter "B" stands for a chromium-molybdenum steel electrode. The letter "C" is a nickel steel electrode and the letter "D" a manganese molybdenum steel electrode. The final digit in the suffix indicates the chemical composition under one of these broad chemical classifications. The exact chemical composition may be obtained from the electrode manufacturer.

The letter "G" is used for all other low alloy electrodes with minimum values of : molybdenum (0.20 percent minimum); chromium (0.30 percent minimum); manganese (1.00 percent minimum); silicon (0.80 percent minimum); nickel (0.50 percent minimum); and vanadium (0.10 percent minimum) specified.

An example of a complete electrode classification is the E8016-B2:

1. E indicates electrode.

2. 80 indicates tensile strength, 80,000 psi.

3. 16 indicates a low hydrogen, potassium covering.

ELECTRODE CLASSIFICATION	END COLOR	SPOT COLOR	GROUP COLOR
E 6010			
6011		Blue	
6012		White	
6013		Brown	
6020		Green	
7010-A1	Blue	White	
7016	Blue	Orange	Green
7018	Black	Orange	Green
8016-B2	White	Black	Green
9016-B3	Brown	Blue	Green
10016	Green	Orange	Green

Fig. 6-33. Table showing the color coding used on grip-end of metallic electrodes.

4. The 1 indicates it is an all-position electrode used with AC or DCRP.

5. The suffix, B2, indicates that the deposited metal chemical composition is a low alloy chromium-molybdenum steel with 1-1/4 percent chromium and 1/2 percent molybdenum.

Most electrode manufacturers now use AWS numbers imprinted on the covering material near the grip end for identification. A few still use the color code. Fig. 6-32 indicates how metal-lic arc welding electrodes are marked with AWS numbers or are color marked. When identification is by color coding, the electrode may be marked in one, two, or three places. These marks are in the form of painted spots, or bands, at the end of the electrode which is clamped in the electrode holder (grip end).

When all three color identifying marks are used, the electrode will have a color painted on the end, on the grip area, and on the flux covering near the grip end of the electrode.

Fig. 6-33 is a table showing the color markings used on various types of electrodes. For information concerning stainless steel electrodes, see Para. 18-18.

6-19. ELECTRODE COVERING ANALYSIS

The AWS classification indicates the main types of metals and coverings. The metals are similar in composition for each classification number, but each manufacturer has its own compounds

RIGHT HAND DIGIT	COVERING COMPOSITIONS	APPLICATION (USE)
0 E—6010	High cellulose with some titanium dioxide ($Ti\ O_2$) magnesium or aluminum silicate, Ferromanganese deoxidizer.	Forms a light slag, and forms a carbon monoxide gas which produces a reducing atmosphere around the arc. It eliminates oxidation. It is used on all mild steel applications (General Welding). Use DCRP.
1 E—6011	Same as "0" with arc stabilizer included (potassium salt).	Same as (0) above except it can be used on DCRP or AC.
2 E—6012	Covering of titania or rutile (sodium) plus ferromanganese and other materials.	It has very little splatter; it is easy to handle the puddle and to fill gaps. Use DCSP or AC.
3 E—6013	Same as "2", but covering contains an ionizing agent (potassium salt).	It is good for low voltage, low current applications. It allows a varying arc length, and makes it easy to maintain an arc. It is particularly useful with small electrodes used in connection with small capacity AC transformer type welders. It is particularly suitable for many farm applications and also for sheet metal work. Use DCSP or AC.
4 E—7014	Same as E—6012 plus iron powder.	Has all the desirable characteristics of the E—6013. May be used where poor fit-up is present. Good wetting and restrike capability. Use with AC or DC either polarity.

Fig. 6-34. Electrode covering compositions and applications.

RIGHT HAND DIGIT	COVERING COMPOSITIONS	APPLICATION (USE)
5 E–7015	Low hydrogen sodium type.	This is a low hydrogen electrode for welding low carbon, alloy steels. Power shovels and other earth moving machinery require this rod. The weld machines or files easily. Use DC, RP only.
6 E–7016	Same as "5" but with potassium salts used for arc stabilizing.	It has the same general application as (5) above except it can be used on either DC, RP or AC.
E–7028	Iron powder (Low Hydrogen) Flat position only.	For low carbon alloy steels, use DC or AC.
8 E–8018	Iron powder plus low hydrogen sodium covering.	Similar to (5) and (6) DC, RP or AC. Heavy covering allows the use of high speed drag welding. AC or DC RP may be used.

Fig. 6-35. Low hydrogen electrode covering Compositions and Applications. These coverings will withstand a high temperature and therefore high currents (amperages) may be used.

for the coverings. Therefore, very few of the electrodes behave exactly the same even though the classification number may be identical.

Some of the more common covering materials are: cellulose, potassium salts, magnesium and aluminum silicates, ferromanganese, sodium silicates, titania, rutile, iron powder, alumina, magnesia, and sodium oxide. See Fig. 6-34 and the electrode manufacturer's specifications for a more complete analysis of coverings and covering properties.

6-20. LOW HYDROGEN ELECTRODES

Hydrogen has harmful effects on alloy steels, causing intergranular cracks called hydrogen embrittlement thus lowering fatigue resistance and strength.

Low hydrogen electrodes deposit a minimum of hydrogen in the weldment. The low hydrogen condition is obtained by using a special covering or (E- - -8). Lime coverings and titania and iron powder coverings are used. The electrode conforms to AWS E-6016, E-7016, and E-7018 specifications and is used on hard-to-weld steels (free machining), high carbon, low alloy, and hardenable steels. The slag is very fluid, but good flat or convex beads are easily obtained. Fig. 6-35 shows the composition of some low hydrogen electrode coverings.

These special coverings contain practically no organic material, and the baking cycle near 600 deg. F. during welding eliminates free moisture. These electrodes may best be used as shown in Fig. 6-36.

The deposited metal has excellent

CURRENT SETTINGS FOR LOW HYDROGEN ELECTRODES

ELECTRODE DIAMETER	AMPS. FLAT	AMPS. VOLT. AND OVERHEAD	VOLTS
1/8	140-150	120-140	22-26
5/32	170-190	160-180	22-26
3/16	190-250	200-220	22-26
7/32	260-320		24-27
1/4	280-350		24-27
5/16	360-450		26-29

Fig. 6-36. Recommended arc machine settings when using ferritic (low hydrogen) electrodes in flat, vertical and overhead positions.

	AS WELDED
Yield Point	60,000 psi
Tensile	72,000 psi
Elongation, % in 2 inches	22%
Charpy V Notch ft. lbs. at -20°F	20

Fig. 6-37. AWS specifications for welds made with E7016, low hydrogen electrodes.

tensile and ductile qualities and is exceptionally clean as may be seen by X-ray inspection. Fig. 6-37 illustrates the physical properties of ferritic electrode welds.

The electrode may be used with either AC or DC RP. The electrodes should be baked at 250 deg. F. before using. If they have been exposed to the atmosphere for an appreciable period, this baking will remove any moisture which may be in the coating. Never exceed a 1/2 in. motion, and practice considerable care during vertical and overhead passes to avoid molten metal flow.

6-21. ELECTRODES (IRON POWDER)

The addition of iron powder to the covering of shielded arc electrodes changes the arc behavior. It greatly increases the amount of metal deposited. Much higher currents can be used to produce faster welds, easier cleaned welds, less spatter, and better bead shapes. The weld results are very noticeable. See Fig. 6-38.

	E7024	E7024	E7018	E7016
Iron Powder Content %	65	50	33	0
Amount Deposited lb./hr.	14	6	3	2
Deposited Efficiency %	190	160	130	80

Fig. 6-38. Effect of using iron powder in electrode coating. Note that E7024 may have iron powder content of either 50 or 65 percent.

The arc obtained is smooth and steady. One may use about 25 percent more current with the EXX18 iron powder electrode, and as much as 50 percent more current with the EXX24 and EXX27 iron powder electrodes, because of their heavy coatings.

The weld puddle is so fluid when the 50 percent iron powder covering is used that this electrode is recommended for downhand (or flat position) welding only. Follow all safety precautions as outlined in this text.

6-22. CARBON ELECTRODES

Carbon electrodes are used for carbon arc welding and carbon arc cutting. These electrodes come in sizes ranging from 1/16 up to 1-in. diameter. Rods may be obtained in 12, 18 and 24-in. lengths. The quality of the rod must be extremely high, as the structure of the carbon must be uniform. The two types of electrodes obtainable are the carbon electrode and the graphite electrode. The graphite has better conductivity and is usually of more uniform quality.

The rod should be inserted in the holder with the end, or the point, of the carbon approximately 10 times the diameter of the rod away from the electrode holder. For example, a 1/2-in. rod will have its end not over 5 in. away from the holder. As the rod is being used, it tends to slowly burn back toward the holder. The rod should be pointed with the taper of the point approximately 6 to 8 times the diameter of the electrode, as shown in Fig. 5-35.

Currents required for different sizes of typical welding electrodes are shown in Fig. 6-39.

More information on carbon electrodes will be found in PAR. 9-3.

ELECTRODE DIAMETER INCHES	WELDING CURRENT		MAXIMUM CURRENT DENSITY AMPS. PER SQ. IN.	POUNDS PER HOUR DEPOSITED
	MIN.	MAX.		
1/8	0	35	2890	. . .
3/16	25	60	2200	. . .
1/4	50	90	1855	. . .
5/16	80	125	1650	. . .
3/8	110	165	1510	. . .
7/16	140	210	1420	1.5
1/2	170	260	1340	2.5
5/8	230	370	1220	4.5
3/4	290	490	1125	6.0
7/8	350	615	1035	. . .
1	400	750	965	. . .

Fig. 6-39. Carbon electrode current requirements. (National Carbon Co.)

6-23. ELECTRODE CARE

It is important that the user follow the electrode manufacturer's recommendations as to ampere settings, base metal preparation, welding technique, welding position, and the like. Electrodes must not be used after being exposed to dampness because the steam generated by the heat of the arc will cause the covering to be "blown" away, and also cause hydrogen inclusions. Questionable electrodes should be "baked" at 250 deg. F. for several hours. Because of the similarity of appearance of electrodes that are much different in welding properties, it is important to label and store them in carefully marked bins. Use masking tape to bind electrodes and label carefully when putting them in storage. Electrodes are costly, and loss of identification may mean loss in time and money.

If an electrode is used beyond its ampere rating, the electrode will overheat; and the covering will crack, thus spoiling the rod. The excess current will also cause considerable splattering of the molten metal.

Electrode coverings vary considerably but their main function is to prevent the atmosphere from reaching the hot liquid metal. The coverings may also contribute certain elements (metallic) to the alloy metal deposits.

6-24. ARC WELDER REMOTE CONTROLS

When an operator works on a multitude of joints or a variety of joints and metals, necessitating changes in current, polarity and electrodes, welding machine adjustments are usually required.

To eliminate the time required to travel back and forth to the welding machine, several manufacturers provide remote control devices which may be kept near the operator for convenient control of the machine.

The small portable remote control panel, shown in Fig. 6-40, provides for voltage and current adjustment, plus a switch to allow remote hand or foot switch current control. Using a panel of this type, the operator may climb into a restricted place, turn on the machine, and adjust it to any polarity and practically any current setting without returning to the machine. The saving in time is important, and better quality welds are produced because

Fig. 6-40. Remote control panel. This type of control panel provides adjustment of arc voltage and current, and the machine may be turned on and off. In addition, a foot control may be used to control the current (amperage from the machine. This control may be mounted on the machine or near the operator.
(Vickers, Inc.)

the machine is more accurately adjusted to the job requirements at all times.

A foot switch, Fig. 6-41, is used for fine adjustment. The machine is set for a certain range of power, then the foot switch provides variations within that range.

6-25. CLEANING EQUIPMENT

It is very important that metals to be welded are clean. It is difficult to weld dirty or corroded surfaces, and, if attempted, the resulting welds will normally be of poor quality. Many types of equipment and tools have been developed for the purpose of cleaning joints and welds. Cleaning may be done by using sand-blasting machinery, rotary wire wheels, and tools such as chipping chisels, hammers, and wire brushes. Nonferrous metals may be chemically cleaned, especially in production welding situations. The amount and size of the welding done usually determines the kind of cleaning apparatus needed, Fig. 6-42.

Slag which covers the finished weld must be removed before the next weld bead is laid, to prevent inclusions in the finished weld. The slag on the final bead must also be removed before the weld can be inspected or painted. This coating may be removed by a rotary wire wheel, or by tapping the scale with a chipping hammer. In either case

Fig. 6-41. Foot operated remote control for changing machine settings.

Fig. 6-42. Combination wire brush and chipping hammer. (Atlas)

suitable eye protection must be provided.

The chipping hammer usually is double ended. One end is shaped like a chisel for general chipping and the other end is shaped like a pick, for reaching into corners, and the like, as shown in Fig. 6-43.

6-26. SHIELDS AND HELMETS

Electric arc welding necessitates the use of special protective devices for skin surface, such as the hands, face, and eyes. The device used to protect the face and eyes may either be mounted and supported on the head (head shield or helmet), Fig. 6-44, or may be held in the operator's hand (hand shield).

The Occupational Safety and Health Act (OSHA) requires the use of a hard hat with the arc welding helmet on construction work. A combination helmet and hard hat is shown in Fig. 6-45. The face shield is usually made of fiber and formed in a shape which covers the front half of the head.

Fig. 6-44. Molded helmet with extra large lens for maximum visibility. (Fibre Metal Products Co.)

Fig. 6-46 illustrates a typical hand shield. An aperture or opening at the level of the eyes provides visibility.

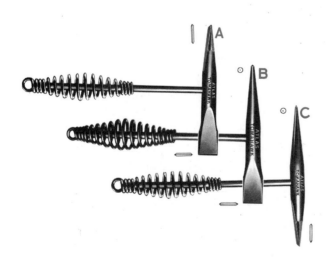

Fig. 6-43. Variety of chipping hammers. Note that in (A) the blades are turned 90 deg. to each other. In (B) and (C) the chipping hammers have a blade at one end, and a pick at the other end.

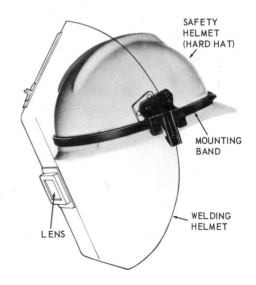

Fig. 6-45. An arc welder's helmet used in connection with a hard hat. (Kedman Co.)

Fig. 6-46. Arc welding hand shield used for welding inspection work, and instructional purposes.

	SHADE NO.
Shielded Metal Arc Welding (ferrous) (covered eledtrodes)	
1/16 – 5/32 in. electrodes	10
3/16 – 1/4 in. electrodes	12
5/16 – 3/8 in. electrodes	14
Gas–Shielded Arc Welding (nonferrous)	
1/16 – 5/32 in. electrodes	11
Gas–Shielded Arc Welding (ferrous)	
1/16 – 5/32 in. electrodes	12
Atomic Hydrogen Welding	10 to 14
Carbon Arc Welding	14

Fig. 6-47. Suggested lens shade numbers for various arc welding applications.

This aperture is approximately 4-1/4 x 2 in., and is provided with at least two glass lenses. The outer lens, which is of double-strength glass, is used to protect the inner and more expensive welding lens from metal spatter and abuse. Many operators also put another clear lens on the face side of the helmet to protect the colored lens from that side. The clear lenses are made of thin plate glass or plastic. Fig. 6-47 shows recommended shade numbers for arc welding.

A good grade of colored arc welding filter lens will remove approximately 99.5 percent of the infrared rays and 99.75 percent of the ultraviolet rays. These figures have been developed by U. S. Bureau of Standards. No. 10 is the common shade used for welding with covered electrodes. Use Shade No. 14 for carbon arc welding and No. 12 for gas shielded arc welding. However, other shade numbers may be obtained and used, following the manufacturer's recommendations. The higher the shade number, the lower the transmission of infrered or ultraviolet rays.

Ultraviolet rays in excess may cause eye pain for 8 to 18 hours after exposure. Infrared light rays tend to injure the sight and every precaution should be taken to shield the eyes from them.

Filter lenses are of such density or shade that the operator cannot see through them until the arc is struck.

Some cover (outer lens) glass is especially treated and resists the adhesion of metallic particles to its surface. Spring clips are generally used to provide a snug fitting of the helmet. Head bands are adjustable.

The helmet, or head shield, has a swing mounting which permits the forward part of the helmet to be lifted above the operator's face, without removing the head band from the head. Some welders who work continuously find that wearing a pair of ordinary welding goggles under the helmet helps to reduce eye strain (goggles eliminate reflected glare around the back of the helmet). Also available are light-weight goggles with a #1 or #2 filter lens. These goggles are called flash goggles and enable the operator to set up work, chip welds (if the lens is tempered) peen, etc., and still have eye protection from flying particles and adjacent arc rays. Some helmets are available into which fresh air is fed by means of a hose to increase the comfort of the operator, see Fig. 6-48.

Special features which may be built into the inside of the helmet include volt-ammeters to help the operator check the machine settings, and varying

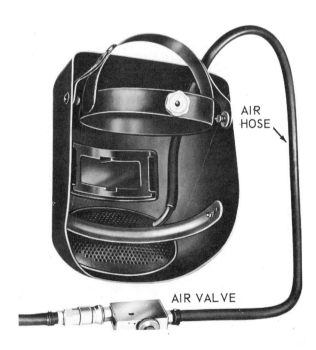

Fig. 6-48. Ventilated arc welding helmet. (Fibre Metal Products Co.)

intensity incandescent bulbs which indicate voltage by the brightness of the light and help the operator check arc length.

6-27. SPECIAL ARC WELDING CLOTHING

While an arc weld is in progress, the molten flux and the metal itself sometimes spatter for a considerable distance around the joint being welded. The operator must, therefore, protect himself carefully from the danger of being burned by these hot particles. Such clothing as gloves, gauntlet sleeves, aprons, and leggings are sometimes necessary, depending upon the type of welding being performed. When performing welds in the overhead position, it is recommended that one wear a jacket or cape to protect the shoulders and arms and a cap or a special hooded arc helmet to protect the head and hair. It is recommended that all of these articles be made of leather (usually chrome leather) with the exception of the leggings and gloves, which are sometimes made of a combination of cloth and asbestos.

It is further recommended that the operator use high-top shoes, meaning shoes that go over the ankle rather than the more common type of oxford. Trousers worn by the welder should not have cuffs, as the cuffs may catch burning particles as they fall. Gloves should be worn to cover the hands and thereby prevent "sunburn."

Another recommendation is that all the clothing worn be carefully inspected to eliminate any place where the metal may catch and burn, such as the open pockets and cuffs in the trousers, etc.

Clothing worn, other than the garments mentioned, should be of heavy material because thin clothing will permit infrared and ultraviolet rays to penetrate to the skin. If the skin is not properly protected, the operator will become "sunburned." Such burns, if they do occur, should be treated as a severe sunburn. If the burns are severe, a physician should be consulted. Easily ignited material such as flammable combs, pens, and the like should not be on one's person while welding.

6-28. TEST YOUR KNOWLEDGE

1. What is a reactance coil?

2. What is the purpose of an exciter?

3. What is an average size motor for driving a 200 amp. generator?

4. Why must the cable be super-flexible?

5. What precautions must be taken when attaching the welding cables?

6. How may a work lead be connected to a welded structure without damaging the surface?

7. Of what material are most electrode-holder handles made?

8. Why are some electrode handles water-cooled?

9. Name two types of arc-welding face protectors.

10. Describe the results if the lens did not fit tightly into the welding helmet frame.

11. What number safety lens is most frequently used in the arc welding shield?

12. When overhead arc welding what special clothing should be worn?

13. List the specifications for some metallic electrodes.

14. Name the minimum and maximum diameters of carbon electrodes for average usage.

15. Carbon electrodes are used for what two purposes?

16. What is the oldest method of generating electricity for arc welding?

17. Why must the generator be kept clean?

18. What special devices may be installed in an arc welding shield to help ventilate the arc shield?

19. When it is desired to carry 150 amperes a round trip distance of 150 feet, what size electrode and ground lead should be used?

20. What is the purpose of a remote control foot switch?

Chapter 7

ALTERNATING CURRENT ARC WELDING

The use of alternating current (AC) arc welding is increasing rapidly. The AC machines now produced are very efficient and easy to handle. Also, electrodes have been developed for AC use which have ionizing compounds in

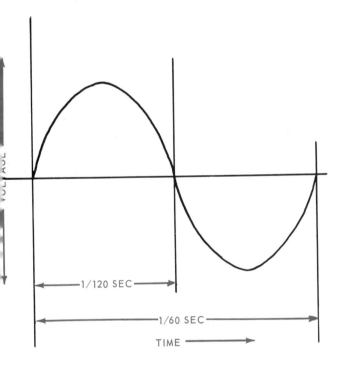

Fig. 7-1. Sine wave form of single-phase alternating current.

the coatings. The ionized arc stream makes it easy to strike and maintain a stable arc.

Another factor which has increased the use of AC arc welding machines is the initial cost. The cost of producing a transformer type AC arc welder is generally less than the cost of either

a motor generator machine or a rectifier type DC machine of equal welding capacity.

Equipment and supplies for AC arc welding are described in CHAPTER 8.

7-1. CHARACTERISTICS OF ALTERNATING CURRENT (AC) ARC WELDING

Most electric current as generated by the electric utility companies is 60 cycle alternating current. With alternating current it is easy to step the voltage up or down (increase or decrease the voltage) by the use of transformers. This alternating form of electricity means that the current reverses its direction of flow 120 times per second. As shown in Fig. 7-1, it requires 1/60 of a second to complete a cycle, or the current flow completes 60 cycles per second hence it is called 60 cycle current.

Most AC arc welders have transformers which step down the voltage, and increase the current for welding purposes. Fig. 7-2 shows what happens at the arc in one cycle of a typical AC transformer type arc welder: The voltage at A and B is zero, then the voltage builds up to a maximum in one direction to point C and then back to zero at point A. The voltage then builds up to maximum in the other direction to point D then back to zero again at point B. This action is repeated at the rate of 60 cycles per second.

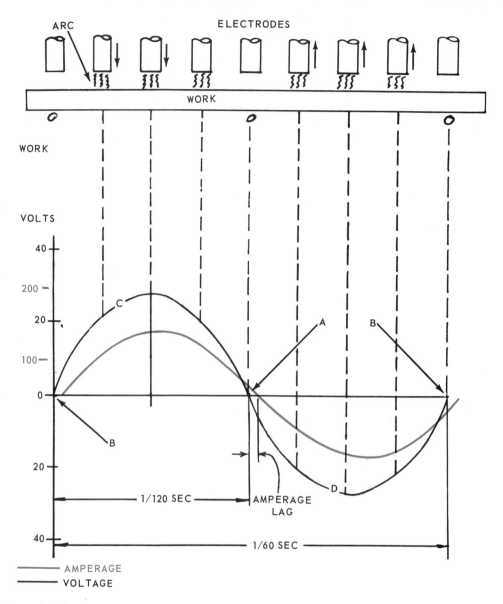

Fig. 7-2. The sine wave curve of alternating current at 60 cycles. At points A and B the voltage value is zero. These two zero values which occur in each cycle (every 1/60 of a second) make it difficult to strike and maintain an AC arc at small current values. Because a certain voltage is required to overcome electron inertia in a circuit there is usually a small lag or lead between the EMF (voltage), and current (amperage). The EMF usually leads the amperage.

Since it is the electrical voltage which enables the current to jump a gap and maintain an arc, it is necessary to reduce the time that the voltage is zero, or to increase the ionization of material in the gap, as the electricity travels across the arc, otherwise the arc would be difficult to maintain. Modern covered AC welding electrodes have ionizing agents in the coverings. This factor is perhaps the greatest reason for the rapid growth of AC welding.

With the current reversing itself, theoretically the same number of electrons would travel in one direction in the arc and then in the opposite direction as the current flow reverses. Therefore 50 percent of the heat is released at the electrode end of the arc and the same amount at the work. In practice, more current travels from the electrode to the work than flows from the work to the electrode. Since the electrical contact area is greater on the work than on the electrode, current will flow easier from the electrode to the work than it will from the work to the electrode (rectification principle). However the difference is small.

Fig. 7-2 also shows the current flow as related to the potential (voltage). The current can be any amount depending on the current used for the size of electrode selected and the welding process. It should be noted that the current lags the EMF (electromotive force) slightly in time because a small amount of time is required for the EMF to overcome electron inertia. Then current begins to flow in the circuit.

7-2. ADVANTAGES OF AC ARC WELDING

The distinct advantage that AC arc welding has is that there is virtually no magnetic blow. The reversal of current flow each 1/120th of a second keeps the effect of the magnetic field to a minimum. The reversal of the field at such a frequency results in a much more stable arc stream. Because arc blow is minimized, quality welds are always possible with AC.

The arc produces good penetration. It is an easy arc to control and to maintain once the arc is started. However, because of the alternating current flow, starting the arc is more difficult than with DC. To overcome this difficulty some machines have a "hot start" circuit built in to provide an extra flow of very high frequency current at the

time the arc is struck. Other machines use capacitors in the secondary (arc) circuit to give high-current surges for the arc-striking or starting periods only.

AC arc welding is usually faster, because larger electrodes and therefore more current can be used due to minimum magnetic blow conditions.

Some of the basic features of AC arc welding are:

1. Good forceful arc.
2. There is an absence of arc blow.
3. Weld arc is easy to hold once it is obtained.
4. A good way to weld aluminum.
5. The most popular application is production welding on heavy gauge steel.

To cause the arc to deposit weld metal at the weldment there are various forces at work.

AC electrode shielding increases ionization and the gases generated tend to direct the globules of metal into the weld puddle.

Surface tension at the surface of the weld tends to hold the metal in place in the weld puddle.

7-3. TYPES OF AC ARC WELDING MACHINES

Several types of AC arc welding machines have been produced and used. The two basic types are:

1. Motor generator type.
2. Transformer type.

A schematic view of these two types is shown in Fig. 7-3.

7-4. GENERATOR TYPE AC ARC WELDING MACHINE

High frequency motor generator AC arc welding machines have been developed and used. Due to the high frequency

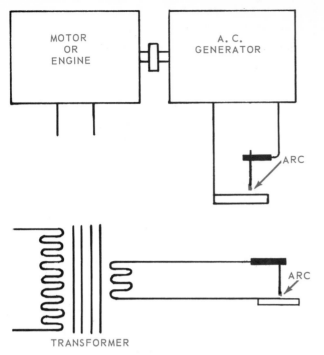

alternating current generated, it is quite easy to strike and maintain an arc with such machines.

Two types of AC arc welding generator units are:

1. Electric motor driven AC generator.

2. Engine driven AC generator.

The electric motor generator is usually driven by a 60-cycle single-phase or three-phase motor at approximately 1750 RPM. The generator may be designed to produce up to 500 cycles per second (CPS). Fig. 7-4 shows a schematic wiring diagram of a high-frequence generator using a three-phase motor drive.

Engine driven AC arc welders are quite commonly used. They have the advantage of being usable beyond the reach of electrical power supply sources.

Fig. 7-3. The two basic types of AC arc welding machines are the AC generator and the transformer.

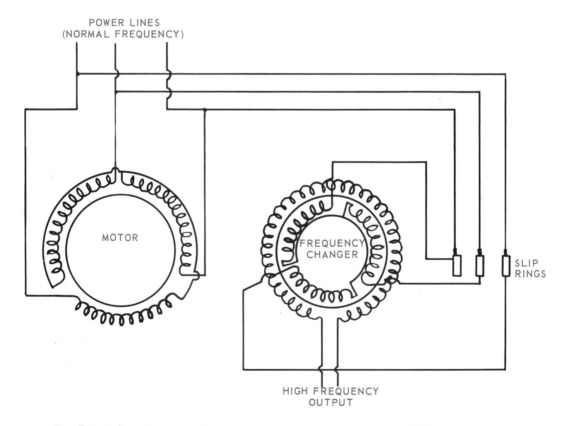

Fig. 7-4. Schematic wiring diagram of a three-phase motor drive, high frequency welding generator.

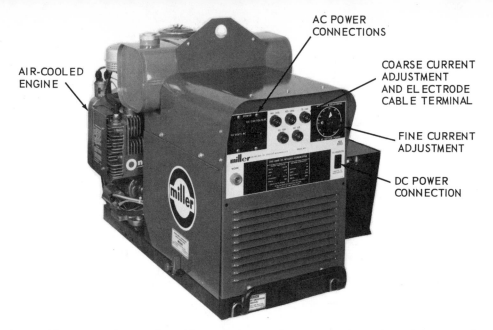

Fig. 7-5. A portable, engine-driven arc welder and electrical power supply machine. It can supply either AC or DC arc welding current and AC or DC power. (Miller Electric Mfg. Co.)

These machines provide the advantage of the AC arc when welding in places where arc blow might make DC arc welding difficult.

Some engine-driven arc welding machines are available as a combination AC and DC unit, Fig. 7-5. This machine can provide an AC arc and a DC straight polarity arc or a DC reverse polarity arc. It also can provide 120-volt AC to operate motors, lights, etc., at 60 cycles Hertz (Hz.) Or, the machine will provide 120-volt DC.

7-5. TRANSFORMER TYPE AC ARC WELDING MACHINE

The transformer type AC welder is most popular of the AC machines. The machine usually uses single-phase, 220 or 440 supply voltage. The welding leads and the arc circuits are similar to those of the DC machines.

There are several types of AC transformer type arc welding machines:

1. Adjustable reactor type.
2. Movable coil type.
3. Movable core type.

A popular AC arc welding machine of the tapped reactor type is shown in Fig. 7-6.

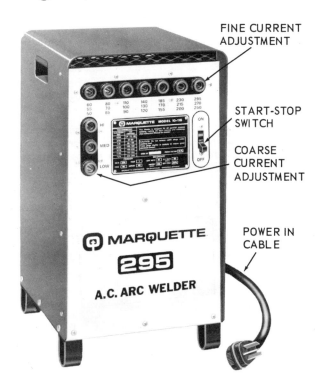

Fig. 7-6. External appearance of an AC arc welder. Electrode lead may be plugged into one of the fine current tap sockets to obtain desired amperage. Ground lead may be connected to one of the three coarse adjustments.

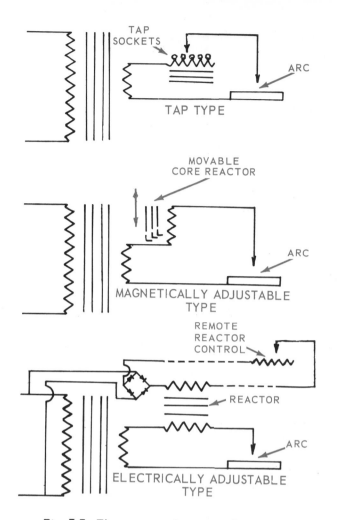

Fig. 7-7. *Three types of reactor adjustments for reactor type AC arc welding machines. The remote reactor control may be located near the operator for convenience.*

reduce the welding current, while a decrease of reactance will permit more welding current to flow.

The electrically adjustable reactor, as shown in Fig. 7-7, is becoming popular because the reactor current adjustment can be located near the operator by using an extension cord. The control may be either adjusted by the operator's hand or foot. In constant voltage machines the windings and the core of the transformer are held constant and the adjustable reactor is used.

If any part of the basic transformer is changed — that is, if the coil or core is movable — the unit is called a constant current type machine as shown in Fig. 7-8.

A stop-start switch is located either on the top, side, or front of the machine. A small electric fan is used to force air through the machine for cooling

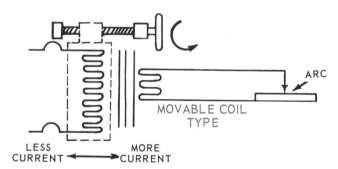

Fig. 7-8. *Two types of AC arc welding machines. One is the movable core type, and one is the movable coil type.*

There are three kinds of reactor adjustments, as shown in Fig. 7-7:

1. Tapped reactor (limited settings).
2. Magnetically adjustable reactor (infinite settings).
3. Electrically adjustable reactor (infinite settings).

In all AC circuits, as the electricity reverses its direction of flow, the changing magnetic field around each wire creates an electrical flow in the opposite direction to the desired flow. This action is called reactance (sometimes called counter electromotive force). An increase of reactance will

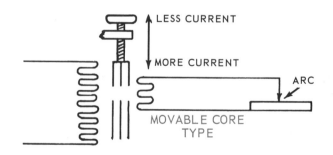

purposes. Fig. 7-9 shows the internal appearance of an AC transformer type welder.

The AC welding transformer consists

arc current flows. The welding transformer is called a step-down transformer. The welding arc current has a lower voltage and a higher amperage

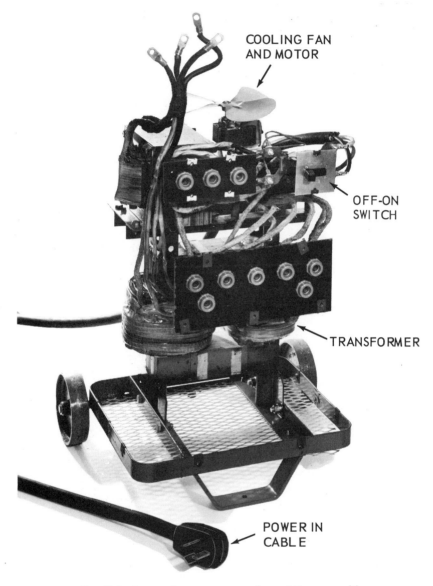

COOLING FAN AND MOTOR

OFF-ON SWITCH

TRANSFORMER

POWER IN CABLE

Fig. 7-9. Internal appearance of an AC arc welder similar to the machine shown in Fig. 7-6. (Marquette Mfg. Co., Applied Power Industries, Inc.)

of two windings, a primary and a secondary. The primary winding is connected to the power source. The secondary winding is connected to the electrode holder and ground and it is through the secondary winding that the welding

or current value than the primary circuit. Fig. 7-10 is a diagram of the electrical circuits in a typical transformer type arc welder.

Some arc welding machines are built to provide either AC arc welding cur-

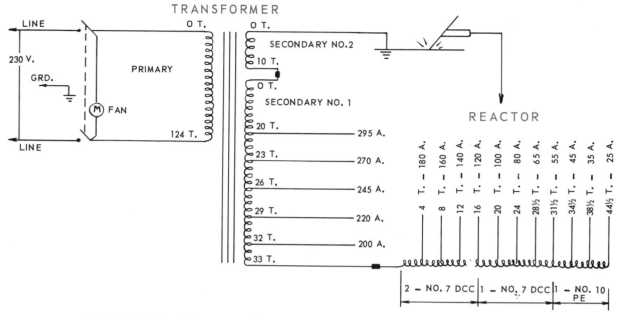

TRANSFORMER

Fig. 7-10. Electrical circuits for a tap type AC transformer. The amperage range of this arc welder is 25 amps. to 295 amps.

rent, or DC arc welding current. These units are described in CHAPTER 6.

7-6. STARTING AC TRANSFORMER ARC WELDER

The AC arc welding machine unit is very easy to start and adjust. A start button is conveniently located near the operator of the machine. Many machines are fitted with a remote switch for convenience. Current settings are easily made with a handwheel using a dial and pointer setting indicator or by using the indicator lead tap adjustment. Fig. 7-11 shows a diagram of a movable primary coil AC arc welding machine. Fig. 7-12 shows the internal construction of the same machine.

It is very important that the complete station be inspected before starting and using the arc welder for:

A. Safety.

B. Efficiency.

Some of the more important inspection items are:

1. Electrode holder, leads, and connections.

A. The electrode holder insulation must be in good condition.

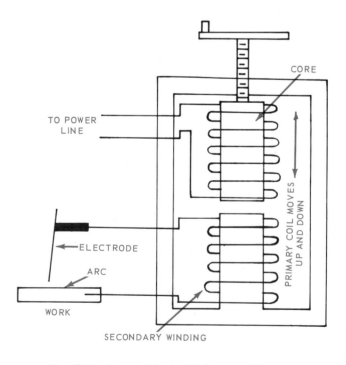

Fig. 7-11. A movable coil AC arc welder.

B. The electrode holder should be hung on an insulated hook.

C. The leads should be well insulated. The cables should be run as close together as possible and should be of minimum length. Separated cables and/or long leads will increase the reactance in the secondary circuit and will reduce the AC machine capacity.

D. The connections should be clean and tight. An ohmmeter can be used to locate excess resistance points in the circuit.

2. The welder should be inspected.

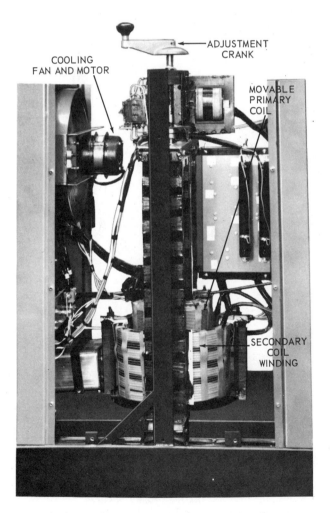

Fig. 7-12. Internal construction of a movable coil AC arc welder.

COOLING FAN AND MOTOR

ADJUSTMENT CRANK

MOVABLE PRIMARY COIL

SECONDARY COIL WINDING

A. The machine ventilating air inlet and outlet must be clean and the openings must be clear of obstructions.

3. The booth walls and curtains must be in good condition.

A. The booth should be clean and dry.

4. The booth ventilation system must be operating and the ventilation openings must be clean and clear of obstruction.

7-7. STRIKING AN ALTERNATING CURRENT ARC

The practice of arc welding with an AC machine is very similar to the DC type. It is a little more difficult to strike the arc with the AC type because of the tendency of the apparatus to break the arc on the change of the cycle (120 times per second in a 60-cycle circuit). As mentioned previously, however, special provisions are made to make up for the no-current interval.

First the machine must be correctly adjusted. It is important that the current setting be matched with the electrode size (diameter) as shown in Fig. 7-13. The current settings vary depending on the thickness of the metal, type of joint, and whether it is a single or multiple pass weld.

The operator will find that the motions recommended in CHAPTER 5, should also be used for starting an AC arc. If the arc breaks continually, regardless of how careful the operator may be, it is probably due to too-low current adjustment of the machine. If the electrode spatters excessively, and if it becomes overheated while welding, the current is excessive. The operator will find that once an arc is maintained for a time, the arc can be maintained quite easily.

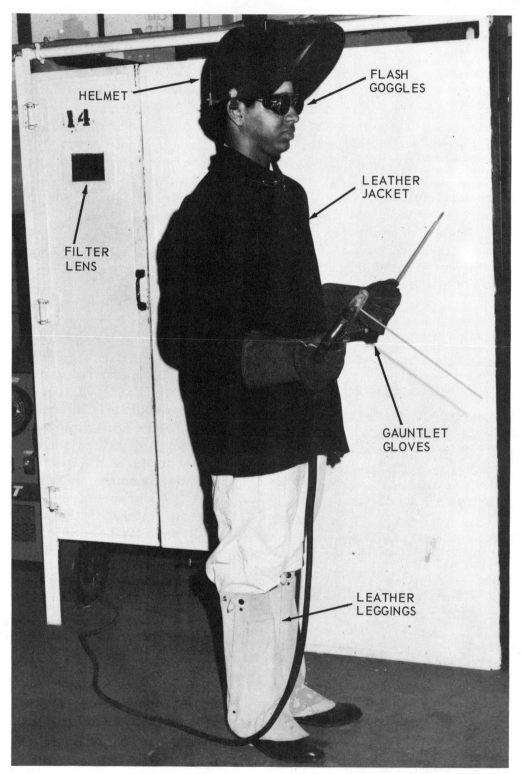

HELMET

14

FILTER
LENS

FLASH
GOGGLES

LEATHER
JACKET

GAUNTLET
GLOVES

LEATHER
LEGGINGS

A welder equipped with flash goggles, leather jacket, gloves, and leather leggings. Note:
The arc welding booth is fitted with a filter lens in order that the instructor may observe
the arc from outside the booth.   (Detroit Board of Education, Detroit Public Schools).

7-8. ALTERNATING CURRENT WELDING PRACTICE

There has been a steady increase in the use of AC arc welders by both manufacturers and jobbing shops. The initial cost of the equipment, the high explained in CHAPTER 5, and the appearance of the weld should be similar to a DC machine weld. The same motions, the same arc length, the same current capacities, and electrode sizes can be used as in DC arc welding. However, AC arc welding can use larger

ELECTRODE DIAMETER	APPROX. METAL THICKNESS SINGLE PASS	CURRENT SETTING MIN.	MAX.
1/16	1/16	20	40
3/32	3/32	30	80
1/8	1/8 - 1/4	50	120
5/32	1/8 - 1/4	75	170
3/16	1/4 - 3/16	100	210
7/32	1/4 - 3/16	120	250
1/4	1/4 - 5/16	160	330
5/16	5/16 - 3/8	200	420

Fig. 7-13. Table of electrode diameters and current settings for AC arc welding. The values are approximate.

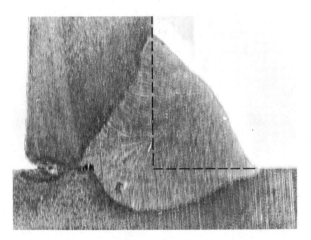

Fig. 7-14. An AC deep fillet weld on a T-joint. (Hobart Brothers Co.)

production rates, and the elimination of magnetic blow are probably the most important reasons. The lack of magnetic arc blow enables a welder to produce better welds in corners, at edges, and in recesses.

The weld should be performed in a flat position, if possible. However, the weld can also be done in vertical and overhead positions. Metal for AC welding should be prepared for welding as electrodes and therefore more current than when DC welding. A longer arc can be drawn when using AC equipment than with DC. A welder operator must be careful to maintain the proper arc length, as it is easy to weld with too long an arc, and thereby increase the chance of a poor quality weld.

One particular advantage of AC welding is its ability to produce excellent deep penetration fillet welds, as shown in Fig. 7-14.

7-9. AC LIGHTLY COATED METALLIC ELECTRODE ARC WELDING

Lightly coated metallic electrodes are economical to use; however, it is more difficult to weld with them and the weld produced is not ordinarily as good as with covered electrodes. Since there is no gaseous shield to keep out the atmosphere, there will be considerable oxidation, and the weld will not be as sound nor as ductile as when shielded electrodes are used. These electrodes are sometimes used where lower strength welds meet the design requirements.

To do an arc welding job using lightly coated electrodes, the following procedures should be followed: The arc should be struck and the electrode held at an angle of about 20 deg. from the vertical in the direction of travel of the weld. It is sometimes necessary to weave a pattern with the arc, in order to obtain width and penetration of the weld. As with all arc welding, the welding arc will be influenced by:

A. Electrode diameter.
B. Current flow (amperes).
C. Arc length.
D. Speed of travel.

In general, it is advisable to use as large an electrode as possible. This increases the welding speed and costs less per pound. Fig. 7-15 is a table of

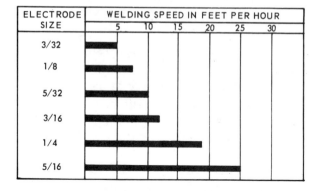

ELECTRODE SIZE	WELDING SPEED IN FEET PER HOUR
	5 10 15 20 25 30
3/32	
1/8	
5/32	
3/16	
1/4	
5/16	

Fig. 7-15. Table which compares the welding speeds using various electrode diameters while making a 3/8-in. mild steel butt weld.

welding speeds for various electrode sizes, if used to make a 3/8-in. butt weld on mild steel.

The amount of current will depend mostly on the size of the electrode. The larger the electrode, the higher the permissible current setting.

The arc length together with the amount of current, governs the amount of heat generated at the arc.

As with DC welding, the sound of the arc (a steady hissing or frying sound) and the appearance of the molten puddle are indications of the quality of the weld being made. A current setting which is too high will overheat the electrode, cause excessive splatter, and the weld will have a poor appearance. A current setting which is too low will give poor penetration (fusion), and greatly reduce the welding speed.

An arc length which is too long will cause excessive splatter, a low, wide, flat bead and it may cause the arc to break frequently or "go out."

Too short an arc will cause a high and narrow bead, a porous weld, and the electrode may stick or freeze to the work.

7-10. AC COVERED ELECTRODE (SHIELDED) METALLIC ARC WELDING

Shielded metallic electrodes have been developed for AC welding. The use of these electrodes helps to make AC welding both easy and reliable. To use these electrodes, strike the arc with the electrode held vertically to the metal and after obtaining the arc tilt the electrode to about an angle of about 20 deg. from the vertical, in the direction of travel of the arc.

The AC shielded electrode, E6011, has a special coating. When the arc is operating, some of the material in the coating ionizes and these ions enable the alternating flow of electricity to easily cross the gap as the current reverses direction. Fig. 7-16 shows two fillet welds on T-joints. One is a typical weld while the other shows complete penetration. The weld with complete penetration is a deep fillet weld made with AC.

The gaseous shield keeps the air from

Fig. 7-16. Conventional fillet weld (A) compared to a deep penetration fillet weld (B) made with an AC arc.

the weld. A higher temperature and more effective heat is developed at the arc with shielded electrodes. This means more rapid heating. The weld metal will remain in a fluid condition longer resulting in less included gases and therefore a more solid weld. Refer to CHAPTER 5, for a description of the effects of electrode diameter, current, arc length, and speed of travel on arc weld characteristics.

7-11. PREPARING THE METAL

As is basic with all work, the materials must be clean, shaped correctly, and firmly mounted. The metal must be known. Mild steel (SAE 1010 - 1020) is recommended for practice exercises. The spark test may be used to determine the type of metal (CHAPTER 24).

The metal edges must be correctly shaped, the metal must be clean (wire brushed, sandblasted, or pickled).

The metal must be correctly positioned with each piece lined up correctly with the other and with the correct gap in the joint (root face).

The metal pieces must be firmly supported to insure against accidental movement while the weld is being made as shown in Fig. 7-17.

After carefully preparing and mounting the metal pieces, the electrode should be selected.

7-12. SELECTING THE ELECTRODE

The electrode must be in good condition. It must be clean, the coating must not be cracked or broken, and it must be dry.

The correct choice of the electrode is very important if good quality, economical welds are to be obtained. A general rule is to use the maximum diameter electrode and the maximum current for the thickness of the metal being welded and still produce good quality welds.

The following variables affect the choice of the electrode:

1. Kind of electrode used; in this case AC.
2. Kind of metal.
3. Thickness of metal.
4. Joint design.

Fig. 7-17. Special shielded screw C clamps being used to hold a metal assembly as it is being arc welded. The shields are to protect the clamping screws from metal spatter. (Adjustable Clamp Co.)

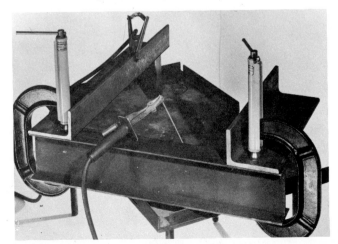

5. Welding procedure; whether one pass or multiple pass.

6. Preheat condition of metal.

Some of the more common alternating current electrodes are identified by AWS standard numbers such as 6011 and 6013 electrodes. See CHAPTERS 5 and 6. These electrodes are able to use a range of current flow. For example, a 1/8-in. diameter electrode can be used for current flows from 80 amperes to 125 amperes. The weld to be made must be made within this current range if this electrode size is to be used.

One must know from past experience or from instructions how much current is to be used for the weld to be made:

1. The thicker the metal, the more current needed.

2. Butt joints and corner joints require less current than T-joints and lap joints.

3. Single pass welds require more current than multiple pass welds.

4. Preheated metal requires less current than ambient temperature metal.

Indications that too much current is being used are excessive splatter of the arc, and poor control of the weld crater. If too much current is being used for the electrode diameter, the electrode coating will not "burn off" evenly. If too little current is used, the arc will be hard to maintain, and the electrode metal will not fuse with the base metal.

The variety of electrode sizes and types is shown by the elaborate storage facilities needed, as shown in Fig. 7-18.

7-13. STARTING THE WELD

The technique described in CHAPTER 5 for starting a weld with DC is also used when welding with AC. Remember that metal is being deposited and parts of the base metal are being melted the instant the arc is created. It is therefore important to strike the arc in exactly the correct spot or the metal may be marred. Fig. 7-19 shows electrode positioning prior to striking the arc. Most welders position the arc end of the electrode just above the exact spot where the weld is to start, then lower the helmet in front of their eyes before actually contacting the metal with the electrode. Start melting and depositing metal on both pieces of

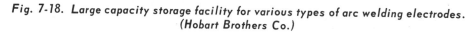

Fig. 7-18. Large capacity storage facility for various types of arc welding electrodes.
(Hobart Brothers Co.)

Fig. 7-19. Chain type clamp being used to securely hold two pieces to be arc welded.

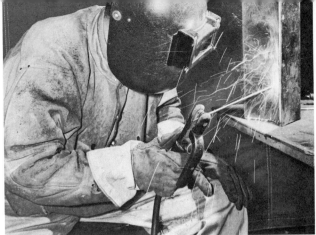

Fig. 7-20. This illustration shows a fillet weld being made in the vertical position. Note the leather cape that the welder is wearing for safety.

metal of the joint as quickly as possible. That is, the arc should be directed until the crater, or puddle, is formed on both sides of the joint at once, before proceeding with the bead.

7-14. RESTARTING THE WELD

Quite often when making a weld the arc is interrupted because the electrode is used up, or for some other reason. When this interruption occurs, care must be taken when restarting the weld that a weak spot is not formed in the weld at this point.

The proper procedure is to first clean the end of the weld by removing the slag and chipping, and/or wire brushing the weld crater. Then, strike the arc at the leading edge of the crater. Move the arc slowly back to the trailing edge of the crater, and finally start welding in the usual fashion and direction.

Fig. 7-20 shows a fillet weld being made on the inside corner joint in a vertical position.

7-15. STOPPING THE WELD

When an arc weld is completed, the weld will tend to have a crater in it, as the electrode is lifted away from the weld. The metal at the end of the weld should be built up to the same level as the rest of the weld, that is, the crater should be filled.

To fill the crater, reverse the electrode movement as the end of the weld is reached. Move the electrode to the trailing edge of the crater, pause until the crater is filled, then lift the electrode slowly until the arc is broken.

7-16. REVIEW OF AC ARC WELDING SAFETY

Always use National Electrical Manufacturers Association (NEMA) approved equipment. Never use homemade or non-approved transformer equipment. It may be dangerous particularly if the primary and secondary winding should become electrically shorted and/or connected.

Alternating current arc welders should be kept dry. Never operate an AC welding machine with the welding cables wrapped around the welder. The magnetic field produced by the welding cables may interfere with the magnetic circuit inside the welder.

The primary coil may be of 120, 200, 220 or 440 volts. The primary circuit should always be installed according to the prevailing electrical code. Follow all precautions described in the text for working near or around electricity.

Burns and flashes are perhaps the greatest hazard to the arc welder. The same gloves, proper clothing, shielding and helmets as described in CHAPTER 6, should be used to provide necessary protection.

7-17. TEST YOUR KNOWLEDGE

1. What has caused the use of AC arc welding to increase in use since arc welding became a common method of welding?

2. What are some of the advantages of using the AC arc?

3. What is meant by the rectifier effect on the AC flow?

4. What are two common types of AC welding machines?

5. How are transformer machines usually cooled?

6. Is the frequency of AC welding generators usually higher or lower than the frequency of transformer type machines?

7. What is the advantage of higher frequency in welding machines?

8. When AC arc welding, what portion of the heat is released at the electrode and what portion at the work?

9. Is it possible for some transformer type AC welding machines to also produce DC? If so, how?

10. What is the purpose of a welding transformer?

11. When arc welding a 3/8-in. thick butt joint on mild steel using a 3/16-in. diameter electrode, what is the approximate welding speed in feet per hour?

12. What is meant by current lag as applied to alternating current?

13. What happens to the welding current as the reactance increases?

14. Why does the welding current change as the tap position is changed in a tap type AC welding machine?

15. Why is more current used as the diameter of the electrode is increased?

Chapter 8

AC ARC WELDING EQUIPMENT AND SUPPLIES

Alternating current for arc welding has certain characteristics which make this type of welding particularly desirable for many welding applications.

8-1. ALTERNATING CURRENT ARC WELDING TRANSFORMERS

The most popular alternating current welding machine uses a transformer to provide the proper electrical characteristics. The welding transformer has an exterior appearance similar to the usual power transformer.

An AC welding machine is shown in Fig. 8-1. Fig. 8-2 shows a complete AC welding outfit.

The machine shown in Fig. 8-1 is of 180 ampere capacity, with a power factor correction to enable it to be used on 3 KVA power line transformers such as used in rural areas.

In general, AC welding transformers are wound to step down the power supply source from 120, 208, 230, 440, or 550 volts, to the required current and potential required for welding. Welding transformers are carefully engineered to produce good arc characteristics with safety. Fig. 8-3 shows the internal construction of the welding machine in Fig. 8-2.

These welding machines are made in accordance with standards established by the National Electrical Manufacturers Association (NEMA). The NEMA standards are concerned with allowable temperature rise in parts of the machine as it is operated at its rated output. The temperature rise maximums are based on the type of insulation material used. Homemade units and welders without NEMA approval should be avoided, as they can be extremely dangerous.

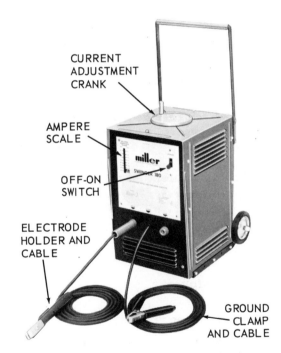

CURRENT ADJUSTMENT CRANK

AMPERE SCALE

OFF-ON SWITCH

ELECTRODE HOLDER AND CABLE

GROUND CLAMP AND CABLE

Fig. 8-1. An AC arc welder with both a tap type and a moving core adjustment. Coarse adjustments are made by moving the electrode plug from the high to the low range tap. Fine adjustments are made by turning the crank to move the core.

AC units come in various capacities such as: 100, 150, 200, 300, 500, 750, and 1000 amperes. These values are output amperes at 40 volts. The units are rated based on a 60 percent duty cycle, that is, if the unit operates 60 percent of the total time (6 minutes out of ten or 60 minutes out of 100, etc.).

winding, which is the arc circuit winding, is the step-down type, which reduces the primary circuit high voltage to an open circuit voltage of 80 volts, and a closed arc circuit voltage between

POWER LEAD

GROUND LEAD

ELECTRODE LEAD

Fig. 8-2. Complete AC arc welder outfit. The AC welder has a high range and low range electrode lead tap. In addition, a crank adjustment is provided with an indicator scale for ampere adjustments from 60 to 100 (high range tap) and 20 to 115 amps. (low range tap). (Miller Electric Mfg. Co.)

It is recommended that the open circuit potential of the arc circuit not exceed 75 to 80 volts as a safety precaution.

An AC welding machine has several main parts:

1. Transformer.
2. Frame.
3. Ventilating system.
4. Shell.
5. Adjustment mechanism.

8-2. THE TRANSFORMER

The transformer consists of three main parts, the primary circuit, the secondary circuit, and the core. The primary circuit generally supplies single-phase current. The secondary

20 and 40 volts. The core consists of sheets of laminated steel of special magnetism properties.

The amount of current produced for welding is varied by several means as explained in CHAPTER 7. These transformers must be specially built to guard against a short between the primary and secondary windings, as this condition can be dangerous. If the primary windings accidentally come in contact with the secondary winding, the primary voltage of 120 or 208 volts might be imposed on the secondary winding and might critically injure the person operating the machine. Also, if some of the primary windings should touch each other electrically, the transformer might become a step-up trans-

TRANSFORMER

COOLING FAN

Fig. 8-3. The internal construction of an AC arc welder. (Miller Electric Mfg. Co.)

former instead of step-down. The higher voltages resulting from the step-up would be dangerous to the person using the welder. Such dangers are virtually nonexistent in professionally made arc welders based on NEMA Standards.

The transformer must be specially constructed to eliminate the need for high open-circuit voltage, as this may also be dangerous or uncomfortable. The method of eliminating the need for high open-circuit voltage is to use a high frequency pilot circuit. This is a type of spark coil which momentarily imposes a high frequency voltage of low current across the arc, making it easier for the operator to start the arc.

These transformers are mounted in a ventilated steel cabinet with suitable controls and meters as desired. Occa-

sionally, the units are mounted on wheels to make them more portable.

8-3. PRIMARY CIRCUIT

The primary circuit carries the input current. The wire for the primary circuit is wound around a special laminated steel core and forms the coil. The magnetic strength of the coil is directly proportional to the number of windings or turns of the coil. These windings must be very carefully insulated from each other and from the magnetic steel core. There are three types of insulation:

Type A, Organic materials (cotton).

Type B, Inorganic materials.

Type C, Combined organic and inorganic materials.

Type B is considered the best, as these insulations can be safely used at higher temperatures. There is also a minimum of moisture effect. Some of these insulations are asbestos, glass, and silicon.

Some of the primary windings have capacitors connected across the primary windings to improve the power factor. NEMA specifies that the machines must have a minimum power factor of 80 percent. This factor means that when the voltage reading and ampere reading are multiplied together

$$V \times A = W$$

that the actual watts must equal 80 percent of the theoretical amount.

Example: 240 V x 30A = 7200 Watts (theoretical)

7200 x .80 = 5760.00 (actual)

or 5.76 KW

The capacitors improve the power factor by discharging and increasing the watts as the sine wave of the input power decreases from maximum and approaches zero.

The primary circuit therefore includes:

1. Stop-start switch (off-on).
2. Overload protection (fuse or circuit breaker).
3. Capacitors in some units.
4. Ventilation fan motor circuit (See PAR. 8-6).
5. Pilot light circuit (in some units).

A movable core AC arc welder wiring diagram is shown in Fig. 8-4. This is the wiring diagram for the machine shown in the three previous figures. The unit has two primary windings, a fan, a movable core transformer and a fixed reactor in the "low" electrode lead. On this diagram, "line" means the power-in leads, FM is the ventilation fan motor, SW is the OFF-ON switch. The primary windings are connected in parallel for 230-volt use. When the two bars are removed and one is put in the center, the two primary windings will

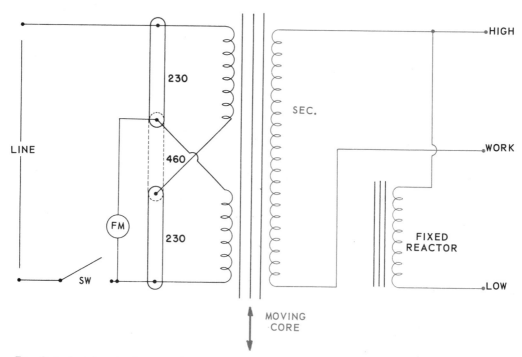

Fig. 8-4. An electrical wiring diagram of a movable core transformer type AC arc welder. (Chemetron Corp.)

be in series for a 460-volt installation. Note that in either case the fan motor will operate at 230 volts.

winding, and the wire is of a larger size (to carry more current).

Its windings must be insulated also;

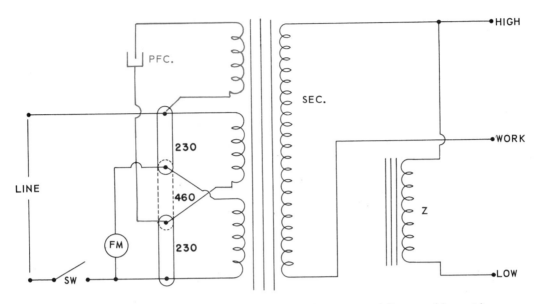

Fig. 8-5. Wiring diagram of movable core transformer type **AC** *arc welder, with a capacitor controlled third primary winding.*

The unit shown in Fig. 8-5 is the same as in Fig. 8-4, except that a capacitor and a third primary winding have been installed in the primary circuit to improve the power factor of the unit, and to improve the performance. The PFC on the diagram means power factor capacitor. As the alternating current builds up and decreases, the capacitor stores electrons during current build up, then releases electrons through the third coil as the original current decreases, thus producing a stronger magnetic field for a longer time and improving the efficiency of the unit.

8-4. SECONDARY CIRCUIT

The secondary winding creates the welding current as the magnetic field created by the primary winding moves back and forth across its windings. It has fewer windings than the primary

and this insulation must be according to NEMA standards.

The leads from the winding are carefully connected to the electrode lead terminal on the welder case, and to the work or ground lead terminal on the welder case. These connections must be mechanically strong, and have minimum electrical resistance.

Some welders have a reactor in the secondary circuit. This reactor consists of a coil of heavy secondary cable and a laminated core made of steel which has special magnetic properties. The reactor may be adjusted to vary the impedance of the reactor coil, and thus vary the amperage output of the machine. The reactor coil produces a voltage which flows in a direction opposite to the normal transformer voltage. By changing the impedance or magnetic strength of the reactor coil, the output of the transformer secondary can be varied.

8-5. WELDING CURRENT ADJUSTMENTS

The current output of the AC arc welder may be adjusted in several different ways:

1. Changing the strength of the reactor winding (secondary circuit).
   A. Changing number of turns in reactor winding.
   B. Changing strength of magnetic field in reactor.
2. Changing strength or position of magnetic field in the main transformer.
   A. Moving coil.
   B. Moving core.

These types are explained in CHAPTER 7.

Some arc welders use taps to in-

The welders with moving elements have a handwheel on the top of the unit, or on the front panel. A pointer and a scale indicates the ampere setting. The machines may be either the main transformer moving coil, moving core type machines, or infinite adjustment reactor type welders.

One type AC welding machine has one adjustable resistance in the reactor winding circuit. The AC is rectified to DC, and a DC winding is used to control the magnetic strength of the reactor. This adjustable control can be remotely located, enabling the operator to more conveniently adjust the machine.

Fig. 8-6 shows a schematic wiring diagram of an AC arc welder with two separate secondary circuit windings

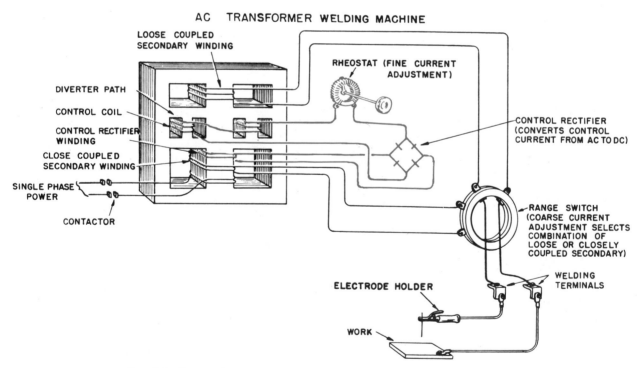

Fig. 8-6. Pictorial wiring diagram of an AC transformer type arc welder.
(Hobart Bros. Co.)

crease or decrease the number of windings or coils in use on the reactor. The taps are usually labeled by the ampere output of that tap. See Fig. 7-10.

for a coarse adjustment and a DC circuit rheostat (resistance) controlled reactor for fine adjustment. The coarse control is mounted on the unit. The

A pipe cutting operation using one cutting torch mounted on a pipe cutting machine.
(H & M Pipe Beveling Machine Co., Inc.)

rheostat control can be either mounted on the unit, or it can be moved to the place of welding for convenience of adjusting.

8-6. COOLING

Most AC welding machines are air cooled. Some of the smaller units have

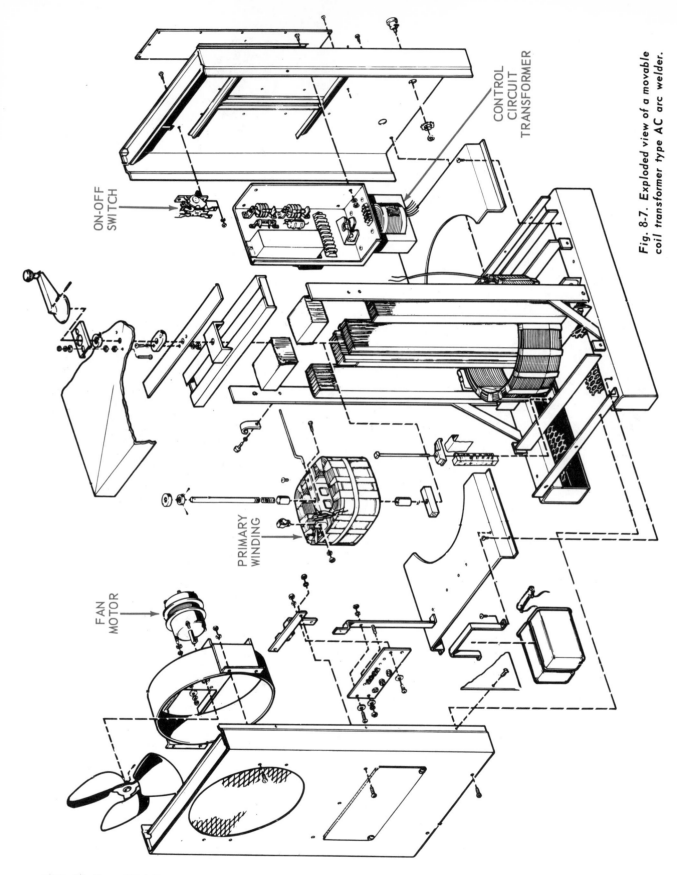

ON-OFF SWITCH

CONTROL CIRCUIT TRANSFORMER

PRIMARY WINDING

FAN MOTOR

Fig. 8-7. Exploded view of a movable coil transformer type AC arc welder.

POLARITY SWITCH — AC SCALE — DC SCALE — OFF-ON SWITCH

WELDING CABLE TERMINALS — INERT GAS CONNECTIONS — CURRENT ADJUSTMENT

POLARITY

Fig. 8-8. Combination AC-DC transformer-rectifier arc welder. The transformer only is used when AC arc welding, while the transformer and rectifiers are used when DC arc welding.

natural or gravity air flow through the mechanisms.

Some machines, particularly the larger sizes, use forced air circulation. An electric motor and fan is connected into the primary circuit and provides forced air circulation automatically when the unit is turned on, and while it is in operation. Fig. 8-7 shows an exploded view of an AC arc welder equipped with a motor-fan forced air circulation cooling system.

Air must get in easily, pass around the mechanism easily, and exit easily. Air passageways must be kept open at all times. The machine should be kept clear of obstructions and should have its ventilation openings uncovered. Periodically, once or twice each year, the casing should be removed (be sure the power is disconnected) and the dust removed from the internal air passages. Compressed air (use goggles) or a vacuum cleaner in combination with a brush, produces excellent results.

8-7. AC-DC ARC WELDING MACHINES

Some welding transformers are made which produce either alternating or direct current by means of a rectifier. With AC-DC transformers, considerable welding flexibility may be obtained with one machine. Settings are made in the same manner as on other transformer welders, but a selection of AC or DC is possible. Fig. 8-8 shows a machine which may be used for AC, DCSP, or DCRP. It has coarse and fine range adjustments for current output. Rectifiers (full wave) change alternating current to direct current. Silicon and selenium material rectifiers are popular for this purpose.

AC-DC machines can be adjusted by using a remote control. Fig. 8-9 shows

Fig. 8-9. Remote current setting device with dial and pointer arrangement. This device will vary the amperes within the coarse range set on the machine.

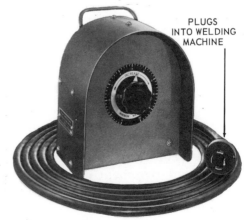

PLUGS INTO WELDING MACHINE

a hand-operated control for a reactance type transformer. This type remote control is also available for foot operation.

Most AC and AC-DC arc welding machines are protected by a thermal overload automatic circuit breaker in the

8-8. MACHINE INSTALLATION

Since these machines are single phase machines, they tend to disturb the electrical power circuit. Because of the low power factor which will be imposed on the power circuit used, it

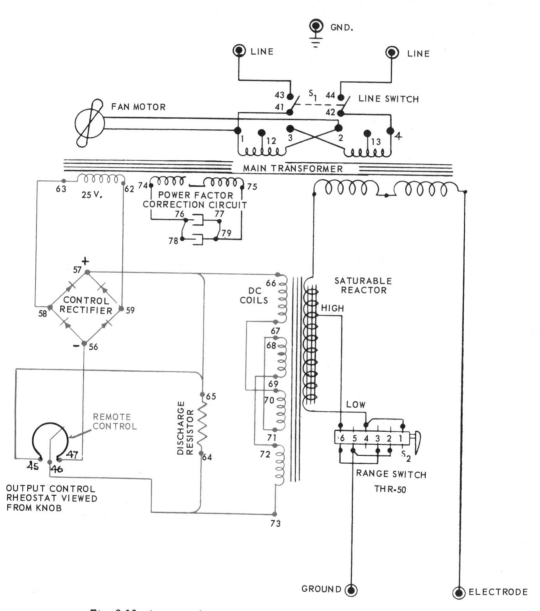

Fig. 8-10. A wiring diagram of an AC welder with remote control.

primary circuit. This protects the winding from overheating in the event of continuous closed secondary circuit or other overloads.

will noticeably affect a single-phase or a three-phase circuit to which it may be connected. Power factor correction capacitors are used to improve the

AC WELDING
LEAD TERMINALS

110 VOLT AC
OUTLETS

DC WELDING
LEAD TERMINALS

220 VOLT AC
OUTLETS

Fig. 8-11. An engine driven combination AC or DC arc welder with outlets for AC auxiliary use.
(Hobart Brothers Co.)

power factor of other electrical machinery on the power line. It is best to consult with the electrical utility company and with an electrical contractor, before purchasing and installing an AC arc welding machine.

Fig. 8-10 is another wiring diagram showing a fan-cooled unit with a coarse range adjustment, a fine range adjustment (DC with rheostat). This machine is a reactor controlled arc welder.

It is recommended when using an alternating current machine, that the arc welding leads be as short as possible, as the electrical resistance of the cables increases the difficulty of striking and maintaining an arc. The cables should also be kept close together to minimize the reactance of the cables as this reactance will reduce the current output of the machine.

8-9. AC ARC WELDING GENERATORS, MOTOR DRIVEN

This type of alternating current, arc welding machine makes use of a motor driven, alternating current generator. The generator is similar to the dynamos or alternators used to generate alternating current for domestic and industrial purposes. Generators of this type, driven by means of a three-phase alternating current motor, have been used. However, this type drive at present, is not in general use.

8-10. AC ARC WELDING GENERATOR, ENGINE DRIVEN

The engine-driven AC arc welding generator is portable and can be used as an emergency electrical power source. This welding machine may be found all over the world. Fig. 8-11 shows a machine which produces AC, DCSP or DCRP for arc welding, and which also may serve as an auxiliary power source 110/220 AC.

Such units are designed to furnish AC arc welding current of capacities varying between 15 to 350 amps. and up.

One model, driven by a 20 HP, 1800 RPM gas engine, will deliver 30 to 350 amps. of welding current, either DC or AC, one KW of DC at 110 V auxiliary power or 10 KW of single-phase (1∅), 115/230 V, 60 cycle AC power.

Some of the units have a built-in high frequency stabilizer unit and can

Fig. 8-12. An engine-driven DC arc welder, generator type, with AC auxiliary outlets for power tools.

be used for inert gas arc welding. A portable, engine-driven DC arc welder with AC auxiliary power is shown in Fig. 8-12.

8-11. AC ARC WELDING ACCESSORIES

The accessories used for AC arc welding are much the same as those used in DC welding. Helmet, gloves, and apron are required. Equipment such as benches, booths and leads are also similar. Small tools - C clamps, chipping hammers, files, chisels, wire brushes, etc. - are the same. Since there is likely to be more metal spatter with AC than with DC, welding solutions are on the market which may be applied to metal surfaces to make spatter easy to remove.

CHAPTER 6 explains in detail the accessories used when arc welding.

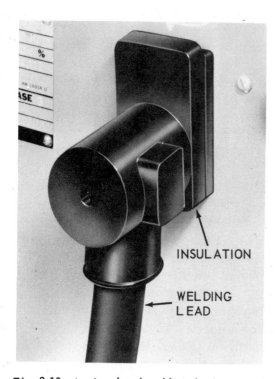

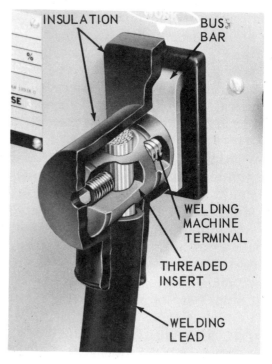

Fig. 8-13. An insulated welding lead terminal. This terminal uses screws to insure a good electrical connection. Left. The complete insulated terminal. Right. Cutaway of the insulated terminal. (Cam-Lok Div., Empire Products, Inc.)

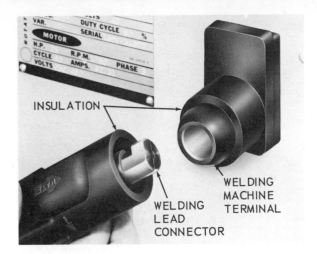

Fig. 8-14. An insulated quick-connect and disconnect
arc welding machine terminal.
(Cam-Lok Div., Empire Products, Inc.)

The Occupational Safety and Health
Act (OSHA) requires insulated welding
cable terminals on welding machines.
An insulated welding terminal is shown
in Fig. 8-13.

Fig. 8-14 illustrates a quick-dis-
connect welding machine terminal of
the insulated type. The insulation on
this terminal is a synthetic rubber
composition which is not affected by
water, oil or grease. The insulator
must cover the entire bus bar, the stud,
the insert and the cable.

8-12. AC ARC WELDING SUPPLIES

Many of the AC arc welding supplies
are the same as those used in DC arc
welding. One major difference is in the
covering of the electrodes for metallic
electrode arc welding. The electrodes
are usually specified as being especially made for AC welding because of the
rate of electrode metal deposit when
using an approximate 50-50 heat distribution. Also, the alternating current
flow necessitates that ionizing elements
be put in the coating for AC arc welding. Some of these electrodes are in
the E 6011 classification, while other
AC electrodes may be found listed in
Figs. 6-30, 6-34 and 6-35. The elec-

trodes are made to the same dimensions
as the DC electrodes. A color code is
used on some electrodes to help the
operator identify the electrodes one
from the other. Some electrodes have
the AWS Classification number stamped
on the coating.

An explanation of electrodes including types, will be found in CHAPTER 6.

8-13. TEST YOUR KNOWLEDGE

1. What type current flows through the
primary coil of the transformer?
2. Name the three main parts of a
transformer.
3. What are the three classes of
insulation?
4. Is there such a condition as Reverse
Polarity AC? If so, why?
5. How are most AC arc welders
cooled?
6. What is a reactor?
7. Under what conditions would it be
necessary to use a generator type AC
arc welder?
8. How does a remote control adjust
the output of an AC transformer type
arc welder?
9. Does an AC arc welding transformer reduce or increase the
potential?
10. Do some AC transformers have two
secondary circuits? If so, why?
11. What is the maximum recommended
potential of the arc welding open
circuit?
12. What other purpose may AC engine
powered arc welders serve?
13. Describe the construction of a
transformer core.
14. Why do AC electrodes have ionizing
agents in their coverings?
15. What is the purpose of the power
factor capacitor?

Using torch in vertical position. U-grooving seams in off shore drilling rig supports.

Chapter 9

ARC, OXY-ARC CUTTING

The intense heat of the electric arc makes it possible to melt and make fluid very small areas of a metal. Therefore, if the fluid metal can be made to flow away either by gravity or by gas pressure, the metal will be removed leaving a cavity or cut.

In metal cutting both carbon electrodes and metal electrodes are used.

9-1. CARBON ARC CUTTING

Metals may be successfully cut by using the carbon electrode arc. The carbon electrode arc leaves a "clean" cut since no foreign metals are introduced at the arc. The cutting current should be 25 to 50 amps. above the welding current for the same thickness

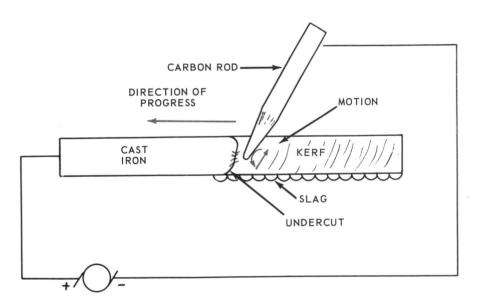

Fig. 9-1. Carbon arc cutting. Note that the carbon electrode is connected to the negative terminal of the welding machine, and the work to the positive terminal.

Many devices have been designed to make this type of metal cutting faster and more accurate. In the case of metal cutting with electrodes, great progress has been made in the development of electrode metals and coverings which produce rapid, smooth and efficient cuts.

of metal. See Fig. 5-4 for a table of current recommendations for DC arc welding.

The carbon electrode should be ground to a very sharp point. During the actual cutting, the carbon electrode should be manipulated in a vertical elliptical movement to undercut the metal, which facilitates the removal of the molten metal. As in oxyacetylene cutting, a crescent motion is recommended. Fig. 9-1 illustrates the relative position of the electrode and the work when cutting cast iron.

The carbon arc method of cutting may be used quite successfully on cast iron because the temperature of the arc is sufficient to melt the oxides formed. It is especially important to undercut the cast iron "kerf" if you desire an even cut. Fig. 9-2 shows a cast iron gear hub being gouged using a carbon electrode with the machine adjusted to approximately 200 amps. DC. Notice the protective equipment the operator uses and the position of the electrode and work. The molten metal must flow away from the gouge or cutting area. Fig. 9-3 is a table of cutting speeds, plate thickness, and

Fig. 9-2. Carbon arc gouging a cast iron gear hub. Note the type electrode holder, and the position of the gear which permits the molten cast iron to flow down and away from the gouge.
(Tweco Products, Inc.)

extension out of the electrode holder to as little as 3 in.

Operating a carbon electrode at extremely high temperatures causes the surface to oxidize and burn away, resulting in a rapid reduction in the electrode diameter. Also as the tempera-

THICKNESS OF PLATE INCHES	CURRENT SETTING AND CARBON DIA.			
	300 AMPS. 1/2 IN. DIA.	500 AMPS. 5/8 IN. DIA.	700 AMPS. 3/4 IN. DIA.	1000 AMPS. 1 IN. DIA.
	SPEED OF CUTTING IN MINUTE PER FOOT			
1/2	3.5	2.0	1.5	1.0
3/4	4.7	3.0	2.0	1.4
1	6.8	4.1	2.9	2.0
1-1/4	9.8	5.6	4.0	2.9
1-1/2	...	8.0	5.8	4.0
1-3/4	...	...	8.0	5.3
2	...	...	...	7.0

Fig. 9-3. Table of recommended electrode sizes, current settings and speed of cutting for carbon arc cutting various thicknesses of steel plate.

current settings for carbon arc cutting.

Because high currents are needed the graphite form of carbon electrode is preferred.

The carbon electrode should not extend more than 6 in. beyond the holder when cutting, to reduce electrical resistance and to reduce the heating effect on the electrode. If the carbon wears away too fast, shorten the electrode

ture increases, the resistance increases.

The standard arc welding generator and other items of arc welding station equipment may be used for carbon arc cutting. DC is always used. It should be adjusted for straight polarity (DCSP).

Due to the very high temperature and the intense arc light, the helmet protector lens should be of a darker

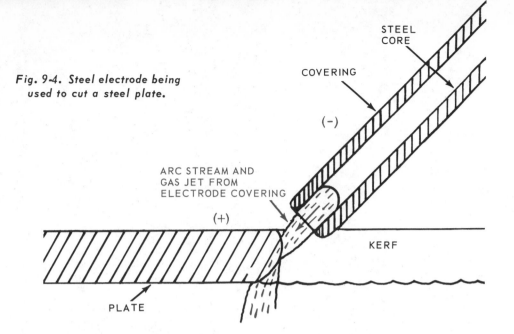

Fig. 9-4. Steel electrode being used to cut a steel plate.

STEEL CORE

COVERING

(-)

ARC STREAM AND GAS JET FROM ELECTRODE COVERING

(+)

KERF

PLATE

shade than for normal arc welding on the same thickness of metal. Shade #12 - #14 is recommended for both carbon arc welding and cutting.

9-2. METAL ELECTRODE ARC CUTTING

Metal may be removed with the electric arc using metal electrodes.

Most cutting electrodes use a covering that disintegrates at a slower rate than the metal center of the electrode. This action creates a deep recess at the arc end of the electrode and produces a jet action that tends to blow the molten metal away, as shown in Fig. 9-4.

Obviously the electrode metal will melt and add metal to the puddle or crater. This extra metal must also be removed. Some of the typical operations that may be performed using this cutting method are shown in Figs. 9-5,

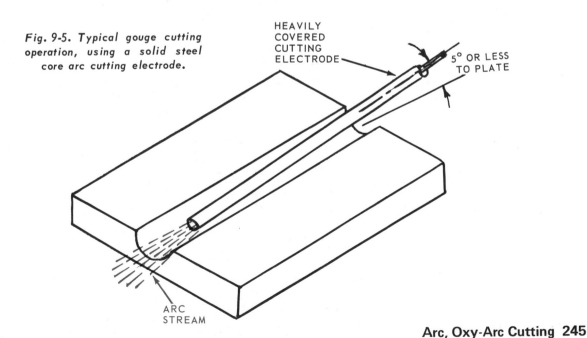

Fig. 9-5. Typical gouge cutting operation, using a solid steel core arc cutting electrode.

HEAVILY COVERED CUTTING ELECTRODE

5° OR LESS TO PLATE

ARC STREAM

Air carbon arc gouging U-grooves in I-beams prior to welding.   (Arcair Co.)

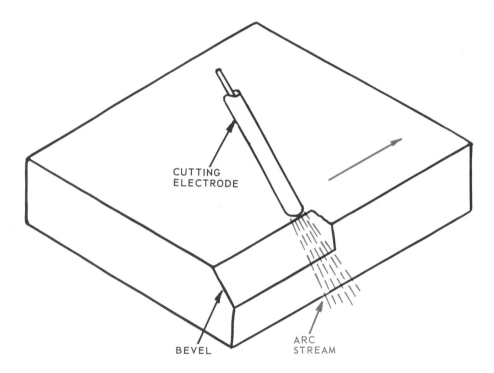

Fig. 9-6. Beveling plate edge, using solid steel core arc cutting electrode.

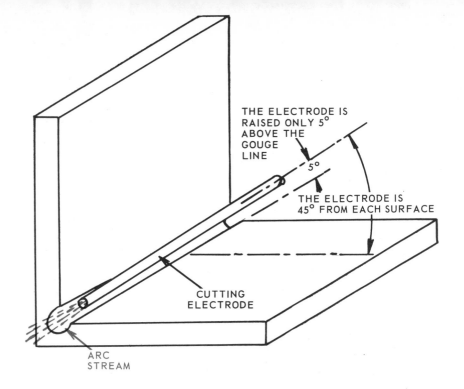

Fig. 9-7. *Gouging an inside corner joint, using solid steel core arc cutting electrode.*

also Figs. 9-6, 9-7, and 9-8.

A current setting, as high as the electrode will take without becoming over-heated to the point of cracking the covering, is recommended. For 1/8-in. electrodes 125 to 300 amps. are used.

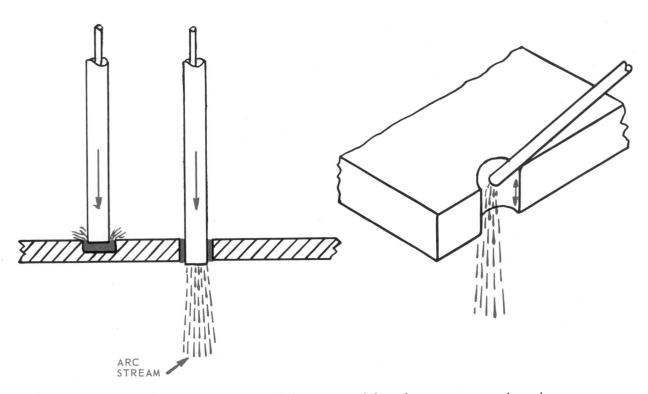

Fig. 9-8. *Piercing hole in steel plate using solid steel core arc cutting electrode.*

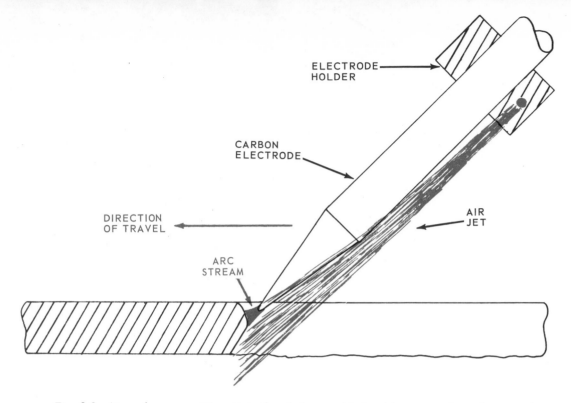

Fig. 9-9. Air carbon-arc cutting. Note the air jet provided to blow away the molten metal.

For 5/32 in. 250 to 375 amps., and for 3/16 in., 300 to 450 amps. are required.

This system is easy to set up. A standard welding station may be used. One needs to use a special cutting electrode for good gouging or cutting results.

Many manufacturers have developed special metal arc cutting electrodes with special coverings which intensify the arc stream for rapid cutting. Such electrodes may be used for cutting stainless steel, copper, aluminum, bronze, nickel, cast iron, manganese, steel or alloy steels.

The system can be used with either AC or with DC straight polarity. A very short arc is recommended. If cutting under water the coating must be waterproof.

9-3. AIR CARBON-ARC CUTTING (AAC)

If the molten metal can be blown away as the electric arc creates the puddle, the metal can be easily gouged, cut, or pierced. A process has been developed to perform this task. Carbon and/or graphite electrodes are used, and compressed air is fed to an orifice built into the electrode holder as shown in Fig. 9-9. A hand valve on the electrode holder controls the air jet.

The air jet blows the molten metal away and usually leaves a surface that is suitable for welding operations without further surface preparation. It operates at air pressures varying between 60 and 100 psig. The use of compressed air is less expensive and because air is principally nitrogen (an inert gas) a high quality cut is usually obtained. Fig. 9-10 shows an air jet electrode holder with the carbon electrode installed.

The electrodes are available in two types:

1. Carbon.
2. Graphite.

Bare carbon and/or graphite electrodes become quite hot over their current carrying length because the carbon offers some resistance to the flow of current. Therefore the heated carbon surface slowly oxidizes and the electrode becomes smaller in diameter. To reduce the temperature and to reduce the oxidation, these electrodes are sometimes copper-coated. Copper-coated electrodes will operate cooler

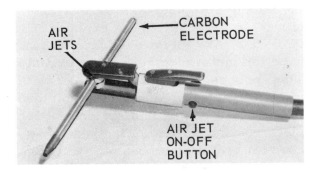

Fig. 9-10. Air carbon-arc cutting torch. Note the two rotating insulated air jet heads for use in interior work in castings. The two air jet heads keep air behind the electrode regardless of the direction the operator may go inside the casting. (Arcair Co.)

since the copper-coating keeps the surface from oxidizing.

The operating procedure is as follows:

1. Adjust machine to correct current for electrode diameter.

2. Start air compressor and adjust regulator to correct air pressure. Use as low air pressure as possible. Use just enough pressure to blow away the molten metal.

3. Insert electrode in holder. Extend carbon electrode 6 in. beyond holder. It is important that electrode point be properly shaped. Fig. 9-11 shows copper-coated carbon electrode being installed in electrode holder.

4. Strike arc, then open air jet valve. Air jet disc can swivel and V groove in disc automatically aligns air jets along electrode at any angle electrode is

adjusted relative to holder.

5. Control arc and speed of travel according to shape and condition of cut desired.

6. Always cut away from operator as molten metal sprays some distance from cutting action. This process may be used to cut metal in all positions such as flat, horizontal, vertical, and overhead.

9-4. AIR CARBON-ARC GOUGING

There are various metalworking applications in which carbon electrode air jet gouging may be used to advantage. Examples are for such jobs as preparing metal for welding, as a metal shaping medium. To gouge, hold the electrode holder so the electrode slopes back from the direction of travel. The air blast is directed along the electrode toward the arc. The depth and contour of the groove will be controlled by the electrode angle and speed of travel. The width of the groove will be controlled by the size of the electrode used and will usually be about 1/8 in. wider than the electrode diameter. Fig. 9-12

Fig. 9-12. Application of air carbon-arc process being used to gouge metal.

shows a casting being gouged using the carbon electrode arc-air process.

The correct position of the electrode holder in relation to the work when cutting or gouging a shallow groove on the surface of metal is a very flat angle. The speed of travel and the current setting also affect the depth of the groove. The slower the movement the deeper the groove. The higher the current setting, the deeper the groove.

Fig. 9-13. A V groove gouged in 2 in. thick carbon steel.

A V groove cut in 2-in. thick mild steel plate is shown in Fig. 9-13. This groove was cut by using an automatic machine to carry and guide the carbon arc-air jet holder. A fillet weld being removed by gouging operations is shown in Fig. 9-14.

9-5. AIR METALLIC-ARC CUTTING

Metal electrode air jet cutting uses the same equipment as carbon electrode-air jet cutting, as explained in PAR. 9-3. Any metal electrode may be used for metal electrode air jet cutting. This process differs from the plain metal electrode arc cutting as explained in PAR. 9-2, in which a special cutting electrode is needed. Electrode sizes used for metal electrode air jet cutting should be selected on the basis of the metal thickness to be cut, and should be a size larger than what would be used for welding the same metal thickness. The current used may be either AC or DC reverse polarity. The current flow should be higher than the current required for welding the same thickness of metal.

The electrode holder is the same as used for the carbon electrode air jet cutting shown in Fig. 9-10.

Hollow carbon electrodes and hollow steel electrodes have both been used to

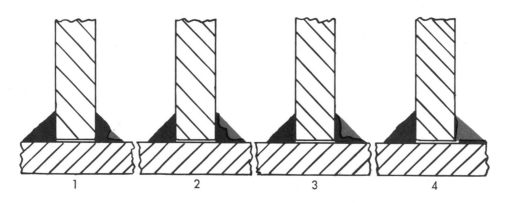

Fig. 9-14. Recommended gouging sequence to be used when removing fillet weld.

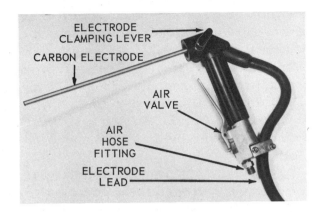

Fig. 9-15. Electrode holder designed for hollow electrodes used when arc-air cutting. Note air hose fitting and air control lever.

cut and gouge metal. Fig. 9-15 illustrates an electrode holder and electrode.

9-6. OXYGEN-ARC CUTTING

This process is a combination of an electric arc and a jet of oxygen. The equipment required may be either an AC or DC welding machine, an oxygen-arc electrode holder, source of oxygen, and suitable oxygen-arc cutting electrodes. The process is mainly used for cutting alloy steels, aluminum, cast iron, etc. which are difficult to cut using the oxygen-acetylene cutting process.

This process usually makes use of a tubular metal covered electrode, and oxygen is fed through the hollow electrode. The electrode is covered if fluxing ingredients are needed for cutting the metal. The covering also enables the operator to hold the electrode end against the base metal and still maintain the arc.

The current settings are approximately the same as for welding. A higher voltage is usually used (28 - 45 volts).

Fig. 9-16 illustrates an electrode holder with the electrode installed as it is used for oxygen-arc cutting. This cutting device is connected to an oxygen source using an oxygen hose, oxygen

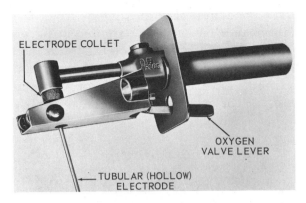

Fig. 9-16. Oxygen-arc cutting electrode and holder. (Arcos Corp.)

regulator, and oxygen cylinder. The electrical lead is connected to a current source as in Fig. 9-17.

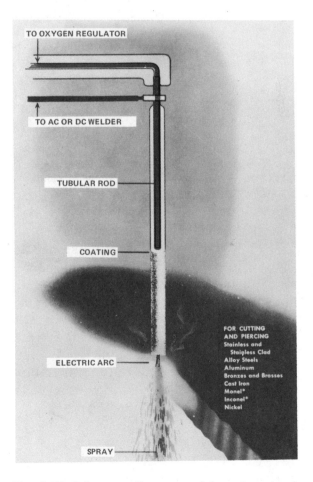

Fig. 9-17. Schematic illustration of the main parts of a tubular electrode and holder as used in oxygen-arc cutting. This process may be used on many different metals such as stainless steel, alloy steel, aluminum, cast iron, etc.

To operate the cutting device, the operator first strikes an arc between the electrode and the metal to be cut. The oxygen valve on the holder is then opened, and the cut begins, as shown in Fig. 9-18. A continuous arc is main-

Fig. 9-18. Tubular steel electrode being used in oxygen-arc cutting job. Note operator's equipment and position of electrode. (Arcos Corp.)

tained throughout the cutting operation. The electrode is gradually consumed as the cut is produced. This cutting process is particularly useful when working on metals which are difficult to cut, such as stainless steel and cast iron.

The oxygen pressure used is the same as described in CHAPTER 3. The pressure is varied as the metal thickness varies and also as the electrode orifice or inside diameter varies.

9-7. REVIEW OF ARC CUTTING SAFETY

Electric arc cutting does not present any great or new hazards. However, due to the fact that some cutting operations may continue for a considerable length of time and at high current rates, precautions must be taken to avoid skin burns or eye damage. The following precautions are recommended:

1. Be sure all surfaces of the skin are shielded from the arc rays. Wear approved gloves, helmet, and approved clothing.

2. Use helmet filter lens a shade or two darker than would be used for welding with the same size electrode.

3. Use completely insulated electrode holders.

4. Stand or work only in dry surroundings.

5. A considerable amount of fumes and gases are liberated during all of the arc cutting operations. The ventilation of the working space must be such that the operator is working in clean, fresh air at all times.

6. Most of the arc cutting processes are accompanied by a great amount of sparking and showers of molten globules of metal. These hot particles may be thrown some distance from the arc. Wear garments which are fire-resistant. Pockets, cuffs, and other clothing crevices must be covered.

7. Be sure that the working area is fireproof. All flammable objects such as wood benches, floors, or cabinets should be removed from the vicinity. Be sure there are no openings in the floor which might allow the sparks to travel to the floor below.

8. Some electric arc cutting stations have a "wet table." The wet table provides for a thin water film under the flame cutting station. The wet film quenches the sparks.

9. When using the oxygen-arc process for cutting any type of tank or container, investigate its former contents. The container should be thoroughly steam cleaned and an inert gas should

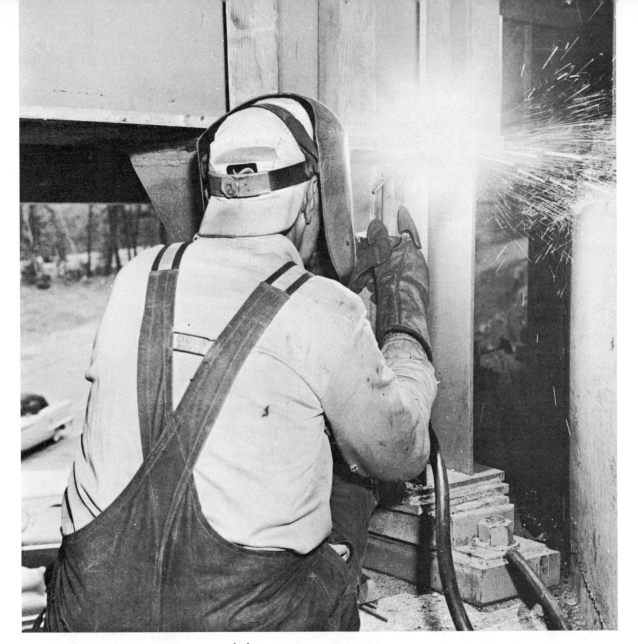

Industry photo--air arc cutting.

flow through the container as it is being cut.

10. Since the oxygen-arc cutting process uses a higher electrical current than is used for welding the same thickness of metal, be sure that the welding machine has ample capacity for the current needed. Most welding machines can operate for a short time at some overload; but if they are operated at an overload for any length of time, they may overheat and burn out.

11. Most oxygen-arc cutting processes are quite noisy. Therefore, such stations should be located where the noise will not be objectionable.

9-8. TEST YOUR KNOWLEDGE

1. How should the carbon electrode end be shaped for arc cutting operations?

2. Should current settings for oxygen-arc cutting be higher, lower, or the

same as when welding using the same diameter electrode?

3. What type electrodes may be used on the air carbon-arc cutting process?

4. How far should a carbon cutting electrode extend from the holder? Why?

5. How do some metal electrodes used for cutting differ from metal electrodes used for welding?

6. What type welding current is suggested for the carbon electrode cutting process?

7. What metals may be cut by the metal electrode arc cutting process?

8. If direct current is used for metal electrode arc cutting, is the current usually straight or reverse polarity?

9. What is one advantage of the air carbon-arc cutting process?

10. What is the condition of the metal surface after being cut by the air carbon-arc cutting process?

11. What is the approximate air pressure required when air carbon-arc cutting?

12. How is the depth of the cut regulated when air carbon-arc gouging?

13. What type of electrode is used with the oxygen-arc cutting process?

14. Name two precautions to be taken when cutting with the electric arc.

15. Why is a darker shade filter lens needed when arc cutting than when arc welding?

Air carbon-arc cutting. This cut is being made on a large casting. Note that the ends of the carbon electrodes are designed to fit together so that there is no waste of carbon electrode material in use.

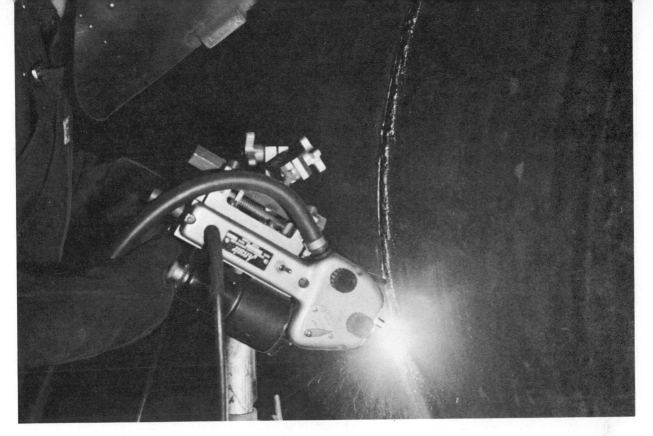

Air carbon-arc cutting machine grooving a pressure vessel wall of 3/4 in. in thickness. The seam has been welded from the inside of the vessel and the external groove is being made in preparation for outside welds.

Chapter 10

ARC CUTTING
EQUIPMENT AND SUPPLIES

The electric arc method of metal cutting is fast becoming a popular cutting process. The use of oxygen, air, and inert gases to remove the metal from the kerf has enabled the operator to produce both improved quantity and quality in arc cuts. Improved equipment and supplies have contributed to the usefulness and efficiency of arc cutting, air-arc cutting, and oxygen-arc cutting.

10-1. CARBON ELECTRODE ARC CUTTING EQUIPMENT

The carbon electrode arc cutting process was the original method used to cut (or melt away) metals.

The complete cutting outfit consists of:

DC or an AC Arc Welding Machine.
Booth with better than normal venti-

lation (1,000 CFM or more).

Electrode lead.

Work lead.

Special carbon electrode holder, as shown in Fig. 10-1.

Carbon electrodes.

Heavy-duty gloves.

Heavy-duty clothes.

Helmet with lens of #12 to #14 shade.

trodes. These two devices are described in CHAPTER 6.

10-2. METALLIC ELECTRODE ARC CUTTING EQUIPMENT

The equipment for metallic electrode arc cutting is identical to that of an arc welding station. All the equipment of a

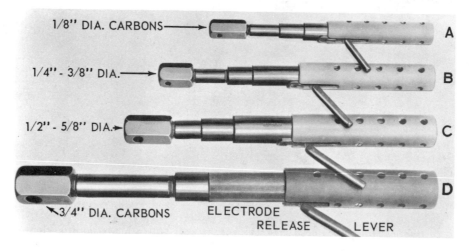

Fig. 10-1. Carbon electrode holders. These holders are designed for: A. 1/8-in. carbon--150 Amps. B. 3/16-in. carbon--200 Amps. C. 3/8-in. carbon--300 Amps. D. 3/4-in. carbon--500 Amps. (Tweco Products)

The high amperage flow at the arc releases a considerable amount of heat. Heavy gloves and clothing must be worn to protect the operator from the intense heat of the carbon arc. The electrode holder often has a shield around the handle to protect the operator's hand from radiant heat. Some carbon electrode holders are water cooled. In using this type of equipment less protection from heat is required.

Because of iron oxide fumes plus other air contaminants, it is very important to have a good ventilation system.

Most of the equipment is standard and has already been described. The two most special devices of a carbon electrode arc cutting station are the carbon electrode holders and the carbon elec-

standard arc welding station is used. The operator need only use the special steel arc cutting electrodes.

10-3. METALLIC ELECTRODES FOR ARC CUTTING

Metallic electrodes for arc cutting are designed to produce a high velocity gas and particle stream formed by the steel core and the covering. This action is created by a deeper cavity in the electrode end, and the resulting jet action of the gases produced by the arc. Using special coatings on the electrodes produces these results. Somewhat the same result may be obtained with ordinary flux covered electrodes by using a higher than normal current for the size of the electrode. Special cutting electrodes are available in standard diameters and lengths. Common sizes which come in 14-in. length, are 3/32, 1/8, 5/32, 3/16, and 1/4 in.

Fig. 10-2 shows the approximate electrode sizes and current settings when using special steel cutting electrodes.

the arc end of the electrode. This deeper cavity creates a stronger jet action which improves the cutting. Water soaked electrodes should be used immediately. Water soaked electrodes should not be used for welding because the moisture produces hydrogen embrittlement in the weld.

METAL THICKNESS	ELECTRODE DIAMETER	AC CURRENT RANGE AMPS.	DCSP AMPS.
1/8	3/32	40-150	75-115
1/8-1	1/8	125-300	150-115
3/4-2	5/32	250-375	170-500
1-3	3/16	300-450	
3, over	1/4	400-650	

Fig. 10-2. A table of suggested uses and approximate current settings for cutting with special steel-cutting electrodes.

When cutting mild steel, covered electrodes similar to these electrodes used in welding may be used; however, a much higher welding current range will be required. Fig. 10-3 lists the

METAL THICKNESS	ELECTRODE DIAMETER	CURRENT RANGE AMPS.
1/8	3/32	100-160
1/8-1	1/8	160-350
3/4-2	5/32	250-400
1-3	3/16	350-500
3, over	1/4	400-650

Fig. 10-3. Approximate current settings and electrode sizes for metallic arc cutting, when using standard welding electrodes.

current values to use when doing various cutting operations with standard covered steel electrodes.

If mild steel electrodes are used for cutting, they will last a little longer if they are soaked in water for a period not to exceed 10 minutes before being used. The moisture in the coating slows the vaporizing of the material thus producing a deeper cup (cavity) at

10-4. AIR CARBON-ARC CUTTING EQUIPMENT

The equipment used for cutting using the electric arc in combination with a high pressure air jet is as follows:

Standard arc welding machine.

Air compressor or compressed air in cylinders (for small applications).

Air carbon-arc torch equipped with an air jet device.

Electrode lead.

Compressed air hose (sometimes combined with the electrode lead, as shown in Fig. 10-4).

Work lead.

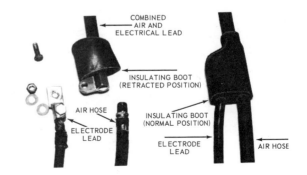

Fig. 10-4. Combination welding lead and air hose assembly. (Arcair Co.)

Fig. 10-5. Automatic air carbon-arc cutting machine. (Arcair Co.)

Carbon electrodes.

The usual gloves, special clothing, booths, ventilation devices, benches, and the like are required. Darker lens (12-14) are usually used in the helmet.

The air carbon-arc process may be used in conjunction with a motorized carriage to cut and gouge metal. Fig. 10-5 shows an air carbon-arc holder mounted on a motorized carriage.

10-5. COMPRESSED AIR SUPPLY

The air is fed to the air carbon-arc torch jet at 60 to 100 psig. This air should contain a minimum of abrasives and moisture. A sufficient amount of air must flow through the jet to remove the molten metal. The size of the air compressor and hose varies with the size of the cutting job. The air compressor is automatically controlled to turn on and off by an air pressure operated electrical switch. The air flow to the hose is kept at a constant pressure by means of a pressure regulator. A hand valve on the torch controls the on and off operation of the air flow.

"Bottled" compressed air (air stored in cylinders) may be used when portability is desired, where accessibility to the job is a problem, and where the cutting project is of small size.

10-6. AIR CARBON-ARC ELECTRODE HOLDER

The air carbon-arc electrode holder (torch) is similar to the standard electrode holder except for the air passageways, the air on-and-off valve, and the air orifice for the air jet, Fig. 10-6.

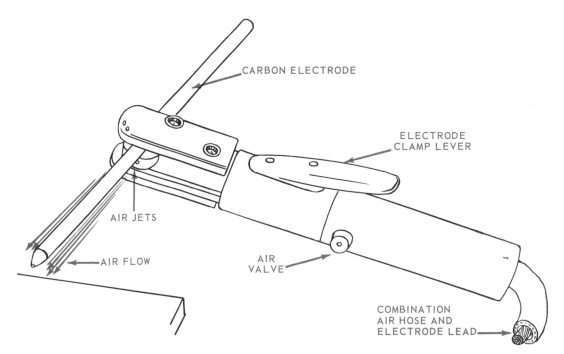

Fig. 10-6. An air carbon-arc electrode holder with carbon electrode installed.

The electrode lead and hose connections are usually made at the handle end of the torch. The air valve is mounted in the handle and is either lever operated. Most of these valves have a lock-open feature to enable the operator to keep the air flowing, and still have a comfortable hand position on the handle.

are copper plated or copper clad to increase the strength, reduce the oxidation of the electrode body, and to reduce the temperature of the electrode.

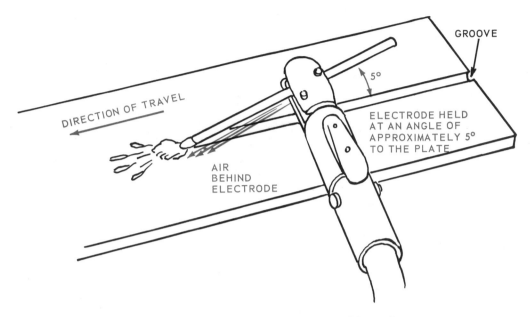

Fig. 10-7. Air carbon-arc electrode holder and carbon electrode being used to gouge a metal plate.

or button operated. Most of these valves have a lock-open feature to enable the operator to keep the air flowing, and still have a comfortable hand position on the handle.

It is very important that the air jet be directed as close as possible to the arc crater. Any misalignment or abuse of the air orifice may produce below average results. Fig. 10-7 shows the electrode holder in position and producing a gouge in a metal plate.

10-7. AIR CARBON-ARC ELECTRODES

These electrodes are obtainable in the carbon form, graphite form and in a mixture of carbon and graphite. The electrodes are usually made of a mixture of carbon and carbon in the graphite form. Some of the carbon electrodes

The copper coating does not add appreciably to the conductivity of the electrode. Fig. 10-8 shows a table of car-

DIA. OF ELECTRODE	AMPERES	
	MIN.	MAX.
5/32	80	150
3/16	110	200
1/4	150	350
5/16	200	450
3/8	300	550
1/2	400	800
5/8	600	1000
3/4	800	1400

Fig. 10-8. Table of current settings for air carbon-arc cutting electrodes.

bon electrode sizes and the approximate current settings for each. The table, Fig. 10-9, shows cutting speeds and results.

Standard carbon electrodes are best

used with DC reverse polarity. Electrodes are also available for use with AC. The AC carbon electrodes are

shown in Fig. 10-10.

Reverse polarity is commonly used when cutting steel, and stainless steel,

"U" GROOVE WIDTH INCHES	DEPTH INCHES	ELECT. DIA. INCHES	AMPERAGE	VOLTS	ELECT. FEED I.P.M.	TRAVEL SPEED I.P.M.	FEET OF GROOVE PER ELECTRODE	FEET OF GROOVE PER MINUTE
3/8	3/32	1/4	270	40	4	54.	10.25	4.5
3/8	1/8	1/4	300	42	4	51.75	9.75	4.5
3/8	3/16	1/4	300	40	6.75	38.25	4.	3.18
3/8	1/4	1/4	320	42	6.25	29.5	2.5	2.05
3/8	3/8	1/4	320	46	3.62	15.	3.	1.25
Deeper Grooves may be made by successive passes with combination of settings:								
3/8	1/8	5/16	320	40	3.	65.5	16.37	5.45
1/2	3/16	5/16	400	46	4.37	46.	7.87	3.87
1/2	1/4	5/16	420	42	3.87	31.25	6.	2.6
1/2	1/2	5/16	540	42	5.62	27.25	3.5	2.25
Deeper Grooves may be made by successive passes with combination of settings:								
1/2	1/8	3/8	540	42	4.25	82.	14.37	6.83
1/2	1/8	3/8	540	42	3.37	65.5	14.5	5.45
17/32	3/16	3/8	540	42	2.62	41.75	14.12	3.48
5/8	1/4	3/8	540	42	3.	29.5	7.37	2.45
9/16	1/2	3/8	540	42	3.25	15.	3.46	1.25
1/2	11/16	3/8	540	42	3.5	12.25	4.29	1.02
For Grooves deeper than 1/2" combinations of settings with multiple passes are recommended:								
11/16	1/8	1/2	800	45	3.06	34.	8.5	2.75
13/16	1/4	1/2	800	45	3.06	22.5	6.37	1.87
13/16	3/8	1/2	800	45	3.06	20.75	4.87	1.75
13/16	1/2	1/2	800	45	3.06	18.5	4.5	1.5
7/8	5/8	1/2	800	45	3.06	15.	3.31	1.25
7/8	3/4	1/2	800	45	3.06	12.5	2.5	1.02
Deeper Grooves may be made by successive passes with a combination of settings:								
15/16	1/8	5/8	900	42	2.5	44.5	19.75	3.62
15/16	1/4	5/8	900	42	2.5	29.5	8.5	2.5
15/16	3/8	5/8	900	42	2.5	20.	5.62	1.85
1	1/2	5/8	900	42	2.5	14.5	3.62	1.25
1	5/8	5/8	900	42	2.5	13.	3.5	1.08
1	3/4	5/8	900	42	2.5	11.	3.15	.93
1	1	5/8	900	42	2.5	10.	2.12	.84
Combination of settings and multiple passes may be used for grooves deeper than 3/4":								

Fig. 10-9. *Table of current settings, speeds, electrode wear, depth of grooves, etc. for air carbon-arc electrode cutting. All figures are for standard DC copper clad electrodes, DCRP. Air pressures throughout are 80-100 psig. A 100 psig pressure is recommended for 1/2 and 5/8-in. electrodes. The figures are for carbon steel, but are valid for other materials with slight variations.*

available in 3/16, 1/4, 3/8, and 1/2 in. diameters. Most carbon electrodes are available in standard 12 in. lengths.

As shown in Fig. 10-8, some of the current ratings are quite high and the arc welding machine must be of ample capacity. Two or more standard machines may be connected in parallel to give the current capacity needed as

while straight polarity is used for nickel alloys. Both DCSP and DCRP are usable on cast iron and copper alloys. Alternating current is also used when carbon electrodes especially made for AC are available.

If one encounters slag problems when cutting with carbon electrodes, it may be better to use metal electrodes. Spe-

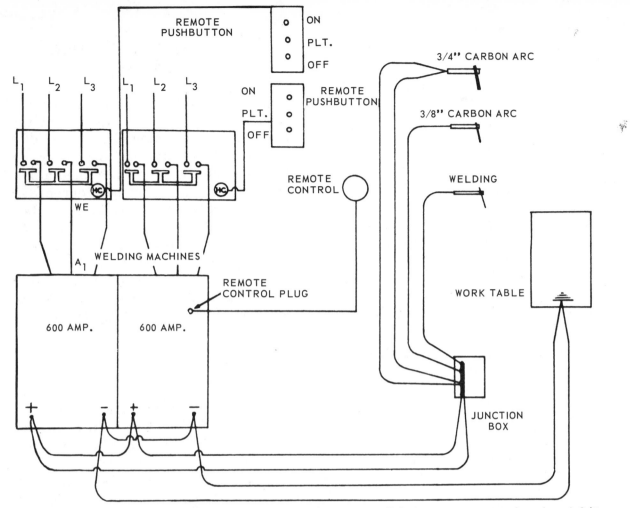

Fig. 10-10. Wiring diagram for connecting two DC arc welders in parallel. An operator can pad-wash with 3/4-in. dia. electrode, reduce amperage and remove fins and defects with 3/8-in. electrodes, then reduce amperage further and weld defects.

cial metal electrodes have been developed to cut such metals as cast iron, copper alloys, or nickel alloys.

10-8. OXYGEN-ARC CUTTING EQUIPMENT

Oxygen-arc cutting equipment is a combination of a gas welding and an arc welding station. It uses the usual gas station equipment as follows:

Oxygen cylinder.
Oxygen regulator.
Oxygen hose.

It uses the following parts of an arc welding station:

Arc welding machine.

Work lead (ground cable).
Electrode lead (cable).
Booth.
Ventilation system.
Bench.

The special equipment needed is:
Special oxygen-arc torch.
Oxygen-arc cutting electrodes.

10-9. OXYGEN-ARC CUTTING ELECTRODE HOLDER

The electrode holder (torch) is designed to hold either ceramic tubular electrodes, carbon tubular electrodes, or metal tubular electrodes. On some models, a special cap equipped with an oxygen orifice is designed to fit over the end of the electrode mounted in the electrode holder. The clamp or collet which holds the electrode also carries the current into the electrode (similar to the standard electrode holder). A soft special gasket seals the opening where the electrode contacts the oxygen orifice cap (neoprene or silicon rubber are common gasket materials).

Some torches use a collet system to hold the electrode. The collet provides the joint to seal against oxygen leakage, and provides an area of electrical contact with the electrodes.

A hand valve is mounted on the electrode holder to control the oxygen flow.

10-10. OXYGEN-ARC CUTTING ELECTRODES

The electrode consists of a steel tube covered with a flux coating. The tube carries the arc current, the inside of the tube is the oxygen passageway, and the covering serves as an insulator, a means to maintain the correct arc length, and to provide fluxing agents to improve the cutting.

These electrodes are available in 3/16 and 5/16 in. outside tube diameters with 1/16 and 1/10 in. inside diameters respectively. These electrodes are made in 14 and 18 in. lengths.

10-11. TEST YOUR KNOWLEDGE

1. When arc cutting, what energy is used to heat the metal to the melting temperature?

2. What action removes the metal from the kerf during the carbon-arc cutting process?

3. How should the operator's hand be protected during a manual carbon-arc cutting operation?

4. Why are some carbon electrodes copper coated?

5. Can air carbon-arc cutting be done using solid electrodes?

6. Is the air turned on before or after the arc is produced?

7. Why should the operator point the electrode away from himself when using the air carbon-arc process?

8. When is it advantageous to use cylinder (bottled) compressed air?

9. What is the position relationship between the air jet and the carbon electrode?

10. Can carbon electrodes be used with AC arc cutting machines?

11. What should be done if slag appears in the kerf when using carbon electrodes for cutting?

12. How is oxygen fed to the kerf during oxygen-arc cutting?

13. What is the purpose of the collet on some oxygen-arc cutting electrode holders?

14. Does the carbon electrode wear during a carbon electrode air carbon-arc jet operation?

15. Are any carbon-arc electrode holders water cooled?

Chapter 11

GAS TUNGSTEN AND
GAS METAL ARC WELDING

In both gas tungsten and gas metal arc welding, an inert gas is fed into the weld. This is done to crowd out the atmospheric air from the weld. Otherwise, oxygen would combine with the molten metals and form oxides which weaken the weld.

Helium, argon, carbon dioxide, or a mixture of these may be used as shielding gases. Substances such as oxygen, hydrogen, nitrogen, and water vapor in the welding atmosphere reduce the quality of the weld.

The gas tungsten arc process is sometimes incorrectly called heliarc welding, because it was developed around the use of helium. Heliarc welding was developed for welding aluminum. It has been found that argon is a better shielding gas than helium. Helium is a very light gas, and when it is heated in the electric arc it becomes even lighter and flows away from the arc. Argon is a heavier gas, and even in a heated condition it remains in the arc and keeps out the atmospheric air. In the beginning, helium was less expensive than argon; but, with the increased use, argon is now generally less expensive.

11-1. GAS ARC WELDING PRINCIPLES

The principle of gas arc welding is quite simple. The electrode holder, often called a torch or gun, is designed to also supply a flow of shielding gas, such as carbon dioxide, helium or

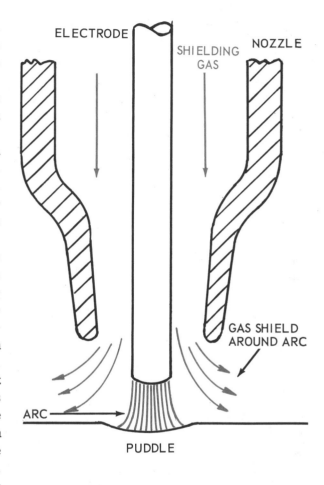

Fig. 11-1. Shielding gas arc welding principle.

argon, which surrounds the electric arc. See Fig. 11-1.

Shielding gas keeps oxygen and other contaminants away from the high temperature molten metal. It also keeps other active elements in the atmosphere away from the molten metal. By eliminating oxidation and other impurities, welds are possible on metals which are impractical, or very difficult, to weld.

The gas arc welding principle may be used either manually, semiautomatically, or completely automatically. The manual and semiautomatic processes are discussed in this chapter, automatic processes in CHAPTER 17.

During gas arc welding the electrode holder flows shielding gas around the electrode. As it flows it pushes the atmospheric air away from the electrode, arc, and weld puddle.

Gas arc welding has three great advantages over the usual forms of arc welding. These advantages are:
1. Faster, minimizing distortion.
2. Cleaner welds.
3. The ability to weld metals previously impossible, or difficult, to weld.

The costs, by percentage, for welding supplies for gas arc welding are approximately:

Tungsten electrodes (when used) 3%
Electric power 5%
Shielding gas 92%

As mentioned before, the gas process is much faster than regular gas welding. Accordingly, the labor saving is considerable, and the productivity of a unit is much greater. Costly cleaning operations are avoided.

11-2. TYPES OF GAS ARC WELDING

Gas arc welding processes include:
1. Gas tungsten arc welding - GTAW (TIG-Tungsten Inert Gas).
2. Gas metal arc welding - GMAW (MIG-Metal Inert Gas).
3. Gas carbon arc welding - GCAW (CIG-Carbon Inert Gas).
4. Gas arc spot welding - GASW.
Three types of current flow are used in gas arc welding:
1. Direct current, straight polarity (DCSP).

2. Direct current, reverse polarity (DCRP).
3. Alternating current.

See Fig. 11-2. Note that the electrons travel in the opposite direction to the

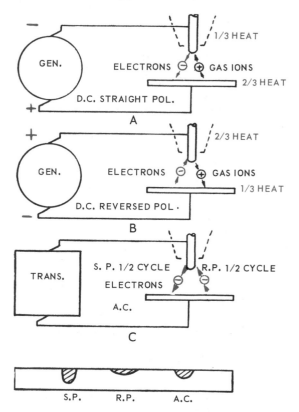

Fig. 11-2. Three basic electrical circuits are used for shielding gas arc welding. A. DC Straight Polarity (DCSP). B. DC Reverse Polarity (DCRP). C. AC. Note direction of travel of electrons. Lower drawing shows relative penetration obtained when each of the three types of welding current is used.

flow of gas ions. Ions are gas molecules temporarily reduced to electrically charged atoms or electrically charged particles other (and smaller) than atoms. These actions are part of the electron theory. The bottom view shows the effect of these energy flows on the melting of the base metal.

When using direct current straight polarity (DCSP), good penetration is obtained because the electron stream flows to the work, concentrating the heat at the work.

When using direct current reverse polarity (DCRP), good cleaning action is obtained but weld penetration is not great since most of the heating effect takes place at the tungsten electrode (anode) or electrode welding wire. This process is best used on thin sections of aluminum, magnesium and other hard-to-weld materials using tungsten electrodes.

When using alternating current with high frequency (ACHF) both good cleaning action and good penetration may be obtained.

The positive portion of the AC sine curve emits positively charged ions. They bombard the surface like sand blasting.

When performing gas metal arc welding, GMAW (MIG), the heat flow from the arc is somewhat different, since the metal electrode is melting and globules of molten metal flow from it to the weld.

Gas metal arc welding is performed by using direct current reverse polarity (DCRP). It should be noted that this arc gives both good cleaning action and fast filler metal deposition rates. A rather high welding current is desirable as this extra current breaks up the globules of molten metal into a fine spray. This increases the rate of transfer and gives better control of the arc, so that it can be directed accurately into the weld joint. This process can also be used with low current flows for thin metal welds and poor fit joints. PAR. 11-12 and 11-13 explain the spray method and dip transfer in more detail.

11-3. GAS TUNGSTEN ARC WELDING PROCESS GTAW (TIG)

This process uses a nonconsumable tungsten electrode. The electrode is mounted in a special electrode holder

Fig. 11-3. *The gas tungsten arc welding torch being used to weld lap joint.* (Dow Metal Products)

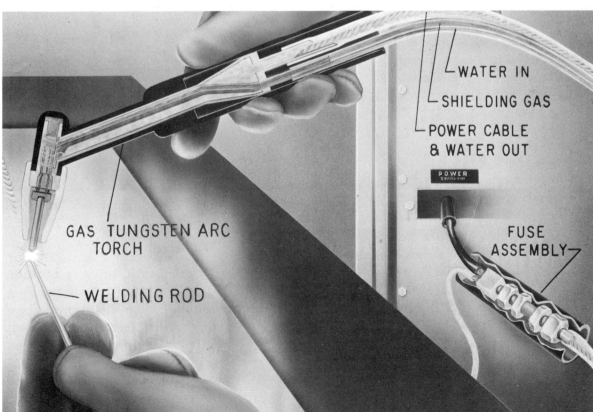

which also is designed to furnish a flow of shielding gas around the electrode and around the arc. Basically, the process consists of striking an arc between

an air-cooled station. If the torch is water-cooled, the station has the following additional equipment: Water Valve, Water Supply Hose, Water Drain

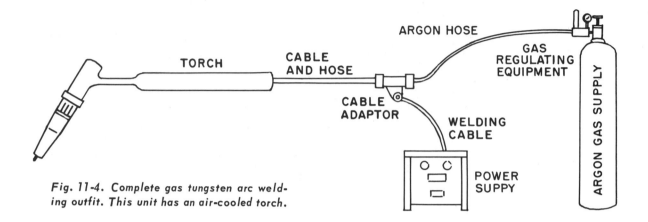

Fig. 11-4. Complete gas tungsten arc welding outfit. This unit has an air-cooled torch.

the base metal and a tungsten electrode in an atmosphere of shielding gas (helium, argon or some shielding gas mixture). CHAPTER 12 describes the various shielding gases and their uses. CHAPTER 28 describes chemical and physical properties of these gases. The filler metal, when needed, is applied in a manner somewhat similar to the method used in gas welding, Fig. 11-3. This process is often known as the TIG (Tungsten Inert Gas) process. However, the correct name is Gas Tungsten Arc Welding (GTAW). See Fig. 11-4 and Fig. 11-5.

11-4. SETTING UP GAS TUNGSTEN ARC WELDING STATION

The complete station consists of: Booth, Ventilating System, Bench, Arc Welding Machine, Shielding Gas Cylinder, Cylinder Regulator and Flow Meter, Shielding Gas Hose, Electrode Lead, Ground Lead, Special Tungsten Electrode Holder (air-cooled) and Gas Shutoff Valve.

The items just referred to are for

Hose, and Tungsten Electrode Holder (water-cooled).

In most cases the water hose, gas hose, and welding lead are all in one

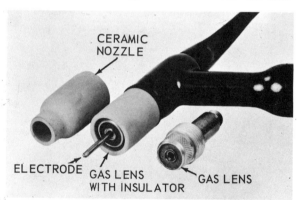

Fig. 11-5. Gas tungsten arc torch. This torch uses several fine-mesh screens (gas lenses) to produce a smoother inert gas flow (reduce turbulence). (Linde Div., Union Carbide Corp.)

jacket and form one lead. Common practice is to flow the outlet water along the electrode lead. This water cooling permits using a smaller diameter lead, which provides a lighter weight torch and greater flexibility, Fig. 11-6.

Fig. 11-7 illustrates a complete

shielding gas tungsten arc welding station in use. Note that the current rate (foot control) and the gas flow were described in previous chapters. It is very important that the operator's eyes and body be protected from the

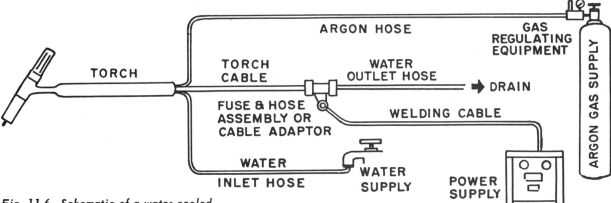

Fig. 11-6. Schematic of a water-cooled gas tungsten arc welding outfit.

(regulator and flow meter) are adjustable to the weld requirements.

Booth and ventilation requirements

intensity of the arc.

The arc welding machine may be either a motor-generator unit or a

Fig. 11-7. Complete gas tungsten arc welding station being used to weld a joint on nose section of aircraft auxiliary fuel tank. (Chemetron Corp.)

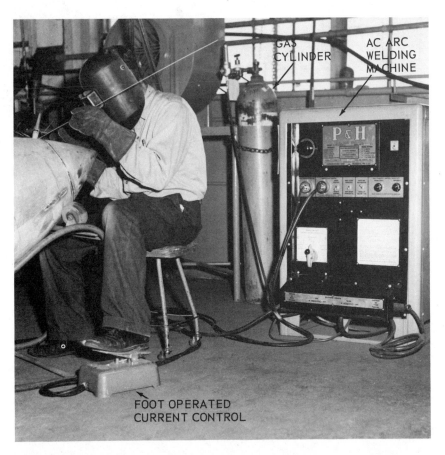

rectifier unit. Most of the welding machines use a high frequency superimposed current in the circuit to aid in starting the arc.

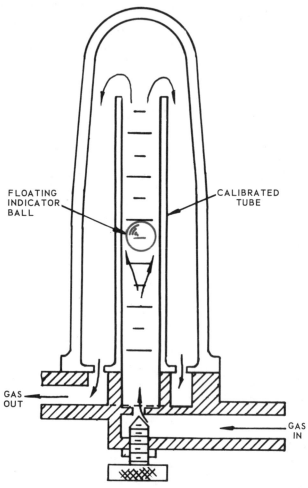

Fig. 11-8. Schematic of a gas flow meter used on gas arc welding outfits. When adjusting this gauge, the top of the ball indicates the gas flow through the meter in cubic feet per hour.

The gas cylinder is very similar to the oxygen cylinder in design, as is the regulator. The same handling and use precautions as described in CHAPTERS 1 and 2 are applicable for these cylinders. However, the consumption of the gas is measured differently. Instead of using psig measurements, the cubic feet per hour measurement is used (quantity of flow). A cross section of a flow meter is shown in Fig. 11-8. The scale on the calibrated tube measures the cfh of the gas flow.

The cables and hose with their connections are very similar to those described previously for other types of welding.

All connections, gas, water and electrical, must be clean and tight.

The only other major difference is the torch. A typical torch and cable assembly is shown in Fig. 11-9.

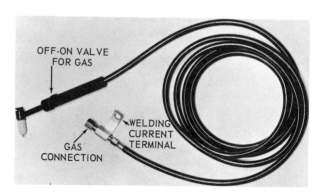

Fig. 11-9. A gas tungsten arc torch and cable. Off-on valve on torch is used to control gas flow. (Airco Welding Products, Div. of Airco, Inc.)

TUNGSTEN DIA.	DIRECT CURRENT			ALTERNATING CURRENT	
	SP HELIUM	RP HELIUM	SP ARGON	HELIUM	ARGON
.040	50		65	MIN.-30	MIN.-40
1/16	50-125	10-20	65-150	20-115	20-60
3/32	125-225	20-35	140-280	100-185	50-100
1/8	200-300	25-50	250-375	150-225	75-175
3/16	250-350	30-75	300-475	200-340	150-240
1/4	300-475	40-125	375	300-445	175-375

Fig. 11-10. Tungsten electrode sizes and the suggested current capacities for each, based on the diameter, the type of gas used, and the type current used.

THICKNESS, INCH	NO. PASSES	CURRENT, AMPERES	TUNGSTEN DIAMETER, INCH	ARGON VOLUME CU. FT./ HR.[1]	CUP OPENING DIAMETER, SIXTEENTHS	FILLER WIRE DIAMETER, INCH
0.051	1	70	1/16	12	6-7	1/8
0.064	1	80	3/32	14	7	1/8
0.081	1	90	3/32	14	7	5/32
0.101	1	120	1/8	15	7-8	5/32
0.125	1	140	1/8	15	7-8	5/32
0.187	1	160	1/8	15	7-8	3/16
0.250	2	220	3/16	20	8	3/16
0.375	2	300	1/4	24	8-10	1/4
0.500[2]	3-4	400	1/4	24	10	1/4
1.000[2]	10-14	500	5/16	30	10-12	5/16
2.000[2]	20-30	600	5/16	30	10-12	5/16

[1] Multiply reading on oxygen flowmeter by 1.5 to convert from liters per minute to cubic feet per hour of argon gas.

[2] Preheat to 400° F.

Fig. 11-11. Table of gas tungsten arc AC settings for welding aluminum when using argon gas.

11-5. STARTING GAS TUNGSTEN ARC WELDING STATION

Before starting a welding station, the station must be carefully checked. All electrical arc circuit connections and gas fittings must be clean and tight. The gas cylinder must be securely mounted to prevent injury to personnel, and to the cylinder.

Adjustments for tungsten electrode amperage, when welding mild or stainless steel, are shown in Fig. 11-10. When welding aluminum, alternating current is recommended and the station adjustments are shown in Fig. 11-11.

Direct current straight polarity (DCSP) is most often used. DCRP can be used but the current flow must be greatly reduced (reduce to approximately 1/5 to 1/8 the DCSP value). See Fig. 11-37.

The amperage for the metal thickness being welded is very similar to standard arc welding practice. The electrode cup size must be varied with the electrode size, and consequently the

current and the gas flow must be adjusted accordingly. Always follow the equipment manufacturer's recommendations. Fig. 11-12 lists some approximate cup inside diameters and amounts of gas flow, as they vary with tungsten electrode diameter.

TUNGSTEN ELECTRODE DIAMETER	NOZZLE SIZE INSIDE DIAMETER	HELIUM GAS FLOW CU. FT./HR.
.040	5/32-3/8	11
1/16	5/16-3/8	15
3/32	3/8-1/2	18
1/8	3/8-1/2	25
5/32	1/2-5/8	32
3/16	5/8	40

Fig. 11-12. Approximate nozzle (cup) inside diameters and gas flow values as related to tungsten electrode diameter.

To start the unit, use the following steps: First, be sure that the gas and water (if water cooled) are flowing or turned on. These fluids are controlled by either manual valves built into the electrode holder or into the torch hanger. Or, the gas flow may be controlled automatically through electric relays and solenoid valves.

Adjust the water flow and the gas flow. The water flow will vary between 12 to 23 gal. per hour. The important measurement is that the water temperature should rise only 10 deg. F. Gas flow rates shown in Fig. 11-13 are typical.

The tungsten electrode must be properly shaped to produce good re-

wise, with the old balled end inside the torch and a new spherical end on the arc end of the electrode, it is necessary to break the electrode to remove it from the collet. The electrode must be straight, and if ground, the point must be concentric (in the center) or the gas flow will be off center from the arc.

To ball the electrode for AC welding, use DCRP and strike an arc for a

KIND	METAL THICKNESS	GAS FLOW CU. FT./HR. ARGON	HELIUM
Steel	.035 - 3/32	8 - 10	20 - 30
Cast Iron	1/4	16	40
Stainless Steel	1/16 - 1/8	11	30
Stainless Steel	3/16 - 1/4	13	32
Copper	1/16 - 1/4	15	38
Magnesium	1/16 - 1/8	10	25

Fig. 11-13. A table of suggested gas flow rates for different metals.

sults. With AC the electrode should have a spherical end. With DCSP the electrode should be ground to a point as shown in Fig. 11-14. Caution: When inserting a used AC tungsten electrode in a collet, be sure the spherical end is at the arc end of the collet. Other-

moment on a piece of carbon or a piece of copper. The ball diameter should be only very slightly larger than the original diameter of the tungsten electrode.

11-6. STRIKING THE ARC

When using alternating current, the unit will need superimposed high frequency. Hold the torch horizontally over the metal starting block or work. Then, very quickly tilt or swing the torch to the upright position with the electrode reaching a point 1/8 in. above the metal. With alternating current and with the superimposed high frequency, the arc will jump this gap.

If DC is used without superimposed high frequency, the tungsten must be touched to the metal and then withdrawn. It is best to touch the tungsten electrode to a tungsten block, or to a used tungsten electrode, as this action will help keep the tungsten electrode clean.

The welding machines designed for

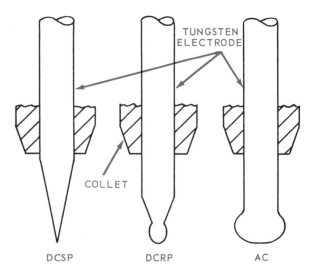

Fig. 11-14. The correct shape of tungsten electrodes for DCSP, DCRP and AC gas tungsten arc welding.

gas tungsten arc welding, whether AC or DC or both, incorporate a built-in high frequency arc starting circuit. This high frequency current may be connected in such a way that it will operate continuously (as in AC welding), or, an electrical relay may be used to disconnect the high frequency circuit, once the arc is started (as in DC welding).

The tungsten electrode torch should be warmed by practicing on a scrap piece of metal, before starting the weld. This practice will warm both the electrode and the nozzle and will give good starting results on the job. Fig. 11-14A shows the electrode holder and the filler wire, welding rod, in the proper operating position for flat position welding.

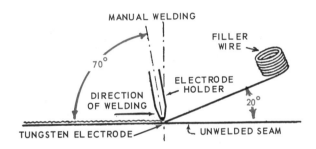

Fig. 11-14A. The correct positions of the tungsten electrode and the filler wire when manual gas tungsten arc welding.

Tungsten electrode should extend out of cup a length equal to inside diameter of cup-about 1/8 in. For fillet welds it should extend a little further. Too long an electrode extension will reduce shielding effect.

11-7. GAS TUNGSTEN ARC WELDING INSTRUCTIONS - GTAW (TIG)

During all arc welding processes, when DCRP and when AC is used, the arc has a cleaning effect on the base metal surface, and therefore these electrical circuits are used especially with aluminum and stainless steel. This action is called the cathodic cleaning effect (work negative polarity all the time or part of the time). Argon gas is recommended if this cleaning effect is desired.

However, the metal must still be cleaned before welding. Aluminum is best cleaned with clean stainless steel wire brushes, wiped, and then chemically cleaned with clean acetone (Caution: Acetone is extremely flammable!) within one hour before welding.

For high quality welding, most steel alloys should be at a temperature of at least 60 deg. F. before welding. Aluminum should be heated to 120 deg. F. before welding (to remove moisture). Between welding passes, the metal should be allowed to cool to approximately 350 deg. F. for steel and approximately 300 deg. F. for aluminum.

Back up shielding gas should be used to protect the root or penetration side of a weld. Before starting to weld, purge the back up space five or six times with spurts of gas, then feed the gas slowly and continuously as weld progresses. Be sure there is a relief opening for the back up gas to prevent pressure buildup if welding a closed structure.

Sometimes the start of a weld will have small cracks. These can be detected by using either a magnifying glass or a dye penetrant. These cracks must be removed by grinding, chiseling, or buffing, and then rewelding. Overheating may also cause cracks, and these, too, must be removed and repaired.

The crater formed at the end of a weld bead should also be carefully inspected. If it has cracks or inclusions, these should be removed and the crater welded again.

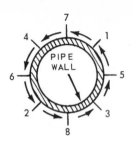

Fig. 11-15. *The recommended sequence of passes when welding pipe. The numbers represent the starting point for each pass, in sequence.*

When welding pipe, for best results weld short passes on alternate sides of the pipe. On pipe less than 6 in. in diameter, each pass should be not over 2 in. in length. For pipe 6 in. or over in diameter, each pass may be 3 in. in length.

Fig. 11-15 shows the proper sequence of passes when welding pipe.

11-8. PREPARING METAL FOR GAS TUNGSTEN ARC WELDING

The preparation of the metal and the design of the joint for gas tungsten arc

Fig. 11-16. *Gas tungsten arc welding a sheet metal assembly. (Miller Electric Mfg. Co.)*

welding is very similar to the preparation and joint design for gas and shielded metal arc welding.

Standard 60 deg. grooves are used for most joints of more than 3/32 in. thickness, although excellent results have been obtained with no edge preparation. The joint should be backed, if possible, to exclude oxygen and other impurities, in addition to promoting

MATERIAL	ALTERNATING CURRENT WITH HIGH-FREQUENCY STABILIZATION	DIRECT CURRENT STRAIGHT POLARITY (DCSP)
Magnesium: up to 1/8 in. thick	X	NR
above 3/16 in. thick	X	NR
castings	X	NR
Aluminum: up to 3/32 in. thick	X	NR
over 3/32 in. thick	X	NR
castings	X	NR
Stainless steel	XX	X
Deoxidized copper	NR	X
Brass alloys	XX	X
Silicon copper	NR	X
Silver	XX	X
Low-carbon steel, 0.015 to 0.030 in.	XX	X
0.030 to 0.125 in.	NR	X
High-carbon steel, 0.015 to 0.035 in.	XX	X
0.030 in. and up	XX	X
Hastelloy alloys	XX	X
Silver cladding	X	NR
Hard-surfacing	X	X
Cast iron	XX	X

CODE: X - Excellent   XX - Good   NR - Not recommended

Fig. 11-17. *Table showing type of welding current recommended for arc welding various metals.*

better fusion through the metal. Other metals, carbon blocks, or a shielding gas, can be used for backing up a weld to keep out the atmosphere when welding such metals as magnesium, titanium, and zirconium.

Flange joints are the best to practice on until one becomes adept in handling the torch. After good welds are made in flange butt joints and in outside corner joints, it is advisable to start

definite speed and the filler metal welding wire, if needed, is fed automatically to the weld puddle. Fig. 11-18 shows the principles of an automatic operation. See Chapter 21.

11-9. GAS TUNGSTEN ARC WELDING PRACTICE

After the arc has been struck and the welding process started, use the small-

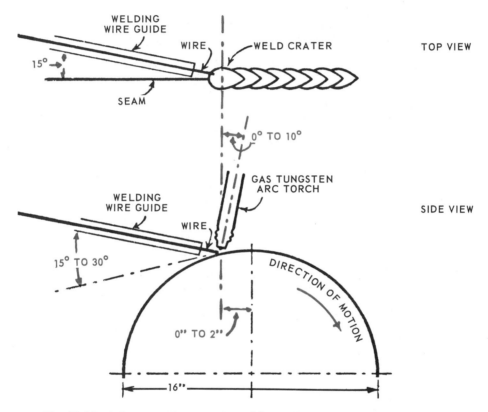

Fig. 11-18. Schematic of automatic welding using gas tungsten arc process.

practicing welding all the joints using filler metal welding rod as shown in Fig. 11-16.

The kind of metal being welded must be known. The type of electric current needed for different metals is shown in Fig. 11-17.

Gas tungsten arc welding can also be done automatically. Either the torch or the metal being welded is moved at a

est circular motion possible, and make a small puddle in the spot where the weld is to begin. The electrode holder should be held at an angle of 60 to 80 deg. with the work. This position is somewhat steeper than when welding with the oxyacetylene torch, and is used to protect the weld puddle with the shielding gas. The molten puddle may be made to move the same as with the oxyacetylene

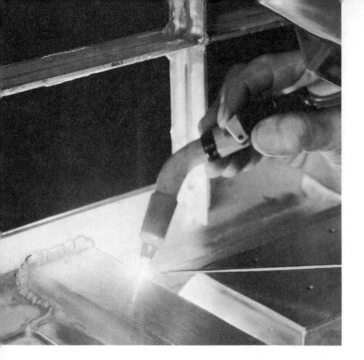

Fig. 11-19. Weld being made using a welding rod and gas tungsten arc. Note relative position of each. (Falstrom Co.)

torch. The motion, if one is used, must be much smaller than the puddle.

If a filler metal welding rod is being used, move the arc to the back edge of the puddle, insert the rod in the front edge of the puddle, and then bring the arc forward to move the puddle a ripple length along the direction of the weld. Hold the welding rod at a very flat angle (approx. 20 deg.) as shown in Fig. 11-19. To stop the arc, lift the electrode holder quickly until the arc is broken

or use the foot control to reduce the current flow. Always keep the gas flowing until the tungsten is cool, or the tungsten will become corroded and will also be consumed too rapidly. This gas flow will also protect the weld puddle until it has cooled sufficiently.

The shielding gas arc torch can be used to produce welds in the flat, horizontal, vertical, and overhead positions. Progress is normally downward when welding in a vertical position.

Welding with a tungsten electrode, Fig. 11-20, will be considerably improved if one remembers the following directions:

1. Keep the tungsten electrode clean and straight.

2. Keep end of electrode in proper condition:

The end of the tungsten electrode should always be kept clean and smooth.

The end of the AC electrode should be rounded. For DCSP, use an electrode ground to a point.

3. Be sure to use the correct size electrode. If the electrode is too small, the end of the tungsten will form into a molten ball larger than the electrode, and this ball may fall into the weld. If much too small, the electrode will melt

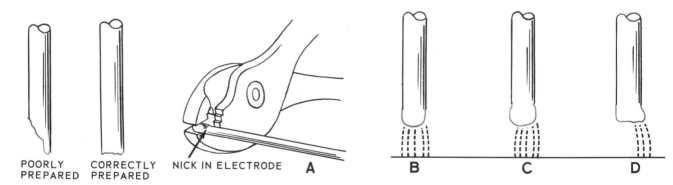

POORLY PREPARED   CORRECTLY PREPARED   NICK IN ELECTRODE   **A**   **B**   **C**   **D**

Fig. 11-20. Preparing and using tungsten electrodes for AC welding. A. Illustrates proper method of breaking tungsten electrode. Electrode is first nicked with grinding wheel and then broken. B. Shows a properly prepared electrode in use. Electrode at C requires more current. D. Shows a jagged or contaminated electrode.

back to the collet or chuck and fuse to the collet. If the electrode is too large, the arc will wander from one side of the electrode to the other.

4. A discolored tungsten electrode usually means that the electrode has been exposed to the air while still very hot.

5. Keep the amount that the electrode extends out of the gas cup to an absolute minimum, even if this practice interferes to some extent with vision of the weld puddle.

6. Remember that the gas connections must be tight or expensive leaks may result. Gas leaks may also permit air to get into the gas lines and cause contamination of the electrode and the weld.

7. Steel welding rods should not be copper coated, as the copper coating will cause spatter and may contaminate the tungsten electrode.

For beginners, it has been found that air-cooled torches are best. These torches can be easily installed. Simply

trode holders are available in capacities up to about 150 amps.

All metals and welding rods should be carefully cleaned prior to welding. The welding should be done as soon as possible after the metal has been cleaned.

Gas tungsten arc welding of aluminum, magnesium, stainless steel, titanium, etc. is described in CHAPTER 18, Special Welding Applications.

11-10. GAS METAL ARC WELDING - GMAW (MIG) - PRINCIPLES

The welding of metals in a shielded atmosphere is also accomplished by using consumable metal electrodes to both maintain the arc and to provide the filler metal. This process is often called Metal Inert Gas Welding (MIG). The metal electrode is fed through the electrode holder and into the arc at the same speed the electrode is melted and deposited in the weld, as shown in Fig. 11-21. A small adjustable speed

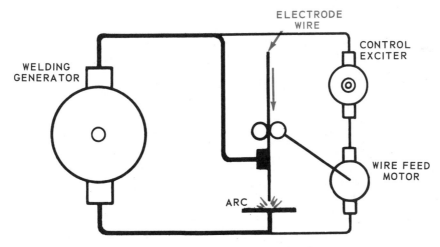

Fig. 11-21. Basic schematic wiring diagram of the gas metal arc process (MIG).

clamp the gas arc welding cable in the DC machine electrode holder, or use quick couplers. Air-cooled elec-

electric motor provides power to remove wire from a spool and feed the wire into the arc. The electrodes are usually supplied in the form of wire wound on special reels or spools. Fig. 11-22 illustrates a typical gas metal one or more of these processes. Such metals as mild steel, stainless steel, aluminum or bronze are often welded

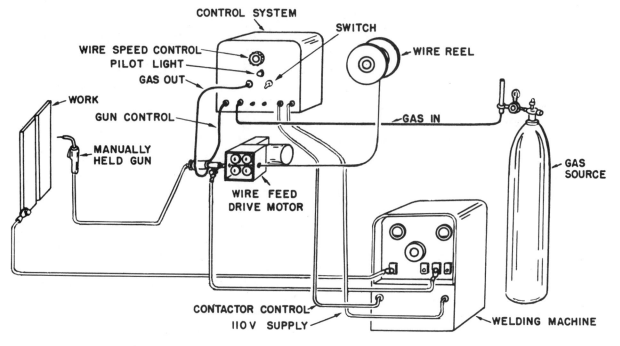

Fig. 11-22. Diagrammatic view of a complete gas metal arc outfit.

arc welding outfit. It should be noted that this electrode holder supplies current for the arc, the wire feed to the arc, and the water-cooled circuit if required. The operation of the gun is controlled by a trigger mechanism on the gun which, through relays, controls the arc current, the water flow, the shielding gas flow, and the electrode wire movement to the arc.

This type equipment is used for several kinds of arc welding:

1. Gas metal arc welding.
   A. Spray arc welding.
   B. Dip transfer arc welding.
2. Gas metal arc welding with magnetized flux.
3. Metal arc welding with flux-cored welding wire.
   A. With shielding gas.
   B. Without shielding gas.

Almost any metal can be welded by using these methods. Hard surfacing can also be performed using this process. See CHAPTER 20.

The three basic types of gas metal arc welding are shown in Fig. 11-23.

11-11. GAS METAL ARC WELDING METHODS

Welding with the Gas Metal Arc process can be done in three ways:

1. The Spray Arc Method.
2. The Short Circuiting Method (Dip-Transfer).
3. The Pulsed Arc Method.

Spray arc means that the metal transfers across the arc as a fine spray of several hundred minute droplets per second. This process is used mainly on normal metal joints and/or good fit-up joints, and for faster welding.

In the short circuit method, the metal transfers across the arc in larger drops

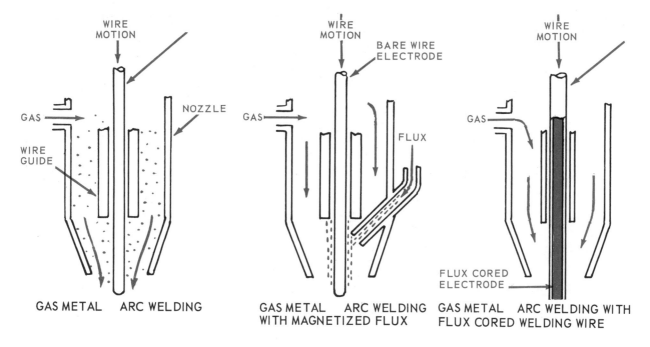

Fig. 11-23. *Three basic types of gas metal arc welding processes.*

at the rate of 100 drops per second or less. The drops are large enough to actually short circuit the electrical flow across the arc (there is no arc gap for a very brief moment of time). There are approximately 100 to 200 drops per second.

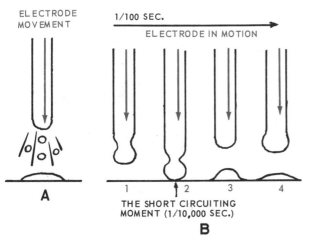

Fig. 11-24. *Two basic methods of metal transfer when using gas metal arc process. Part A shows "spray" of metal across the arc (spray transfer). Part B shows dip transfer method. The four steps in the sequence of the dip transfer arc are: 1. Electrode metal melting. 2. Short circuit interval. 3. Filler metal deposited on base metal. 4. Restarting of same cycle.*

This process is used for welding thin metal joints and for poor fit-up joints.

Fig. 11-24 illustrates two of these basic types of metal transfer.

11-12. SPRAY-ARC METHOD

The spray-arc method consists simply of using a heavy current flow for a certain size electrode wire. At a certain level of current (ampere) flow the electrode wire will divide into very small droplets of welding wire metal, and will travel forcibly across the arc. Fig. 11-25 lists the minimum current

ELECTRODE WIRE DIAMETER, INCHES	WELDING CURRENT AMPERES (DCRP)
0.030	150
0.035 (1/32)	175
0.045 (3/64)	200
0.0625 (1/16)	275
0.09375 (3/32)	350

Fig. 11-25. *Minimum current requirements for spray transfer arc welding when using gas metal-arc process. The shielding gas is usually argon with 5 percent oxygen.*

Gas Tungsten, Gas Metal Arc Welding 277

values needed to achieve spray-arc results with different size electrode wires.

Current values below these values are used for the dip-transfer method in the spray-arc method, but the current values are much lower, as shown in Fig. 11-26.

The dip-transfer method is used with several different gases or combi-

ELECTRODE WIRE DIAMETER, INCHES	APPROXIMATE WELDING CURRENT (DCRP)	
	MINIMUM	MAXIMUM
0.030	50	150
0.035	75	175
0.045	100	225

Fig. 11-26. Current values for dip transfer (short-arc) arc welding. These values are used when the shielding gas is 75 percent argon and 25 percent carbon dioxide.

(short-arc). When carbon dioxide gas is used, the change from spray-arc to dip transfer arc is not as definite. For example, with carbon dioxide gas, globular transfer results until the current flow is somewhat higher, and it requires about 500 amps. to obtain spray transfer with 1/16-in. diameter electrode wire (welding metal).

Several different shielding gases have been used in the spray-arc method:

Argon-carbon dioxide mixture.

Carbon dioxide.

Argon-oxygen mixture.

These gases influence the shape of the weld, the amount of current needed, the splatter, and the like. Carbon dioxide tends to promote spatter and it uses more current, but welding speeds are higher and penetration is deeper.

Argon tends to reduce spatter and enables a reduction in current, producing a neater appearance of the weld.

The standard arc welding machine (DCRP) can be used for spray transfer gas metal arc welding, because no current recovery is needed and the arc does not short even for very brief moments of time.

11-13. THE DIP-TRANSFER METHOD

The same electrode wires are used for the dip-transfer method as are used

nation of gases as practiced with the spray-transfer method.

The conventional arc welding machine is impractical for this method, as the short circuit current flow is too low to reestablish the arc at each short circuit moment. Special power supply units are used. These machines must be constant potential type, or constant potential type with slope control.

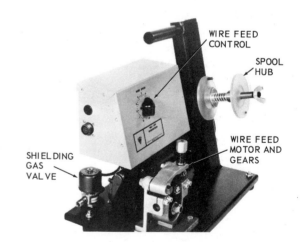

Fig. 11-27. Gas metal arc wire feed unit complete with a spool hub, wire feed control, wire drive motor and shielding gas control valve.
(Chemetron Corp.)

The speed of travel is greater with the spray-arc method. For example, when welding .200-in. thickness metal, the dip-transfer process will weld at approximately 10 in. per minute, while spray-arc process will weld approximately 22 in. per minute.

11-14. GAS METAL ARC WELDING STATION

The complete equipment for a gas metal arc welding station consists of the following:

Arc welding machine. Constant voltage (constant potential) machines are preferred. DC machines may be used depending on the application.

Booth.

Bench.

Ventilation system.

Combination gas, water, wire and cable control wire lead to the gun.

Shielding gas cylinder.

Cylinder regulator and flow meter.

Gas metal arc gun.

Ground lead and ground clamp.

Remote current control.

Electrode wire feed unit.

Reel of electrode wire.

Water control.

This welding system is semiautomatic since the consumable metal electrode is automatically fed to the electrode holder (torch) at a rate equal to its rate of consumption, and the electrode holder is manually operated, positioned and moved. The equipment requires a welding wire feed mechanism in addition to the water, gas and current flow controls. Fig. 11-27 shows the welding wire feed unit, the spool hub and the shielding gas control.

11-15. SETTING UP GAS METAL ARC WELDING STATION

A gas consumable electrode arc welding station should be assembled by an experienced person. It is very necessary that the local codes on Building and Safety and Industrial Hygiene be followed. The manufacturers supply installation instructions which should be carefully followed.

The booth should provide for good lighting, adequate ventilation, fireproofing, and electrical shockproofing. The welding power supply machine should be mounted in a dry place. It should be level, and the primary circuit cables and control should be installed according to the local electrical code. The welding circuit electrical leads should be kept to a minimum length, and should be placed where they cannot be injured or abused by personnel or by stock or equipment movement.

The shielding gas cylinders should be placed near the arc welding machine, and should be securely fastened to prevent accidental tipping or abuse.

11-16. STARTING GAS METAL ARC WELDING STATION

To start the unit, first check the station to be sure all parts are properly connected. Then proceed with the following steps.

1. Make visible inspection of hose and leads for wear, etc.

2. Inspect gas, water and electrical connections.

3. Examine booth, gas cylinder, and welding machine for cleanliness, correct position and mounting.

4. Turn ventilating system on, and check air flow.

5. Turn on main water valve.

6. Turn on shielding gas system (observe all safety precautions).

7. Adjust welding machine for wire size, type of joint, etc. (follow manufacturer's recommendations).

8. Test electrode wire feed mechanism by operating without striking arc.

　　A. Nozzle must be clean.

　　B. Electrode wire tube must be

clean. Tube should be cleaned and lubricated each time a new reel of wire is installed.

C. Electrode wire feed rollers must be clean. The rollers must be free of chips and dirt.

D. Gas flow and water flow should be checked at this time.

9. Cut excessive electrode wire extending out of nozzle away, using diagonal pliers. Correct extension should be about 1/2 to 3/4 in.

10. Position electrode holder with electrode touching metal where weld is to start.

11. Lower helmet (wear gloves, and proper clothing).

12. Press starter button on torch handle and proceed with the welding.

11-17. GAS METAL ARC WELDING PRACTICE

Using the proper welding wire, the proper settings, the correct gases or gas mixtures, this type of welding can be successfully done on most commercial metals. Such welding can be done in all positions.

Fig. 11-28 illustrates a gas metal arc weld being done in the flat position on a thin metal butt joint. Notice the gas shielding, the direction of travel,

and the position of the electrode in respect to the weld.

Fig. 11-29 shows a thicker gauge butt joint being welded in the flat posi-

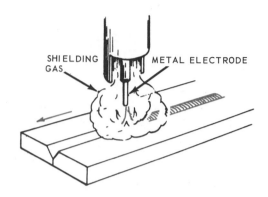

Fig. 11-29. A steel plate being welded in flat position using gas metal arc process. Notice V-groove preparation for butt joint.

tion. The metal edges are V-grooved to aid complete penetration. This weld too is being made in the flat position.

Horizontal, vertical, and overhead position welds can also be made. Fig. 11-30 shows a thin metal butt joint

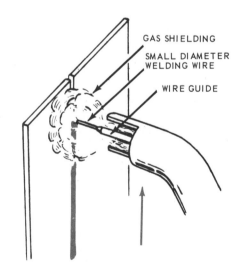

Fig. 11-30. Gas metal arc welding butt joint in vertical position.

being welded in the vertical position. Notice that this weld is being made by progressing upward.

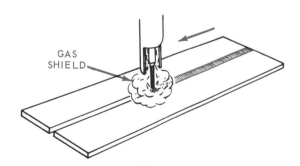

Fig. 11-28. Thin metal butt joint being welded in flat position using gas metal arc process. Notice gas shielding and electrode position in relation to base metal. (Hobart Bros.)

Carbon dioxide is increasing in popularity as a shielding gas. This process is shown in Fig. 11-31. Note how the carbon dioxide is reduced to carbon monoxide and atomic oxygen at the arc stream. The carbon monoxide immediately recombines with the oxygen to form carbon dioxide in the outer shielding gas envelope.

When using a consumable electrode, first practice on a scrap of metal similar to the metal to be welded.

A good technique to use to start the arc in the correct place when consumable electrodes are used, is to touch the electrode end to the proper place on the metal with the helmet up and with the welding current shut off. Next, lower the helmet over your face, then press the start button on the torch.

If the start button is pressed before properly positioning the electrode holder or torch, the wire will feed out of the holder until an arc contact is made. This excessive length of electrode wire will quickly melt away when the arc is struck, causing an unnecessary use of the wire, and it is detrimental to the appearance of the weld. A steady crackling or hissing sound of the arc is a good indicator of correct arc length. It is important that the wire size, the wire feed speed, and the current setting be properly related.

For the same size wire, if it is desirable to use more current, the wire feed speed should also be increased. When changing to a larger diameter wire, either the current must be increased or the wire feed speed must be reduced, or both.

These relationships are very important and the manufacturer's recommendations should be followed.

When consumable electrode welding mild steel of .060 thickness, it is recommended that about 130 amps. of DC reverse polarity be used with .035 dia. wire. This weld can be made at a speed of 26 in. per minute using an argon flow of 17 cu. ft. per hour.

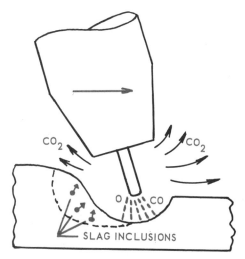

Fig. 11-31. Weld action with carbon dioxide used as the shielding gas.

To obtain high-quality steel welds, weld should be made on "killed" steel only. These steels are fully deoxidized as they are made. Rimmed steels cause porosity because there are no deoxidizing elements present in the metal. See CHAPTER 24 for definitions of these steels. Most of the electrode wires for mild steel welding have deoxidizing elements in the welding wire metal, such as manganese, silicon, aluminum, titanium, and zirconium.

11-18. GAS METAL SHORT ARC WELDING

The dip transfer method of gas metal arc welding is now further improved for certain applications by using an interrupted short arc circuit. Some welding machines have been developed to supply an interrupted or pulsed arc welding current. The interruption or pulse can be regulated to control the frequency of the dip transfer metal from

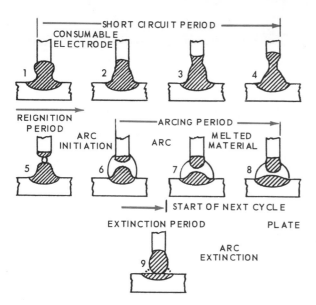

Fig. 11-31A. Diagrams cover the nine-step sequence of a gas metal arc welding process, using constant current flow. Steps 1 through 9 show the electrode behavior: Period 1–4. Short circuit. 5. Reignition. 6–8. Arcing. 9. Arc extinction.

the metal electrode. Fig. 11-31A shows a constant current flow. With pulsed arc, the welding current is shut off for a portion of the dip transfer period as a method of puddle and weld control. Fig. 11-31B illustrates the characteristics of the pulsed arc.

Constant voltage is used to provide the general current flow while a drooping voltage source is used to provide

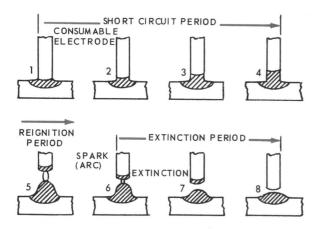

Fig. 11-31B. An eight-step sequence of a gas metal arc welding process, using pulse arc: Period 1–4. Short circuit. 5. Reignition and arc. 6–8. Extinction.

the pulse current (silicon rectifiers are used). See Paragraph 12-14. This pulse current is variable from 0 to about 2000 cycles per second (cps). The pulse current duration is adjustable from .005 sec. to a steady DC.

The arc penetration can be controlled independently of the melting rate of the electrode. The penetration can be increased or decreased by proper current setting and pulse duration.

11-19. GAS TUNGSTEN PULSE ARC WELDING

A welding circuit using a pulsed electrical flow is available for use with a tungsten electrode to both control the cleaning action of the metal surface and also to control the penetration. With this circuit, welds may be made on clean metal with good penetration and narrow beads.

To better understand the control of the arc core, one should know that the core of the arc is a plasma. This plasma concentration is affected by the current flow. Different current patterns can be obtained by changing the ratio of the DCSP and DCRP during each cycle. Fig. 11-31C shows one pattern used. The DCSP or the DCRP each vary from 0 to 100 percent, or the machine can operate as an AC unit with 50 percent SP and 50 percent RP.

The plasma core shape is also affected by the shielding gas used, by the mixture of gases used, and by the volume of gas flow. There is also a jet action of the plasma core which increases as the current (amperage) intensity increases. Therefore, a constriction of this core by shaping the electrode will influence the jet action effect. (An extreme case of this action is the plasma arc torch. See Chapter 17.) The length of the arc also controls the arc core diameter.

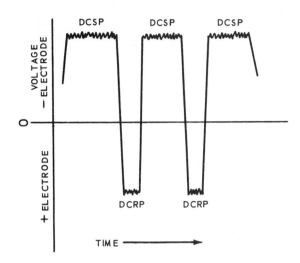

By properly adjusting all these variables, the depth of the penetration and the shape of the weld bead can be controlled. Fig. 11-31D shows a gas tungsten arc welding machine with filler metal feed.

11-20. GAS TUNGSTEN ARC WELDING TECHNIQUES

One must observe the following basic operations if quality welds are to be obtained: Striking the arc with ACHF: Hold the torch horizontal. Revolve it into a vertical position with the electrode approximately 1/8 in. above the metal -- the HF will help the arc jump the gap. Striking the arc with DC: Touch the electrode to the work (or to an adjacent carbon or copper surface).

When starting to weld, first move the torch in a small circular motion until a fluid puddle is formed. Then tilt the top of the torch 15 deg. away from the direction of travel. No torch motion is necessary except along the weld seam.

Avoid touching the metal parts of a water-cooled torch to the base metal or bench. If this is done, the water jacket may be burned through and a leak developed.

Add filler rod by moving the torch (arc) to the rear of the puddle and insert the correct size clean filler rod in the front of the puddle. After some metal (the amount melted determines the crown of the weld) has melted from the filler rod, remove the rod just a short distance, but keep it inside the shielding gas envelope, and move the torch to the front or leading edge of the puddle. The frequency of repeating this sequence determines the number of ripples and size of ripples in the weld.

Lap joint welds should be made by keeping the electrode as near the edge of the top piece as possible and still melting some of the lower piece. Do not move the torch ahead of the puddle at any time.

The correct arc length of 1/16 to 1/8 in. is very important. If the electrode is too close to the base metal causing a short arc, the weld will become contaminated with tungsten. If the electrode is too far from the base metal, it will cause a wandering arc, and the gas shielding will be reduced.

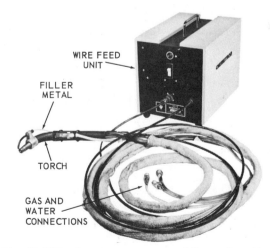

Fig. 11-31D. Gas tungsten arc welding automatic wire feeding machine. Torch is water-cooled. Wire feed system is simple to operate. (Chemetron)

With DCRP, one must use larger size tungsten electrodes and lower current settings, as shown in Fig. 11-31E.

WELDING STEEL IN FLAT POSITION

TUNGSTEN ELECTRODE DIAMETER	AMPERES		
	DCSP	DCRP *	AC
1/16	70–120	5–10	60–90
3/32	200–275	15–20	115–160
1/8	275–375	20–25	180–210
5/32	375–450	25–30	210–240
3/16	450–525	30–35	240–270

* CAUTION --
be sure to reduce the amperage when using DCRP.

Fig. 11-31E. A table showing the current carrying capacity of tungsten electrodes when using the three types of welding current.

11-21. GAS METAL ARC WELDING USING FLUXED WIRE

Flux has long been used as an agent to improve soldering, brazing and welding. It is used also to improve the flame and arc cutting processes. It is natural that it is used in the gas metal arc welding process and in some automatic and semiautomatic metal arc welding processes.

Flux may be used in three ways:

1. As a suspension fine powder mix in the shielding gas.

2. As a flux powder coating on the metal electrode as it leaves the torch nozzle. Magnetism is used to hold the flux to the wire. The flux contains iron particles which are attracted to the magnetized metal electrode.

3. As a flux inside a tubular metal electrode.

Fig. 11-32 illustrates the fundamentals of the magnetic flux process.

The flux-cored process uses 1/8-in. diameter and larger electrodes. The arc and arc behavior is very similar to the stick electrode arc welding process as explained in CHAPTERS 5 and 6.

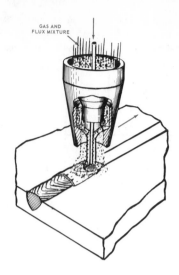

Fig. 11-32. Principle of combined magnetized flux and gas metal arc process.

This flux-cored process may be used with carbon dioxide gas, gas mixtures, or used without any shielding gas. The process is shown in Fig. 11-33.

The flux-cored process is growing in popularity. It is now used for more than 20 percent of arc welding. Some flux-cored electrode welding still uses CO_2 shielding, but the use of flux-cored wire alone is increasing. In many cases, the flux-cored wire alone produces welds equal to or better than the original metal, and its use eliminates the need for the gas shield equipment and cost of the gas.

11-22. GAS CARBON ARC WELDING PROCESS

Gas arc welding may also be done using a carbon electrode with the arc surrounded by a shielding gas. A DC straight polarity current supply is used. The gas carbon arc process is excellent for teaching purposes as the arc is easy to start. It is a good process to use when welding thin metals, for the same reason. Fig. 11-34 shows how the carbon electrode should be prepared for welding. The operating cur-

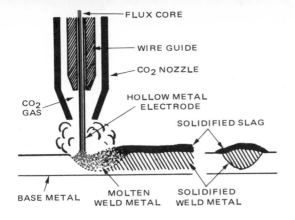

Fig. 11-33. Basic design of flux-cored wire and inert gas metal arc process. Note that the electrode wire is hollow, with a flux core.

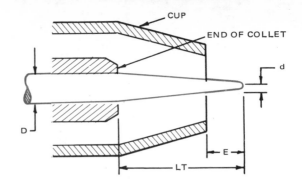

Fig. 11-34. Proper method of preparing end of carbon electrode for use in inert gas arc welding. D. Carbon electrode diameter. d. Operating diameter (same as tungsten diameter). LT. Length of taper. E. Extension of the carbon electrode beyond nozzle.

rent (DCSP) and carbon electrode diameter recommendations are shown in the table, Fig. 11-35.

11-23. GAS SPOT WELDING

A special application of the gas arc welding system is the shielding gas spot weld. By correct timing and using a special torch, this method is used to melt one member of an assembly and fuse it to the underneath member in a "spot" as shown in Fig. 11-36. This process eliminates the need to reach both sides of an assembly with the welding device.

All of the usual consumable electrode equipment is used. In addition, a timer is used to control the time of current flow, size of the spot, and the amount of metal deposited. The timers usually control the spot (or button) time between 3 cycles (1/20 of a second) for approximately 1/16-in. thickness metal to 360 cycles (6 seconds) for approximately 3/16-in. thickness metal.

Some GTAW systems are used for this type of spot welding. A shield minimizes the need for a face helmet and colored lens. A special application of this type of torch and process is the cutting or piercing of holes in a thin metal (up to 1/8-in. thickness). The arc is started, and after a set time inter-

val a high-pressure shielding gas is pressed against the heated and molten metal. This high-pressure gas will blow or force the molten metal through the plate leaving a hole which is clean, accurate in size and shape.

CARBON DIAMETER	AMPERAGE (DCSP)
1/16	50 - 125
3/32	125 - 225
1/8	200 - 300
3/16	250 - 350
1/4	300 - 475

Fig. 11-35. Table of current settings recommended for various diameter carbon electrodes when used in gas arc welding.

11-24. REVIEW OF SAFETY IN GAS ARC WELDING

The operator should use a dark shade lens (#12 to #14) when gas arc welding because the arc is more exposed, and is therefore very intense. The operator must also be very careful to protect the skin from the arc rays.

The machine must be kept in good condition, for best results.

Repair water leaks immediately as wet equipment and/or a wet floor increases the chance of electrical shock. The primary circuit of the machine, and the high-frequency circuit, should

Fig. 11-36. Gas spot welding torch being used to fabricate pulleys. Concealed arc minimizes need for special protection clothing.
(A. O. Smith Corp.)

The influence of the ultraviolet rays from the arc tends to change oxygen into ozone. The amount of ozone is very small but good ventilation is still very much recommended. Ventilation is also needed to minimize the small amounts of oxides of nitrogen, carbon dioxide, carbon monoxide, and decomposed gases such as trichloromethylene.

Because the gas shielded arc is not enclosed in opaque gases, the ultraviolet rays are approximately twice as strong as with the flux shielded arc. One should wear #2 flash goggles, with side shields, under a helmet fitted with #12 lens. All skin surfaces must be covered. Light-colored clothing on the upper part of the body will allow too much light reflection behind the helmet. Cotton clothing will rapidly deteriorate if exposed to the arc rays. Leather and/or wool clothing is recommended.

be disconnected while working on the equipment.

The gas arc welding processes require that the operator and those in the vicinity follow all of the usual arc welding safety practices. In addition there are the typical welding precautions which must be observed.

11-25. TEST YOUR KNOWLEDGE

1. Of what material is the nonconsumable gas arc welding electrode usually made?

2. Why is the shielding gas kept flowing after the welding arc is broken?

3. Should one use lighter or darker lens when performing gas arc welding as compared to coated steel electrode arc welding?

4. List four shielding gases used in arc welding processes.

5. How far should the 1/16-in. dia. tungsten electrode extend past the end of the gas cup?

6. Is flux necessary when gas arc welding aluminum? Why?

7. Is it possible to automatically weld with gas arc welding?

8. What is the popular welding rod angle when using welding rod while gas arc welding?

9. What is gas arc spot welding?

10. How is the gas arc welding electrode holder or torch cooled?

11. Name a gas mixture and list some of the properties of the gases.

12. What is a flow meter?

13. How is a consumable metal electrode fed to the arc?

14. Why is it recommended that a tungsten electrode arc be started on a scrap tungsten surface?

15. What is the appearance of the arc end of a correctly used tungsten electrode?

16. What would happen if the tungsten electrode were bent off center?

17. What is the disadvantage of wearing cotton clothing when using gas shielded arc welding?

18. Can consumable metal electrodes be used when gas arc spot welding?

19. What holds a tungsten electrode in place in the torch?

20. Of what materials are the gas nozzles made?

Chapter 12
GAS TUNGSTEN AND GAS METAL ARC WELDING EQUIPMENT AND SUPPLIES

Fundamentally the gas arc welding process consists of performing a weld in which the arc is surrounded by a shielding gas such as helium, argon, or carbon dioxide. This gas keeps atmospheric air away from the weld, and prevents elements in the air from combining with molten metal during the welding process.

This chapter discusses the various items of equipment and supplies used in gas arc welding.

12-1. GAS ARC WELDING STATION EQUIPMENT

Complete gas arc welding station consists of:

1. Shielding gas supply cylinder.

This gas may be helium, argon, carbon dioxide, or a mixture of gases.

2. Pressure regulator.
3. Flowmeter.
4. Hose.
5. Arc welding machine either DC or AC usually with high frequency stabilization.
6. Cable (electrode holder lead). Combination cable, gas, electrical and water (if a water-cooled electrode holder is used).
7. Electrode holder (torch).
8. Ground lead.

This equipment, together with a booth ventilation, benches, lights and the like, makes up a complete station.

Fig. 12-1 illustrates a typical manually operated gas arc welding station

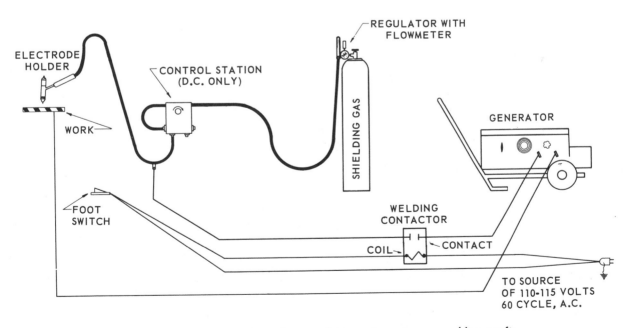

Fig. 12-1. Diagrammatic view of a complete gas tungsten arc welding outfit. (Airco Welding Products, Div. of Airco, Inc.)

suitable for gas tungsten arc welding - GTAW (TIG). The electrode holder is air-cooled, so it has no water valves or water hose. This station would be suitable for either helium or argon gas tungsten arc welding.

12-2. GASES FOR GAS ARC WELDING

Three common shielding gases are used in the gas arc welding process.

The most popular shielding gases used are:

1. Helium.
2. Argon.
3. Carbon dioxide.
4. Mixtures of the above gases in a variety of combinations.
5. Mixtures of any of the above gases with nitrogen and/or oxygen.

Both helium and argon are monatomic, meaning that the molecule and the atom are the same, whereas most gases have molecules that consist of two or more atoms. Neon, krypton, xenon, and radon are also monatomic, but they are more expensive at the present time, and more difficult to use.

Fig. 12-2 shows different weld forms

This gas was used extensively during World War II to weld aluminum and aircraft parts.

Argon gas is also used in the gas tungsten arc welding process. It is available from manufacturers of oxygen in most cases. Argon is heavier and diffuses more slowly than helium; therefore about one third as much argon is needed by volume as compared to helium. The equipment required for both gases is very similar.

Argon is less expensive than helium per cubic foot, but helium is much lighter than argon and more of the gas is needed because it floats away rapidly when heated by the arc. Helium enables deep penetration welds and permits fast welding.

Argon, as mentioned before, gives a stable arc. It is less sensitive to arc length than helium, and provides a smoother arc. It is good as a starting arc-gas and for use where the arc length may vary during the welding operation.

Carbon dioxide is the least expensive of the gases used. It is quite sensitive to arc length.

Carbon dioxide was first used in the

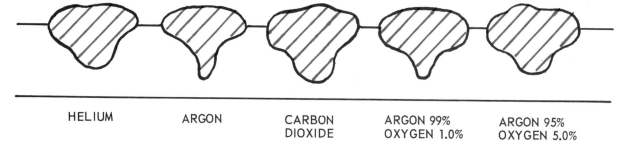

| HELIUM | ARGON | CARBON DIOXIDE | ARGON 99% OXYGEN 1.0% | ARGON 95% OXYGEN 5.0% |

Fig. 12-2. End views of gas arc beads made on steel plate. These drawings show the effects of various shielding gases and shielding gas mixtures on the cross sectional shape of the weld bead.

obtained as influenced by the various shielding gases.

Helium was the first gas to be used and it is satisfactory; however, the gas is costly and it has some arc resistance.

fully automatic machines only. It is now used in both the automatic and the semi-automatic gas metal arc processes.

Carbon dioxide provides good shielding, however a constant arc length must be maintained; consequently, carbon dioxide is most used where the arc length is automatically controlled, as in semiautomatic and automatic welding, for example.

Nitrogen is used as mixing gas for other shielding gases, when welding nonferrous metals. Mixtures of these gases give good results. See Fig. 12-3 for a table of suggested gas mixtures.

Gas mixtures such as 20 percent carbon dioxide and 80 percent argon may be used when welding low alloy steels. A 10 percent oxygen, 90 percent carbon dioxide may be successfully used on steel plates using the gas metal arc process. A mixture of 20 percent nitrogen and 80 percent argon may be used on stainless steel. Also a 5 percent nitrogen, 95 percent argon has been used on stainless steel. These mixtures, and others, are used only after considerable experimenting in reference to the metals used and the type of joint desired.

Hydrogen is sometimes mixed with argon (10 to 15 percent hydrogen) and is used when welding stainless steel. Hydrogen and nitrogen mixtures have also been used as back-up gases when welding stainless steel. Back-up gases are used to keep the atmospheric gases away from the penetration side of a weld.

Information on the chemical and physical properties of these gases will be found in CHAPTER 28.

Names such as Heliarc, Heliweld, and Argonarc have sometimes been used to identify gas arc welding processes.

Fluxed electrodes and other fluxing devices have been developed which, when used with carbon dioxide in gas arc welding, provide a stable arc and the arc length becomes less critical. Many of the devices are now successfully used in manual applications.

The machines are usually called semiautomatic, since the rate of flow of electrode wire, shielding gas and arc current are automatically controlled. The operator controls only the movement of the electrode holder (gun).

12-3. HELIUM

Helium has a fast voltage change as the arc length changes, and this voltage change is used to operate electronic controls. Therefore, helium works well on automatic gas tungsten arc machines. It requires somewhat higher arc voltages than argon (approximately 40 percent more). Because of its higher resistance a more intense heat is generated in the arc. This intense heat produces good penetration and fast welding.

Helium is a by-product of the natural gas industry. Its production and sale is controlled by the U. S. Bureau of Mines. It is found in substantial quantities in gas fields. For successful welding, helium must be pure (99.99 percent) and must be very dry.

Helium is obtainable in steel cylinders of the same style and size as oxygen cylinders. These cylinders must conform to ICC regulations. It is compressed to a pressure of more than 2000 psig at 70 deg. F. Because it is the second lightest gas, only hydrogen is lighter, it has a tendency to rise from the weld very rapidly and for this reason, more cubic feet per hour are needed to maintain an inert atmosphere around the arc.

12-4. ARGON

Argon is one of the gases in the mixture of gases that make up the atmospheric air. It has been used for many

Argon is one of the most popular gases for manual shielding gas arc welding. It offers less resistance to the arc current than helium. This condition permits small changes in the arc length without breaking the arc. The less in-tense arc temperature is also advantageous for welding thin metal and when welding unlike metals. Fig. 12-4 lists various metals and welding currents to be used when argon is the shielding gas.

Argon is contained in standard cylinders. These cylinders must conform to

METALS	PERCENTAGES IN GAS MIXTURES						METHOD		POLARITY
	ARGON	HELIUM	CO_2	O_2	H_2	N_2	TIG (GTAW)	MIG (GMAW)	
Aluminum Alloys	100							*	DCRP
	100						*		ACHF
		100					*		DCSP
	25	75						*	
Aluminum Bronze	100						*	*	DCSP
Copper	25	75					*	*	DCSP
	100							*	DCSP
		100					*		
	95					5			
Magnesium	100							*	DCRP
	100						*		ACHF
Nickel	100						*	*	
	20	80							
		100					* A		DCSP
Silicon Bronze	100						*	*	DCSP
			100					*	ACHF
Steel mild	100						*		DCSP
	75		25					*	DCRP
		100					* A		
	100						*		
Low alloy	98			2				*	
	97			3				*	
	95			5				*	
	80				20				
	80		20					*	
Stainless	99			1				*	
	95			5				*	
	80				20		*		DCSP
	100						*		DCSP
		100					* A		
Titanium	100							*	
	100						*		
		100					* A		
Dissimilar Metals	100						*		
Back-up Gas					5	80			

Fig. 12-3. Table of suggested shielding gases and gas mixtures with their recommended applications.
A. Indicates automatic application.

ICC (Interstate Commerce Commission) specifications. The cylinders are very similar to those containing oxygen. Argon should be at least 99.8 percent pure (the rest is nitrogen) and the gas should be absolutely dry.

The three common sizes of cylinders are:

AS - 78 cu. ft.
S - 150 cu. ft.
T - 330 cu. ft.

Large consumers (over 6000 cu. ft. per month) may obtain argon in its liquid form in 2900 cu. ft. cylinders.

Argon in liquid form is stored in thermos bottle-like cylinders under a pressure of 235 psig; however, the liquid argon at this pressure will slowly evaporate at normal ambient temperatures. Argon evaporates very slowly to maintain the 235 psig pressure. If argon is being used continuously from the cylinder, the amount drawn off will consume the vapors which evaporate keeping the cylinder pressure at the proper level. In the event the vaporized argon is not used as rapidly as it boils from the liquid, a safety valve allows it to escape and a constant pressure of 235 psig is maintained on the liquid argon. The cylinders hold approximately 260 lbs. of liquid argon (or equivalent to 2900 cu. ft. of gaseous argon).

Argon is obtained as a by-product from the liquefaction of air during the production of oxygen. Argon is more expensive per cubic foot than most other gases used in the shielding gas arc welding, but the higher cost is somewhat offset by the fact that less of the gas is needed for most welding applications. Its main advantage is its ability to carry a stable arc, and at the same time exclude the atmosphere.

12-5. CARBON DIOXIDE (CO$_2$)

Carbon dioxide is much used in the gas metal arc welding of mild steel. An advantage of CO_2 is its low cost - - about one tenth that of argon gas.

Unlike the monatomic gases, the CO_2 reduces to carbon monoxide and oxygen in the arc. However, the gases return to CO_2 as they cool. The oxides and gas inclusions leave the weld metal before the metal solidifies. A higher current (about 25 percent) is used with CO_2 causing more agitation of the weld puddle, making it easier for the entrapped gases to rise to the surface of the metal, thus reducing weld porosity.

Direct current, reverse polarity is generally used.

Because some carbon monoxide is liberated during this process, the welding station should be thoroughly ventilated. This toxic gas should not be allowed to collect near the operator.

There is also some ozone generated. This is an additional reason for providing good ventilation. It is specially important to keep these weld gases from reaching the operator's face.

The carbon dioxide used must be moisture free or the hydrogen generated will cause weld porosity and brittleness.

Carbon dioxide is furnished in liquid form in 50 lb. cylinders. These cylinders are approximately 9 in. dia. 51 in. high and weigh 105 lbs. when empty. Each pound of liquid CO_2 will furnish 8.7 cu. ft. of gas (equals 435 cu. ft. per cylinder). It must boil (change to a gas) as it is being used. This action is dependent on the room temperature; for example: one cylinder can only furnish about 35 cfh when the cylinder is in a 70 deg. F. room. Sometimes two or more cylinders must be connected in parallel to furnish enough gas. The pressure in the cylinder when liquid is present is 835 psig at 70 deg. F. As the liquid boils it cools, and as the gas passes through the regulator, the expansion of the gas causes further cooling. If moisture is present, it may condense and freeze in the regulator, causing blocking of the gas passage. Excessive moisture may also be indicated by erratic flowmeter operation. It is recommended that CO_2 with a -20 deg. F. dew point or lower be used. Since CO_2 has a high arc resistance, and because even very small arc length changes produce a wild arc and splatter, a very short and constant arc length must be maintained. It is recommended that constant voltage arc (CV) or rising arc voltage (RV) machines be used with CO_2.

Carbon dioxide is also obtainable in gas form and in solid form (dry ice); however, the cylinder liquid form is chiefly used for gas arc welding purposes.

Because carbon dioxide is 50 percent heavier than air, its ability to shield the arc is quite satisfactory.

Due to its rather high electrical resistance and therefore rather critical arc length, CO_2 gas metal-arc welding lends itself particularly well to automatic machine welding; however, many successful manual applications are also in regular use.

12-6. GAS MIXTURES

A very successful gas mixture for nonferrous welding is a mixture of argon and hydrogen. This mixture produces excellent results on nickel bearing metals such as stainless steel, monel and similar alloys. It is recommended for automatic welding. Since the process produces an arc of great intensity, high weld speeds are made possible.

The addition of small amounts of oxygen (1 to 5 percent) to argon, helps produce arc stability when using a reverse polarity DC welding circuit.

This mixture is not used in gas tungsten arc welding because the oxygen will damage the tungsten electrode. The resultant welds have less porosity when this mixture is used. The mixture has helped make the metal arc process suitable for welding mild steels.

Argon and helium have been used in various mixtures. The mixtures tend to give the weld the advantages of each gas. For example: if, for a given application, the helium arc is too hot and the argon arc is too cool, a mixture of the two will produce the correct balance.

These gases are easily mixed in any proportion desired. Ratios of 50 - 50 percent of each to 14 percent of one and 86 percent of the other have been used. A mixture of 75 percent helium and 25 percent argon is a good all-purpose mixture.

Stainless steel has been successfully welded using argon gas around the electrode and CO_2 as an outside shield. This combination is sometimes called the dual gas shielded method. The amount of argon is reduced by about two thirds, using this method.

Considerable research has been done with different mixtures of gases and, as a result, there are now several mixtures, each recommended for one or more welding applications. (See Fig. 12-3 for a table of the commonly used shielding gases and gas mixtures along with their applications.)

It should be noted that some of the recommended gas mixtures contain small amounts of oxygen. A small amount of oxygen is desirable in gas mixtures used for gas metal arc welding - GMAW (MIG). The metal oxide formed at the tip of the molten metal electrode makes the molten metal more fluid and breaks the surface tension of the molten globules.

12-7. SHIELDING GAS EQUIPMENT

Shielding gases must be safely stored, safely reduced in pressure, the volume must be accurately controlled, and the gas delivered in pure dry condition to the torch nozzle. The gas is usually obtained in cylinders except for very large consumers.

The complete gas arc welding outfit therefore has: gas cylinders, regulators, flowmeters, on-and-off valves, accessories, power sources, torches, torch nozzle and electrodes.

12-8. SHIELDING GAS CYLINDERS

The gases described in the previous paragraphs can be obtained in cylinders of various sizes. These cylinders are similar to the oxygen cylinders described in CHAPTER 2. They are manufactured to Interstate Commerce Commission (ICC) specifications.

Some gases are stored in cylinders as a gas; some are stored as liquids. The quantity of gas in the cylinder is determined:

1. By the high-pressure gauge or volume scale if stored as a gas.

2. By weight if stored as a liquid.

The gases - with the exception of hydrogen - are not flammable and therefore combustion and/or support of combustion is not a problem. Still, the high-pressures in the full cylinders make it necessary to handle the cylinders with care.

The cap should be securely threaded over the cylinder valve whenever the

MATERIAL	POWER SOURCE	WELD CHARACTERISTICS
Aluminum, any thickness	ACHF *	Good arc starting, clean weld, low gas consumption.
Aluminum Bronze	DCSP	Mainly for surfacing.
Magnesium 1/16 and over	ACHF *	Clean weld, fluid puddle, low gas consumption.
Mild Steel 0-1/8	DCSP	Clean weld, ease of manipulation. Good puddle control flat position.
Low Alloy Steel	DCSP	Same as mild steel.
Stainless Steel to 14 ga.	DCSP	Controlled penetration on thin material.
Titanium, thin ga.	DCSP ACHF	Clean weld, good metal transfer.
Nickel, Copper Alloys	DCSP ACHF	Good control.
Silicon Bronze	DCSP	Arc length sensitivity is reduced.

* AC High Frequency is used because the reverse polarity
half of the cycle tends to break the oxide surface conditions.

Fig. 12-4. Welding characteristics of various metals using argon gas in the manual gas tungsten arc process.

cylinders are being moved, or when they are in storage. The cylinder should be fastened to a wall or non-tippable object when in use. It should be placed where it is virtually impossible to accidentally injure it with an arc or cutting torch. The cylinders should be stored in an upright position and used in an upright position.

12-9. SHIELDING GAS REGULATORS

Shielding gas regulators are designed to perform the same as the oxygen, acetylene, and hydrogen regulators described in CHAPTER 2.

They have either a gauge or a pressure indicator to show the cylinder pressure. Some of them have only a flowmeter gauge on the gas delivery side. This regulator has a constant outlet pressure to the flowmeter of approximately 50 psig. Fig. 12-5 shows

Fig. 12-5. An argon gas cylinder with the pressure regulator and flowmeter attached. Note that the regulator pressure is preset and seldom needs adjustment. (National Welding Equip. Co.)

a flowmeter and regulator mounted on a cylinder. The fitting used to connect the regulator to the cylinder varies with the kind of gas. These regulators will deliver gas flows up to 60 cu.

ft./hr. The flowmeter scales are accurate only if the gas entering them is at approximately 50 psig. If higher inlet pressures are used, the gas flow rate will be higher than the actual reading, and the reverse is true if the inlet pressure is lower than 50 psig. It is therefore important to use accurately adjusted regulators. Fig. 12-6 illustrates a two-stage regulator for argon gas. The gauge is a high-pres-

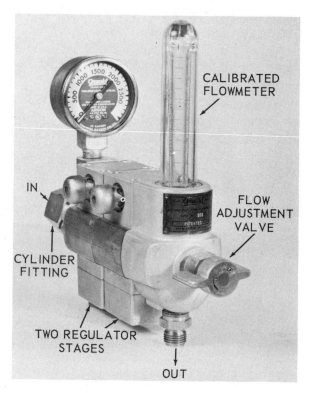

Fig. 12-6. Combination two-stage regulator and flowmeter. Regulator is preset to provide 20 psig pressure to flowmeter. (Linde Div., Union Carbide Corp.)

sure gauge and is used to indicate the pressure in the cylinder.

12-10. FLOWMETERS

The amount of gas around the arc can best be measured by the volume of gas coming out of the nozzle, rather than the pressure of the gas. There-

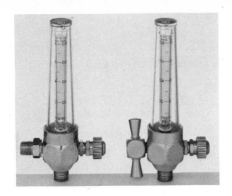

Fig. 12-7. Flowmeter for argon gas. The meter on the left has a finger-tightened regulator fitting.

fore, the shielding gas system is usually equipped with a flowmeter that is calibrated in cubic feet per hour. Fig. 12-7 illustrates two floating ball-type flowmeters. They are identical except for the regulator fittings.

In one type of flowmeter a tapered plastic or glass tube contains a loosely fitted ball. As the gas flows up the tube, it passes around the ball and lifts the ball. The more gas that moves up the tube, the higher the ball is lifted. Fig. 12-8 shows a cross section of the flowmeter shown in Fig. 12-7.

The tube and return gas housing are either clear plastic or glass. Some have a metal protecting cover. Joint between plastic tubes and body must be gastight. Scale on the inner tube is usually calibrated from 0 cfh to 60 cfh. Scale is usually read by aligning the top of the ball with cfh desired.

For an accurate reading, it is important that this type instrument be mounted in a vertical position. Any slant will cause an inaccurate reading.

Because gas densities vary, it is necessary to use different flowmeters for the different gases.

Some universal type flowmeters are available with a 0-100 scale on the flow tube. An accompanying chart or table will indicate the flow of gases (in cfh) in various densities.

Gas volume is related to pressure if the orifices and gas passages remain constant in size and shape. Therefore a gas pressure gauge can be used as a

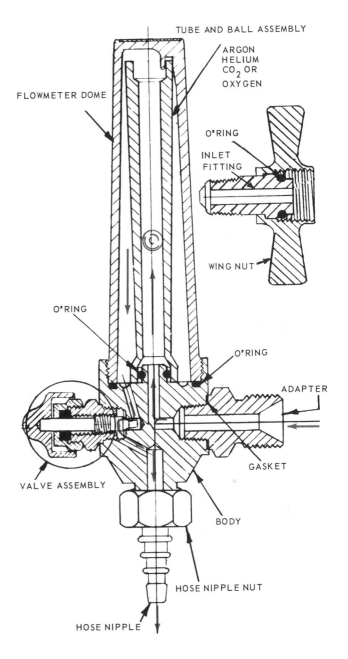

Fig. 12-8. Cross section of a floating ball type flowmeter.

volume gauge when the scale is calibrated as shown in Fig. 12-9. The single-stage regulator in the illustration has a plastic adjusting screw,

Gas Arc Welding Equipment 295

a rubber (fabric reinforced) diaphragm, the brass body is drop-forged, while the cap is die cast. Flowmeter outlets are usually equipped with a male threaded fitting (9/16 in. - 18 RH).

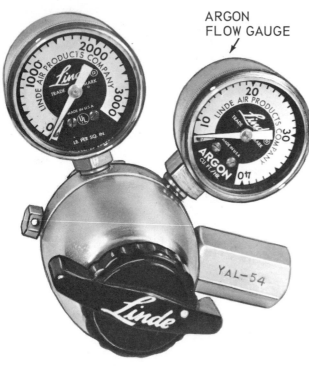

Fig. 12-9. Argon gas regulator with a Bourdon type gauge for measuring gas flow. Flowmeter gauge at right reads in cubic feet per hour (cfh).

12-11. ON AND OFF VALVES

Most flowmeters have a needle valve to turn on the gas flow, shut off the gas flow, and control the volume of the gas flow. Because this valve controls the volume, and because the pressure to the valve is constant, the valve orifice and the needle must be accurate and in good condition at all times. Any abuse of the needle and/or seat will result in errat- ic volume feeds. Also if the packing around the needle or if the needle threads are abused, the needle will not stay in adjustment. It is necessary that this valve be handled carefully. It should

not be forced. FINGERTIP HANDLING ONLY is recommended.

12-12. ACCESSORIES

When water-cooled torches are used, it is important that water flow be main- tained. If the water flow ceases or diminishes, the torch and the cable (welding lead) may quickly overheat and be ruined. To protect the equipment, a safety switch is sometimes placed in the water circuit. If the water pressure decreases, an electrical switch opens and shuts off the electrical power source. Fig. 12-10 shows a flow safety device.

Fig. 12-10. Water flow safety control. If the water flow decreases, the current is shut off until the flow of water resumes.

Some torches use fuse protection for the water-cooled torches. The fuse will interrupt the current flow when and if

Fig. 12-11. Fuse and hose assembly. This device protects against water flow failure and/or excessive current.

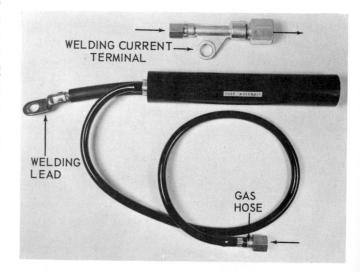

Fig. 12-12. Fuse and hose assembly with insulator cover removed. (Linde Div., Union Carbide Corp.)

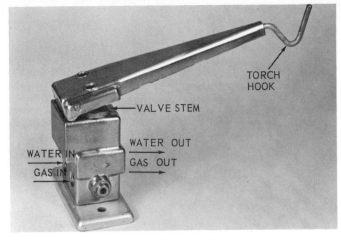

Fig. 12-13. Shielding gas and water flow economizer.

water flow is stopped, or decreases to a dangerous minimum, causing the fuse link to overheat, Fig. 12-11. The construction of this safety device, with the casing removed, is shown in Fig. 12-12.

To economize on water flow and/or gas flow, a lever-operated shutoff valve is sometimes used. Fig. 12-13 shows this type economizer. The valve operating lever acts as the torch holder. When the torch is not in use, it is hung on this lever and its weight closes the valves. When the torch is removed, the valves automatically open, allowing the water and gas flow to start.

Ground clamps as used to fasten the ground lead to the article being welded are similar to those shown in CHAPTER 2.

12-13. GAS ARC WELDING POWER SOURCES

Practically all of electrical power sources described in the previous chapters may be used to provide the welding current for gas arc welding, i.e. direct current, both straight and reverse polarity, and alternating current. All types of machines have also been used such as:

Motor-Generator (DC only).

Transformer (AC only).

Rectifier (DC only).

Combination-transformer and rectifier (DC or AC).

It has been found that each type of current and each type of machine has advantages for particular welding applications.

The motor-generator unit, the transformer unit, and the rectifier unit are described in CHAPTERS 6 and 8.

The units using a nonconsumable electrode (tungsten or carbon) are standard machines except the units usually have a high-frequency AC current in the welding circuit during starting, to permit starting the arc without difficulty.

The units using consumable elec-

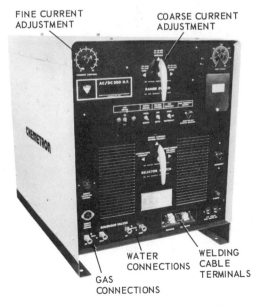

Fig. 12-14. AC-DC arc welder designed for gas arc welding. (Chemetron Corp.)

trodes (MIG) and especially those using CO_2 gas, need current sources with special properties. These units must be electrically designed to provide a constant voltage even though the current flow value changes, and some installations, such as the dip transfer process, even require a rising voltage across the arc as the current flow increases.

A combination DC-AC arc welder constructed especially for gas arc welding is shown in Fig. 12-14. A DC rectifier arc welder with a remote foot operated switch and current adjustment combination is shown in Fig. 12-15. The remote adjustments are also

Fig. 12-15. Gas arc welding machine with a remote foot operated off-on switch and current control.

available in hand-operated models, as shown in Fig. 12-16. These remote adjustments enable the operator to make current adjustments while welding. They also enable the operator to start and stop the machine without leaving the work.

Fig. 12-16. Hand-operated remote current adjuster. (Airco Welding Products, Div. of Airco, Inc.)

12-14. DROOPING VOLTAGE MACHINES

This type electrical power source is so built that as the current increases, the voltage decreases. In other words when the current is zero, the voltage is highest, and when the current is maximum the voltage is lowest as shown in Fig. 12-17.

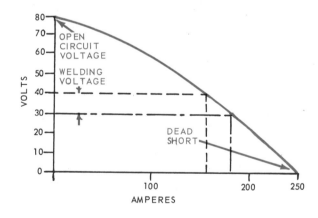

Fig. 12-17. Typical volt-ampere curve for a drooping voltage arc welding machine.

This operation can be either AC or DC, and is the most common machine on the market. It performs satisfactorily for stick electrode, tungsten electrode, and carbon electrode units.

12-15. CONSTANT VOLTAGE MACHINES

With the introduction of power-fed electrodes to the arc zone, it was found necessary to use an electrical power source with a different behavior pattern in volt-amp. characteristic as explained in PAR. 12-13. The unit must be designed to provide a wide variety of current values at the arc, without necessitating a voltage change, Fig. 12-18.

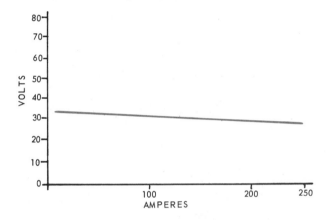

Fig. 12-18. Typical volt-ampere curve for constant voltage arc welding machine.

The operator adjusts the unit to the voltage across the arc desired and the machine will maintain this voltage over a large current adjustment.

A DC rectifier arc welder with constant voltage characteristics is shown in Fig. 12-19. This machine has a 3-phase primary power input. Note the output voltmeter, ammeter and the voltage adjustments.

12-16. RISING VOLTAGE MACHINES

The rising voltage machine was also developed especially for consumable electrode gas arc welding machines. Its design is based on having the machine deliver more current as the

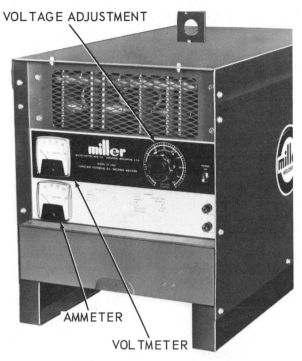

Fig. 12-19. Constant voltage arc welding machine. (Miller Electric Mfg. Co.)

voltage rises to enable the machine to weld satisfactorily as the wire size increases. The machine adjustment is commonly labeled as arc length adjustment rather than voltage or amperes. Fig. 12-20 shows the current-voltage relationship for the range of the machine.

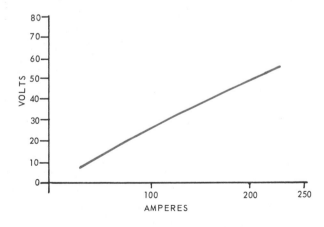

Fig. 12-20. Typical volt-ampere curve for a rising voltage arc welding machine.

Gas Arc Welding Equipment 299

12-17. ELECTRODE HOLDERS (TORCHES)

Although this item is basically an electrode holder, trade practice has

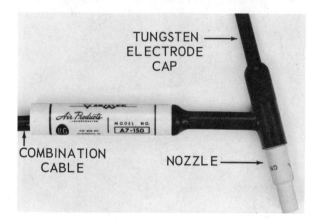

Fig. 12-21. *Air-cooled gas tungsten arc torch with a 130-amp. capacity. Uses long tungsten electrodes and ceramic nozzles. (KGM Div. Air Products)*

(tungsten or carbon) as shown in Fig. 12-21.

2. Consumable electrode torch as shown in Fig. 12-22.

The main parts of the nonconsumable electrode torch are:

1. Handle.
2. Cup or nozzle.

Fig. 12-22. *Gas metal arc torch. The trigger switch starts and stops the current, water, and wire feed. Note the metal nozzle used on this torch.*
(Auto Arc-Weld)

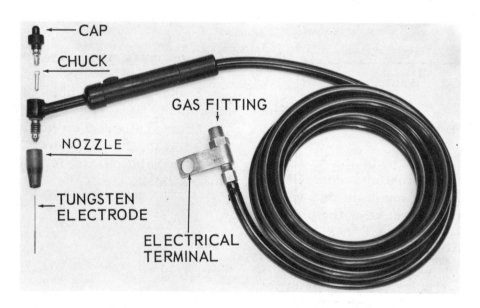

Fig. 12-23. *Air-cooled, gas tungsten arc welding torch. Short cap and electrode permit use of this torch in difficult-to-reach places.*
(Airco Welding Products, Div. of Airco, Inc.)

established the name "torch" for this combination electrode holder and gas dispenser.

The torches are of two basic types:

1. A nonconsumable electrode torch

3. Chuck.
4. Electrode lead connections.
5. Hose connections.
6. Electrode.

The torches may also be classed as

either air-cooled or water-cooled. The air-cooled models as shown in Fig. 12-23, are used for light duty, intermittent service. The stub electrode is used for welding in close, confined quarters. The water-cooled models are used when higher currents are desired and/or when the torch is used for longer continuous periods of time. The latter type is shown in Fig. 12-24. The details

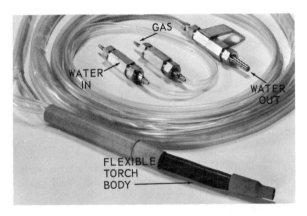

Fig. 12-24. Complete torch, hose, and electrode lead assembly. This is a flexible body torch and it is water-cooled.

of this torch and its connections are shown in Fig. 12-25. Note that the torch

body is flexible to enable positioning the electrode at different angles to the torch handle.

These torches must be sturdily constructed, leakproof (both for the gas and for water, if water-cooled). Leaks of either fluid may contaminate the weld and, in the case of the inert gas, will be costly.

Plastic insulating materials are used to cover the torch. The electrical connections must have a minimum resistance, both for economy and to minimize heating of the torch. The electrical connection to either the tungsten electrode or to the consumable electrode wire must be excellent.

The complete torch unit should be as light in weight and as well balanced as possible.

12-18. TUNGSTEN ELECTRODE TORCH – GTAW (TIG)

Tungsten electrode torches come in many sizes, capacities, and shapes. Two basic models are available:

1. For short-length tungsten elec-

Fig. 12-25. Exploded view of gas tungsten arc torch and hose assembly. This torch has a flexible body which can be bent to suit the job. (Falstrom Co.)

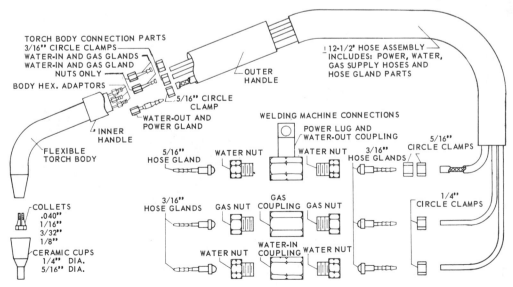

Fig. 12-26. Stub length tungsten electrode torch.

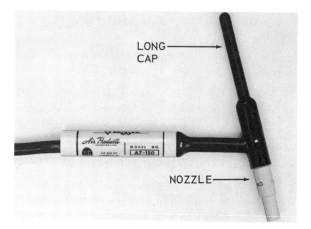

LONG CAP

NOZZLE

Fig. 12-27. Full length tungsten electrode torch.

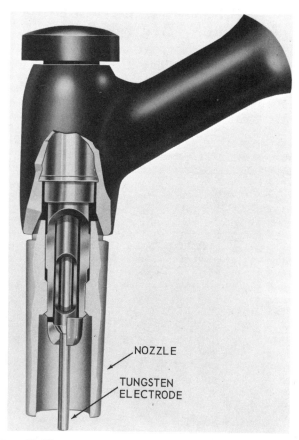

NOZZLE

TUNGSTEN ELECTRODE

Fig. 12-28. Cross-sectional view of a gas tungsten arc welding torch. (Linde Div., Union Carbide Corp.)

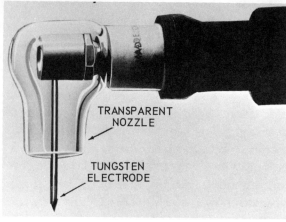

TRANSPARENT NOZZLE

TUNGSTEN ELECTRODE

Fig. 12-29. Stub electrode gas tungsten arc torch with transparent nozzle. Arc end of tungsten electrode has been prepared for DC welding. (Tec Torch, Inc.)

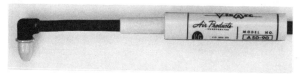

Fig. 12-30. Stub electrode gas tungsten arc torch for welding in confined positions. Note the small head. (KGM Equip. Co.)

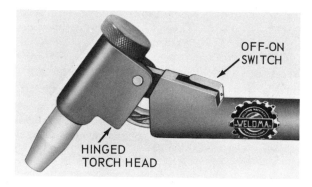

OFF-ON SWITCH

HINGED TORCH HEAD

Fig. 12-31. Tungsten electrode torch has solenoid valve that controls flow of shielding gas.

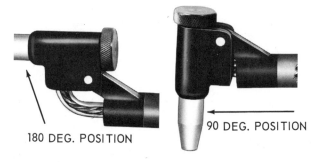

180 DEG. POSITION

90 DEG. POSITION

Fig. 12-32. Hinged torch head shown in two positions. This torch is air-cooled and has a 100 amp. capacity.

trodes (stub) as shown in Fig. 12-26.

2. For tungsten electrodes of 6 to 7 in. in length as shown in Fig. 12-27.

The chuck used to hold the tungsten electrode is usually made of steel. Some are operated by fingertip pressure (with gas cup removed), some need a chuck wrench to operate them, and some are designed as shown in Fig. 12-28. A stub tungsten electrode torch with a transparent cup or nozzle is shown in Fig. 12-29. A similar type of torch which has an opaque cup or nozzle is shown in Fig. 12-30.

Some of the torches have a manually operated switch on the torch to operate the gas solenoid valve, as shown in Fig. 12-31. The valve should be opened before the arc is struck, and it should be kept open for a few moments

Fig. 12-33. Pen-size gas tungsten arc torch. In use, operator would wear gloves. (Argopen)

commercially in the field.

Gas tungsten arc torches are also made for automatic welding. This type torch has an appearance similar to the automatic gas cutting torch as shown in Fig. 12-34. Most of these torches are

GAS HOSE

HEIGHT ADJUSTMENT RACK

WATER IN

Fig. 12-34. Gas tungsten arc torch designed for automatic welding. (Tec Torch Co., Inc.)

after the arc is broken (until the tungsten electrode cools) to prevent oxidation of the tungsten electrode.

A short electrode adjustable angle torch is shown adjusted to two different positions in Fig. 12-32.

The gas, electrical, and water connections must be clean and tight.

The torches vary in size according to the amount of current they are to handle. Torches are available as air-cooled models up to 200 amp. capacity. Torches of the higher amperage capacities are available in water-cooled models. Fig. 12-33 shows one of the smallest air-cooled models available

water-cooled. A cross section of this type of torch is shown in Fig. 12-35.

12-19. NOZZLES

Nozzles used on gas tungsten arc torches are usually made of ceramic materials, steel (chrome-plated materials), plastic materials and glass (pyrex) materials.

Nozzles must be fastened to the torch body with a leakproof joint. They must have the correct gas flow opening, and they must be centered around the electrode to provide a concentric gas flow. Nozzles are also called "cups."

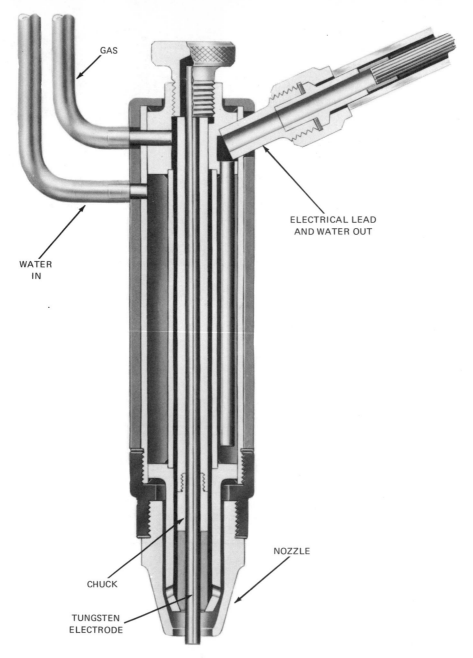

GAS

ELECTRICAL LEAD
AND WATER OUT

WATER
IN

NOZZLE

CHUCK

TUNGSTEN
ELECTRODE

*Fig. 12-35. Cross section of a water-cooled gas tungsten arc torch designed for
automatic welding. This torch is designed with a capacity of up to 500 amps.
(Weldma Co.)*

Ceramic cups are usable up to approximately 250-300 amps. flow. Above this level, water-cooled metal cups are needed. Fig. 12-36 illustrates several different ceramic cups.

Plastic type cups are made of a transparent plastic of high temperature stability. One make is Vycor, which is stable to temperatures above 3000 deg. F.

Glass cups (Pyrex) have been used to some extent.

The metal cups are heavy-duty cups. They can be designed with built-in water cooling passages. They are usually chrome-plated to reflect heat radiation and to prevent accumulations of weld splatter.

12-20. TUNGSTEN ELECTRODES

Tungsten electrodes are more popular than carbon electrodes in nonconsumable electrode shielding gas arc welding applications.

Four types of tungsten electrodes are used:

1. Pure tungsten AWS EWP.

2. Tungsten with 1 percent zirconium dioxide (ZrO_2) AWS EWZr.

Fig. 12-36. Ceramic nozzles for gas tungsten arc welding torches. (Diamonite Products Mfg. Co.)

A gas metal arc welding outfit. A light duty torch is being used. The wire drive unit is in the cabinet on top of the constant potential DC welder. (A. O. Smith Corp.)

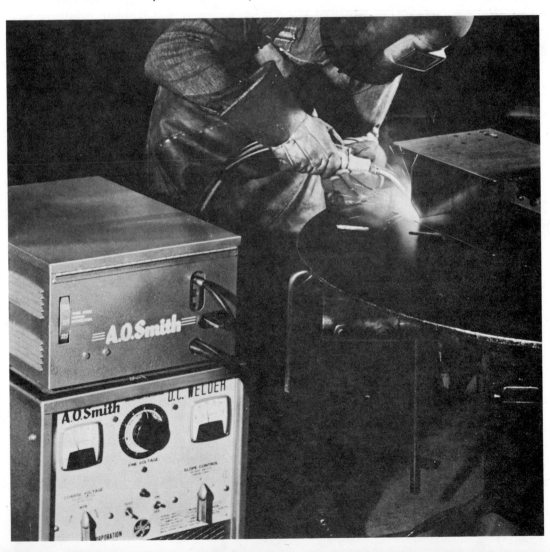

3. Tungsten with 1 percent thorium dioxide (ThO_2) AWS EWTh-1.

4. Tungsten with 2 percent thorium dioxide (ThO_2) AWS EWTh-2.

The pure tungsten is least expensive and it is satisfactory for most welding applications, but is consumed faster.

Electrodes with zirconium dioxide content last longer and provide a more stable arc. They are used frequently with AC high-frequency stabilized current. The thorium dioxide content enables easier arc starting, provides a more stable arc, reduces weld contamination. The 2 percent thoriated electrode has a longer life than the 1 percent electrode. There is also available a pure tungsten electrode with stripes of thorium dioxide.

One company uses a color code to identify the four types of electrodes. Green is for the pure tungsten, brown

These electrodes come in a range of lengths 3 to 24 in. Some popular lengths are 5, 6, 7, 8, 9, 12 and 18 in. The most popular lengths are 3 in. and 7 in.

The diameters of the electrodes vary from .010 to .250 in. dia. Some popular sizes are .040, 1/16 (.063), 3/32 (.093), 1/8 (.125), 5/32 (.156), and 3/16 (.188).

12-21. CARE OF TUNGSTEN ELECTRODES

The electrode must be straight. If it is off center in the cup, the weld may become contaminated. As the electrode is used, it becomes brittle up to 3/8 in. back from the arc end. If one uses a pair of pliers and grasps the arc end of the electrode, the electrode can be broken off easily to form a new clean arc end.

The tungsten must be clean. It must have good electrical contact with the collet, and it should be adjusted to ex-

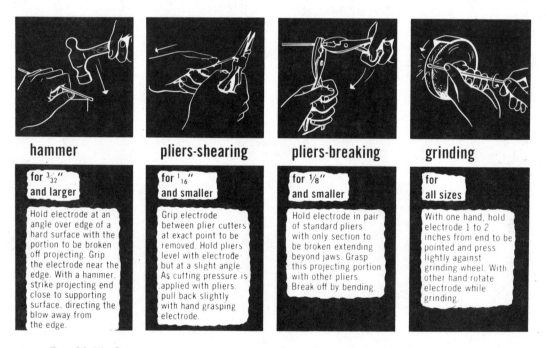

Fig. 12-37. Proper ways to retip tungsten electrodes. (Sylvania Electric Products)

is for the zirconium, yellow is for the one percent thoriated, and red is for the two percent thoriated electrodes.

tend 1/8 in. beyond the end of the cup. It is extremely important that the helium or argon hose and all the connections be tight because air or moisture in the gas would be harmful to the weld.

Fig. 12-37 shows the proper way to break and/or condition the rod for best results. If the retipping is not properly done, the tungsten electrode may shatter and/or split.

Conditioning the electrode end for

Fig. 12-38. *A special grinder used to grind points on tungsten electrodes in preparation for DCSP arc welding. Note that both the electrode and grinding wheel are rotated to obtain a uniformly ground point.* (Weldma Co.)

DC welding is very important. Fig. 12-38 illustrates a grinding machine especially designed to grind the electrode end safely and accurately.

12-22. CONSUMABLE ELECTRODE (WIRE) UNIT - GMAW (MIG)

The consumable wire torch is constructed much the same as the tungsten electrode model. The major differences are the tube built into the torch to carry the consumable wire and the controls needed by the operator.

One model has the wire spools in an automatic feeding machine and the consumable wire is fed through a flexible tube up to the arc point. Variable speed rollers push the wire through the tube. Fig. 12-39 shows a view of such a machine which may be used for either semiautomatic or automatic gas metal arc welding.

Another model mounts a small spool of wire in the torch and also incorporates the drive motor in the torch. This type unit eliminates pushing the

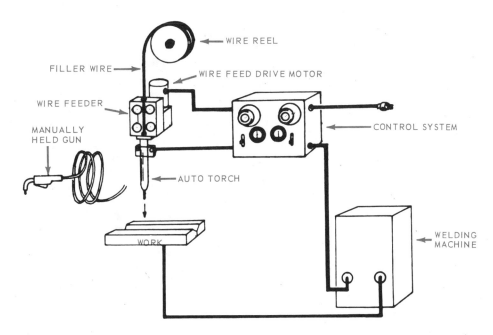

Fig. 12-39. *Schematic drawing of a combination manual and automatic gas metal arc welding outfit.*

wire through the long tubes. A special version of this type torch is to mount the feed motor in the torch and pull the electrode wire through its guide tube to the torch.

The complete station is illustrated in Fig. 12-40. Note the electrode wire reel, the lubricator mounted near the electrode wire feed rollers. The valve in the gas line is operated electrically. The construction of a compact, semiautomatic gas metal arc welding unit is shown in detail in Fig. 12-41. The instrument and control panel has a

Fig. 12-40. *Diagrammatic view of a semiautomatic gas metal arc welding outfit.* (Arcos Corp.)

CONTROLS GROUND LEAD

AC – DC WELDING MACHINE

ENLARGED VIEW OF WELDING NOZZLE

FLOW METER

HEATER

ELECTROVALVE

WORK LEAD

WORK

CO$_2$ GAS

Fig. 12-41. *A gas metal arc unit, showing wire reel, wire feed rollers, leads, controls and torch.* (Arcos Corp.)

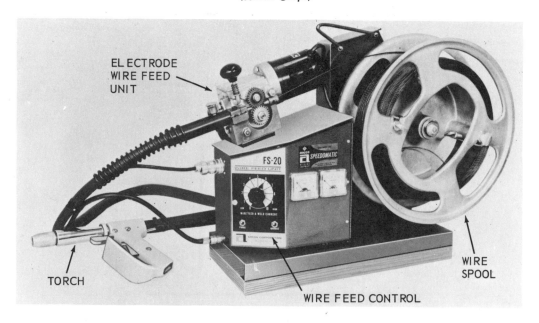

ELECTRODE WIRE FEED UNIT

FS-20

SPEEDOMATIC

TORCH

WIRE FEED CONTROL

WIRE SPOOL

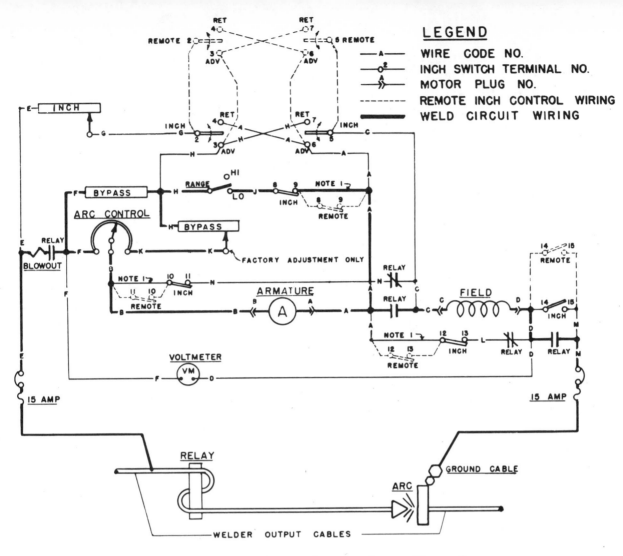

Fig. 12-42. Wiring diagram of gas metal arc welding outfit.

hi-low switch (current), a gas purge switch and an electrode wire advance switch. Note the electrode wire reel in the rack, the feed wire wheels and the number of leads going to the torch. Fig. 12-42 shows a wiring diagram of a semiautomatic gas metal arc welding machine. The wiring diagram includes the electrical connections of the control and wire feed units.

12-23. GAS METAL ARC TORCH

A gas metal arc torch with its leads and hoses is shown in Fig. 12-43. The nozzle is made of metal. The cables and hose are kept together by using tape.

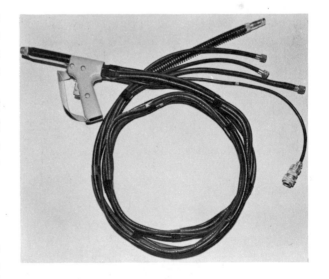

Fig. 12-43. Assembled torch, leads, and hose for gas metal arc welding outfit. (Auto Arc-Weld Mfg. Co.)

This torch has a starting switch with a guard to prevent accidental starting and also to reflect the arc radiation from the operator's hand. The torches are available in many different styles, some of which are shown in Fig. 12-44.

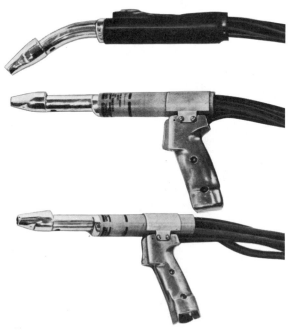

Fig. 12-44. Three styles of gas metal arc torches. (Linde Div., Union Carbide Corp.)

The torch nozzle must be kept clean both inside and outside and the tube through which the electrode wire passes should be cleaned each time the electrode reel is changed. Fig. 12-45 illustrates the details of a gas metal arc torch or gun. Note the knurled nozzle for ease of assembly. The pin

Fig. 12-45. Gas metal arc torch. (Arcos Corp.)

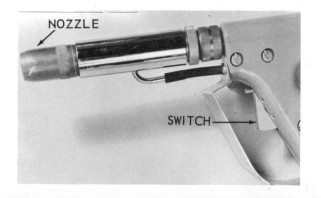

switches are used to "inch" the electrode wire forward or backward, while the lower switch is used to start the welding process. Most machines are wired to start and operate the wire feed rollers only when the arc is actually operating.

12-24. CONSUMABLE ELECTRODE WIRE

Welding wire used for consumable electrode gas welding is similar in many ways to the filler rod used in gas welding. The wire contents must match the base metal being welded and must have deoxidizers included.

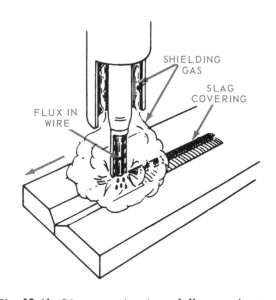

Fig. 12-46. Diagrammatic view of flux-cored wire in combination with gas shielding, as used in gas metal arc welding process.

The wire comes in a variety of diameters, a variety of compositions, and is usually furnished in 1, 2.5, 12, 12.5, 25, 50 and 60 pound spools, and 250, 500, and 750 pound reels or drums.

Some of the more common diameters are: .025, .030, .035, .045, 1/32, 3/64, 1/16, 5/64, 3/32, 5/32 in.

Another process which uses flux-

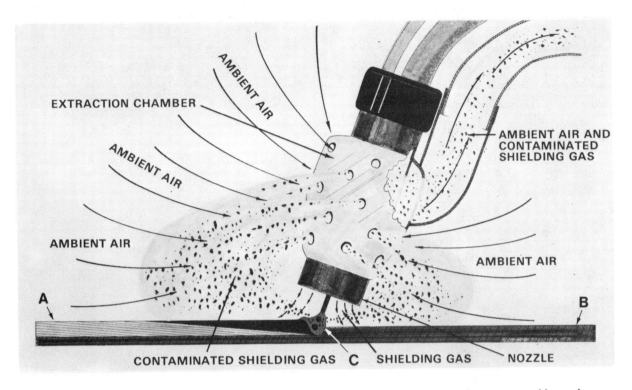

Fig. 12-47. A gas metal arc welding torch equipped with a smoke extracting device. A. Weld metal. B. Base metal. C. Puddle. (Bernard Co., Div. of Dover Corp.)

cored welding wire is available. The wire is a small tube filled with welding flux. Outside diameter (OD) is 1/8 or 3/16 in. Wire is formed from strip metal and powdered flux is trapped in tube as tube is formed. Some flux-cored wires are used with CO_2; other wires operate without CO_2.

When used with carbon dioxide, gas protects metal in arc and metal in puddle. The flux deoxidizes metal, scavenges impurities, and forms a slag to protect weld metal once metal leaves the gas shield, Fig. 12-46. The arc has less spatter than when cored wire is used without gas.

Solid wire may also be used with a combination of gas and flux. This process uses the principle of magnetism to hold the flux on the electrode. The iron powder flux adheres to the wire due to the magnetic field surrounding the electrode. This magnetic field is caused by the welding current flowing through it. The powder flux is fed to the torch cup by gravity and is carried to the nozzle by the flow of carbon dioxide gas.

12-25. SMOKE EXTRACTING TORCH

Another torch (gun) design is one with an exhaust system built into the torch. The extracting unit collects the welding fumes and some air before these fumes can reach the operator's face. See Fig. 12-47. When properly operated, this fume-removing system does not interfere with the flux and gas protection.

Fig. 12-48 shows a gas metal arc welding torch. Note the vacuum placed around the torch nozzle. The vacuum nozzle removes between 30 and 60 cubic

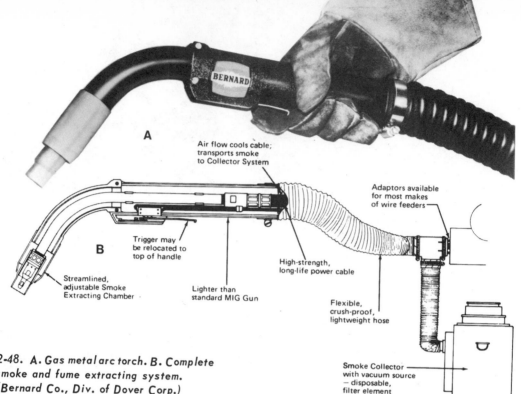

Fig. 12-48. A. Gas metal arc torch. B. Complete smoke and fume extracting system. (Bernard Co., Div. of Dover Corp.)

feet per minute. The fume-laden or smoke-laden air is exhausted to the outside, or it can be filtered, Fig. 12-49.

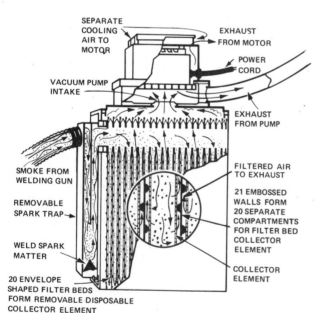

Fig. 12-49. A vacuum pump and filtering cabinet. This unit is used to remove dirt from the gases coming from a gas metal arc torch equipped with a fume extractor. (Bernard Co., Div. of Dover Corp.)

12-26. WATER-COOLING EQUIPMENT

When currents of over 250-350 amps. are needed, the heat generated necessitates that the torch be water-cooled. The water is carried to the torch, and then to the drain through flexible hose. In most cases, solenoid valves are used to control the water flow.

If a water source is not available, or if the available water contains impurities, portable water-cooling equipment is often used, as shown in Fig. 12-50. The internal construction of the recirculating water-cooler is shown in Fig. 12-51.

12-27. SHIELDING GAS ATMOSPHERE CHAMBERS

In addition to using gases as a flowing gas shield around the arc, much arc welding is done in a shielding gas chamber.

Two types are:

1. Small units where the operator

Fig. 12-50. Recirculating water system for use on water-cooled gas tungsten arc and gas metal arc outfits. (Bernard Co., Div. of Dover Corp.)

stands outside the chamber with arm and glove manipulators to handle the electrode holder.

2. Large units with gas locks. The chamber is full of shielding gas and the operators use gastight suits similar to diving outfits.

Fig. 12-52 shows a small reach-in gas atmosphere welding chamber. The arm reach-in openings are just below the observation window. This unit has a vacuum pump and a refrigerating unit to cool and dry the shielding gas atmosphere (the gas atmosphere must be as dry as possible), and a recirculating system to filter the shielding gas.

A flexible bag type of shielding gas atmosphere welding chamber is shown

Fig. 12-51. Cross section view of water circulating unit used on water-cooled welding machines. (Bernard Co., Div. of Dover Corp.)

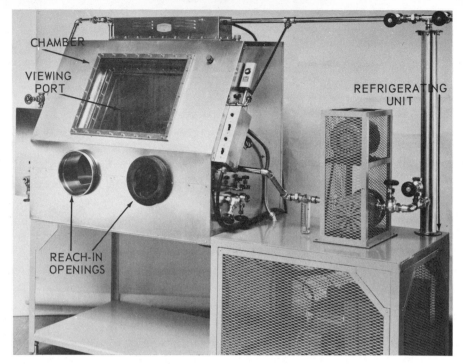

Fig. 12-52. Shielding gas atmosphere cabinet. Metals which are difficult to weld in normal atmosphere may be welded in this chamber in a shielding gas atmosphere. (S. Blickman, Inc.)

in Fig. 12-53. The flexible bag is inflated with shielding gas; therefore, no air is in the bag. The bag is designed to permit use by three operators at the same time.

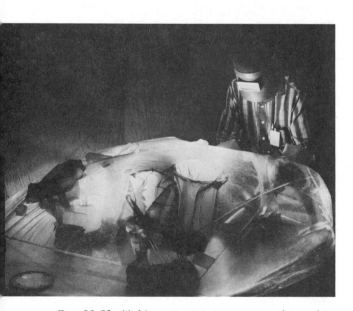

Fig. 12-53. Welding titanium in an atmosphere of inert gas within an inflated plastic container. (Boeing Co.)

12-28. ARC SPOT EQUIPMENT

The gas tungsten arc welding station and the gas metal arc station can also be used for arc spot welding, and for arc hole piercing.

The standard cup or nozzle is replaced with a special nozzle, and a timer is placed in the electrical circuit.

For spot welding, the gun is placed over the spot on the lapped metals to be welded together. The arc is started and then at the correct interval of time (this time varies with the metal thickness and the amount of current) current from the welding machine is switched off. Fig. 12-54 shows a gun for gas metal arc spot welding.

For arc hole piercing, the gun is modified to start an arc in a shielding atmosphere provided by a special cup or nozzle pressed against the metal to be pierced. At the end of a certain time sequence, a timer operating a solenoid valve controlling a high-pressure gas supply releases the gas pressure on the molten spot and the pressure forces the molten metal through the plate forming a fairly accurate size and shape hole in the metal.

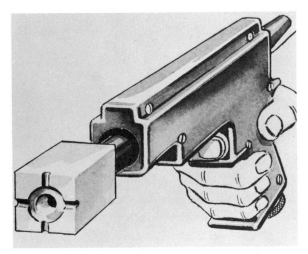

Fig. 12-54. A gun type gas metal arc spot welder. Sketch shows arc end of ceramic nozzle and how gun is held. In operation, arc is completely covered by ceramic nozzle. (Tec Torch Co., Inc.)

12-29. CARBON ELECTRODES

Carbon electrodes may be used to weld metals in a gas atmosphere. These electrodes are not used as frequently or as much as tungsten electrodes. The same specifications of the carbon rods, as listed in CHAPTER 6, are used for gas carbon arc welding.

12-30. HELMETS

Because of the clearer atmosphere around the arc, the operator must use arc welding lenses in a darker shade to reduce eye fatigue and possible eye damage. Most helmets for gas arc welding use a clear cover glass, a filter lens, and sometimes a clear cover lens. It is very important to you that all these lenses be used and that they

be clean. It is also recommended that flash goggles of approximately #2 shade be worn under the helmet. See PAR. 6-26.

12-31. PROTECTIVE CLOTHING

Asbestos (CAUTION: See Par. 2-34.) and leather clothing are recommended. Wool is satisfactory. Cotton does not provide sufficient protection, and it deteriorates rapidly under infrared and ultraviolet rays. Always wear dark clothing to reduce reflection of light behind the helmet.

The clothing should be without cuffs or open pockets as these will collect sparks.

12-32. MAINTENANCE OF GAS METAL ARC WELDING EQUIPMENT

The gas metal arc welding mechanism requires good maintenance, which includes: 1. Welding current source. 2. Gas equipment. 3. Wire roll equipment. 4. Wire moving equipment. 5. Leads. 6. Welding gun.

Previous paragraphs in this chapter have explained some of the more important maintenance operations. Some other maintenance steps one must follow regularly are:

If a wire spool is used, keep the wire on the wire spool clean. Keep the spool covered except when replacing it.

The wire spool must move freely on its shaft.

The wire must not have any sharp bends or kinks.

The wire feed rollers should be cleaned each time a new spool of wire is installed.

The wire should be lightly lubricated before the wire enters the rollers. See Fig. 12-55.

The welding cable should be as

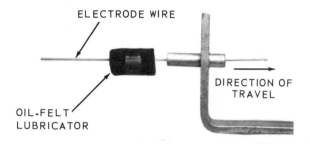

Fig. 12-55. *The lubricated felt pad both cleans and lubricates the wire before the wire reaches the feed motor rollers. Friction drag is reduced as much as 50 percent.*

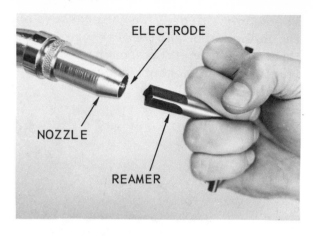

Fig. 12-56. *A nozzle reamer used to keep the nozzle clean and free from metal splatter.*

straight as possible to improve the wire feed.

The feed rolls should be in line with each other and in line with the inlet and outlet bushing.

Always electrically ground the welding machine. Always clean the area to which a ground is to be attached.

Ground clamp surface must be clean.

The gun nozzle should be kept as clean and as free from metal spatter as possible. See Fig. 12-56.

Follow all safety rules.

12-33. TEST YOUR KNOWLEDGE

1. What is a shielding gas?

2. What does a flowmeter measure?

3. Why must the tungsten electrode be straight?

4. How is the tungsten electrode held in the torch?

5. Of what materials are torch cups made?

6. Are some arc welding machines made that will increase the ampere flow as the arc potential increases?

7. Under the same welding conditions, would one use more cu. ft./hr. of helium or argon?

8. What is the allowable moisture content of shielding gases for arc welding?

9. How is argon stored in cylinders?

10. What method is used to manufacture oxygen?

11. Which is more expensive per cubic foot, argon or carbon dioxide?

12. Is oxygen ever added to shielding gas for arc welding purposes? Why?

13. When is high-frequency stabilization necessary?

14. Are water-cooled gas arc torches used? Why?

15. What is a magnetic flux shielding gas welding system?

16. Explain the dip transfer method.

17. When is a pointed tungsten electrode used?

18. Is nitrogen used in the shielding gas mixture when welding steel? Why?

19. How does the CO_2 molecule differ from the argon and helium molecules?

20. As the inlet pressure to a flowmeter increases, how does this increase affect the flowmeter reading?

21. In Fig. 12-29, for what type of current flow has the electrode been prepared?

Chapter 13

ELECTRIC
RESISTANCE WELDING

All resistance welding is based upon the fundamental principle that when an electrical current is sent through metal, the resistance of the metal to this electrical flow heats the metal. By applying sufficient current, the resulting high temperature may produce fusion temperatures and make welding possible.

The term electric resistance welding includes a variety of welding applications, and it is described under a variety of names such as Spot Welding, Shot Welding, Gun Welding, Flash Welding, Stud Welding, Spike Welding, Upset Welding, Press Welding and others. Some of these names are official American Welding Society nomenclature, others are common usage "shop" terms.

Resistance welding has a number of advantages. It is fast, there is very little warpage of the metal, the process can be accurately controlled, and the welds are consistently uniform. This type of welding is particularly well suited to all forms of automatic production.

13-1. PRINCIPLES OF ELECTRIC RESISTANCE WELDING

When an electrical current is passed through two pieces of metal that are touching, the local high resistance

Spot welder being used to fabricate a part from sheet metal. (ITT Holub Industries)

generates or produces a high temperature. If enough current is used, the metals will become plastic, and then molten. If the two pieces are pressed together while their surfaces are plastic, or molten, the pieces will fuse into one piece. Fig. 13-1 illustrates the parts of an elementary resistance spot welder. Because the two pieces cannot be in perfect contact, the parts of the metals that form the contacting surfaces offer the highest resistance and therefore these surfaces heat up first and to the highest temperature. If the metal is pressed together while plas-

TOP
ELECTRODE →

POWER
CORD

SHEET METAL

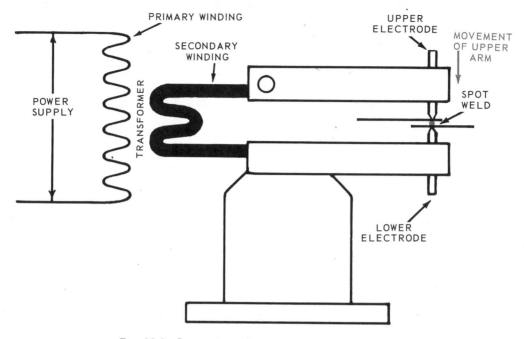

Fig. 13-1. Parts of a typical resistance spot welder.

tic, and if the parts are held together after the current is shut off and the metal cools to a solid condition, a good weld may be obtained, and the metal will retain most of its physical properties because of the rapidity with which such operations are accomplished. If the joint is clean, the physical properties of the weld will be as good as with any other welding method.

A resistance welding machine is fundamentally an electric transformer operating from an alternating current circuit. In order that the resistance welder may perform the welding operation, it must produce a very high current (amperes) at a relatively low voltage. This requirement means that the primary circuit will have many turns in the transformer, while the secondary winding will ordinarily have only one turn. Sometimes two or three turns are used if the secondary leads are long, such as in gun welders. Fig. 13-2

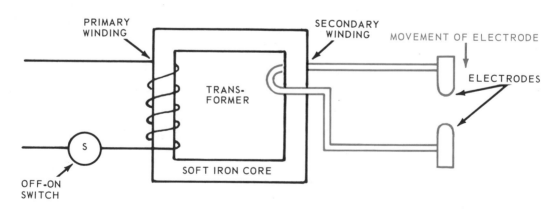

Fig. 13-2. Basic resistance spot welder electrical circuit. Note that the primary winding has many more turns than the secondary winding. This is a step-down transformer.

illustrates an elementary resistance welder electric circuit.

13-2. FUNDAMENTAL RESISTANCE WELDING PRACTICE

The correct application of resistance welding depends on the proper application and control of the following variables:
1. Current.
2. Pressure.
3. Time.
4. Electrode contact area.

All resistance welding requires enough current to heat the metal being welded to its plastic or molten state (usually the plastic state).

The desired welding current may be set on the current control on the machine as shown in Fig. 13-3. There are

pressure on the electrodes which holds the metal tightly while the welding current is flowing is called Weld Pressure. After the metal reaches its plastic state the weld current is usually turned off and the metal is then squeezed and forged together by the Forge Pressure. The metal is held under this pressure for a short time to allow the weld to solidify. After the weld is completed, the pressure is released from the electrodes and the part is removed.

Four different timing periods must be set up on the automatic machine or manually controlled on a manual machine during one welding cycle:
1. Weld time.
2. Squeeze time.
3. Hold time.
4. Off time.

The period of time during which the

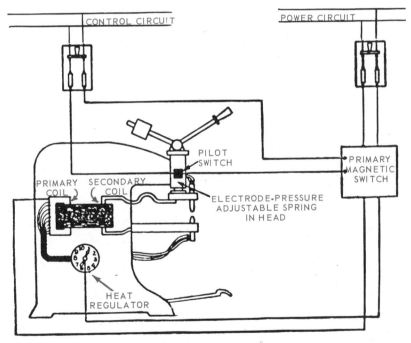

Fig. 13-3. Schematic drawing and wiring diagram of a spot welding machine. (Electric Controller and Mfg. Co.)

two pressures which are required in making a resistance weld. They are the weld pressure and forge pressure. The

welding current is flowing is known as Weld Time. Squeeze Time is the time during which the metal is forged and welded by the forge pressure. The cooling period is known as Hold Time and the period of time from the release of the electrodes from the work after cooling to the start of the next weld cycle is known as the Off Time of the machine.

The size of the weld is controlled by the surface area at the contact face of the electrodes. This area can be varied by choosing sets of welding electrodes used to give the area of contact desired. The welds can be done very rapidly and the resultant weld is clean and strong. The electrodes. must be made of a high conductivity metal and also a wear-resisting metal, usually a copper and beryllium alloy. The metal being welded must be clean, the electrodes must be clean and correctly shaped. The operator should protect himself from flying sparks by wearing flash goggles, gloves, and coveralls. Some of the automatic machines have a casing surrounding the weld, with window inserts to enable the operator to safely observe the welding operation.

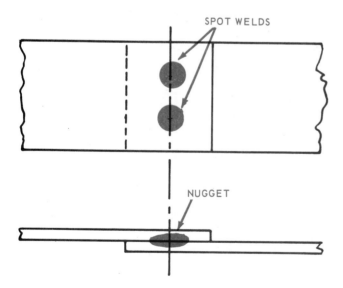

Fig. 13-4. Two pieces of metal, lapped, and spot welded in two places. Note the nugget where the fusion of the two pieces takes place.

Automatic resistance welding machines are accurately set up by specially trained technicians and engineers, who are usually trained by the manufacturers of the equipment.

Manually operated machines depend to a great extent on the operator for the control of the variables. The size of the electrode contact areas is of considerable importance, and specifications are provided by the manufacturer giving electrode sizes, currents, time, and pressures recommended for varying thicknesses and kinds of metal.

Special problems such as welding two pieces of different thickness together, or welding sheet steel to solid rod, etc., require considerable ingenuity on the part of the setup man or the operator.

13-3. TYPES OF ELECTRIC RESISTANCE WELDING

There are several types of resistance welding equipment based on the above principle. Some of the more common types are:

1. Spot Welding.
2. Gun Welding.
3. Shot Welding.
4. Upset Welding.
5. Flash Welding.
6. Seam Welding.
7. Projection Welding.
8. Spike Welding.
9. Stud Welding.
10. Metal Foil Welding.
11. Metal Fiber Welding.
12. Percussion Welding.
13. High Frequency Resistance Welding.

All of these operations are fundamentally the same, but the preparation of the metal may be different, and the construction of the machine may be different.

13-4. SPOT WELDING PRINCIPLES

One way to join sheet metal parts is to drill holes and either rivet or use machine or sheet metal screws to fasten the parts together. Another way to join sheet metal parts is to spot weld them together.

Spot welding is resistance welding two pieces of metal together with a small nugget of fused metal. It consists of lapping two pieces of metal and clamping the joint between two electrodes. A current is then passed between the two electrodes; the resistance of the sheet metal between the electrodes heats the metal at the clamped spot, and the clamping action results in a fusion of the metals at that particular spot. Fig. 13-4 shows a diagram of spot welding in both plan and cross section. This method is popular in manufacturing processes, either as a preliminary step to other forms of welding, or as a final procedure.

To operate a manual spot welder, the operator must clean the electrodes, turn on the cooling water, set the variable transformer to the correct cur-

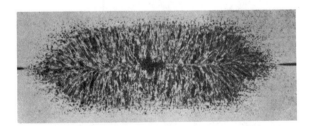

Fig. 13-5. A 10X macrograph of a spot weld nugget in cross section. The material is mild steel .094 in. thick. (Taylor-Winfield Corp.)

rent, and turn on the power. Place the metals being welded on the stationary electrode. By pressing the foot lever, you press the movable electrode against the metal. Pressing the foot pedal a little further will close the electrical

Fig. 13-6. A pneumatically powered spot welder. The timing of the weld is electronically controlled. (Acro Welder Mfg. Co.)

switch manipulated by lever. Spring on lever and its tension adjustment control the pressure during welding. When operator estimates proper time elapsed, current is discontinued by pressing foot lever to bottom which opens lever electrical switch. Also, maintaining pressure after current is shut off enables molten nugget to cool and fuse before clamping pressure is released. Fig. 13-5 shows cross section of spot weld in mild steel.

Spot welders have been built with several operating devices such as mechanical-electrical electrode movements, pneumatically powered electrodes, and hydraulically powered electrodes. The electrical circuits have been controlled by both vacuum tube circuitry and solid state circuitry. Fig. 13-6 illustrates a bench model spot welder, pneumatically powered and electronically controlled.

Many automatic machines have been developed using this method of welding. Spot welders have been made in many

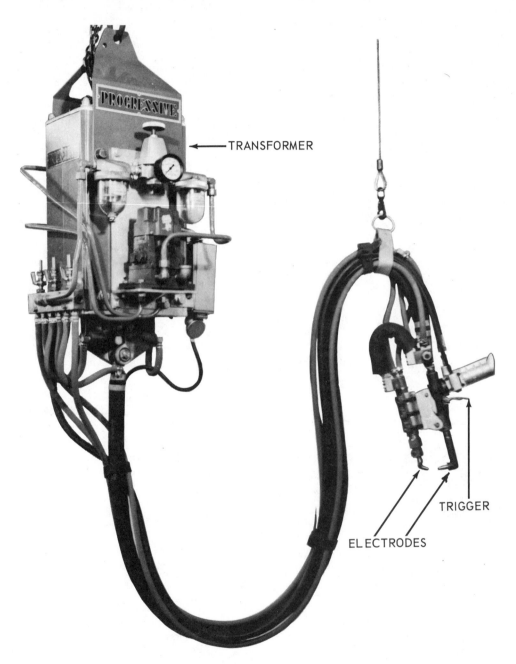

Fig. 13-7. Gun welding machine. Note that the transformer on this welder is supported by a rail overhead. (Progressive Machinery Corp.)

combinations of electrode arrangements, etc. However, most of the machines of over one set of electrodes are machines designed for special production purposes. Automobile body production is a good example of the resistance spot welding method.

13-5. GUN WELDING

A gun welder is a portable form of resistance welder which is designed to enable the spot welder machine to be moved easily from place to place and then positioned to fit irregular surfaces. Fig. 13-7 shows a typical gun welder. "Gun" welders are popular in production sheet metal fabricating shops, such as in automobile body fabrication. The electrodes are usually operated either hydraulically or pneumatically. The timing, pressure, and amount of current are usually specified by the engineer and technician. The operator has only to select the places to weld, put the nonmoving electrode against the metal, pull the switch (trigger) and the automatic controls do the rest. The operator must keep the electrodes in good condition. The most common method is to occasionally replace the electrode tips.

Some portable welders are of the seam welder principle (see PAR. 13-9).

13-6. SPOT WELDING ALUMINUM AND STAINLESS STEEL: SHOT WELDING

Shot welding is an unofficial (shop) term applied to special spot welding where a carefully controlled amount of current is used for a very brief interval. Shot welding is necessary when spot welding aluminum, its alloys, and stainless steel. The timing is so short an interval that only electronic controls

can do a satisfactory job of timing the current flows. The basic principle is to heat and cool the metal as fast as possible, to minimize heat treating and oxidation abuses. Shot welding is usually performed on press type, pneumatically operated and electronically timed resistance (spot) welders.

13-7. UPSET WELDING PRINCIPLES

Upset welding consists of clamping two pieces of metal to be welded together in separate electrode jaws. The two metals are then brought together as shown in Fig. 13-8. In upset welding the

Fig. 13-8. Basic principle of upset welding. The diameter of the rod is increased in the area of the weld.

two pieces are touched together, heavy current is passed from one piece to the other, and the resistance to the electrical flow heats the faces to fusion temperature. The two pieces are pressed together (upset) under pressure while the current is flowing and after the current is turned off. The metals fuse together and upon cooling become one piece.

This process, by its upsetting action mixes the two metals intimately, tends to push the impurities, if any, out of the weld and reduces the heated zone to a minimum. Fig. 13-9 shows a method of clamping the parts and an electrical wiring diagram.

13-8. FLASH WELDING

Flash welding is a form of butt welding which uses considerable pressure

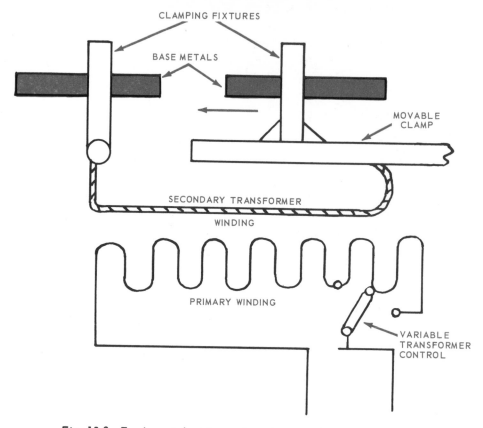

Fig. 13-9. Fundamental wiring and mechanism of an upset welder.

to join the parts together, when the plastic temperature of the metal is reached. This method of welding is used extensively in production work, particularly in welding rods and pipes together.

The two pieces to be welded are held by clamps in a special flash welding machine. The parts are brought together and the resistance to the current flow heats the contacting surfaces. As soon as the metal has been brought to its melting temperature, the current is shut off and the pieces are rapidly brought together under considerable pressure. When this action takes place the squeezed molten metal gives off sparks or a flash, hence the name. As the metal is heated to its plastic state, the pieces are forced together under high pressure which forces fused metal

and slag out of the joint making a good solid weld. Fig. 13-10 shows two pieces of metal before and after being

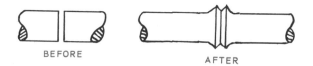

BEFORE   AFTER

Fig. 13-10. Metal before and after it has been flash welded.

"flash" welded together, while Fig. 13-11 is a schematic diagram of a flash welder.

13-9. SEAM WELDING

Most seam welding produces a continuous or intermittent seam weld near the edge of two overlapped metals, by using two roller electrodes. As these

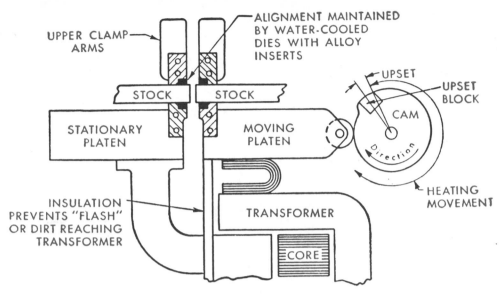

Fig. 13-11. Schematic drawing of a flash welder. Note how the heated metals are pushed together with considerable pressure by the upset block.
(P. R. Mallory and Co.)

rollers travel over the metal, the current passing between them heats the two pieces of metal to the fusion point. This might be called a seam, spot weld method. Fig. 13-12 shows a schematic

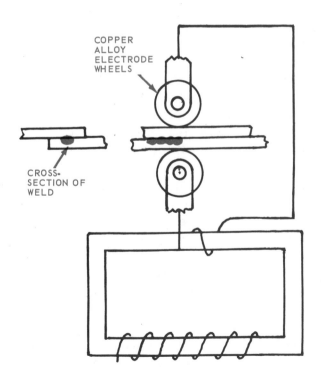

Fig. 13-12. Schematic drawing of metal being seam welded.

diagram of a typical seam welder in operation. The spots can be timed to overlap or timed to have spaces between the spots. The main object of the overlapped spots is to produce gas and liquid leakproof lap joints. Two types of seam welds are shown in Fig. 13-13.

The intermittent overlap spot weld technique is used for metals that are critical to heat treatment.

Another form of seam welding is called butt seam welding. This technique is often used to butt weld a longitudinal seam in pipe. The two rollers are mounted at an angle and as they press on the pipe the pipe edges are forced together and the electric current brings the metal to the plastic-molten condition and fusion takes place.

13-10. PROJECTION WELDING

Projection welding consists of forming slight projections or bumps on one piece of metal. The projections are accurately formed in precise locations on the metal by a special set of dies. After the projections are formed, the raised portions on one piece are pressed into contact with another piece, while at the same time a heavy current is passed through the two pieces. When these raised portions touch the second sheet of steel, as they are clamped by

Macrographs of a projection weld in various stages of completion is shown in Fig. 13-15. Each successive photograph illustrates the effect of the number of cycles of the electrical flow through the weld.

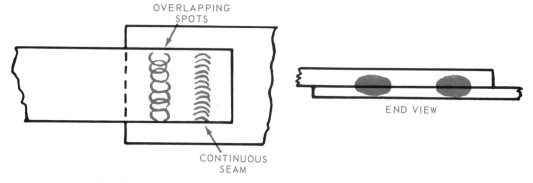

Fig. 13-13. Sketch showing metal that has been seam welded. A seam weld is made with wheel-type electrodes. If the current to the electrodes is turned off and on, a series of overlapping spots form the seam. An uninterrupted flow of current to the electrodes will form a continuous seam.

the electrodes in a projection welder and the current is applied, current flow at the points heats and fuses the two pieces together.

An advantage of this method is that it locates the welds at certain desired points. Also, several spots may be welded at the same time. However, the tooling required makes this method practical only when high production is planned. Fig. 13-14 shows a typical projection welding setup.

Fig. 13-14. Schematic drawing of metal before and after it is projection welded. Some of the depression formed when the projection is stamped remains on the top surface of the finished weld.

13-11. SPIKE WELDING

This type of resistance welding is a special application of the maximum energy, minimum time method of resistance welding. A common method is to store large amounts of current in capacitors, then discharge these capacitors through the electrodes and through the metals to be welded. Electronic controls keep the current flow to an extremely small interval of time. The welding, therefore, takes place so fast that warpage is usually eliminated and contamination is also eliminated if the proper sequence of operation is followed. This process enables metals of different thicknesses to be welded together successfully. It also enables successful welding of almost any metal or metal alloy. The method has been particularly useful for fabricating missile components, where accurate alignment of the assembly is of paramount importance.

A spike welding machine looks very

similar to the standard spot welding machine.

13-12. METAL FOIL WELDING

Metal foil welding is a patented process. While it is a resistance welding electrodes and the joint is covered both top and bottom with a thin foil (approximately .010 in.) of metal of the same material as the metal to be joined.

The foil tends to concentrate the welding current in the immediate area of the weld joint. The weld formed has

Stage 1, Cold Set Down—No Current—Original Projection Height of .050 decreases .010 caused by .008 indentation and .002 projection collapse. Sheet separation .040.

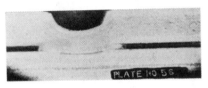

Stage 2, 1-Cycle Weld Time—Pressure Weld Forming with Tensile Shear Strength of 1050 lbs.—Sheet separation .010.

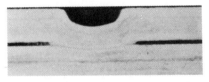

2-Cycle Weld Time—Strength 1750 lbs.—Sheet separation of .007.

Stage 3, 4-Cycle Weld Time—Strength 2300 lbs.—Sheet separation .005.

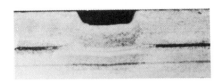

Stage 4, 6-Cycle Weld Time—Strength 2600 lbs.—Sheet separation .005.

Stage 5, 10-Cycle Weld Time—Strength 3125 lbs.—Sheet separation .002. This is last stage of welding without nugget formation.

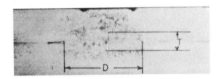

Stage 6, 12-Cycle Weld Time—Strength 3100 lbs.—Sheet separation .001. Nugget formation has started. Nugget diameter .150.

8-Cycle Weld Time—Strength 3000 lbs.—Sheet separation .002.

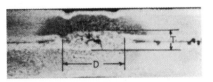

Stage 7, 16-Cycle Weld Time—Strength 3700 lbs.—Sheet separation .0. Nugget diameter .160. This is recommended weld time.

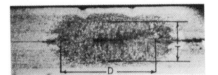

Stage 8, 20-Cycle Weld Time—Strength 4000 lbs.—Nugget diameter .300. This last stage indicates increase in nugget size with increased weld time.

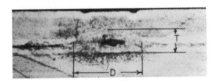

14-Cycle Weld Time—Strength 3200 lbs.—Sheet separation .0. Nugget diameter .160.

Fig. 13-15. Samples of projection welds as weld progresses from 0 cycles to 20 cycles (20 cycles - 1/3 second) weld time. (Taylor-Winfield Corp.)

process, it has many features which make it unique. As shown in Fig. 13-16, sheet steel may be butt welded with this process. The sheets to be butted together are placed between wheel type a slightly raised bead as compared to a slight indentation from the usual resistance welding operation. Fig. 13-17 shows the weld in its various stages of formation. This process is

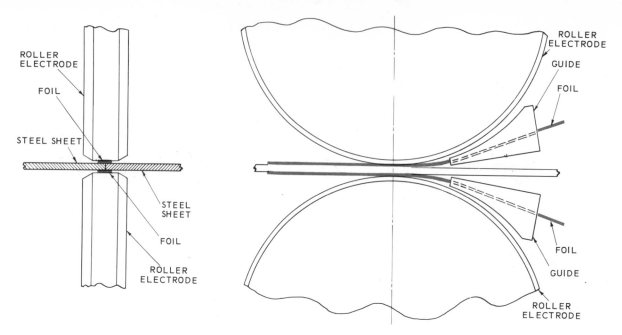

Fig. 13-16. Principle of foil welding. Note butt joint, and that the foil serves as
filler and reinforcing material.

used in joining sheets for automobile body construction and similar uses because this type seam may be finished by the usual metal finishing processes and therefore no solder or other "filler" material is required.

An automatic metal foil resistance welding machine for welding automobile roof panels is shown in Fig. 13-18.

13-13. METAL FIBER WELDING

A metal fiber weld is usually a lap joint resistance weld. Metal fiber sheets are formed by "felting" very tiny filaments of the metal to be used to produce a finished sheet much like a thick sheet of felt cloth. A strip of this felted fiber is then placed between the two pieces of metal to be joined. Fig. 13-19 shows a cross section of the metal fiber and welding electrodes preparatory to performing a metal fiber resistance weld.

Metal fiber welding is applicable to a variety of resistance welding jobs. The metal fiber used may be impregnated with various metals such as copper, brazing metals, silver or an alloy. Such fibers may be used to join two different metals together. Copper may be joined to stainless steel and other similar joints may be made.

When resistance welding, using fiber metal, less electrode pressure is required as the fiber offers a greater resistance to the flow of current than the solid metal so it heats first, and to a higher temperature. This high temperature condition at the surfaces of the metal to be joined greatly assists in forming a good resistance weld. Also, since a lower electrode pressure is required, less indentation of the metal will be made.

13-14. PERCUSSION WELDING

Percussion welding is a form of resistance welding in which the heating of the metals to be welded occurs over their entire surface area by an electrical discharge. The weld is made by applying force rapidly (percussively) immediately after the electrical discharge occurs between the adjoining surfaces.

The parts to be joined are held apart by a small projection (nib), or one part is moved toward the other. After the complete surface areas of both parts are heated to a welding temperature, an impact forces the metals together to complete the weld.

Power supplies for percussion welding are of four types as follows:

1. Low voltage, capacitive storage.
2. High voltage, capacitive storage.
3. Electromagnetic or inductive storage.
4. Low voltage AC that uses a transformer to furnish the welding voltage.

The arc is initiated by high frequency (HF) imposed on the DC welding current. The welding force is applied through pneumatic cylinders, electromagnets, springs, or gravity (falling weights). The metal after welding will be slightly shorter (burn-off) than before welding, but this burn-off rate can be ignored when the current is supplied by capacitor discharge.

13-15. HIGH FREQUENCY RESISTANCE WELDING

When extremely high frequency current flows through a conductor it flows at or near the surface of the metal and not through the entire thickness.

This phenomenon of high frequency

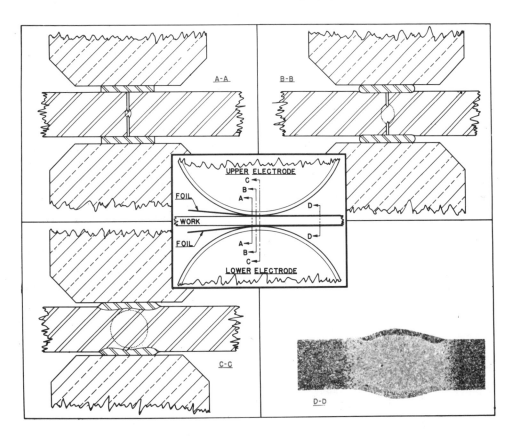

Fig. 13-17. Fusion action during foil butt welding process: AA—Start of weld. BB—Nugget is approximately one half developed. CC—Weld approximately complete. DD—This is an actual sample cross section of a completed weld etched and macrophotographed.

Fig. 13-18. *An automatic metal foil resistance welding machine used to weld a seam on automobile roof panels. The panels are clamped in the fixtures, and the seam welder is moved toward the fixture. When the full length of the seam has been welded, the seam welder is moved back to the starting position. (Precision Welder, Flexopress Corp.)*

electricity has been used to weld sections as thin as .004 in.

Low amperage current with frequencies as high as 450,000 cycles per second are used. This high frequency heats

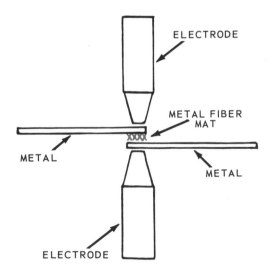

Fig. 13-19. *Setup for a metal fiber resistance weld. Diagram showing the principle of fiber metal spot welding. The metal fiber mat is placed between the base metals before they are welded.*

the surface of the metal and when pressure is applied to force the metal parts together, an excellent fusion weld occurs at the surfaces of the parts.

Copper has been welded to steel and alloy steel to mild steel by this method. Exotic metals can also be welded in an inert gas atmosphere.

The electronic equipment needed for this high frequency resistance method of welding is very complicated and extensive special training is needed to become a competent technician for maintaining, adjusting, and servicing the machines.

13-16. SPOT WELDING SETUP

Physical distortions may be due to misalignment of tips, original malformation of the pieces, improper clamping, improper welding technique causing buckle, as when operator works from both ends to the center on a long sheet, instead of from the center, out to both ends. Reasons for physical distortion must be traced immediately and corrected. See Fig. 13-20.

Blow-holes and cavities are a common occurrence from improperly set controls, too short squeeze time, improper positioning of parts with reference to the tips. See Fig. 13-21. Other physical conditions such as dirt, burrs, etc., on the surface of the metal result in a quick, excessive concentration of welding current. Insufficient capacity to absorb or conduct such current to surrounding metal causes the blow-holes.

Tips which have parallel faces, but whose centers are not aligned, have drastically reduced net effective tip areas. See Fig. 13-22.

When surfaces of welding tip are not parallel, pressure and current are confined to a fraction of the proper area. See Fig. 13-23.

13-17. SPOT WELDING ALUMINUM

The electrodes collect more metal when spot welding aluminum than when spot welding mild steel. Tip maintenance is required after about 500 spots, whereas with steel the tip requires cleaning after approximately 6000 spots. The aluminum surfaces to be spot welded should be cleaned of aluminum oxides immediately before the weld process.

13-18. REVIEW OF SAFETY IN RESISTANCE WELDING

Resistance welding and resistance welding equipment, if properly handled, is one of the safest types of welding. The greatest dangers are: (1) Flying sparks; (2) Electrical shock; (3) Hot or sharp objects; (4) Moving machinery.

It is recommended that all operators of resistance welding equipment wear face shields or flash goggles. There may be some flying sparks or "flash" thrown from the joint being welded. Protective clothing is also necessary.

The welding voltage across the electrodes is very low. The primary circuit

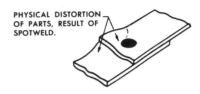

Fig. 13-20. Sheet metal warping due to improper spot weld.

Fig. 13-21. Faulty spot weld which results in a blow hole in sheet metal.

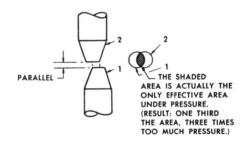

Fig. 13-22. Effect of using electrodes which are off-center.

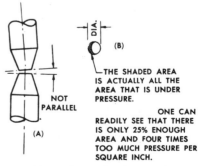

Fig. 13-23. Welding electrode tips are in contact but not parallel.

wiring should be handled only by qualified electricians. All resistance welders should be grounded.

Naturally, any surface being welded will be very hot. Operators should always wear leather or asbestos gloves if it is necessary to handle the materials being welded. Gloves will also protect the hands against sharp or ragged sheet metal edges.

Most modern resistance welders are power operated. For this reason, the electrode pressures and forces may be quite high. There is danger of injury if the operator's hand or fingers should accidentally be caught between the electrodes as they come together at the work. Safety devices are available and should be used to insure that no part of the operator's body can be in a danger area at the time the machine starts to operate.

13-19. TEST YOUR KNOWLEDGE

1. What kind of current is used in a resistance welder?

2. How is the current changed by the transformer?

3. Are resistance welding machines used for automatic welding?

4. What is the difference between an upset welder and a flash welder?

5. What metals are used for electrodes on spot welders?

6. What metal alloyed with copper tends to make the electrode harder?

7. How much current is passed through a spot weld?

8. What is shot welding?

9. What are the four greatest dangers when resistance welding?

10. Name the four variables of resistance welding.

11. Where is electric resistance welding most commonly used?

12. What is spot welding?

13. Name the principal parts of an electric resistance welder.

14. Is pressure used when spot welding? If so, can you explain why?

15. Why do most gun welders have more secondary turns than spot welders?

16. What is hold time?

17. What usually happens if one attempts to spot weld corroded or dirty metal?

18. Can a butt joint be welded using the seam welding machine?

19. What causes the "flash" in a flash welder?

20. May projection welding be done using roller electrodes?

21. What is the basic principle of a resistance welder?

22. What is the source of the welding current in most resistance welding machines?

23. What is a popular spot welder electrode material?

24. What are the three variables of resistance welding?

25. Can you describe the physical condition of the spot weld during the time the weld is being formed?

26. What is the most popular joint used when spot welding?

27. What resistance welding machine is used for making butt joints?

28. What type of resistance welding machine can make a continuous weld?

29. What type of resistance welding machine requires indentations in the metal being welded?

30. What is the most important safety precaution to be taken when working on a spot welder?

Chapter 14
RESISTANCE WELDING
EQUIPMENT AND SUPPLIES

Designing and constructing resistance welding machines involves a combination of electrical design, machine structures, mechanisms, and controls. The machines vary from simple mechanisms to exceptionally complicated units.

Accurate control of all three of the resistance welding variables is essential if good resistance welds are to result.

The Resistance Welder Manufacturer's Association (RWMA) and the National Electric Manufacturer's Association (NEMA) have developed standards for resistance welding machines, and these standards have enabled the manufacturers to produce machines of known rated capacity and recognized durability.

The supplies required in resistance welding are chiefly electrodes and electronic control elements.

Resistance welding equipment is classified in several ways. One way is according to the type of joint which it is capable of producing, such as a lap joint and a butt joint. Equipment for lap welding is usually designed to produce either spot, seam, or projection welds. Equipment for butt welding is designed to produce either flash or upset welds.

Another classification is according to the type of electrical power used:
1. Single- phase direct energy.
2. Three- phase direct energy.
3. Stored energy.
Of the three types, the direct energy machines which use electricity directly from the power line without storing it are the most popular.

14-1. ELECTRIC RESISTANCE WELDING MACHINES

In design, most electric resistance welding machines are quite similar. However, the manner in which the metal to be welded is held, and the appearance and position of the welding electrodes varies from machine to machine.

The methods used to actuate the operating mechanisms of the machine and the methods used to control the various time and welding factors also vary.

Generally these machines may be divided into three main groups:
1. Manual control of time and electrode pressures.
2. Manual control of welding time with hydraulic or pneumatic pressure devices. Electric solenoids are used to control both the fluid flow, the electrode force and the electricity flow.
3. Automatic control of welding time with hydraulic, pneumatic, or electric solenoids being used to supply and control electrode forces and electrical flow.

The various parts of the manual machine are:
A. Frame.
B. Transformer.
C. Welding arms or electrodes.
D. Manipulating mechanisms.
E. Control switches.

The transformer is of special construction and usually has several taps or adjustments. It may be either air-cooled or water-cooled. The secondary winding usually consists of one loop or of several parallel loops. The ends of these loops are soldered, bolted, or brazed to the electrode arms of the machine.

The welding arms and operating mechanisms are different for each type of electric-resistance welder. However, each one usually uses foot power to press the parts together; or a foot-operated switch in connection with electro-magnets, pneumatic cylinders, or hydraulic cylinders to perform this work. The operating mechanism also operates the welding current switch. As the mechanism acts, the switch is simultaneously turned on and off at the correct moment for corresponding positions of the clamping arms or electrodes. The control switches are manual, primary circuit switches, and are designed to close at the proper time, and to give the correct amount of current (weld time). The National Electric Code and the local electrical utility require that all code regulations be observed and that power factor and other information be considered when connecting these machines onto the power line.

The welder switch usually has large copper contacts which may be easily cleaned or replaced. They are adjustable both for wear and for correlating their timing with the movement of the welding electrodes or points.

The various parts of the automatic machine are similar to the manually operated machine, with the exception that the movement of the electrode arms and the contacting of the switches is all performed automatically. Automatically timed electro-magnets, hy-

draulic pressure, or air pressure do all of the operations. Once a machine is timed correctly for a certain operation, the above device eliminates the human element, and the machine will repeatedly produce almost identical welds.

14-2. TRANSFORMERS

The electric resistance type of welding machine uses a transformer which is designed and built to use the voltage and current provided by the electric utility companies. Neither the current flow nor voltage supplied is suitable for welding needs.

This transformer, then, must deliver a higher amperage and a lower voltage of electrical energy for use at the welding electrodes or welding points of the machine. This type of transformer is called a "step-down" transformer since it reduces the voltage supplied at the electrical outlet.

For example, assuming 100 percent efficiency and 100 percent power factor, a step-down transformer will transform 20 amps. at 115 V. to 100 amps. at 23 V., or 1,000 amps. at 2.3 V. or 10,000 amps. at .23 V. The resistance welding transformer usually delivers secondary welding current in the range of 10,000 to 100,000 amps. The secondary circuit voltage is usually about 10 V. open circuit and may decrease to less than 1 V. when current flows during welding.

Most of these transformers are intermittent in operation. That is, they deliver current for only short intervals of time.

The amount of time the transformer delivers current, in ratio to the off time, is called the duty cycle. Thus, if a transformer delivers current 3 seconds out of each minute (60 seconds)

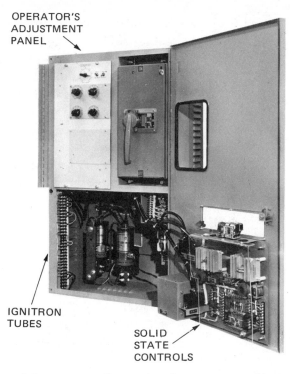

OPERATOR'S
ADJUSTMENT
PANEL

IGNITRON
TUBES

SOLID
STATE
CONTROLS

A solid state controller used with resistance welders.

the length of the duty cycle is 5 percent:

$$3 \div 60 = .05$$

$$= \frac{5}{100} \text{ or 5 percent}$$

An interesting design problem is introduced in the design of resistance welders in that the primary windings, the core, and secondary windings tend to move in relation to each other as the current flows and is then interrupted. A rigid design is needed to prevent insulation wear which would soon lead to short circuiting and destruction of the transformer.

The secondary wiring is usually made of cast copper or rolled copper bars. These bars are sometimes water-cooled--either by welding or brazing copper tubes to the bars or by using

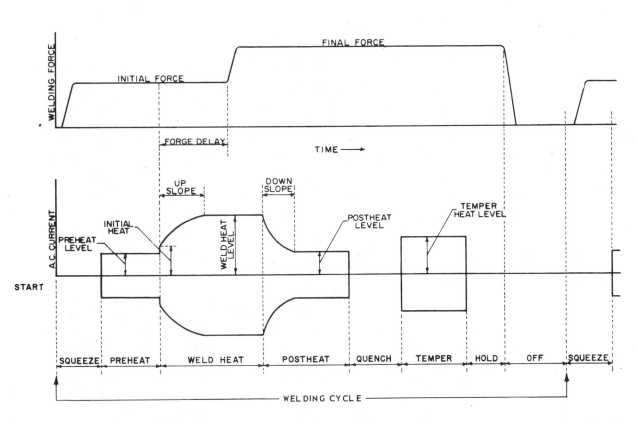

Solid state electronically controlled resistance welding cycle. Note how the squeeze force and the heat level vary through the welding cycle.

Resistance Welding Equipment 335

hollow bars to carry the cooling water as shown in Fig. 14-1. Smaller units may be air-cooled.

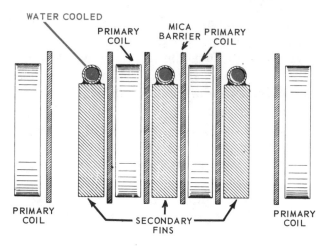

Fig. 14-1. Schematic cross section of a resistance welding transformer. Note that the secondary circuit fins or bars have water-cooling coils brazed to them.

The laminated core of the transformers in resistance welders is usually made of 4 percent silicon transformer iron.

The insulation of the primary windings must be such that accidental grounds cannot occur. Fig. 14-2 illustrates a typical primary winding. Great care is taken to insulate the ribbon type primary windings from each other and from the frame. A secondary winding which consists of three parallel loops is shown in Fig. 14-3. This design is still considered as one winding even though the electricity divides into three parallel paths.

Fig. 14-4 shows a schematic wiring diagram for a resistance welding transformer. This machine provides for eight current ranges. The range is dependent on the ratio of the number of turns in the primary circuit to the number of turns in the secondary. A wiring diagram for a resistance welder with a high-low tap and an eight-step primary winding is shown in Fig. 14-5.

The transformers vary in capacity and will suit varied welding needs.

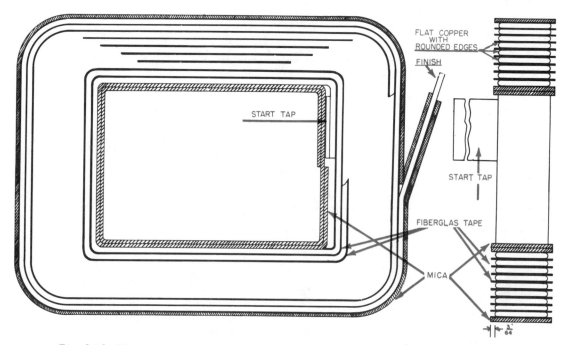

Fig. 14-2. The construction of the primary winding of a typical resistance welding transformer. The conductor is flat or ribbon shaped and the turns are insulated from each other by fiber glass. (Taylor-Winfield Corp.)

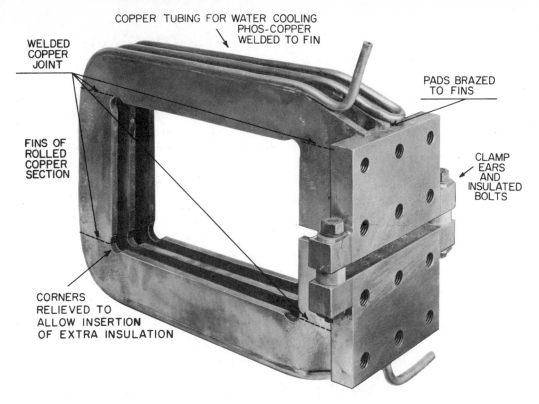

Fig. 14-3. A typical secondary winding design for a resistance welding transformer. In this case three bars are connected in parallel to increase current carrying capacity and still retain the one turn electrical function.

Fig. 14-4. Schematic wiring diagram which shows adjustable *primary windings* of a resistance welding transformer. This unit has eight taps (adjustments) in the primary circuit.

The capacity is usually listed as the KVA (kilovolt-ampere) rating. These

Fig. 14-5. Resistance welding transformer schematic wiring *diagram*. This unit has a high-low tap plus eight regular taps in the primary winding for purposes of current adjustment.

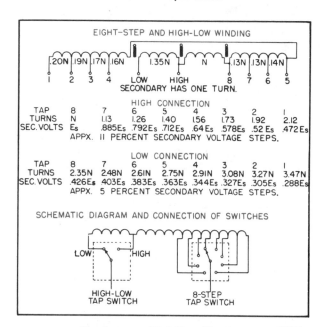

transformers are rated on a different basis than most transformers. The KVA is the input or primary circuit rating. For example, with a 230 volt input and 43.48 amps. flowing, the machine would be rated at 10 KVA. (230 x 43.48 = 10,000.40.

$$\frac{10,000}{1,000} = 10 \text{ KVA.})$$

A welding transformer is usually rated at 50 percent duty cycle, which means that it should be used in actually creating welds 30 seconds out of each 60 seconds, 10 seconds out of 20 seconds, or any segment of time where the idle time at least equals the weld time.

Duty cycles vary all the way from 1 percent to 100 percent, with a normal rating of 50 percent. If the duty cycle is high, the KVA setting must be smaller to insure safe transformer temperatures. The manufacturer usually specifies the KVA rating based on the 50 percent duty cycle.

The welding transformer is rated on a specified input and not on the output like other transformers. Since the output of the transformer will vary with the conductivity of the metal being welded, the electrode circuit, and other factors, the welder manufacturer rates only the primary input capacity of the transformer.

When a machine is used at its rated KVA, the temperatures of the transformer windings will stay within safe limits for the insulation used. When the KVA rating or duty cycle is exceeded, the transformer efficiency decreases rapidly as the temperature rises. There is also the possibility of an insulation breakdown occurring.

Small spot welders may be rated in VA (volt-amperes). This is done when the KVA rating would be a decimal as in the case where 300VA = .3KVA.

Some of the large resistance welding machines use 3-phase transformers.

When it becomes necessary to increase the total metal thickness being welded, remember that the current required to weld this new thickness will also increase. The increase in required current is not, however, in direct ratio to the increase in the thickness of the metal.

The required welding current increase is the square root of the ratio of the new total metal thickness to the original total metal thickness. For example, if the thickness doubles, the required current is 41 percent more ($\sqrt{\text{double thickness}} = \sqrt{2} = 1.41$, because 1 or 100 percent is the original current, the additional .41 = 41 percent increase).

The primary current flow influences the electrical power service. For example, a 100 KVA unit may draw as much as 1000 amps. The input electrical service must therefore be carefully considered. For example, a 10 percent drop in line voltage will cause a 20 to 30 percent reduction in weld heat available. Thus automatic machines will produce substandard welds if the line voltage varies.

14-3. SPOT WELDING ELECTRODES

Resistance welding electrodes conduct the current to the surfaces of the metals to be welded. There are certain requirements which these electrodes must possess. They must:

1. Be a good conductor of electricity.

2. Be a good conductor of heat.

3. Have good mechanical strength and hardness.

4. Have a minimum tendency to alloy (combine) with the metals being welded.

Pure copper possesses good electrical and thermal properties; however, it is rather soft and does not wear well.

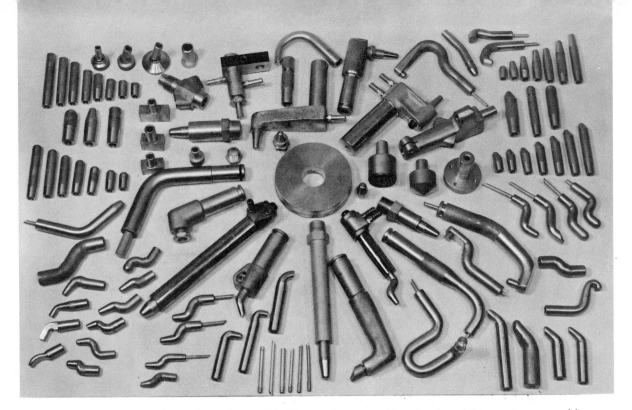

*Fig. 14-6. Variety of electrodes, electrode holders and seam welder wheels used in resistance welding.
(Hercules Welding Products Co.)*

Also it tends to soften with heat.

The electrode must be a good conductor of electricity in order that the current may flow to the work piece without overheating the electrode. It must be a good conductor of heat so that the heat from the high temperatures generated at the point of contact at the weldment may be conducted away from the point without causing it to become overheated. Typical spot welding electrodes are shown in Fig. 14-6.

The Resistance Welder Manufacturer's Association (RWMA) and the Resistance Welding Alloy Association (RWAA) recognize two groups and several classes of materials used for resistance welder electrodes. The groups are Group A and Group B.

Group A are copper-base alloys having good electrical and thermal (heat) conductivity and with improved hardness and wear qualities.

Group B are refractory metal alloys. The electrical and thermal properties of this group are not as good as the alloys in Group A. However, they have extremely high melting temperatures, and high compressive strength and wear resistance. These materials are usually sintered mixtures of tungsten and copper.

When ordering or purchasing resistance welder electrodes the following data may be referred to as a guide:

GROUP A, COPPER-BASE ALLOYS
Characteristics And Use

Class 1. General purpose--recommended for spot welding aluminum alloys, magnesium alloys, galvanized iron, brass or bronze. The electrode material is a copper-cadmium alloy.

Class 2. For high production spot and seam welding, clean mild steel, low alloy steels, stainless steels, nickel alloys and monel metal. The electrode material is a copper-chromium alloy.

Class 3. Higher strength than Class 2. Recommended for projection welding electrodes, flash and butt welding electrodes. Recommended for stainless steel. The electrode material is a copper-zirconium alloy.

Class 4. This is a hard-high strength alloy. This alloy is recommended for use as an electrode material for special application when the pressures are extremely high and the wear is severe and electrode heating not excessive.

Class 5. This alloy is used chiefly as a casting and has high mechanical strength and moderate electrical conductivity. It is used chiefly for electrode holders.

GROUP B, REFRACTORY METAL COMPOSITIONS

Characteristics And Use

Class 10. This electrode material has high electrical conductivity and is somewhat malleable. It is recom-

obtainable in various sizes and forms. These electrodes must be kept clean and correctly shaped if they are to produce good results. A number of different spot welding electrode shapes are shown in Fig. 14-8.

The electrodes used in the other types of resistance welding equipment are described in their respective Paragraphs.

14-4. ELECTRODE HOLDER

The resistance welder tips are held by electrode holders. The holders in turn are clamped into the ends of the movable spot welder arm and the stationary arm. Most of the tip holders are water-cooled.

The electrode holders are the ma-

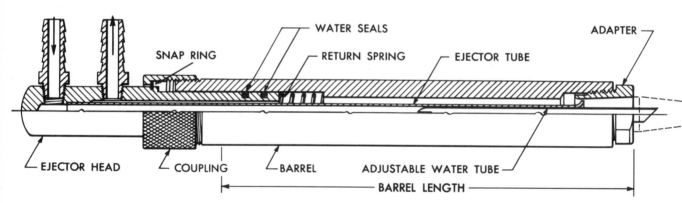

Fig. 14-7. Resistance welding water-cooled electrode holder.
(Tuffaloy Products, Inc.)

mended for facings for projection welding electrodes and flash and butt welding electrodes.

Class 11. This material is harder than the material in Class 10.

Class 12, 13, 19. These are special materials particularly for welding metals having a higher electrical conductivity. In general, special setup and machines are required for this type of resistance welding.

Resistance welding electrodes are

chine arms or supports which hold the electrode (tips) in proper position, carry the welding current, and provide the tips with water-cooling. Fig. 14-7 shows a typical electrode holder and tip.

It should be noted that on most of the spot welding machines the electrode holders are adjustable for length and position. In general, the electrode holders should be adjusted to the shortest length at which the weld metal may be easily inserted. Electrode holders are

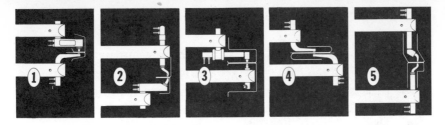

Fig. 14-8. *Offset electrode holders, electrode designs, and how they are used.*

made of a copper alloy which provides good current carrying qualities and rigidity.

14-5. MECHANISMS

The operation of resistance welding machines involves levers, cams, gears, racks, screws, etc.

The spot welder arm is normally lever operated. The upset welder movable electrodes are sometimes cam operated. The flash welder sometimes uses the gear-and-rack mechanism. The motor driven parts of the automatic spot welder often use the screw mechanism.

14-6. CONTROLS

Controls for resistance welding machines vary from simple hand adjustments and switches, to precise electronic controls (SEE CHAPTER 21, FOR MORE INFORMATION ON AUTOMATIC CONTROLS).

The variables of resistance welding can all be controlled manually or automatically. These variables are:

1. Current.
2. Pressure.
3. Time.

The current may be manually adjusted by the primary steps on the transformer.

The pressure of the electrodes on the metal being welded can be controlled by the operator's pressure on the levers, semiautomatically controlled by the use of adjustable springs, or automatically controlled by hydraulic or pneumatic pressure and the size of the cylinders.

The time that the current flows and the time the pressure is imposed can be manually controlled on simple machines. The more automatic machines use electronic timers that function on the basis of the number of AC cycles that pass through the circuit such as:

1. Pressure applied for 20 cycles. (Squeeze Time)
2. Current flows for 6 cycles. (Weld Time)
3. Current off-pressure maintained for 15 cycles. (Hold Time)

14-7. SPOT WELDING EQUIPMENT

Spot welding machines are the most common of the resistance welding machines. Spot welders are made in a great variety of sizes from small bench units used to spot weld such items as costume jewelry and electron tube components as shown in Figs. 14-9 and 14-10, to mammoth machines to spot

Fig. 14-9. *Small bench model spot welder with 3 KVA capacity. The cabinet in the background contains the electronic controls. The overall height is only 16 in. Note that the electrode tips are directly clamped to the arms.   (Eisler Eng. Co.)*

ELECTRODES

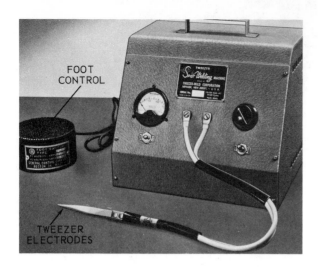

Fig. 14-10. Spot welder for fine precision work. The electrodes are of the tweezer design.

machines using many electrode sets automatically sequence operated.

The equipment basically consists of the same main parts as all resistance welding equipment.

The basic purpose of the spot welder is to spot weld (form a fused nugget) between two lapped pieces of metal. Fig. 14-13 shows a table of sheet metal thicknesses and the KVA needed to successfully spot weld the material together.

The dimensions of the operating area are important in that the size of the assembly which may be welded is controlled by the throat depth. The size of the sheets (curves and attached parts) the machine can handle is controlled by the horn spacing, as shown in Figs. 14-14 and 14-15.

weld hundreds of spots on large sheet metal productions such as automobile bodies, refrigerator cabinets, and the

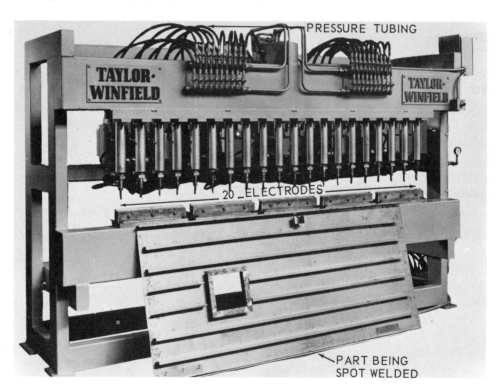

Fig. 14-11. Large multiple electrode, automatic spot welder.

like as shown in Figs. 14-11 and 14-12.

The machines vary from one set of electrodes manually operated to large

The force imposed by the electrodes against the metal is controlled by variable compression spring forces,

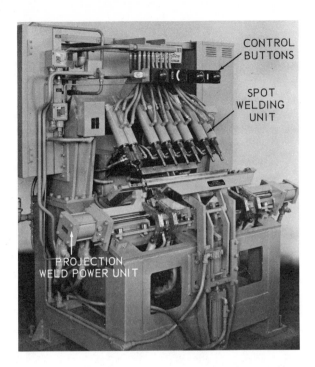

Fig. 14-12. Combination spot welding and projection welding machine with four 55 KVA transformers. The input power is 440 V., 60 cycle. The spot welding guns lower, weld, and retract in 1-1/2 seconds. There is one projection welding unit at each end. The units advance, weld, and retract in 1-1/2 seconds. There are dual controls. The operator manually loads and unloads the parts. Note the pneumatic, hydraulic, and electric devices. The necessary parts are water-cooled using water valves and electric timers to control the water flow. (Resistance Welder Corp.)

hydraulic forces, pneumatic forces, or magnetic forces. The forces are varied from a few ounces to hundreds of

THICKNESS OF EACH OF TWO PIECES		
GA.	IN.	KVA
16	.063	12
14	.078	15
13	.093	17
11	.125	20
10	.141	28
9	.156	35
7	.187	35
5	.218	50
3	.250	50

Fig. 14-13. Table of approximate KVA settings for spot welding various thicknesses of steel.

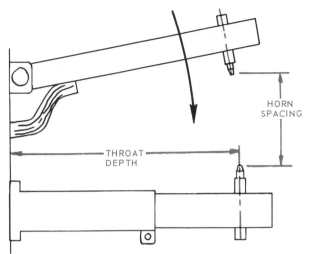

Fig. 14-14. Throat and horn openings of a spot welder. Such a unit has a lever type top arm operated by a motor driven cam or a mechanical linkage to a foot pedal.

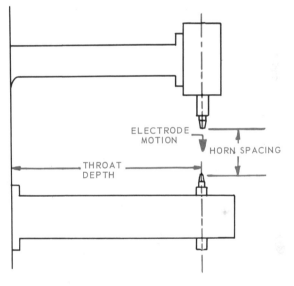

Fig. 14-15. Throat and horn openings of spot welder with a pneumatic, hydraulic or magnetic top electrode mechanism.

pounds in the different size spot welders. In a particular machine, forces may be varied from as low as 5 lbs. to as high as 100 lbs. Fig. 14-16 shows a schematic of a spring pressure mechanism.

Small portable spot welders are also used to spot weld sheet metal. These units are light enough so they may be

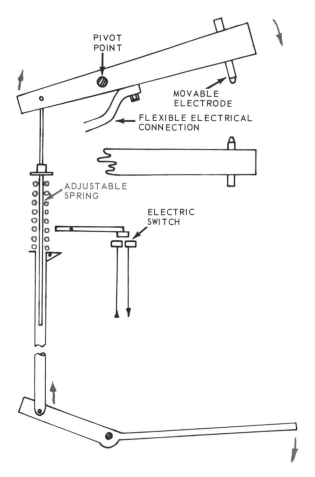

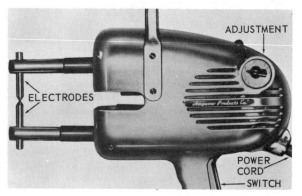

Fig. 14-16. *Adjustable compression spring is used to control the electrode force against the work.*

Another application of spot welding is the high rate discharge machine. This process is sometimes called "spike" welding. Most spike welders are basically spot welders with added electrical characteristics. There are several types of resistance welders developed to do this work:

A. Auto-transformer type.
B. Electro-magnetic type.
C. Electro-static type (capacitor).

The capacitor type is most used. Its principle of operation depends upon electrical condenser action. An accurate timing device provides for charging a condenser and then directing the

Fig. 14-17. *Small portable spot welder. The current flow is electronically timed. The squeeze force is approximately 1000 lbs. (Ampower Products, Inc.)*

carried around a large sheet metal fabrications job to make spot welds on irregular edges and in hard-to-reach places. Fig. 14-17 illustrates a portable spot welder, while Fig. 14-18 shows the wiring diagram of the same welder.

Shot welding is a term used in the trade for a process which is identical with spot welding, with the exception that in shot welding, a very high current is used for extremely short periods of time (1/100 second or less). The accurate time control enables this equipment to be used extensively for spot welding of stainless steels, aluminum, magnesium and other metals and metal alloys.

discharge through spot welder points and the metal to be welded. The control or timing device makes use of vacuum tubes or semi-conductors, and is popularly known as electronic control. Fig. 14-19 shows a spike welder with water-cooled electrodes and a water-cooled transformer.

The use of the capacitors to store current enables the welder to use a smaller KVA transformer because the off-time part of the cycle of the machine is used to electrically load the capacitors or condensers.

The electronic switching and timing makes possible minimum distortion

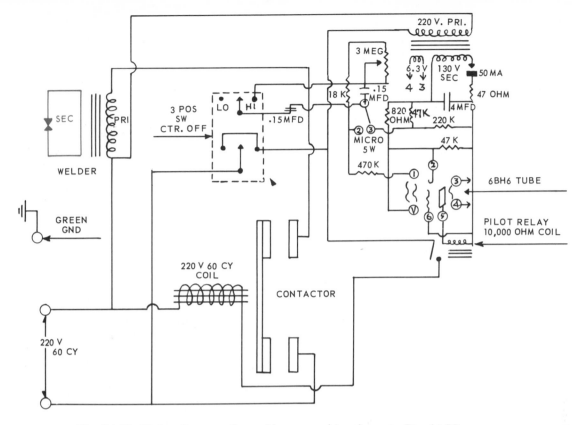

Fig. 14-18. Wiring diagram of portable spot welder shown in Fig. 14-17.

Fig. 14-19. A spike welder. This welder is pneumatically operated. The cabinet on the right houses the electronic controls called coaxial ignition spike controls. (Weldex Div., Metal Craft Co.)

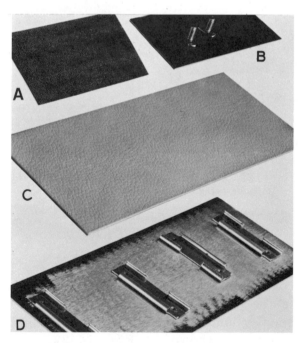

Fig. 14-20. Two samples of small parts welded by spike weld process. This process does not mar or blemish the other side of the metal. A. Top of coated metal. B. Same piece on the other side with two small screws spike welded to it. C. Top of coated metal. D. Underside of same metal with the pieces spike welded to it.

Resistance Welding Equipment 345

and minimum corrosion. Spot welds may be made as fast as 500 spots per minute.

These machines deliver a "spike" discharge of as much as 100,000 amps. for approximately a millisecond (1/1,000 of a second).

These machines successfully spot weld parts to vinyl-coated and lacquer-coated metal without discoloring the coatings. Figs. 14-20 and 21 illustrate some typical finished parts welded by that process. This process is particularly useful for spot welding aluminum and zinc-coated metals. Also it is often used for welding small parts where too much heat may be conducted to areas outside the weld joint and cause distortion, discoloration, or other damage.

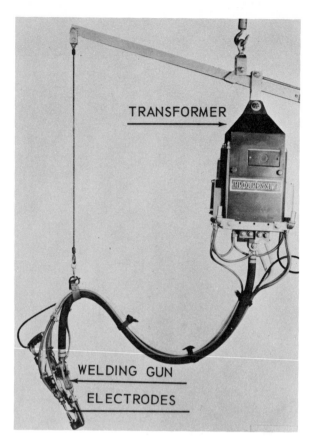

Fig. 14-22. Suspended type portable gun welder.

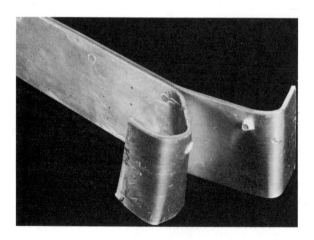

Fig. 14-21. Test sample of an aluminum spot weld made using the spike welding process.

14-8. PORTABLE WELDING MACHINES

A special application of spot welding is the use of a portable spot welder, usually supported from an overhead track, as shown in Fig. 14-22. The electrode arms of the welder are operated by pneumatic or hydraulic forces. Note that the transformer is mounted on the track to minimize the bulkiness of the gun. The gun-type spot welder usually requires a higher secondary voltage because of the increased length of the secondary leads. These transformers may have two or three turns in the secondary. The operator merely puts the fixed electrode on the spot to be welded, and presses a trigger. Air or hydraulic pressure presses the moving electrode into place. The electrical current is usually electronically timed. This welder can produce a great number of spot welds on an irregular shaped article very quickly. Most portable welding machines operate on a 25 percent duty cycle. This tool is used extensively in automobile body manufacturing. The electrode holders, electrodes, and operating cylinders for a portable gun welder are shown in Fig. 14-23.

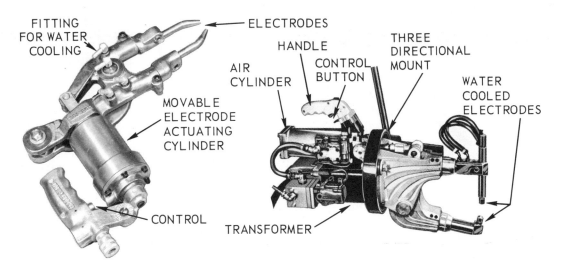

Fig. 14-23. Gun welders showing electrode holders, electrodes, and actuating cylinders. Left. Gun welder which would be connected to a remote transformer. (Progressive Machinery Corp.) Right. Portable gun welder which has a built-in transformer. The lower electrode is pneumatically operated. Note the mounting which permits electrode movement into any position. (Falstrom Co.)

14-9. SEAM WELDING MACHINES

A seam welding machine is a special form of a resistance welder, usually one which uses two rollers as electrodes. These rollers press the two pieces of metal together and roll slowly along the seam. Either a continuous current is passed between the rollers welding the two pieces with a continual weld or current is passed through the metal at timed intervals to produce either overlapping weld nuggets or weld nuggets spaced at intervals. Fig. 14-24 shows a wheel-type seam welder. The machine must be carefully

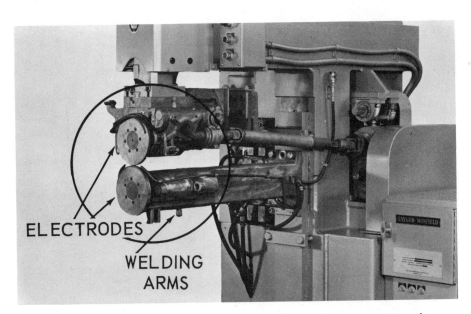

Fig. 14-24. Seam welding machine. Electrodes on this machine are positioned transversely to the welding arms. In this position the welded seam moves in a direction perpendicular to the welding arms.

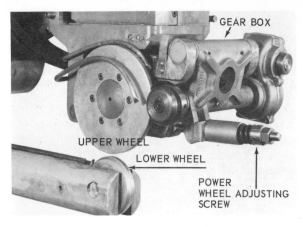

Fig. 14-25. *This illustration shows the electrodes positioned longitudinally to the welding arms. The wheels are positioned so the welded seam travels parallel to the welding arms.*

timed to secure good results, cooling of the rollers must be continuous, and pressure must be uniform. Both the metal and the rollers must be kept clean.

These machines are also made as portable units similar to the gun welders. Some seam welders are used to butt weld the longitudinal seam of welded pipe. Examples of seam welding wheels are shown in Fig. 14-25.

Seam welders are made with the axis of the electrode wheels positioned parallel to the front of the machine (longitudinal setup), perpendicular to the front of the machine (circular or transverse setup), or the machine may be built so that the wheels can be positioned at any desired angle (universal). Fig. 14-26 shows a universal seam welding machine.

The details of the upper seam welder wheel operating mechanism are for this

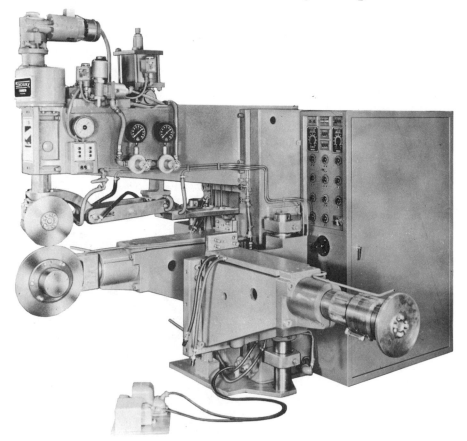

Fig. 14-26. *Universal seam welder. Note that the electrode wheels are set up for longitudinal welding. The arm to the right has the electrode wheel set for seam welding in the transverse position. When this arm is moved into position under the upper wheel, the upper wheel must be rotated 90 deg. (Sciaky Bros., Inc.)*

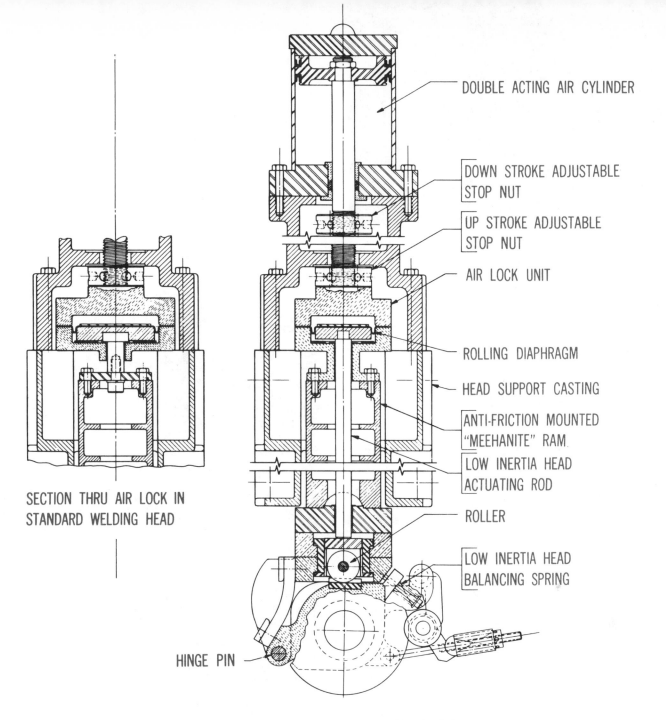

DOUBLE ACTING AIR CYLINDER

DOWN STROKE ADJUSTABLE STOP NUT

UP STROKE ADJUSTABLE STOP NUT

AIR LOCK UNIT

ROLLING DIAPHRAGM

HEAD SUPPORT CASTING

ANTI-FRICTION MOUNTED "MEEHANITE" RAM

LOW INERTIA HEAD ACTUATING ROD

ROLLER

LOW INERTIA HEAD BALANCING SPRING

HINGE PIN

SECTION THRU AIR LOCK IN STANDARD WELDING HEAD

TRANSVERSE SECTION THRU AIR CYLINDER, AIR LOCK AND LOW INERTIA HEAD

Fig. 14-27. Pneumatically operated upper electrode of a universal seam welder. It is used to raise and lower the upper roller and to develop the correct welding pressure. (Taylor-Winfield Corp.)

type of machine shown in Fig. 14-27. Occasionally seam welds are made on parts which will not allow a continuous seam. In this case a shoe type electrode (wheel segment) may be used as shown in Fig. 14-28. The shoe electrode

Fig. 14-28. Shoe electrode used to weld a seam close to a part which is in the path of what would otherwise be a continuous seam.

allows a seam weld to be made closer to a part which lies across the seam than would be possible with a full electrode wheel.

14-10. PROJECTION WELDING EQUIPMENT

Another type of resistance welding uses a machine which welds together two pieces of metal, one having small projections pressed into it. When the two pieces are pressed between the electrodes and the current turned on, the projections are fused into the mating piece. This method speeds the welding, but requires the extra operation of pressing the projections into the metal.

Dies used to force the metals together may be special electrodes, rollers, or special dies. Fig. 14-29 illustrates a resistance welding machine which would be suitable for making projection welds on surfaces of large areas.

14-11. UPSET WELDING MACHINES

Upset welding is often called butt welding. However, butt welding merely means that the pieces are welded together in a butted position. Therefore a butt weld may be made by any of the usual welding techniques.

Upset welding is a resistance welding process where combining of the metals is produced simultaneously over the entire area of abutting surfaces, or progressively along a joint. The heat is obtained from resistance to the flow of electric current through the area of contact of the surfaces. Pressure is applied before heating is started and is maintained during the heating period.

The machine uses the same type of transformer as the spot welder, but the

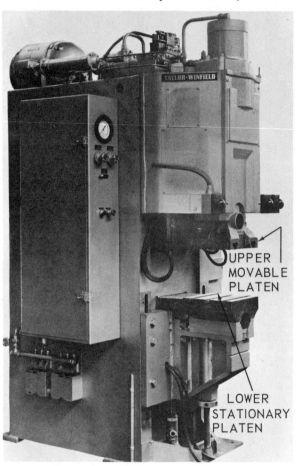

Fig. 14-29. A projection welder. The dies are fastened to the upper movable platen and to the lower adjustable platen.

electrodes in this case are vises; one movable, and one fixed. The metal must be clean for satisfactory work. Fig. 14-30 shows the top section of an upset welder. Fig. 14-31 shows a complete

UPPER CLAMPING DIES IN THE OPEN POSITION

Fig. 14-30. An upset welder. The completed weld is shown mounted in the open clamping dies.

Fig. 14-31. Complete butt welder. Note the drive mechanism on the right and the foot controls.

automatically operated upset welding machine. Fig. 14-32 shows the power required and the time needed for butt welding 1/2-in. square bar steel stock.

Another use of electric resistance heating is to pass a high current through a piece of metal located in a die. After a few seconds, the temperature of the metal is such as to permit the die to form the metal easily into almost any shape desired. Upset welding machines are used to heat and then upset rods and bars. Special fixtures are needed to upset other type assembles.

14-12. FLASH WELDING EQUIPMENT

Flash welding is a special form of butt welding. It is set up much the same

KW	TIME IN SECONDS	DISTANCE BETWEEN GRIPS IN IN.
19	3	0.79
14	6	1.6
12	9	2.4
10	11	3.2
8	14	3.9
7.5	17	4.5
7	20	5.5

FOR 1/2-IN. SQ. STEEL BARS

Fig. 14-32. Butt welding energy and time table.

Fig. 14-33. Flash welder in operation.
(Airmatic/Beckett-Harcum)

as the upset-welding machine, because the metal is held in the dies in such a way that the metal ends touch together under light pressure. When the welding current is applied, an arc or flashing action takes place between the ends at the pieces to be welded together. As the flashing proceeds, metal is melted away requiring one piece of metal to be moved toward the other in order to maintain the flashing action. As the ends of the two pieces attain welding temperature, the upset force is applied completing the weld. This action forges the metal and forces out impurities during the flashing action. An illustration of the mechanism for doing this is shown in Fig. 14-33.

Fig. 14-34 shows a table of Flash Welding Data. Note that better than a

ton and a half of pressure is used to upset a one square inch cross section, and that 60 KVA are recommended.

14-13. SPECIAL RESISTANCE WELDING MACHINES

There have been numerous specially designed resistance welding machines. These machines have consisted of some special application of one or more of the basic type machines.

Two of the more popular models were the cross wire welding machine, shown in Fig. 14-35, and forge welding machines.

Cross wire resistance welding machines use special single electrodes or multiple electrodes which hold the wire or rod in correct alignment. The current passes through the diameters of the wires where they cross, and the resistance at the contact spot creates enough heating to cause fusing of the metal after the current ceases flowing, to produce a strong welded joint.

Forge welding is very similar to the cross wire welding and butt welding. After the current flow has heated the metals to a plastic and/or flow temperature, the metal is upset by a movement of the electrodes (press action) and the parts to be welded are forced together.

CROSS SECTION SQ. IN.	AVERAGE WELDS PER HOUR	KVA REQUIRED	PUSH-UP PRESSURE
1/8	600	5	75
1/4	500	10	150
3/8	300	13	350
1/2	400	25	800
5/8	300	30	1225
3/4	200	30	1750
1	150	60	3150
1-1/4	100	80	5000
1-1/2	75	125	7000
1-3/4	60	200	9600

Fig. 14-34. Table of flash welding data.

Fig. 14-35. Welding crossed wires, with a resistance welding machine.

14-14. CARE OF RESISTANCE WELDING EQUIPMENT

There are several main areas of resistance welder maintenance:

1. Mechanical.
2. Electrical.
3. Hydraulic.
4. Pneumatic.
5. Electronic.
6. Water-cooling.

The mechanical maintenance consists of:

1. Lubrication.
2. Checking moving parts for wear and alignment.
3. Checking the force applied by the electrodes against the base metal being welded. Fig. 14-36 shows a convenient tool for checking spot welder electrode forces. These forces may be obtained by compression springs, hydraulics, pneumatics or by magnetism. The amount of the force is very important if good results are to be obtained. Fig. 14-37 shows a force gauge in use.

CHAPTER 26 contains additional information relative to resistance welding machine maintenance.

4. Checking mechanical safety devices such as guards, machine mounting, etc.

The maintenance manuals for each machine should be consulted to determine the exact specifications to be used on each particular machine.

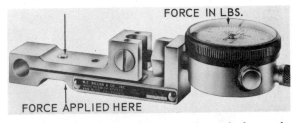

Fig. 14-36. Gauge for measuring electrode force of a resistance welder. (W. C. Dillon & Co.)

Fig. 14-37. Electrode force measuring gauge in use.

Electrical maintenance consists of:

1. Checking primary potential cycles and current.

2. Checking secondary potential

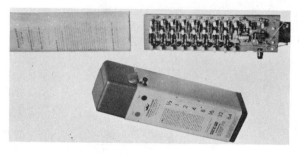

Fig. 14-38. Instrument to measure weld time in cycles. (Instrument Control Co.)

cycles, and current. Figs. 14-38 and 14-39 illustrate a test instrument for measuring weld time. An instrument

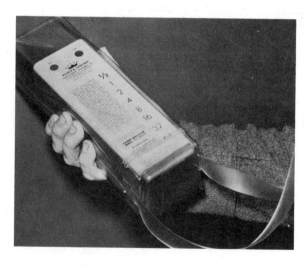

Fig. 14-39. Cycle counter instrument in use. No electrical connection is necessary when using this instrument.

to measure resistance welding secondary current is shown in Fig. 14-40.

3. Removing, cleaning and installing the electrodes. An electrode removal tool is shown in Fig. 14-41. Use a rawhide or lead hammer in setup aligning operations, or when striking ejector buttons to remove tips. Do not use steel hammers, wrenches or chisels to re-

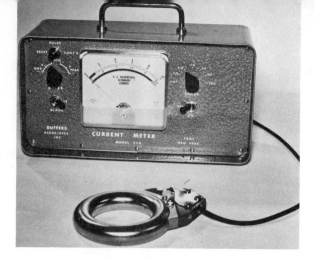

Fig. 14-40. An instrument designed to measure the secondary current in a welding circuit. (Duffers Associates, Inc.)

move tips. Both the tip and holder may be damaged. Keep the tip and holder tapers clean and free from foreign deposits, to allow easy passage of the welding current. Any corrosion in the circuit will cause welding difficulties. A thin film of castor oil or graphite grease will facilitate tip removal and

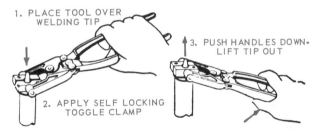

Fig. 14-41. Tool for removing tapered electrode tips.

is harmless. Dress the tip welding faces regularly to insure sound welds, as shown in Fig. 14-42. Neglect may result in poor and uneven welds.

4. Checking, cleaning, installing of switches and relays.

5. Loose shunts are a frequent source of trouble. All electrical connections should be checked daily. Either a voltmeter or an ohmmeter may be used.

The hydraulic maintenance consists of:

1. Checking pressure.

2. Checking quantity and condition of hydraulic fluid.

3. Checking hydraulic lines and connections.

4. Checking hydraulic valves.

5. Checking hydraulic pumps and cylinders.

Pneumatic maintenance is similar to the hydraulic maintenance procedures.

Electronic maintenance is usually done by people who make a specialty of this work. However some maintenance items that can be handled on the premises are:

1. Checking the tubes.

2. Checking the circuit.

Tube checking instrumentation, oscillographs, and oscilloscopes are needed to make a complete analysis of the functioning of the electronic equipment.

The water-cooling circuit is an important component of a resistance welder. Thermometers, pressure gauges, and volume measurements are necessary for checkup purposes.

An adequate supply of cool water should be provided to each welding machine, at a minimum of 30 lbs. line pressure. Do not circulate hot water from one machine to another.

Be certain that the water inlet hose is connected to the holder inlet, so that water first passes through the center cooling tube. Do not use holders with leaky heads or tapers.

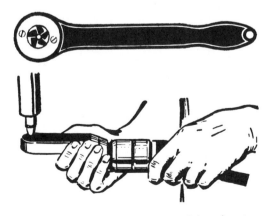

Fig. 14-42. Hand tool and power tool for cleaning and dressing electrodes. (Mallory Metallurgical Co.)

Use holders which will pass at least two gallons of water per minute. Do not allow holders or water lines to become clogged with deposits.

Use holders with free working extension tubes. Be positive that the extension cooling tube in the holder reaches to within 1/4 in. of the bottom of the tip water hole.

White lead or other foreign materials should not be used in an attempt to seal a leaking taper as this may create electrical resistance.

Always provide tips, dies, and seam welder wheels with adequate water cooling. Adequate cooling will produce better welds and longer electrode life.

14-15. RESISTANCE WELDING ACCESSORIES

A resistance welding machine operator should wear gloves, and use pliers to handle metal parts to be welded. Goggles should be worn for eye protection. If the operator cleans and changes the electrodes, he will need proper tools.

If various kinds of articles are to be welded, a number of different shaped tips or electrodes will be required.

14-16. TEST YOUR KNOWLEDGE

1. List the five (5) main parts of a resistance welding machine.

2. What is the difference between an electrode holder and an electrode tip?

3. Which transformer winding is adjustable?

4. What do the letters KVA mean?

5. List the various types of forces used to move electrode holders and electrode tips.

6. Why do the transformer windings tend to move or jump as the power is turned on and off?

7. How is a transformer cooled?

8. How much is a transformer in use, if it has a 25 percent duty cycle?

9. Of what material is the secondary winding made?

10. How many winding turns are most common in the secondary winding?

11. What is a common electrode tip alloy?

12. If the welding time is three cycles, what is the actual time in seconds?

13. What is horn spacing in a spot welder?

14. Why is one form of resistance welding called upset welding?

15. When is the pressure applied to the metal being welded during the upset welding process?

16. Can one produce weld nuggets using a seam welder?

17. Can a butt joint be welded using a seam welder?

18. What type resistance welding results in minimum heating, warpage, and corrosion?

19. Can the secondary current be measured?

20. What time factors must be set on a fully automatic resistance welding machine?

Chapter 15
SOLDERING

Soldering is a term applied to fastening two metals, either like or unlike, together with another metal entirely different from either or both of the base metals. Soldering fastens two metals together without melting either one of them. The theory of soldering is that, using clean surfaces, the binding or joining metal, upon becoming molten, adheres to the parent metal by means of molecular attraction. The molecules of solder entwine with the parent metal molecules and form a very strong bond. This process is called adhesion. In some cases the metals in the solder may form a surface alloy with one or both of the parent metals.

In soldering, the joining metal melts and flows at temperatures less than 800 deg. F. If the joining metal melts and flows above 800 deg. F., the process is called brazing. Brazing is described in CHAPTER 16. This method of identifying soldering and brazing was established by the American Welding Society.

15-1. SOLDERING PRINCIPLES

Soldering is used where a leakproof joint, neatness, a low resistance electrical joint, and sanitation are desired. The joint produced by means of soldering is not as strong as a brazed or welded joint, and in many cases a mechanical joint is used together with the solder seam. The solder commonly used is a lead and tin alloy. Other solders contain such metals as antimony and/or bis-

muth in the alloy. The proportions of the three metals in the alloy are varied, producing a variety of properties. Different solders are produced to meet various needs. There are different solders for soldering tin, copper, brass, bronze, sheet iron, and sheet steel. Soldered joints have excellent heat conductivity and electrical conductivity. However the joint is usually of less strength than the metals being joined and the assembly must be kept at a lower operating temperature.

15-2. SOLDER ALLOYS

The properties of six common soft-solder alloys are shown in Fig. 15-1. The No. 1 solder is usually called 50-50 solder while the others are called 60-40, 70-30, and 95-5. The percentage of tin is listed first.

Solder starts to soften at the melting temperature and will flow freely at the flow temperature. The 65-35 and 70-30 solders stay plastic (between the melting and flow temperature) over a wide temperature range. These solders are used for body soldering and for wiping lead joints in plumbing work.

The 62 percent tin, 38 percent lead alloy has the lowest melting temperature--361 deg. F. This is the eutectic alloy (alloy with lowest melting point possible).

Fig. 15-2 is a graphical picture of alloy combinations of tin and lead.

The biggest advantage of soldering is

minimum warpage and minimum disturbance of heat treatment of the parent metals being joined. Some solders ning or spreading action of the solder. Bismuth is used for the lower temperature alloys and silver is used for the

NO.	TIN PERCENTAGE	LEAD PERCENTAGE	MELTING TEMP. DEG. F.*	FLOW TEMP. DEG. F.	USE
	62	38	361	361	General
	60	40	361	370	General
1	50	50	361	420	General
2	40	60	361	460	General
4	30	70	361	500	General
	5	95	570	595	High Temp. Solder

*Notice that in the first five alloys part of the alloy melts at the eutectic temperature (361 deg. F.) and the entire alloy becomes a liquid at the flow temperatures (liquidus line) for each alloy.

Fig. 15-1. Table of some common tin-lead solders.

have small amounts of other metals included in the alloy to produce special properties. The three other metals often added in small amounts to lead-tin solders are:

1. Antimony.
2. Bismuth.
3. Silver.

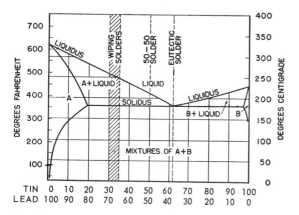

Fig. 15-2. Graph of melting and flow temperatures of alloys of tin and lead.

Antimony is added (up to 2 percent) to increase the strength of the solder. However, this solder when used on zinc, cadmium, or galvanized metals produces a brittle solder and therefore a weak joint.

Bismuth and silver improve the tin-higher temperature alloys. Three bismuth alloys are shown in Fig. 15-3.

Another group of solders are the lead-silver solders using 2.25 percent to 2.5 percent silver. These solders are a satisfactory substitute for tin-lead solders. The 2.5 percent silver alloy melts at 580 deg. F. and flows at 585 deg. F.

An alloy of 95 percent tin and 5 percent antimony is used when higher strength joints are desired. It has been used for soldering joints on cooling coils and for soldering copper to cast-iron joints. It flows at 450 deg. F. (232 deg. C).

A solder for high temperature applications is made of 95 percent cadmium and 5 percent silver. This alloy melts at 640 deg. F. (338 deg. C) and flows at 740 deg. F. (393 deg. C).

Zinc-base die castings have been successfully soldered with an alloy of 82.5 percent cadmium and 17.5 percent zinc. This flows at 508 deg. F. (264 deg. C).

Solders containing indium are used where corrosion difficulties may otherwise be encountered. A typical indium solder consists of 25 percent indium,

37.5 percent tin, and 37.5 percent lead which melts at 274 deg. F, and flows at 358 deg. F.

15-3. SOLDER FLUXES

Joints to be soldered must be chemically clean because the presence of dirt, or of oxidation, will hinder good soldering. The chemicals to be used for cleaning depend upon the kind of metals to be soldered together and the kind of solder to be used.

These chemicals, called fluxes if properly used, produce and maintain a chemically clean surface during the soldering operation. Some fluxes add alloying elements while others actually increase the fluidity or reduce the surface tension to promote more rapid

alcohol to form a more applicable paste. Clear shellac mixed with flux is an excellent way to keep the flux at the joint and on the adjoining surfaces, even under severe conditions.

The most common flux is a mixture of zinc chloride (71 percent) and ammonium chloride (29 percent). This mixture is an "acid" flux and has corroding tendencies. It must, therefore, be thoroughly cleaned from the joint after the soldering operation. Some fluxes contain only zinc chloride (with perhaps just a trace of ammonium chloride or stannous chloride). These fluxes are usually dissolved in water although some are in paste form (using a petroleum jelly).

In action, these fluxes, as they are heated, first lose their liquid if there

NO.	CONTENTS OF ALLOY, PERCENTAGE			FLOW TEMPERATURE DEG. F.
	LEAD	TIN	BISMUTH	
1	25	25	50	266
2	50	37.5	12.5	374
3	25	50	25	338

Fig. 15-3. Table of bismuth, lead, tin soldering alloys showing flow temperature.

flow of the metal. Fluxes are available in powder form, liquid form, paste form, and solid form.

Flux excludes or keeps away the atmosphere (mainly oxygen) from the molten metal. It dissolves any oxides that may form due to air contacting the molten or hot metal. It is of vital importance not to blow or vaporize the flux away before it can perform its job. A soldering copper at the correct temperature and large enough to furnish enough heat, or a large soft flame, which will quickly heat the joint (more time means more oxides) and will not blow excessively, is desirable. Some powder fluxes may be mixed with water or

is one, then melt and partially decompose to form a hydrochloric acid (HCL). All fluxes have some moisture content. The first boiling of a flux as it is heated is vaporizing of the moisture. The acid dissolves the oxides and permits the solder metal to adhere to the parent metal. The flux fuses as it cools. The fused flux covers the solder to prevent or minimize further oxidation.

Certain solder joints must be soldered using noncorrosive fluxes. Such joints as electrical and electronic connections are included in this class. Rosin is the flux used. It comes from tars found in pine trees. It is the residue formed after distilling the tar. Rosin does have the handicap of operating slowly and spreading slowly. To overcome this action, activators (1 to 2 percent) are sometimes mixed with the rosin. These activators are usually a hydrochloride compound. Most of these activators are slightly corrosive and good cleaning is recommended. Rosin is commonly mixed with ethyl alcohol to create what consistency (thickness of mixture) is desired. It may also be mixed with petroleum jelly when a paste mix is desired.

Articles coated with tin and other soft alloys can best be soldered when a rosin-dissolved-in-alcohol flux is used. When soldering copper and iron, sal ammoniac (ammonium chloride) is an effective flux. Sheet metals, copper, brass, etc., are best soldered when a zinc chloride flux is used, while hydrochloric acid (muriatic acid) with an excess of zinc is best for galvanized iron. Flux for soldering cast iron usually consists of zinc chloride added to tallow and heated to a brown color. A solution of zinc chloride may be used for the same purpose.

In using all soldering fluxes, the instructions supplied by the manufacturer should be carefully followed.

15-4. SOLDERING PROCEDURES

The five general methods of providing heat for soldering are:

1. Soldering copper.
2. Torch.
3. Dip bath.
4. Furnace.
5. Carbon arc.

Before describing soldering procedures in detail, it should be pointed out that several things must be done in order to produce successful soldering. These are:

1. Metals to be soldered together must be chemically clean. All the oxides, grease, and dirt must be removed.

2. Metals to be soldered together must be heated.

3. Metals to be soldered must be firmly supported during the soldering operation.

4. The proper flux must be used. This flux must be fresh and it must be as chemically pure (CP) as possible.

5. The solder should be melted only by the heat in the metals to be soldered together.

6. The soldering operation should be done as quickly as possible.

7. An excess of solder is useless and unsightly.

8. The solder flux should be removed from the joint as thoroughly and as soon as possible after the soldering operation is completed.

The metals may be cleaned chemically if they are then thoroughly rinsed and dried. The metals may also be cleaned by filing, or wire brushing. Use only clean tools, clean steel wool, and/or clean stainless steel wool. The metals should be heated just above the flow temperature of the solder. Clean heat should be used.

If either piece of parent metal moves while the metal is cooling from its flow temperature to its solidification temperature, the solder will probably contain cracks and will fail. It is therefore necessary to firmly support the metals with a fixture, clamps, etc., to make sure they do not move while the soldering operation is in process.

Be sure the main flux container is sealed when not in use in order to keep the flux clean. Remove only that quantity of flux needed for the particular job. Apply the flux with a clean brush or paddle. Brushes and paddles should be thoroughly washed each day. Water-base fluxes should be used immediately.

Fig. 15-4. Kit of various special paste solders. These compounds contain both a powdered soldering metal and a flux. A syringe is provided for applying the mixture to the soldered joint. (Fusion, Inc.)

With most paste type fluxes, you may wait as long as an hour before using the flux. The soldered joint should be completed as soon as possible once the flux is heated, as any delay will cause flux salts to form.

One method used to properly proportion the flux to the joint to be brazed or soldered and to insure a correct distribution of the filler metal all along the joint, is to use a paste made of the flux and of the filler metal in a powder form. Such a paste is shown in Fig. 15-4. This is obtainable in a variety of solders and brazing materials mixed with the proper fluxes. These fluxes are applied by using dispensers as shown in Fig. 15-5.

An interesting method of soldering electrical wire joints is shown in Fig. 15-6. The operator is using a twin-carbon electrode holder, and the arc between these electrodes is providing enough radiant heat to heat the wire joint sufficiently to permit soldering. The operator is holding a spool of wire solder in his right hand. Note that only the heat from the electrical wire is used to melt the solder filler metal wire.

15-5. TINNING

Adhering a very thin layer or film of solder to a metal surface is called tinning. The word tinning is from an old sheet metal term. Tin cans are actually not tin cans but are made of steel with an extremely thin coating of tin or a tin alloy on the surface of the steel. Copper wire is also frequently tinned as it is manufactured. In all soldering operations, the solder tins the surfaces as the process travels

Fig. 15-5. Solder paste flux applicator. The paste may be measured in quantities from .001 oz. to 1 oz. and deposited under a given pressure at the joint by the paste applicator nozzle. To complete the soldering operation, heat must be applied. (Fusion, Inc.)

Fig. 15-6. Electrical joint being soldered using a twin-carbon electrode arc as the source of heat.

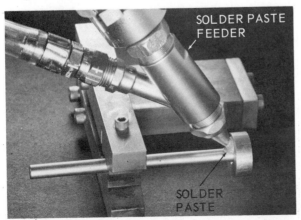

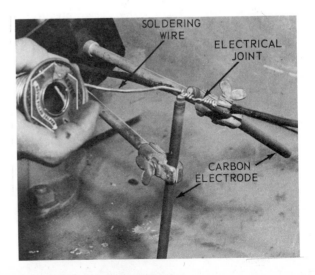

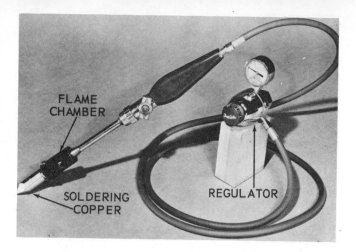

Fig. 15-7. Air-acetylene torch equipped with a soldering copper attachment.

square or octagonal solid copper bar with a four-sided tapered point. Soldering coppers come in several sizes and shapes. One pound to sixteen pound coppers are available. Shapes available include pointed, flat bottom, blunt pointed, tapered flat bottom, and hatchet. The efficiency of transmission of the heat from the copper to the work makes copper ideal for the purpose. Also, copper is easily tinned, or coated with solder, so that the molten solder will adhere to it, making the handling of the solder less difficult. The copper may be heated by a gas flame, by a blowtorch, or by an electrical resistance heating element. Fig. 15-7 shows an air-acetylene flame heating a soldering copper, while Fig. 15-8 shows soldering coppers being heated by the twin-carbon arc method.

along joint. Some operations can be tinned before parts are assembled and joint then soldered if corrosive flux is needed for tinning. Noncorrosive flux can be used where final assembly is difficult to clean.

Fig. 15-8. Twin-carbon electric arc being used to heat soldering coppers in a specially built fixture. (Lincoln Elec. Co.)

15-6. SOLDERING COPPER METHOD

The soldering copper method of soldering is the oldest and still a popular method for doing certain types of lead-tin alloy soldering. This tool is often called a soldering iron because it "irons" the solder along a seam. A soldering copper usually consists of a

Advantages of the soldering copper are that it produces a concentrated heat, and the copper is not likely to be heated to such a high temperature that it will injure the metals to be soldered together, or the solder. If the soldering copper is heated to too high a temperature the tinning will burn off the copper and the copper will blacken. The soldering copper also acts as a means of spreading, or smoothing (ironing) the solder at the same time it is melting and adhering the solder to the metals, as shown in Fig. 15-9. A soldering

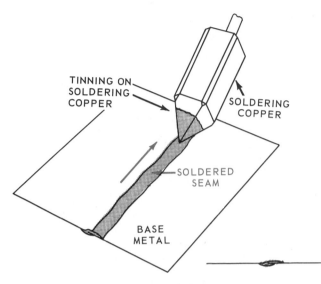

copper has the disadvantage of requiring reheating quite frequently. However, electrically heated soldering copper and internal flame heated coppers do not have this problem. The tip of the soldering copper must be kept clean and tinned at all times.

The process of cleaning and solder coating the soldering copper is called "tinning." To "tin" a soldering copper, it is necessary to file the point with a clean file to a smooth coppery finish without leaving any dirt or pits, and then to clean the point chemically by dipping it in a cleaning compound, or to apply the thin coat of solder to the tip in the presence of sal ammoniac.

One of the best applications of the soldering copper is to use it for "sweating" a soldered joint. By this term is meant that the two metals to be soldered together are lapped at their joint with

a previously applied film of solder on the two surfaces that come in contact. The two edges to be lapped together are previously tinned by using the soldering copper. These edges are then lapped, and the copper is slowly moved along the seam, permitting the heat from the soldering copper to penetrate through the metal to the solder and fusing the solder films together, as shown in Fig. 15-10. The resultant joint is strong and neat. It is very important that the two metals fit together snugly. This method is recommended when a high quality, leakproof joint is desired and for difficult to solder joints.

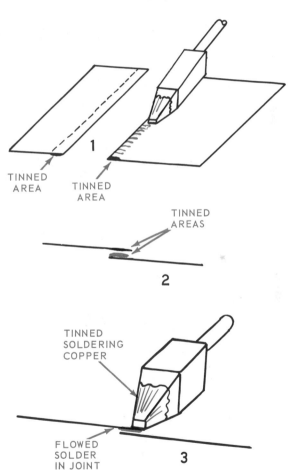

Fig. 15-10. *Steps required to make a "sweated" soldered joint. 1. Solder is applied in a thin film (tinning). 2. Surfaces are lapped to form the joint. 3. Soldering copper is applied and moved along the joint to flow the solder on to the previously tinned surfaces.*

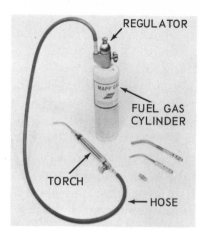

Fig. 15-11. An air-fuel gas soldering and brazing out-fit. Regulator is attached to a fuel gas cylinder. (Airco Welding Products, Div. of Airco, Inc.)

15-7. SOLDERING TORCH (TORCH SOLDERING) METHOD

Soldering torches provide a fast and flexible method for providing heat for soldering. Several types of soldering torches are:

1. Gasoline blowtorch.
2. Natural gas-oxygen torch.
3. Natural gas-compressed air torch.
4. Air-fuel gas torch as shown in Fig. 15-11.
5. Compressed air-acetylene torch, as shown in Fig. 15-12.

Fig. 15-12. Compressed air-acetylene torch.

6. Oxyacetylene torch.
7. Propane torch (with fuel cylinder).

In order to solder satisfactorily with a torch, the flame must heat the metals to be soldered; the flame must be clean so that the surfaces heated will not be corroded by the flame gases; the flame

heat must be concentrated; and the amount of heat must be easily adjusted. Fig. 15-13 shows a variety of tips used for soldering with the air-acetylene flame.

The general method of torch soldering is as follows:

1. Clean the surface.
2. Heat the surface with the torch (if possible direct the torch flame on the metal a short distance away from the spot to be soldered to avoid oxidizing the place where the solder is to adhere).

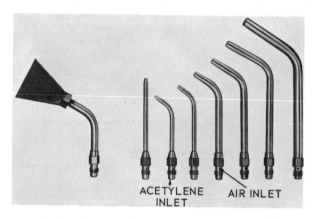

Fig. 15-13. Variety of tips used with an air-acetylene torch. More heat may be obtained by using a larger tip size; however, each torch tip flame operates at the same temperature. (Linde Div., Union Carbide Corp.)

3. Apply a small quantity of solder at the same time drawing the torch away from the metal (the metal only should melt the solder).
4. Smooth the surface. Do not hold the torch too close, but keep the inner cone of the flame from 1/2 to 3 in. away from the metal to avoid over-heating the metal and the solder. Always keep the torch moving to help prevent overheating any part of the joint. If the solder does not adhere, stop the soldering operation, thoroughly reclean the metal and then repeat the soldering operation.

A common torch soldering fault is to use too much solder, which does not

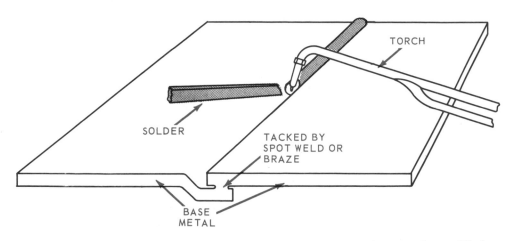

Fig. 15-14. This illustration shows the cavity of an offset lap joint being filled by using the *torch soldering* method to form a smooth upper surface for final metal finishing.

strengthen the joint and which is wasteful. If too much solder is used, a neat looking joint may still be obtained by wiping off the excess molten solder, using a clean thick cloth. The solder should be wiped away while it is at the flowing temperature rather than at the melting temperature. It is very important to avoid overheating the metal.

A growing use for torch soldering is to build up irregular surfaces on manufactured articles by applying solder to secure a smooth finish on the finished article. This method is used extensively in automotive body manufacturing and in body repair work. The irregular surface is mechanically cleaned and it is then chemically cleaned with a weak acid. A wood paddle is sometimes used to help apply the solder to the torch-heated surfaces. Fig. 15-14 shows a metal joint being filled in by the torch soldering method. The solder is then "dressed" by filing and sanding to match the sheet metal surface.

Refrigeration, air conditioning, and plumbing industries are now using soldered joints at an increasing rate. Instead of threading the joints, a special fitting is made into which the pipe is inserted as shown in Fig. 15-15. Here,

too, the joint must be mechanically cleaned, fluxed, the tubing firmly supported, and the soldering done as quickly as possible. The flux is usually put on the outside of the tubing only to prevent an excess of flux inside the tubing and fitting. A weaving action of the torch flame will heat the metal evenly. This is important. It prevents overheated spots. This type joint may be soldered successfully with the tubing and the fittings in any position. A change of color of the secondary flame usually indicates that the tubing and fitting are

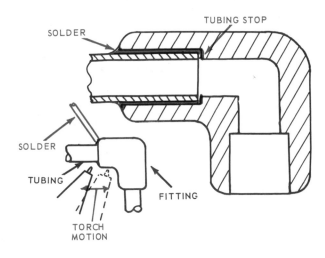

Fig. 15-15. Soldered tubing or pipe joint. Solder is drawn into the joint by capillary action.

overheated. This type joint utilizes the principle that when two surfaces are placed close together, a liquid placed in any one spot between them will quickly spread to all parts of the space between the two surfaces. The joint needs small clearance to provide space for solder to flow into the joint.

Upon heating the pieces and applying solder to the joint, solder fills in the space by capillary attraction, making a strong and neat joint. The joints may be either soldered or brazed. See CHAPTER 16 for brazing procedures.

To solder cast iron, it must first be filed, machine blasted, or shot blasted to remove the oxide skin. The mechanically cleaned surface must then be degreased and cleaned in a special molten flux bath that will remove the surface graphite.

15-8. DIP BATH METHOD

This method consists of melting a quantity of solder in a tank or pot. The solder is protected by means of a hood or chemical covering, such as powdered charcoal, to prevent oxidation of the solder. The articles to be soldered are dipped in a flux bath, and then in the solder bath. This method is a labor-saving device to either solder coat surfaces to make them rustproof, or to fasten the various parts of an assembly together.

The articles to be soldered together are usually assembled and then acid cleaned, "pickled," after which they are thoroughly washed and dried before being dipped into the solder bath. The articles are lifted out of the bath; and excess solder is allowed to drain from the surfaces.

An important precaution to be observed with this method is that under no circumstances should any articles which have the slightest amount of moisture on them be put into the bath. A small amount of moisture will produce instant high-temperature steam, causing an explosion of the bath with possible injuries to workers and destruction to property.

Many solder joints are made automatically. The parts to be soldered are mounted in fixtures, the flux and solder are fed automatically. Heating is automatic and the soldered joint is cooled before the parts are ejected from the fixtures. See Fig. 15-16. Some installations use preplaced solder rings or solder foil instead of automatic feeding of the solder filler wire. Precleaning and postcleaning are sometimes a part of the automatic process.

15-9. ELECTRIC SOLDERING IRONS

The electric type soldering iron maintains a uniform heat, and is desirable for many soldering jobs. A 200-watt iron is suitable for most sheet metal work.

The soldering gun is used mostly for electric/electronics work.

15-10. STAINLESS STEEL SOLDERING

Many kinds of stainless steels are now being used. Some of the more common ones are:

1. 200 series (approximately 17 percent chromium, 5 percent nickel).

2. 301 to 308 series (approximately 18 percent chromium, 8 percent nickel).

3. 309-314 series (approximately 23 percent chromium, 14-18 percent nickel).

4. 315-347 series (approximately 18 percent chromium, 12 percent nickel).

Properties of stainless steels are described in CHAPTERS 18 and 24.

Stainless steel joints can be either soldered, brazed, or welded. Soldering is used often to make low strength leak-proof joints, and to eliminate cracks and crevices.

A strong flux is needed to promote adhesion of the solders as the chromium surface film on the stainless steel resists ordinary fluxes. A corrosive flux is used. It is very important that the flux be completely removed after soldering. A warm water bath with perhaps some mechanical brushing, will do a satisfactory job of removing the flux.

A 50-50 solder makes a good joint but its color does not match the color of stainless steel. Special solders which match stainless steel color (usually with no lead content) are available.

Solder for stainless steel comes in rods 1/8 dia. by 15 in. long.

15-11. STAINLESS STEEL SOLDERING FLUXES

One way to solder stainless steel successfully is to first clean the metal with a hydrochloric acid (50 percent) and water (50 percent) mixture. Only experienced persons should mix the acid (always add the acid slowly to the water--not water to acid). Wear goggles, face guards, rubber-lined clothing and rubber gloves when preparing this

Fig. 15-16. Automatic soldering operation using a pressurized solder and flux mixture. (Fusion, Inc.)

cleaning fluid. Be sure the mixture temperature is moderate and stays near the ambient (room) temperature.

After the acid has been on the stainless steel surface for approximately four minutes, rinse, apply a zinc chloride flux and carefully heat the metal; then apply solder as usual. Great care is needed as stainless steel is difficult to solder. One will find that the solder will adhere easier and better if the stainless steel surface is scratched using a small scrubbing motion with the solder filler rod. This action seems to mechanically remove any residual film which prevents the solder from adhering to the surface.

15-12. ALUMINUM SOLDERING ALLOYS

One of the most difficult soldering tasks is to solder articles made of aluminum. At the present time commercially pure aluminum and aluminum alloys of not more than 1 percent manganese or 1 percent magnesium can be soldered most successfully. Heat treated aluminum and clad aluminum lose some of their physical properties when soldered. It is recommended, if possible, to weld any aluminum fractures rather than attempt to solder them. Lead solders are most difficult to use for soldering aluminum due to the reluctance of the aluminum and lead to adhere to each other. Scratching of the aluminum surface with the solder does help. Special solders prepared for aluminum soldering are recommended. Some of these are as follows:

1. 60 percent tin, 37 percent zinc, and 3 percent copper.

2. 30 percent tin, and 70 percent zinc.

3. 40 percent cadmium, and 60 percent zinc.

These solders melt and flow at approxi-

mately 400 to 450 deg. F. (well below the "hot short" temperature of aluminum-see Chapter 16).

There is an increasing number of patented and/or registered alloys which can be successfully used to solder aluminum. The manufacturer's recommendations should be carefully followed.

To obtain the best results, the parts soldered together should be held firmly until the solder cools to near room temperature.

The aluminum solder alloys are available in both wire and bar form.

15-13. ALUMINUM SOLDERING PROCEDURE

The same basic procedure is necessary for soldering aluminum as for any other metal, see PAR. 15-4. The metal, the flux, and the filler rod must be clean. A good support for the parts while they are being joined is important.

Usually, the joint is heated with a carburizing flame. A rather large torch movement is used. The rod (solder) should be melted only by the heat from the metal to be joined. The joint may be built to the desired fillet or build-up by rubbing the rod on the heated surface. The rubbing action helps break the oxide film on the metals to be joined. Flux is not always necessary, but a good aluminum flux usually improves the filler metal flow.

Fluxes are now available that permit soldering without using the rubbing technique.

The joint must be cleaned with a clean wire brush or clean steel wool before and after to remove dirt and grease. It is important to remove all traces of the flux after soldering, by washing and/or scrubbing, or otherwise the residual flux may corrode the metal.

If there is any chance of exposure to moisture which may produce galvanic action, the joint should be painted or varnished as a further protection.

When using a soldering copper to solder aluminum, the copper may be tinned with a tin-lead solder.

Brazing aluminum is covered in CHAPTER 16.

15-14. DIE CAST SOLDERING

Some zinc die castings can be soldered. However, it is a difficult operation. The solder used is usually an alloy of 82.5 percent cadmium and 17.5 percent zinc. This alloy melts (flows) at 508 deg. F. (264 deg. C). Lacquer, paint, or chromium plated surfacing must first be removed before soldering die castings.

Die castings can sometimes be soldered without flux. One can use a soldering copper by rubbing the copper with vigor into the die cast surface. Die castings are also successfully soldered by first coating the die casting with nickel and then soldering with a tin-lead solder.

15-15. SOLDERING MISCELLANEOUS METALS

Copper alloys such as aluminum bearing copper alloys, silicon alloys, and beryllium alloys can be soldered. The oxides have a high melting temperature (refractory properties) and must be mechanically removed (grinding). A strong flux is needed. One satisfactory flux consists of:

Concentrated hydrochloric acid (50 percent), 25 percent solution of zinc chloride in water (50 percent).

One of the best solders to use is a high-tin solder.

Galvanized iron can also be soldered.

After cleaning the metal, use a flux made of ammonium chloride and zinc chloride, then use tin-lead solder which is free of antimony. The antimony may mix with the zinc coating and produce a brittle and gritty solder mixture. To avoid damage to the zinc (galvanized) coating, remove the flux as quickly as possible after soldering.

15-16. TESTING AND INSPECTING SOLDERED JOINTS

Soldered joints are usually tested in two ways:

1. For being leakproof (where it is applicable).
2. For being immune to humidity.

The tightness of the joint can be tested hydrostatically using water.

The joint is tested as to its corrosion resistant powers by putting the joint in a humidity cabinet where it is kept at a temperature of 100 deg. F. and at approximately 100 percent humidity for at least 72 hours. A good joint will reveal no evidence of corrosion under low magnification inspection.

Visual inspection of the joint will usually show such defects as poor adhesion, incomplete soldering, too much solder, overheating, dirt inclusion, etc.

15-17. REVIEW OF SOLDERING SAFETY PRACTICE

Soldering, brazing, or welding with or on alloys containing cadmium or beryllium can be extremely hazardous.

Fumes from cadmium or beryllium compounds are extremely toxic. Several deaths have been reported from inhaling cadmium oxide fumes.

Skin contact with cadmium and beryllium should also be avoided.

An expert in industrial hygiene should

be consulted whenever cadmium or beryllium compounds are to be used or when repairs are to be made on parts containing these metals.

Fluxes containing flouride compound are also toxic.

Good ventilation is essential when soldering or brazing and the operator should always observe good safety practices.

The most common hazard when soldering is exposure of the skin, eyes, and clothing to acid fluxes. Always work in a way that flux will not be spilled on the skin or clothing. Surfaces accidentally contacting acid fluxes should be washed immediately in clear water.

Heating soldering coppers sometimes presents a fire or heating hazard if an open flame is used. Be sure flammable material is kept away from the heating flames.

Be sure that there are no flammable fumes such as gasoline, acetylene or other flammable gasses present where soldering is being performed.

Perhaps the greatest hazard one meets in soldering is the attempt to solder gasoline or other fuel tanks. The job should never be attempted in the usual school shop.

15-18. TEST YOUR KNOWLEDGE

1. What is soldering?
2. Name the different types of soldering.
3. What sources of heating are used for soldering?
4. What is the purpose of the flux?
5. Why are oxides detrimental to soldering?
6. What alloy is used for soldering a silicon-copper alloy?
7. What is the indication when the soldering copper is becoming too hot?
8. What is the melting temperature of 50-50 solder?
9. May a welding torch be used for soldering?
10. What is the flow temperature of 50-50 solder?
11. What temperature level does the American Welding Society define as the difference between soldering and brazing?
12. Does a soldered joint have good resistance to heat flow?
13. What solder action does a small addition of silver cause?
14. Can an oxidized surface be soldered?
15. Of what use is zinc chloride when soldering?
16. Name the main chemical in a non-corrosive flux.
17. What evaporates first when a flux is heated?
18. What chemical in a cast iron surface must be removed before it can be successfully soldered?
19. Why must the parts of a joint be firmly supported during the soldering operation?
20. What is the difference between adhesion and cohesion?
21. Why is the surface sometimes scraped under the layer of molten solder?
22. In what condition should a wire brush be, before it is used to clean metal for soldering?
23. What type joint uses capillary action to distribute the solder in the joint?
24. Can aluminum be soldered?

Chapter 16

BRAZING
BRAZE WELDING

Both brazing and braze welding are metal joining processes which are performed at temperatures above 800 deg. F., as compared to soldering which is performed at temperatures below 800 deg. F.

The American Welding Society defines these processes as follows:

BRAZING - "A group of welding processes wherein coalescence is produced by heating to suitable temperatures above 800 deg. F., and by using a nonferrous filler metal having a melting point below that of the base metals. The filler metal is distributed between the closely fitted surfaces of the joint by capillary attraction."

BRAZE WELDING - "A method of welding whereby a groove, fillet, plug or slot weld is made using a nonferrous filler metal, having a melting point below that of the base metals, but above 800 deg. F. The filler metal is not distributed in the joint by capillary attraction."

Brazing has been used for centuries. Blacksmiths, jewelers, armorers, and other tradesmen have used the process on large and small articles since before recorded history. This joining method has grown steadily both in volume and popularity. It is an important industrial process as well as a jewelry making and repair process. The art of brazing has become more of a science as the knowledge of chemistry, physics, and metallurgy has increased.

The usual terms Brazing and Braze Welding imply the use of a nonferrous alloy as a filler metal which consists chiefly of copper and zinc or tin.

Brass is an alloy consisting chiefly of copper and zinc. Bronze is an alloy consisting chiefly of copper and tin. Most rods used in both brazing and braze welding are brass alloys rather than bronze. The brands which are called bronze usually contain a small percent (about 1 percent) of tin.

Special alloys containing either silver or aluminum will be explained in later paragraphs.

16-1. BRAZING AND BRAZE WELDING PRINCIPLES

Brazing is an adhesion process in which the metals being joined are heated but not melted; the brazing filler metal melts and flows at temperatures above 800 deg. F. (427 deg. C.).

A brazed joint is stronger than a soldered joint and in certain instances is as strong as a welded joint. It is used where mechanical strength and pressure proof joints are desired. Brazing and braze welding are superior to welding in some applications, as they do not affect the heat treatment of the original metals as much as welding.

Brazing and braze welding warp the original metals less, and it is possible to join dissimilar metals. Examples: steel tubing may be brazed to cast iron, copper tubing brazed to steel, and tool steel brazed to low carbon steel.

The joints and the material being brazed must be specially designed for the purpose. Fig. 16-1 shows several different types of joints suitable for brazing.

In braze welding, joint designs as used for oxy-gas or arc welding are satisfactory. Such joint designs are discussed in CHAPTERS 1 and 5. When brazing, poor fit and alignment result in poor joints and in inefficient use of brazing filler metal. See Fig. 16-2.

It is important to provide good ventilation when brazing or braze welding, because fumes from the heated fluxes and vaporized filler metals (such as zinc and cadmium) may affect the respiratory system, eyes, or skin.

16-2. BRAZING AND BRAZE WELD-ING FILLER METAL ALLOYS

Many filler metal alloys are available for use in brazing and braze welding. Some of them are:

struction. Brazing is included in the methods for joining these exotic metals.

Fig. 16-3 shows various brazing filler metal alloys, alloying elements, and their melting and flow temperatures.

16-3. BRAZING AND BRAZE WELDING FLUXES

The American Welding Society defines a flux as, "Material used to prevent, dissolve or facilitate removal of oxides and other undesirable substances."

A brazing flux must be of a composition that keeps both the brazing filler metal and the metals being joined clean during the joining operation. The fluxes must be chemically pure. Some manufacturers mark their fluxes C.P. (chemically pure).However, most fluxes are chemically pure even though they may not be marked C.P.

Borax is a flux which has been used over the greatest period of time with the copper brazing filler metal alloys.

Borax or boric acid (Borax plus

BRAZING ALLOY NAME	AWS-ASTM CLASSIFICATION
1. Copper and zinc alloys (brass)	*BCuZn
2. Copper	BCu
3. Nickel and Chromium Alloys	BNiCr
4. Copper and Phosphorus Alloys	BCuP
5. Silver Alloys	BAg
6. Copper and Gold Alloys	BCuAu
7. Aluminum and Silicon Alloys	BALSi
8. Magnesium Alloys	BMg

*In each of these filler metal alloys the "B" stands for Brazing Alloy.

As more of the earth metals are refined and used commercially, brazing filler metal alloys are being developed so these metals may be joined by brazing and braze welding. Such metals as palladium, titanium, beryllium zirconium (the exotic group), are now being used in missile and satellite con-

water) is a common base for brazing fluxes. A popular mixture is 75 percent borax (powdered form) and 25 percent boric acid (liquid form) mixed to form a paste. Other ratios of these two chemicals are also used to form more solid or more fluid fluxes as may be required. These mixtures range from 75

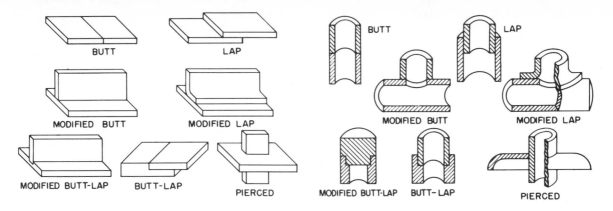

Fig. 16-1. Joints designed to produce good brazing results. (Handy and Harman)

percent to 25 percent borax with the remainder being boric acid. Some of the commercial fluxes also contain small amounts of phosphorus and halogen salts (halogen means any one of the

specially trained in its use as it must be kept dry, away from acids, nitrates, and other oxidizing agents.

When selecting a flux for brazing or braze welding all of the variables in

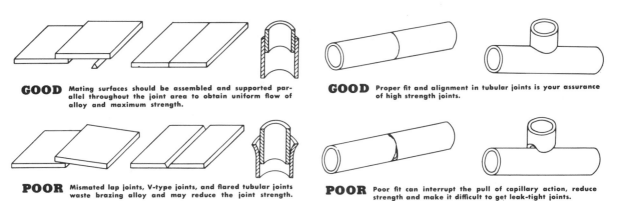

GOOD Mating surfaces should be assembled and supported parallel throughout the joint area to obtain uniform flow of alloy and maximum strength.

GOOD Proper fit and alignment in tubular joints is your assurance of high strength joints.

POOR Mismated lap joints, V-type joints, and flared tubular joints waste brazing alloy and may reduce the joint strength.

POOR Poor fit can interrupt the pull of capillary action, reduce strength and make it difficult to get leak-tight joints.

Fig. 16-2. Some well-designed joints which have been prepared for brazing, and some poorly-designed joints shown for comparison.

iodines, bromine, fluorine, chlorine, and astatine chemical elements).

Alkaline bifluoride is used as a flux for brazing or braze welding stainless steel, silicon bronzes, aluminum or beryllium copper alloys. As most of the fumes from these fluxes are harmful to health, good ventilation is very necessary.

A special flux made of sodium cyanide salts is excellent when silver brazing tungsten to copper. THE FUMES ARE VERY DANGEROUS. AVOID BREATHING THE FUMES AND DO NOT LET THE FLUX CONTACT THE SKIN. This flux should only be handled by people

each application must be considered. The variables include:

1. Base metal or metals used.

2. Brazing filler metal used.

3. Source of heat used: oxyacetylene, carbon arc, or electric induction heating.

16-4. METHODS OF APPLYING BRAZING AND BRAZE WELDING FLUXES

Various methods may be used to apply fluxes during a brazing or braze welding operation.

The fluxes are most commonly sup-

Filler Metal Classification	Nominal Composition												Temperature, ° F	
	Cu	Ag	P	Zn	Cd	Au	Ni	Al	Cr	Si	B	Other	Solidus	Liquidus
BAlSi-1	..	..	..	..	..	..	..	95	..	5	..	..	1070	1165
-2	..	..	..	..	..	..	..	92.5	..	7.5	..	..	1070	1135
-3	4	..	..	..	..	..	..	86	..	10	..	..	970	1085
-4	..	..	..	..	..	..	..	88	..	12	..	..	1070	1080
BCuP-1	95	..	5	..	..	..	..	..	..	..	..	..	1305	1650
-2	93	..	7	..	..	..	..	..	..	..	..	..	1305	1485
-3	89	5	6	..	..	..	..	..	..	..	..	..	1195	1500
-4	87	6	7	..	..	..	..	..	..	..	..	..	1185	1380
-5	80	15	5	..	..	..	..	..	..	..	..	..	1185	1300
BCuAu-1	62.5	..	..	..	..	37.5	..	..	..	..	..	..	1755	1815
-2	20	..	..	..	..	80	..	..	..	..	..	..	1620	1630
BCu	99	..	..	..	..	..	..	..	..	..	..	..	1980	1980
BCuZn-1	60	..	..	40	..	..	..	..	..	..	..	..	1650	1660
-2	57	..	..	42	..	..	..	..	..	..	..	Sn-1	1630	1650
-3	56	..	..	40	..	..	1	..	..	..	..	Sn-1, Fe-1 Mn-1	1590	1630
-4	52.5	..	..	47.5	..	..	..	..	..	..	..	...	1570	1595
-5	51.5	..	..	45	..	..	..	..	..	..	..	Sn-3.5	1585	1610
-6	48	..	..	42	..	..	10	..	..	..	..	...	1690	1715
-7	47	..	..	41.5	..	..	10.5	..	..	..	..	Ag-1	1685	1710
BMg	..	..	..	2	..	..	..	9	..	..	..	Mg-89	1110	1120–1160
BNiCr-1*	..	..	..	..	..	..	..	..	16.5	4	3.8	C-0.8	1760	1875
-1a*	..	..	..	..	..	..	..	..	16.5	4	3.8	C-0.2	1760	2000
-2*	..	..	..	..	..	..	..	..	7	4	3	..	1750	1825
-3*	..	..	..	..	..	..	..	..	..	4	2.6	..	1800	1875
-4*	..	..	..	..	..	..	..	..	19	10	..	..	2030	2080
BAg-1	15	45	..	16	24	..	..	..	..	..	..	..	1125	1145
-1a	15.5	50	..	16.5	18	..	..	..	..	..	..	..	1160	1175
-2	26	35	..	24	18	..	..	..	..	..	..	..	1125	1295
-3	15.5	50	..	15.5	16	..	3	..	..	..	..	..	1170	1270
-4	30	40	..	28	..	..	2	..	..	..	..	..	1220	1435
-5	30	45	..	25	..	..	..	..	..	..	..	..	1230	1370
-6	34	50	..	16	..	..	..	..	..	..	..	1250	1425	..
-7	22	56	..	17	..	..	..	..	..	..	..	Sn-15	1145	1205
-8	28	72	..	..	..	..	..	..	..	..	..	..	1435	1435
-9	20	65	..	15	..	..	..	..	..	..	..	..	1235	1325
-10	20	70	..	10	..	..	..	..	..	..	..	..	1275	1360
-11	22	75	..	3	..	..	..	..	..	..	..	..	1365	1450
-12*, †	..	85	..	..	..	..	..	..	..	..	..	Mn-15	1760	1780
-13*	40	54	..	5	..	..	1	..	..	..	..	..	1340	1575
-14*	52.5	25	..	22.5	..	..	..	..	..	..	..	..	1250	1575
-15*	45	20	..	30	5	..	..	..	..	..	..	..	1140	1500
-16*	45	20	..	35	..	..	..	..	..	..	..	..	1315	1500
-17*	7	92.5	..	..	..	..	..	..	..	..	..	Li-0.2–0 5	1400	1635

* Proposed—not included in current AWS specification A5.8–56. † Formerly BAgMn.
Solidus—Melting Temperature. Liquidus—Flow Temperature.

Fig. 16-3. Table of brazing filler metal alloys showing their compositions, melting and flow temperatures.

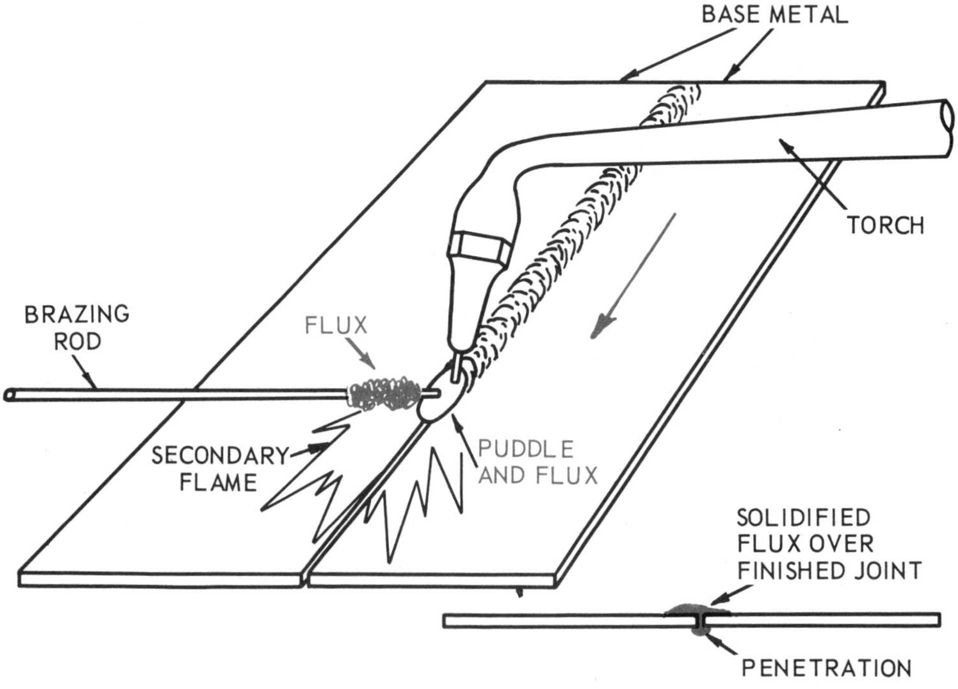

Fig. 16-4. A butt joint being braze welded using the oxyacetylene flame as the heat source. Flux has been applied to the brazing rod.

plied in a powder form. If the brazing rod is heated on the end for a distance of 2 - 3 in. and is then placed in the

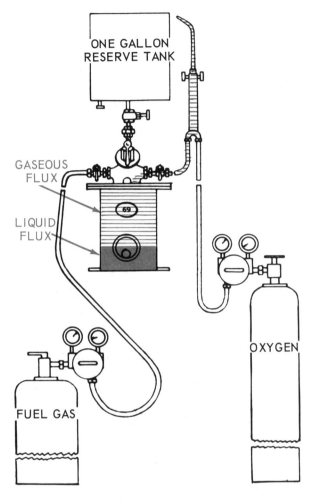

Fig. 16-5. Oxygen-fuel gas brazing station with gas fluxing unit in fuel gas supply line. (Gasflux Co.)

flux container, the flux will adhere or stick to the brazing rod. Another method used to apply powdered flux to a joint is to heat the base metal slightly and sprinkle the powdered flux over the joint. The powdered flux will partly melt and adhere to the base metal. Fig. 16-4 shows the flux being applied to the brazing rod first, then brought to the joint.

Frequently, the flux is mixed with clean water (distilled water is best) to

form a paste. This paste is then painted on the brazing rod, or on the base metals, or both. Whenever the fluxes are painted on, an inexpensive, clean, small brush is used (commonly called an acid brush).

Brazing rods with a preapplied flux coating or covering are available. Tubular brazing rods with a flux core have also been used.

A method which eliminates the extra handling of separate flux is to feed the flux into the fuel gas, and have the flux brought to the joint being brazed or braze welded along with the flame gases, as shown in Fig. 16-5. The flux is usually dissolved in alcohol. The installation requires two separate lengths of fuel gas hose with fittings. The fluxing equipment, complete with the reserve flux tank, is shown in Fig. 16-6.

The fuel gas is fed through a container of liquid flux, and a controlled amount of the flux mixes with the fuel gas and is fed to the operation through the torch

Fig. 16-6. A gas fluxing unit. (Gasflux Co.)

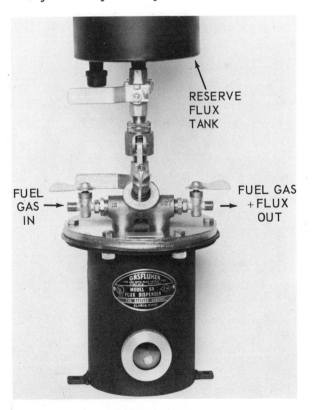

tip. This method not only eliminates the separate operation of adding flux, but assures a continuous flow of flux of the correct amount, and results in an excellent and clean joint.

Powdered braze metals can also be injected into the welding flame. By this method the heated powder particles are protected from oxidation as they are transferred to the surface of the base metal. Fig. 16-7 shows a hopper type powder brazing torch.

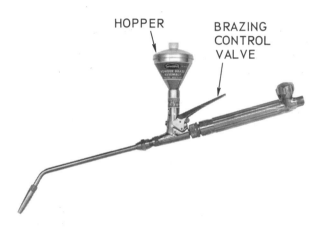

Fig. 16-7. A special oxyacetylene torch with a hopper feed for ultra fine powder fluxes and/or metals. The torch can be used for brazing, metal surfacing, and welding.

Another method is to thoroughly mix powdered brazing filler metal with the flux in the proper proportions to form a paste. This combination is then added to the joint. This mixture may be hand fed using flexible plastic bottles or it may be gun fed (either manually or power). Fig. 16-8 shows a pneumatically powered feed applicator for paste type fluxes.

16-5. BRAZING AND BRAZE WELDING PROCEDURES

Heat source for brazing and braze welding may be:
1. A molten bath of brazing metal alloy or a molten salt bath into which the assembled joint to be brazed is dipped.
2. Torch heating with:
   A. Oxyacetylene.
   B. Air-acetylene.
   C. Oxyhydrogen.
   D. Oxy-propane.
   E. Oxy-natural gas.
3. Controlled atmosphere furnaces.
4. Electric resistance heating.
5. Carbon arc.
6. Induction heating.
7. Block brazing--where the joint is heated by applying electrically or flame-heated blocks to the joint.
8. Flow brazing--where the joint is heated to the brazing temperature by flowing molten brazing metal over the joint.

Braze welding usually requires more heat than brazing, and since braze welding is similar to welding, the operation is usually done with an air or oxy-fuel gas torch or carbon arc torch.

Brazing and braze welding procedures are quite similar, the main difference being in the joint design and the quantity of brazing rod applied to the joint.

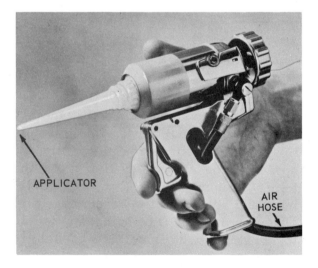

Fig. 16-8. A pneumatically powered flux and powdered metal applicator. (Wall Colmonoy Corp.)

Some good pointers applicable to both brazing and braze welding follow:

1. Metals to be joined must be mechanically and chemically clean.

2. Two pieces to be brazed together must be fitted properly: that is, the metals should not be spaced too far apart or forced together. The braze welded joint must be prepared in the same manner as any oxy-gas or arc welded joint.

3. The two pieces must be firmly supported during the brazing and cooling operations. Any movement of either part while the joining metal is molten or plastic will weaken the joint.

4. The metals to be joined must be heated to a temperature slightly above the melting temperature of the brazing filler metal, but below the melting temperature of the metal being brazed or braze welded.

5. Clean, fresh flux must be applied as the operation proceeds to reduce oxidation and to float the oxides to the surface.

6. A heat source listed previously is required to obtain a high enough temperature to obtain a good joint.

The torch flame (oxyacetylene) is usually adjusted to a neutral flame. A reducing flame will produce an exceptionally neat looking joint, but strength will be sacrificed. An oxidizing flame will produce a strong joint with rough looking surface. A neutral flame will give the best results under ordinary conditions.

Carbon arc brazing or braze welding is done with a two-carbon electrode holder and with AC, Fig. 16-9.

7. As in steel welding, the brazing filler metal must penetrate to the other side of the joint. This penetration must be such that the joining metal covers 100 percent of the joint surfaces and adheres to these two surfaces. A min-

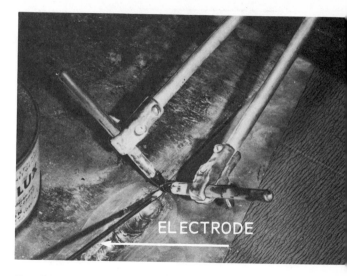

Fig. 16-9. A carbon-arc twin electrode torch used for brazing. The arc is drawn between the two electrodes and the radiated heat is used for brazing. (Lincoln Electric Co.)

imum amount of joining material should be used.

8. The braze must be cooled properly to obtain the desired properties in both the original metal and the joining metal. Heat treatment of metals is covered in CHAPTER 25.

16-6. JOINT DESIGN FOR BRAZING AND BRAZE WELDING

Basically, brazed joints are either a lap joint or a butt joint, Fig. 16-10. Braze welded joints have the same design as a gas or arc welded joint. However, there are many varieties of each type of joint. It is important to remember that the smaller the brazing filler metal thickness when brazing, the stronger the joint. However, the space between the metals being joined must be enough to allow the brazing metal to flow through the joint. The brazing alloy will not flow between two surfaces that are pressed together (press fits, force fits, clamped together and the like). In a poorly designed brazing joint, the flux may be trapped or the metals may

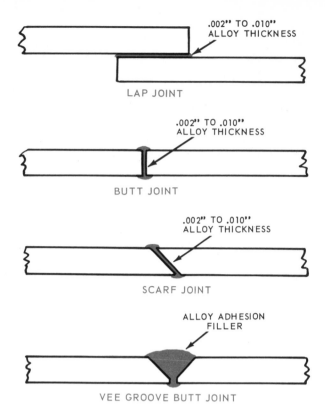

Fig. 16-10. Four types of common joint designs for brazing. The V-groove butt joint requires more filler metal and is commonly called braze welding.

expand and reduce the space between the base metals. The joint must be vented to allow the brazing filler metal to flow freely in the joint.

16-7. COPPER AND ZINC ALLOYS (BRASS)

One of the principal metals used in brazing filler metal alloys is copper with zinc being used as an alloying metal. This family of alloys melts at temperatures between 1570 and 1690 deg. F., and becomes fluid at temperatures between 1595 and 1715 deg. F.

Fig. 16-11 lists some copper-zinc alloys, their composition, use, melting, and flow temperatures.

The most common proportion of copper and zinc in brazing filler metals is 57 percent copper, 42 percent zinc, and 1 percent tin. This alloy melts at 1630 deg. F., and becomes fluid at 1650 deg. F.

An unusual alloy is Tobin Bronze. This alloy is a patented mixture of copper and zinc with just a little tin in it.

16-8. FLUXES FOR COPPER AND ZINC ALLOYS

Flux required depends on the brazing filler metal alloy and kinds of metals to be brazed or braze welded together. Borax or a mixture of borax and boric

POPULAR COPPER BRAZING ALLOYS

	Copper %	Zinc %	Tin %	Fe %	Mn %	Si %	Ni %	P %	Use	Trade Name	Melting Temp °F	Flow Temp °F
Brass Brazing Alloy	60	40							Copper Nickel Alloy Steel	All State 41	1650	1660
Naval Brass	60	39.25	.75						Copper Steel Nickel Alloys		1630	1650
Tobin Bronze	59	40.5	.50						Steel Cast Iron	Anaconda 481	1625	
Manganese Bronze	58.5	39.25	1.0	1.0	.25				Steel	Oxweld #10 Anaconda 984	1590	1630
Low Fuming Bronze	57.5	40.48	.9	1.0	.03	.09			Cast Iron Steel	Anaconda 997	1598	
	52.	48.									1570	1595
	50	50.									1585	1610
Nickel Silver	55-65	27-17					18		Steel Nickel Alloys Cast Iron	Anaconda 828		
	48	42					10		Steel Nickel Alloys		1690	1715
Copper Silicon	98.25					.25	1.5		Steel to Copper	Anaconda 943	1981	
Phosphor Bronze	98.2		1.5					.3	Copper Alloys	Anaconda 903 All-State 21	1922	

Fig. 16-11. A table of copper alloy brazing filler metals. Note that all of these alloys melt at a lower temperature than the melting temperature of copper which is 1981 deg. F.

acid will prove satisfactory in most of the copper and zinc alloy brazing filler metal applications. Many other fluxes are also available which provide satisfactory results.

16-9. PROCEDURE FOR BRAZING OR BRAZE WELDING WITH COPPER AND ZINC ALLOYS

Prepare pieces to be braze welded just as they are prepared for steel welding. That is, provide for contraction and expansion. If the metal is thicker than 8-gauge, it must be grooved or chamfered to permit adequate penetration. The joint to be brazed must be designed and fitted so that a close fit of the two base metals is obtained. The surfaces to which the brazing filler metal is to adhere must be mechanically cleaned using clean file or abrasive paper. An emery wheel or emery cloth is not recommended due to the possibility of imbedding abrasive particles and oil in the metal.

Adjust the torch for the flame desired, using the same size tip as would be used for welding the same thickness of the base metal. The flames used for brazing or braze welding operations when LP fuel gas is used are shown in Fig. P-15B. When brazing, the flux is usually applied to the joint, while in braze welding the flux is often applied to the brazing rod. To apply flux to the brazing rod, warm the first two or three inches of the brazing filler metal rod with the torch flame, and then dip the heated end of the rod into the flux container. This action will coat the brazing rod with semi-fused flux 1/8 to 1/4-in. in thickness. Now apply the torch to the metals to be brazed or braze welded, heating the metals at the joint (each one equally) to a dull cherry red. The width of the bead when braze welding will be

determined by how wide a portion of the metal is heated to cherry red. The brazing filler metal will not flow over the surface unless the surface is at the brazing filler metal flow temperature.

The width of the braze weld bead should be a little wider than a steel weld on the same thickness of metal. While heating the metal, the brazing filler metal rod should be kept near the torch flame to maintain a fairly high rod temperature as shown in Fig. 16-4. After the metals have been heated, bring the brazing rod (flux-coated portion) into contact with the cherry red metals, meanwhile maintaining the torch motion. The brazing rod will quickly melt and flow over or between the parent metals. Do not overheat the brazing rod. Keep it away from the inner cone.

When braze welding, the bead should proceed along the joint just as in the welding process except the procedure should be faster. The torch flame is not held as close to the metal as in steel welding (approximately double the distance). The width of the bead may be controlled by raising and lowering the torch flame. When the flame is held close to the metal a wide bead will result. By drawing the torch away, the metal cools slightly and the bead will not be so wide.

The finished braze or braze weld should have the appearance of adequate fusion with the base metal. The brazing filler metal should penetrate through the joint and appear underneath. A white deposit on the outside of the brazed or braze welded joint indicates an overheated joint, also the color of the braze filler metal in the joint will indicate if it has been overheated. The best looking brazed or braze welded joint will show a color exactly similar to the brazing filler metal used. If the brazing filler metal is heated to an excessive temperature, some of the zinc will be burned out leaving a coppery appearance. If an oxidizing flame is used, the brazing filler metal will have a red color due to the oxidation of the copper.

16-10. SILVER BRAZING ALLOYS

In the fabrication of jewelry, small precision articles, and instruments which require strong joints, various silver alloys are used as the joining metal. The original use of silver alloys was in jewelry manufacturing, but at present there are but few industries

performing the brazing varies little from the brazing previously described.

The metals which form the alloys of silver are gold, copper, cadmium, and zinc. The best grades of silver alloys are the ones which are formed partly of gold. This type of silver alloy is used mostly in jewelry work and precision instrument work. A practical silver alloy consists of 45 percent silver, 15 percent copper, 16 percent zinc, and 24 percent cadmium. Another common silver alloy consists of 15 percent silver, 80 percent copper, and 5 percent phosphorus. Other silver alloys are shown in Fig. 16-12.

ASTM Spec #B-73-29	Silver	Cu	Percent Zinc	Cadmium	Melts °F	Flows °F	Color
	9	53	38		1450	1565	
1	10	52		.05	1510	1600	Yellow
	*15	80		(5% Phos)	1185	1300	Gray
2	20	45	35.	.05	1430	1500	Yellow
3	20	45	30	.05	1430	1500	Yellow
	30	38	32		1370	1410	
	**35	26	21	18	1125	1295	Almost white
	40				1135	1205	Almost white
4	45	30	25		1250	1370	Almost white
	**45	15	16	24	1125	1145	Almost white
5	50	34	16		1280	1425	Almost white
	**50	15.5	16.5	18	1160	1175	Almost white
	**50	15.5	15.5	16 (3% Ni)	1195	1270	White
6	65	20	15		1280	1325	White
7	70	20	10		1335	1390	White
8	80	16	4		1360	1490	White

*—A special alloy containing phosphorous and used only on nonferrous metals.
**—Some special alloys of silver using a fairly high cadmium content.

Fig. 16-12. A table of silver brazing filler metal alloys. NOTE: When using brazing alloys which contain cadmium, the work area must be well ventilated.

which do not have some industrial applications for this method of brazing.

Silver brazing is often called silver soldering. However, the correct term is silver brazing.

Many alloys of silver have been developed. Each has a different melting point, flow characteristics, strength, and color. Some silver alloys melt at relatively low temperatures while others melt at fairly high temperatures.

Due to the cost of silver alloys, they are most frequently used for brazing operations rather than for braze welding. The application or the method of

Many new silver brazing alloys are on the market. Each alloy has its own particular fields of application.

These alloys vary in melting temperature and in flow temperature. The term "melting" temperature means the temperature at which the alloy starts to melt, while "flow" temperature means the temperature which the alloy must reach so all of the metal alloy is liquid. These temperatures may vary with some alloys from 1125 deg. F. melting to 1295 deg. F. flow.

Wide temperature difference between melting temperature and its final higher flow temperature presents some difficulties. While applying one of these latter alloys, the lower melting temperature metals (constituents) may flow into the cracks and leave the higher temperature metals or alloy behind.

This action causes a change in color and strength, and it also causes difficulty in flowing the remaining alloy onto the base metal. It is best therefore, to heat the alloy quickly, first to minimize oxidation, and second to prevent alloy separation. The separation feature of the wide temperature range alloys is an advantage in poor fit-up joints and in fillet joints, as the higher temperature alloys parts will bridge the gaps.

Cadmium and zinc are used as alloying metals in silver brazing alloys, because they have the peculiar ability to "wet" or flow and alloy with iron.

They also lower the alloy melting and flowing temperatures.

There is some danger of producing harmful fumes from the zinc and cadmium, if alloys containing these metals are violently overheated and the metals vaporize. The work area should be well-ventilated.

Silver brazed joints are very strong if properly made. When joining stainless steel butt joints by silver brazing, the tensile strength varies as shown in Fig. 16-13.

Silver brazing is one of the best meth-

STRENGTH OF SILVER BRAZED BUTT JOINT

Thickness of alloy in joint in inches	Tensile Strength lbs. per sq. in.
.002	133,000
.003	115,000
.006	90,000
.009	83,000
.012	76,000
.015 app. 1/64	70,000

Fig. 16-13. A table showing how the strength of a silver brazed joint is affected by varying thicknesses of silver brazing filler metal in the joint.

Fig. 16-14. A microphotograph of a silver alloy brazed steel joint. The brazing filler metal in the joint is Easy-Flo which is a 45 percent Si, 15 percent Cu, 16 percent Zn, and 24 percent Ca alloy.

ods used to connect parts in a leakproof manner and to provide maximum strength. These joints are strong and will stand up under the severe conditions. The excellent adhesion qualities are shown in Fig. 16-14.

When joining copper, brass, and bronze parts, a copper, silver, and phosphorus alloy may be used. This alloy is less expensive than most silver brazing alloys. No flux is necessary on copper, but brass (copper-zinc) is usually brazed using a flux. This type of alloy is not used on steel or iron alloys. The silver content is approximately 15 percent. The alloy flows at 1300 deg. F. The strength of this joint is shown in Fig. 16-15. The absence of flux when brazing copper improves the visibility during the brazing operation.

16-11. FLUXES FOR SILVER BRAZING ALLOYS

The fluxes used when silver brazing consist of various mixtures of boric acid, borates, fluorides, fluoroborates and the like. These fluxes are usable as dry powders, paste, or in the molten condition. They must be clean. The

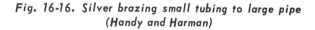

Fig. 16-15. A microphotograph of a copper and brass part joined by a 15 percent silver plus copper and phosphorus brazing alloy.

Fig. 16-16. Silver brazing small tubing to large pipe (Handy and Harman)

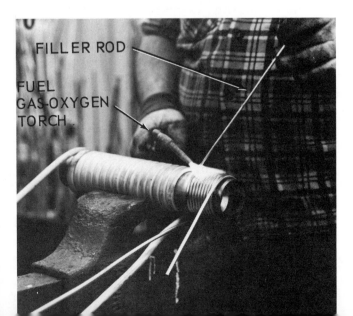

paste is made by using water or alcohol, to form a paste suitable to the purpose. These liquids must also be clean. Distilled water is preferred because tap waters often contain chemicals that might contaminate the flux.

It is important that the flux wet the surface to be brazed. This means that a low surface tension flux is needed. The liquid flux must be displaced by the molten silver alloy as the brazing proceeds. A combination powdered silver alloy mixed with a paste flux is available.

16-12. PROCEDURES FOR BRAZING WITH SILVER ALLOYS

Silver brazing can be easily done, if the correct procedure is followed. The points to be remembered are:

A. Clean the joints mechanically (use clean materials or tools).

B. Fit the joint closely and support the joint.

C. Apply the proper flux.

D. Heat to the correct temperature.

E. Apply the silver brazing material.

F. Cool the joint.

G. Clean the joint thoroughly.

An oxyacetylene torch is an excellent heat source for silver brazing, as shown in Fig. 16-16, although any liquefied petroleum fuel in conjuction with oxygen is acceptable. There are various silver alloys on the market. Silver brazing may be done by a number of methods such as:

1. Sweat method.

2. City gas and blowpipe method.

3. Welding torch method.

Fluxes used for silver brazing must be clean, chemically pure and fresh. Chlorides are popularly used as silver brazing fluxes. Borax made into a paste with water may also be used successfully as a silver brazing flux.

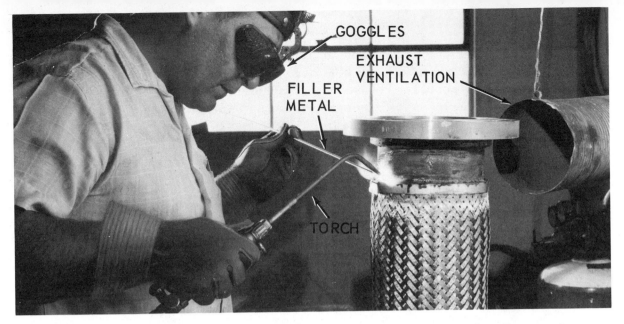

Fig. 16-17. *Silver brazing a flexible tube to a flange, using an oxyacetylene flame as the source of heat.* (Handy and Harman)

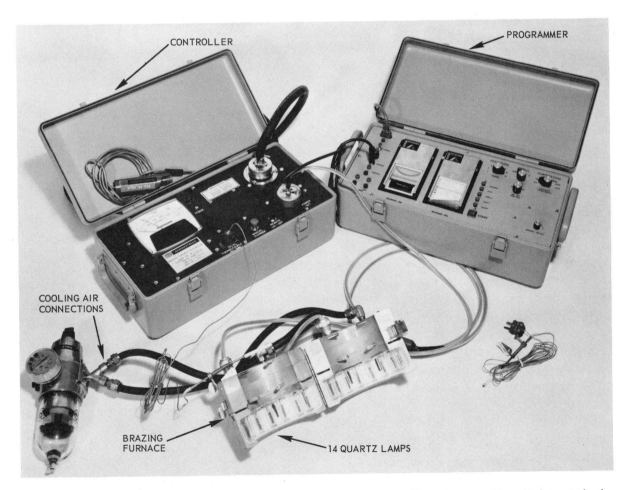

A quartz lamp heating device which may be used for brazing operations. Note the controller which controls the temperatures, the programmer which determines the heating time cycle, and the device for cooling the lamps.

The brazing filler metals having a 35 to 45 percent silver content are becoming increasingly popular. The parts to be brazed must be made to fit accurately and must be thoroughly cleaned. Any external surface should be cleaned with clean steel wool to remove dirt. Other types of cleaners may leave abrasives or an oil film. Internal circular surfaces can be cleaned with clean wire brushes, with clean steel wool rolled on a rod, or by using a clean drill.

The parts must have contacting surfaces of sufficient size, such as a tubing sliding into a fitting (not a drive fit) to get a strong fit. The contacting surfaces need not be very large, usually three

to bubble at about 600 deg. F. Finally it will turn into a clear liquid at about 1100 deg. F. This is just short of the brazing temperature. The clear appearance of the flux will indicate the time to start adding the filler metal. Fig. 16-17 shows a flexible tube being brazed to a flange. The 45 percent alloy melts at 1120 deg. F. and flows at 1140 deg. F.

A large tip is recommended for heating the joint, as the extra heat permits a shorter brazing time, thus reducing the time for oxides to form. The basics of a brazing operation when using a flame as the source of heat are shown in Fig. 16-18. The joint should be kept

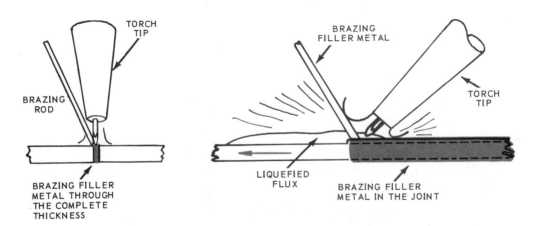

Fig. 16-18. Basic flame method of brazing a butt joint.

times the thickness of the thinnest piece of metal in the joint. If the parts are dented or are out-of-round, this fault must be corrected before the brazing is done. It is important to support the parts securely during the operation, as no movement should take place while the alloy is flowing and while it is solidifying.

The joint must be heated carefully. Behavior of the flux will indicate the temperature changes in the joint. First, the flux will dry out as moisture (water) boils away at 212 deg. F. Then the flux will turn milky and start

covered with the flame during the whole operation to prevent air from getting to the joint. The flame should not blow either the flux or the molten metal. Use

Fig. 16-19. A large assembly being silver brazed using the oxyacetylene flame as the source of heat. (Handy and Harman)

a slight feather (reducing flame) on the inner cone for best joint appearance. Fig. 16-19 shows a large assembly being silver brazed using an oxyacetylene flame as the source of heat.

It is necessary to heat both pieces evenly. If a thick piece is being joined to a thin piece, the heat must be applied more to the thick piece. Keep the torch in motion while the alloy is being added or local "hot" spots may develop and result in a poor joint.

A popular way to apply silver brazing filler metal when making a silver braze on a tubing joint is to use silver alloy rings as shown in Fig. 16-20. This is

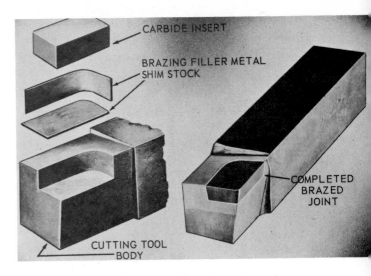

Fig. 16-21. A machining tool bit showing how the carbide insert is brazed to the tool bit body using preplaced brazing filler metal shims.

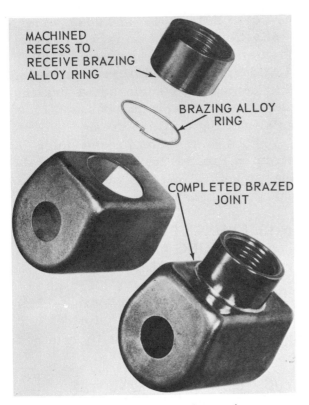

Fig. 16-20. Silver brazed joints designed to use preplaced silver alloy rings. The alloy forms almost perfect fillets and no further finishing is necessary.

a practical way to add silver alloy which is economical when used on a production basis.

A method of brazing using preplaced brazing shims is shown in Fig. 16-21.

It is necessary to thoroughly clean the completed silver brazed joint by washing in water or scrubbing. Any flux left on the metals will tend to corrode them and also hinder any painting or plating operations.

The joint may be cooled quickly or slowly. Cooling with water is permissible. Water quenching may also serve to wash the flux from the joint.

Visual inspection of the joint will quickly reveal any places where the braze metal did not adhere. But it is best to watch for this adherence and make any corrections during the brazing operation.

16-13. BRAZING STAINLESS STEEL

Stainless steel may be more easily silver brazed than welded. A silver brazed joint is strong if properly designed and made. The metal does not warp to any great extent because of the relatively low silver brazing temperatures, and a good color match can usually be obtained. With stainless steel, a 45 percent silver alloy filler rod produces a good silver brazed joint.

Fig. 16-22 shows the adhesion properties of a silver brazing alloy to both stainless steel and copper.

Fig. 16-22. A microphotograph of a silver brazed stainless steel and copper joint.

16-14. ALUMINUM BRAZING AND BRAZE WELDING ALLOYS

Some aluminum alloys may be successfully brazed or braze welded. The brazing filler metal is an aluminum alloy usually with a high silicon content which melts at a lower temperature than the parent metal. Fig. 16-23 shows some of the aluminum brazing filler metal alloys. Brazing or braze welding is usually confined to the 1100 and the 3003 (has some manganese) series aluminum alloys (non-heat treatable or "soft" alloys) and also to the 6061, 6062, 6063, and 6951 series aluminum alloys.

Aluminum castings of the A612, C612, 4043, and 356 series may also be brazed or braze welded. Some sheet aluminum is manufactured with a brazing alloy on its surface (brazing alloy clad) as shown in Fig. 16-24. See CHAPTER 24 for more information on aluminum and its alloys.

Some aluminum alloys have melting temperatures below that of the brazing alloys and therefore cannot be brazed or braze welded. Some of these alloys are the 2011, 2014, 2017, and 2025 series.

The aluminum brazing filler metal flow temperatures range from 1030 to 1190 deg. F. as shown in Fig. 16-25. This comparison scale shows brazing alloy melting temperatures and base metal melting temperatures.

A good aluminum brazed joint is shown in Fig. 16-26. Note the crystalline structure of the metal in this macrophotograph of the etched sample.

16-15. FLUXES FOR ALUMINUM BRAZING AND BRAZE WELDING ALLOYS

Aluminum brazing and braze welding fluxes consist mainly of chlorides and fluorides. This flux is usually obtained

FILLER METAL	COMPOSITION			TEMPERATURE DEG. F.		
CLASSIFICATION	Si	Al	Cu	SOLIDUS	LIQUIDUS	BRAZING TEMP.
BALSi-1	5.0	95.0	..	1070	1165	1150-1185
BALSi-2	7.5	92.5	..	1070	1135	1120-1140
BALSi-3	10.0	86.0	4	970	1085	1060-1185
BALSi-4	12.0	88.0	..	1070	1080	1090-1185

In the classification BALSi, the B means brazing; the AL means aluminum; and the Si means silicon. Solidus is the temperature at which the metal begins to melt. Liquidus is the point at which the metal will begin to flow.

Fig. 16-23. A table of aluminum brazing filler metal alloy compositions and temperatures. (American Welding Society)

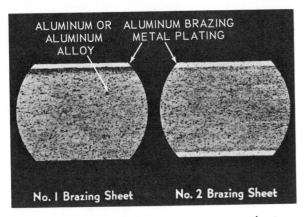

Fig. 16-24. Aluminum sheets showing an aluminum brazing metal plating on one side (No. 1) and on both sides (No. 2).

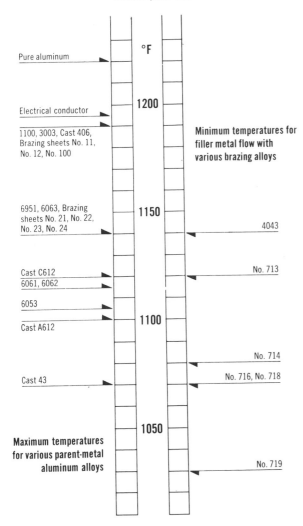

Fig. 16-25. A temperature scale comparing the melting temperatures for various aluminum alloys with the melting temperatures of various aluminum brazing filler metal alloys.

in the powder form. It can be applied as a powder, a paste, or a liquid. As for all fluxes, these fluxes must be clean. Some of the manufacturers have

Fig. 16-26. A 15x microphotograph of an aluminum brazed T-joint. Note that little or no parent metal has melted.

developed special fluxes for aluminum brazing and braze welding which produce excellent results. A small addition of alkali bifluorides appears to improve flux for aluminum brazing. Most commercial aluminum brazing fluxes include this ingredient.

16-16. PROCEDURES FOR BRAZING AND BRAZE WELDING WITH ALUMINUM ALLOYS

Aluminum alloys can be brazed or braze welded using a number of the methods of heating. Dip brazing and furnace brazing are used when enough production warrants the cost. The carbon arc, electric resistance, block brazing, and flow brazing methods are not used for aluminum brazing.

Practically all of the common joints have been successfully brazed or braze

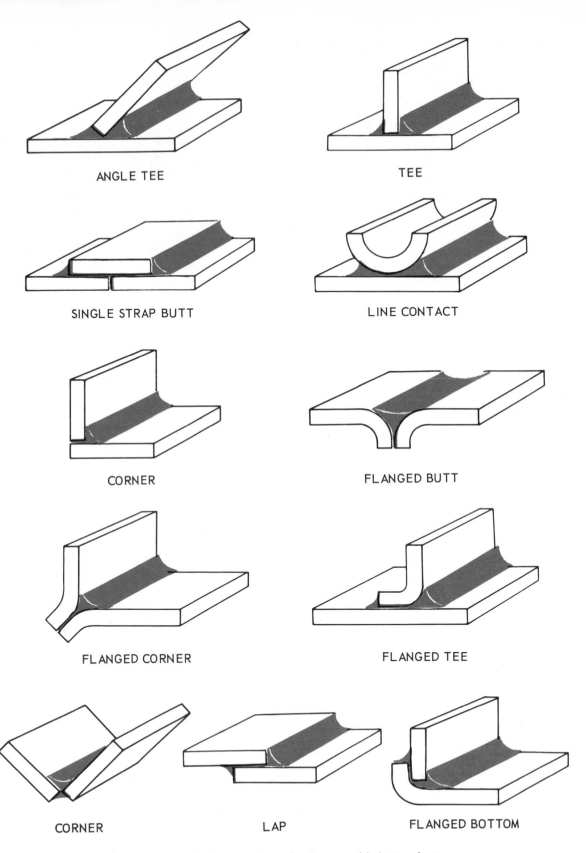

ANGLE TEE

TEE

SINGLE STRAP BUTT

LINE CONTACT

CORNER

FLANGED BUTT

FLANGED CORNER

FLANGED TEE

CORNER

LAP

FLANGED BOTTOM

Fig. 16-27. Typical aluminum brazed or braze welded joint designs.
(Aluminum Co. of America)

welded. Figs. 16-27, 16-28, and 16-29 show typical examples.

In the torch method of aluminum brazing and braze welding, both the oxyhydrogen and the oxyacetylene flames

2. Apply a fresh, chemically clean flux, specially compounded for aluminum brazing or braze welding, along the joint and on the brazing rod (flux may be in either powder or paste form).

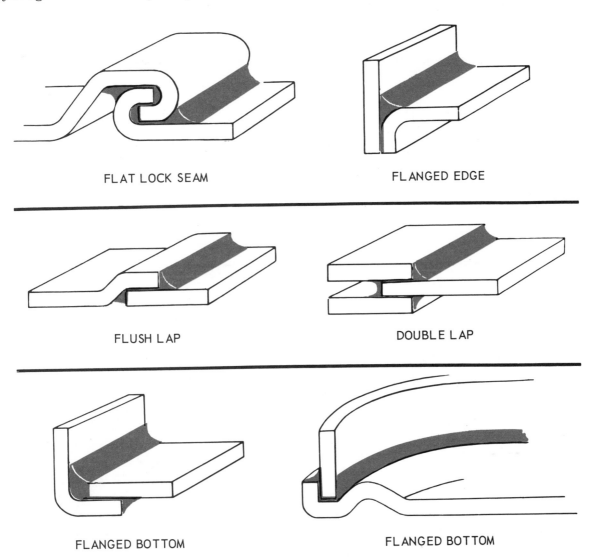

FLAT LOCK SEAM

FLANGED EDGE

FLUSH LAP

DOUBLE LAP

FLANGED BOTTOM

FLANGED BOTTOM

Fig. 16-28. Typical aluminum brazed or braze welded joints designed for use on containers.

may be used. In either case, a reducing flame with a 1 to 2-in. intermediate cone should be used. The following procedure produces good results:

1. Clean and degrease the surface. Braze or braze weld the joint as quickly as possible after cleaning the metal, while the surface is free of oxides.

3. Heat the joint and as the flux chemicals turn liquid, start to apply the brazing rod. The rod will melt and flow into the joint and through the crevices to produce a neat brazed or braze welded joint. Hold the torch so the inner flame cone is 1 to 2 in. away from the joint.

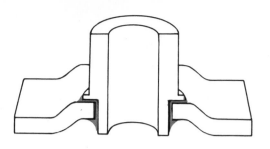

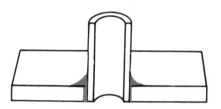 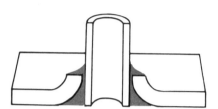

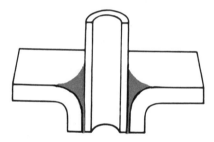 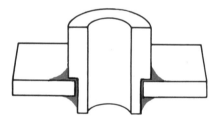

Fig. 16-29. Aluminum sheet and tube joints which have been designed for brazing.

4. Cool the finished braze or braze welded joint. Do not disturb the parts until the temperature has dropped to below 900 deg. F.

5. Clean the flux from the joint. Use hot water, then a concentrated nitric acid solution. After this, wash again in boiling water. Any flux left on the metal will seriously corrode it.

DANGER - THE HANDLING AND USE OF THESE ACIDS SHOULD ONLY BE UNDERTAKEN BY AN EXPERIENCED PERSON WHO HAS BEEN GIVEN THOROUGH TRAINING IN THE USE OF DANGEROUS ACIDS.

The finished weld should show 100 percent adhesion and the surface should be smooth and clean. The brazing filler metal should penetrate, with good adhesion, the full thickness of the joint.

16-17. BRAZING AND BRAZE WELDING MAGNESIUM ALLOYS

Successful brazing filler metal alloys have been developed for brazing and braze welding magnesium, as shown in Fig. 16-30. Fluxes for brazing and braze welding with magnesium alloys are usually made of chlorides and fluorides. The flux is normally a powder but is generally used as a paste or in electric resistance method. However, the torch, furnace and dip bath methods are the most popular.

| FILLER METAL | COMPOSITION | | | | TEMPERATURE | |
CLASSIFICATION	Mg	Al	Zn	MELT	FLOW	BRAZING TEMPERATURE
BMg	89	9	2	770	1110	1120–1160
AZ92A	89	9	2	830	1110	1130–1160
AZ125X	83	12	5	770	1050	1080–1130

Fig. 16-30. Table of magnesium brazing filler metal alloy properties.

dip bath form. A brazed magnesium T-joint is shown in Fig. 16-31.

Brazing of magnesium alloys can be done with the source of heat being:

1. Torch.
2. Furnace. A beryllium addition to

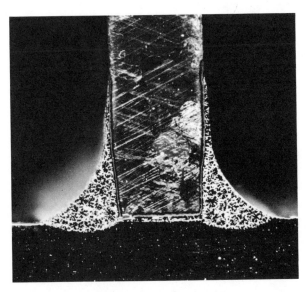

Fig. 16-31. A magnesium brazed joint. The sample has been etched; the macrophotograph is 8x.
(Dow Metal Products Co.)

the brazing alloy is needed when furnace brazing magnesium to prevent ignition of the metal.

3. Dip Bath. Examples of dip bath brazed magnesium assemblies are shown in Fig. 16-32.

The methods listed in PAR. 16-5 have been successfully used for brazing and braze welding magnesium except the

Heating must be very carefully controlled. The brazing temperature is between 1080 and 1160 deg. F. The melting temperature of one of the base metals is 1200 deg. F. Other base metals which may be brazed or braze welded melt at 1050 and 1160 deg. F.

To obtain a good brazed or braze welded joint:

1. Clean the metal thoroughly (both the base metal and the brazing rod).

    A. Clean mechanically--use clean tools and materials.

    B. Clean with acid by dipping parts for two minutes in a room temperature bath (24 oz. chromic acid, 5-1/3 oz. ferric nitrate, and .47 oz. potassium fluoride, plus enough water to make one gallon). Then rinse in running water. Finish with hot water rinse to aid in drying. This operation should be performed only by a well trained, experienced person.

2. Apply paste flux to the surfaces to be joined, and to the brazing rod.

3. Parts to be brazed should have a .004 - .010-in. clearance at the joint. Parts to be brazed or braze welded should be firmly supported until they cool.

4. Adjust the oxyacetylene or air-acetylene torch to a neutral flame.

5. Heat the joint evenly (same technique as for aluminum brazing).

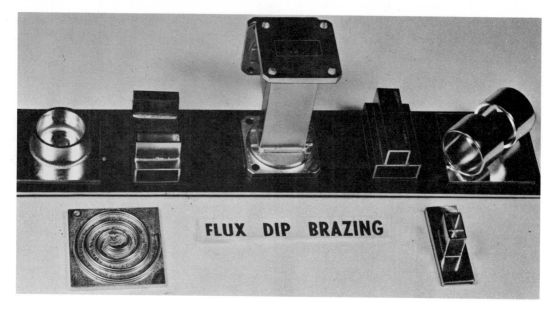

Fig. 16-32. Some typical magnesium structures as brazed by the flux dip brazing method.

6. Apply the brazing rod at the correct time. Flux behavior gives the best signal. Watch closely for full brazing filler metal flow.

7. Allow to cool to a solid state before moving the assembly out of its supports.

It is very important that the flux be completely removed. Hot water and scrubbing are popular methods.

A good quality magnesium brazed joint is shown in Fig. 16-33.

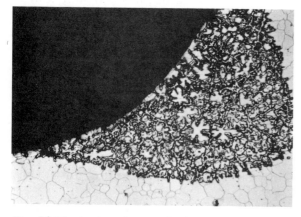

Fig. 16-33. A microphotograph of a magnesium brazed inside corner joint (65x). Note how the brazing alloy has penetrated the grain boundaries of the parent metal.

16-18. BRAZING AND BRAZE WELDING CAST AND MALLEABLE IRON

Cast iron can be braze welded quite satisfactorily. The cast iron surface should be processed much the same as for welding, except that the surface should not be ground. Instead, it should be filed or machined. Grinding smears the graphite particles over the grains of iron and results in poor adhesion between the braze filler metal and the iron. The braze welded cast iron joint may be reinforced the same as for cast iron welding, as described in CHAPTER 18. For best results preheat the casting to a temperature between 400 and 600 deg. F. Cast iron brazing flux must be used. Brazing rod for cast iron braze welding should be a high temperature rod (melting temperature approximately 1700 deg. F). Such rods are brass, usually high in copper content with some nickel added. Nickel has the property of making the joint metal adhere better to the iron. The high temperature used also assists in removing the graphite from the joint and helps the brazing filler metal adhere to the base metal. Fig. 16-34 shows a braze welding operation on a V-groove butt joint on cast iron. In braze welding cast iron, a tip which will give a high heat output with low fuel gas pressures (high heat soft flame) should be used. A high velocity flame (small tip, high pressure) tends to blow the flux away from the joint, and a poor braze weld will result.

Braze welding is the only satisfactory repair for broken malleable castings. Welding of malleable iron is not recommended. The procedure is the same as for braze welding cast iron, except that no preheating is necessary. For both cast iron and malleable iron the braze welded joint should be cooled

slowly by placing it in a box filled with sand or powdered asbestos. The slow cooling will eliminate formation of white cast iron which is hard and brittle. There are special cleaning processes for cast iron which make it possible to braze cast iron parts successfully. See Fig. 16-35.

16-19. CONTROLLED ATMOSPHERE FURNACE

A method of fabricating articles, and one that has become very popular in the last few years for the fabrication of small articles, is the controlled atmosphere brazing furnace. The method originated in 1907, but only recently has it become very popular. It consists of a typical furnace, usually heated on the inside by electrical resistance coils

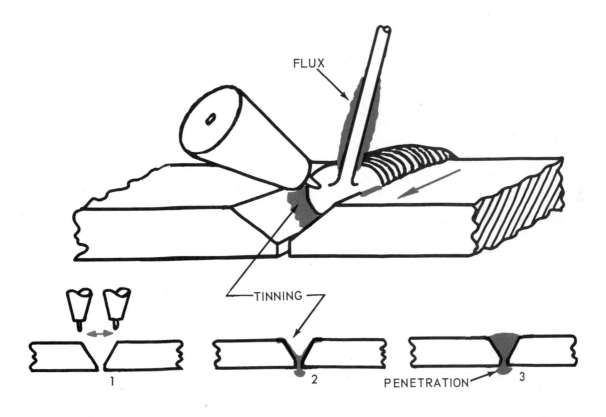

Fig. 16-34. A schematic showing the braze welding of cast iron. The backhand method is being used on this joint.

or partly combusted fuel gases. The inside of the furnace is usually kept as nearly gas-tight as possible, and little or no air is allowed to enter. Instead, the gas fed into the furnace is a low pressure mixture of air and fuel gas. Sufficient temperature is maintained inside the furnace to melt copper, brass, bronze, and silver alloys without melting the iron and steel. Recent developments in aluminum alloys enable the brazing of aluminum and its alloys in the reducing atmosphere fur-

Fig. 16-35. A microphotograph of steel being joined to cast iron by the silver brazing process. The cast iron was cleaned by a special process prior to the brazing operation. (Kolene Corp.)

nace. The operation of the furnace is as follows:

If two articles such as the two ends of a small steel cylinder are to be fastened together, they are first cleaned and then assembled in their final form with a small wire of the alloy, which is used to join them, placed inside the assembled joint. The assembly is then placed in the controlled atmosphere furnace.

Upon being subjected to the high temperature in the reducing atmosphere furnace, the brazing filler metal wire melts and joins the two surfaces together tightly. The article is then removed from the furnace and cooled slowly in a neutral atmosphere. When finished, the joint is approximately as strong as a weld. The method is well adapted to production work, as it insures a high quality joint, and the reducing atmosphere in the furnace simultaneously cleans the complete article, eliminating one manufacturing step. The reducing atmosphere of partly burned fuel gases eliminates any oxidation of the metals during the brazing operation.

Furnaces which are filled with an inert gas have also been used for brazing operations. Brazing has also been done in vacuum furnaces.

16-20. HEAT RESISTANT BRAZED JOINTS

There are several brazing applications in situations where the brazed joint must retain most of its physical properties when exposed to relatively high temperatures (2000 deg. F.). These special brazing filler metal alloys must also have good corrosion resistance.

These alloys are usually made of either a nickel-chromium alloy or a silver-manganese alloy, as shown in Fig. 16-36.

The nickel-chromium alloy has good corrosion resistance. It is used on jet engine blade joints, on stainless steel

3. When acids are used, these acids should be handled only by persons thoroughly trained for this type work.

FILLER METAL CLASSIFICATION	COMPOSITION								TEMPERATURE		
	Ni	Cr	Fe	Si	C	B	Ag	Mn	MELT	FLOW	BRAZING
BNiCR	70	16	*	*	*	4			1850	1950	2000–2150
BAgMn							85	15	1760	1780	1780–2100

* The total of these three elements is approximately 10 percent maximum.

Fig. 16-36. A table of brazing filler metals intended for use in high temperature and/or corrosive conditions. In the American Welding Society Classification BNCr and BAgMn the B means brazing; N-Nickel; Cr-Chromium; Ag-Silver; Mn-Manganese.

joints, and on low carbon steel joints. The brazing is usually done in a furnace.

The silver-manganese brazing filler metal is used mainly for joining stainless steel and high nickel alloys. This alloy is also usually applied in a furnace.

16-21. REVIEW OF BRAZING AND BRAZE WELDING SAFETY PRACTICES

Brazing and braze welding operations are safe to perform if reasonable safety precautions are carefully followed. The normal precautions as explained in previous chapters about the safe use of gas welding and arc welding equipment, must also be followed when brazing or braze welding.

The special precautions which are needed when brazing or braze welding are:

1. Ventilation must be excellent to eliminate the health hazards which are presented by toxic metal and flux fumes so often present when brazing or braze welding.

2. Many of the fluxes used are harmful to the skin and care should be taken in handling them so that no direct contact is made. If the fluxes do come into contact with the skin, the skin should be washed thoroughly by soap and water.

4. Some brazing filler metal alloys contain cadmium. When molten, and especially if overheated, these alloys emit cadmium oxide fumes to the atmosphere. Cadmium oxide fumes are very dangerous if inhaled. The limit value for cadmium oxide fumes is 0.1 milligrams per cubic meter of air for daily eight-hour exposures. This value represents the maximum tolerance under which workers may be exposed without adverse effects.

Cadmium fumes have no odor, and a lethal dose need not be sufficiently irritating to cause discomfort until after the worker has absorbed sufficient quantities to be in immediate danger of his life. Symptoms of headache, fever, irritation of the throat, vomiting, nausea, chills, weakness, and diarrhea generally may not appear until some hours after exposure. The primary injury is to the respiratory passages.

Most states have industrial hygiene personnel, usually in the health department or in the department of labor. If there is any question concerning the amount of cadmium oxide fumes in the work area, the industrial hygiene department should be asked to take air samples in the work area and evaluate the concentrations of cadmium oxide fumes present. Recommendations may then be made to correct situations that are potentially hazardous.

Eating or storing of lunches should not be permitted in the work area, and workers should wash both hands and faces before eating, smoking, or leaving from work. Workers brazing with cadmium-bearing alloys should be made aware of the hazard involved and trained to take precautionary measures relative to the environment of the particular job.

Alloys free of cadmium are available and should be used wherever possible. Workers should be trained to recognize the brazing alloys which contain cadmium.

An increase in temperature over the molten state accelerates the quantity of the fumes produced. For this reason, the use of an oxyacetylene flame as the heating source, due to its higher temperatures, may produce higher concentrations of cadmium oxide fumes than will the flames produced by air-acetylene, air-natural gas, or oxygen LP gas.

Obviously, it is of great importance that the work space used in brazing operations be thoroughly ventilated.

5. Another brazing operation which must be handled carefully is the brazing of beryllium. Oxides from this metal are also very dangerous if inhaled. Thorough ventilation is a must when heating and working with this metal.

16-22. TEST YOUR KNOWLEDGE

1. What is brazing?
2. Explain the difference between brazing and braze welding.
3. List the main constituents of brass.
4. What are some of the chemicals used in copper alloy brazing fluxes?
5. Name the constituents of a typical 45 percent silver brazing alloy.
6. Why is silver brazing used so extensively for fabricating stainless steel food and liquid containers?
7. What is meant by C.P.?
8. To what temperature must steel be heated when it is being copper alloy brazed?
9. Name one typical application for each of the following: copper alloy brazing, silver brazing, and aluminum brazing.
10. What might cause a coppery appearance in some brazed joints?
11. What should be the appearance of the flux before the operator adds brazing rod during a silver brazing operation?
12. Is it possible to make a silver brazed joint for the same cost as a welded joint? Explain.
13. What is the most popular brazing filler metal alloy used when brazing copper?
14. What is the flow temperature of the BALSi-2 aluminum brazing alloy?
15. What are the constituents of magnesium brazing filler metal alloy?
16. Why must there be clearance between the faces of the metals to be brazed together?
17. Why should the metals being joined be firmly mounted?
18. Name one heat resistant brazing filler metal alloy.
19. How do the graphite flakes in gray cast iron affect the adhesion?
20. Why must heating be carefully controlled when brazing magnesium?

Chapter 17

SPECIAL
WELDING PROCESSES

The general classification of welding includes various methods that have been developed for fusing metals together. Some of these methods are highly specialized. Some are patented, and these may be used only with the

adjacent surfaces and produce intermingling of the molecules (cohesion) is a welding process. Many other processes have been developed. Some are difficult to relate to the three main methods. These special processes dif-

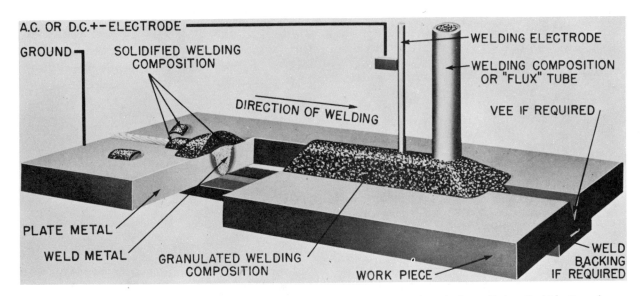

Fig. 17-1. Diagrammatic view of a submerged arc weld in progress. (Linde Div., Union Carbide Corp.)

permission of the patent owners. Others are special developments of gas flame welding, arc welding, resistance welding, and combinations of these methods.

17-1. SPECIAL WELDING PROCESSES

The three main types of welding are gas welding, arc welding, and resistance welding. Each of these three methods has several related processes. However, any method that can melt

fer mainly in the method used to agitate the metal molecules so they will mix. Some of the processes which may be considered special are:

ARC RELATED PROCESSES
Submerged Arc Welding
Electroslag and Electrogas
Stud Welding
Arc Spot Welding
Metal Arc Underwater Welding
Underwater Shielded Metal Arc
  Welding
Atomic-Hydrogen Welding

Plasma Arc Welding
GAS RELATED PROCESS
  Self-Generating Oxyhydrogen
    Welding
OTHERS
  Thermit Welding
  Forge Welding (Blacksmith Welding)
  Cold Welding
  Ultrasonic Welding
  Electron Beam Welding
  Friction Welding
  Explosive Welding
  Laser Welding

17-2. SUBMERGED ARC WELDING

Submerged arc welding is a welding process that has grown rapidly in popularity. It has some outstanding advantages. It is very fast. There is no visible arc, no splatter, and the welds are of high quality. See Fig. 17-1.

The method involves striking the arc between a consumable electrode and the joint while the arc is buried in a granular flux (such as titanium oxide-silicate).

Some of the submerged arc machines are able to produce single pass welds for butt joints up to 3 in. in thickness, plug welds up to 1-1/2 in. in thickness, and fillet welds with up to 3/8 in.

Electrodes up to 1/2-in. diameter may be used. The current may be as high as 4000 amps, and it may be either AC or DC. A typical submerged arc welding machine as shown in Fig. 17-2, has a power operated carriage with a universal variable-speed motor to control the travel rate from 7 to 210 in. per minute. A hopper feeds the granular flux to the joint just ahead of the arc. As the arc is struck and the bead is formed, the heat generated

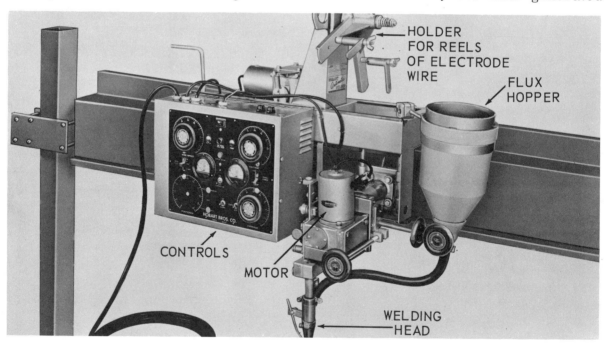

Fig. 17-2. Submerged arc welding outfit.   (Hobart Bros. Co.)

melts the adjacent flux granules, and a portion of the flux fuses and covers the weld with an airtight slag that protects the weld until it cools. This slag may be readily removed. The unused gran-

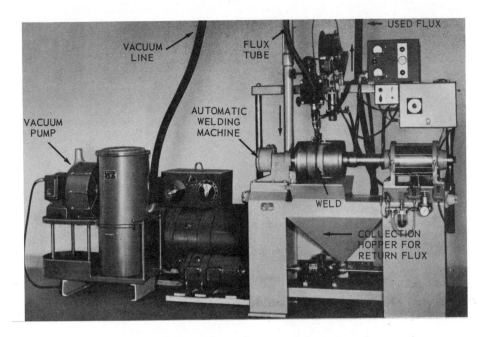

Fig. 17-3. Machine for recovering unused flux from a submerged arc welding machine. (Invincible Vacuum Corp.)

ules can be used again. Special equipment is used on some applications to pick up the unused flux and to feed it back into the hopper, Fig. 17-3.

Since the arc is submerged and therefore not visible to the operator, the correct setting is indicated by ammeter reading, and the correct arc length is indicated by voltmeter reading. The position of the arc is important, and a guide light shown in Fig. 17-4, has been developed which helps the operator align the arc to the groove.

This method is excellent for production jobs which require welds on heavy materials. The carriage may be used on a standard track, a template track, or the metal being welded may be mounted on a carriage.

The electrode is usually furnished in coils. A variable-speed electric motor controls the electrode feed. The controls for the process are mounted on the machine. Only a few adjustments are required on the machine to handle a variety of jobs.

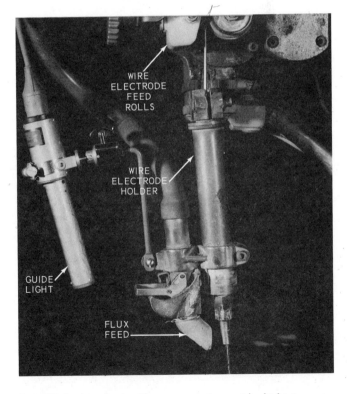

Fig. 17-4. A submerged arc operation guide light is mounted in line with the electrode and the line of the carriage or work movement, to help the operator determine if the submerged arc is centered in the weld groove. (Jimmie Jones Co.)

Special Welding Processes 399

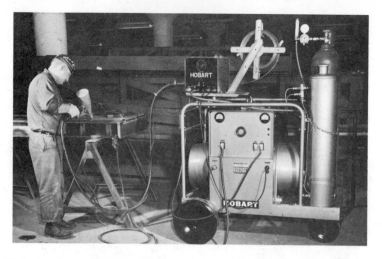

Fig. 17-5. A semiautomatic submerged arc welding station. (Hobart Bros. Co.)

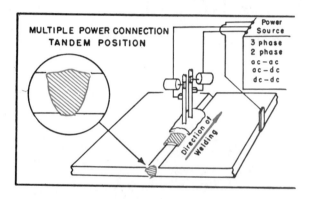

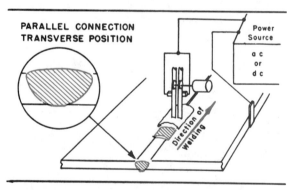

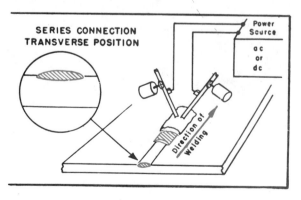

Submerged arc welding may also be done manually. See Fig. 17-5. Note the small hopper for granular flux built into the electrode holder. A combination electrode feed wire and cable is attached to the electrode holder. The operator guides the electrode holder and is able to make fast welds on either straight or irregular joints.

Submerged arc welding may be done with more than one metal electrode. The use of more than one metal electrode may be desirable when the finished weld must be wide or higher welding speeds are desired. See Fig. 17-6 for schematic drawings of three different methods of multiple electrode welding.

It has been found that when welding a horizontal joint the most practical position for the electrode is the 2 to 3 o'clock position. Many industrially made parts are welded in fixtures which will allow the electrode to weld in the 3 o'clock position for greatest weld efficiency, shown in Fig. 17-7.

See Fig. 17-8 for a comparison of weld metal deposition rates using single and multiple electrode machines.

Fig. 17-6. Three different ways to use multiple electrodes in submerged arc welding.

3 O'CLOCK WELDS

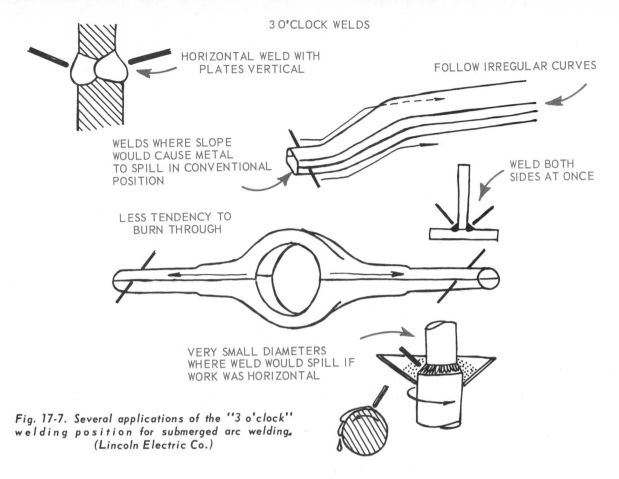

HORIZONTAL WELD WITH PLATES VERTICAL

FOLLOW IRREGULAR CURVES

WELDS WHERE SLOPE WOULD CAUSE METAL TO SPILL IN CONVENTIONAL POSITION

WELD BOTH SIDES AT ONCE

LESS TENDENCY TO BURN THROUGH

VERY SMALL DIAMETERS WHERE WELD WOULD SPILL IF WORK WAS HORIZONTAL

Fig. 17-7. Several applications of the "3 o'clock" welding position for submerged arc welding. (Lincoln Electric Co.)

Fig. 17-8. A comparison of weld deposition rates for single and multiple electrode submerged arc welding.

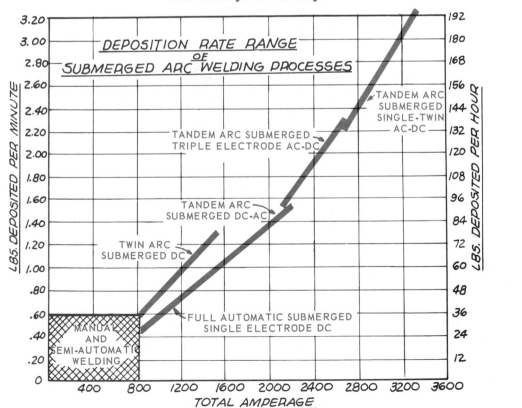

DEPOSITION RATE RANGE OF SUBMERGED ARC WELDING PROCESSES

LBS. DEPOSITED PER MINUTE

LBS. DEPOSITED PER HOUR

TOTAL AMPERAGE

TANDEM ARC SUBMERGED SINGLE-TWIN AC-DC

TANDEM ARC SUBMERGED TRIPLE ELECTRODE AC-DC

TANDEM ARC SUBMERGED DC-AC

TWIN ARC SUBMERGED DC

FULL AUTOMATIC SUBMERGED SINGLE ELECTRODE DC

MANUAL AND SEMI-AUTOMATIC WELDING

17-3. ELECTROSLAG AND ELECTROGAS WELDING

A method has been developed (Electroslag) to weld seams of very thick sections or joints. The process elim-

one pass. Butt joints, T-joints, corner joints, and most other joints can be made with this process.

In this process there is no arc, as the resistance to electrical flow through the flux creates the heat necessary to

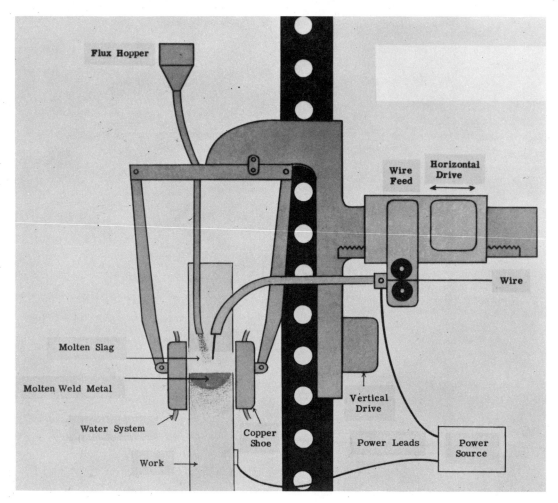

Fig. 17-9. Electroslag welding operation showing the flux hopper, electrode wire feed, shoes, and the rail on which everything moves as it moves up the weld joint. (Arcos Corp.)

inates the need for multiple passes and for bevel, V, U, or J grooves.

The process consists of putting the joint in a vertical position, using one to three electrodes, shoes (molds) on each face of the plates, and powdered flux. As the weld is made the molds move up with the welding head. Fig. 17-9 shows an electroslag weld in progress. The weld is completed in

melt the metal. The electrodes used are either solid wire or flux cored. The process is fast and requires no edge preparation of the metal. More than one electrode may be used. This permits a thick joint to be welded faster. Figs. 17-10, 17-11, and 17-12 show the electroslag process as it is used to weld a thick-walled pipe joint.

The extreme heat produced by the

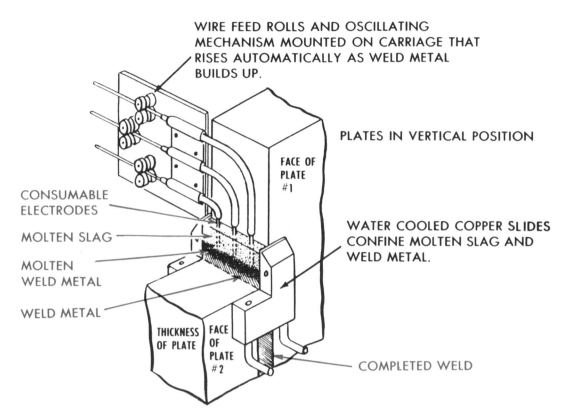

WIRE FEED ROLLS AND OSCILLATING
MECHANISM MOUNTED ON CARRIAGE THAT
RISES AUTOMATICALLY AS WELD METAL
BUILDS UP.

PLATES IN VERTICAL POSITION

FACE OF PLATE #1

CONSUMABLE ELECTRODES

MOLTEN SLAG

MOLTEN WELD METAL

WELD METAL

WATER COOLED COPPER SLIDES CONFINE MOLTEN SLAG AND WELD METAL.

THICKNESS OF PLATE

FACE OF PLATE #2

COMPLETED WELD

Fig. 17-10. A three-wire electroslag welding operation shown diagrammatically. The molten slag floats above the weld metal and helps prevent oxidation.

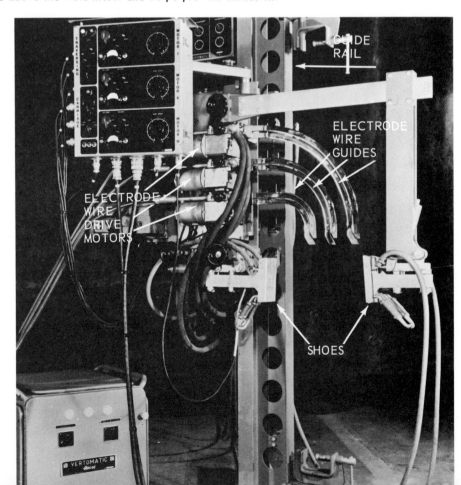

GUIDE RAIL

ELECTRODE WIRE GUIDES

ELECTRODE WIRE DRIVE MOTORS

SHOES

Fig. 17-11. The electroslag welding station for three electrode feed. Notice the water-cooled shoes and the fact that each of the wire feed mechanisms has its own drive motor.

GIRTH SEAM WELDING

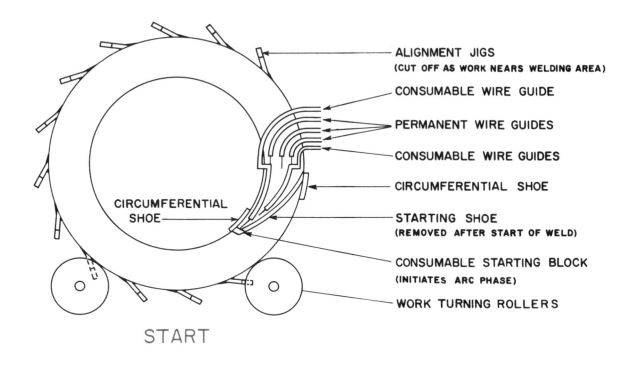

ALIGNMENT JIGS
(CUT OFF AS WORK NEARS WELDING AREA)

CONSUMABLE WIRE GUIDE

PERMANENT WIRE GUIDES

CONSUMABLE WIRE GUIDES

CIRCUMFERENTIAL SHOE

CIRCUMFERENTIAL SHOE

STARTING SHOE
(REMOVED AFTER START OF WELD)

CONSUMABLE STARTING BLOCK
(INITIATES ARC PHASE)

WORK TURNING ROLLERS

START

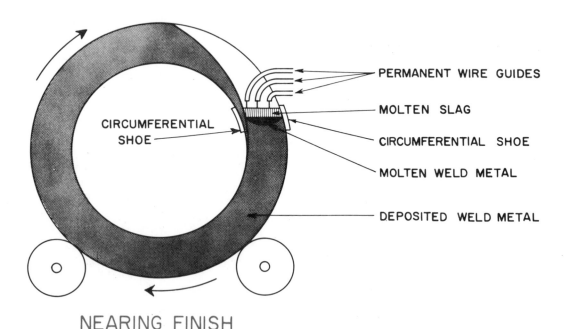

PERMANENT WIRE GUIDES

MOLTEN SLAG

CIRCUMFERENTIAL SHOE

CIRCUMFERENTIAL SHOE

MOLTEN WELD METAL

DEPOSITED WELD METAL

NEARING FINISH

Fig. 17-12. The use of the electroslag process for welding butt joints in cylindrical objects. In this case two consumable wire guides are used at the beginning of the weld.

molten slag and metal in the weld arc causes the base metal to melt away from the original joint gap as shown in Fig. 17-13.

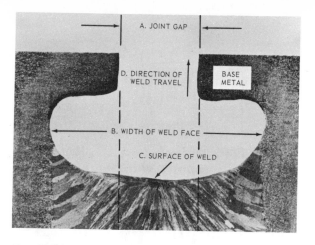

Fig. 17-13. A macrograph (1.5x) of a manganese-molybdenum steel joint welded by the electroslag process. Note the way the base metal melts away from the edges of the original joint gap.

The process of electrogas welding is similar to the electroslag process, except in this process a shielding gas is used. Flux cored wire is automatically fed to the molten weld pocket. An electric arc is continuously maintained between the electrode and the weld puddle. Fig. 17-14 illustrates a typical setup for electrogas welding.

17-4. STUD WELDING

Stud welding is a special welding development that quickly and efficiently welds studs and other fastening devices

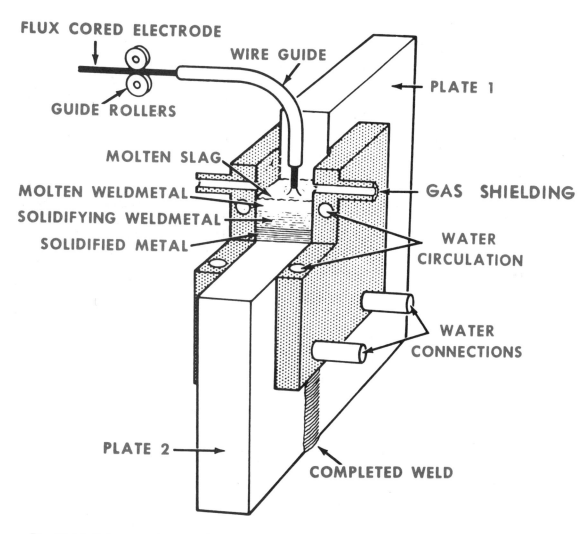

Fig. 17-14. Schematic drawing of an electrogas weld in progress. The shoes are water-cooled and are moved up as the weld proceeds. A shielding gas protects the molten weld metal from oxidation.

to plates and other surfaces. The process permits fastening an assembly device to a structure, and fastening different parts to this structure without piercing the metal, Fig. 17-15.

Fig. 17-15. Stud welded to a steel plate. Note the depth of fusion in this application as shown in the stud which has been sectioned. (Gregory)

Stud welding eliminates drilling or punching holes in the main structure and saves the work of mechanically fastening an object to the main structure, using bolts, rivets, or screws. Fig. 17-16 illustrates the process of stud welding.

A special collet on the gun holds the stud (1), and an arc is struck between the stud and the main plate (2). At the end of an automatically timed interval, the molten end is forced against the molten metal pool in the plate (3). A ceramic ferrule or collar around the stud holds the molten metal in place and helps form a good fillet. The flux on the end of the stud aids the arc control, and enables the operator to make stud welds in any position. By using appropriate jigs and fixtures, the studs can be placed to an accuracy tolerance

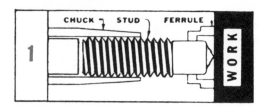

Fluxed end of stud is placed in contact with work.

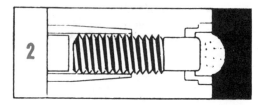

Stud is automatically retracted to produce an arc.

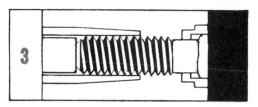

Stud is plunged into pool of molten metal.

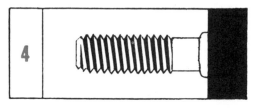

Operation completed — stud is welded to work.

Fig. 17-16. Steps required in the operation of stud welding. (Nelson Stud Welding)

as low as .005 in. both as to position and height (4). The equipment required in a complete stud welding station is shown in Fig. 17-17. Fig. 17-18 shows the details of a stud welding gun.

Fig. 17-19 shows a stud being welded to a cover plate.

Aluminum as well as steel can be welded using the stud welding process. In aluminum welding, a shielding gas is fed into the chamber formed by the fer-

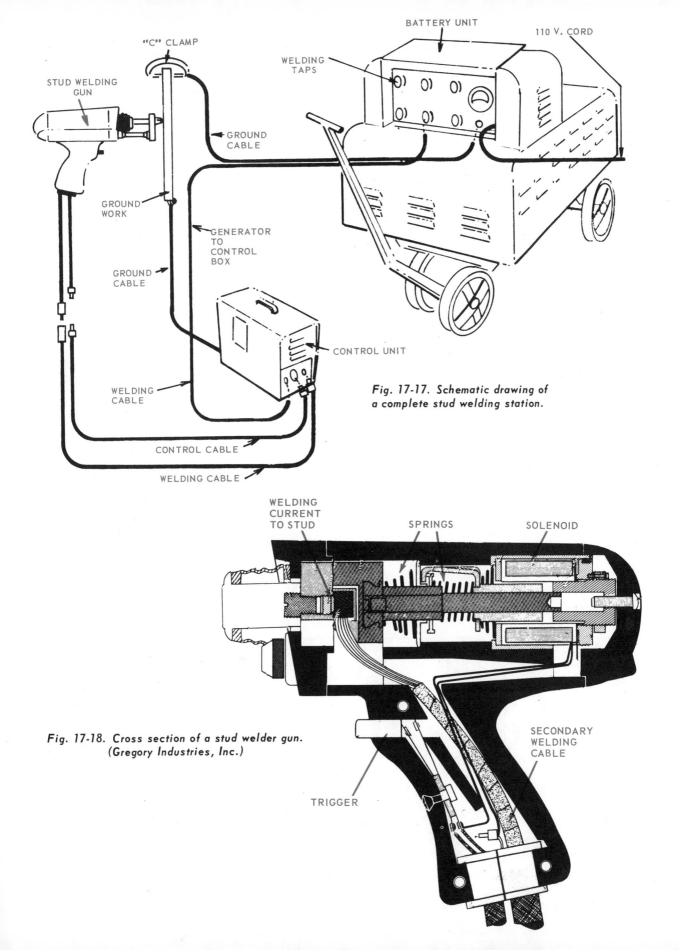

"C" CLAMP

BATTERY UNIT

110 V. CORD

WELDING TAPS

STUD WELDING GUN

GROUND CABLE

GROUND WORK

GENERATOR TO CONTROL BOX

GROUND CABLE

CONTROL UNIT

WELDING CABLE

Fig. 17-17. Schematic drawing of a complete stud welding station.

CONTROL CABLE

WELDING CABLE

WELDING CURRENT TO STUD

SPRINGS

SOLENOID

SECONDARY WELDING CABLE

Fig. 17-18. Cross section of a stud welder gun. (Gregory Industries, Inc.)

TRIGGER

Fig. 17-19. Stud being welded to a plate.
(Nelson Stud Welding)

rule, the stud and the base metal. Fig. 17-20 illustrates the operations required when stud welding on an alumi-

num plate. The gun is similar to the regular gun except a shielding gas feed is built into the gun as shown in Fig. 17-21.

The stud welding gun operates in the following manner: The stud is loaded into the collet and spring pressure from the gun holds it in place against the work. When the operating trigger is depressed, current flows through the solenoid and through the secondary welding circuit to the stud. The solenoid overcomes the spring tension in the gun and pulls the stud away from the work a preset distance. As the stud is pulled away from the work, an arc is struck between the stud and the base metal. The weld current and solenoid current are interrupted by the welding timer after a preset time, and the springs in the gun force the molten stud end into the molten pool in the base metal and hold it there until fusion takes place.

17-5. STUD WELDING EQUIPMENT

Stud welding equipment consists of:
1. Power source.
2. Cables.
3. Gun.
4. Studs.

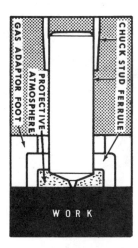
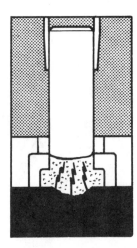
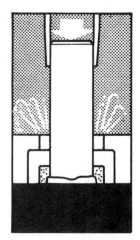
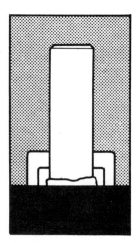

Fig. 17-20. Operations required when welding a stud to an aluminum plate.

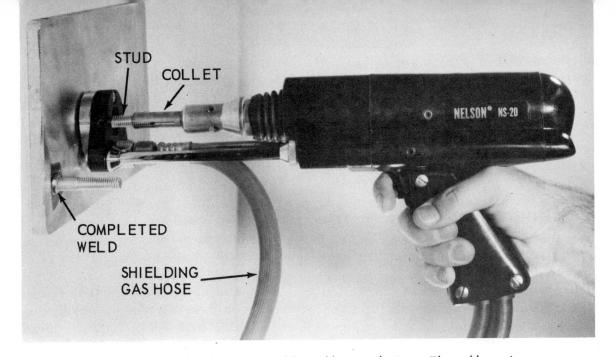

Fig. 17-21. Stud welding gun used for welding on aluminum. The weld area is protected by shielding gas fed into a chamber around the stud.
(Nelson Stud Welding)

The power source is a welding transformer adapted to stud welding use. The cables are made in the same manner as standard welding cables. The gun is of a design that ensures correct alignment, and has the following main parts: device to hold stud, trigger to start current flow, and timing device that withdraws the stud at the proper time to create the arc, move the stud to the correct position, and hold it until the metals cool to the solid state.

The studs are available in a variety of shapes, as may be seen in Fig. 17-22. Some are designed for threading, some for riveting, and some operate as nails.

17-6. ARC SPOT WELDING

The metallic arc welding process may be used to produce welds between two sheets of metal or between a sheet of metal and a plate or structural metal member.

The process consists of striking an arc, holding this arc in one place until the top sheet of metal melts through

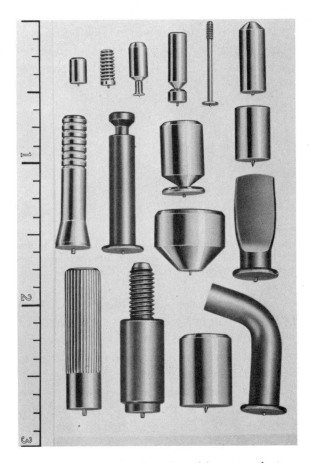

Fig. 17-22. A variety of studs and fastening devices designed for stud welding. (Omark Industries)

and fuses with a molten portion of the sheet or structural member underneath. This process is also described in CHAPTER 11.

The arc time is usually automatically controlled. The electrode is usually tungsten. The electrode holder usually has an insulator which properly spaces the tungsten electrode from the metal. A trigger moves the electrode up to the sheet metal. When the electrode is withdrawn, the arc is created and the timer is started.

The process is also known as button welding. It is used extensively for automobile body assembly, and for assembling sheet metal to structural steel.

17-7. UNDERWATER SHIELDED METAL-ARC WELDING

Metal arc underwater welding is done under water using a well-insulated electrode holder and special covered electrodes, Fig. 17-23.

Fig. 17-23. Underwater welding electrode holder. Note that the holder is well-insulated with no bare metal areas exposed. (Craftsweld Equipment Corp.)

The electrode covering is protected from damage by the water by special waterproof outer coatings. About 10 percent higher current setting is used for welding under water than is used when welding the same object in air. Because of the problem of rapid heat loss in water, stringer beads should be used rather than wide weaving beads. Also, a close arc and DCSP is recommended.

Due to the poor visibility under water, a #4 - #8 welding lens is recommended for use on the mask or helmet. When wearing the metal helmet for deep dives, the operator must be careful not to ground the helmet to any part of the welding circuit.

It is recommended that a telephone communication system be used and that the arc welding current be turned on from above water only after receiving the diver's orders when he is actually welding.

In underwater welding, fillet welds are recommended. When properly made such welds will develop approximately 80 percent of the tensile strength and 50 percent of the ductility of similar welds made above water. One should take extensive diving and welding training before attempting metal arc underwater welding for safety, and to acquire the necessary skills.

17-8. ATOMIC HYDROGEN WELDING

The atomic hydrogen process is a unique method of welding, which combines gas welding with electric arc welding. A special electrode holder is used having two tungsten electrodes and a hydrogen gas outlet, as shown in Fig. 17-24. An electric arc is passed between the two electrodes, at the same time that a stream of hydrogen gas is

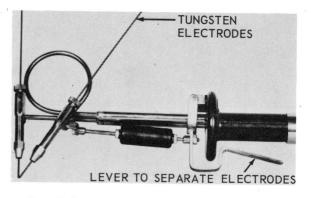

TUNGSTEN ELECTRODES

LEVER TO SEPARATE ELECTRODES

Fig. 17-24. Atomic hydrogen electrode holder.

Welding a T-joint using gas metal arc welding (GMAW) equipment. Note the protective gloves and helmet. A protective metal shield helps keep the right hand at a comfortable temperature. (Lincoln Electric Co.)

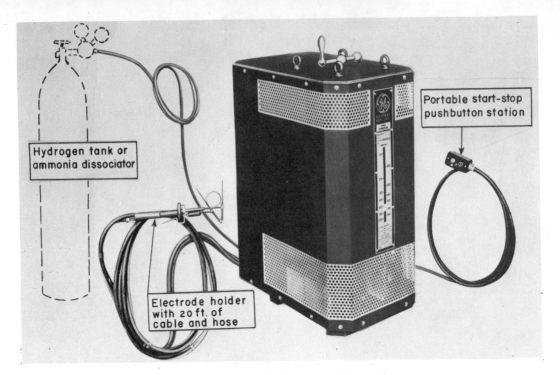

Fig. 17-25. Atomic hydrogen welding outfit.

ejected from the hydrogen nozzle and directed between the two electrodes. The electric arc breaks down the molecular hydrogen into atomic hydrogen. When the electrodes are held near the base metal, the atomic hydrogen stream contacts or impinges upon the metal to be welded. In touching the relatively cold metal, the atomic hydrogen recombines into molecular hydrogen, liberating considerable heat. This heat melts the metals to be welded and creates a molten puddle into which welding rods may be added.

Usually AC energy is used, which, instead of passing through the metal being welded, passes from one tungsten electrode to another. The temperature produced is approximately 7500 deg. F. The molecular hydrogen, as it blankets the metal, also prevents oxidation as it forms steam vapor when it combines with the oxygen in the air.

The method has two outstanding advantages:

1. The arc and, therefore, the amount of heat liberated are consistent at all times regardless of the work being welded.

2. The hydrogen provides a reducing gas atmosphere under which the fusion takes place. This eliminates any oxidation (burning) of the weld metal, producing greater density and more ductility.

The operator uses a rod to supply any additional metal needed. The heat input, atomic hydrogen stream, and shielding gas atmosphere are easily controlled, thus allowing this process to be used for welding applications such as tool and die repair where alloy control and heat input are important.

17-9. ATOMIC HYDROGEN WELDING EQUIPMENT

Equipment used in the atomic hydrogen welding may be divided into three sections:

1. Electrical and the gas supply.
2. Electrode holder.
3. Operator's equipment.

The electrical supply is usually

alternating current. It is transformed from higher voltage down to a voltage considered safe to be handled by the operator. The transformer used for this purpose is usually a constant potential type; that is, the voltage used at the holder is kept constant. The amperage is varied by means of an adjustment mounted on the transformer body. The transformer proper is usually air cooled.

The hydrogen is furnished in 200 cu. ft. cylinders under a pressure of approx. 2,000 psig. A regulator is used to reduce this pressure to 2 or 3 lbs. per sq. in. at the electrode holder.

The electrode holder is equipped with two tungsten electrodes. The electrode ends form a gap just beyond the hydrogen orifice. The arc is obtained by touching the tungsten electrode ends and then separating them the correct distance as shown in Fig. 17-25. The operator's equipment is standard and similar to that which the usual arc welder uses.

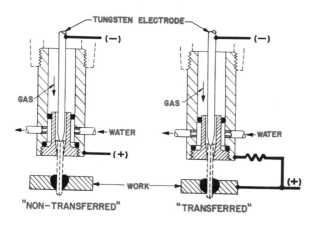

Fig. 17-26. *Schematic drawing of two types of plasma arc welding torches. Left. No electrical flow in the plasma stream between the torch and the work. Right. An electrical potential between the electrode and the work.*

The head shield lens used is usually a shade or two darker because the arc is not shielded by flux vapors.

17-10. PLASMA ARC WELDING

The term plasma as used in physics means a stream of ionized particles. When a person sees a streak of lightning in the sky he is actually seeing the gases of the atmosphere ionized and heated to incandescence. The electric arc is capable of ionizing both solids and gases. Substances are usually thought of as existing in three states either as solids, liquids or gases. Ionized substances are sometimes thought of as being in the fourth state of matter.

Plasma arc cutting originated in 1955 as an aluminum cutting process. Surfacing with plasma arc was used in 1960, and finally successful welds were made in 1963.

The plasma torch provides an electric arc between a tungsten electrode and a water-cooled copper nozzle. Gases such as helium or hydrogen are forced thru the arc and nozzle with the result that they are heated and become ionized and the stream from the nozzle is therefore a plasma stream of ionized particles.

This plasma torch may be used for both welding and cutting. A schematic of such a torch is shown in Fig. 17-26. Note that the tungsten electrode is negative (cathode) and the torch housing is positive (anode) DCSP. Two types of plasma torches are in general use:

1. Transferred arc torch.
2. Nontransferred arc torch.

The transferred arc type electrical connections are used for welding and cutting while the nontransferred arc process is used for plating applications (metals, ceramics, and some plastics). Top view, Fig. 17-27 shows a weld being made on stainless steel using a plasma arc torch. Note that DCSP is used. The temperature of the plasma ranges from 6,000 deg. F. to 100,000 deg. F.

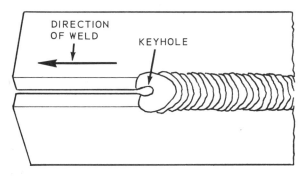

Fig. 17-27. Plasma torch (nontransferred arc) being used to weld stainless steel joint. Top. Notice how filler metal is fed into molten puddle. (Linde Div., Union Carbide Corp.) Bottom. Partially completed plasma arc weld showing "keyhole" at leading edge of puddle. Size of penetration.

Advantages of this system are that it has the directional control of a gas flame, a gas shield at the metal surface, along with the high temperature of the electric arc.

This torch can use almost any gas, for example air, argon, helium, hydrogen, nitrogen, or most any gas mixture. Most popular for cutting gases are argon-hydrogen, nitrogen-hydrogen. The hydrogen content varies between 10 and 35 percent. Argon may be used as a "backing bar" gas (root gas).

In plasma arc welding, the torch flame gives off a very penetrating and disturbing whine and ear muffs are needed to reduce the discomfort when working with or around the plasma arc.

The plasma arc voltage is usually between 27 and 31 volts when welding. The following table shows typical current and gas settings:

THICKNESS	AMPERES	TORCH SPEED	ARGON CU.FT./HR.
1/8 in.	125–160	10 ipm	45
1/4 in.	185–210	15 ipm	68
3/8 in.	210–250	18 ipm	80

The torch is held almost vertical to the base metal surface when welding.

Because the ion impingement has the ability to remove oxide films from the base metal surface, plasma arc welding of aluminum and other space-age metals is very practical. Direct current pulses from 6 to 60 cycles per second have been used with good success. The slower pulses seem to aid metal control in the puddle. In some situations when the metal being welded has a tendency to oxidize (aluminum, etc.) rapidly, the plasma torch is first passed quickly along the joint using DCRP to clean the metal (cathode cleaning) without fusing the metal, then the joint is welded immediately.

A plasma torch gives the best results when operated automatically.

The plasma arc has a tendency to produce enlarged root openings at the leading edge of the puddle (keyhole). Too much plasma jet action causes oversize keyholes, and the current and gas flow must be reduced to prevent this action. See bottom view of Fig. 17-27.

17-11. SELF-GENERATING OXYHYDROGEN GAS WELDING

The production of oxygen gas and hydrogen gas by the electrolysis of water has been done for many years. The burning of hydrogen in air or in pure oxygen has been used as a welding flame or gas welding process for several decades.

However, only recently have these two processes been combined. Distilled water is changed into its two gases by electrolysis. These gases are then fed to a torch and burned as shown in Fig. 17-28.

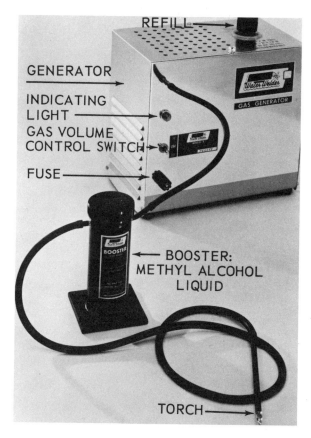

Fig. 17-28. *Self-generating gas welding station. The oxygen-hydrogen generator has two gas volume capacity positions which are controlled by a switch.*

The unit operates on 115 volts AC. An AC to DC converter provides power to the electrolytic reactor. The mixed gases travel along a single hose to the torch. The unit also has a booster which contains methyl alcohol. As the gas mixture bubbles through the booster, alcohol vapor is mixed with the gas and it also burns in the torch flame. The alcohol vapor provides a slightly reducing flame for soldering, brazing or welding. Fig. 17-29 shows a self-generating gas welding station in use without the methyl alcohol booster.

You should observe all the safety precautions as recommended in the previous chapters on oxy-fuel gas welding, soldering and brazing.

17-12. THERMIT WELDING

Thermit welding is based upon the fundamental chemical principle that aluminum is a more chemically active metal than iron. The process consists of mixing iron oxide and aluminum, both in a powder form, placing them in a hopper above a mold which surrounds the joint area, and then bringing this mixture to a temperature of 2000 to 2500 deg. F. The aluminum will then combine chemically with the oxygen molecules in the iron oxide producing a temperature of approximately 5000 deg. F., an aluminum slag (aluminum oxide), and a liquid iron. This high-temperature, molten iron is heavier than the aluminum oxide and it settles to the bottom of the mold which permits it to come into contact with the steel joint. It will melt the surface of the steel to be welded and fuse with it. This method may be used to weld both large and small steel parts together. It is very popular where large sections are to be welded together. Preheating is often used if the size of the pieces to be welded is large.

Fig. 17-29. *Self-generating gas welding station being used to fuse thermocouple wires in missile temperature sensing system. Methyl alcohol booster is not being used in this case.* (Henes Mfg. Co.)

Thermit welding may be roughly compared to a foundry casting operation, with the one difference that the metal being poured is of a considerably higher temperature than metal melted in a furnace.

Some common applications of thermit welding are: To weld railroad rails together, new teeth on large gears, large fractured crankshafts, sections of castings where size prevents their being cast in one piece; to repair large steel structures that are made only on special order and would be very costly to replace. Thermit welding has been applied successfully in almost every industry.

Thermit welding is also used for welding pipe; however, in this application, the thermit does not mix with the pipe metal, but merely furnishes the heat to melt the pipe ends, which are then butted together as they melt.

17-13. PROCEDURE FOR THERMIT WELDING

As thermit welding is a specialized form of casting, molds are necessary to control the flow of the liquid iron, and to shape it while it is in a liquid form. The metal parts to be welded together are firmly and accurately set up for welding. In preparing the parts to be welded, the joint is usually machined to provide a V gap all around the joint to enable the molten metal to gain access to all parts. One method is to use a wax pattern placed in this joint and built up to the form that the molten metal is to take as it flows into the joint. A sand mold is built around the wax pattern and the work to be welded. Both pouring gates and risers must be provided in this mold just as in regular molding. Vent holes are necessary on the larger jobs. During the preheating of the metal to be welded, the wax used as a pattern burns away, leaving the correctly shaped cavity to receive the thermit processed metal.

A second process uses a permanent mold. Fig. 17-30 shows a section of a crucible mold assembled around a rail joint to be welded.

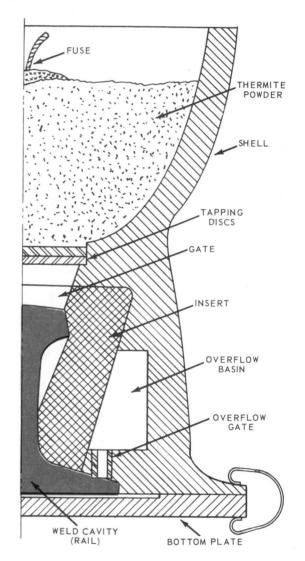

Fig. 17-30. Half of a cross sectional drawing of a set-up for thermit welding showing a crucible and the mold designed for welding a railroad rail butt joint. (Exomet, Inc.)

On most applications a funnel-shaped container, constructed of the same materials as the mold, is built above the mold. This funnel contains the aluminum and the iron oxide in their original powder form in sufficient quantity to provide enough iron for the weld. The usual procedure is to place this mixture above the weld in order to allow the highly super-heated iron to flow into the weld by gravity through the pouring gate. Some molds have a preheating gate, which provides for preheating the metal just prior to the pouring of the iron. This preheat also insures that all moisture is removed. As in all casting work, any moisture present when molten metal is poured may cause a sudden creation of steam and an explosion. An ignition powder, a special powder which burns at a high temperature, must be used in order to start the ignition of the thermit mixture. The aluminum and iron oxide mixture, as mentioned previously, will ignite only after it has been brought to the temperature of approximately 2000 deg. F. Once the mixture starts to burn, it is self-propagating. The chemical action should be allowed to go to completion before the pouring is started. The process is considered safe because of the very high temperature necessary to ignite the thermit mixture. This eliminates chance of accidental ignition of the mixture.

After the weld is completed and the metal has been allowed to cool slowly, the mold should be removed and the weld cleaned.

After the removal of the mold, the weld may need trimming because of the excess metal clinging to the weld, such as the pouring gate metal, the riser metal, and the metal in the vent holes.

The molding sand used in thermit welding is usually of a special mixture of silica sand and plastic clay. The pouring gate is closed with a thin sheet

of material which will melt only when the molten iron contacts it.

Before the mold is placed around the joint to be welded, it is necessary to clean the fracture until all surfaces are bright, and clean. It is especially important to remove any oil, grease, or water from the metal being welded as the presence of any or all of these materials, and the vaporizing of them may build up a dangerous pressure and may cause the mold to burst.

The thermit process is also used for welding cables for electrical conductors such as shown in Fig. 6-22.

17-14. FORGE WELDING (BLACKSMITH WELDING)

The oldest form of fusing two pieces of metal together is forge welding. This type of welding requires considerable skill on the part of the operator, and is usually limited to the joining together of pieces of solid steel stock. The two pieces of metal to be welded together are heated in a blacksmith forge such as shown in Fig. 17-31.

Fig. 17-31. Blacksmith forge such as used to heat metal for hammer welding.
(Champion Blower and Forge Co., Inc.)

This forge uses a banked fire of charcoal or coal with an air blast from below the coals to produce the heat necessary for forge welding. The metal to be welded must have coals under it in order that most of the oxygen from the air blast will have been consumed by the burning coals before the gases reach the iron which is being heated. Otherwise, the iron will become burned, or oxidized, and a poor weld will result. The metals to be joined are heated, in the area where the weld is to be made, to a white heat just short of the rapid oxidizing or burning point, and are then placed together so that the surfaces may be forced one against the other suddenly by the impact of a hammer. A blacksmith usually swages (enlarges) the ends to be joined. The pressure of the hammer blows, and the extra heat produced from the hammering, fuses the two pieces together. See Fig. 17-32. The energy from the shock of the hammer blows forces the oxide from between the surfaces of the metals, and produces a relatively clean area of fusion.

This same type of welding may be done using power hammers, fixtures, and presses. It is often called forge welding and/or hammer welding.

A weld of this nature, when done correctly, has every quality of the original metal. However, because of the skill necessary to produce a successful joint, and the relative ease with which other processes accomplish the same task, this type of welding has virtually been replaced by more modern welding processes.

17-15. COLD WELDING

The General Electric Company, Ltd. of England, announced a room temperature pressure welding process in

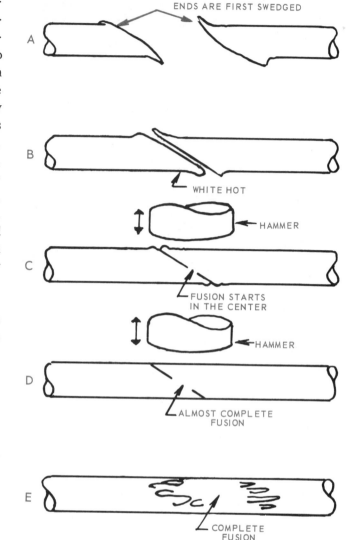

Fig. 17-32. A diagrammatic illustration of the steps followed when hammer welding.

1949. This process works best on aluminum and its alloys, copper, alloys of cadmium, nickel, lead, zinc, etc. This process is now available in the United States through the Koldweld Div. of Kelsey-Hayes Co. No heat is needed. The cleaned metal is forced together under considerable pressure, and the ductility of the metals produces a true fusion condition. Aluminum when welded by this process, has revealed a tensile strength of 22,000 psi. When pressed together, enough

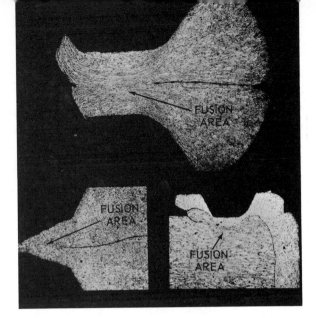

Fig. 17-33. Sample of a cold welded part. The metals
are aluminum to aluminum (1003).
(Koldweld Div., Kelsey-Hayes Co.)

pressure must be applied to reduce
the original thickness of the metal to
about 1/4 of the original thickness.
Fig. 17-33 shows three typical cold
welds.

The metals to be joined must be very
carefully prepared. The oxides and
other contaminants must be completely
removed. An abrasive wire wheel, turning at high speed, has been found to be
very satisfactory, as it not only removes the oxides but throws the particles clear of the metal. This method is
considered to be better than chemical
methods because of residual solvents.
Aluminum, copper, magnesium and
stainless steels, have been successfully welded by this process.

The theory of the process is that the
pressure along the molecular boundary
at the surface causes a fusion only a
few molecules deep. Temperatures at
the surface generally rise to about 600
deg. F. during the process, but never
up to the melting temperature of the
metals being joined.

The design of the tool for imposing
the pressure on the metals is very im-

portant. The tool must also be designed
to compensate for varying hardness of
the metals. Thus the tool must have
twice as much contact against the softer
aluminum when cold welding it to copper. The welds obtained can either be
the straight, ring, or seam design.
Fig. 17-34 shows a hand operated weld-

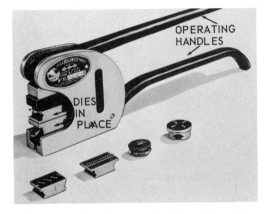

Fig. 17-34. Hand-operated cold weld machine.

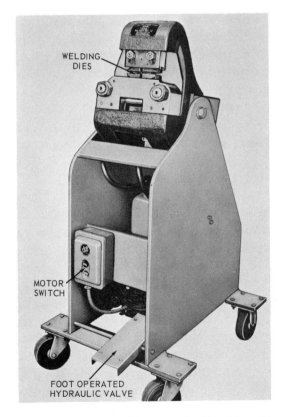

Fig. 17-35. Hydraulically powered cold weld machine.

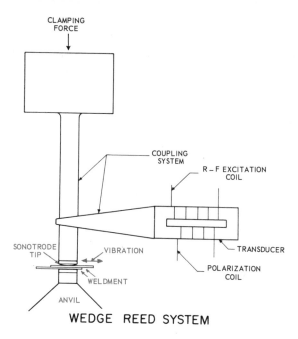

WEDGE REED SYSTEM

Fig. 17-36. Wedge-reed principle of ultrasonic welding. (Sonobond Corp.)

All types of equipment contain one or more transducers which convert high-frequency electrical power into mechanical vibration at the same frequency. (This is usually between 15,000 and 75,000 Hertz or cycles per second.) Beside the transducers there is a coupling system which transmits the mechanical vibration to the welding tip and thus into the metals being joined. The high-frequency electrical power used to drive the transducer is produced by a frequency converter.

Ultrasonic welding has several advantages. There is no heat distortion nor does the weld metal become brittle. No fluxes or filler metals are required. Many types of metals can be joined to themselves or to other metals and the equipment can be operated by semi-skilled personnel with minimal training.

For spot welding, the metals are clamped together and transverse or lateral vibration is imposed on the assembly, as shown in Figs. 17-36 and 17-37. Two different sizes of spot welders are seen in Fig. 17-38.

With the seam welder in Fig. 17-39, the metals to be joined are drawn between two counter-rotating rollers, and vibration is introduced through the upper roller. A continuous leaktight weld seam is produced between the two metals.

ing tool. It can weld up to a combined thickness of .080-in. aluminum or .060-in. copper.

The tools can be obtained either hand, pneumatically, or hydraulically powered. Fig. 17-35 shows a hydraulically powered unit which can butt weld up to 3/4-in. (round) or up to 3/4-in. strip copper. It can also perform lap welds.

17-16. ULTRASONIC WELDING

In ultrasonic welding, metals are clamped together under pressure and high-frequency vibrations are introduced into these metals through a welding tip or sonotrode. The vibrations (usually at a frequency above those one can hear) break up the surface films and cause the solid metals to bond tightly together. This happens without the use of heat and without melting the metal. Careful cleaning of the metals is not necessary. The low clamping pressure reduces deformation.

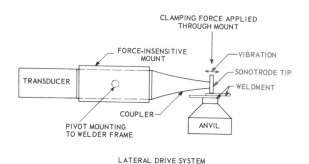

LATERAL DRIVE SYSTEM

Fig. 17-37. Lateral drive principle of ultrasonic welding.

Fig. 17-39. A 100-watt ultrasonic seam welder for making continuous seam welds in thin metal foil. (Sonobond Corp.)

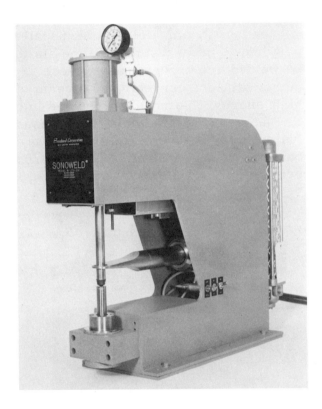

Fig. 17-38. Top. This low-power ultrasonic spot welder is equipped with frequency converter and foot switch for welding electronic devices. Bottom. A table model ultrasonic spot welder operating at a 500-watt input. (Sonobond Corp.)

With ultrasonic ring welders, such as that in Fig. 17-40, the welding tip vibrates in a circular motion like a cookie cutter. It produces a complete circular weld usually in less than a second.

Ultrasonic welding is used for making many types of electrical and electronic

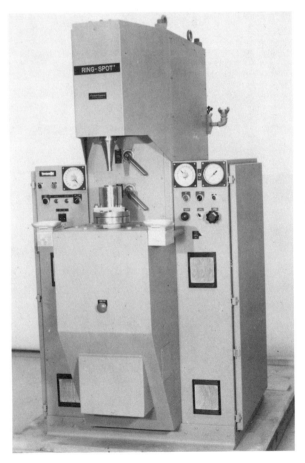

Fig. 17-40. Ultrasonic ring welder for welding metal covers to small hat-shaped containers.

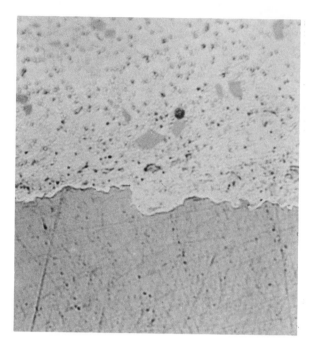

Fig. 17-41. Photograph (magnified) of an ultrasonic weld joining aluminum to copper.

connections, for fabricating honeycomb sandwich structure, for splicing aluminum foil in rolling mills, for making leaktight packages, and many other applications. Its greatest use is in welding aluminum to itself and to other metals. Fig. 17-41 shows a photomacrograph of a section of an ultrasonic weld joining aluminum to copper.

17-17. ELECTRON BEAM WELDING

Electron beam welding is usually performed in a vacuum. The metals to be welded are brought rather close together while a concentrated stream of high energy electrons is directed into the gap between the metals. This causes fusion to take place. Fig. P-22 shows a basic four-color diagram of an electron beam welding system.

Fig. 17-42. Complete electron beam welding machine. The weld is being made within an enclosed chamber. The operator views the weld through a special optical system.
(Hamilton Standard Div., United Aircraft Corp.)

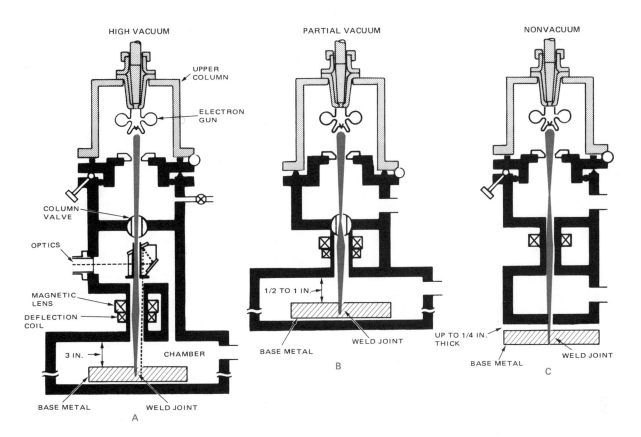

Fig. 17-43. Three types of electron beam welding systems: A. Weld being made under high vacuum. B. Weld being made under partial vacuum. C. Weld being made under atmospheric pressure. (Hamilton Standard Div., United Aircraft Corp.)

Fig. 17-42 shows a complete electron beam welding station. The equipment consists of a vacuum chamber, vacuum pumping system, electronic equipment, electronic controls, electronic gun, a gun and/or carriage moving mechanisms.

Fig. 17-43 shows three electron beam welding systems. In part A, under high vacuum, the top surface of the metal being welded is usually 3 to 4 in. below the top of the weld chamber. Under partial vacuum, part B, Fig. 17-43, the top surface of the metal being welded is usually 1/2 to 1 in. below the top of the weld chamber. In part C, at atmospheric pressure, the top surface of the material being welded is usually not over 1/4 in. away from the vacuum chamber. In this case, the electron

beam extends through a hole in the bottom vacuum chamber and, as in the case of Views A and B, the gun is under a high vacuum. The vacuum is maintained by a powerful vacuum pump. This vacuum pump removes the air which has entered the vacuum chamber through the opening on the bottom. In part C is shown a weld being made without the use of vacuum.

The electrons are emitted or released by the heated low-voltage filament. This filament is, in a way, very much like a light bulb. The cathode, which is the negative pole, and the anode which is the positive pole, send out a concentrated beam. The effect is to propel the electrons along the direction of the beam.

The focus coil (a doughnut shaped

Special Welding Processes 423

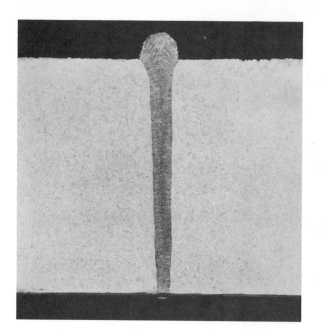

Fig. 17-44. Cross section of electron beam weld. Note how narrow weld is in relation to thickness of base metal. (Hamilton Standard Div. United Aircraft)

Fig. 17-45. Electron beam weld superimposed over a tungsten electrode metal arc shielding gas weld. Material is .250-in. stainless steel. Notice how narrow the electron beam weld is in comparison with the TIG weld.

electromagnet) concentrates or spreads the electron beam according to the user's welding needs.

The electron beam position can be controlled by deflection. Low frequency rotation of the beam is used to weld circular patches in thin materials. High frequency rotation will increase the stirring action in a puddle. The beam can be offset, circular, or square and the beam width can be adjusted as the welding operation requires.

To provide the low voltage AC, the high voltage DC (approximately 30,000V DC), and the low voltage DC, with all their controls, requires rather extensive electronic apparatus. It requires approximately 260 milliamps of current at 30,000V to weld a 1/2-in. thickness of stainless steel at 34 in. per minute.

Some units have the gun mounted on a wall of the vacuum chamber. In others the gun is movable within the chamber. Any gases in the chamber would interfere with the electron flow.

The welds are as shown in Figs. 17-44 and 45. Notice the narrow width of the weld metal, compared to the thickness of the weld.

17-18. FRICTION WELDING

Friction generates heat. If two surfaces are rubbed together, enough heat can be generated and the temperature can be raised to the level where the parts subjected to the friction may be fused together.

Therefore, if a joint is mounted in a device with one surface stationary and the other joint surface revolved, under pressure, the joint surfaces can be raised to a fusion temperature. The process is similar to electrical resistance butt welding except friction is the source of heat.

The process produces efficient joints and is especially adaptable to pipe, tubing, or solid round rods. Fig. 17-46 shows the various stages of a friction weld being made on a 1-in. dia. carbon steel rod. To perform this weld, 1,500 rpm is used at 7,500 psi heating pressure, then a 70,000 psi forging pressure is applied. The resulting weld is produced in 15 seconds.

Various metals can be welded using this process, such as:

1. Carbon steel.
2. Stainless steel.
3. Stainless to carbon steel.
4. Tool steel.
5. Copper.
6. Aluminum.

Fig. 17-47. Modern friction (inertia) welding machine. (Production Technology, Inc., a Caterpillar Subsidiary)

7. Alloy steels.
8. Alloy steels to carbon steel.
9. Titanium.

Most friction welding is used to weld ferrous parts. However, the electrical, chemical, nuclear, and marine industries use the process to join dissimilar (unlike) metals. The process is well suited for joining dissimilar materials (aluminum to stainless or mild steel, aluminum to titanium, copper to aluminum) in production. No protective atmosphere is required. This saves a significant amount of both time and materials.

The machines are somewhat similar in appearance to a lathe as shown in Fig. 17-47. An electric motor of 15 to 75 HP creates the rotation while hydraulic units are usually used to produce the heating (friction) pressure and the forging pressure. The apparatus is of greatest advantage on repetitive operations after the welding cycle has been established.

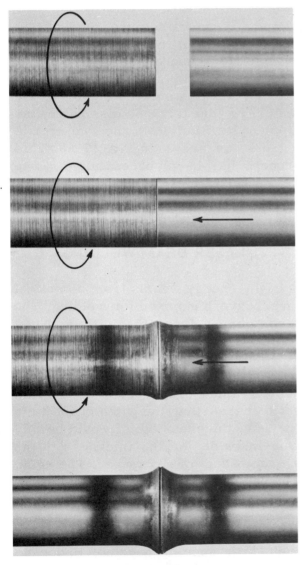

Fig. 17-46. Steps in friction welding two pieces of 1-in. diameter carbon steel. The rod at the left is rotated at high speed while the rod on the right is forced against the rotating rod. Friction creates enough heat to perform a sound weld. (Hamilton Standard Div. United Aircraft)

Special Welding Processes 425

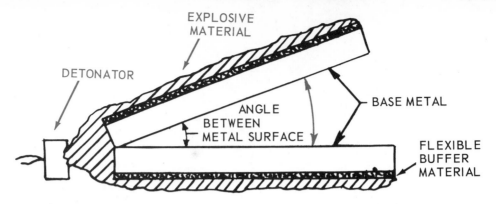

Fig. 17-48. Schematic drawing of a setup for explosive welding showing how the explosive material is placed on the base metal.

17-19. INERTIA WELDING

Inertia welding is a form of friction welding. In this process a movable part revolves against a stationary part. The energy for doing the welding comes from a flywheel spinning at high speed. A small amount of energy is required to give the flywheel a high velocity. During the time that the parts to be welded are being inserted in the machine, the flywheel will gain enough speed so that, when the clutch is engaged and the parts are brought together, they will have been brought up to a welding temperature. Inertia welding is used for high-volume runs, principally cylindrical parts such as axles, drill pipe, and the like.

17-20. EXPLOSIVE WELDING

This process is very dangerous. It should be performed only by explosive experts, and in specially designed chambers or water-filled chambers. Special permits must be obtained from local, state and federal authorities before this type of work can be done. Also, special training must be taken by the people involved.

Metals have been successfully formed, welded and/or work-hardened using the energies derived from an ex-plosion. When metals are welded by an explosive force, the metals are placed at an angle to each other, and the explosion forces the plates together at high velocity causing surface ripples in the metal. As the force is dissipated, the ripples lock or weld the two metals together as shown in Fig. 17-48. This process has been successfully used to weld steel to steel, aluminum to aluminum, copper to steel, stainless steel to molybdenum, and many other metals.

17-21. LASER BEAM WELDING

Light energy beams from an optical laser have been used for welding. The term comes from the initials, Light Amplification by Stimulated Emission of Radiation. The laser method can weld at a 200 to 1 ratio (the weld can be 200 times deeper than wide). This ratio means that the width of the weld for 1-in. thickness metal would be only .005-in. wide, or the thickness of the average sheet of newspaper. The welds can be made in either an inert atmosphere, or in the normal atmosphere.

There are no dangers of X-ray radiation but the operator must protect his eyes with special goggles. Under no circumstances should the operator put any part of his or her body in the way of the beam.

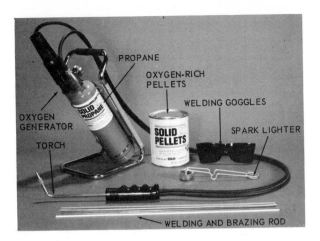

Fig. 17-49. An oxygen-propane welding system. It produces oxygen by heating an oxygen-bearing pellet. (Cleanweld Products, Inc.)

The operation of the laser consists of firing a brilliant light (capacitor discharges into Xenon tubes or almost instant ignition of aluminum or magnesium foil or wire). Usually this intense light is directed into the ruby by using parabolic mirrors. The electrons in the ruby are stimulated by this light source. These electrons return to their normal energy level releasing energy of a constant wave length. This accumulation collects inside the ruby, rebounds off one end of the ruby (mirrored) and the energy leaves the ruby at the other end with a pulse of very short duration (1/10,000 sec.) but very concentrated. The Xenon light and the ruby retain a considerable amount of heat and must be cooled between pulses of the laser beam. For example, the laser may operate 1/10,000 of second, and the rest of the second is a cooling period.

17-22. SOLID PELLET OXYGEN-FUEL GAS WELDING

This convenient oxygen-fuel gas welding system uses oxygen-rich solid pellets. It has been specially developed to use with a propane cylinder as the fuel source.

The advantage of this system, Fig. 17-49, is that it does not require an oxygen cylinder. The propane cylinder and a supply of pellets is all that is required to do a welding, brazing or cutting job anywhere.

The torch is designed to use different sizes of pellets, depending on the torch tip size. For larger welding jobs, the torch uses a larger tip and a larger pellet.

The manufacturers of this system supply a very thorough guide for its operation and use. If the guide is carefully followed, satisfactory welds may be made and the operation of the system is quite safe. If all safety instructions are not carefully followed, there may be some danger, since oxygen is produced within the torch handle.

17-23. REVIEW OF SAFETY

In submerged arc welding, the flux remaining over the completed weld acts as a heat insulator. The weld, as a result, remains very hot for some time after it is completed.

The usual eye protection and protective clothing should be worn when doing electroslag or electrogas welding.

There is some flash and sparking from stud welding operations. Eyes, face and hands must be protected from this sparking. Similar protection is necessary when performing arc spot welding.

The underwater welder should be in touch with the surface operator by means of a telephone. Precautions must be taken if a metal helmet is worn so that no part of it comes in contact with the welding circuit.

Plasma arc welding and cutting produce intense sound waves. In addition to eye, face and body protection, it is usually necessary to wear ear muffs during such welding or cutting operations.

Oxygen and hydrogen gas produced in a self-generating unit may form a very explosive mixture. Allow the gases to come together only at the torch flame.

A considerable quantity of molten metal is formed in the crucible of the thermit welding process. When this molten metal is flowed into parts to be joined, the metal must be carefully managed so that it does not come in contact with moist or wet surfaces. Protect eyes, face, hands and body.

There is usually considerable sparking during hammering together of parts to be joined in forge welding. Face, eyes and hands must be adequately protected.

Friction welding generates considerable sparking. Protective clothing, helmets and gloves should be worn.

Handling explosives is always dangerous. They must be carefully stored, applied and detonated. Follow manufacturer's instructions.

Laser welding uses light energy. The beam of light from the laser welding equipment must never strike any object except the parts to be welded.

Solid oxygen pellet welding systems require careful handling. The pellet burns at a very high temperature. A partially burned pellet may be kept for future use; however, handling a burning pellet requires great care. The usual eye protection, clothing and gloves must be worn. Since this system can be used in so many locations, surrounding areas must be cleared of all combustible material. The flame can produce temperatures of more than 5000 deg. F.

17-24. TEST YOUR KNOWLEDGE

1. What are the two metals used in the thermit steel welding mixture?

2. What kinds of metals are best welded with the thermit process?

3. How are underwater welding electrodes protected for use under water?

4. Where does the heat come from originally which is used to melt the parent metal in the atomic-hydrogen welding process?

5. What kind of metal may be blacksmith-welded?

6. Can steel pipe be thermit welded?

7. Of what materials are the atomic-hydrogen electrodes made?

8. What protects the metal and the arc during a submerged arc weld?

9. How does one determine if the correct arc length is being used in submerged arc welding?

10. What energy is expended during cold welding?

11. Does the metal temperature rise during cold welding?

12. Are the parts to be welded under pressure during ultrasonic welding?

13. Why is most electron beam welding done in a vacuum?

14. What means are used to focus the electron beam?

15. What is plasma?

16. Why are some plasma systems called transfer arc systems?

17. In what direction does the weld travel during electroslag welding?

18. What type electrode is used in electroslag welding?

19. How many pressure stages or steps are used during friction welding?

20. Is stud welding an arc process or a resistance welding process?

21. Is stud welding done using a gas shielding medium?

22. How many carbon electrodes may be used when carbon arc welding?

23. Do some welding processes use a flux-cored electrode?

24. May metals be welded using the energy from an explosion?

25. Is it possible to use the electron beam for cutting metals?

Chapter 18
SPECIAL
WELDING APPLICATIONS

Practically all metals can be successfully welded. Earlier chapters of this text covered the welding of mild (low carbon) steel. This chapter is devoted to metals and alloys which are not as common as mild steel, and for which special welding instruction may be needed. This includes also pipe and tube welding.

Pipes and tubes are mediums used to carry fluids from one point to another. The term fluid includes substances in either the liquid or gaseous form, or both. The term "pipe" usually refers to cylinders of hollow metal or other material, of substantial wall thickness. Pipes are made by casting, extruding (seamless), or by rolling flat stock into cylinder form and welding the seam. The thickness of a pipe is usually such that the pipe may be threaded. The term "tube" usually refers to hollow cylinders having a thinner wall than pipe which are of either seamless or seamed construction. The wall thickness of tubing is usually too thin for threading. Therefore, the joints must be soldered, brazed, or welded.

The types of plastics in commercial use and the quantity of plastics is rapidly increasing. Some types of plastics may be welded.

Most of these special applications necessitate preheat treatment and postheat treatment. CHAPTER 25 describes heat treatment processes.

18-1. CLASSIFICATIONS OF METALS

The two large classifications of metals are:

1. Ferrous metals.
2. Nonferrous metals.

Ferrous metals are those bearing a substantial iron content. The growth and popularity of special forms of ferrous metals has necessitated the development of many new techniques of welding these metals. The use of low carbon alloy steels for high strength, lightweight construction, and the use of stainless steels for applications where appearance and corrosion resistance are important, has brought about many special welding problems.

The coating on metals must also be considered when these metals are being welded. Galvanized steel (zinc coated steel) is a very popular metal and it can be successfully welded.

Welding finds an important use in the fabrication and repair of cast iron articles. Cast iron is normally a brittle metal. Under the welding flame, or the electric arc, it behaves differently from most of the metals previously described.

Dissimilar metals must frequently be welded together.

The nonferrous metals consist of all metals not composed of iron or steel. Many metals are classified as nonferrous. Some of the more popular of the nonferrous metals are:

1. Copper and its alloys.
2. Aluminum and its alloys.
3. Lead and its alloys.
4. Zinc and its alloys.

There are also many new metals being developed for commercial use. Some of these metals are:

1. Titanium.
2. Beryllium.
3. Zirconium.

18-2. ALLOY STEELS

Alloy steels have been produced and have been used for many decades. New alloys are constantly being developed. Each commercially produced alloy has special properties. Some are more corrosion resistant, some are stronger, some are tougher. A variety of properties have been developed to meet existing needs.

In welding these steels they must be accurately identified so the proper welding procedures may be used. The correct preheat procedure, the proper type of welding, the welding sequence, the postheat procedure and the like, all must be known and must be followed carefully for best results.

Some of the more popular alloy steels are:

1. Low carbon alloy steels.
2. Chromium steels.
3. Nickel steels.
4. High nickel-chromium steels.
5. Low carbon molybdenum steels.
6. Tool and die steels.
7. Galvanized steels.
8. Stainless steels.

For low temperature use, the steel alloys shown in Fig. 18-1 are recommended. These steel alloys retain enough strength and ductility to permit their use at these low temperatures. Welds in these metals for use at these low temperatures must have similar properties. See CHAPTER 28 for more information on steels and steel alloys.

18-3. LOW CARBON ALLOY STEELS

Many articles which were formerly made of cast iron or cast steel are now being fabricated of rolled steel. Many of these applications are found to be much more desirable, especially when a reduction in weight is possible. Low carbon, alloy steels (alloy steels under .30 percent carbon) are 10 to 30 percent stronger than the straight, carbon steel, and are desirable where a saving in weight is important. An example of this is a steam shovel bucket. Generally speaking, each pound that is saved in structure means an extra pound of material may be handled with the same power. One company was able to convert a 5-ton capacity, steam shovel bucket into a 6-ton carrying capacity by changing from a cast design to a fabricated low carbon, steel alloy design.

These alloy steels are slightly more expensive than the straight carbon steel, but by reducing the overhead (lost power) they are preferred. The method of welding the low carbon alloy steels is similar to that of straight carbon steel.

When oxyacetylene welding, flux should be used to counteract the oxidation of alloying elements, and filler rods of a composition corresponding to the base metal must be used.

When arc welding these metals use covered electrodes. Use an electrode recommended for the steel being welded. When gas metal arc welding these metals, the electrode wire must match the base metal. The expansion coefficient of these steels slightly exceeds that of straight carbon steel: however,

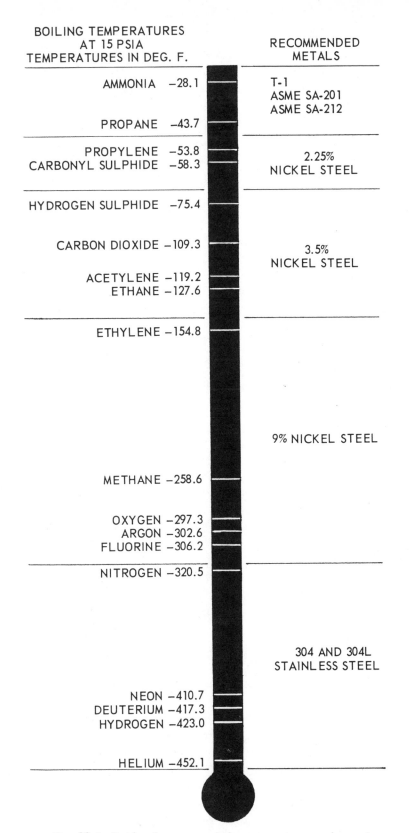

Fig. 18-1. Table of recommended construction metals used to contain low temperature fluids. (Welding Engineer)

no special precautions need be taken in reference to warping, except in extreme cases. For the best results, one should anneal the complete structure after all the welding has been completed. The main alloying elements of these low carbon steels are usually copper and nickel. The amounts of these are .1 to 1.0 percent copper and .2 to 2.0 percent nickel, as shown in Fig. 18-2.

always be annealed as quickly as possible after welding. If the alloy has over 18 percent chromium, a postheat treatment is recommended. A postheat of 1500 to 1600 deg. F. and then water quench, is sometimes used to increase hardness and corrosion resistance. Some of these steels occasionally have a little molybdenum or tungsten added to increase the strength of the steel especially at the higher temperatures.

TYPE	C	Mn	Si	Cu	Ni	Mo	Cr
Cor-Ten	.10	.20	.70	.40	....	...	1.0
Man-Ten	.35	1.50	.22	.13	....	.20	....
Sil-Ten	.30	.70	.20	.13	....	...	....
Yoloy	.15	....	...	1.0	2.0	...	....
RDS-1	.12	1.00	...	1.0	.75	.20	....
RDS-1A	.30	1.00	...	1.0	.75	.20	....
Hi-Steel	.12	.60	.30	1.1	.55	...	....
HT-50	.12	.50	...	.55	.55	.15	....
AW-70-90A	.25	.75	.25	.55	.25	...	....
AW-70-90B	.25	.75	.25	.55	.25	...	.25
Jal-Ten	.35	1.50	.30	.40	....	...	.25
Gr City	.14	.80	.18	.28	....	...	.12
Gr City 2	.25	1.40	.18	.28	....	...	.12
Centralloy	.15	.75	.50	.50	.25	...	.25
Konik	...	....	...	.20	.35	...	.12

Fig. 18-2. Table of low carbon alloy steels.

18-4. CHROMIUM STEEL

Some steels which are corrosion resistant, but still are not classed as stainless steels, contain chromium as their principal alloying metal. There are four main chromium alloys:

1. 12 percent Chromium.
2. 16 percent Chromium.
3. 18 percent Chromium.
4. 28 percent Chromium.

The group containing 14 to 16 percent chromium are corrosion resistant, especially to sulphides, and are also applicable to cold working. The carbon content of these steels varies between .10 to .25 percent. These metals are best welded when preheated to 300 to 500 deg. F. Welds on these metals of 16 percent chromium or less should

These metals are usually arc welded. The current settings are approximately the same as for similar conditions when welding mild steel.

18-5. NICKEL-CHROMIUM STEELS

The high nickel, low chromium steels are better known as Inconel or Nichrome steels. These metals contain from 72 to 80 percent nickel, from 10 to 15 percent chromium. The remainder is principally iron.

Inconel is usually arc welded using a E3N12 electrode. A preheat and/or postheat is not required.

Ni-chrome may be arc welded with DC reverse polarity using an E4N12 electrode. No preheat, stress relieving, stabilizing, or annealing is required.

18-6. CHROME-NICKEL-MOLYBDENUM STEELS

These steels contain approximately 10 to 15 percent nickel, 16 to 20 percent chromium, and 2 to 4 percent molybdenum. For example:

1. 316 has 14 percent nickel, 18 percent chromium, and 3 percent molybdenum.

2. 317 has 15 percent nickel, 20 percent chromium, and 4 percent molybdenum.

A preheat treatment is not required. A postheat treatment is not required. The arc welding is usually done with DC reverse polarity.

18-7. LOW CARBON MOLYBDENUM STEEL

In high-pressure, piping work, and in some other welded structures which operate at a high temperature, low carbon, molybdenum steels are used. The molybdenum content is approximately .5 percent. It is best to preheat the metal before welding. Otherwise, the welding operation is much the same as for mild steel. It should be annealed after the metal has been welded.

18-8. CHROME-MOLY STEELS

These steels contain a small amount of molybdenum and a medium amount of chromium. For example:

1. Type 502 or 5 Chrome contains 5 percent chromium and .5 percent molybdenum.

2. Type 505 or 9 Chrome contains 9 percent chromium and 1 percent molybdenum.

These metals are used where creep strength is needed, and where oxidation resistance is needed at high temperatures; for example, with high temperature steam, 1200 deg. F.

A preheat of 300 to 500 deg. F. is recommended. The arc welding is done using DC reverse polarity. The stress relieving is done at 1350 to 1450 deg. F. with air cooling. Annealing is done by heating to 1550 to 1600 deg. F., furnace cool to 1100 deg. F., then air cool. If hardness is desired, furnace cool to 600 deg. F., then air cool.

18-9. TOOL AND DIE STEEL WELDING

The use of stamping, drawing, and forging presses is a growing trend in modern production manufacturing. The dies and tools used in these machines are capable of producing thousands of duplicate parts. Obviously these stamping, drawing, and forging tools and dies must be strong and must resist wear. They must be made very accurately, and they must hold their accuracy. These tools are usually made of a high carbon alloy steel. They are accurately shaped, heat treated, and ground. If a part of the tool or die wears or breaks, the production of a duplicate one is slow and expensive. Tool and die welding makes it possible to reclaim many of these tools by rebuilding the worn surface or replacing the broken part, as shown in Fig. 18-3. Another exam-

Fig. 18-3. Water hardening tool steel prepared, welded, and ground. (Welding Equipment and Supply Co.)

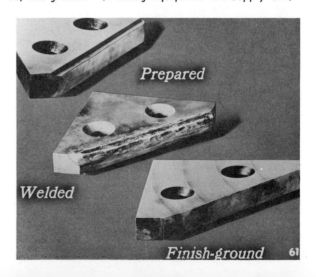

Fig. 18-4. A boring mill cutter. A. Before welding. B. After welding.

ple of salvaging a valuable tool by welding is shown in Fig. 18-4.

18-10. TYPES OF TOOL STEELS

There are many different alloys and trade brands of tool steel. However, as far as the welder is concerned, it is only necessary for him to select the electrode or welding rod to match the heat treatment required in repairing a particular tool. There are four general heat treatment classifications which may be used as the basis for selection:

1. Water hardening tool steel.
2. Oil hardening tool steel.
3. Air hardening tool steel.
4. Hot working tool steel (ni-chrome steel base).

18-11. TOOL STEEL WELDING PROCEDURE

To weld tool steels the surface must be clean and occasionally shaped (ground) to best receive the electrode deposit. It is also good practice to preheat the tool or die. It is very important to deposit a minimum of metal and then peen the metal to relieve the shrinkage stresses. Finally, the job should be heat treated (hardened and drawn) according to the original specifications of the tool or die.

Some of the basic steps to be observed when arc welding tool steels are:

1. Know the type of tool steel. Obtain this data from the drawings or from the manufacturer.

2. Use an electrode recommended for the type of steel to be welded. Refer to electrode manufacturers' catalogs.

3. Make the proper joint preparation.

4. Preheat as shown in Fig. 18-5.

5. Weld using reverse polarity (never exceed the maximum draw temperature as specified by the manufacturer of the metal).

6. Heat treat the weld.

Tool steel electrodes are available in sizes ranging from 1/16 to 3/16-in. diameter.

The electrode deposit must contact 1/8-in. of surface width of the base metal. For large deposits a 1/8-in. thickness deposit is recommended except on draw and forming dies. Fig. 18-6 shows welding of tool steel. Note the overlapping of the weld beads.

Weld in a slightly uphill direction, if possible, as this method produces an even buildup and permits the slag to float to the rear of the crater. It is desirable to peen to relieve stresses in the deposit after each pass. A postheat treatment is necessary for maximum

Fig. 18-5. Preheating furnace used to preheat tool and die steels.

results. Use temperature indicating colors and pyrometer cones to accurately determine the temperature. Large pieces require more time to complete the heat treatment than small pieces. In general, use one hour per 1/8-in. thickness being treated. Size determines time.

The weld quality depends upon several factors:

1. Type of steel.
2. Amount that the base metal mixes with the weld deposit.
3. Rate of cooling.
4. Preheat treatment.
5. Technique of welding.
6. Heat treatment after welding.

18-12. GALVANIZED STEEL

A popular way to protect mild steel against corrosion is to plate it with zinc. The zinc is deposited electrically or the steel is dipped in molten zinc (galvanized).

When joining pieces of galvanized steel, it is essential that the galvanizing not be destroyed.

Braze welding is a common method of joining galvanized steel. The metal is heated only to the melting temperature of the brazing rod and the galvanized coating is affected very little. This method may be used on pipe that is galvanized inside as well as outside. The inner coating is very important; it must not be destroyed. Carbon arc braze welding (the arc is struck between the two carbon electrodes only) or gas tungsten arc are the most satisfactory sources of heat, rather than torch heating, as the faster processes reduce the damage to the galvanized coating.

Galvanized steel is also arc welded. The temperatures reached usually result in some destruction of the galvanized coating in the areas immediately adjacent the weld seam. This protection must be replaced. Special galvanizing metal rods are available for this purpose.

18-13. STAINLESS STEELS

There are a large number of different steels known as stainless steels. All contain varying amounts of chromium or a combination of chromium and nickel. These steels are corrosion resistant, retain a good clean appearance, and have good physical properties. The American Iron and Steel

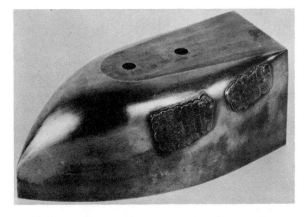

Fig. 18-6. Welding tool steel. Note welding technique which shows overlapping of the weld beads.

AUSTENITIC STAINLESS STEELS (CHROMIUM AND NICKEL ALLOYS)

A. I. S. I. SERIES	% Cr	% Ni	% C	% OTHERS
Type 301	17	7	.08-.20	....
302	18	8	.08-.20	....
304	18.5	8.5	.08 max.	....
316	17	12	.10 max.	Mo 2.5
317	19	14	.10 max.	Mo 3.5
347	18.5	10	.08 max.	Cb 10 x C min.
321	18.5	10	.10 max.	Ti 4 x C
308	20	11	.08 max.	Mn 2 max.
309	24	13	.20 max.	....
310	25	20-	.25 max.	....
318	17	12	.10 max.	Mo 2.5

Fig. 18-7. Table of compositions of austenitic stainless steels.

Institute has classified these alloys by number.

There are three general classifications of Stainless Steels:
1. Austenitic, Fig. 18-7.
2. Martensitic, Fig. 18-8.
3. Ferritic, Fig. 18-9.

See Paragraphs 24 and 25 in CHAPTER 24, for further information concerning the iron-carbon properties of these metals.

that retain the existence of austenite at room temperature are called austenitic steels. Stainless steels of the 300 series (chromium and nickel) are a good example of austenitic steels. These steels do not harden by heat treatment, but rather by cold working. Fig. 18-7 shows a table of the composition of austenitic stainless steels.

As a result of research dating back to about 1935, chromium-nickel steel

MARTENSITIC STAINLESS STEELS
(STRAIGHT CHROMIUM OF LESS THAN 14%)

A. I. S. I. SERIES	% Cr	% C	% OTHERS
410	12	.15 max.	....
416	13	.15 max.	P. S. or Se .07 min.
431	16	.20 max.	ZR or MO .06 max.
501	5	.10 max.	MO .45-.65
502	5	.10 max.	MO .45-.65

Fig. 18-8. Table of compositions of martensitic stainless steels.

18-14. AUSTENITE STAINLESS STEEL

Austenite is a physical condition in plain carbon steel that exists at temperatures above 1200 to 1300 deg. F. It is a solid solution of Fe_3C in iron.

It has been found by adding chromium and nickel to the steel that this austenite form can be retained in the metal as it cools down to room temperatures, if the metal is cooled quickly. Therefore, any steels with alloying elements

electrodes are now used to weld high tensile steels (low alloy) without a preheat or postheat being used. During World War II there was a tremendous increase in the use of this technique,

FERRITIC STAINLESS STEELS
(STRAIGHT CHROMIUM OF 14% OR MORE)

A. I. S. I. SERIES	% Cr	% C
430	14-18	.12 max.
446	26	.35 max.

Fig. 18-9. Table of compositions of ferritic stainless steels.

with the result that armor plate and the like were very successfully welded with stainless steel (austenitic) electrodes.

These austenitic steels are also non-magnetic, or, are very weakly magnetic. A popular use is for the bright sheet metal parts of household appliances.

18-15. MARTENSITIC STAINLESS STEEL

The martensitic stainless steels have less than 14 percent chromium content and a varying carbon content. See Fig. 18-8. These steels are good for tableware, instruments, ball bearings, etc. They may be hardened by heat treatment. They are also magnetic.

18-16. FERRITIC STAINLESS STEEL

Ferritic stainless steels have a high chromium content and less than .15 percent carbon. They have a ferrite grain structure at room temperature and are not hardenable by heat treatment, Fig. 18-9.

Gas welding stainless steel of the austenitic group is practical, and fairly easy. The metal being welded must be carefully cleaned (stainless steel wool or a clean stainless steel wire brush).

18-17. STAINLESS STEEL GAS WELDING

To successfully gas weld stainless steel, the metal and the welding rod must be first chemically or mechanically cleaned (a clean stainless steel wool is a good cleaner).

Both the surfaces of the metals being joined and the rod must be flux coated using a small clean brush. A common flux contains 1/2 pound zinc chloride,

1/3 oz. hydrochloric acid, 16 oz. water, 1/2 oz. potassium dichromite. Because of the relatively high amounts of the alloy metals, most of which have a high coefficient of expansion, these metals warp and buckle more noticeably than straight carbon steels. An oversize tip with a soft flame may be used to minimize warpage. Clamp the pieces to be joined carefully in a fixture. Some welds may best be performed by first spot welding the pieces of metal together. Hold the torch (neutral flame or very slightly carburizing) at a 45 to 60 deg. angle. As soon as a small puddle is formed, feed the welding rod to the puddle. When that spot is welded, move forward slightly, and repeat. Always leave the welding rod in or near the puddle (front edge). Never lift the welding rod away or the oxides formed on the rod will injure the quality of the weld. Always remove the torch by pulling it away slowly so that the flame may keep the last portion of the weld protected from the atmosphere until it has cooled somewhat. The control of the puddle in stainless steel is much more critical than with mild steel. A little too much heat and the alloy may disintegrate. Fig. 18-10 shows stainless steel being gas welded using the backhand technique.

The completed weld should be straight, even in width, slightly crowned, and evenly penetrated. You must be careful not to oxidize, or overheat stainless steel, as this seriously discolors it and ruins the properties of the metal. You will usually find that after finishing a welding operation, the metal is discolored. However, if correctly done, the discoloration will be only on the surface. The slight discoloration of the weld, and the metal adjacent to the weld, is easily removed by buffing and polishing.

When stainless steel was first welded using stainless steel filler metal, the welds sometimes had intergranular cracking (chromium carbide). To stop this action columbium and/or titanium

steel manufacturer and the electrode manufacturer.

The American Welding Society has prepared identification numbers for a variety of stainless steel welding elec-

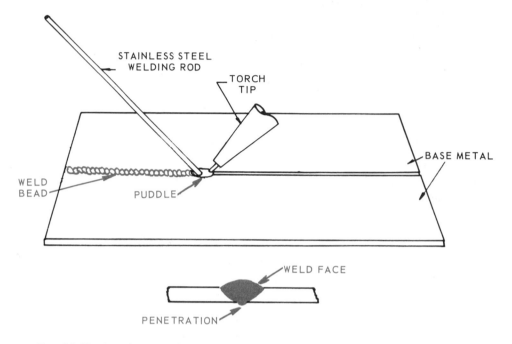

Fig. 18-10. Stainless steel joint being gas welded using backhand technique.

is added to the stainless steel. Titanium is proving the most popular. (American Iron and Steel Institute Series 321 and 347).

18-18. STAINLESS STEEL ARC WELDING

It is easier to arc weld stainless steel than to gas weld it. All arc welding methods such as manual arc welding, automatic arc welding, atomic-hydrogen arc welding, and shielding gas arc welding may be easily and effectively used in welding this type of steel. The consumable electrode must match the type of stainless steel being welded. To obtain good, consistent results, the operator must carefully follow recommendations made by both the stainless

trodes. For example, E 308-15 is for AISI 301 to 308 series, and is used in all positions as indicated by the number 1 in the AWS designation, EXXX-IX, and has a lime coating as indicated by the number 5 in the AWS designation EXXX-X5 which calls for DC reversed polarity. Because there are at least seven main classifications for the austenitic steels, great care should be used when selecting the electrode to be used. The current settings are slightly lower, but the electrode sizes are very similar to the choices for mild steel welding. DC reversed or AC energy is used. EXXX-15 electrodes are used with DC reversed polarity. The electrodes must always be kept dry and clean. The electrode coating must be free from cracks and chips. The welding procedure is almost identical with mild steel practice. The same electrode position, crater appearance, electrode motion or weave should be used. It is advantageous to use as short an arc as possible.

Gas tungsten arc welding is particularly desirable with stainless steel. Strong, good looking welds may be obtained with a minimum of discoloration and warpage. The metal behaves normally under the arc. Both the gas tungsten arc and the gas metal arc processes are suitable. When using the gas tungsten arc method, the welding rod is added in a manner similar to gas welding.

Since the use of flux is not necessary with shielding gas arc welding, no after cleaning is necessary.

18-19. MARAGING NICKEL STEELS

These steels are ultra low carbon .03 percent or 3 point carbon (low carbon steels are usually .10 to .20 percent or 10 to 20 point). The term maraging is the result of combining two words-- martensite and aging. The maraging steels owe their properties to the transformation of austenite to martensite, and then aging the martensite under precisely controlled conditions.

There are three grades of this steel:
1. 18 percent nickel.
2. 20 percent nickel.
3. 25 percent nickel.
These steels have tensile strengths of about 250,000 psi. The composition is explained in CHAPTER 24.

18-20. WELDING MARAGING STEELS

These steels are weldable using covered electrodes or using shielding gas arc welding. The filler metal should be of about the same composition as the base metal. Preheat is not re-quired but postheat is necessary with the 20 and 25 percent nickel alloys in order to develop the full properties of the metals.

General information on heat treatment follows, but the exact procedures and temperatures used will vary with the alloy, and the manufacturer's recommendations should be followed:

1. No preheat is required. However, if the temperature of the metal is at freezing or below, it is best to preheat the weld joint area to 70 deg. F.

2. Anneal at approximately 1500 deg. F.

3. Ausage (austenite age) at approximately 1300 deg. F.

4. Cold work.

5. Refrigerate to -100 deg. F.

6. Marage (martensite age) at approximately 900 deg. F.

7. Cool in air.

18-21. PROPERTIES OF CAST IRON

Cast iron is a casting made of iron and carbon with between 2.25 and 4 percent carbon. For certain purposes, alloy metals may be added to the cast iron. For example, nickel makes the casting more dense; nickel and chromium make the cast iron rust resistant. Phosphorus makes the metal pour more easily (low surface tension), and cast more accurately. Cast iron is used extensively for heavy machine parts. Its melting temperature is approximately 2600 deg. F.

18-22. TYPES OF CAST IRON

There are three principal types of cast iron:
1. White cast iron.
2. Gray cast iron.
3. Malleable cast iron or malleable iron.

White cast iron is a casting that has been cooled rapidly (chilled) after it has been poured, or cooled rapidly after being heated above its critical temperature. The name, white cast iron, is given due to the appearance of the fracture. The metal is extremely hard and is very difficult to machine. Most welds on any type of cast iron will tend to have a white cast iron structure unless they are heat treated after welding or cooled slowly after welding.

Gray cast iron is cast iron that has been cooled very slowly from its critical temperature (in sand, asbestos, or in a furnace). The name is derived from the gray appearance of the fracture. The gray is the result of graphite flakes in a matrix of white iron and iron carbide. This metal is easy to machine. Usually a welder heat treats all cast iron welds, that is, allows them to cool slowly, in order to produce gray cast iron.

Malleable cast iron or malleable iron, is a white cast iron that is heated to 1400 deg. F. for 24 hours per each inch of thickness and then cooled slowly. This heat treatment permits the release of carbon from the iron and allows it to form small nodules, or spheres of carbon in the matrix of low-carbon iron. The surface of the casting is affected first. The length of the heat treatment determines the depth of the change. Most malleable castings have the heat treatment extend only 1/8 to 1/4 in. into the metal. This heat treatment helps to make the casting stronger and more resistant to shock, fatigue, and vibration. Welding malleable iron will destroy this heat treatment and turn the metal into white or gray cast iron. Therefore the welding of malleable casting is generally not recommended. Malleable castings may be brazed, or braze welded, quite satis-

factorily. See CHAPTER 16 for instructions concerning braze welding cast iron.

All cast iron weld joints tend to become white cast iron; heat treatments previously administered are destroyed, and the casting must again be heat treated to bring back its original properties. Occasionally even this precaution will not be sufficient if the welder has allowed oxide spots to remain in the weld (poor procedure or improper flux). There is no satisfactory remedy for hard spots except to reweld them and remove the oxide inclusions.

18-23. NODULAR CAST IRON

Casting cast iron is an economical way to make metal parts. However, in its ordinary forms, white or gray, it is a brittle and fairly weak metal because the free carbon or graphite is in flake form which causes easy cracking.

Since 1948 a new form of cast iron has become available. It is called Nodular, Ductile, or Spheroidal cast iron. The process, which is patented, produces this ductile result by adding a small amount of magnesium to the cast iron. The graphite then becomes nodules or little spheres in the metal. This metal has much greater ductility and is stronger.

This ductile iron can also be obtained alloyed with silicon (3-6 percent) if heat resistant properties are desired, or with nickel (8-35 percent) and chromium (5 percent), if corrosion resistance is desired, plus heat resistance. Fig. 24-10 lists some of the properties of these ductile irons.

Ductile iron has been satisfactorily welded by arc welding using a 60 percent nickel, 40 percent iron electrode.

Step or intermittent welding of several passes is recommended as this process minimizes stresses. Preheat and post-heat improves the weld especially for machining, (600 F. preheat and 900 F. postheat).

18-24. PREPARING CAST IRON FOR WELDING

Cast iron may be prepared for welding much the same as steel. Thin pieces should be cleaned by grinding and then filing, whereas thick sections, 1/4 in. or more, should be chamfered (beveled)

narily obtained. The methods used are:

1. Insert studs in the edges to be joined, which when welded, add materially to the strength of the weld.

2. Cut notches in the cast iron surfaces to be welded. See Fig. 18-12.

18-26. PREHEATING CAST IRON

Most cast iron objects (except some simple structures where expansion and contraction are freely permitted without producing undue stresses and strains) must be preheated before welding, to prevent cracking of the metal

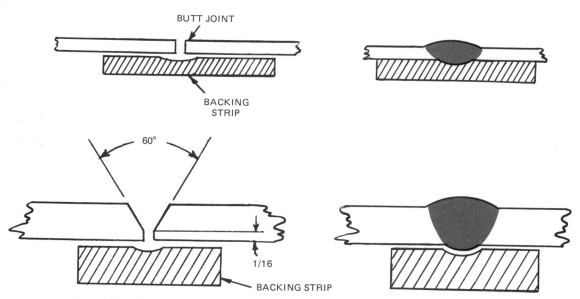

Fig. 18-11. Preparing cast iron joints for welding and cross sections of finished welds.

at a 60 deg. angle, leaving a 1/16-in. blunt edge. Fig. 18-11 shows a cast iron joint prepared for welding, also a cross section of the finished weld.

18-25. METHODS OF STRENGTHENING A CAST IRON WELD

Occasionally the welder may prepare the cast iron seam in special ways to obtain a stronger joint than ordi-

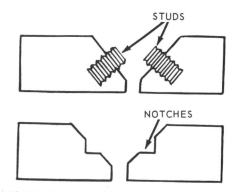

Fig. 18-12. Methods of strengthening cast iron weld joint.

as the weld cools. This is due to the brittleness of the metal and to the fact that most cast iron welds are on complicated frames and structures, i. e., cast iron wheels or frames. This preheating also permits faster welding, and assures closer alignment of the structure after the weld is completed and has cooled. Preheating is done in a furnace. The structure to be welded should be heated to a temperature of between 1500 and 2000 deg. F. (a dull cherry red). If possible, the actual welding should be done in the preheating furnace with the preheater operating. Fire bricks are usually built around the structure to be welded to retain the heat and permit quicker and more economical preheating. If a furnace is not available and preheating is necessary, this may be done with a torch, oxyacetylene, city gas-air, city gas-oxygen, gasoline-air, propane-oxygen, and the like. By knowing what members to preheat, most of the stresses and strains can be eliminated, Fig. 18-13. If the casting is flat, it is best to weld it on both sides to prevent warping.

18-27. GAS WELDING CAST IRON

The tip size used for cast iron should be similar to the size used for steel of the same thickness. A neutral flame should be used along with a cast iron welding rod of the proper size. Cast iron welding rods are available in 1/8 in. diameter x 18 in. in length, or 1/4 in. square x 24 in. in length. Oxides must be cleaned from the parent metal. Use a clean, sharp file. Grinding or sanding may leave abrasive particles that interfere with the welding. The torch should be held at a 60 deg. angle with the seam. The inner cone must not touch the metal. The flux should be added by coating the welding rod with

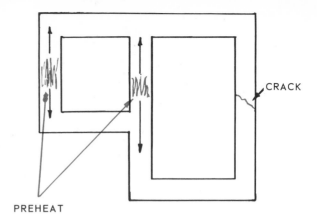

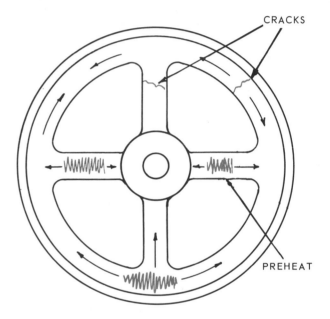

Fig. 18-13. Local preheating of structural cast iron bodies.

flux and adding it to the weld. The flux must have the correct constituents; it must be fresh, clean, and be moisture free. In cast iron welding, the molten puddle is not very fluid. Therefore, it is important that gas pockets and oxides be worked to the surface of the weld. This may be done by stirring the molten puddle with the filler rod. The torch may be moved either forward or backward, usually forward for thin sections, and backward for thick sections. The weld must have thorough penetration. A slight crown is preferred. An oscillating torch and filler rod motion is

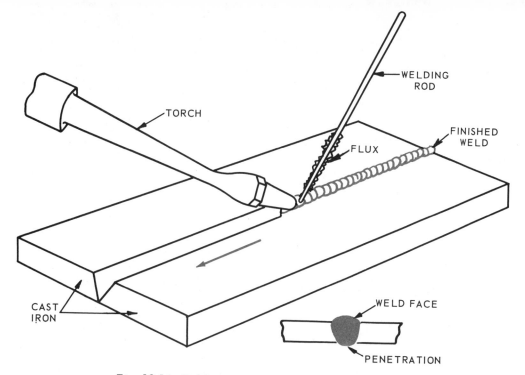

Fig. 18-14. Welding cast iron by oxyacetylene process.

usually used, although no ripples are likely to appear on the surface of a weld as shown in Fig. 18-14.

18-28. ARC WELDING CAST IRON

Cast iron may be welded with a carbon arc, or with a metallic arc.

When using a carbon arc, the manipulation is much the same as with the oxyacetylene method shown in Fig. 18-15. If a metallic arc is used, special covered steel electrodes are necessary--usually steel with a nickel content.

Steel studs are commonly used to

Fig. 18-15. Welding cast iron by carbon arc process.

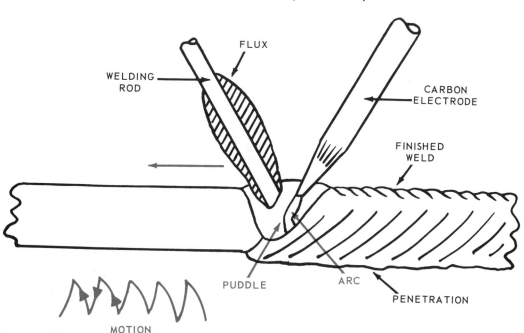

make a stronger weld. The same motion, electrode size, arc length, and the like, are used in cast iron welding as are used with the same thickness in steel welding. Fig. 18-16 shows, in cross section, an arc weld on cast iron, in progress.

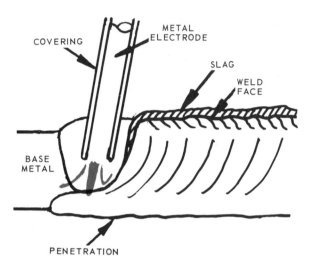

Fig. 18-16. Welding cast iron by shielded metal arc process.

18-29. INSPECTING CAST IRON WELDS

The finished cast iron weld should be slightly crowned without undercutting. It should have penetration. No pits or bubbles should appear on the upper or underneath (penetration side) surface of the metal. The width of the weld should be approximately three times the thickness of the metal.

18-30. HEAT TREATING CAST IRON WELD

If a cast iron weld is allowed to cool in the air and is cooled quickly, it will consist mainly of white cast iron, or at least a large number of these spots will exist. If cooled very slowly in sand, or in a brick housing, most of the weld will be gray cast iron, but

some hard spots may still be included in the weld. To secure the best results, the whole structure should be heated to its critical temperature and then cooled very slowly. This latter practice is sometimes more expensive than the welding operation itself, but it is necessary to insure a good weld.

18-31. TESTING CAST IRON WELDS

Cast iron welds may be tested by inspection of the outer surface. However, the beginner should break sample welds along the length of the seam, and inspect the fracture for gas pockets, cracks, burned spots, and hard spots. A good weld will have no such spots.

18-32. WELDING DISSIMILAR METALS (FERROUS BASE)

There are many iron and steel fabrications which require welding metals together even though their compositions are different. Welding stainless steels to mild steels, and low-alloy steels to high-alloy steels are typical examples.

It is very important that the properties of each metal be known. The welding rod or electrode composition should be close to the same composition as the metal which has the least amount of alloying metals. When in doubt or if the information needed is unknown, it is advisable to weld with filler metal of 25 percent Cr and 20 percent Ni.

It is always best to preheat the joint, even as low as 200 deg. F. to 300 deg. F. to produce better welding results. Practice on sample pieces, if possible.

18-33. ALUMINUM

Aluminum is one of the more common and popular metals. It is present in the earth in great quantities. Since it first became commercially available, it has grown in use, many alloys have been developed, and there are many uses for this metal.

It is available in all standard shapes and forms, and it is shaped and formed by all the standard methods. Parts made of aluminum are joined by all the conventional methods. Growing in popularity is the joining of aluminum parts by welding.

18-34. TYPES OF ALUMINUM

Aluminum is available in its commercially pure form. It is also available as it is alloyed with many other metals. It is also available in clad and anodized forms.

The metal identification has been standardized by code numbers. CHAPTER 23 explains how aluminum is manufactured, how it is alloyed, identified and shaped. Fig. 18-17 lists alloys and welding filler metals recommended.

The three digit code number (xxx) are castings while the four digit code numbers (xxxx) are wrought metals. The xxx series have seven main classifications (0XX through 7XX).

The four digit code number (xxxx) has 8 main classifications ranging from (EC through 7XXX).

The three digit series has letter prefixes (example B 214) to indicate slight changes in the alloying elements.

18-35. PREPARING ALUMINUM FOR WELDING

Aluminum is used commercially in two principal forms, cast aluminum, and drawn or rolled aluminum. The welding procedure will be governed by the form, cast or drawn. Certain characteristics of aluminum make it rather difficult to weld:

1. The ease with which the aluminum oxidizes at high temperatures.
2. The melting of aluminum before it changes color.
3. The oxide melting at a much higher temperature than the metal.
4. The oxide is heavier than the metal (more dense).

However, despite these difficulties aluminum welds can be made which are just as strong and ductile as the original metal. The filler metal, if used, should be of proper aluminum composition.

In welding this type of joint, the welder must obtain the same welding results as in steel welding:

1. Good fusion.
2. Good penetration.
3. Straight weld.
4. Build-up over the seam.
5. Clean appearance.

The metal must be clean. It can be mechanically cleaned (clean stainless steel wire brush or clean stainless steel wool). It can be cleaned chemically (dipped in cleaning solution and then rinsed).

The metal must be solidly supported before, during and immediately after the weld is made.

The welder must know the composition of the metals being joined and should check the American Welding Society's recommendations for the best welding filler metal to use.

The edge preparation should conform to recommended practices.

The structure section weldments should be preheated to approximately 600 - 800 deg. F.

18-36. OXYACETYLENE WELDING WROUGHT ALUMINUM

Gas welding aluminum can be done successfully. However, if shielding gas arc welding facilities are available, the

Legend

Filler alloys are rated on the following characteristics:

Symbol	Characteristics
W	Ease of Welding
S	Strength of Welded Joint ("as welded" condition) (Rating applies particularly to fillet welds. All filler alloys rated will develop presently specified minimum strengths in butt welds.)
C	Corrosion Resistance in Continuous or Alternate Immersion in Fresh or Salt Water
T	Suitability for Service at Sustained Temperatures above 150°F
M	Color Match After Anodizing
D	Ductility (Rating is based on free bend elongation of the weld.)

A, B, C and D are relative ratings in decreasing order of merit. The ratings have relative meaning only within a given block. Combinations having no rating are not usually recommended.

Filler Alloy 5154 may be used in place of 5254.

Once the parent alloys have been determined, the governing factor in choosing a filler alloy then becomes the desired characteristic in the weld area.

Example: If one parent alloy is 6061 and the other 5052, and if ease of welding (W) is the desired characteristic, trace the W sector in the 5052 block (along the left margin) horizontally to the 6061 block (appearing at the top) and note that 4043 (rated A) is the filler alloy recommended.

Fig. 18-17. List of *aluminum and aluminum alloys showing the welding filler metal to be used.*
(Alcoa)

*Rating does not cover these alloys when heat-treated after welding.

latter process is recommended.

Flux is always used when gas welding aluminum. Both the metals being joined and the welding rod, if used, are flux coated. The operator usually mixes the flux with water, (3 to 1 ratio) to form a paste. The edges to be welded together are then heated, and the paste applied to both edges with a brush. If the inside of the fabricated part cannot be cleaned, you should not put flux on the metal, only on the welding rod. In any case the welding rod should also be coated with this flux.

Due to the high heat conductivity of

reducing or slightly carburizing flame should be used.

The appearance of the metal changes very little when the parent metal is becoming molten. Light blue color welding lens should be used. You can then see the light gray color of the metal just as it is melting. You may also "feel" the surface of the metal with the welding rod by lightly touching the metal with the rod as the metal is heated. The aluminum may be solid one instant, and then without any apparent change in appearance, melt and sag. The operator will feel this change

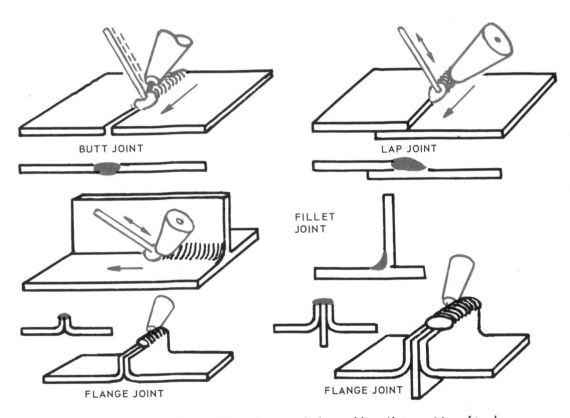

BUTT JOINT

LAP JOINT

FILLET JOINT

FLANGE JOINT

FLANGE JOINT

Fig. 18-18. Aluminum joints shown during and after welding. Note position of torch and welding rod. No welding rod is required when welding flange joint on thin metal.

aluminum, a larger torch tip is needed for a corresponding thickness of aluminum than for steel welding. Because the weld puddle does not emit sparks a large tip may be used and the flame reduced without the torch "popping." A

coming since the metal surface will feel soft or elastic under the welding rod just as the metal melts. The instant that this "feel" indicates that the metal is about to melt the welding rod should not be withdrawn from the metal, but the tip of the welding rod should be allowed to melt with the metal. If the welding rod is withdrawn at the instant of melting, there is danger (particularly with thin metal) of the metal breaking away and part coming away with the welding rod leaving a hole in the work being welded. However, with a little practice and by learning the knack of scraping the surface with the welding rod one is soon able to obtain a strong, neat looking aluminum weld. Fig. 18-18 shows various aluminum joints being welded, and after welding.

The commercially pure aluminum (1100 or 3003) are usually welded with 1100 welding rod. The 5050, 5052 and 6061 alloys are welded using 4043 welding rod.

Flange welding is also a very popular type of aluminum weld seam. Welding rod may or may not be used. Occasionally three pieces are welded in this way. One must be careful when welding three pieces to concentrate the heat on the middle piece as it requires the greatest amount of heat. A larger than normal tip may be used when welding aluminum, as it may then be cut down (gas flow reduced) without the danger of backfire. To insure fusion all the way across the thickness of the metal when flange welding, the weld should bulge a little (it should be a little thicker than the total thickness of the two or three sheets). Very little torch motion, if any at all, is used. Aluminum sections up to 1 in. in thickness have been gas welded. Fig. 18-19 shows an aluminum plate being gas welded.

The flux used for welding aluminum contains chlorides and occasionally fluorides. The fumes from the flux are irritating and aluminum should be welded only in well ventilated places. This flux is also irritating to the skin and

harmful to clothing, and must be carefully handled. Keep the flux in airtight containers when in storage. The flux should be mixed with pure water to a paste consistency, and added to the filler rod only with a clean brush. Never contaminate the flux with dirt, rust, or dust. Keep the brush in the flux bottle when not in use; never lay if on the bench or this will ruin the flux. It is important to use good, fresh, aluminum flux at all times. The finished weld, in addition to all the usual appearances of a good weld, should be of a color similar to the welding rod material with a bright, shiny surface. If the metal is overheated and oxidized, the color becomes darker and the metal has a dead white appearance with a rough surface.

A slightly carburizing flame may be used for aluminum welding to insure that the metals will not be oxidized. An excessive amount of carburizing flame will give the weld a dirty appearance. Flux should be washed from a weld with water or better with a sulphuric

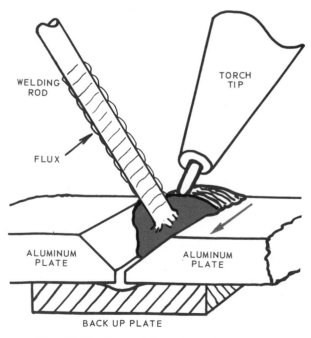

Fig. 18-19. Gas welding plate aluminum.

acid water solution as soon as possible, as flux left on the weld will have a corrosive effect. A student will find it very convenient to use backing material when first learning how to weld sheet aluminum. When welding aluminum, nickel, lead and some other metals one must, if possible, use holding jigs and back up surfaces, because these metals become very weak just prior to their melting temperatures. This weakness is called "hot shortness." See PARA. 18-80 and PARA. 18-81.

18-37. OXYHYDROGEN WELDING WROUGHT ALUMINUM

Oxyhydrogen flame has a temperature of 4100 deg. F. Because it is very clean, it can be used to weld aluminum.

The opening of an oxyhydrogen station is similar to the steps followed when opening an oxyacetylene station. The hydrogen cylinder is built similar to an oxygen cylinder except that the regulator attaching fitting has left-hand threads. The average cylinder holds 195 cu. ft. of hydrogen at 2000 psi.

The hydrogen torch releases 436 BTU per cu. ft. in comparison to the 1,640 BTU released per cu. ft. of acetylene. This necessitates the use of a larger tip orifice than with the oxyacetylene flame for a given metal thickness.

To adjust the oxyhydrogen flame, open the hydrogen torch valve about one half turn, light the gas, adjust the regulator until the burning gas (almost colorless) just starts to hiss and become turbulent, or rough, then open the oxygen torch valve and adjust the regulator until an inner cone is just visible.

If the end of the tip is dirty or coated with aluminum flux, the flame will have an orange color and the inner cone will not be visible. Clean the end of the tip with fine grain polishing paper and ream the orifice very lightly to remove the impurities causing the orange color. The hydrogen flame is almost colorless and the neutral flame is hard to ascertain even under good working conditions. A black background helps one to see the flame more clearly.

Another way of determining the length of the inner cone is to slowly move the torch up to a cold metal surface. When a small black dot appears in the center of the point of contact of the flame with the metal, note how far the torch tip end is from the metal. This distance is the length of the inner cone.

When butt welding aluminum, hold the torch at a 45 to 60 deg. angle and dip the welding rod in the center of the puddle with an up and down motion. The inner cone of the flame must not touch the aluminum base metal or the filler rod. Do not stir the puddle or push or pull the welding rod through the puddle.

To make a lap weld hold the tip at an angle of almost 90 deg. to the line of weld, and tilt it up about 60 deg. from the surface of the weld. When adding the welding rod in this case push the puddle back toward the trailing edge of the puddle with the welding rod. Keep the welding rod at the edge of the upper piece of aluminum at all times to prevent overheating.

When fillet welding, hold the tip equally between the two pieces and at a 60 deg. angle to the line of weld. Push the puddle toward the back edge of the puddle with the welding rod and at the same time add the welding rod to the puddle.

18-38. ARC WELDING WROUGHT ALUMINUM

Aluminum may also be welded with the electric arc. Heavily covered electrodes (5 percent silicon) used with reversed polarity (electrode position)

and a short arc (20V) are recommended. The electrode coverings must be
kept dry. Good ventilation is very important. Metal of less than 1/8-in. in
thickness is difficult to arc weld and
great care must be taken. The metal
should be backed if possible.

Arc welding aluminum is about three
times as fast as welding steel. A slower
progress will result in too much buildup
of electrode metal or a melt through
of the base metal.

If the arc is interrupted, remove all
the flux before attempting to strike the
arc again, or the weld will have flux
inclusions. Restart the arc about 1/2 in.
back along the weld to produce enough
preheat and buildup. After the weld is
completed immediately remove the
flux, wash the weld with a 5 percent nitric or 10 percent sulphuric acid solution, then wash with warm water. This
thorough cleaning is necessary to prevent corrosion of the aluminum.

18-39. OXYACETYLENE WELDING CAST ALUMINUM

Occasionally an operator encounters
a project which involves the welding of
cast aluminum articles. Cast aluminum
welding is used in foundry repair, general repair, and in production. As in
sheet aluminum welding, a neutral
flame or a slightly reducing flame is
recommended.

Use a fresh, chemically-pure flux.
The same preparations should be followed for cast aluminum as for wrought
aluminum. To prevent the sagging at
the joint while welding, use carbon,
copper, or steel blocks to back up the
weld.

A very valuable aid for cast aluminum
welding is a steel paddle which may be
used to:

1. Stir up the molten metal.

2. Remove the oxidized aluminum
from the weld.

3. Smooth out the surface of the weld.

These paddles are usually made from
1/4-in. steel welding rods flattened at
one end into a flat spoonlike shape,
Fig. 18-20. When welding large and/or

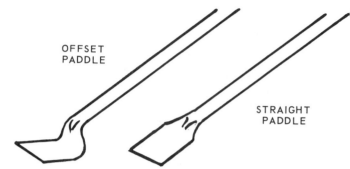

OFFSET
PADDLE

STRAIGHT
PADDLE

Fig. 18-20. Steel paddles used for puddle control when
gas welding cast aluminum.

thick section aluminum castings the
castings should be preheated to 500 -
600 deg. F. This temperature may be
easily checked by heating the casting
until it will char a soft pine stick touched
to the surface. See PAR. 18-11 for
details of construction and operation of
a preheating furnace. Thicker sections
should be chamfered as is done with
steel. In addition, the beveled edges
should be notched with a chisel or hacksaw (every 1/8 in.). When this is done,
the edges will quickly reach the welding
temperature without heating the adjacent metal too much.

18-40. ARC WELDING CAST ALUMINUM

Arc welding cast aluminum is done
in much the same manner as arc welding wrought aluminum. Both sand castings and permanent mold castings are
weldable. Aluminum die castings are
not usually weldable because of certain
ingredients in the aluminum alloys.
Castings that have been impregnated

ASTM ALLOY	ASTM FILLER ROD ALLOY	ORIGINAL METAL STRENGTH TS 1000 PSI	WELDED STRENGTH 1000 PSI
AZ 31B-H24	AZ-61A	42	37
AZ 61A-F	AZ-61A	45	40
ZK 21A-F	AZ-61A	38	32
ZE 10A-H24	AZ-61A	38	33
HK 31A-H24	EZ-33A	38	31
HM 21A-T8	EZ-33A	34	31
HM 31A-T5	EZ-33A	44	28

Fig. 18-21. Table of weldable magnesium alloys.

to insure tightness are not weldable.

Castings that have been heat treated lose the heat treatment properties when welded. If a casting requires heat treatment, it should be welded prior to heat treatment. One should determine the preheat treatment and the postheat treatment prior to attempting a weld on a casting.

It is important to minimize the area of heating during the welding operation.

18-41. WELDING ALUMINUM ALLOYS

All but the 1100 series of aluminums are aluminum alloys. Some of the weldable aluminum alloys are as shown in Fig. 18-17.

All of the aluminum welding instructions in the previous paragraphs apply to the welding of aluminum alloys.

18-42. WELDING MAGNESIUM

Magnesium and its alloys are often confused with aluminum and welders often try unsuccessfully to use aluminum welding techniques. The metal is shaped into various wrought forms and is also cast in sand and in permanent molds.

There are several magnesium alloys on the market. These metals can usually be arc welded and resistance welded. Gas welding is recommended only as

an emergency repair. Fig. 18-21 lists the weldable magnesium alloys.

Magnesium oxidizes very rapidly when heated to its melting point. In fact, when small shavings are thus heated, the magnesium will burn spontaneously and leave a white ash. This burning and white ash is one way to identify the metal. BE CAREFUL TO USE ONLY A VERY SMALL AMOUNT OF SCRAPINGS OR SHAVINGS.

The most popular magnesium alloys are those with aluminum and zinc added. These are the AZ series (A for aluminum and Z for zinc).

The high zinc alloys are weldable by the spot and seam processes. These alloys are the ZH and ZK series. The usual arc or gas processes are not recommended for welding these alloys.

18-43. USE OF CARBON PASTE IN WELDING DIE CASTINGS (WHITE METAL)

Die castings are metal-alloy castings (sometimes called white metal) cast in iron and steel molds (dies) under pressure or by gravity. The same die (or mold) may be used many times and the finished article requires very little machining or finishing. Die castings may be accurate to .001 inch. Die casting alloys are usually alloys of high zinc, high aluminum, or high magnesium content. These castings are brittle and break easily, but because of the ease of manufacture, many articles are now being made this way. Failure due to brittleness produces considerable demand for repair welding of these castings.

To weld a die casting successfully, you should know the constituent metals in the alloy. The zinc casting is heaviest, the magnesium casting lightest, while the aluminum casting is in between in weight. Zinc die castings are the most common. The zinc die castings melt at about 725 deg. F. while the magnesium and aluminum castings melt at about 1100 to 1200 deg. F. Zinc alloy die castings are very difficult metals to weld because of their low melting temperature and high rate of oxidation. Some typical die casting alloys are shown in Fig. 18-22.

during and after welding. Fig. 18-23 shows a die casting placed in a temporary mold being welded. Fig. 18-24 shows carbon paste and carbon plates used to form molds for controlling weld metal. A heavy carburizing flame should be used and the usual welding procedure followed. A very small tip is recommended. You must be careful because the metal melts before it changes color. Use the welding rod to break the surface oxides as welding metal is being added. The welder may use a steel or brass paddle to smooth the surface of the weld and to remove oxide inclusions.

Another successful method of repairing die castings is to use a soldering copper to melt the metal. An oxyacetylene torch is used to keep the body of the copper at a red heat while the point

METAL	ALLOY NO. 1	ALLOY NO. 2 *	ALLOY NO. 3*
Zinc, percent	85		
Aluminum, percent	3		
Copper, percent	4	1	4-15
Tin, percent	8	4	70-90
Antimony, percent		15	Trace
Lead, percent		80	
Melting Temperature, deg. F.	852	550	675

* Soldering is recommended for repairing these alloys rather than welding.

Fig. 18-22. Table of three die casting alloys.

The welding rod should be of the same composition as the original metal, if possible; although one may purchase die casting welding rods for general repair.

Die casting is prepared for welding just as other metals. It must be chamfered if there is a thick section. It must be thoroughly cleaned. Any plating must be ground away in the area to be welded. It must be backed up with carbon paste or blocks if possible. The parts must be firmly supported before,

of the copper heats the die casting and fuses the two pieces together. The torch flame is not put on the die casting at all unless the casting requires preheating. This method is especially successful when repairing small sections.

18-44. WELDING COPPER AND COPPER ALLOYS

Copper and most of its alloys can be welded. There are two kinds of copper:

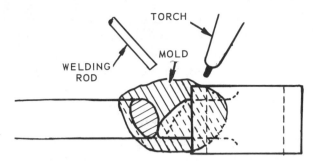

Fig. 18-23. Preparing a die casting for welding by forming a mold of carbon or asbestos paste around the fracture.

1. Electrolytic - 99.90 percent copper minimum, Oxygen .04 percent.

2. Deoxidized - 99.50 percent copper minimum, Phosphorus .015 to .040 percent.

Copper has a high specific heat and consequently it heats at about half the rate of aluminum. Both copper and aluminum have higher heat conductivity than steel.

Some copper alloys are:

1. Copper and Zinc (Brass)

Gilding - 94 to 96 percent Copper - Zinc Remainder.

Commercial Bronze - 89 to 91 percent Copper - Zinc Remainder.

Red Brass - 84 to 86 percent Copper - Zinc Remainder.

Low Brass - 80 percent Copper - Zinc Remainder.

Cartridge Brass - 70 percent Copper - Zinc Remainder.

Yellow Brass - 65 percent Copper - Zinc Remainder.

Muntz Metal - 61 percent Copper - Zinc Remainder.

Admiralty - 71.5 percent Copper - 1.1 percent Tin - Zinc Remainder.

Naval Brass - 61 percent Copper - 0.75 percent Tin - Zinc Remainder.

Manganese Bronze - 58.5 percent Copper - 1.00 percent Tin - 1.4

percent Iron - 0.5 percent Manganese (Maximum) - Zinc Remainder.

Aluminum Brass - 77.5 percent Copper - Aluminum 2.2 percent Zinc Remainder.

2. Copper and Tin (Bronze)

Grade A - .19 percent Phosphorus - 94.0 percent Copper - 3.6 percent Tin.

Grade C - .15 percent Phosphorus - 90.5 percent Copper - 8.0 per-

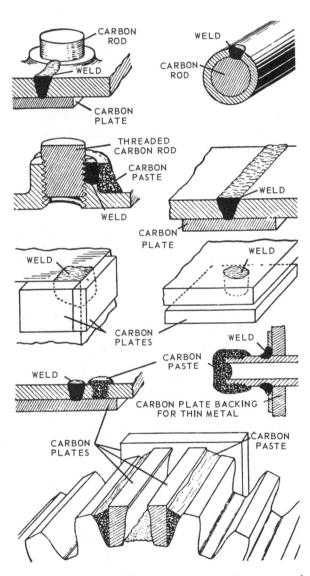

Fig. 18-24. Carbon block and paste backing as used in typical welding applications.

cent Tin.

Grade D - .15 percent Phosphorus - 88.5 percent Copper - 10.0 percent Tin.

Grade E - .25 percent Phosphorus - 95.5 percent Copper - 1.25 percent Tin.

Alloys of copper are used a great deal, mainly because of their ductility, which enables them to be worked easily and shaped into many complicated patterns. These alloys are used in both cast and wrought forms. They are resistant to certain kinds of corrosion, and are good conductors of heat and electricity. Copper alloys may be recognized by their characteristic red or yellow color.

Pure deoxidized copper is comparatively easy to weld, while some alloys of copper are difficult to weld. In order to test a specimen to determine if it may be easily welded, quickly heat a sample of the metal with a torch to the molten state. If the puddle remains quiet, clear, and shiny, it indicates the metal is comparatively pure copper, and that it will be easy to weld. However, if the puddle boils vigorously and gives off a quantity of gaseous fumes, this indicates ingredients in the metal that make it difficult to weld.

Annealed deoxidized copper has a tensile strength of 30,000 - 35,000 lbs. per sq. in. and may be welded with the oxyacetylene process to produce full strength welds.

The tensile strength of annealed electrolytic copper is 30,000 to 35,000 lbs. per sq. in. When welding electrolytic copper the cuprous oxide redistributes itself in the area just outside the weld and weakens the zone next to the weld. It is difficult to obtain a weld with a strength greater than 70 - 87 percent of the annealed base metal.

When using the oxyacetylene flame to weld electrolytic copper the hydrogen in the flame penetrates the metal just back of the line of fusion and reduces the tensile strength in this area by about 50 percent.

Also if the metal sample is brittle and breaks easily, this is an indication of alloying elements or impurities. The alloying elements which make welding copper alloys most difficult are: bismuth, antimony, and arsenic. Phosphorus in small quantities makes welding copper easier.

18-45. OXYACETYLENE WELDING COPPER

Copper and its alloys may be welded with either the oxyacetylene or the electric arc process, and the general procedure is similar to that followed in welding steel. No flux is needed when welding pure copper. However, in welding copper alloys a flux is required.

The physical properties of copper require that certain changes be made in the welding procedure. Welds on copper in its commercial state produce poor welds usually because of the cuprous oxide dissolved in it.

Commercial copper, after welding, can be strengthened by mechanically working it. The weakest point of a copper weld is always next to the joint. Always use specially prepared welding rod for copper welding in order to obtain deoxidized copper.

One problem encountered in welding copper is the difficulty experienced in eliminating blow holes and gas bubbles in the finished weld. This may be partially remedied by the use of small quantities of phosphorus in the welding rod. When welding copper, it must be remembered that because of the rapid conduction of heat a tip larger than normal must be used.

18-46. ELECTRIC ARC WELDING COPPER

Copper may be arc welded successfully using the "long" (high voltage) carbon arc. Set the welding machine on straight polarity and with a voltage of 40 to 60 volts. A copper weld should always be backed by a steel plate during the welding operation.

Shielded gas arc welding of copper is also successful. Either the tungsten electrode, with welding rod, or the metal electrode method can be used.

Nitrogen gas may be used with good results. Argon gas may also be used. Nitrogen gas does cause some spatter but its use allows thicker one pass welds. A mixture of nitrogen gas and argon gas has proven very successful, the ratios vary between 80 percent argon 20 percent nitrogen and 85 percent argon and 15 percent nitrogen.

The consumable electrode is usually a boron alloy of deoxidized copper. When welding thick copper plates, twin consumable electrode heads, side by side with a U-bevel edge preparation, have given good results. The drooping curve type generator is recommended for copper arc welding. Carbon electrode arc welding may also be successfully used to weld copper.

18-47. OXYACETYLENE WELDING BRASS

Brass consists of an alloy of copper and zinc, the proportion of the two metals being variable (10 to 40 percent zinc). Occasionally other metals are added to the alloy. These, however, do not usually affect the welding procedure. The alloy has a much brighter appearance (more yellow) than pure copper. It is a very common alloy.

While not as ductile as copper, it does have a higher tensile strength. Its resistance to certain corrosive action is better than that of pure copper.

Brass is more difficult to weld than copper, because some of the zinc, under the high temperature that is required to melt the copper in the alloy will vaporize, forming irritating fumes which destroy the proportions of the metals in the alloy. The welder should make certain there is excellent ventilation and the welder must avoid breathing the zinc dust and zinc oxide fumes.

When welding brass the proportions of the two metals should be known, and the welding rod used should be of similar alloy. If the color of the parent metal and of the welding rod is the same, this color will usually indicate that the alloys are approximately the same. It is more important to use good flux in welding brass than in welding copper. Fresh, chemically-clean borax paste may be satisfactorily used. The torch flame should be slightly oxidizing to reduce the zinc fumes, and to reduce the tendency for gas pockets to form. A carburizing flame has a tendency to form gas pockets, and to permit the weld to accumulate too much width, wasting heat.

One of the most common applications of brass welding is the use of brass welding rod to fuse two different steels together or to join unlike metals together. This practice of using brass as a joining metal is sometimes called hard soldering. The correct term however, is brazing. Brazing is described in detail in CHAPTER 16. A good flux is necessary when brazing. Chlorides are the principal ingredients for brazing fluxes.

Carbon blocks will be found a handy accessory as backing material, especially when there is danger of the weld penetrating too much, Fig. 18-24.

18-48. ARC WELDING BRASS

Special electrodes are available for arc welding brass. The metal is prepared in a manner similar to the preparation of the metal for oxyacetylene welding. Reversed polarity is commonly used and a slightly higher current (10 to 20 percent) is used than for steel of the same thickness. The welds should be backed with steel or carbon plate and should be preheated. Typical electrode motions are recommended. A short arc should be maintained. Weld in a flat position, if possible. The weld should show a clean, slightly crowned appearance after the slag has been removed by chipping.

The gas tungsten arc welding process may be successfully used for welding brass. The metal may also be welded using a double electrode carbon arc with only reflected heat melting the metal.

18-49. OXYACETYLENE WELDING BRONZE

Bronze is an alloy of copper and tin in various proportions. Some of the bronze alloys may also have lead content. Bronze is very resistant to corrosion. Its resistance is greater than that of the brass alloys. Also it is easier to weld than brass alloys, since the tin does not have the tendency to separate from the copper in the molten state.

An oxidizing flame should be used for this type of welding to eliminate fumes as much as possible.

The three metals, copper, tin, lead become liquid without appreciably changing their color, which is one of the reasons the beginner experiences difficulty when welding them. However, a little practice soon overcomes this difficulty.

The edges to be welded together should be chamfered so the total angle will be approximately 90 deg. Backing material is recommended. Welding rods of the same alloys as the parent metal should be used when welding both brass and bronze.

18-50. ARC WELDING BRONZE

Metallic bronze electrodes, heavily covered (coated) are used to weld bronze. This process is rather expensive and is not as popular as the oxyacetylene method. Use reversed polarity with a short arc and back the weld with a steel or carbon plate. As in brass welding, use a slightly higher current than is used for steel of the same thickness.

Bronze can also be welded using the gas tungsten arc welding process.

The carbon-arc process is also usable for welding bronze.

18-51. WELDING COPPER - NICKEL ALLOYS

The cupro-nickel alloys are very corrosion resistant and have a good hot strength character. These alloys have many uses.

Cupro-nickel alloys have been popular for years in marine service where salt water and hot condenser temperatures are involved. They are also used in many kinds of food processing equipment.

The alloys vary from 2.5 percent nickel to 30 percent with the 10 percent nickel and the 30 percent nickel alloys being most common.

These alloys can be soldered, brazed and welded.

Gas arc welding is usually preferred to oxyacetylene welding. Either the tungsten electrode, or the consumable electrode methods are used with helium

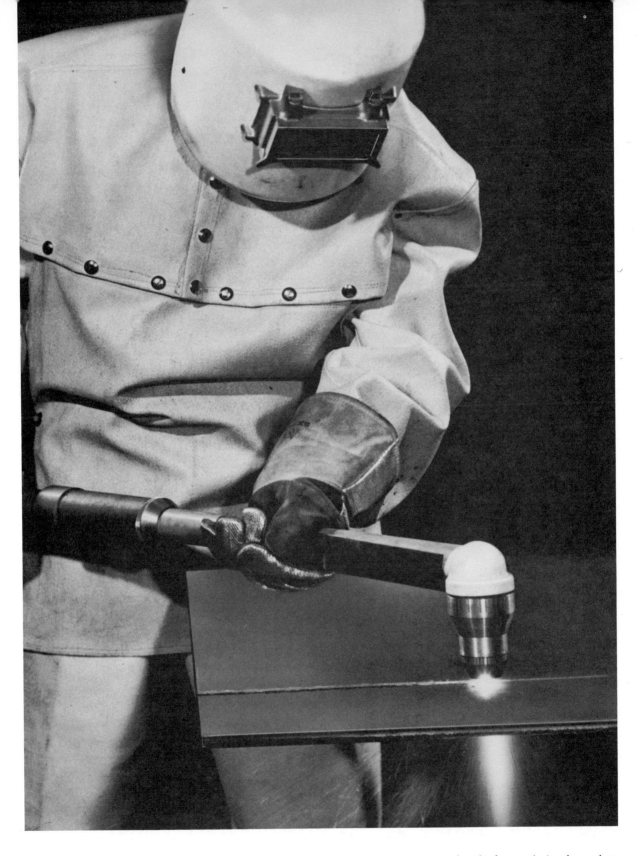

A hand held plasma-arc cutting torch. This torch may be used to cut any metal at high speed. Arc heated air is used as the operating gas. Note the protective clothing worn by the operator. (Thermal Dynamics Corp.)

or argon gas. If a welding rod is used, it should be of the same composition as the base metal.

18-52. WELDING LEAD

Lead is a ductile metal having a high resistance to corrosion by certain chemicals. It is used as a metal for tanks, cylinders, and as lining material for containers used to hold corrosive chemicals. The most common use for lead welding is in connecting the parts of the lead-acid (automobile) storage battery.

The procedure to follow when welding lead is simple. The metal edges are cleaned. Then, using an oxyacetylene torch and a cast welding rod, the edges are fused together by an operation similar to steel welding. The lead does not change color upon becoming molten, but this offers no serious difficulties. The joint should be backed, if possible, to prevent sagging of the metal upon securing penetration. Inasmuch as lead melts at approximately 620 deg. F., the welder will be likely to overheat the weld at first. Fig. 18-25 shows a lead weld in progress.

Lead can also be welded using the gas tungsten arc welding process.

LEAD FUMES ARE VERY DANGEROUS-DO NOT OVERHEAT-USE GOOD VENTILATION.

18-53. WELDING TITANIUM

Titanium, a newer commercial metal, is one of the most corrosion resistant metals. It is one of the more expensive structural metals, and it is difficult to shape. The metal is used where considerable corrosion resistance is required particularly in missile, rocket, space craft, and airplane parts.

Titanium is weldable, this process being one of the most common assembly

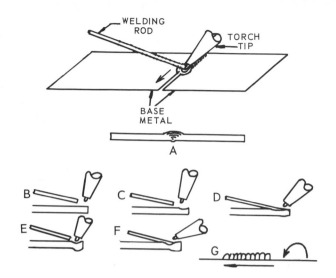

Fig. 18-25. Lead butt joint being gas welded.

processes. The parts must be carefully designed and fixtured to obtain successful welds. Copper chill bars are used in the fixtures to minimize heat spread and to cool the weld quickly.

Titanium will readily absorb oxygen, nitrogen and hydrogen as it is heated, for which reason the shielding gas arc welding processes are recommended for welding this metal.

There are several titanium alloys. Some with good weldability are: RC 55, Ti 75, RC 70, RS 70, Ti 100, RC-A 110 AT.

One with fair welding characteristics is RS 110 BX.

The metal must be thoroughly cleaned prior to welding. After weld completion the weld must be kept in the shielding gas atmosphere until it has cooled to 1000 deg. F. or below.

When tungsten electrode welding, one should keep the end of the welding rod in the gas atmosphere during the whole weld operation. The shielding gas should be fed to both sides of the weld.

18-54. WELDING BERYLLIUM

Beryllium is also one of the newer commercial metals, the use of which has been promoted by demands of the space age. It is lighter than aluminum and has a fairly high melting temperature. Its tensile strength is approximately 55,000 psi. It is used for such items as nose cones, brake discs and reentry units. It is a very expensive metal. It can be resistance welded and it can be gas arc welded. AC with high frequency is usually used in conjunction with tungsten electrodes. Beryllium welding rod is used. The shielding gas used should be a mixture of 50 percent argon and 50 percent helium. Approximately 55 amperes are used for .050-in. stock up to approximately 160 amps. for .250-in. stock. About 30 cfm gas flow has been found adequate.

18-55. WELDING ZIRCONIUM

Zirconium is a rare metal now being used commercially. Its unique properties make it useful for astronautic applications and for new chemical and nuclear energy uses.

Zirconium is weldable using the shielding gas tungsten arc process. The metal must be both chemically and mechanically cleaned. It must be very carefully fixtured. The weld is best done in a 100 percent shielding gas atmosphere.

18-56. WELDING DISSIMILAR METALS

It is possible to weld together dissimilar nonferrous metals such as: Aluminum to copper, Aluminium to lead, Copper to lead, and many other combinations.

Steel and some of the stainless steels can be welded to such metals as aluminum, copper, and others.

In most cases, arc welding and oxyacetylene welding have not proven as successful as resistance welding, friction welding and/or cold welding.

18-57. PLASTICS

Plastics are used for a large variety of component parts in manufactured articles. These plastics have a great variation of chemical compositions and physical characteristics. However, regardless of their properties or form, plastics all fall into one of two groups-- the thermoplastic or the thermosetting types.

The thermoplastic group become soft and capable of being formed when heated. These plastics harden when cooled. Thermoplastics may be reheated and reformed repeatedly.

These plastics fall into the thermoplastic group classification: Acrylic, Cellulosic, Acetal resin, Nylon, Vinyl, Polyethylene, Polystyrene, Polyvinylidene, Polypropylene, Polycarbonate, Polyfluorocarbon.

The thermosetting plastics, once formed and cooled, cannot be reheated and reformed.

Thermosetting plastcs include: Aminoplastic, Phenolic, Polyester, Silicone, Alkyd, Epoxy, Casein, Allylic.

Fig. 18-26. Cross section of plastic butt joint welded on both sides.

18-58. PLASTIC WELDING PRINCIPLES

Thermoplastics may be welded in the same manner as metal. All joint designs may be used and successful welds are possible in various positions. The joints must be cleaned carefully if the welds are to be of high quality.

The welding rod must be of the same composition as the plastic being welded. Fig. 18-26 shows a butt joint and Fig. 18-27 a fillet weld on plastic

Fig. 18-27. Cross section of a plastic T-joint welded on both sides. (Laramy Products Co.)

using plastic welding rods. Plastic welding rods may be obtained in round, oval, triangular, or flat-strip forms.

The welders choice of welding rod shape is affected by the shape of the joint, the thickness of the plastics to be welded, and the welding equipment to be used.

As an example, if a thick bevel weld is to be made, a triangular-shaped rod may be used. Using a triangular-shaped rod will partially eliminate the multiple passes required with a round rod.

18-59. PLASTICS WELDING EQUIPMENT

Heat for plastic welding may be supplied by a heated gas (either compressed air or shielding gas), which is heated as it passes through the welding torch and then is directed onto the surface of the joint. No flame touches the joint.

Electrically heated coils may be used to heat the gas as it flows through the torch. Fig. 18-28 shows a complete plastic welding station. Fig. 18-29 illustrates a welding torch used for plastic welding. This torch uses an electric coil element to heat the welding gas.

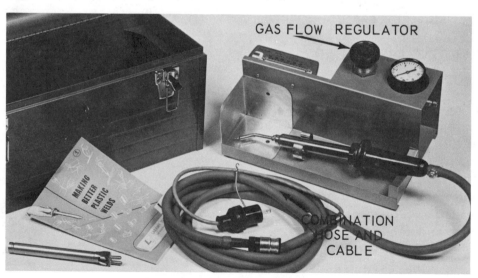

Fig. 18-28. Complete plastic welding outfit. (Laramy Products Co.)

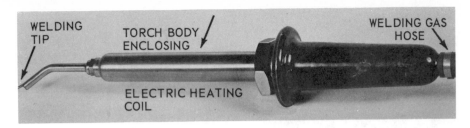

Fig. 18-29. Welding torch, designed for welding plastics, uses electric heating coil to heat welding gas. (Laramy Products Co., Inc.)

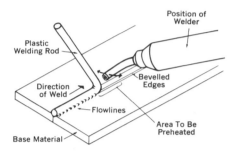

Fig. 18-30. Correct procedure for hand welding a plastics material butt joint. Note position of plastics filler rod and torch tip. (Kamweld Products Co., Inc.)

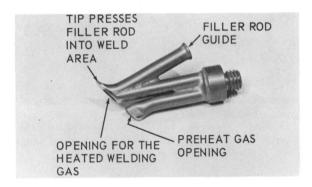

Fig. 18-31. Speed welding tip for welding plastics. (Laramy Products Co., Inc.)

The torch is held in one hand and the plastic welding rod is fed to the weld area with the other hand. Fig. 18-30 illustrates hand welding of a butt joint on plastic material. The plastics industry uses the word "welder" rather than "torch."

Fig. 18-31 shows a tip in which the plastic rod is fed through one section of the tip while the area ahead of the rod is heated to fusion temperature. An electric welding torch is shown in Fig. 18-32, with a speed welding tip

Fig. 18-32. Speed welding torch being used to weld a plastic butt joint.

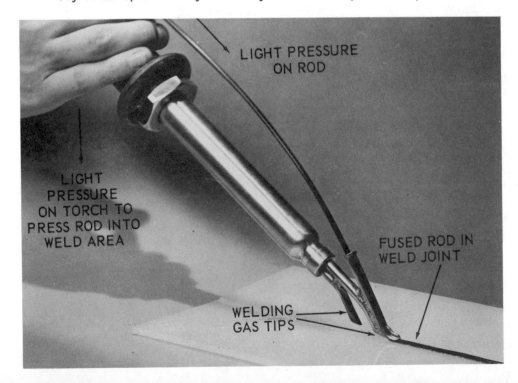

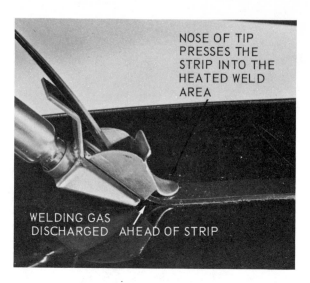

Fig. 18-33. Plastic butt joint being welded using plastic strip as joining material. Note special shape of strip guide and of hot gas tip.

installed. A light uniform pressure on the plastic rod is required while light downward pressure is exerted on the torch to press the rod into the heated weld area.

If the weld seam warrants it, the welding rod may be in a strip form. The strip is applied in a manner similar to the rod shape using a special speed welding tip, Fig. 18-33.

The welding gas may be shielding gas, or compressed air. To regulate the welding gas temperature, decrease the welding gas volume to increase the temperature; and increase the welding gas volume to decrease its temperature. This is possible because, with the gas flowing slowly over the heating element, it picks up more heat and, with the gas flowing rapidly over the heating element, it picks up less heat.

To change the heating capacity of the torch, the tip size is changed. The torch heats the welding gas to 450 - 800 deg. F. This heated gas is directed to the weld area where it heats the joint and the welding rod to a soft state. While both the joint and rod are soft, (plastic) they are forced together by

hand with about 3 lbs. of pressure. The surfaces of the joint and welding rod bond. When they cool the weld is accomplished.

Four things are necessary to accomplish a good weld on plastics:
1. Correct welding temperature.
2. Correct pressure on welding rod.
3. Correct welding rod angle.
4. Correct welding speed.

18-60. TYPES OF PIPES

The common practice for many years has been to connect sections of pipe to the proper fitting, or to each other, by means of the Briggs Standard Pipe Thread. These threads, being of a tapered construction, are self-sealing and make a tight joint. However, the cutting of the threads on the different sections of the pipe necessitates considerable labor; also the cutting of the pipe to accurate lengths requires skill. The threads become more difficult to make as the pipe becomes larger.

The size of the pipe is measured according to its inside diameter, and the thickness of wall usually varies according to the diameter of the pipe. Generally, the larger the pipe, the greater the wall thickness. There are several standard types of metal pipe:
1. Cast iron.
2. Wrought iron.
3. Single strength, low-carbon steel.
4. Double strength, low-carbon steel.
5. Steel alloy pipe.
6. Copper and brass pipe.

Cast iron pipe is seamless pipe formed in a mold. The properties of cast iron are such that the sections of the pipe cannot be fastened together by means of threads, but must be connected by the shoulder type of construction. This limits the use of cast iron pipe to low pressure installations, such as drainage work and the like. Fig. 18-34 shows three types of mechanical pipe butt joints.

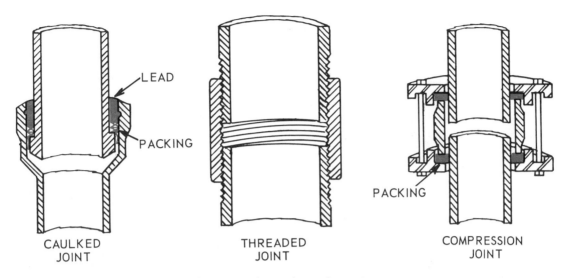

Fig. 18-34. Mechanical pipe butt joints.

Wrought iron pipe is made of an iron with practically no carbon present, but slag is usually found in the material. Wrought iron pipe may be threaded, but it will not stand up under excessive pressure because the pipe is made by rolling flat stock into a cylindrical shape; and the seams are either press welded or fusion welded.

Mild steel and double strength pipe may either be seamed or seamless pipe. The seamed pipe is fabricated just as wrought iron pipe, but the seamless pipe is fabricated from solid bar stock, and the pipe is drawn through special dies. Seamless pipe has considerably more strength and can be used for much higher pressure work than seamed pipe. The double strength pipe uses the same carbon steel, but the wall thickness is greater. This pipe is used for higher pressure pipe lines.

Stainless steel and other noncorrosive steel alloys are used whenever corrosive chemicals are carried in a piping system. During the last few years copper and brass pipe have come into extensive use for water and refrigerant lines. This pipe is usually connected by means of the soldered joint, using "streamline fittings." (See CHAPTERS 15 and 16). Fig. 18-35 shows various types of welded, brazed and soldered pipe butt joints.

Plastic pipe and tubes are also being used where highly corrosive chemicals are being transported at low pressure.

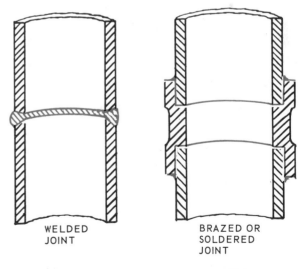

Fig. 18-35. Cross sections of pipe and tube butt joints joined by welding, brazing, and soldering.

Plastic pipes are also used in low pressure systems which are laid in water or in the ground since the plastic used is free from corrosion.

Aluminum alloy pipe and tubing is also finding wide use. Aluminum tubing may be soldered or brazed relatively easily with the fluxes recently developed. Aluminum pipes can also be welded using automatic machines.

18-61. TYPES OF TUBES

Tubing is thin wall pipe. Some tubing is flexible and because the wall is thin it cannot be threaded. One of the most popular metals used in the construction of tubing is copper. Steel and aluminum are also used as tubing materials. In the fabrication of automatic machinery, in refrigerating systems, in automobiles, and in many other applications, tubing is found to be the most satisfactory method of transmitting fluids. Tubing is available in the flexible (soft) form and in the rigid (hard) form. The size of tubing is measured by its outside diameter (OD). When the direction of the run is changed, soft tubing may be easily bent to conform with the change in direction, whereas hard tubing and pipe necessitates the use of appropriate fittings. Tubing is obtainable in different strengths, and is also manufactured in both the seamed and seamless construction.

Seamless tubing is made by piercing a piece of metal and drawing the metal through a die and over a mandrel or by forcing a pierced piece of metal over a mandrel by means of rollers as shown in Fig. 18-36.

Seamed tubing is made by:

1. Rolling flat stock into a cylinder and welding the seam.

2. Rolling thin flat stock into a two metal layer cylinder and then furnace-

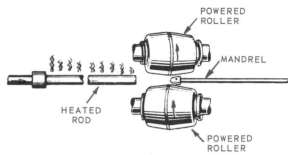

START OF OPERATION

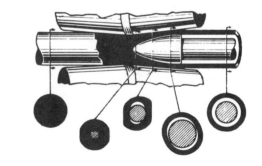

SEAMLESS TUBING BEING FORMED

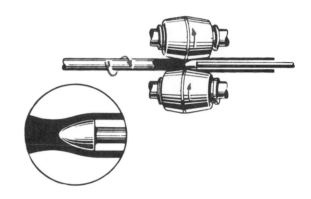

VARIOUS STEPS IN THE OPERATION

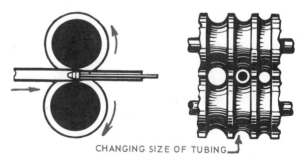

CHANGING SIZE OF TUBING

Fig. 18-36. Forming seamless tubing from a solid rod. (Michigan Seamless Tube Co.)

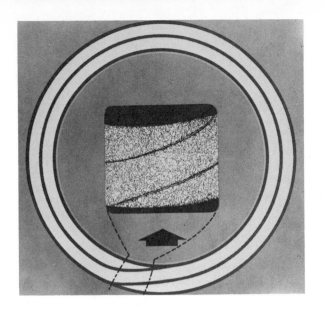

Fig. 18-37. Cross section of seamed tubing which has been furnace brazed. Note macrograph of brazed tubing at center. (Bundy Tubing Div., Bundy Corp.)

brazing the lapped surfaces as shown in Fig. 18-37. A brazed joint is shown in Fig. 18-38.

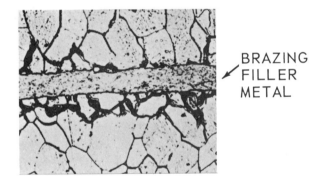

BRAZING FILLER METAL

Fig. 18-38. Microphotograph showing adhesion in brazed tubing joint.

A recent development in the plumbing, heating, and refrigeration industries is the use of copper pipe. Copper pipe is nonflexible, heavy-duty piping, having a wall thickness somewhat less than steel piping. Its size is based on its internal diameter. Copper pipe is used extensively in plumbing work. However, its test pressure of approximately 800 psi prohibits its use in very high pressure work. The joints made

in copper pipe are either soldered or silver brazed as the thickness of the copper makes it impractical to thread the copper pipe.

The aircraft industry uses special, seamless, steel tubing of special alloy. Aircraft tubing is seldom bent and is joined by welding.

Tubing standards have been established by the American Society of Mechanical Engineers, the American Society for Testing Materials, the Society of Automotive Engineers and the U. S. Government (MIL specifications).

Tubing is available in various cross section shapes:

1. Round.
2. Square.
3. Rectangular.
4. Hexagonal.
5. Octagonal.
6. Streamline.
7. Oval.
8. Irregular shapes.

18-62. METHODS OF JOINING PIPE

Steel pipes may be joined together, or joined to fittings by:

1. Pipe threads.
2. Flange fittings.
3. Welded joints.

The method of joining by threads has been previously explained.

The flange method incorporates the use of threads, but the final connection of the two sections is by means of flange bolts, which clamp the flanges together with a gasket between the two flanges. This construction is popular in threaded pipe installations where it is thought that dismantling might occasionally be necessary.

The welding of pipe joints is a popular means of making a pipe installation. The advantages of welding piping are neatness, compactness, rapidity,

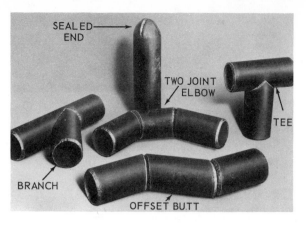

Fig. 18-39. A comparison of pipe joint designs. (Airco Welding Products, Div. of Airco, Inc.)

and low cost. A variety of welding joints are shown in Fig. 18-39. Pipe welding may be performed by using either the oxyacetylene or the arc method. Manual and automatic types of arc welding are used more often since the arc method is less expensive and faster.

When pipe is butt welded in a production plant, the flash welding technique or the friction welding technique is often used.

18-63. METHODS OF JOINING TUBING

Tubing joints may be made by: 1. Compression fitting. 2. Flared fitting. 3. Quick couplers. 4. Soldering. 5. Brazing. 6. Welding.

Compression fittings and flared fittings are used on small copper tubing connections. These fittings are commonly used with the seamed type of tubing (automobiles) or low-pressure systems, whereas the flared connection is used where seamless tubing is used in high-pressure systems such as some refrigeration systems.

Quick couplers are fittings that are spring loaded, gasketed assemblies that enable quick assembly and dismantling of tubing systems without loss of the fluid. Fig. 18-40 shows a quick coupler fitting.

Soldered connections utilize special fittings, having receptacles accurately sized for the tubing insertion. The solder is admitted between the tubing and the fitting, where it forms a thin film and strongly binds the two pieces of metal together. The soldered connection may be used for seamed tubing, seamless tubing, or for copper pipe.

Brazing and welding of copper tubing, or copper pipe, is very rarely done; but when it is performed, no special difficulties are encountered. Brazing of copper tubing and pipe occurs when the metal is to be attached to a different metal, or where corrosion is possible. Where copper pipe is to be attached to a copper fitting, or to another section of copper pipe, the joint may be welded.

18-64. CODE REQUIREMENTS

Special applications of welding piping, steel tubing, copper tubing, and copper pipe in certain localities are covered by a welding code or codes necessitating that certain requirements be met before the welds may be used. These requirements usually concern high-pressure work, including steam and air, or where corrosive fluids are carried. In aircraft work the Federal Aeronautics Administration (FAA) maintains a crew of inspectors who judge whether the welding of the aircraft tubing is sufficiently well done

Fig. 18-40. Quick couple used for tubing joints which need easy and fast connecting and disconnecting such as in hydraulic, pneumatic, water, or refrigerant lines. (Aeroquip Corp.)

Fig. 18-41. Pipe joint fittings. Such fittings are commercially made in various sizes and thicknesses of steels and steel alloys.   (Tube Turns-Div. of Chemetron Corp.)

to permit its use. Various insurance organizations require that all welding which is done on properties they have insured must be done by men who have passed qualification tests prepared by experts.

18-65. PIPE JOINTS

As mentioned in the preceding paragraph, piping must be carefully pre-

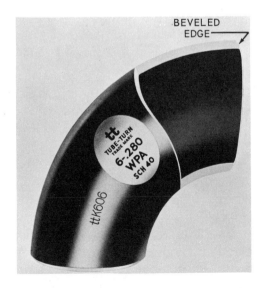

Fig. 18-42. Commercially fabricated pipe elbow fitting designed for welded joints. The 6-.280 means 6-in. pipe with a .280 wall thickness.

fabricated to insure strong and neat welds. Many companies now carry in stock, elbows, adapters, tees, etc., which are made especially for use in pipe welding. The ready-to-use weld fittings with chamfered edges which are available to the pipe welding trade include tees, 90 and 45 deg. elbows, welding neck flanges, concentric reducers, lateral nipples, straight nipples, eccentric reducers, saddle-caps, reducing tees, and 180 deg. return bends, Fig. 18-41. These fittings may be obtained in various sizes, and are stocked by wholesale pipe establishments. You must be sure to specify the proper kind and size of fitting when ordering. Fig. 18-42 illustrates a method used to identify a particular fitting.

When preparing pipe for welding, the contour of the chamfer is important. The angle of the chamfer is recommended to be approximately 22-1/2 deg., 30 deg., or 45 deg. from the vertical. The depth of the chamfer depends somewhat on the thickness of the metal. Under no circumstances is it to extend through the full thickness of the metal, but instead should extend to within approximately 1/16 in. of the inside of the pipe to insure adequate penetration as shown in Fig. 18-43. A method

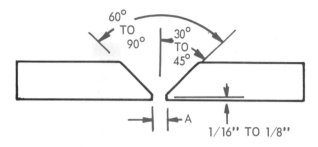

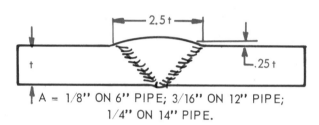

Fig. 18-43. Pipe joint preparation. The bevel angles, root and spacing dimensions vary with different applications.

for preparing a butt pipe joint for welding is illustrated in Fig. 18-44. This

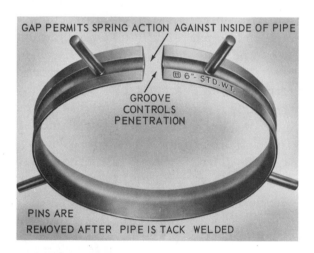

Fig. 18-44. Backing ring for pipe butt joints. This device aligns the pipe and also helps control the penetration. The ring is for 6-in. ID pipe. (Tube Turns, Div. Chemetron Corp.)

shows a backing ring which spaces the pipes and helps to control penetration.

The beveling of the ends of a pipe

may be done with an oxyacetylene torch, by machining, or by grinding. A hand operated oxyacetylene beveling machine is illustrated in Fig. 18-45. A

Fig. 18-45. Hand-operated pipe beveling machine. Note the hand crank and gearing and angle adjustment for flame cutting head. The pipe being cut is 22 in. in diameter.

motorized oxyacetylene pipe beveling machine with a device for cutting shapes such as saddles, is shown in Fig. 18-46.

18-66. LAYOUT OF PIPE AND TUBE JOINTS

Cutting of pieces of pipe to fit one another when the pipe is placed at various angles necessitates knowing the irregular curves of the contact.

Many times, a pipe welder is forced to fabricate or manufacture his own fittings, or to prepare special joints when connecting pipes. A fundamental knowledge of geometry (especially descriptive) may be used to good advantage for this type of layout. Several practical methods have been developed, however, which solve most of the problems encountered.

Paper or sheet metal templets laid out on a drafting board and cut out for

MOTOR

TORCH
ANGLE
ADJUSTMENT

TEMPLATE
FOLLOWER

SHAPE
CUTTING
TEMPLATE

Fig. 18-46. Motor-driven pipe beveling machine with shape cutting attachment.
(H and M Pipe Beveling Machine Co.)

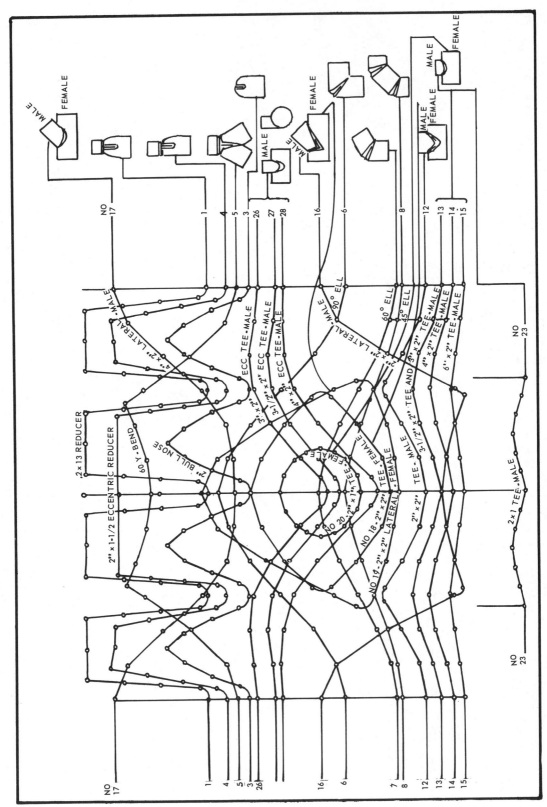

Fig. 18-47. Geometric curves for 2-in. dia. pipe joints. (Airco Welding Products, Div. of Airco, Inc.)

shop use are in common use, as shown in Figs. 18-47 and 18-48. Special devices are also used to determine the on one pipe and sliding it around the pipe while keeping the soapstone in line with the center line of the pipe to

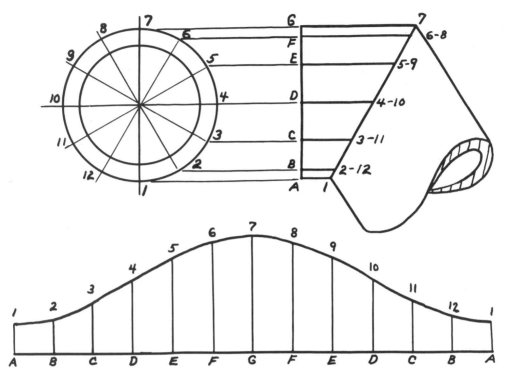

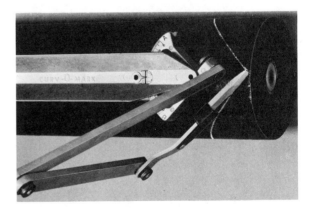

Fig. 18-48. Preparing paper or metal template (layout) for 60 deg. angle pipe joint.

shape of the curves, as shown in Fig. 18-49. Another method used is placing the two or more pipes in their proper alignment and laying a soapstone pencil

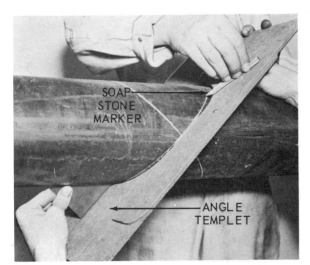

Fig. 18-49. Method used to determine shape of end of pipe for T-joint. (Linde Div., Union Carbide Corp.)

Fig. 18-50. Special fixture used to mark pipe for cutting.

mark both the opening shape and the shape of the pipe end. Special mechanisms may be purchased which aid in making correct pipe layouts.

Many special mechanisms have been developed to aid in producing accurate layouts quickly and easily. Fig. 18-50 illustrates a fixture for marking pipe at an angle. Note the protractor and the freedom of the arm to travel all around the pipe. The fixture uses magnets to hold it in place.

For irregular joints and for final assembly, manual welding is the only method available.

As in all welding, pipes must be firmly held and the pieces accurately

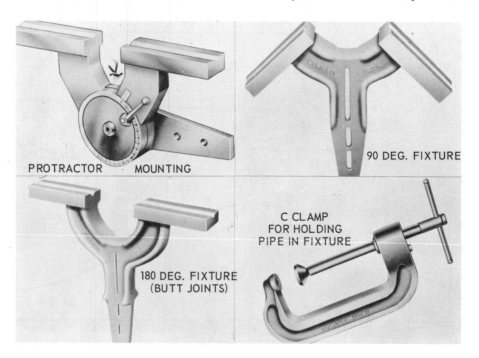

PROTRACTOR   MOUNTING

90 DEG. FIXTURE

180 DEG. FIXTURE
(BUTT JOINTS)

C CLAMP
FOR HOLDING
PIPE IN FIXTURE

Fig. 18-51. Fixtures for holding pipes while they are being welded.
(Wales Strippit, Inc.)

18-67. PIPE WELDING

The method to use when welding pipe is dependent on the kind of pipe, size, and location of welds. Many types of joints are encountered in pipe welding. These joints include Butt and T-Joints in all positions, Elbow Pipe Connections, and the adapting of different sizes of pipe. Pipe welding of a more simple nature, butt joints for example, may be done on automatic machines. These machines are of two types:

1. Where the flame or arc travels around the pipe and welds the metals together.

2. More commonly, the pipe is rolled and a fixed position arc or flame performs the welding.

aligned and spaced if the welds are to be successful. Fig. 18-51 shows a fixture for holding two pipes at any desired angle to each other as long as the pipes are in the same plane. The fixture itself can be clamped or bolted to a bench or some other support. If a fixture is not available, V-blocks or short pieces of angle iron can be assembled to serve as fixtures. Other devices used to help accurately lay out pipe and align pipe are shown in Figs. 18-52 and 18-53.

One difficulty encountered with all pipe and tube welding is the alignment or positioning of the pipe or tube after the welding is completed. One method is to clamp the pipe or tube in a fixture while welding, and allow it to cool be-

Fig. 18-52. Centering device to make center punch mark at any angle (spirit level and protactor). It locates pipe in any position and at any angle of downward slope. (Contour Sales Corp.)

fore removing the clamps. Another method allows for the contraction of the joint weld metal by providing approximately a 1/4-in. movement of the pipe for each 24-in. of length. The preset out of alignment can be adjusted until the correct results are obtained. Fig. 18-54 shows pipe mounted in a fixture or mounted out of line with the intent that the cooling weld metal will pull the pipe and/or tube into correct alignment.

Standards have been established which determine the construction of the weld bead for certain size pipes. The requirements are based on the fact that these welded joints must be at least as strong as the original pipe.

Pipe greater than 1/8-in. wall thickness must be specially prepared for welding (chamfered); special joints must have special contours cut into the

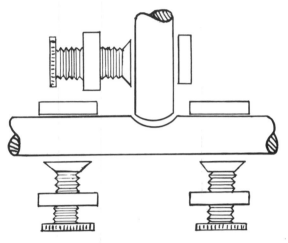

CLAMPING FIXTURE

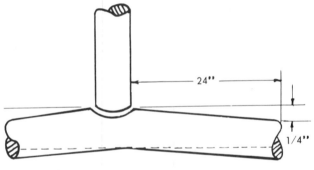

PRE-BENT PIPE

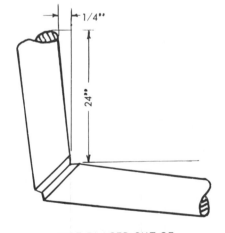

PIPE PLACED OUT OF LINE BEFORE WELDING

Fig. 18-54. Methods used to prevent pipe from being out of line after welding.

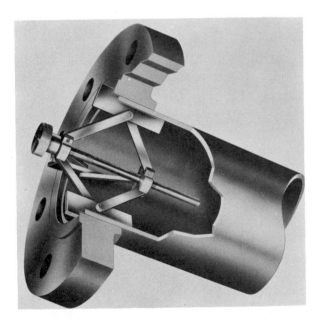

Fig. 18-53. Device for centering flanges and for accurately aligning flanges 90 deg. to axis of pipe.

Special Welding Applications 473

pipe to permit accurate fitting. Prefabrication of the pipe is performed by using metal-cutting tools or a cutting torch. See CHAPTER 3. The actual welding of the joint may be accomplished either by the arc method, or by the flame method. There are many new covered electrodes available which are designed for pipe welding.

In an attempt to obtain more uniform welds on pipe, automatic welding equipment is being used extensively. Special machines have been developed for automatically welding pipe on cross country pipe installations.

18-68. OXYACETYLENE WELDING PIPE JOINTS

When welding with the oxyacetylene torch, three important things must be observed before the final welding operation may be performed:

1. The pipes must be accurately cut to fit one another.

2. The metal must be beveled, or chamfered, to insure accurate penetration and maximum strength. The chamfering may be done with a cutting torch. Some high-pressure lines specify, however, that the bevels must be machined or chipped.

3. The joint must be carefully mount-

Fig. 18-55. Fixture for aligning and holding pipe as it is being butt welded. (Jewel Mfg. Co.)

ed in the proper alignment of the piping during the welding operation. Figs. 18-55 and 18-56 show two fixtures used to insure correct alignment of the pipe.

Fig. 18-56. Special clamp device for positioning pipe T-joint and holding pipe during welding operation.

If aligning fixtures or if back-up rings are not being used, the pipe should be tacked in at least three places around the periphery. When the actual weld is being done, these tacks should be rewelded; also the end of the weld should run over (overlap) the beginning of the weld about 1 in.

After the preparations have been completed, the weld proper is performed by one of two methods:

1. The easiest and best way, if feasible, is to roll the pipe or turn it gradually, as the welding progresses so the part welded is uppermost at all times (position welding) as illustrated in Fig. 18-57.

2. The pipe may be held stationary, and the welding done by starting at the bottom and working up both sides of the pipe. This method is shown in Fig. 18-58.

Of the two, the first usually insures a better weld. However, only in rare cases may this type of welding be utilized for the complete job. Usually both methods are employed on one assembly of a pipe installation.

The size of torch tip to be used is determined by the thickness of the pipe wall, and is practically the same as for

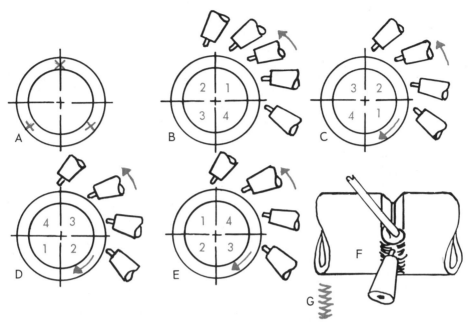

Fig. 18-57. Rolling method of welding pipe butt joints: A. Three tacks 120 deg. apart. B. First part welded. C. Pipe rolled and second quarter welded. D. Pipe rolled and third quarter welded. E. Pipe rolled and fourth quarter welded. F. Position of torch and welding rod. G. Weaving motion used.

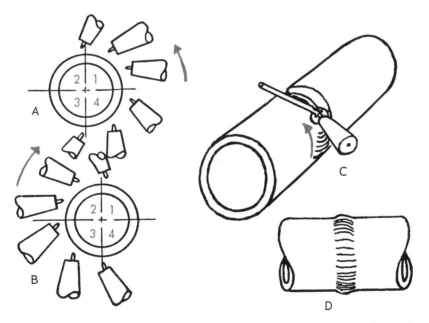

Fig. 18-58. Fixed method of welding pipe butt joints: A. Starting underneath and progressing to the top of one side. B. Starting underneath and progressing to top of other side. C. Position of torch and welding rod. D. Appearance of finished weld.

the same thickness of flat sheet or plate. The weld proper may be done either by using a forward weld, or as many companies recommend, the backward weld may be used. One procedure recommends that the backward method of welding be used in combination with the reducing flame and a special welding rod. It is extremely important that penetration be secured when welding a pipe. This root pass is of extreme importance. Excessive penetration will restrict and turbulate the fluid flow. Insufficient penetration will leave cracks and crevices on the inside surface of the pipe and may cause failures especially in high-pressure pipe. The existence of either of these faults may necessitate a reject of the joint and it will have to be redone. Back-up rings, such as shown in Fig. 18-44, should be used to promote good root passes.

On T, angle joints, and cluster joints, one should use a sequence welding technique, mainly to prevent weld metal contraction from pulling the pipe out of line. Fig. 18-59 illustrates one satisfactory sequence for a T-pipe assembly. If possible do all the welding in a downhand position. The weld should also be crowned or "built-up" a definite amount. Experience has shown that a well penetrated weld, sufficiently "built-up," has more strength than the original metal, and will withstand more wear and corrosion than the original metal. The metal must be well fused and there must be no undercutting.

There are many occasions when used pipe must be welded. If the pipe has carried flammables, the inside of the pipe must be cleaned and the pipe must be purged free of air (O_2) during the welding. This same technique is also used if the inside of new pipe is to be kept as clean as possible during welding operations or to prevent oxidation of the weld metal. Fig. 18-60 shows a method of providing shielding gas during a welding operation.

If the piping is for high-pressure work, the joint must be annealed. Spe-

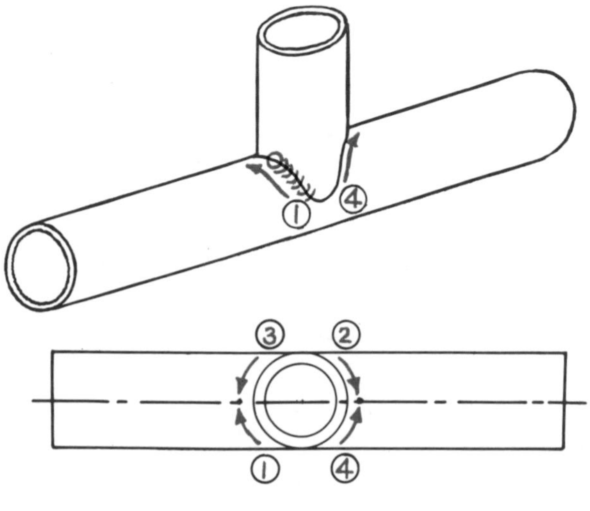

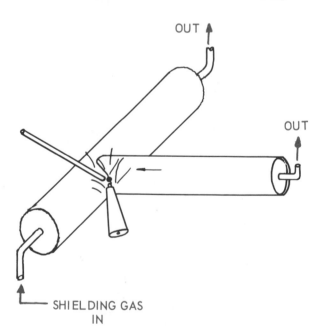

Fig. 18-59. Method of welding a 90 deg. T-pipe assembly with one pipe in vertical position. Note: Recommended sequence is to prevent contraction from pulling pipes out of alignment as welds cool.

Fig. 18-60. Shielding gas being used to purge flammables from the inside of the pipe during welding operation. This technique may also be used to keep inside of pipe clean during welding operation.

Fig. 18-61. Pipe and an elbow being assembled prior to arc welding joint. Note backing ring used here for alignment and to control penetration.
(Tube Turns, Div. Chemetron Corp.)

cial, high-frequency, induction heating devices are now available for joint pre-heating and annealing.

18-69. ARC WELDING PIPE JOINTS

Arc welding can be satisfactorily done on pipe. The operator must be adept in flat, horizontal, vertical, and overhead welding on flat plate joints before attempting to become an expert on pipe welding. Puddle control by using the correct electrode sizes, correct current settings, arc length, and elec-trode motion is very important. It is very important that the operator see the difference between the molten slag and the molten metal. As one arc welds a pipe joint, the same joint may one moment be in the flat position and then change to the horizontal, vertical, or overhead position as the weld pro-gresses along the seam.

Fig. 18-61 illustrates an elbow being fitted to a pipe with a backing ring used to align the fitting to the pipe.

18-70. SHIELDING GAS ARC WELDING PIPE JOINTS

Shielding gas arc welding of pipe joints gives excellent results. Some of the more exacting jobs put shielding gas inside the pipe during the welding to re-duce contamination at the penetration side of the weld as shown in Fig. 18-60.

It is possible to obtain even and positive penetration using the shielding gas tungsten electrode method. Some high-pressure pipe welding procedure in-structions require that the root pass be made using this method. This extreme-ly clean method of welding, plus the in-herent surface tension of the molten metal, produces smooth inner pipe sur-faces and perfect fusion.

The consumable electrode shielding gas welding method may also be used with good results.

18-71. TUBE WELDING

Tube welding is similar to thin sheet steel welding except, as in pipe weld-ing, the weld joint is a three dimen-sional curve. Also, because the root of the weld is not accessible for welding, and because the inner surface is in con-tact with flowing fluids, the penetration standards are high. Fig. 18-62 shows two tube welding faults which must be

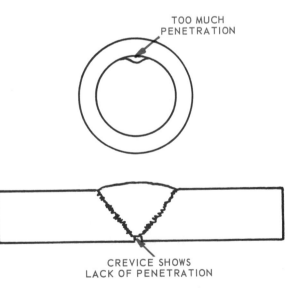

Fig. 18-62. A tube joint weld showing two common penetration defects, either not enough penetration or too much penetration.

corrected before the tubing can be used. These faults can be located by X-ray or by hypersonic devices.

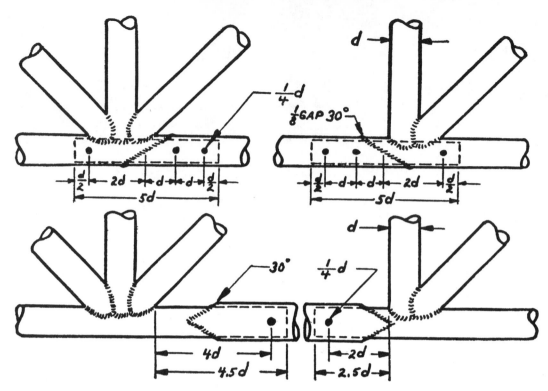

Fig. 18-63. Aircraft tube longeron repair welds. The letter "d" indicates outside diameter of tubing. Dimensions used are to be in accordance with repair standards found in Civil Aeronautics Manual No. 18, prepared by Federal Aeronautics Administration.

Tube materials commonly welded are:
1. Low carbon steel.
2. Chrome-molybdenum steel.
3. Stainless steel.
4. Aluminum.

18-72. OXYACETYLENE WELDING TUBE JOINTS

Control of the weld puddle as a tube weld is being made is very important. The thickness of the tube wall determines the weld technique to be used. If the tube wall is thin, best results can be obtained by coordinating the torch and welding rod motion as follows:

1. Heat the base metal edges to the plastic temperature (melting temperature, not flow temperature).

2. Move the preheated welding rod into the weld area. As this action is taken, lift the torch flame enough so the flame is concentrated on the welding rod. Melt some of the welding rod and allow it to settle on the weld (only a very small amount).

3. Remove the welding rod a short distance, lower the flame, and fuse the added metal to the two tube edges.

You should first become expert on flat surface welding of this thickness metal in all positions, before attempting tube welding. The complex three dimensional joints require expert torch and welding rod handling.

18-73. ARC WELDING TUBE JOINTS

Consumable covered electrode welding of tube joints has been successfully done, but it is a weld that requires considerable skill especially on tubing with thin walls. Low amperages must be used and the weld puddle kept small. Expert manipulation of the electrode is necessary to produce high quality welds.

DC welding using high ionized covered electrodes of 1/16 and 3/32-in. diameter is quite common. The electrode material and the welding procedure varies with tube metals and tube wall thicknesses.

The same tacking procedures, aligning procedures, and welding sequences are used for tube joints as for pipe joints.

18-74. SHIELDING GAS ARC WELDING TUBE JOINTS

Both the GTAW (TIG) method and the GMAW (MIG) methods of shielding gas arc welding may be used for tube welding. Using these methods, it is possible to produce quality welds on all weldable tube joints.

Current settings, gas flows, and the like, are based on wall thickness of the tubing and depend on the OD of the tubing and other heat absorbing metals and fixtures in the weld area.

Frequently the same kind of shielding gas used for welding is also passed through the inside of the tube or tubes as shown in Fig. 18-60. The gas supply is kept at a low pressure to economize on gas consumption and to prevent forcing the plastic and molten metal out of position.

18-75. STAINLESS STEEL TUBE WELDING

Refer to Paragraphs 18-13 through 18-18 for instructions on the welding of various types of stainless steel sheets. These instructions also apply to stainless steel tubing. It is impor-

tant to remember that the coefficient of expansion of stainless steel is more than that of mild steel. Therefore, more care must be taken to tack or fixture stainless steel tubing.

Stainless steel tubes may be welded using the oxyacetylene flame, stick electrode arc welding, or shielding gas arc welding processes.

18-76. ALUMINUM TUBE WELDING

Aluminum tubes are available using most of the aluminum and aluminum alloys. Some of the alloys are weldable and virtually all of the more common welding methods may be used to weld the weldable aluminums.

It is important to clean the metal shortly before welding and to use clean welding rods.

When shielding gas arc welding aluminum tubes, the technique of flowing shielding gas through the inside of the tube is more necessary with this metal than with carbon steels.

18-77. AIRCRAFT TUBE WELDING

A special division of tube welding has been created by the aircraft industry. The Federal Aircraft Administration (FAA) requires that one be a licensed airplane mechanic before being allowed to weld aircraft structures. Special alloy tubes are used in the fabrication of the fuselages, the empennage (tail), landing gears, and the wing sections of two and four passenger airplanes. This tubing is always of seamless construction, and is made of a very high quality steel with some strengthening, alloying metal added. Such steels as chrome-molybdenum (1 percent chromium, 8-1/4 percent molybdenum), and the like, are quite commonly used for this type of work. The tube has a relatively thin wall as compared to the ordinary steel pipe. Good fusing through the thickness of the metal is required with an absolute minimum of visible sag penetration.

Because a maximum of strength must be obtained when welding the joints of aircraft tubing, special fabrication de-

if the aircraft mechanic doing the welding is also certified to do the welding using these processes. The certification is usually obtained by actual performance tests under the auspices of FAA officials.

Fish-tail fitting is a strengthening, or repair joint, where the larger tube

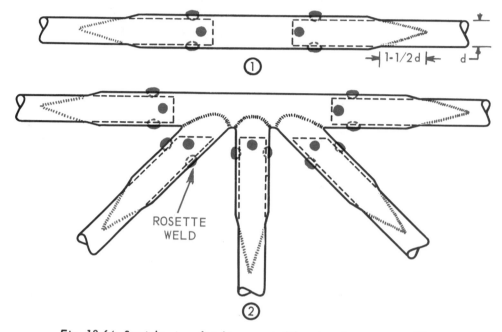

Fig. 18-64. Straight aircraft tubing repair (1) and a cluster repair (2).

signs have been developed. The most common types of special aircraft tubing welded joints are:

1. Fish-tail are:
2. Telescope tubing weld.
3. Gusset weld.
4. 90 deg. T-weld.
5. 45 deg. T-weld.
6. Cluster tube weld.

Fig. 18-63 shows samples if aircraft tubing repairs using the welding process.

Oxyacetylene welding of aircraft tubing is common. However, the tubing may also be welded using the covered electrode arc welding process or the shielding gas arc welding processes if procedure is certified by the FAA and

is cut into a V-shape and then slipped over the small tube which it is to strengthen. The weld is a lap joint; the heated portion of the tube, where the weld occurs, is spread over a considerable length of tubing, distributing the load along a greater length of the tubing. Fig. 18-64 shows an injured tube repair (1) and a cluster repair (2). The rosette welds are plug welds done by drilling the outer tube only before assembly and then welding the inner tube to the outer tube.

The telescope weld is used to obtain a stronger joint. If a simple butt joint weld were used, the joint might fail under bending; but by telescoping the tubing at the joint, additional strength and stiffness is obtained, thus reducing the chance of failure.

The cluster type joint is used where three or more tubes come together, at an angle, with one another. The tubing ends are then welded together where they come into contact, forming an extremely strong joint to better withstand tension, vibration, compression, and bending loads. Fig. 18-65 shows a type of reinforcement weld used on aircraft structures.

Aircraft tubing welds that are going into service are often inspected by the X-ray method. However, many companies also periodically test samples of the welds as made by their welding operators. These samples are tested to destruction to determine both the quality of the weld, and the welder's ability.

One of the greatest problems encountered in aircraft tubing welding is the corrosion which takes place inside of the tubing at the point where a weld has been made. To prevent this corrosion it is a common practice to coat the inside of the tubing with hot linseed oil or some other approved preservative, which prevents rusting or oxidation of the metal at the weld. If you are interested in specializing in aircraft welding, the Civil Aeronautics Manual #18, which is prepared by the Federal Aeronautics Administration, contains official information. It is available from the Superintendent of Documents, U.S. Government Printing Office, Washington, D.C.

18-78. HEAT TREATING PIPE AND TUBE WELDED JOINTS

Because the weld metal and the metal immediately adjacent to a weld has all the properties of a casting, mechanical working of this metal and heat treatment of this metal is sometimes necessary to obtain the maximum physical properties desired. Manual and automatic peening of the weld are methods of mechanically working the metal.

Heat treating a pipe or tube joint presents special problems. The assembly is usually too awkward to put into a furnace. Gas flame heating has been

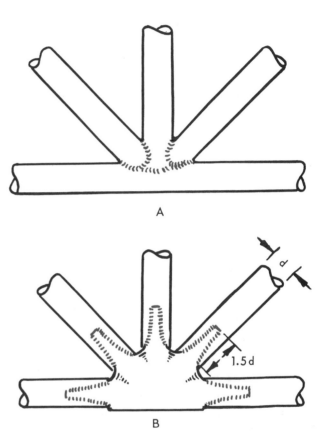

Fig. 18-65. Aircraft tubing repair weld: First the tubes are welded to each other (A), then a repair reinforcement is welded to this assembly (B).

used but temperature accuracy along the weld is difficult. Two methods which have been successfully used are electric heating methods:

1. Resistance wire heating.
2. Induction heating.

Resistance heating for preheating, or heating during the welding operation, or postheating consists of wrapping resistance wire around the pipes or tubes at the proper places and heating the pipe, controlling the temperature with a thermostat. Special units have been developed that are easily installed (quick clamp and toggle method). CHAPTER 25 describes methods used to heat metals for heat treatment.

Preheating is usually up to 600 deg. F.

Postheating is usually up to 1300 deg. F.

Concurrent heating is usually up to 600 deg. F.

The time the weld is held at the specified temperature and the rate of cooling are important factors when postheating welds. Some alloys require a two-step postheat treatment. You should always check with the manufacturer of the metal and the supplier of the metal, to ascertain the exact preheat and postheat treatments recommended.

18-79. INSPECTING PIPE AND TUBE WELDS

It is important to know whether a completed pipe or tube weld has sufficient strength. CHAPTER 22 explains inspection tests and destructive tests. The welding procedure must be carefully established and checked by inspection and testing facilities for adequacy. The operator must be checked to make certain he can conform to the accepted procedure. In determining the ability of an individual welder, standard procedures have been developed for testing sample welds produced by a welder. However, on the final fabrication, or welding, of pipe joints it is impossible to actually determine the strength of the weld, as the ability of a welder may vary with fatigue and interest. Therefore, methods have been devised to determine the strength of the joint without destroying it. One example of this

type of testing is usually called visual inspecting. Under the heading of visual inspection, the following qualities of the weld are observed:

1. Clean bead.
2. Constant width of bead.
3. Fusion of the added metal.
4. Height of the bead.
5. Presence of pits and spots in the weld.

The method used to inspect a weld for sufficient height (build-up) of the weld, and for the width of the bead, is to use a templet which is a steel form that fits on the pipe with a slight cavity on one edge. The cavity is of the shape, size, and contour of the correct size weld.

Before working on high-pressure fluid lines, the welder must pass an examination to determine whether the welds are of good enough quality to withstand the stresses in the high-pressure lines. It is only after passing this examination with a good grade that an operator may work in a plant doing welded pipe fabrication.

Practically all pipe welds are tested either under gas pressure (air, nitrogen, or carbon dioxide), under water pressure, or by using a halogen gas. The water or hydrostatic pressure imposed is usually twice the amount the weld is expected to be under in actual use. The water method (hydrostatic) determines the strength of the complete assembly.

The gas pressure method is used mainly to detect leaks. A soap solution is placed around the joint and a leak is indicated by the appearance of bubbles. When testing pipe-joint welds during the gas pressure test, it is important that the pipe immediately adjacent to the weld be pounded vigorously with a hammer. This pounding is done to dislodge any small scale (oxide) particles which might temporarily close a leak. Do not pound on the weld itself, as the peening action might temporarily close small pin holes which will eventually give trouble.

The halogen gas leak test is becoming popular. This test will locate smaller leaks than any other known method. A halogen gas, such as Refrigerant 12 or Refrigerant 22, is put into the pipe or tube under pressure. Any leak is detected by a very sensitive electronic ionizing sniffer. It is indicated by a meter reading or by a buzzer sound. This method is an adaptation of leak detecting methods used in the Air Conditioning and Refrigeration industry. It will signal a leak of minute size.

18-80. HOT SHORTNESS

Some metals, such as cast iron and aluminum, when heated to the melting point for welding, become extremely weak in the weld puddle area. During this time the metal has a tendency to suddenly fall away from the weld puddle, leaving a hole in the metal. This tendency is called "hot shortness."

18-81. USE OF A BACKING STRIP

Metal which extends beyond the back side (side opposite the bead) of a weld is called penetration. When it is necessary to control the shape and size of this penetration, a backing strip is often used. See Fig. 18-11.

Backing strips are used to control the molten metal in the welding puddle and reduce the possibility of "hot shortness" failure occurring (see Para. 18-80).

A groove is often machined into the metal backing strip. This groove insures that the penetration has the desired shape and size when it cools.

Preformed backing strips can be purchased in roll form. These pre-formed strips are taped onto the metal.

When round parts such as pipes are welded, a backing ring may be used. (See Fig. 18-44.) Backing strips and rings are usually removed after welding is completed. The backing strips and rings are made of metals which conduct heat readily and of dissimilar metals so that they are not welded into the joint.

18-82. REVIEW OF SAFETY IN SPECIAL WELDING APPLICATIONS

All of the safety rules explained in the previous chapters also apply to pipe and tube welding.

Of particular importance to the pipe and tube welding operator are:

1. The piping and tubing must be absolutely free of flammable liquids or fluids to prevent an explosion. If the pipes or tubes have been used, the lines should be steam cleaned, and a shielding gas should flow through the piping system while the weld is being made.

2. The welding of pipes and tubes requires that the operator be in an awkward position or climb into almost inaccessible places. A firm footing, excellent ventilation, and a quick escape route are essential.

3. Because pipes and tubes are frequently installed in occupied buildings, a thorough investigation must be made to locate and remove any and all flammables from near the place of work. Fire fighting equipment must be handy and a fire protection person should be standing by during the welding and/or cutting. The area must be checked continuously for several hours after the work is completed.

One must carefully ventilate work stations where toxic fumes and particles exist. Because lead is a very toxic substance in particle or vapor form, one must have excellent ventilation. One must also wash thoroughly and change clothes after welding and/or handling lead materials. When using nitrogen for testing, one must always install a pressure relief valve in the test line. It should be adjusted to 5-10 psi above the testing pressure.

It requires considerable skill to weld pipe correctly. A pipe welder's ability should be periodically checked by checking and testing samples to destruction. Pipe welding, when properly done, is faster and stronger than the threaded assembly method. Pipe welds should be vigorously inspected and tested. Penetration and buildup are absolutely essential to a successful weld.

Many pipe and tube welding procedures are subjected to code regulations. One should always inquire of the Local, County, and State Building and Safety Departments before accepting and undertaking a structural, pipe or tube welding project.

18-83. TEST YOUR KNOWLEDGE

1. What are two alloy elements included in low carbon, alloy steels?

2. Why is columbium added to stainless steel filler metal?

3. List three different types of stainless steels.

4. How much iron is contained in high-nickel, low-chromium steel?

5. Why must special precautions be taken to prevent the oxidation of stainless steel during welding?

6. What is the principal advantage of a low carbon, alloy steel?

7. What is the range of the chromium content in chromium steels?

8. What two purposes does a copper backing plate serve when welding stainless steel?

9. What is the approximate, maxi-

mum, carbon content of a low carbon alloy?

10. Is a low carbon alloy steel stronger than a straight, carbon steel of the same carbon content? Explain.

11. What welding method is used with Chrom - Nickel - Molybdenum steels?

12. Why is tungsten sometimes added to a chromium steel?

13. List four types of tool steels.

14. What polarity is recommended when arc welding tool steel?

15. Describe galvanized steel. Of what is this steel made?

16. If a stainless steel weld is discolored when the welding is completed, does this mean the weld is worthless in reference to its appearance?

17. Is stainless steel used solely for its enduring bright finish, or are there other reasons for the name "Stainless?"

18. What organization has developed a stainless-steel, standard listing?

19. What is the molybdenum content of chrome-moly steels?

20. Which stainless steels are magnetic?

21. What is the purpose of titanium in stainless steel welding rods?

22. Is it possible to arc weld stainless steel?

23. What institute standardized the types of stainless steel?

24. Why is a flux necessary when gas welding stainless steel?

25. How may one arc weld stainless steel without using a flux or a covered electrode?

26. What is cast iron?

27. How may white cast iron be turned into gray cast iron?

28. What is the best method of repairing a fractured malleable iron casting?

29. Why should cast iron structures be preheated before welding?

30. May the backhand welding method be used for welding cast iron?

31. Explain the purpose of steel studs when welding thick cast iron fractures.

32. Is cast iron more fluid or less fluid than steel when melted by a torch?

33. What is the melting temperature of cast iron?

34. Does cast iron contain more carbon than steel?

35. Is malleable iron more vibration resistant than white cast iron?

36. Why is the "hot shortness" temperature of aluminum important?

37. List the constituents of the average aluminum weld flux.

38. Is penetration necessary when welding a sheet-aluminum butt weld?

39. What is the principal difference between sheet aluminum and cast aluminum?

40. What two main characteristics of aluminum make it difficult to weld?

41. What are the alloy metals of brass?

42. What are the alloy metals of bronze?

43. Is a flux necessary in welding copper?

44. Does one use a larger tip for welding 1/4-in. copper plate than for 1/4-in. steel plate?

45. What happens if brass is overheated?

46. Describe the type of flame used when lead welding.

47. How are the oxides eliminated during the bronze welding process?

48. Why must titanium be welded in an inert atmosphere?

49. What shielding gases are used when welding beryllium?

50. What gas is used when torch welding plastics?

51. Why should the ends of a pipe being joined be spaced apart prior to welding?

52. What are the two methods of performing the gas welding of a pipe?

53. What is the difference between tubing and pipe?

54. May tubing be welded?

55. Of what is aircraft tubing usually made?

56. What technique may be used to compensate for contraction in a pipe joint?

57. Explain the air pressure method of testing pipe welds.

58. Describe the halogen gas leak test.

59. What method of welding pipe is also called "position" welding?

60. Should a pipe joint be tacked before welding?

61. Sketch several tubing cross sectional shapes.

62. May a pipe be arc welded?

63. Should a pipe weld be peened before testing for leaks? Why?

64. May a pipe always be chamfered by oxyacetylene cutting?

65. Of what materials are pipe cutting templates made?

66. Should the chamfer extend through the thickness of the pipe?

67. Is a carefully welded pipe joint considered as strong as the original metal?

68. Explain the initials, OD, and ID.

69. Why is some pipe made of plastic material?

70. Describe two methods used to fabricate seamed pipe and tube.

71. How are flanged fittings fastened to pipe?

72. What is the difference between a procedure qualification and an operator qualification?

73. Describe the purpose of a backing ring.

74. How is the underside of a pipe joint weld kept free of contaminants?

75. What agency controls aircraft welding standards?

76. Why is pipe generally rotated for welding when possible?

77. What effect on the weld does under-

cutting have?

78. When welding thin wall tubing, what type edge preparation is recommended?

79. What should be done with the tack welds as a weld proceeds?

80. When welding a pipe in quarters, should one weld over the top before rotating the pipe to the new position?

81. At the end of a pipe weld, why is it necessary to remelt and reweld some of the beginning weld?

82. Is it possible to use the backhand welding technique when gas welding pipe?

83. How do the number of tack welds vary with the size of the pipe?

84. Generally what shape does the bead contour have when pipe welding?

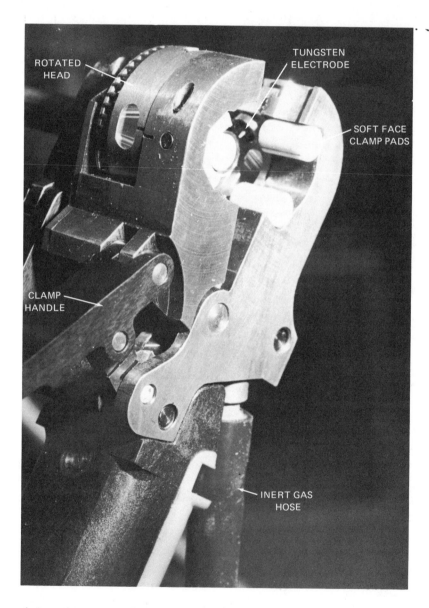

Automatic tungsten inert gas arc welder for welding tubing. When the jaws are closed, the tubing is clamped. Inert gas surrounds the tubing within the clamp, and the tungsten electrode holder is rotated in a complete circle around the tubing. The weld is completed without the addition of filler wire.

Chapter 19

SPECIAL CUTTING PROCESSES

Some examples of metal-cutting processes which are considered special and will be discussed in this chapter are: cutting metal under water in ship salvaging operations, cutting metal sections several feet thick with the oxygen lance, and cutting through concrete walls with powder cutting equipment. Other examples of special processes are shielding gas arc cutting of nonferrous metals, plasma-arc cutting and laser beam cutting.

19-1. OXYGEN LANCE CUTTING PRINCIPLES

The principle of oxyacetylene cutting, as discussed in CHAPTER 3, is that once the metal is heated and the oxygen for cutting is turned on, the metal becomes fuel and the oxygen supports the oxidation (burning) of the metal. With the ordinary oxyacetylene torch and tip, the preheating and the oxygen blast for cutting is combined in one torch. The thickness or depth to which the oxygen can cut is limited to about 24 in. with the conventional hand-held torch. By using a long piece of low alloy steel tubing to carry oxygen and a long handled preheating torch, it is possible to penetrate to the limit of the length of the tube to oxidize and cut steel, or about 8 ft.

The oxygen lance is used for such operations as cutting risers from large castings, cutting through thick pieces of steel in scrapping operations and/or opening gates in large furnaces. A lance is available which consists of a tube lined with alloys. In the presence of oxygen, it burns at a very high temperature. It will cut metals, concrete or stone.

19-2. OXYGEN LANCE CUTTING EQUIPMENT

The equipment required for the oxygen lance operation is:

1. Complete oxy-fuel gas welding or cutting outfit to provide the preheat for lancing. For extremely thick sections a special long preheating torch is required to provide a constant preheat temperature. Fig. 19-1 shows an oxygen lance in use.

2. One or more oxygen cylinders. For large volume cutting, several cylinders may be manifolded to the regulator.

3. A heavy-duty oxygen regulator.

4. Oxygen hose of an appropriate length complete with fittings.

5. Length of 1/8 or 1/4-in. internal diameter pipe long enough to penetrate the desired thickness and sufficiently long to allow for consumption of the tubing as the cutting progresses.

6. Safety equipment will be needed. This equipment includes goggles, helmet, leather jacket, leggings, apron and gloves.

Considerable sparking accompanies oxygen lance cutting. The operation, as well as the equipment and surroundings, must be protected against these hot particles.

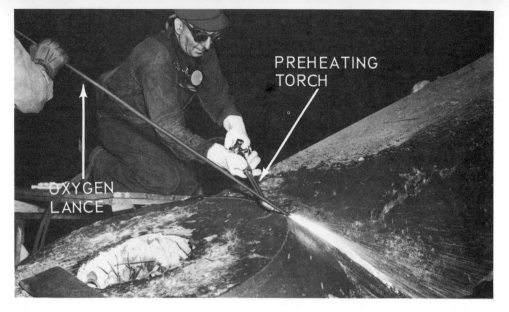

Fig. 19-1. Oxygen lance, with aid of regular cutting torch, being used to cut riser from steel casing. (Linde Div., Union Carbide Corp.)

19-3. OXYGEN LANCE CUTTING PROCEDURE

To start a cut with the oxygen lance, it is first necessary to preheat the edge of the piece with a conventional heavy-duty welding or cutting torch. After the metal reaches a cherry red or white hot temperature, the oxygen for the lance may be turned on and the pre-heating torch may be removed. On ex-tremely thick pieces, the preheat tem-perature is maintained with a special long torch. Generally the cut is started at an angle and gradually brought to a vertical position by manipulating the lance. The cut is much easier to start and maintain using this principle. Fig. 19-2 is a schematic drawing showing the oxygen lance principle.

Lancing may also be used to pierce holes through pieces up to 8-ft. thick. This makes it possible to pierce holes in machinery which would be expensive if not impossible to machine.

As the pipe extends down into the cut, it is subjected to excessive tempera-ture in the presence of oxygen, and the end of the pipe will also be burned

away. However, the end of the cutting lance has been known to pierce the cen-ter of a freight car axle shaft through its length without burning through the

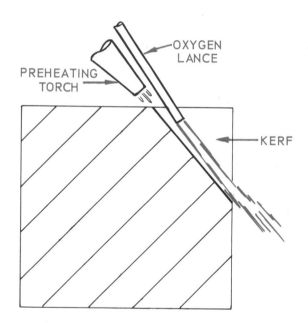

Fig. 19-2. Schematic of oxygen lance being used to cut thick metal.

side of the shaft. Considerable expense is eliminated, when cutting large steel or cast iron sections, by using this method.

Fig. 19-3 shows operators at work using a standard cutting torch and an oxygen lance, to cut a large casting.

19-4. OXYGEN-FUEL GAS UNDER-WATER CUTTING PRINCIPLE

The method of cutting ferrous metals underwater is much the same as the method used on the surface with one exception. In order to keep the flame lighted at extreme depths, the cutting tip and combustion area is enveloped by compressed air which keeps the water away from the combustion zone.

The use of acetylene below 15 ft. is not recommended since acetylene is not safe to use at pressures over 15 psig. Hydrogen should be used as the fuel gas below a depth of 15 ft. A proprietory fuel mixture called MAPP gas, manufactured by the Dow Chemical Co., may be used for this purpose as it is stable under high pressures. Oxyhydrogen flame has been used to a depth of 200 ft.

The underwater torch has a special high-pressure air system, the purpose of which is to keep the water away and maintain an air pocket at the point of cutting as shown in Fig. 19-4. Special training in underwater welding and cutting techniques is required and the operator should also be a qualified diver.

Gas torches are usually constructed as compactly as possible. The compressed air sheathing around the tip is adjustable to enable the operator to hold the torch head against the metal being cut. Slots in the sheathing permit escape of the products of combustion and air. The tip must have exceptional preheating capacity to overcome excessive heat removal by conductivity and convection, to overcome corrosion on the surface of the metal,

Fig. 19-3. Large casting being cut using standard cutting torch and oxygen lance.

and to save the operator's time since he cannot remain at any great depth for long intervals. The air pressure, and therefore the amount of air, is controlled by a separate air valve designed to be a part of the cutting torch. Fig. 19-5 illustrates a typical underwater oxyhydrogen cutting torch.

19-5. OXYGEN-FUEL GAS UNDERWATER CUTTING EQUIPMENT

The equipment needed for underwater cutting is the same as for surface cutting with these exceptions. A special torch is needed which has a special slotted cylinder surrounding the cutting tip, and an additional connection for compressed air, as shown in Fig. 19-4. A source of compressed air and air hose will also be needed.

The shielding cylinder around the tip which is filled with compressed air keeps water out, and the slots in the shield allow excess air to escape. Be-

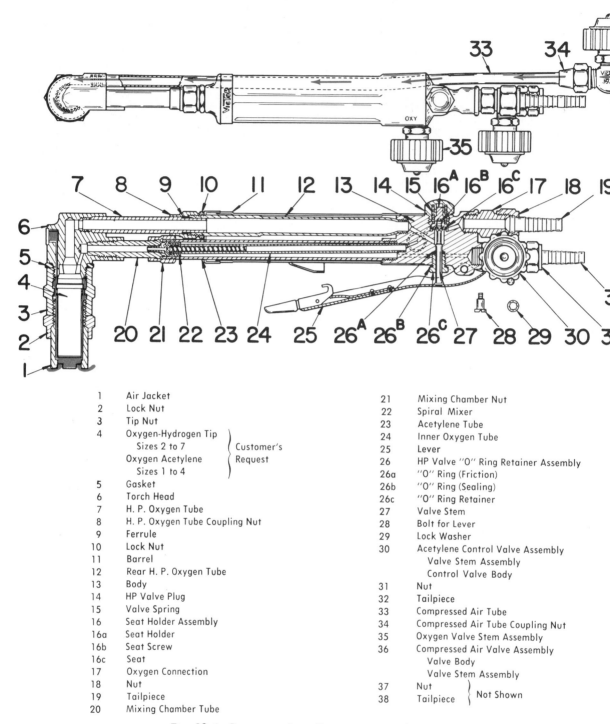

1	Air Jacket	21	Mixing Chamber Nut
2	Lock Nut	22	Spiral Mixer
3	Tip Nut	23	Acetylene Tube
4	Oxygen-Hydrogen Tip	24	Inner Oxygen Tube
	Sizes 2 to 7 } Customer's	25	Lever
	Oxygen Acetylene } Request	26	HP Valve "O" Ring Retainer Assembly
	Sizes 1 to 4	26a	"O" Ring (Friction)
5	Gasket	26b	"O" Ring (Sealing)
6	Torch Head	26c	"O" Ring Retainer
7	H. P. Oxygen Tube	27	Valve Stem
8	H. P. Oxygen Tube Coupling Nut	28	Bolt for Lever
9	Ferrule	29	Lock Washer
10	Lock Nut	30	Acetylene Control Valve Assembly
11	Barrel		Valve Stem Assembly
12	Rear H. P. Oxygen Tube		Control Valve Body
13	Body	31	Nut
14	HP Valve Plug	32	Tailpiece
15	Valve Spring	33	Compressed Air Tube
16	Seat Holder Assembly	34	Compressed Air Tube Coupling Nut
16a	Seat Holder	35	Oxygen Valve Stem Assembly
16b	Seat Screw	36	Compressed Air Valve Assembly
16c	Seat		Valve Body
17	Oxygen Connection		Valve Stem Assembly
18	Nut	37	Nut } Not Shown
19	Tailpiece	38	Tailpiece
20	Mixing Chamber Tube		

Fig. 19-4. Compressed air flow in typical underwater
torch and names of torch parts. (Victor Equipment Co.)

cause of the heat conductivity of the
water, the preheating orifices (holes)
in the underwater torch tip must be
much larger than for similar cutting

torches used only on the surface.

The torch has four hand valves:

1. Acetylene or hydrogen valve.
2. Preheat oxygen valve.

3. Cutting oxygen valve (lever operated).

4. Compressed air valve.

Vision is generally very poor under water, so the operator must know the position of the valves well enough to operate them skillfully without being able to see them.

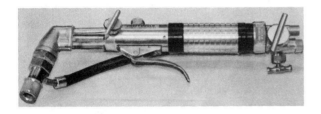

Fig. 19-5. Oxyhydrogen underwater cutting torch, Ellsberg model. (Craftsweld Equip. Corp.)

19-6. OXYGEN-FUEL GAS UNDERWATER CUTTING PROCEDURES

The torch is lighted with an electric sparking device after the oxygen, fuel gas, and compressed air are all turned on and adjusted. A diver who is thoroughly familiar with surface cutting will not find underwater cutting too difficult.

To start a cut, the operator holds the torch against the metal and watches the color of the flames around the torch. When a bright yellow flame is seen the preheat temperature has been reached. The cut is then started by turning on the cutting oxygen and proceeding in the same manner as on the surface. Because of the pressure created by the compressed air within the shielding cylinder on the torch an extra effort must be made by the operator to keep the torch against the work. The air tends to push the torch away from the metal being cut. A distinct rumbling sound is heard when the torch is burning. When the sound stops, the diver knows the flame is out.

19-7. OXYGEN-ARC UNDERWATER CUTTING PRINCIPLES

There are two basic principles of cutting metal underwater with an arc. These are the metallic arc and the oxygen-arc methods. In the oxygen-arc process, an arc is struck on the metal to be cut, using a hollow steel electrode which has high-pressure oxygen flowing through it, as shown in Fig. 19-6.

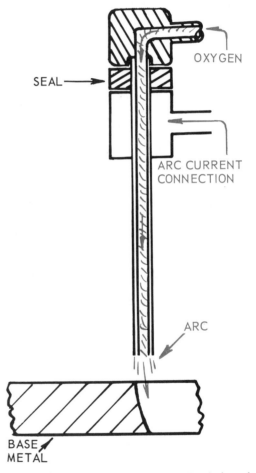

Fig. 19-6. Schematic cross section of tubular electrode and oxygen-arc electrode holder for underwater cutting.

The arc heats the metal and the oxygen jet quickly oxidizes the metal to perform the cut.

Oxygen-arc underwater cutting has proven to be very successful and is widely used in industry. The electrode is usually 5/16-in. tubular steel, flux-covered and waterproofed. A special torch (electrode holder) fully insulated without taping, is shown in Fig. 19-7.

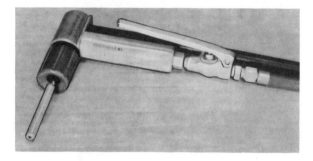

Fig. 19-7. Arc-oxygen underwater cutting torch, patented Model A. (Craftsweld Equip. Corp.)

This system works well because the moment the arc is struck a very high temperature spot is produced on the metal. With the oxygen turned on, the cutting starts instantly.

The operator then draws the cutting electrode along the line of cut while holding an arc. This method of cutting is usable at all depths. The electrode is consumed quite rapidly. The welding current is usually shut off when electrodes are changed. All metals may be cut with this system including steels, alloy steels, cast iron, and all the nonferrous metals.

Using direct current, straight polarity (electrode negative), is the general practice. Fig. 19-8 illustrates the internal construction of an arc-oxygen underwater cutting torch.

Tubular (hollow) carbon electrodes and ceramic (silicon carbide) electrodes have also been used but their brittleness has resulted in their almost complete replacement by steel electrodes.

In the metallic arc process solid electrodes are used. The process can be used to cut both ferrous and nonferrous metal under 1/4-in. in thick-

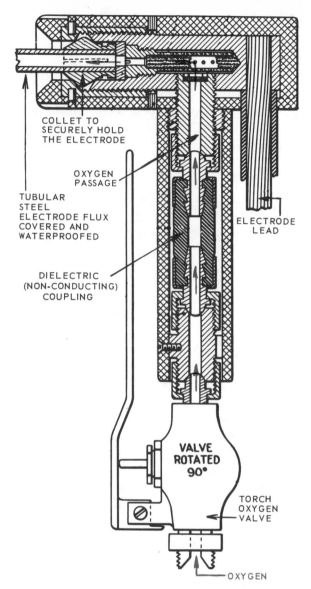

Fig. 19-8. Cross sectional sketch of an arc-oxygen underwater cutting torch. (Craftsweld Equip. Corp.)

ness. The base metal is melted by the process and falls from the kerf by gravity. The resulting cut is quite wide and is not as smooth as a cut obtained using the oxygen-arc process.

19-8. OXYGEN-ARC UNDERWATER CUTTING EQUIPMENT

The equipment used in an oxygen-arc underwater cutting outfit uses some standard arc welding equipment and some standard oxygen equipment.

The standard arc welding equipment used is an arc welding machine (DC preferred).

The standard oxygen equipment used includes oxygen cylinders and manifold oxygen regulator, extra-heavy oxygen hose, welding lens 4-6 shades lighter than required on the surface.

The underwater welder is serviced from a tender ship on the surface. Some special equipment needed includes: oxygen-arc underwater cutting outfit; electrical off and on switch for the welding leads on the tender; ground lead (waterproofed); electrode lead (waterproofed); insulated oxygen-arc torch (waterproofed); waterproofed, covered, tubular electrodes, and rubber gloves.

19-9. OXYGEN-ARC UNDERWATER CUTTING ELECTRODES

Electrodes used in underwater cutting come in three types:

1. Carbon electrodes (hollow; some with metal tubes built in).

2. Ceramic electrodes (hollow).

3. Steel electrodes (tubular) and covered.

The carbon electrode if used presents some difficulties. The 12-in. electrode lasts about 30 min. The electrode holder is returned to the surface for the insertion of the replacement electrode. The torch has to be thoroughly retaped each time to make it waterproof. Some of the carbon electrodes are rectangular in cross section (3/8 x 3/4 in.). These electrodes are usually 12-in. long.

Ceramic electrodes are usually 1/2-in. diameter and 8-in. long. They are made of a material such as silicon carbide. Ceramic electrodes last about 10 min. in operation and are replaceable under water. The electrode is

brittle and must be handled carefully.

Tubular steel electrodes, which are very popular, are available in a length of 14 in. and diameter of 5/16 in. OD. The oxygen orifice is usually 1/8 in. in diameter. These electrodes last only a few minutes but may be replaced under water.

19-10. OXYGEN-ARC UNDERWATER CUTTING PROCEDURE

With the electric generator and the oxygen turned on, the operator strikes the arc as one would on the surface. The operator then depresses the oxygen valve handle to start the cut. The technique used for underwater arc cutting is to drag the electrode along the line to be cut. A constant arc length is maintained by this method due to the thickness of the flux covering.

Skill is required to move the electrode at a constant speed. The operator must move the electrode slowly enough to insure the cutting of the complete thickness of the metal. If movement is too slow, however, the cut may be stopped due to the lack of proper preheating of the advance metal. This will cause too many electrodes and too much oxygen to be consumed during the cutting operation.

19-11. GAS METAL ARC AND GAS TUNGSTEN ARC CUTTING PRINCIPLES

Aluminum, stainless steel, chrome steel, nickel, copper, monel, inconel have been successfully cut by means of the gas metal arc cutting (GMAC) process and the gas tungsten arc cutting (GTAC) process. Satisfactory severing cuts also have been made on titanium, magnesium, and most of the

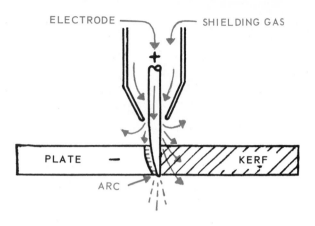

Fig. 19-9. Basic operation of shielding gas cutting process. Notice use of reverse polarity and extension of steel electrode through thickness of metal being cut.

exotic metals.

The principle is the same for both of these gas arc (GMAC and GTAC) processes. An arc is struck between the electrode wire and the work. As the metal melts it is blown away by the flow of shielding gas. The shielding gas flow must extend through the full thickness of the metal being cut to keep the surface of the cut from oxidizing. Fig. 19-9 shows this cutting process.

The effectiveness of this cutting process is due to the scavenging action of the gas velocity. Sufficient gas must flow to prevent oxidation. The use of excess flow is wasteful and will not speed the cutting process.

DC reverse polarity is usually used with shielding gas metal arc equipment because of its penetrating powers.

Argon is the most popular shielding gas, but helium, nitrogen, and carbon dioxide have also been used. Air and oxygen have been used when oxidation does not interfere with the cutting.

Automatic equipment works best as wire feed and torch speed are coordinated. Cuts are made using the gas metal arc (GMAC) process. Cuts produced in aluminum, copper, brass, and many other metals, using this process

are frequently smooth enough so they do not require machining or grinding prior to any subsequent welding.

Manual cuts may be made, but at a slower rate and the results are of poorer quality.

19-12. SHIELDING GAS ARC CUTTING EQUIPMENT

Equipment required for cutting of metals using gas tungsten arc - GTAC (TIG) and gas metal arc cutting - GMAC (MIG) is the same as that required for welding with these same processes. However, in all cases the equipment should be heavy-duty to make it completely versatile. Fig. 19-10 shows a complete shielding gas metal arc cut-

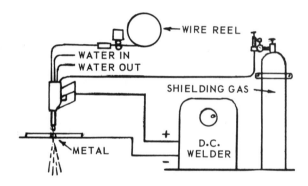

Fig. 19-10. Schematic of shielding gas metal arc cutting outfit.

ting station. The DC power source must be of higher capacity than normal, since the required amperage for some metals, such as 1/4-in. monel, may be as high as 1000 amps.

19-13. SHIELDING GAS ARC CUTTING PROCEDURE

When cutting with gas metal arc, the amperage and shielding gas volume required should be carefully set in accordance with the equipment manufacturer's manuals. Reverse polarity

Fig. 19-11. Large powder-cutting torch
being used to cut riser from casting.

should be used to obtain the required arc force. The water cooling should be adequate to cool the torch, particularly if high ampere flows are required. Since a good cut is only made when the wire is extending just to the bottom surface of the metal, the speed of the wire feed and speed of the cut must be carefully set. Cutting speeds of 140 in. per minute (IPM) on 1/4-in. aluminum and 85 IPM on 1/4-in. monel have been obtained using automatic shielding gas arc cutting equipment.

19-14. POWDER CUTTING

Powder cutting was introduced in 1943 for cutting stainless steel. The process has since been used to cut alloy steels, cast iron, bronzes, nickel, and aluminum. Most materials may be cut with the powder cutting technique-- even reinforced concrete.

In powder cutting, iron powder is introduced into a standard oxyacetylene cutting torch flame. The heat of combustion of the iron powder increases the total heat of the flame.

When cutting stainless steel with an oxyacetylene torch, a refractory chromium oxide is formed on the surface of the metal in the cutting zone. This oxide insulates the base metal preventing further oxidation by the cutting oxygen. By introducing iron powder into the oxyacetylene flame, insulating chromium oxides are kept from forming.

The cut is able to proceed unhindered. The high-in-iron particles oxidize or burn in the oxygen stream at the torch tip producing a very high temperature and concentrated heat. The powder also produces a reducing action on the metal, and a high velocity gouging action. Fig. 19-11 shows a large casting being cut by this process.

Cutting speeds comparable to oxyacetylene cutting of low carbon steels may be obtained on stainless steels using the powder cutting torch.

When cutting nonferrous metal, it is helpful to add a more active ingredient than iron powder to the oxyacetylene flame, so about 10-30 percent aluminum powder is used with iron powder.

Ferrous metals up to 5-ft. thick have been cut with machine cutting powder torches.

Reinforced concrete up to 18-in. thick may be cut with speeds of 1 - 2 1/2 in. per minutes and walls 12-ft. thick may be cut with a powder cutting lance, as shown in Fig. 19-12.

Several devices have been used to feed the iron powder to the flame:

1. Introduction of the powder into the oxygen orifice of the tip.

2. Introduction of the powder into a separate orifice in the tip.

3. Introduction of the powder into the preheating orifices of the tip.

If the oxygen orifice method is used, the oxygen pressure must be less than the air or nitrogen pressure feeding the

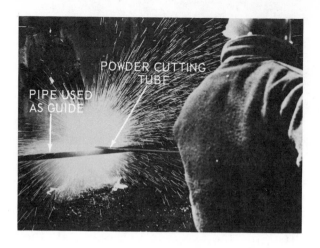

Fig. 19-12. Cutting concrete using the powder cutting process.

iron powder to the tip. If a separate orifice is used, there is a delay in ignition of the iron powder and the tip must therefore be held 3 or 4 in. away from the work.

The extremely high temperature that results from the ignition of the powder practically eliminates the need of preheating the metal prior to the cutting action, and much time is saved. The torch also has a considerable penetrating action or carry over. The cutting easily jumps slag pockets and carries over from one plate to another.

19-15. FLUX CUTTING

In the usual oxyacetylene cutting process, during the heat of combustion, oxides are formed. Occasionally a flux powder is used rather than an iron or aluminum powder in cutting. These fluxes have a reducing reaction on oxides which tend to form. In other words, the fluxes cut down the amount of these undesirable formations.

Fig. 19-13 is a schematic of a flux cutting station. Note that a separate supply line feeds the flux into the work.

In the flux cutting method, a flux is fed into the oxygen stream. This causes a reducing action on the alloy metal oxides, permitting their easy removal from the kerf. The carry-over action of the flux action enables the cutting of multiple layers of metal plates without tight clamping.

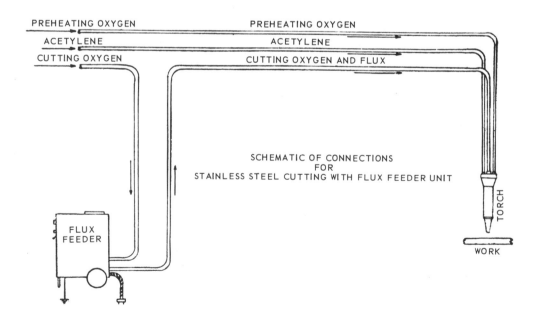

Fig. 19-13. Schematic of a flux cutting outfit. (Airco Welding Products, Div. of Airco, Inc.)

19-16. POWDER CUTTING EQUIPMENT

Equipment required for powder cutting includes a cutting torch or lance, oxygen and acetylene supply, and a method of introducing the powder.

Some special powder cutting torches have an additional tube to carry the powder to the torch head where it is added to the preheating flames. Such a torch is shown in Fig. 19-14. The oxygen and powder tubes on the torch are both opened when the oxygen control lever is depressed. Normally there is a slight delay in the opening of the oxygen valve, thus allowing the iron oxide to coat the surface before the oxygen reaches the surface. This delay helps to start the cut faster particularly on nonferrous metals.

Powder cutting adapters are made which may be attached to the conventional machine-type torch as shown in Fig. 19-15. Powder is carried to the adapter through tubing, and escapes through multiple openings around the conventional cutting tip. It is melted by the preheat flames and carried to the molten metal by the cutting oxygen stream. Since the powder is directed into the oxygen stream from all sides, the multiple opening attachment is suitable for form cutting and straight cutting. When powder cutting nonferrous metals, or when cutting extra thick ferrous metals, an attachment with a single powder tube is used. The powder is carried to a point adjacent to the conventional tip where it is discharged into the flame. Since more powder may be discharged from the single tube opening than can be released from the small openings in the multiple-opening attachment, it is usually considered to be better suited for thick sections than the multiple opening attachment. The single tube attachment works best on straight cuts and is not as well suited

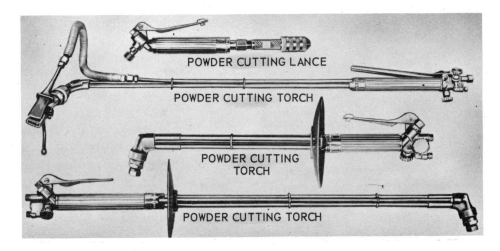

Fig. 19-14. Several types of powder cutting torches. (Linde Div., Union Carbide Corp.)

POWDER CUTTING LANCE

POWDER CUTTING TORCH

POWDER CUTTING TORCH

POWDER CUTTING TORCH

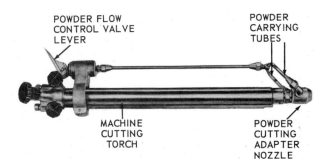

POWDER FLOW CONTROL VALVE LEVER

POWDER CARRYING TUBES

MACHINE CUTTING TORCH

POWDER CUTTING ADAPTER NOZZLE

Fig. 19-15. Machine cutting torch with powder cutting nozzle adapter, powder valve, and manifold tubing.

Fig. 19-16. *Pressurized powder dispenser for use with powder cutting equipment. Notice pressure regulator used to control pressure for powder feed. Note also that only air or nitrogen are to be used to pressurize the powder.*

for shape cutting as the multiple-opening attachment.

Two types of powder dispensers are presently in use. In one, an enclosed hopper is used as shown in Fig. 19-16. The hopper is filled and then pressurized. Powder is then fed under constant pressure to the powder-cutting torch or adapter whenever the powder cutting control valve on the torch is opened. Air or nitrogen may be used to supply the pressure on the powder dispenser. CAUTION -- never use oxygen as it may form an explosive mixture.

The second type of dispenser uses vibration action to keep the powder agitated and moving along a trough. The powder falls from the vibrating trough into a hopper and is fed to the torch by air pressure. Vibrating type dispensers are used where a more constant and accurately adjusted powder flow is required.

When the thickness of metal to be cut is greater than 18 in., it is advisable to use the powder cutting lance. A special designed mixing handle is used to which long lengths of hollow iron pipes are attached as shown in Fig. 19-17. Oxygen and powder are mixed and carried to the cutting area through the hollow pipe. The pipe is slowly consumed during the operation. Reinforced concrete walls up to 12 ft. thick may be cut using the powder lance.

19-17. POWDER CUTTING PROCEDURES

The techniques and torch angles used for powder cutting are approximately the same as those used for oxyacetylene cutting. However, where an oscillating motion is sometimes required when cutting cast iron with an oxyacetylene torch to overcome the insulating effects of the graphite particles on the metal, no torch motion is required when powder cutting cast iron.

To allow the powder to reach its ignition temperature before it hits the work, the torch should be held about 1-1/2 in. above the work.

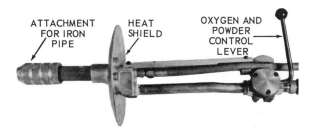

Fig. 19-17. *Powder cutting lance handle. Consumable black iron pipe is attached to the handle. Clean cuts in reinforced concrete over 12 ft. thick can be made with 70 percent iron and 30 percent aluminum powder. (Linde Div., Union Carbide Corp.)*

A new technique for starting a cut faster on carbon steels has evolved because of powder cutting. This technique is known as a "flying start."

To use this method, the oxyacetylene

Hand-held plasma arc cutting torch. This torch is being used to cut 1/4 in. stainless steel at high speed.
(Thermal Dynamics Corp.)

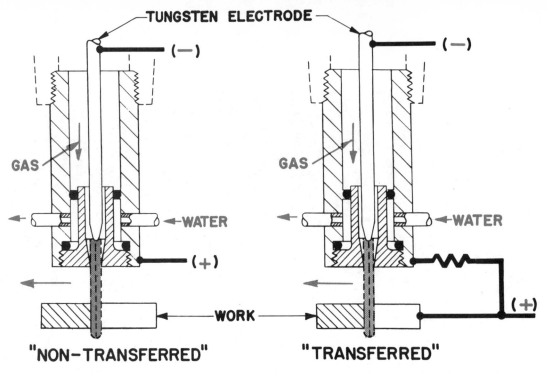

Fig. 19-18. Schematics showing principle of operation of nontransferred arc, and transferred arc plasma cutting processes.

torch is equipped with a powder-cutting attachment. Iron oxide or a mixture of iron oxide and aluminum oxide powder is used. The powder and oxygen are turned on at the start of the cut for a few seconds and the powder is then turned off. Due to the additional heat created by the combustion of the oxide powder, the cut is started almost immediately. Prior to the use of this method 15 - 25 percent of the cutting time was spent bringing the base metal up to a temperature where it would oxidize or be cut.

19-18. PLASMA-ARC CUTTING PRINCIPLE (PAC)

Plasma arc cutting uses the principle of passing an electric arc through a quantity of gas and using a restricted outlet for this electric arc heated gas to flow through. The electric arc heats the gas to such a high temperature that it turns into what physicists call "plasma." Plasma is the fourth state of matter - - not a gas, liquid, or solid. As plasma is formed, a great deal of heat (heat needed to form the plasma) is put into the gas and it is this heat (released as the plasma changes back to a gas) plus the extremely high sensible heat, which heats the base metal to the melting point. Using the plasma process, temperatures as high as 60,000 deg. F. have been reached.

DCSP is used when cutting. Clean, narrow kerfs are possible at high speeds. Several ports are sometimes used in the torch. Only the center port has the arc plasma, while the surrounding ports provide a shielding gas protection. Gas flow may be as high as 250 cfh.

Usually a 25 percent hydrogen - 75 percent argon mixture is used for cutting. The operation is usually started with 100 percent argon; and, after starting, the hydrogen is added.

As the gas is heated in the nozzle it tries to expand, but it is restricted

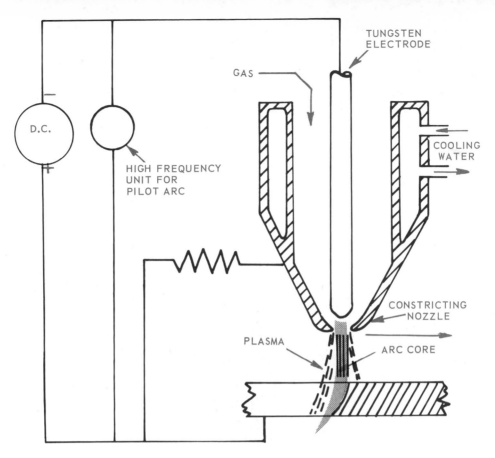

Fig. 19-19. Cross section schematic of torch nozzle used in transfer arc plasma cutting process.

by the nozzle, which acts as a venturi. As it passes through the nozzle its speed is increased to several thousand feet per second. It is this high velocity of the gas which removes the molten metal from the work and creates the kerf. The plasma formed from a shielding gas, and the shielding gas, shields the sides of the kerf from contamination and oxidation.

Two types of plasma arc cutting processes are shown in Fig. 19-18. They are:

1. Transfer arc.
2. Nontransfer arc.

In some plasma arc cutting processes the transferred arc principle is used. The work is attached so that it becomes a part of the electrical circuit, and an arc is created between the electrode and the work in addition to the arc be-

tween the electrode and the holder. Figs. 19-19 and 19-20 show schematics of the transferred arc process. Straight polarity is used to create most of the heat on the work.

In the nontransfer arc process, the electrical circuitry is different from that of the transfer arc process. The arc is struck between the electrode and the torch nozzle. The work is not part of the electrical circuit. A schematic of this system is shown in Fig. 19-18 at the left.

Of the two methods, the transferred arc creates the greatest amount of heat and is used when cutting. The plasma arc cutting method may be used to cut almost all types of metals with great success and economy because of the extreme temperatures and the shielding qualities of the gases used. Speeds of

300 in. per minute (IPM) have been obtained on 1/4-in. aluminum plate and 50 IPM on 1-in. aluminum plate. Speeds of 100 IPM have been made on 1/4-in. stainless steel plate and 30 IPM on

be made over the entire length of the rotating metal.

The plasma transfer arc process is more often used for surfacing operations. See CHAPTER 20.

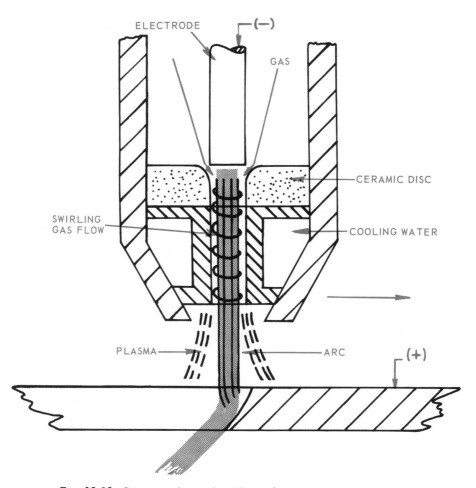

Fig. 19-20. Diagram of transferred-arc plasma cutting process.

1-in. stainless steel plate. All of these cuts may be made dross (slag) free. Underwater cuts have also been made by some operators using this process.

The plasma-arc process has also been used to "machine" inconel and stainless steel metals which are extremely hard on ordinary cutting tools. The metal is held and rotated in a lathe and the plasma-arc torch is held in a special fixture on the traversing tool holder. In this way, accurate cuts may

19-19. PLASMA-ARC CUTTING EQUIPMENT AND SUPPLIES

Plasma-arc cutting may be done with automatic equipment, or manually as shown in Fig. 19-21. The equipment, with the exception of the power source, is the same as that used for plasma-arc welding. See CHAPTER 17. The manual type torch for heavy-duty cutting is usually equipped with a heat shield to protect the operator's hands from the intense heat. It is necessary for heavy-duty cutting up to 5 in., to

to 1-1/2-in. aluminum and 1-in. stainless steel. A high frequency unit is required to start the pilot arc at the be-

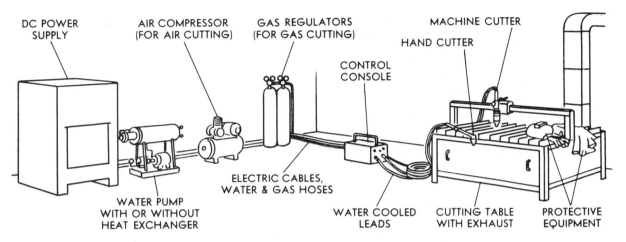

Fig. 19-21. Diagrammatic sketch of the equipment required for a plasma-arc cutting station. (Thermal Dynamics Corp.)

have a power source capable of supplying DC power of 700 amps. and at 170 volts. A power source of 400 amps. and 80V under load may be used to cut up

ginning of a cut. Cooling water pumps and a shielding gas or gases under pressure are also required. For heavy-duty cutting, cooling water pumps are required to circulate enough water through the torch to prevent damage from the extreme temperatures of the plasma-arc as shown in Fig. 19-22.

The torches may be either manually operated or machine operated. Fig. 19-23 shows a manual torch. The ma-

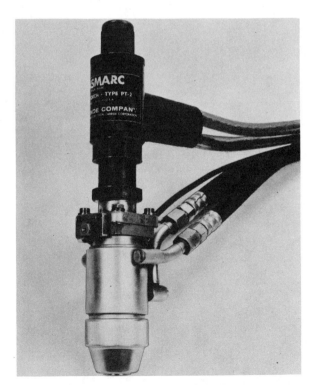

Fig. 19-22. A water-cooled plasma-arc torch. (Linde Div., Union Carbide Corp.)

Fig. 19-23. Manual plasma-arc torch (400 ampere) cuts metal up to 2 in. thick. (Thermal Dynamics Corp.)

chine-operated torches are usually designed with a gear and rack adjustment similar to a oxyacetylene machine cutting torch as shown in Fig. 19-24. A 200 KW plasma torch designed for mounting on a motorized carriage is shown in Fig. 19-25.

Common shielding gases used for gas tungsten arc cutting are nitrogen and

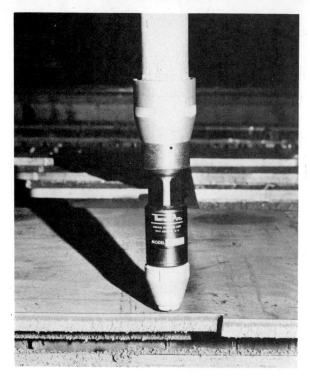

Fig. 19-24. Machine mounted 400 ampere plasma-arc cutting torch for use on automatic cutting machine. (Thermal Dynamics Corp.)

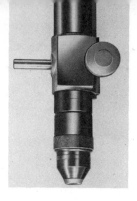

Fig. 19-25. A 200 KW plasma-arc cutting torch designed for use on an automatic cutting machine. (Thermal Dynamics Corp.)

varying mixtures of argon and hydrogen. Air has also been used as the plasma gas with acceptable results.

19-20. PLASMA-ARC CUTTING PROCEDURE

The first step required in making a plasma-arc cut is to adjust the power supply and gas settings. The cooling water must also be turned on, as a water flow safety device will prevent the starter arc circuit from functioning until the water is flowing.

To start the arc, the operator must press the starter button or trigger. The control box will then create a high frequency current to start the pilot arc. At the same time the plasma gas begins to flow and the DC power supply is turned on. After the plasma arc is established through the ionized starter arc, the starter arc is turned off, and the cut may proceed as shown in the photograph, Fig. 19-26.

In the transfer arc process, the arc will go out when the cut is completed because there is no longer a complete circuit through the base metal. When the arc goes out, the gas will also stop flowing and if an automatic carriage is being used, it will stop also, due to the wiring of the circuits. In the nontransfer process, the operating switch must be opened when the end of the cut is reached.

Square edge kerfs are possible with this process, but at a slight decrease in cutting speeds. Fig. 19-27 shows a machine-operated plasma arc torch in action. At high-speed settings both sides of the kerf will be beveled.

Fig. 19-26. Complete plasma-arc cutting system used for manual cutting 6 in. diameter flanges from 3/16 in. type 304 stainless steel. (Thermal Dynamics Corp.)

Fig. 19-27. A 200 KW plasma-arc torch being used to cut a metal plate.
(Thermal Dynamics Corp.)

Plasma-arc cutting may be done in any position, and it will cut virtually any metal. This ability makes it a very useful and versatile process. Fig. 19-28 shows the quality of a cut obtained with a compressed air plasma arc torch.

Fig. 19-28. Mild steel plate 1-in. thick, cut with a plasma-arc torch using compressed air as the gas.

19-21. OXYGEN-PROPANE CUTTING

Propane may also be used as the fuel gas when oxygen cutting. Propane

pletely burn one volume of propane and 1.7 volumes of acetylene are required to equal the heat of one volume of propane, the propane requires about three times as much oxygen as acetylene does to produce the same quantity of heat.

Equipment used for oxy-propane gas cutting is slightly different from that used for the oxyacetylene method as shown in Fig. 19-29. Regulators require special tank attachment fittings and torch tips vary in design from the oxyacetylene cutting tips. Oxy-propane gas cutting is competitive in cost to oxyacetylene gas cutting in areas where

Fig. 19-29. Oxygen-propane gas cutting station. (Phillips Petroleum Co.)

is obtained in liquid form and is available in 20 and 100 lb. net weight capacity cylinders. Propane gas has 2,520 Btu's of heat per cu. ft. as compared to acetylene with 1,475 Btu's per cu. ft. This heat-producing difference means that 1.7 cu. ft. of acetylene must be burned to equal the amount of heat produced by burning one cu. ft. of propane. The oxyacetylene flame produces a flame temperature of 5,900 deg. F. and oxy-propane produces a flame temperature of 5,300 deg. F. Since five volumes of oxygen are required to com-

oxygen is produced cheaply, or acetylene is relatively expensive.

Fig. P-2D shows the colors of oxygen-propane cutting flames.

19-22. LASER BEAM CUTTING

CHAPTER 17 describes the fundamental principles of the laser beam. This beam can also be used for cutting and piercing metals. The advantage of the laser beam is that it can be operated to cut or pierce metals in locations in a metal structure very difficult to reach by any other method. It also performs the operation with a minimum of heat distortion to the base metal, due to the rapid rate at which it operates.

19-23. REVIEW OF SAFETY IN SPECIAL CUTTING PROCESSES

In this chapter, one is dealing with very high temperatures, high gas pressures, high currents, and high voltages.

The operator must be particularly alert for the types of hazards which these conditions may produce.

Burns, flying sparks and electrical shock are ever-present hazards.

Reread the review of safety for CHAPTERS 3 and 9. In addition, be sure to observe the following precautions.

1. Clothing, face, eye, and hand protection must be provided. The illustration of oxygen lance cutting shows an example of an operator being well protected for cutting operation.

2. Due to the rather high oxygen pressures carried in cutting hoses, be sure all hoses used are in good condition and that all fittings are tight.

3. Avoid cutting on tanks, cylinders or other containers which may have, at some time, held flammable materials.

4. If it becomes necessary to cut into such a container with any kind of a flame or arc cutting device, first prepare the container as follows:

   A. Using live steam under pressure, continuously steam out the container for at least a half hour until the entire container is heated to the boiling temperature before beginning the cutting operation.

   B. If it is an edge or dome that is to be cut, fill the container with water up to the point where the cut is to be made.

   C. In some cases after washing out with live steam, the operator may wish to flow argon or nitrogen through the container for some time before the cut is to be made in order to replace all of the air in the container with this inert gas. Argon is preferred to helium for this purpose as it is heavier and will not flow away as readily as helium. Never gas or arc weld or cut on a container which may have an air or oxygen-fuel gas mixture in it, as it may explode!

   D. Do not use carbon tetrachloride for displacing air in a container to be cut or welded, as the fumes from this substance are very toxic.

5. Be sure that the area in which all cutting operations are performed is well ventilated. There will be both minute metal particles and metal oxides produced in the cutting operations.

6. When performing underwater cutting operations, be sure approved diving equipment is provided and, if possible, provide a two-way telephone communicating system between the underwater operator and the surface staff assisting him.

7. With most cutting operations there is considerable sparking and throwing of molten slag. Therefore, the area in which the cutting is performed must be of fireproof construction.

8. Always have an approved fire extinguisher at hand when performing cutting operations.

9. If performing heavy oxyacetylene cutting operations for a considerable length of time, be sure to manifold

two or more acetylene cylinders together as, particularly in cold weather, the acetylene cylinders cannot provide a heavy flow of acetylene over a long time because of low vapor pressures.

10. Due to the great amount of splatter when flame and arc cutting, be sure the eyes are well protected with goggles or a helmet, even though rays from the flame or arc may not seem to require such protection.

11. Never use oxygen with metal powder when powder cutting. Oxygen should not be used for pressurizing any operating or test procedure. Use carbon dioxide, nitrogen, helium or argon. Use a pressure relief valve adjusted to the maximum pressure to be used.

19-24. TEST YOUR KNOWLEDGE

1. Will metal burn?

2. Is the oxygen lance consumed as it is used for cutting?

3. Are regular cutting torches used in combination with the oxygen lance?

4. Of what use is compressed air during underwater oxy-fuel gas cutting?

5. Describe powder cutting.

6. How is the flux introduced into the kerf when flux cutting?

7. How deep under water may one safely perform oxyacetylene cutting?

8. May one cut under water with an arc-oxygen torch?

9. What type electrode is used in arc-oxygen cutting?

10. How accurately can one cut with an automatic machine?

11. What is plasma?

12. What purpose does iron powder serve in a powder-cutting operation?

13. How is the oxyacetylene flame ignited under water?

14. What is the purpose of the tubular underwater electrode?

15. Is hydrogen usually used as a fuel gas for underwater cutting operations?

16. What is the advantage of using a shielding gas for cutting purposes?

17. Describe the transfer arc plasma torch principle.

18. Can compressed air be used in a plasma arc torch?

19. Can concrete be cut with a gas flame cutting torch?

20. What is the temperature of the plasma arc torch flame?

Chapter 20
METAL SURFACING

Metal surfacing is a special application of cohesion and adhesion which applies a thin coat of a different material to the surface of the base metal. The new material may be a metal, a metal alloy, a ceramic, or a plastic.

If, when the metal is applied, the base metal (or its surface) does not melt, the process is called plating. This is a type of adhesive bonding. Some examples of the process are galvanizing, tin plating and silver plating.

When metal is applied and the base metal surface melts, the method is known as cohesion. Many hard surfacing alloys are added to base metal using this method.

Many of today's metal surfacing materials are applied by thermal spraying. This, of course, is one of the cohesion methods. Thermal spraying includes electric arc spraying, flame spraying, and plasma spraying.

20-1. METAL SURFACING

The process of metal surfacing may be defined as the process by which a layer of metal is bonded or fused to the surface of a base material, usually metal. When the applied metal is a hard metal, the process is called hard surfacing.

In the 1930's the railroads began surfacing areas on their equipment which received abrasive wear and an attempt to decrease the "down-time" on the equipment. Concrete, stone, and many synthetic material surfaces have been metal sprayed to give them a better appearance, also physical and chemical protection.

Today, since new equipment costs are constantly rising, more and more industries are looking with favor on metal surfacing processes to extend the service life of their equipment.

20-2. METAL SURFACING PRINCIPLES

By applying a layer of the correct alloy to the surface of a piece of equipment, the part may be made more resistant to corrosion by chemicals, to wear and abrasion from contact with abrasive materials, to cracks or breakage due to shock loads.

Most steels and steel alloys may be hard surfaced with the exception of some high vanadium steels and high-speed tool steels.

Some of the advantages of surfacing a part with a desired metal alloy are:

1. Certain dimensions may be maintained under adverse conditions of abrasion, corrosion, or impact shocks.

2. The service life of a part may be greatly increased.

3. Production costs may be lowered since less expensive low alloy materials may be used and only the areas

of high wear need be coated with the more expensive alloy required to stand up under the service conditions. In corrosive conditions of chemical production, low alloy steel surfaces which contact the corrosive gases or liquids may be surfaced with stainless steel to resist the corrosive action. The resulting cost per square foot would be appreciably lower than if 100 percent stainless steel was used.

4. Costs would be lowered also, since fewer replacement parts would need to be carried in stock because of the increased service life of each part.

20-3. DETERMINING JOB REQUIREMENTS

Many surfacing treatments may improve the surface's resistance to corrosion, impact breaks, or abrasive wear, but no single surface treatment will give the maximum resistance to all of these types of deterioration at the same time.

When consideration is given to the surfacing of a particular part, four factors should be considered:

1. The nature and cause of the wear problem (loss of surface material).

2. The metal surfacing material needed to reduce this wearing condition.

3. The metal surfacing process which will most economically apply the selected surfacing material.

4. The proper technique to be used for depositing the surfacing material.

20-4. THE NATURE OF WEAR PROBLEMS

As mentioned previously, wear, or the deterioration of metals may be caused by chemical action. The gradual consumption of the metal may be by external rusting or corrosion, or may be by chemical etching of the parts (pipes or tanks) carrying certain chemicals.

It is possible to combat chemical wear by putting a layer of material on the surface which will resist oxidation and chemical etching. Such materials as nickel, stainless steel, lead, and zinc are possibilities depending on the chemical process involved.

Parts may also wear due to constant impacts with hard materials. Road grader blades, tool and die surfaces, power shovel buckets, and engine valve faces are examples of impact wear which change the original dimensions of a part by pounding it out of shape or through breakage.

To reduce impact wear or breakage, a material which has greater toughness should be used. Harder surfacing material will prevent deformity due to pounding on a softer surface.

Another cause of wear is contact with abrasive materials. Bull dozer blades, rock crushing rollers and housings, and rock quarry conveyors are examples of abrasive wear.

Hard materials such as tungsten carbide or ceramics (chromium and aluminum oxide compounds) may be used to reduce abrasive wear.

20-5. HARDNESS DETERMINATION

Various characteristics of metals and their importance are explained in CHAPTER 24.

In the metal surfacing field, the property usually given the most attention is hardness. By making a part from a low carbon steel alloy, the part will remain tough and will resist shock load which might otherwise cause it to fracture, but being tough it will normally be comparatively soft and subject to wear. To remedy this weakness the wear surfaces may be coated with a harder material such as tungsten carbide.

Hardness is the property of a metal to resist the action of cutting tools (or materials). It is important that the hardness of the surfacing material be known in order to insure the requirements of a particular job. The three most common methods of measuring hardness are the Brinell scale, the Rockwell scale and the Shore Scleroscope. Test procedures for these methods are given in CHAPTER 22.

A reasonably accurate test for hardness may be conducted by using a mill file, if testing machines are not available:

1. If metal is removed easily--100 Brinell; 60 Rockwell B; for example, low carbon steel.

2. If metal is readily cut with moderate pressure exerted--200 Brinell; 15 Rockwell C; for example, medium carbon steel.

3. If metal is difficult to cut, though possible--300 Brinell; 30 Rockwell C; for example, high alloy steel.

4. If metal is cut with great pressure only--400 Brinell; 40 Rockwell C; for example, tool steel.

5. If metal is nearly impossible to cut--500 Brinell; 50 Rockwell C; for example, tool steel.

6. If metal cannot be cut--600 Brinell; 60 Rockwell C: for example, hardened tool steel.

20-6. SELECTION OF HARD FACING MATERIAL

Hard surfacing materials may be grouped into three general classifications: Ferrous Alloys, Nonferrous Alloys, and Diamond Substitutes.

The ferrous alloy rods and electrodes have an iron base and are alloyed with chrome, manganese, nickel, molybde-

num, silicon, boron, and zirconium. The low alloy ferrous materials may contain less than 20 percent of the alloying element and the high alloy materials may contain between 20 and 50 percent of the alloying elements. Chromium, tungsten, molybdenum, and cobalt are alloying elements in nonferrous rods and electrodes. The so called nonferrous rods may contain small quantities of iron.

Diamond substitutes consist of carbides of tungsten, boron, aluminum oxides, tantalum, titanium, and borides of chromium. Hard surfacing materials in this classification are made in rod and powdered form as well as in insert form. The rod form may have the surfacing material contained within a hollow alloy steel as shown in Fig. 20-1,

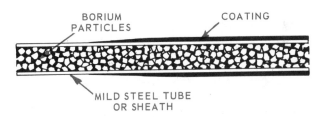

Fig. 20-1. Composite structure of a hard-surfacing electrode.

or the material may be uniformly distributed in a cast rod form. Powdered forms of diamond substitutes may be in the form of minutely crushed particles or particles suspended in a paste. In either case the material is placed on the base metal and it sinks into the metal as the base metal is puddled.

Inserts may be diamond substitutes cast into a desired shape. These are placed in position and brazed or welded to the base metal.

When selecting a hard surfacing material, consideration should be given to the ability to resurface it later. Some molybdenum, cobalt, and austenisticmanganese steel electrode applications cannot be resurfaced later because the original deposit may crack.

20-7. PROCESSES USED TO APPLY SURFACING MATERIALS

The following welding processes may be used to apply surfacing materials:

1. Gas welding - oxyacetylene is the most popular.
2. Inert gas tungsten arc welding.
3. Atomic-hydrogen arc welding.
4. Manual metal-arc welding.
5. Metal spraying.
6. Plasma arc welding.

20-8. SELECTION OF HARD SURFACING PROCESS

The oxyacetylene process for depositing hard surfacing material has many desirable features:

1. The oxyacetylene process is used where a surface is to be applied which will require a minimum of final surface finishing.
2. The carburizing or reducing flame, normally used to apply hard surfacing material, adds carbon to the surface and improves the abrasive resistance of the metal.
3. Preheating and postheating may have to be carefully controlled with the oxyacetylene flame to prevent cracking of the surfacing material.
4. Nonferrous metals are more easily applied by this method.
5. Oxyacetylene is well adapted to fusing hard surfacing materials.

When a flawless surface deposit is required, the inert gas tungsten arc process is frequently used. However the metal being hard surfaced must be very clean. This requires additional equipment and labor costs. In addition, the rate of deposit is comparatively slow. But because of the high quality of the deposit, the inert process is being used more and more in the aircraft and rocket fields for surfacing new parts subjected to some form of extreme wear.

The atomic hydrogen process may be used whenever a thin layer of surfacing material is to be placed on thick pieces. One application is in building up larger dies which become broken or require changes.

The manual metal arc process is often used. Larger volumes of surfacing material may be deposited. This process has these advantages:

1. The equipment is highly portable and is common to most welding shops.
2. Surfacing materials may be laid on more easily in various positions and locations due to the great variety of electrodes available.
3. Great varieties of surfacing materials may be laid because of the large variety of electrodes available.
4. Austenitic stainless steels, austenitic manganese steels, and nickel-chromium-ferritic steels may be surfaced with the metallic arc process.
5. All factors taken into account, most types of normal hard surfacing may be done economically with the metallic arc process.

Automatic and semiautomatic processes normally will give a higher quality of surfacing deposit, but the processes are not as portable as the manual process, nor are they as adaptable to position welding. Automatic and semiautomatic processes are best used in the down-hand position. The advantages of speed and accuracy of the deposit are best utilized in hard surfacing of long straight or curved surfaces in new production, or for repairs when parts can be dismantled and brought to the automatic machines.

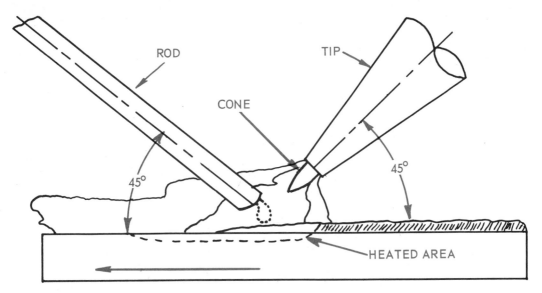

Fig. 20-2. Hard-surfacing material being applied with the use of an oxyacetylene torch. Note that the puddle of molten metal is larger than the puddle used when making a weld. The outer flame area preheats the base metal in advance of the deposited surfacing metal. A carburizing flame is used when hard-surfacing.

In the figure, labels read: ROD, TIP, CONE, 45°, 45°, HEATED AREA.

Metal spraying may be used to deposit almost all virgin metals and many alloys to other metal surfaces. The deposit is usually thin, and the thickness may be held to close tolerances. The surfaces being treated must be cleaned and slightly roughened. Sandblasting may be used for this purpose. In some instances multiple passes may be made to build up a surface with hard-surfacing material without danger of cracking the previous layers.

Virtually any material may be bonded to any other with the plasma-arc process. Ceramics may be sprayed on the surfaces of metals to increase their resistance to corrosion from heat and/or chemicals. Stainless steel may be sprayed on low alloy steel to increase resistance to corrosion and abrasive wear.

20-9. OXYACETYLENE SURFACING PROCESS

To obtain the best results, the surface should be cleaned and the metal preheated to eliminate warpage. A tip, one to two sizes larger than used with the same rod when welding, should be used when hard surfacing. A slightly reducing (carburizing) flame is desired since any carbon added to the surface will aid in the hard surfacing process and the reducing flame will reduce the oxides on the surface of the work. The angles for the torch and rod are the same as for welding, as shown in Fig. 20-2. The metal should be brought up to its sweating temperature, approximately white heat or 2200 deg. F., before the rod is brought down to the work as when brazing. This step is shown in Fig. 20-3. The preheat temperature is important and it is advisable to use a temperature indicating crayon, or some other means for determining the temperature. The process of hard surfacing is very similar to brazing as it applies to the welder's techniques and manipulation. Fig. 20-4 shows steel being hard surfaced using the forehand process, while Fig. 20-5 shows the backhand technique.

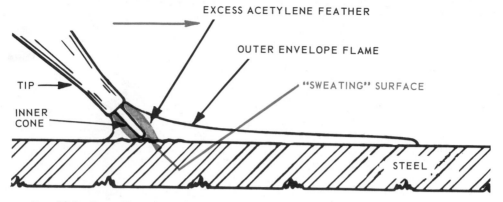

Fig. 20-3. Type flame and tip position to use when producing a "sweating" condition prior to hard-surfacing. (Haynes Stellite Co.)

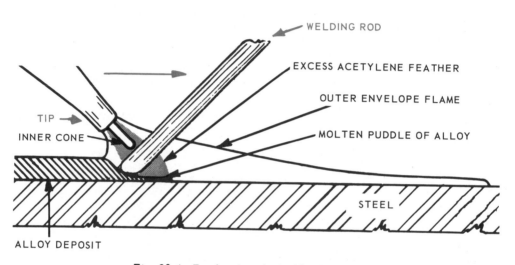

Fig. 20-4. Forehand method of hard-surfacing.

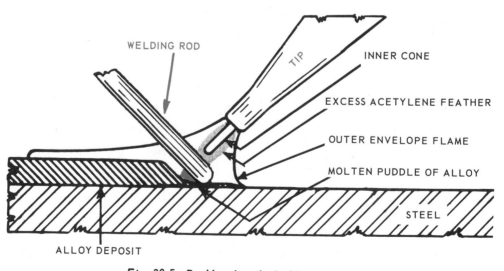

Fig. 20-5. Backhand method of hard-surfacing.

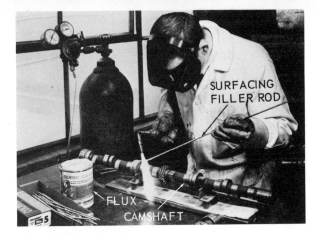

Fig. 20-6. Hard-surfacing engine cam shaft.
(Wall Colmonoy Corp.)

The techniques used for holding the torch and the hard surfacing rod are shown in Figs. 20-6 and 20-7.

When hard surfacing cast iron, the metal will not sweat as it does with steel, and the surfacing material will not flow as easily. To obtain a good job, the rod must be used to break through the surface crust of the iron by rubbing the heated surface with the rod.

20-10. MANUAL METAL ARC SURFACING PROCESS

Direct current is normally used with reverse polarity and with a higher current setting than is normal for the rod diameter being used. The high current will permit a long arc which will preheat the metal ahead of the arc, and keep the surfacing material molten long enough for the impurities to float to the surface. An angle of about 45 deg. is correct for the electrode. This position provides postheating of the deposited areas behind the arc. A long arc is necessary to spread the heat over a large area and thus prevent deep penetration, localized heating and warpage.

Preheating is not necessary before applying surfacing materials to all metals, but with alloy steels it is advisable.

Fig. 20-7. Helix type conveyer being rebuilt using hard-surfacing rod and oxyacetylene flame.
(Wall Colmonoy Corp.)

When in doubt about preheating the following guide rules may be helpful:

1. Except for heavy sections, low and mild carbon steels do not require preheating.

2. Preheat high carbon and medium carbon high alloy steels.

3. 12-14 percent manganese steel parts should be preheated to 200 deg. F. to relieve stresses. However, the part temperature should never exceed 500 deg. F.

4. Cast iron should be preheated to 500-700 deg. F. at a slow and uniform rate.

5. When in doubt preheat to 500-700 deg. F., but be certain the part being preheated is not a manganese steel.

If narrow beads are desired, an oscillating motion should be used in the direction of travel with no sideways motion. A good technique is to lay a bead about 1-in. long and then start the next stroke of the same length 3/4 in. back on the first stroke and extend it 1/4 in. past the first stroke. The third stroke should start 3/4 in. back on the second stroke and extend 1/4 in. beyond the second stroke. This procedure is continuously repeated until the desired length bead is laid.

Wider beads require a circular mo-

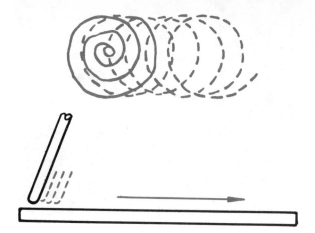

Fig. 20-8. *Circular motion being used to obtain bead widths from 3/4 to 1-1/4 in.*

until the desired width of bead is obtained. The operator then continues to make circular motions and extend each circle about 1/4 in. beyond the last circle in the direction of the desired path each time a circular motion is made as shown in Fig. 20-8.

The recommended height of a deposit is 1/16 to 1/4 in. and the recommended bead width is 3/4 to 1-1/4 in.

When laying a second bead, the first and second bead should overlap 1/4 to 1/3 the width of the first bead. This process will insure a fairly uniform surface to the deposited metal as shown in Fig. 20-9.

If more than one layer of surfacing

tion. A circular motion is started at the center and continued in spirals

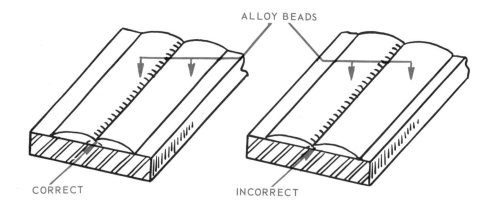

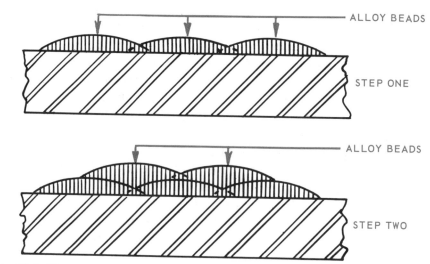

Fig. 20-9. *The proper way to overlap adjacent hard-surfacing beads.*

material is required the beads of the first layer should be cleaned before laying the second layer. This will minimize slag inclusion. Each bead should be made in the same manner as described for laying beads from 3/4 to 1-1/4-in. wide.

An interesting technique for laying hard surfacing beads is to lay the beads in a basket weave pattern. This leaves low spots between the beads. If the part being hard surfaced is used in abrasive material such as sand, the abrasive material will build up in the low spots between the beads and act as an added protection for the base metal as shown in Fig. 20-10.

Fig. 20-11. A lime crusher hammer being hard-surfaced using the DCRP arc process.

Fig. 20-10. Basket weave pattern for laying beads. The abrasive material will build up between the beads and help protect the base metal.
(Stoody Co.)

It should be noted here that small beads made with small diameter rods cool fastest and mix the least with the base metal. Wide beads made with high current and large electrodes cool slowly and mix with the base metal to a greater extent.

The welder must decide how wide a bead, how fast the metal should cool, and how harmful the mixing of base metal with the surfacing material will be to the finished job, before he selects

a rod and procedure for a particular job.

In many situations better results may be obtained by using superimposed layers of different hard surfacing alloys. Fig. 20-11 shows a lime crusher being hard surfaced, after preheat to about 300 deg. F., with an iron base alloy having a nickel and manganese content. After hard surfacing, a layer of iron base alloy containing chromium, molybdenum, silicon and carbon is welded over the first layer.

An application of the twin carbon arc method is shown in Fig. 20-12. The percussion end of the chisel is being repaired by resurfacing.

Fig. 20-12. A twin-carbon arc being used as the source of heat to surface a cold chisel.
(Lincoln Electric Co.)

The automatic and semiautomatic methods of hard surfacing are the same as for welding except that a different type of wire is used. See CHAPTER 21 for automatic welding procedures and equipment.

20-11. METAL SPRAYING PRINCIPLES

The metal spraying process may be used to spray pure or alloyed metals

Fig. 20-13. Metal surfacing a shaft using a metal spray torch which has a built-in wire feed motor. The shaft is being built up to the required diameter.

Fig. 20-14. A metal spray torch with a built-in wire feed mechanism being used to surface a crankshaft journal. The torch is mounted on the lathe cross-feed for better spray control. (Metallizing Co. of America, Inc.)

onto a surface which requires a coating buildup. The sprayed coating may be used "as sprayed," or it may be fused to the base metal by reheating with an oxyacetylene torch, or by using electric induction heating coils.

Worn or inaccurately machined parts may be repaired by having the surfaces built up to the acceptable dimensional limits by the metal spraying process. In general, this process is workable with most inorganic materials that can be melted without decomposition.

Flame sprayed coatings, for example, can be used to build up scored surfaces on crankshaft journals, so they can be machined to accept standard size main bearing inserts.

Fig 20-13 shows a shaft being metal surfaced, using a wire feed torch held in the operator's hand. A similar operation, with the torch attached to the lathe cross feed mechanism (which insures even metal spraying) is shown in Fig. 20-14. Occasionally, in original production, surfaces are metal sprayed with a harder alloy material to reduce wear on these surfaces. Fig. 20-15

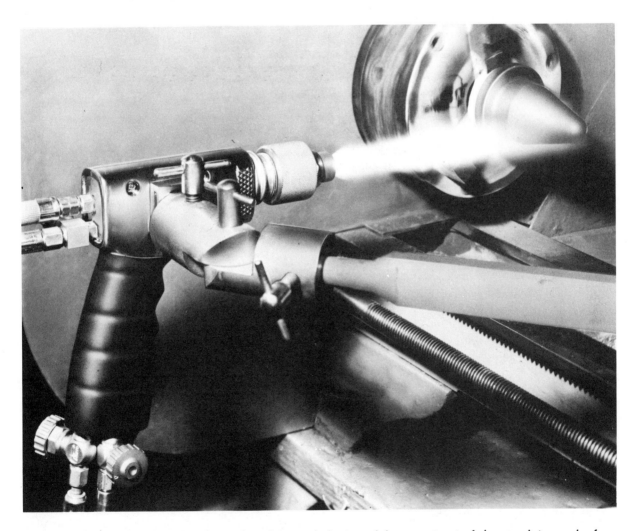

Fig. 20-15. Metal surfacing a part as it revolves in lathe. Metal for spraying is fed to torch in powder form. Air pressure feeds powder metal to torch. This metal surfacing process is followed up with an oxyacetylene flame fusing operation. (Wall Colmonoy Corp.)

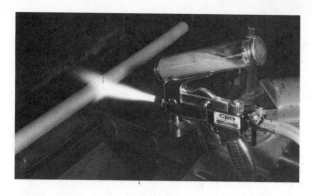

Fig. 20-16. Metal *spraying a shaft. The powder is gravity fed to the torch from the container mounted on the torch. This torch is equipped with an electric vibrator to improve powder flow to the torch.*

shows a part being metal sprayed with a torch which is supplied with air pressure fed metal powder. Another method used to supply the powdered metal for metal spraying is shown in Fig. 20-16. The powder is fed to the flame by gravity from the hopper mounted on top of the torch. Some close fitting parts of jet engines, for instance, require airtight seals which run at extremely high temperatures. The contacting surfaces of these parts are metal sprayed with a soft metal. As the rotating parts expand

the parts to be joined (capillary flow) is difficult to obtain. To insure that the brazing material will cover the entire surface to be joined, the surfaces may be first metal sprayed with brazing metal, assembled, and then heated in an oven to complete the brazing operation.

Metal spraying may also be used to build up a wear or corrosive resistant surface on a part subjected to much wear. The surface produced by metal spraying normally requires the least finishing of all the surfacing processes. Metal spraying is often used to refinish worn shaft bearing journals.

20-12. METAL SPRAYING

Metal spraying is a process in which metal, in a wire or powdered form, is fed through a special torch (spray gun), where it is melted in an oxygen-fuel gas flame and atomized by high-pressure gases which carry the atomized particles to the cleaned and prepared surface. Fig. 20-17 illustrates a metal

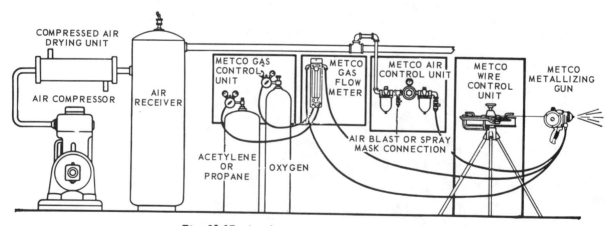

Fig. 20-17. A schematic of the equipment used in a typical wire feed metal spray outfit.

and contract, the soft metals contract and form to the rotating parts to produce the required airtight seal.

In the production brazing of certain alloys, flow of brazing material between

spraying outfit used with wire feed spray metal.

The equipment necessary for wire feed metal spraying is:

A. Air Compressor--Capacity of at

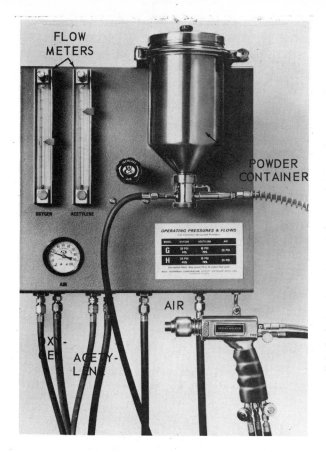

Fig. 20-18. An outfit for spraying powdered metals. The metal powder is fed from the hopper to the torch under air pressure. (Wall Colmonoy Corp.)

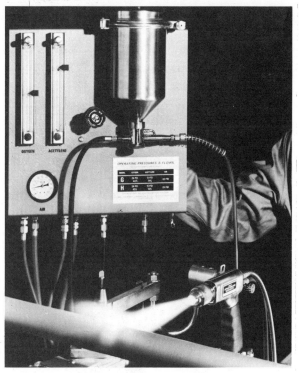

Fig. 20-19. Powdered metal is applied by spraying it on a revolving shaft. (Wall Colmonoy Corp.)

least 30 cu. ft. per minute of free air required.

B. Compressed Air Drying Unit--If there is an excess amount of moisture in the air supply, it is generally advisable to use an air drying unit, since any amount of moisture will interfere with the metallizing bond.

C. Air Receiver Tank -- This unit smooths out the flow of compressed air to the compressed air regulator by compensating for compressor pumping pulsations.

D. Fuel Gas and Oxygen Regulators-- Two-stage regulators recommended.

E. Fuel Gas and Oxygen Cylinders-- Acetylene or propane used as fuel gas.

F. Gas Flow Meters -- Twin tube standard gas flow meters keep constant gas volume flow to gun nozzle.

G. Compressed Air Control--A sin-gle-stage regulator may be used to control air pressure. Air filters are normally used to further purify air, particularly if air supply is also used to supply air to operator's mask.

H. Wire Feed Control--This unit will straighten wire as it comes from coil and feeds it at uniform rate and uniform tension to gun nozzle.

I. Metal Spraying Gun--Metal spraying gun mixes fuel gas, oxygen, and the desired alloy wire. Some guns contain mechanisms to draw wire from wire feed control coils.

Some of the equipment needed to spray metal powder is shown in Fig. 20-18. Equipment not shown includes an air compressor, oxygen and acetylene cylinders complete with regulators. The normal air compressor accessories are used to insure dry air and a constant air supply. The torch with its oxygen and acetylene supply system is shown in Fig. 20-19. This outfit operates with 16 psi oxygen and 10 psi acetylene pressure. The air pressure used is usually 35 psi.

20-13. METAL SPRAYING TORCH

The metal spraying torch must feed oxygen, fuel gas, high-pressure air, and the wire or powder to the nozzle in correct amounts in order that the metal spraying process may be successful.

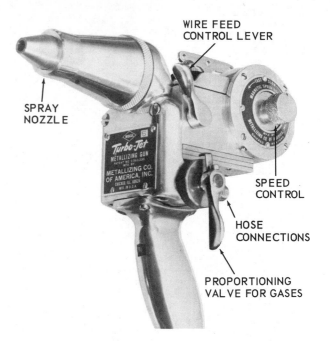

Fig. 20-21. Wire feed metal spray torch. (Metallizing Co. of America, Inc.)

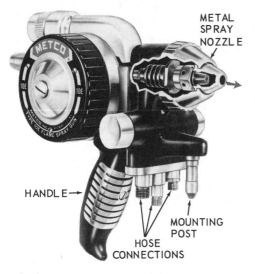

Fig. 20-20. A wire feed metal spray torch. The wire feed is air turbine powered. (Metco, Inc.)

Fig. 20-20 shows a typical metal spraying torch for use with wire surfacing materials.

Fig. 20-21 shows the location and names of the essential parts of a typical wire type metal spraying torch. The basic operation of this torch is shown in Fig. 20-22. The high temperature

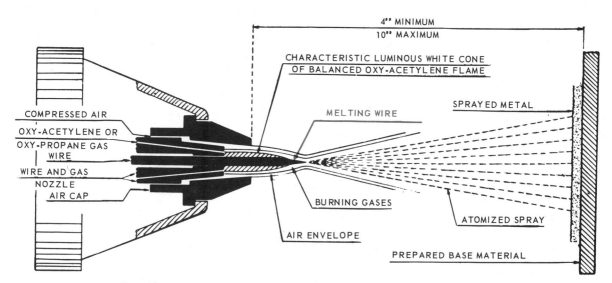

Fig. 20-22. Schematic cross section of a wire feed metal spray torch.

flame completely surrounds the wire as it is fed through the torch tip. The compressed air atomizes the molten wire metal, and sprays this metal against the material to be surfaced. A wire feed type metal spraying torch which uses an electric motor to move the metal spraying wire is shown in Fig. 20-23.

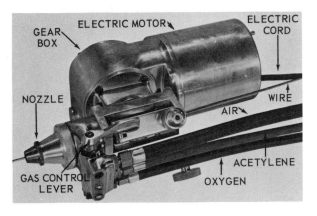

Fig. 20-23. Wire feed metal spray torch with an electric motor driving the wire feed. Air pressure is used to spray the metal.
(Metallizing Co. of America, Inc.)

20-14. STARTING AND ADJUSTING METAL SPRAYING STATION

The safety precautions recommended for use with all gas welding and cutting equipment apply also to the use of the metal spraying torch. Check all hoses, connections and gauges for leaks before using the equipment. The procedures for adjusting and lighting the wire feed type torch are as follows:

1. Open the air valve and adjust the air pressure regulator to pressure as shown in Fig. 20-24.

2. Close the air valve.

3. With the drive roll knob loose, insert the wire into the rear wire guide and through the gun to the nozzle.

4. Tighten the drive roll knob until the wire begins to feed.

5. Adjust the speed control ring which controls the speed of the turbine in the air motor. When the speed control ring is moved clockwise the wire will speed up and when the control ring is turned counterclockwise it will slow down.

6. Close the valve handle.

7. Open the valve handle to the run position and adjust the fuel gas and oxygen pressures to the values shown in Fig. 20-24, and close the valve handle.

20-15. LIGHTING METAL SPRAY TORCH

Generally, the wire feed metal spraying torch is lighted using the following procedure:

1. Open the valve handle and wait three seconds.

2. Close the valve handle to the 45 deg. position until a click is heard. This is the lighting position. Light the gases at the nozzle with a spark lighter and IMMEDIATELY open the valve

METAL	WIRE SIZE INCHES	AIR CAP SIZE	AIR PRESS PSIG	LIGHTING PRESSURE OXYGEN PSIG	LIGHTING PRESSURE ACETYLENE PSIG	FLOWMETER READINGS OXYGEN CU. FT./HR.	FLOWMETER READINGS ACETYLENE CU. FT./HR.	SPEED SQ. FT./HR. .001 IN. AVERAGE THICKNESS
Metco Aluminum	1/8	J	55	36	15	32	32	676.8
Metco Copper	1/8	H	55	35	15	32	32	349.0
Metco Nickel	1/8	H	55	36	15	32	31	225.2
Spraybond Wire	1/8	H	50	38	15	33	33	136.6
Metco-Weld H	1/8	H	40	38	15	35	37	93.5

Fig. 20-24. Table of metal spraying data as applied to a Metco 4E Metal Spraying Torch.

handle to the running (vertically down) position.

3. After lighting the torch, adjust the fuel gas and oxygen flow regulator according to Fig. 20-24.

4. Set the speed ring to obtain the highest wire speed which will still allow the wire to melt off to a certain desired point without spattering. The wire should usually extend about 1/2 in. in front of the air cap; however, this distance will vary with the type of coating and diameter of the metal being sprayed.

20-16. SURFACE PREPARATION

Aluminum and magnesium castings which have been in contact with oil or grease must be chemically cleaned to remove the oil which penetrates into the metal, and then grit blasted to further clean and roughen the the surface. When grit blasting, the moisture in the compressed air lines must be held to an absolute minimum, since any amount of moisture on the surface will reduce the bonding strength of the sprayed metal. Fig. 20-25 shows a gear being sandblasted prior to a metal spraying operation.

Other methods of surface preparations are (a) rough threading, (b) grooving, (c) electric, and (d) metallic spray bonding.

Fig. 20-25. Preparing to clean a gear by sand blasting. A special enclosed chamber is used.

On cylindrical surfaces which need to be built up, the worn area should be turned on a lathe to provide a space to deposit the built up metal before any of the preparations on the surface are used as shown in Fig. 20-26. Shafts

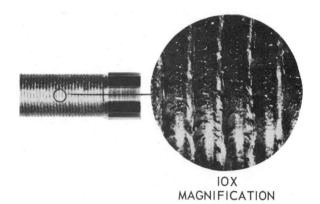

IOX
MAGNIFICATION

Fig. 20-26. Appearance of steel stock after machine cutting the surface prior to metal spraying. The 10-X magnification shows the roughness needed for good spraying results.

may be rough threaded after being turned down to provide a surface rough enough to insure a good bond between the shaft and the metal spray material.

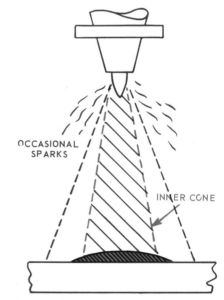

OCCASIONAL SPARKS

INNER CONE

Fig. 20-27. Schematic showing that approximately 95 percent of the sprayed metal is in the inner cone, and is deposited on the surface being sprayed.

Fig. 20-28. An automated metal spraying operation.
The electrically powered wire feed is electronically
controlled. (Metco, Inc.)

For heavy buildups of hard surfacing materials, the surface of the metal should be grooved with a chamfering electrode or carbon arc electrode. The recommended size of the grooves is 1/4-in. deep and 5/16-in. wide. The grooves should never be connected so that each groove may more effectivly break up stress lines which may develop in the surfacing material.

When a surface is too hard to roughen by grit blasting or machining, electric bonding may be employed. In this method, a nickel electrode is normally used to fuse small irregular particles to the surface which is to be metallized.

A layer of molybdenum alloy may be sprayed uniformly onto the surface to produce a surface rough enough to receive a metallizing spray. This method of surface preparation is called metallic spray bonding.

Metallic spray bonding may be used on hard or soft surfaces, but should not be used on copper or copper alloys.

20-17. SPRAYING

When spraying, the gun cap (end of the spraying nozzle) should be held from 5 to 8 in. from the surface being surfaced. The gun should be held perpendicular to the surface being sprayed, to insure an even coating.

In the case of thin parts or small diameter shafts, the parts may have to be cooled by an air stream to prevent overheating. The width of the spray cone varies with the wire diameter, the type of metal being sprayed, and the spraying speed (wire feed rate). The spray cone which is visible is considerably larger than the actual cone of spray metal due to the sparks which travel around the actual cone of metal.

About 95 percent of the metal being sprayed is in a smaller inner cone not visible when spraying, as indicated in Fig. 20-27.

Different types of nozzles are used for acetylene and propane gases and these should not be interchanged. If the propane nozzle, with the cupped end, is used with acetylene gas, back-firing may occur.

A completely automated flame spraying installation in operation is shown in Fig. 20-28. The torch is electronically controlled.

Another process used to hard surface metals is by means of a powdered surfacing material carried to the work by an inert gas, compressed air, or gravity. The powder in the pressure feed units is carried to the torch by a third hose from a hopper where the powdered material is stored under pressure. Fig. 20-29 shows a gravity feed powder metal spray gun. When this torch is used with powdered ceramics, an electric vibrator is used to help the ceramic powder flow. The powder metal type of torch is shown in action in Fig. 20-30.

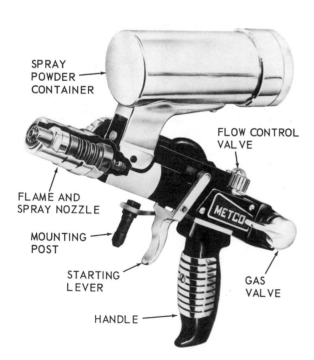

SPRAY POWDER CONTAINER

FLOW CONTROL VALVE

FLAME AND SPRAY NOZZLE

MOUNTING POST

STARTING LEVER

GAS VALVE

HANDLE

METCO

Fig. 20-29. Gravity-feed metal spraying torch. The metal to be sprayed is in the powder form and supplied in the container mounted on top of the torch.

Fig. 20-30. Gravity feed metal spraying torch in operation.

This process is used for bond surfacing, ceramic surfacing and other metal surfacing. One advantage of this type torch is that the spray powders may contain fluxes when necessary. The surface of the metal being hard surfaced is sometimes preheated by using an oxyacetylene flame.

An interesting application of flame spray surfacing is using a metal surfacing operation as a base surface for a protective coating of some type of sealer. In those situations where the sealer would ordinarily have difficulty adhering to the metal it is to protect, an intermediate metal spray operation enables the protective sealer to adhere for much greater periods of time. Fig. 20-31 shows a magnified cross section of such an intermediate application.

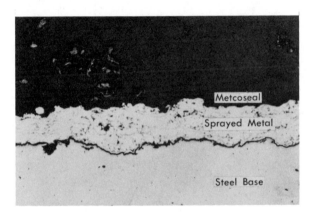

Fig. 20-31. Macrograph of a steel surface, showing the metal sprayed on to the base metal surface to form a base for a protective sealing surface.

20-18. PLASMA ARC PROCESS OF METAL SPRAYING

The plasma arc or jet is the result of forcing gas, such as nitrogen, hydrogen, argon and the like into an enclosed electric arc. The gas is heated to such a high temperature by the electric arc that its molecules become ionized atoms containing a great deal of energy. Plasma is considered by physicists as a fourth state of matter, i. e. neither gas, liquid, or solid. CHAPTER 17 explains the process in detail.

Temperatures of 3000 - 30,000 deg. F. may be obtained with plasma arc equipment. This process can be very successfully used for surfacing by spraying. In general, any inorganic material which will not decompose may be sprayed. Examples of materials being applied to a base metal are: All ferrous metals, ceramics, tungsten, tungsten carbides, tantalum, zirconium diboride, platinum, columbium, hafnium, vanadium carbides and others.

Coating densities are up to 98 percent of theoretical density and the plasma arc application is easy to control. Pure tungsten and tungsten carbides may also be applied to almost any base material with up to 95 percent densities.

A typical plasma arc torch is shown in Fig. 20-32. The principle of the process is:

Using direct current, an arc is formed between the internal electrode and the nozzle and the plasma or gas is ionized within the nozzle. The material to be sprayed is carried to the arc area in the nozzle, by the same gas used for forming the plasma, where the metal is melted and atomized. The atomized surfacing material is then carried to the surface to be sprayed by the high velocity of the plasma jet. Speeds of 20,000 feet per second may

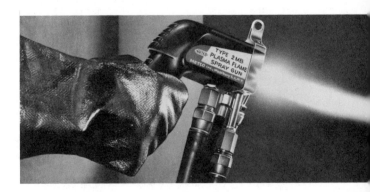

Fig. 20-32. Plasma-arc metal spraying torch in operation.

be reached with the plasma jet, but normally much lower speeds are required. Fig. 20-33 shows a schematic of a plasma arc metal spraying torch.

Nitrogen with from 5 - 10 percent hydrogen gas is generally used for the

A complete plasma arc station would usually include:

1. Plasma-jet spray gun.

2. Control unit with an on-off switch, starter arc controls, gas flow controls, ammeter and voltmeter.

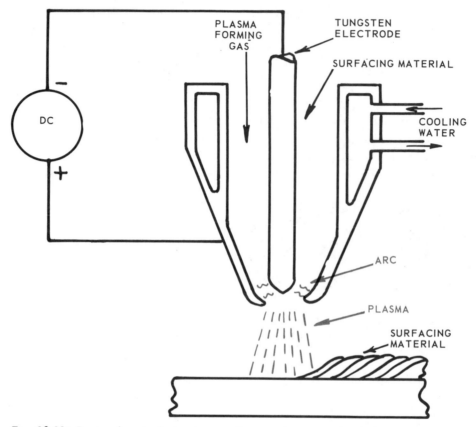

Fig. 20-33. Basic electrical circuit, gas flow, surfacing material flow, and cooling of plasma-arc metal spraying torch, shown schematically.

plasma gas, with pure nitrogen being used as the carrier gas to propel the powdered surfacing material to the melting zone within the nozzle.

The entire plasma jet nozzle is usually water cooled. Distilled water is often used in the cooling circuit to prevent mineral deposits and thereby lengthen the life of the equipment. Fig. 20-34 shows a cross section of a plasma arc metal spraying torch. Fig. 20-35 shows the external view of a similar torch. The cooling circuit, gas circuit, and powder feed are shown.

3. Heat exchanger in which city water is used to cool the distilled water circulating in the gun. Water flow controls are also included in this unit.

4. Powder control unit--This unit controls the pressure and quantity of the powder fed to the gun.

5. Power supply unit--The current supplied to the electrodes is controlled and stabilized by this unit.

6. Gas and water hoses and electrical cables.

7. Safety equipment--Face shield, heat resistant and reflecting clothing,

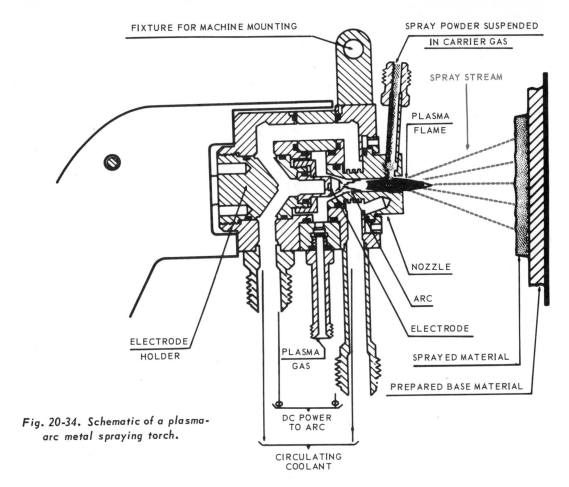

Fig. 20-34. Schematic of a plasma-arc metal spraying torch.

Labels in figure:
FIXTURE FOR MACHINE MOUNTING
SPRAY POWDER SUSPENDED IN CARRIER GAS
SPRAY STREAM
PLASMA FLAME
NOZZLE
ARC
ELECTRODE
SPRAYED MATERIAL
PREPARED BASE MATERIAL
ELECTRODE HOLDER
PLASMA GAS
DC POWER TO ARC
CIRCULATING COOLANT

gloves, and ear plugs to protect ears of the operator from the intense noise caused by the plasma jet.

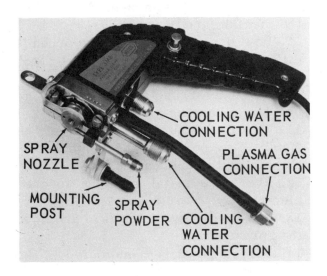

Labels in figure:
SPRAY NOZZLE
MOUNTING POST
SPRAY POWDER
COOLING WATER CONNECTION
PLASMA GAS CONNECTION
COOLING WATER CONNECTION

Fig. 20-35. An external view of plasma-arc metal spraying torch similar to one shown in Fig. 20-34. (Metco, Inc.)

8. Water wash spray booth is used to protect other workers from over spray of both flame and material.

An interesting application of the plasma arc metal spraying process is used when making rocket parts such as nozzles. The parts are exposed to extremes of temperature and erosion wear due to the high velocities and temperatures to which they are exposed. Rocket nozzles are made from zirconium, platinum, or other rare and expensive metals, too rare in fact to be machined to shape where the machining chips would have to be salvaged and remelted.

In the plasma arc process, the desired metal is sprayed into an aluminum or brass form (mandrel) to the desired thickness and then the mandrel is removed from the shell of rare metal by dissolving it with acid.

20-19. ELECTRIC ARC SPRAY SURFACING

Another process used to spray metals is the electric arc spray method. This method consists of maintaining an arc between two metal wires which are automatically fed to the arc position in the spray gun by electric drive mechanisms similar to those used in gas metal-arc welding. The metal melted by the arc is then sprayed onto the surface by using an air jet or an inert gas jet directed across the arc and at the surface to be sprayed.

20-20. FINISHING HARD-SURFACED OBJECTS

Cobalt base alloys may be machined with carbide cutting tools, but the cutting tool angles must be properly ground for the hard-surfacing material.

Grinding is an economical way to finish iron-base hard-surfaced alloys. High cutting speeds must be avoided since they may overheat the surface and cause surface cracks to form.

Nickel based materials should be machined for best results.

Tungsten carbide surfaces must be ground since the surface is too hard to be machined economically.

20-21. REVIEW OF METAL SURFACING SAFETY

All of the safety precautions specified for gas welding, arc welding, welding on containers and the like, also apply to metal surfacing operations.

Excellent ventilation is essential at all times because of the fluxes used, and because of the toxic effects of some of the alloys in the surfacing materials.

Protective clothing should be kept clean and in good condition. Your eyes should be protected at all times.

Ear plugs are necessary when using the plasma arc metal spraying torch as the high plasma velocities create a damaging sound pitch and intensity.

You must be constantly on the alert to avoid spraying metal on flammables, or on a person.

20-22. TEST YOUR KNOWLEDGE

1. List several materials which are used as surfacing material for metals.
2. Describe the type bond between the added material and the base metal.
3. What are the advantages of hard-surfacing versus making the article out of 100 percent hard material?
4. What happens if the base metal is melted too much during a hard-surfacing operation?
5. Is metal spraying a fusion or adhesion process?
6. What hard-surfacing process uses compressed air?
7. Of what material is the surfacing rod made when the oxyacetylene method of hard-surfacing is used?
8. May a metal be hard-surfaced using the electric arc process?
9. How does the plasma arc process differ from the oxygen-fuel gas torch process?
10. May metal surfacing be used to build up a bearing journal?
11. May metal surfacing materials be finished by grinding the surface? By machining the surface?
12. Are some metal spraying torches electrically power driven?
13. How does the compressed air move the spray wire?
14. How is powdered metal fed to the torch flame?
15. May ceramics be sprayed on a surface using the plasma arc torch?

Chapter 21
AUTOMATIC WELDING, CUTTING EQUIPMENT

Today, virtually every industry and business is moving toward automation and cybernetics (mechanical-electrical communications system). The welding industry is converting various hand welding operations to either semi-automatic or fully automatic processes. These changes are aimed at:

1. Increasing production.
2. Improving quality.
3. Lowering costs.

Serious students of welding should become familiar with automatic equipment and controls.

Various control units, power devices, timers, gauges, and instruments are needed in automated processes. Automatic welding has many forms and can be adapted to perform almost any welding task.

One who sets up, adjusts and services the machines is usually a skilled manual welder. In addition, this person also knows metals, common welding processes and most manufacturing processes.

21-1. PRINCIPLES OF AUTOMATIC WELDING

Automatic metallic arc welding has several advantages over the hand welding processes and may be used as an example of the advantages and principles of other forms of automatic welding.

Some advantages of automatic arc welding over manual arc welding are:

1. Low electrode stub loss due to the continuous feed from a reel of welding wire.

2. Relief from the labor of concentrating on the arc length, speed, and other variables which must be controlled for a good quality weld.

3. Much higher currents may be maintained with any given electrode size as compared to hand welding.

4. Weld height, width, fusion and penetration will be uniform once the automatic controls are adjusted correctly.

5. Weld rates are higher in automatic welding than in hand welding. See Fig. 21-1.

The advantages mentioned can only be obtained however, with correctly fitting weldment components. In hand welding, the operator can slow down or speed up his weld to compensate for poor fits and alignments, but in automatic welding, the machine once adjusted, will continue along a seam at a constant rate. This action might cause poor welds in poorly fitted sections of a joint. However, some of the modern automatic welders have solved the arc length problem. As the arc length (voltage) varies the speed of the welding wire feed speed, is changed instantly to compensate, by electronic controls.

Included in the various types of automatic welding equipment are variable speed drive motors, gas controls, coolant fluid controls, timers, current controls, sequence controls, and instruments for testing and inspecting various circuits in these machines. Not all of these components are included on each welder, but most welders have several of these units operating in conjunction with one another to form a more automated unit.

In some automatic operations, when the electric motor is used for the purpose of moving wire, the motor operates through a gear box. Knurled drive wheels press against the wire and move the wire at the desired feet per minute. The type drive motor varies according to the type welding current source. If the source is the drooping voltage type or the rising voltage type,

MANUAL WELDING

METAL THICKNESS	ELECTRODE SIZE	AMPS.	IN. PER MIN.
1/4 BUTT	1/8	150	18
1/2 BUTT	5/32	200	7
3/4 BUTT	1/4	250	3
1/4 FILLET	1/8	150	14
5/16 FILLET	5/32	200	11
1/2 FILLET	1/4	250	5

AUTOMATIC WELDING

METAL THICKNESS	ELECTRODE SIZE	AMPS.	IN. PER MIN.
1/4 BUTT	.040	300	70
1/2 BUTT	.060	400	36
3/4 BUTT	·060	500	22
1/4 FILLET	.040	300	40
5/16 FILLET	.040	350	28
1/2 FILLET	.060	400	15

Fig. 21-1. Comparison of approximate manual and automatic arc welding speeds.

21-2. ADJUSTABLE SPEED DRIVE MOTORS

Variable speed drive motors may be used in automatic equipment to:

1. Move a torch or electrode along the work.

2. Move the work under the torch.

3. Move the electrode at a controlled rate.

Large coils of welding wire are used in automatic welders to enable continuous welding operation. The wire is fed into the weld area by adjustable speed drive motors.

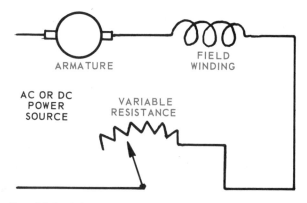

Fig. 21-2. Schematic wiring circuit of an adjustable speed motor. A manually-operated variable resistance, in series with the motor, is used to control the speed. The universal motor may be connected into an AC or DC circuit.

the arc length must be very accurately controlled. With these types of welding current machines, the motor is usually connected into the welding transformer electrical circuit. As a change in arc length occurs due to poor fits or for some other reason, the voltage changes in the welding transformer and a signal is sent to the wire drive motor to speed up or slow down the speed of the wire feed as required.

An arc voltage rise will speed up the motors while an arc voltage decrease will slow them down. Two types of motor speed controls are commonly used:

1. A make and break governor.
2. An electronic tube or transistor control.

In gas metal-arc welding, when a constant potential current source is used, an adjustable speed drive motor is used to feed wire to the electrode holder at a rate adjusted by the operator on the control box. Fig. 21-2 shows a wiring diagram of an adjustable speed motor.

One of the more common variable speed drive motors is the low voltage series wound universal motor. This motor varies its speed directly with the voltage. As the voltage increases, the motor speed increases. When these motors are electrically connected to the arc circuit, as the arc voltage increases (a longer arc) the motor will speed up and feed electrode wire faster. Then, as the arc length decreases, and therefore the arc voltage decreases, the motor speed will decrease and less wire per unit of time will be fed to the weld. The motor, therefore, responds to the arc voltage and will maintain any preset arc voltage desired within the capability of the machine.

The wire feed motors must have controls to enable the operator to manually control the wire feed for loading purposes; also to reverse the wire movement, that is, to inch the wire either forward or backward.

The operator may also control the wire feed to the arc area by using an on-off switch in the holder to stop the wire feed. The operator has full manual control of the torch movement along the joint.

21-3. GAS CONTROLS

Gas controls for gas welding or inert gas arc welding consist of pressure regulators, flowmeters, and gas flow control valves. The pressure regulators used are the same in most respects as those used in oxyacetylene welding. See CHAPTER 2.

The flowmeter is described in CHAPTERS 11 and 12. It is used to measure the quantity of gas needed.

For applications of gas tungsten-arc welding (TIG) or gas metal-arc welding (MIG), a valve in the gas line may be controlled by a solenoid. It is held open by the solenoid while current is flowing in the coil. When the operator pushes the start button or trigger, the solenoid is electrically energized. The magnetism opens the valve and gas begins to flow immediately. The solenoid circuit can be in parallel to the arc circuit. When the arc is struck, current will flow in the secondary windings of the welder, and some current will flow in a parallel circuit to operate a relay. This relay controls the flow of electricity to energize the solenoid. Fig. 21-3 shows a simplified solenoid valve circuit.

When the arc is broken, the gas should continue to flow for a short time to protect the electrode from contamination while it cools. This feature is sometimes provided by a bimetal strip in the solenoid valve. While current is flowing, this bimetal strip is

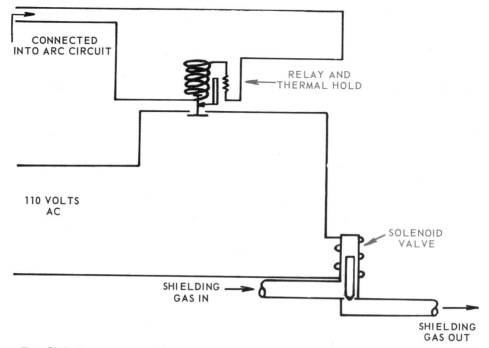

Fig. 21-3. Relay-controlled solenoid valve used to control shielding gas flow to a gas arc welding torch.

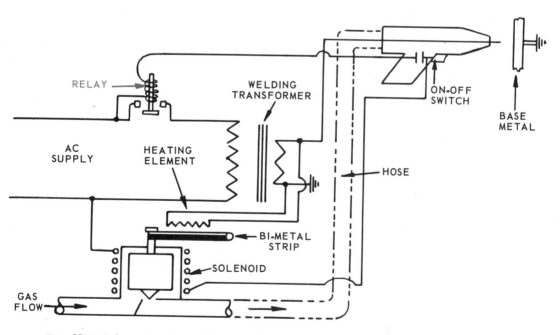

Fig. 21-4. Schematic wiring diagram of a shielding tungsten arc welding outfit. This unit has a relay to regulate the shielding gas flow.

heated, which locks the solenoid valve open. After the arc is broken and current no longer flows in the solenoid or heating coil of the bimetal strip, the solenoid valve is held up or open by the bimetal unit until it cools. As the bimetal strip cools, it will bend to its unlocked position. The solenoid core will drop, shutting off the gas flow as shown in Fig. 21-4. Some devices use an electronic tube to control the solenoid electrical flow. The delayed closing of the solenoid is obtained due to the continued flow through the tube until it cools after the arc shunt current has stopped flowing.

In some less-automated units, a mechanical gas flow control valve is used. The electrode holder is hung on a hook which is connected to a needle valve. When the holder is picked up, the spring-loaded needle valve opens, permitting inert gas to flow. When welding is completed and the holder is replaced on the hook, the gas flow is stopped. The same device may be used to stop and start the flow of the electrode holder cooling water. See CHAPTER 12.

Another gas used frequently in automatic welding operations is compressed air. Air under pressure is used to operate air cylinders which move parts of the machinery, move stock, and/or clamp stock.

Compressed air flow to these devices is controlled by automatic valves. A popular type valve for this purpose is the solenoid operated valve, as shown in Fig. 21-5.

In all fluid piping systems, it is important to know at all times whether the fluid is flowing. This may be done by using a flow indicator. One such device uses a reed inserted in the fluid piping as shown in Fig. 21-6.

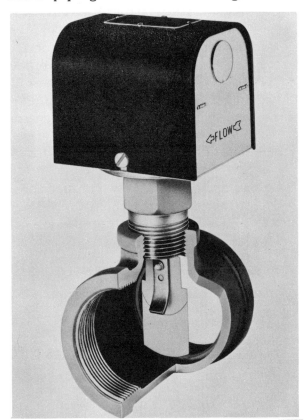

Fig. 21-6. *Fluid flow switch. This switch may be wired to turn on a warning light and-or turn off the electrical power to the welder in case the water flow stops or becomes insufficient for cooling.* (McDonnell and Miller, Inc.)

When the fluid is moving, the reed will bend and its bending movement is electrically connected either to a safety switch or to an indicator.

21-4. WATER CONTROLS

Water or some other coolant is required for cooling transformers, electrodes, and electrode holders of many automatic and semiautomatic welders. Such parts as the transformers and the electrodes in resistance welders, and the electrode holders in gas tungsten-

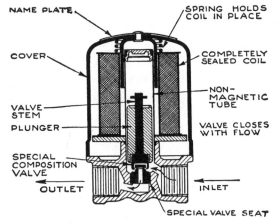

Fig. 21-5. *Solenoid valve used to control shielded gas flow.* (A-P Controls Corp.)

arc and gas metal-arc welders, are frequently water-cooled. A well-controlled supply of coolant reduces the wear and increases the efficiency of these machines.

Solenoid valves, pressure regulators, volume controls, and safety relays are often used to control coolant flow. Pressure regulators are set at a few pounds less than the supply water pressure, since the supply pressure may vary and the machine must have a constant pressure and a constant volume flow to obtain uniform cooling.

Safety relays are electrically connected to the pressure and volume controls, and also to a shut-off switch on the welder. If the coolant is not flowing or is flowing at a rate insufficient for proper cooling, the electrical circuit is opened and the machine is automatically shut off, Fig. 21-7.

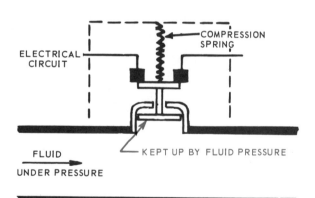

Fig. 21-7. Safety switch for water circuits. If the water pressure decreases, the electrical circuit will open.

In the more complex machines, water flows only when current is flowing in the secondary circuit and welding is being done. This procedure reduces the water flow and therefore reduces the cost of water usage. This control is obtained by wiring the water-flow solenoid valve into the secondary circuit

rather than into the primary circuit. The circuit in this case uses a relay operated by the secondary circuit to open and close a circuit to operate the solenoid valves.

Coolant lines must be cleaned on a scheduled basis to insure proper cooling. Calcium and sulphate deposits may form in the coolant lines and seriously affect the cooling of the machine. Mechanical or chemical cleaning may be used to clean the water lines.

21-5. THREE-WAY VALVES FOR HYDRAULIC OR PNEUMATIC SYSTEMS

Three-way solenoid valves are often used to open and close the operating passages of a hydraulically or pneumatically operated cylinder on an automatic welder. Fig. 21-8 shows a three-way valve.

This valve is normally in the closed position. The passage of hydraulic pressure from port A on the inlet, to port C on the outlet side of the valve, is blocked by the main valve.

When the solenoid is energized, the solenoid pilot valve is moved down. A passage is then completed from port D through port E to the top of the main valve. This pressure on top of the main valve forces it down, thus opening the passage from A to C and to the operating cylinder connected to passage C.

When the solenoid is de-energized, the solenoid pilot valve is moved up by spring pressure blocking the passage from port D to port E. Pressure is relieved from above the main valve by passing through port D, F, G, and to the release port at B. When pressure is relieved from the top of the main valve, it is moved up by pressure from below, thus closing the passage from A to C, and opening the passage from

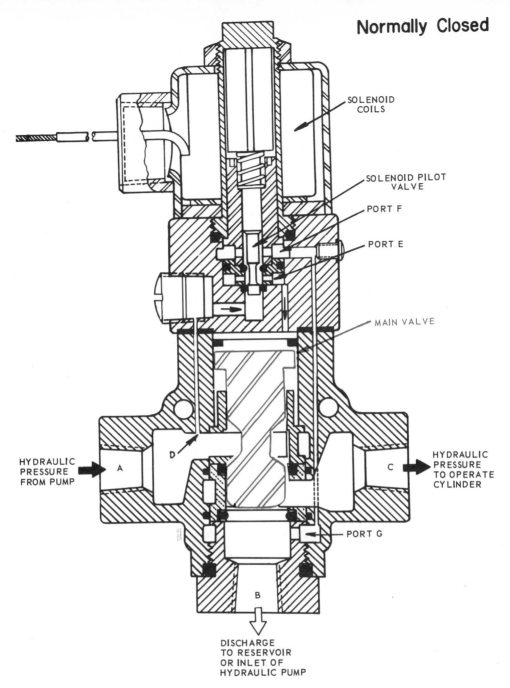

Normally Closed

SOLENOID
COILS

SOLENOID PILOT
VALVE

PORT F

PORT E

MAIN VALVE

HYDRAULIC
PRESSURE
FROM PUMP

A

D

C

HYDRAULIC
PRESSURE
TO OPERATE
CYLINDER

PORT G

B

DISCHARGE
TO RESERVOIR
OR INLET OF
HYDRAULIC PUMP

Fig. 21-8. Three-way solenoid valve for controlling hydraulic fluid flow.
(Airmatic/Beckett-Harcum)

C to B allowing the fluid in the C passages to discharge through opening B.

21-6. TIMERS

Timers for controlling the operation phases of the welding cycle may be of a mechanical nature, such as a clock or cam. They may be of an electronic nature, such as a vacuum tube, gas tube, or transistor.

The clock or chronograph can be used to make electrical contacts at specific time intervals to operate a machine. It is not practical to use this type of timer for time intervals of one second or less.

Cams rotated by constant speed electric motors may be used to make and break electrical contacts to energize various circuits within the welder.

Where timing must be measured in cycles, that is in parts of the 60 cycles

Fifteen hundred interruptions per minute are possible with a thyratron control. Hot cathode mercury vapor tubes may be used with these control tubes and a synchronous timer and a series transformer used to control the firing of the welding transformers. See Fig. 21-9. The control tubes and syn-

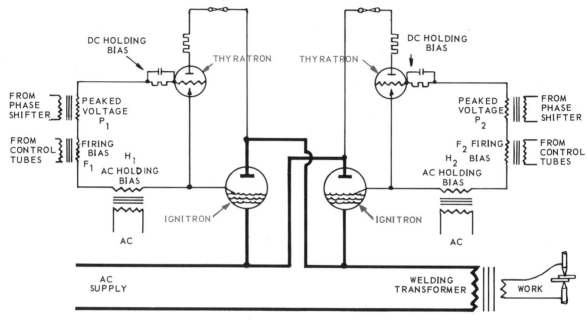

Fig. 21-9. Schematic wiring diagram of spot welder electrical circuit using two ignitron tubes and two thyratron tubes.

per second time, tubes or transistors are generally used. The simplest form of this type of timer is a three electrode tube, a condenser, which is kept charged between cycles to a definite value, and a variable resistor through which the condenser discharges to give the timing period. Times of 3 cycles or 3/60 of a second are possible using this method. This time may be varied by changing the resistance in the circuit using the variable resistance.

The exact timing required for aluminum and stainless steel spot welding methods requires the use of the thyratron controls in conjunction with ignitron controls.

chronous timer cause the mercury tube or thyratron tube to be either conducting or nonconducting. A series transformer provides the high voltage for the thyratron tubes through its secondary windings.

When the thyratron tubes are made conducting, the secondary of the series transformer is short circuited. The full voltage is then applied to the welding transformer primary and the weld is made. When the tubes are made nonconducting, they act as a valve in the circuit, and the line voltage to the welding transformer is insufficient to perform a weld. This interval is then the "off" period in the welding cycle.

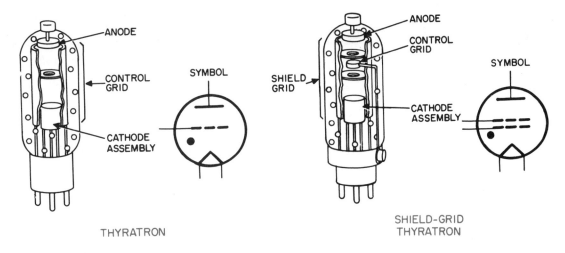

THYRATRON

SHIELD-GRID
THYRATRON

Fig. 21-10. Schematic of thyratron control tubes. These tubes are connected in
the circuit which sends current through the mercury pool of the ignitron tube.

A schematic of two types of thyratron tubes is shown in Fig. 21-10. The tube is shown in Fig. 21-11. Such tubes serve as timing devices for the ignitron tubes.

Fig. 21-11. A thyratron tube.

21-7. CURRENT CONTROLS

The ignitron control operates in a similar but more complex electronic method than the thyratron tube control. This tube permits large current flow. When the mercury is vaporized by the thyratron ignition circuit current, the vaporized mercury allows large amounts of current to flow from the cathode to the anode. The ignitron tube may be made of glass or steel depending on the current requirements of the welder.

Ignitron tubes come in five sizes: A, B, C, D, and E. These tubes are water-cooled. Sizes B, C, D, and E have a built-in water jacket, while size A has a clamp-around water jacket. Tube capacities are shown in Fig. 21-12.

Tube temperature should not exceed 125 deg. F. Some of the tubes have

IGNITRON TUBE MODEL	AMPERE CAPACITY BASED ON APPROX. 20% DUTY CYCLE	COOLING WATER GALS/MIN. MINIMUM
A	200	1-1/2
B	600	1-1/2
C	1300	1-1/2
D	3000	3
E	8000	5

Fig. 21-12. Table of ignitron tube properties. Capacities fluctuate
depending on many variables. Size tube to be used should be carefully
calculated based on Resistance Welder Manufacturers Association
(R.W.M.A.) standards.

thermostatic protection. If the tube temperature reaches 125 deg. F., the thermostat opens, shutting down the system. The thermostat points close again at 105 deg. F. Fig. 21-13 shows

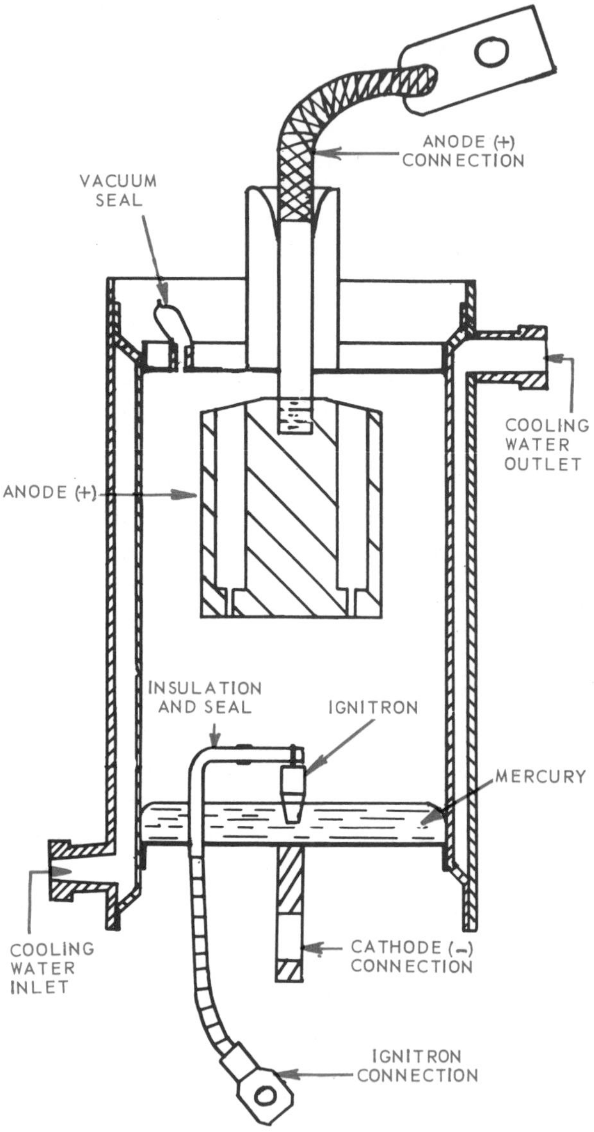

VACUUM SEAL

ANODE (+) CONNECTION

ANODE (+)

COOLING WATER OUTLET

INSULATION AND SEAL

IGNITRON

MERCURY

COOLING WATER INLET

CATHODE (−) CONNECTION

IGNITRON CONNECTION

Fig. 21-13. Main parts of an ignitron tube. When ignitron fires mercury vaporizes. The vaporized mercury completes the circuit between the cathode and anode. (General Electric Co.)

a schematic of an ignitron tube showing the cooling water jacket.

Some units use the combination of an ignitron mounted thermostat and a solenoid valve in the water lines, to control

the water flow. Such a device is used to economize on water. If the temperature of the tube lowers to approximately 85 deg. F., the water circuit is closed. When the temperature of the tube rises to 105 deg. F., the thermostat contacts will close, and the solenoid water valve will be energized allowing cooling water to flow. Fig. 21-14 illustrates an igni-

Fig. 21-14. Ignitron tube thermostat.

tron tube thermostat, while Fig. 21-15 shows a thermostat mounted on the tube.

It is recommended that the cooling water-in temperature be 90 deg. F. maximum and 50 deg. F. minimum. The pressure drop in the circuit is usually 15 to 20 psi. The exhaust water should

Fig. 21-15. Ignitron tube with a thermostat mounted on the tube.

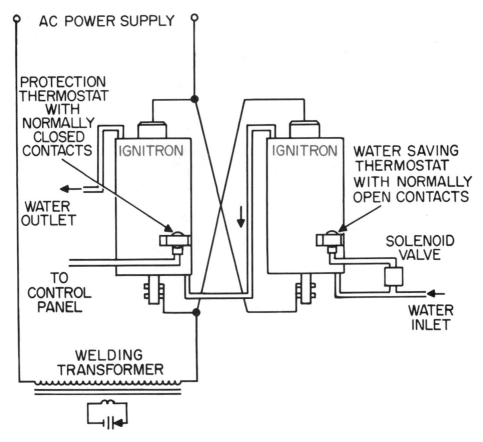

Fig. 21-16. Simplified circuit using two ignitrons to enable use of both halves of alternating current cycle. (National Electronics, Inc.)

be to an open drain (1) to provide an air gap protection against reverse syphoning and (2) to allow a visual check to be sure water is flowing. These tubes have a decreasing amperage capacity as the duty cycle increases.

The ignitron tube fires for only one half the alternating current cycle ($\frac{1}{120}$ sec.). Two tubes must be used to obtain current flow for both halves of the cycle as shown in Fig. 21-16. A pictorial view of the two-tube combination is shown in Fig. 21-17.

The tubes are fired by a peaked voltage from a transformer supplied with current from a phase shifter. The phase can be shifted over 100 deg., the exact amount depending on the power of the welder.

Ignitron tubes are connected in reverse parallel. This arrangement allows one tube to flow one half of the cycle and the other tube the other one half of the cycle. The tubes control the current flow in the primary windings of the transformer, and are connected in series with the primary windings of the welding transformer.

The ignitron tubes are fired by the thyratron tubes in the shunt circuit. The grids of the thyratron tubes are controlled by four different voltages in series:

A. DC holder bias voltage.
B. Peaked voltage.
C. Firing bias voltage.
D. AC holder bias voltage.

Wiring of the various adjustments is shown in Fig. 21-9. AC and DC holder

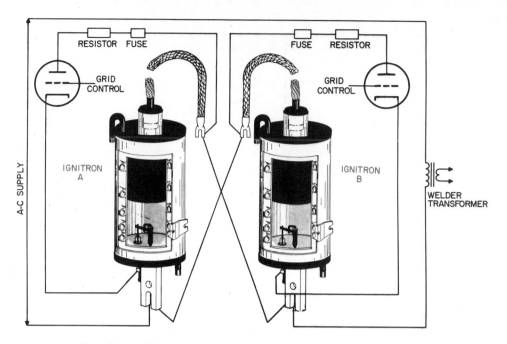

Fig. 21-17. Pictorial view of full-wave ignitron installation.

bias keeps the thyratron tubes nonconducting, since the peaked voltage is insufficient to overcome the biases.

When the control tubes energize the firing bias, the voltage from the firing transformer overcomes the balance in the parallel circuit and causes the thyratron tubes to become conductors. Current then flows to the ignitron tube. The ignitron tubes fire and immediately short circuit the thyratron tubes, making them conducting again. The first ignitron tube now permits current to flow to the primary of the welding transformer and the transformer secondary winding produces the low-voltage high-current electrical flow which produces the weld.

The number of cycles or time is determined by the control tubes. The average current value is set by the heat control phase shifter. The effect of the phase on the cycle is shown in Fig. 21-18. These two controls decide the firing point of the thyratron tubes and therefore the action of ignitron tubes.

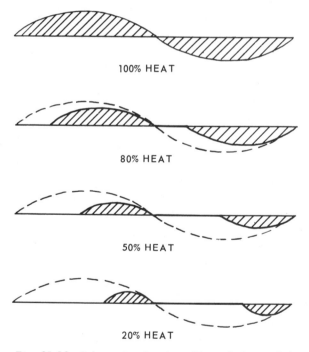

100% HEAT

80% HEAT

50% HEAT

20% HEAT

Fig. 21-18. Schematic showing effect of phase shifting on the AC cycle.

A complete cabinet installation of solid state circuitry, ignitron tubes, control, and main switch is shown in Fig. 21-19. The same cabinet with the door open is shown in Fig. 21-20.

Fig. 21-19. *Resistance welder electrical control cabinet.* *(Robotron Corp.)*

21-8. SEQUENCE CONTROLS

Where many automatic welding machines are operating in one shop, careful consideration must be given to the current load on the AC supply lines. If all the machines are operated at the same time, the power circuits might become seriously overloaded, causing malfunctions in all of the electrical equipment in that shop and in the neighborhood. Therefore, some control over the sequence of operation of the various machines is necessary to prevent such overloads.

In a single, high-current machine where multiple electrodes are used, the electrodes may be sequenced (timed

Fig. 21-20. *Resistance welder electrical cabinet showing tubes and wiring.* *(Robotron Corp.)*

to be used in a certain order) by means of cams. The cams rotate when the machine is energized and force the electrodes into contact in a preset order according to the placement and shape of the cam lobes.

The electrodes of a multiple electrode welder may be sequenced by trolled pneumatic circuit, and the solenoid will be energized when it receives current from the distributor.

Electronic timers may also be used to sequence the functions of a resistance welder. Thyratron or ignitron tube timers in conjunction with relays may be used to actuate solenoids con-

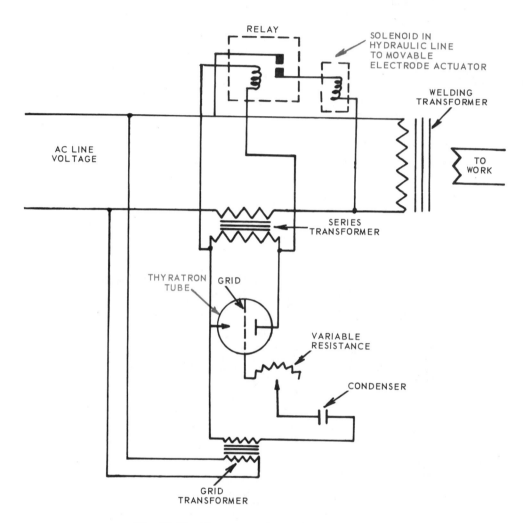

Fig. 21-21. Thyratron-tube controlled sequence circuit connected to hydraulic circuit.

means of a rotating current distributor. As the timer rotates, it makes and breaks contacts to various electrode solenoid circuits. Each electrode will be forced against the metal to be welded by means of this solenoid valve controlling either hydraulic, pneumatic or electromagnetism circuits. By shifting the phase slightly on each of a number of welders, no two welders would operate at exactly the same time, as shown in Fig. 21-21.

21-9. INSTRUMENTS FOR TESTS AND INSPECTIONS

Expensive highly automated equipment necessitates instrumentation either attached to the machine or readily available in the shop. The instruments are used to determine how the various electrical, electronic, cooling, pneumatic, and hydraulic circuits are functioning.

Gauge-type thermometers may be used to indicate the temperature of the coolant at the inlet and outlet, and to check the efficiency of the cooling system. Flowmeters may be used to check the volume of coolant flow.

An oscilloscope with a visual representation on a cathode tube screen may be used to check the various electronic and electrical circuits for proper functioning.

When resistance welding aluminum and other nonferrous metals, timing and pressure application is a critical factor. The duration of current flow is measured in terms of a part of the 60 cycle/second input current. Cycle counters have been employed to determine if the current flow lasts for 1 cycle of the 60 cycle period or 1/60 of a second. These counters are electronic units which enable the operator or the setup technician to adjust the weld current time to precise needs.

When resistance welding ferrous materials, usually a mechanical pressure gauge on the hydraulic or pneumatic system provides sufficient accuracy. However, when the pressures must be more exact, and the true force of the electrode against the metal must be known, transistorized strain gauges are used to measure the electrode force during the weld and forge periods.

To keep accurate control of variables such as weld pressure, forge pressure, weld current, weld time, and forge time, a permanent record of the variables may be made by using a recording oscillograph. As many as 12 variables can be recorded on the graph paper at one time. The variations can be detected and changed immediately as they occur when this device is used.

21-10. AUTOMATIC GAS WELDING

Oxyacetylene and oxyhydrogen welding equipment may be adapted to automatic welding by using an adjustable speed drive motor to propel the welding torch along the joint to be welded. Some joints may be designed to use the base metal as the filler metal for this type operation. Some automatic gas welding is done where the welding rod may also be fed into the puddle at a constant rate by using an adjustable-speed drive motor on the filler wire roll.

21-11. AUTOMATIC ARC WELDING

Inert gas-metal arc welding systems such as gas tungsten-arc (TIG) and gas metal-arc (MIG), have been designed to take advantage of the high welding rates, uniformity of weld, and the savings in gas consumption which are possible with automatic equipment. The automatic TIG and MIG welder uses welding rod in the coil form. The wire is fed to the electrode holder at a predetermined rate by adjustable drive motors as shown in Fig. 21-22. The electrode holder may be moved across the work, or the work moved under the electrode at a constant speed, by other adjustable-speed drive motors.

Arc length may be controlled automatically.

Aluminum, stainless steel, magnesium, titanium, and other previously hard to weld metals have been successfully welded with automatic equipment. As in manual welding, the base metal welding rod is needed, this rod may be fed to the weld automatically.

Multiple arc welding may be used with any of the above mentioned pro-

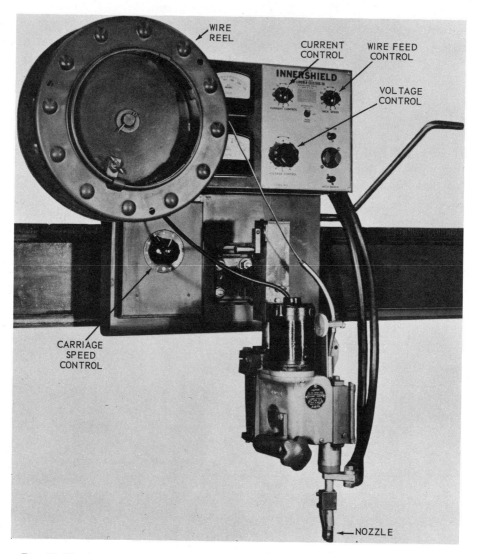

Fig. 21-22. Automatic welding head for flux cored wire. (Lincoln Electric Co.)

must be thoroughly cleaned to insure sound welds.

The carbon arc may be used in automatic equipment by moving the arc over the joint at a uniform rate with an adjustable speed drive motor. The flange is melted down rapidly due to the extreme temperature of the carbon arc and weld rates of 400 ft/hr are possible with no welding rod required. When

cesses by mounting several welding heads above the work, so that several weld beads may be made in one joint simultaneously, or, several welds may be made at the same time as shown in Fig. 21-23.

Studs may be welded to metal fabrications at an extremely fast rate using a semiautomatic or automatic gun-type resistance welder. A stud,

Fig. 21-23. Two welds may be made at the same time with this automatic gas metal arc welding equipment. (Cecil C. Peck Co.)

spike, or rod is loaded into the gun together with a ceramic ferrule which is used to limit the weld size and insulate the weld area. When the gun is triggered, the stud is pulled up off the work slightly and the arc is formed. After a suitable time, the stud is forced into the molten pool and fusion takes place.

The automatic plasma-arc process uses adjustable speed drive motors to move the work under the welding head, or to move the welding head. Gas pressure regulators and flow meters are used to control the inert gas. Adjustable speed motors are used to feed the electrode into the arc area. See CHAPTER 17.

21-12. AUTOMATIC RESISTANCE WELDING

Spot, seam, projection, butt, and flash-type resistance welds may be made automatically or semiautomatically. CHAPTERS 13 and 14 describe these machines in detail. In a fully automatic machine, after making the proper adjustments for weld and forge pressures, weld, hold and off time, coolant flow rate, and welding current, the operator needs only to load and unload the parts to be welded. However, even this work can be done mechanically.

Any number of welds may be made at one time using projection welding. The roller electrode of the seam welder could have an intermittent current sent to it to produce a long line of spot welds. Welds of many types can be made on the automatic machine by using programming tapes which signal electronically when the weld action should start and stop, when the weld movement should start and stop, when the current should increase and decrease, and in what direction the weld operation should

travel. Tape controlled automatic welding is growing in popularity.

Three operating devices for automatically operating resistance welding electrodes are shown in Fig. 21-24.

21-13. AUTOMATIC BRAZING

Brazing also has been automated in a number of ways. Vapor-flux has been in use with the oxyacetylene torch for years. Flux is mixed by bubbling acetylene gas through a liquid flux held in a closed container. The container is connected between the regulator and torch. In this manner a constant supply of flux is fed to the work with the acetylene gas. This eliminates the need to stop the welding process to flux the joint or the brazing rod.

When soft soldering or brazing is to be used to connect two parts together, a special solder form (ring or shim) may be designed to fill the joint to be soldered or brazed. The pieces of the fabrications, together with the solder forms, may be placed into a gas fired or electric induction furnace and heated. When the correct temperature is reached, the joining materials will tin and braze the joint.

21-14. AUTOMATIC CUTTING

Carbon arc cutting may be automated in the same manner as the arc processes described previously.

Inert gas fed through a hollow metallic electrode may be used to cut nonferrous metals by means of the heat developed in the arc and the pressure of the gas through the hollow electrode. This process can be automated by using drive motors, pressure regulators, flowmeters, and other equipment similar to that used on the inert arc process.

Cutting heads may be mounted on

tracer-type machines so that when an operator moves a stylus or follower around a form to be cut, the cutting head at some distance from the operator, will cut the same shape as that of the template used. Several cutting heads may be used and several parts cut at the same time.

Adjustable speed drive motors have been used to move a cutting head along a preshaped track to cut a shape in the same form as the track. Such motors are used to move a cutting head around a pipe to chamfer the edge in preparation for a butt weld. These forms of automation have been used for years.

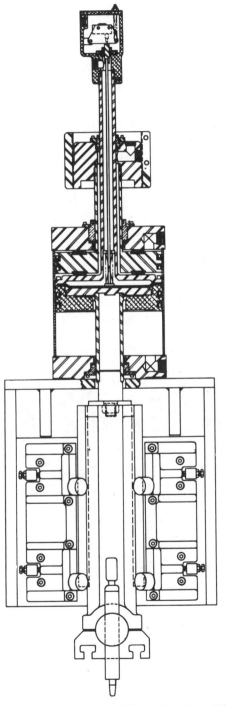

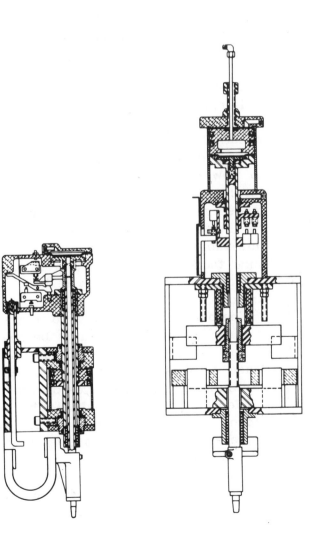

Fig. 21-24. Three different models of resistance welding electrode heads. The middle head uses magnetic force, while the others use fluid pressures to create forces. Note the use of diaphragms and complete switch.
(Acro Welder Mfg. Co.)

21-15. AUTOMATIC METAL SURFACING

The processes of metal surfacing may be automated by passing the part under the surfacing device at a uniform rate of speed or moving the device automatically over the surface. In this way a uniform thickness of a desired metal may be bonded to the base metal.

The plasma-arc may also be automated to apply a controlled thickness of surfacing material to a surface. The part to be surfaced is moved under the arc by drive motors and the surfacing metal is fed to the surface at a uniform rate. The speed of the part or surfacing metal wire feed may be automatically changed, to change the thickness of the surface applied.

21-16. ELECTRICAL LEADS

The electrical leads in automatic welding must be carefully planned and placed. Welding circuit connections must be as close to the weld as pos-

Fig. 21-25. *Rotating ground attachment which may be used to insure good electrical connections when work moves during welding operation.*
(Lenco, Inc.)

sible. If the article being welded rotates or moves, the electrical connections to this article can be solved by using a rotating ground attachment as shown in Fig. 21-25.

21-17. WELD SENSING SYSTEM

Using sensors which never touch the metal, an electronic system controls the electrode or arc torch distance above the metal. This constant distance is carefully maintained even as the electrode or torch passes over rough surfaces, tacks, and spatters. This electronic device is accurate to ±0.015 in.

21-18. REVIEW OF SAFETY PRACTICES IN AUTOMATIC WELDING

All of the safety practices discussed in previous chapters relative to gas welding, arc welding, resistance welding and the like also apply to automatic welding.

In addition one must be very alert to power driven devices. Moving torches and/or welding heads may move against an operator. The electrical leads and gas hoses also move and this movement must be checked to be sure these devices do not catch on obstructions or move into a dangerous area.

The pressure devices used to create the forces necessary must be maintained in good condition or sudden release of pressures by broken lines or parts may injure ones eyes or body.

Electrical cabinets should be kept closed to prevent objects falling into the electrical devices causing dangerous shorts and sparks.

One should manually operate an automatic system through its full length of travel or operating cycle to make certain that there are no obstructions to prevent accurate and safe operation during the automatic operation.

21-19. TEST YOUR KNOWLEDGE

1. Does automatic welding involve different welding process than manual welding?

2. Does automatic welding mean automatically moving and controlling the metal to be welded, automatically moving and controlling of the welding method, or both?

3. List some advantages of automatic welding.

4. List some disadvantages of automatic welding.

5. What basic system is commonly used to maintain a constant arc length?

6. Name some energy mediums used to produce motion.

7. List several methods used to guide a flame or an arc along a predetermined path.

8. Why should the water-cooling system of a resistance welder be electrically connected to the resistance welding circuit?

9. Why is it common to say that electronics are the brains of automation while pneumatics and hydraulic mechanisms are the muscles of automation?

10. Describe the principle and operation of a solenoid valve.

11. Why are pressure regulators used on some pneumatic piping?

12. How does a bimetal strip protect the electrode of some gas metal-arc outfits?

13. What is the most common reason for installing a solenoid valve in a water circuit?

14. Why are timers used on some resistance welding units?

15. What is a thyratron?

16. What vapor enables current to flow in an ignitron?

17. How is resistance welding electrode force measured?

18. What is a popular method of applying solder in an automatic soldering operation?

19. Why is an oscilloscope needed to check or inspect some automatic welding operations?

20. What is meant by the term "firing bias"?

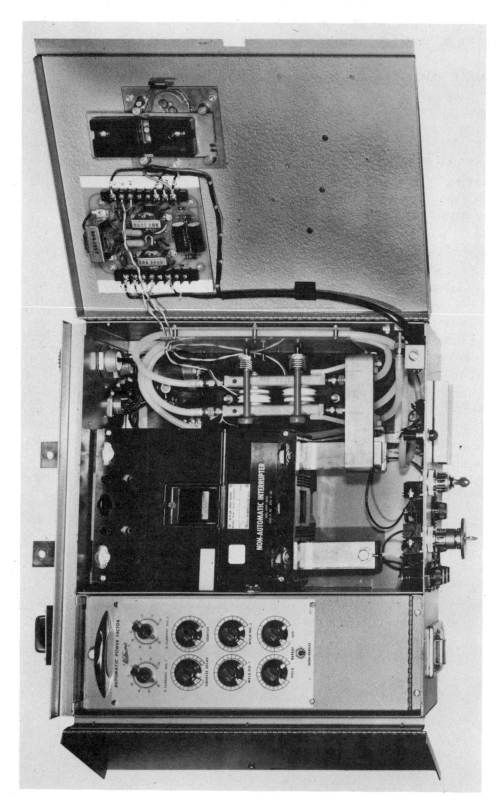

Solid state resistance welding timer showing solid state controls, wiring, and water-cooled 1200-ampere rectifiers. (Weltronic Co.)

Positioning X-ray tube head for testing aircraft landing gear.
(Sperry Div., Automation Products, Inc.)

Chapter 22
INSPECTING AND TESTING WELDS

A finished weld is not always as good or as bad as it may appear to be on its surface. Due to the increasingly high standards of production, reliable methods of testing and inspecting welds are required.

The methods used to determine the quality of a weld may be broken down into two general classifications:

1. Nondestructive tests.
2. Destructive tests.

The welding method used, the shape

of the article, and the kind of metal influence the type of testing or inspection required.

Spot welds and seam welds have inspection limitations.

A corner joint or T-joint is difficult to test by some methods.

Pipes and containers lend themselves to pressure testing.

In this chapter methods of testing welds will be shown and explained.

22-1. NONDESTRUCTIVE TESTS

The method of testing a weld without destroying its usefulness as a finished product (nondestructive testing) is by far the fastest and least expensive in terms of a finished product.

Methods which fall under this classification include:
1. Visual inspection.
2. Magnetic particle inspections.
3. Liquid penetrant inspection.
4. Ultrasonic inspection.
5. X-ray inspection.
6. Eddy current inspections.
7. Mass spectrometer detection.
8. Air pressure leak tests.
9. Halogen gas leak tests.

Test method numbers 2, 3, 4, and 5 are the more popular methods used in the industry. Standards of inspection have been established for these four types.

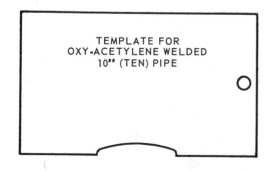

Fig. 22-1. Template for testing bead contour of welds.

22-2. DESTRUCTIVE TESTS

Certain types of weldments must be cut up and even ground up to determine the various physical properties. When the weld is destroyed or damaged beyond use, the test is a destructive test.

Several destructive tests are:
1. The tensile test.
2. Chemical analysis.
3. Bend test.
4. Microscopic test.
5. Macroscopic test.
6. Hardness test.
7. Charpy test.
8. Hydrostatic test to destruction.
9. Peel test.

22-3. VISUAL INSPECTION

A weld which is not required to have a high physical strength may be inspected for cracks, inclusions, contour, and certain other qualities visually. This type of inspection is subjective in nature, and usually has no definite and rigidly held limits for acceptability. A template may be used to check the contour of the weld bead. Figs. 22-1 and 22-2 show such templates. Using the visual inspection method, an inspection may compare a finished weld with an accepted standard, and pass or reject a weld by comparison method only.

This test is effective only when appearance is the most important quality of the weld.

22-4. MAGNETIC PARTICLE INSPECTION

This method is most effective in checking a weld for surface or near surface flaws. It is used only on materials which can be magnetized. A liquid solution containing very tiny magnetic particles is painted or sprayed onto the surfaces being checked and the metal is then subjected to a strong magnetic field. These particles are colored red or black and are suspended in a fine oil vehicle. The choice of color is dependent on which color, black or red gives the best color contrast. Any lack of continuity at or near the surface of the metal when magnetized creates a local north and south magnetic pole, and attracts the

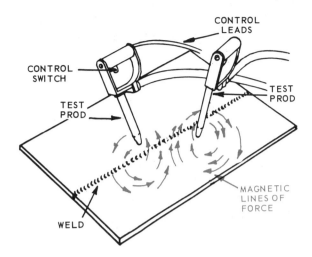

Fig. 22-3. Magnetic field is created around a weld as current is passed through the weld between two test prods. (Magnaflux Corp.)

Two methods are used to create the magnetic field:

1. Passing current through the object being inspected by means of test prods.

2. Placing a powerful electromagnet or permanent magnet against the object to allow the magnetic field to pass through the material being inspected.

The magnetic field flux lines should

Fig. 22-4. Magnetic powder and electrical test prods being used to locate surface defects in a weld. One of the test prods used here is magnetically attached to one end of the weld. This frees one of the inspector's hands and allows him to apply the magnetic powder at the same time. (Magnaflux Corp.)

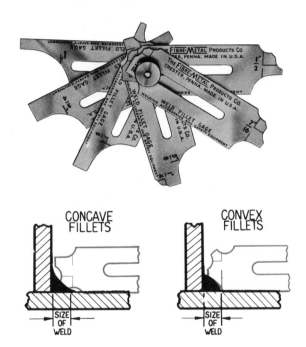

Fig. 22-2. Above. Gauges for measuring and inspecting both convex and concave fillet welds from 1/4 to 1 in. Below. Blades of gauge used to check concave and convex weld contours.
(Fibre Metal Products Co.)

metallic particles in the solution used. When the magnetic field is removed, the inspector will fine a concentration of magnetic particles in the area of every flaw. If imperfections are found they are ground or chipped away, the part is rewelded, and then retested. Fig. 22-3 shows the magnetic field which is created around the weld area being tested.

be as perpendicular to the crack as possible. Because the location of the crack is not known, the article should be magnetized twice; one inspection being made at 90 deg. to the other. Fig. 22-4 shows magnetic powder being used to detect flaws in a weld which has been magnetized by electrical test prods.

A kit which includes an electromagnet used to create a magnetic field in the metal is shown in Fig. 22-5. The use of this method is shown in Fig. 22-6.

Fig. 22-5. *Kit for magnetic inspection using an electromagnet.*

Fig. 22-6. *An electromagnet being used to inspect a cylinder block for cracks by means of the magnetic particle test. (Magnaflux Corp.)*

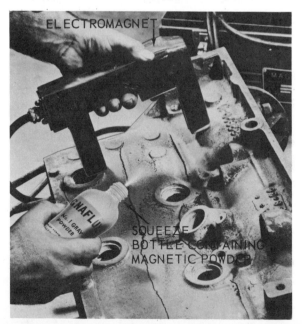

The magnetic field can also be produced by using a permanent magnet as shown in Fig. 22-7. The electro and

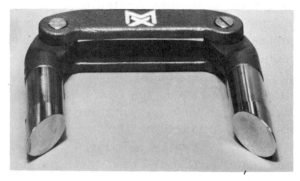

Fig. 22-7. *Permanent magnet used to magnetize metal which is to be inspected with magnetic particles.*

permanent magnet systems are convenient, and are also used where an electrical spark may be dangerous. Another type of electromagnet, shown in Fig. 22-8, is used when inspecting pipes and shafts.

Fig. 22-8. *Ring type or torus shaped electromagnet used to magnetize pipes or shafts to be inspected with magnetic particles.*

The material to be tested should be as clean and bright as possible prior to the test.

The magnetic test for cracks is also very popular as a checking device in many service trades (aircraft, automotive, truck, marine, and the like).

Locating cracks in axles, shafts, gears, crankshafts, and landing gear parts before performance failure occurs is of considerable importance.

22-5. LIQUID PENETRANT INSPECTION

The liquid penetration inspection method uses colored liquid dyes and fluorescent liquid penetrants to check for surface flaws. This system can be used to detect surface flaws in metals, plastics, ceramics, and glass. This method will not detect subsurface flaws.

The liquid dye penetrant is sprayed onto the clean surface being inspected. After allowing a short time for the liquid to penetrate the excess amount of dye is removed with a cleaner, and the surface is washed with water and allowed to dry. After the surface is thoroughly dry, a developer is sprayed on the surface which brings out the color in the dye penetrant that has penetrated into any cracks or pin holes. Surface flaws will show up as shown in Fig. 22-9.

Fluorescent liquids are used in a manner similar to dye penetrants. A fluorescent liquid is applied to the surface being inspected. After a short time, the excess fluorescent liquid is removed with a cleaner and the surface is washed and dried. A blacklight source is then brought up to the surface. All areas where the fluorescent liquid has penetrated will show up clearly under the blacklight, as shown in Fig. 22-9.

The dye, the cleaner, and the developer are available in aerosal spray cans for convenience. Some solvents used in the cleaners and developers contain high percentages of chlorine, a known health hazard, to make the liquids nonflammable. Solvents and developers containing chlorine should be used with great care.

Because the penetration ability of the dye varies according to the materials being tested, and because the penetrant activity varies with the temperature, it is important to allow sufficient time (this varies from 3 to 60 minutes) to permit accurate inspection. At room temperatures, the time usually recommended is from 3 to 10 minutes.

22-6. ULTRASONIC INSPECTION

A relatively new method of inspecting welds is to use high frequency sound waves. This testing technique can detect internal flaws as well as surface flaws. The principle involved is the same as the echo ranging principle used to find submarines under water. A high frequency sound wave (ultrasonic wave) is sent into the metal for very short periods (1 to 3 microseconds). Then the wave is stopped. The same unit which was used to send the sound wave then acts as a receiver to listen to the ultrasonic wave as it is reflected through the metal. The sound again

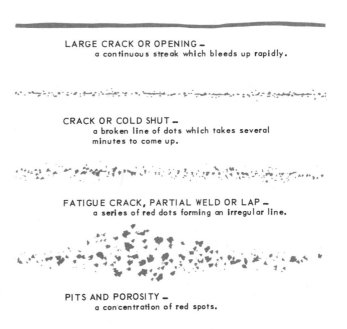

LARGE CRACK OR OPENING —
a continuous streak which bleeds up rapidly.

CRACK OR COLD SHUT —
a broken line of dots which takes several minutes to come up.

FATIGUE CRACK, PARTIAL WELD OR LAP —
a series of red dots forming an irregular line.

PITS AND POROSITY —
a concentration of red spots.

Fig. 22-9. Indications of surface flaws found by liquid dye or fluorescent penetrant method of inspection.

flows, stops, and its reflected wave is picked up by the transceiver (transmitter-receiver unit). This cycle is repeated from 1/2 to 5 million times per second. Each wave is visually represented on an oscilloscope. The oscilloscope is calibrated to pick up only flaws of a size which would be considered harmful. The oscilloscope wave pattern is also calibrated to show the distance between the searching unit and any flaw found, as shown in Fig. 22-10.

of this system is that no extra materials are needed, and the parts are not damaged.

22-7. X-RAY INSPECTION

Welds may be checked for internal flaws by means of the X-ray. X-ray is a wave of energy which will pass through most materials and reproduce their image on film (radiography), on a fluorescent screen (fluoroscopy), or on a

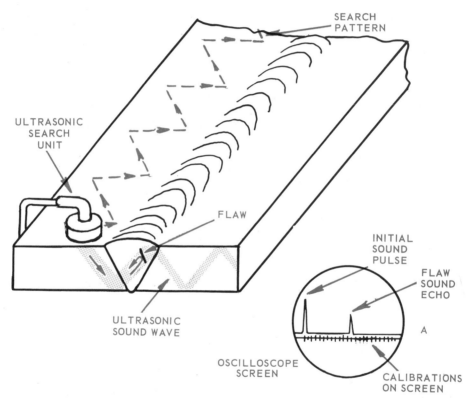

Fig. 22-10. Schematic drawing showing the path followed by the ultrasonic search unit and path of sound waves as they move through the metal being tested. The initial, also echo indications, are shown on the sound oscilloscope in a manner similar to that shown at A. The distance between peaks is an indication of how far the flow is from the searching unit head.

A complete ultrasonic test unit, and its use is shown in Fig. 22-11. This system is fast and results are determined at the moment of testing. However, the operator must have some training, to insure consistent interpretation of results. Another advantage

television screen for viewing at a remote spot (TVX). Fig. 22-12 illustrates diagrammatically the principle of the fluoroscopic system of examining a pipe weld. The radioactive energy may be produced electronically in an X-ray machine, or by means of radio-

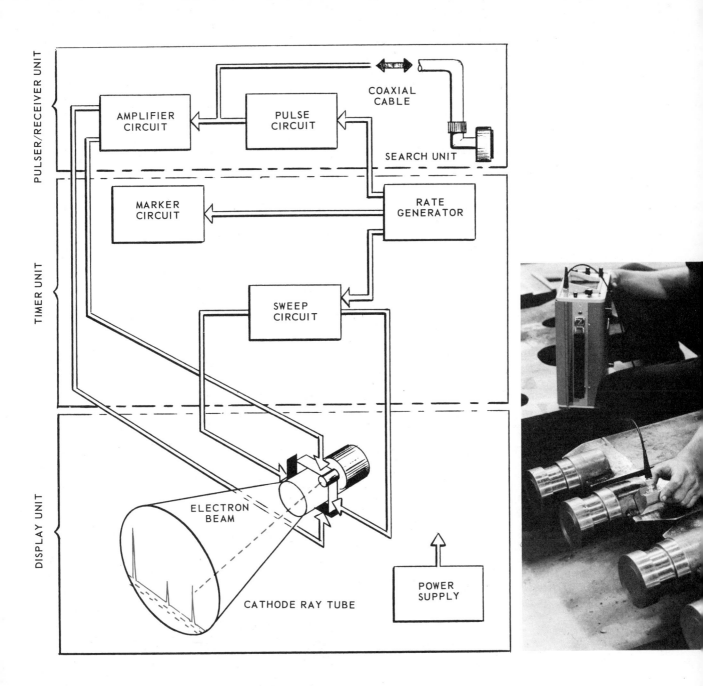

Fig. 22-11. *Left. Block diagram shows component functions of an ultrasonic testing unit. Ultrasonic sound is sent and received by search unit. Signal is displayed on cathode ray tube. Pattern indicates any defects. Right. Ultrasonic unit in use testing weld on equalizer bar for railroad car.* (Sperry Div., Automation Industries, Inc.)

active isotopes. Equipment using radio-active isotopes is portable and may be used to check welds made in the field. Portable X-ray equipment is shown in Fig. 22-13. Certain safety precautions must be observed while using radio-active materials, to prevent illness and even death from overexposure. Radiographers inspect welds made before any further work is done on the weldments. A radiograph, if made (and this action is often required), is a perma-

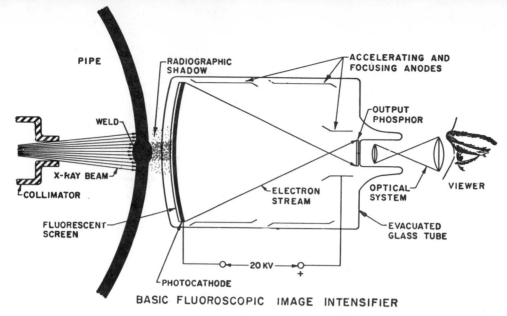

BASIC FLUOROSCOPIC IMAGE INTENSIFIER

Fig. 22-12. Diagrammatic explanation of a fluoroscopic examination of a pipe weld.
(National Tube Div., U. S. Steel)

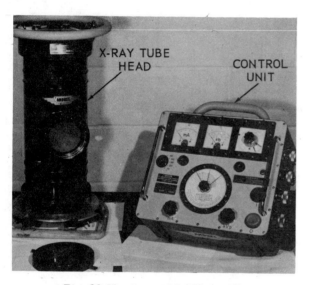

Fig. 22-13. A portable X-ray unit.
(Diano Corp.)

nent record of welds made on critical construction such as pipelines, ships, aircraft, and the like. Fig. 22-14 shows equipment being readied for taking a radiograph.

Several popular radioactive isotopes are:

Cesium 137
Cobalt 60
Iridium 192
Samarium 153
Thulium 170

These isotopes give off gamma rays which are about 1/4-in. in size. Because they are small and need no electrical source, and because they give off rays in all directions, they are more flexible in use than standard X-ray equipment.

Isotopes must be handled with great caution. Only specially trained personnel should handle them.

The energies needed to X-ray vary considerably. Equipment is available from 50 KV (K = 1000, V = volts) to 24,000 KV (2,400,000 Volts). 140 KV will

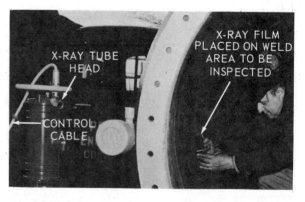

Fig. 22-14. Portable X-ray equipment being set up to inspect a weld on a large pipe section. The operator will move a safe distance away from the X-ray head when he operates the controls which release the radio active energy through the metal to the sensitive film.

X-ray 2-in. steel while 24,000 KV will X-ray 20-in. steel. Fig. 22-15 illustrates the principle of radiography.

which the flaw occurs cannot be determined with an X-ray made from only one direction.

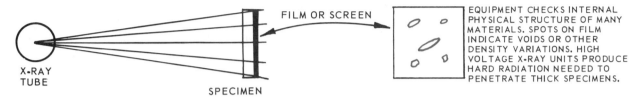

Fig. 22-15. Schematic of X-ray photography of a weld.
(Philips Electronic Instruments)

Fig. 22-16. X-Ray machine being set up by an operator to radiograph the welded seams of a spherical steel electronic gear housing.
(Baldwin-Lima-Hamilton Corp.)

Fig. 22-16 shows an X-ray machine being set up to radiograph a welded seam.

Flaws in a weld usually are easily seen in an X-ray radiograph as shown in Fig. 22-17. However, the depth at

Fig. 22-17. An X-Ray (radiograph) of a weld in 1-1/2-in. steel plate. Note how the undersurface crack at the left shows up in the X-Ray photograph.
(General Electric X-Ray Corp.)

The equipment required when inspecting by means of X-rays depends on:

1. Kind of material.
2. Thickness of material.
3. Accessibility of part to be tested.
4. The geometry of part to be tested (for example, a flat plate is easier to inspect than a pipe cluster).

22-8. EDDY CURRENT INSPECTION

When an AC coil is brought up close to a conductive metal, it will induce eddy currents in the object. These eddy currents produce their own magnetic field which opposes the field of the AC coil, increasing the impedance (resistance) of the AC coil. Coil impedance can be measured. When a flaw passes under the AC coil, the eddy currents vary in the metal. This in turn changes the impedance of the coil which is wired to flash a warning light, and the location of the flaw can be found. This method is presently being used to check thin wall tubing at the rate of 200 fpm.

22-9. INSPECTING WELDS USING PNEUMATIC OR HYDRAULIC PRESSURE

A common method of testing pressure-vessel welds (tanks and pipe lines) to determine if leaks are present, is to use gas or air pressure. Carbon-dioxide gas pressure is well suited for the purpose because of its nonexplosive properties when in contact with oils or greases. A small pressure should be built up in the vessel or pipe (25 to 100 psi), and a soap and water solution put on the outside of each weld. Leaks will be indicated by the formation of bubbles. The ability of the vessel to hold pressure is also an indication of its tightness. The vessel to be inspected may be pressurized and the pressure noted on a gauge. After 24 hours the gauge is again checked. Any drop in pressure indicates a leak. This test is easily applied and is safe since the pressures used are usually under 100 psig.

Another test for pressure vessels is to coat the surface with a lime solution. After the lime has dried, pressure is built up in the vessel. Where the lime flakes from the metal, a flaw is indicated as being present. Hydraulic pressure, using water as the fluid, is the usual medium used in this test. This test may also be used to reveal the weakest portion of a welded vessel if enough pressure is created without destroying the vessel.

To inspect pressure vessels for leaks, water is still a popular method, but water is not a reliable check for extremely small leaks.

By pressurizing a vessel with chlorine, fluorine, helium, or other non-oxygenated gases, and placing a pickup mass spectrometer tube on a suspected leakage area, a flow of one part of these gases in 1,000,000 parts of air may be detected to indicate a minute leak in the pressure vessel.

22-10. BEND TESTS

A popular method of testing a weld which does not require elaborate equipment, is the destructive bend test. The method is fast and shows most weld faults quite accurately. A sample specimen may be tested to destruction to determine:

1. Physical condition of the weld, and thus check on the weld procedure.

2. To determine the welder's qualifications.

This method is particularly appropriate where large numbers of identical pieces are to be fabricated. The usual method of testing is to take one piece out of a predetermined quantity and test it to destruction. The destructive test may show up such qualities as tensile strength, ductility, fusion, penetration, and crystalline structure.

The equipment used for a test of this kind depends on the shape and type of articles to be tested: a common method is to clamp the piece to be tested in a vise and by means of a bending bar bend the metal at the welded joint. Fig. 22-18 illustrates this test.

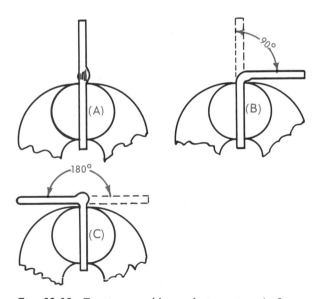

Fig. 22-18. Testing a weld sample in a vise: A. Sample before bending. B. Sample after bending 90 deg. (note the weld is closed in on itself, to test the penetration). C. Sample after bending it back 180 deg.

This method of bending quickly gives the approximate strength of the weld, while the stretching of the metal determines to some extent its ductility. Any cracking of the metal will show false fusion or defective penetration.

Many fixtures have been devised to help test weld specimens to destruction. After the weld has been broken, the appearance of the fracture will show the crystalline structure. Large crystals will usually indicate wrong welding procedure, or poor heat treatment after welding, while small crystals will usually indicate a good weld.

The guided root or face bend test, as described in CHAPTER 27, is a bend test made under specified conditions to determine certain physical properties of a weld. The radius of the bend made on the test sample must be 3/4 in. to meet AWS test standards. As the bend is made, the metal is guided or controlled at all times by the bending fixture so that all samples are bent in exactly the same manner. Fig. 22-19 illustrates a hydraulically operated guided bend tester. Free bend

tests are also made on specially prepared weld samples as described in CHAPTER 27.

22-11. IMPACT TESTS

It has been found that it is possible for a weld to test sound by a variety of tests, but fail under a rapidly applied load (impact). The impact test may be made by either the Izod or Charpy method. These methods are similar, but the shape and position of the notch varies.

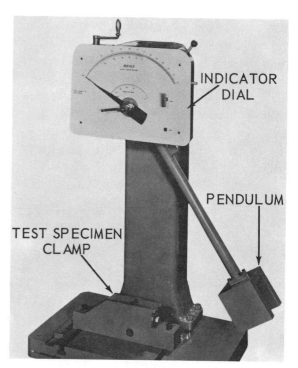

Fig. 22-20. A Charpy impact testing machine. In this tester the pendulum is lifted and dropped against the test specimen which is held in the clamp. The impact force is registered by the dial indicator.
(Riehle Testing Machines Div., Ametek, Inc.)

A test piece is notched in a specified manner and clamped in the jaws of an impact testing machine. A heavy pendulum is lifted to a given height and then dropped against the notched specimen to determine what impact force it can withstand, Figs. 22-20 and 22-21. The

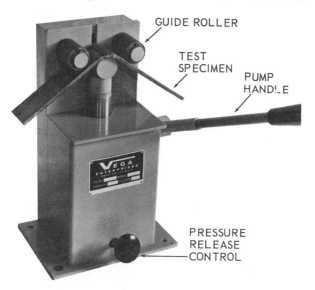

Fig. 22-19. A hydraulically-operated guided bend test machine. (Vega Enterprises)

higher the weight is lifted before it is dropped, the greater the force on the notched specimen.

ical properties of sample welds or base metal specimens. Occasionally, the testing is performed by a metallurgi-

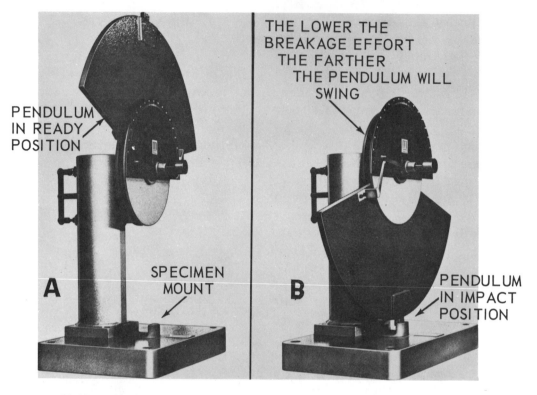

Fig. 22-21. A Charpy impact testing machine. A, shows the pendulum in the ready position. B, shows the pendulum in the impact position. (Physmet Corp., Div. of Manlabs, Inc.)

The indicator needle on the testing machine shows what force was exerted on the test specimen. The impact force is increased until the test piece finally fails. This test determines the ultimate impact strength of the weld specimen.

22-12. LABORATORY METHODS OF TESTING WELDS

To determine weld strength, most companies that do welding to any extent have established laboratories for scientifically determining the exact characteristics of welds. These laboratories are equipped with modern equipment, which can be used to determine the complete physical and chem-

cal department, and in some cases a part of the shop is set aside for this purpose.

Some items to be determined in a laboratory test are:
1. Tensile strength.
2. Ductility.
3. Hardness.
4. Microstructure.
5. Macrostructure (deep etch).
6. Chemical constituents.

The conditions under which the specimens are tested are kept identical, meaning that the specimens are all of a standard size. The length need not be the same, but the cross section area must be the same. Samples should also be taken from identical positions from the large weld which is being tested.

The Society of Automotive Engineers, the American Society of Mechanical Engineers, and American Society of Testing Materials, have adopted standards for laboratory tests of metals. These standards involve specifications of various physical qualities to be maintained for different types of welds, and specifications for various test specimens or samples. Some sample test pieces are shown in Fig. 22-22. This is done in order to have a means of comparing the standards of one company with those of another, and to enable standard testing machines to be built.

Fig. 22-22. Samples of weld test specimens tested to destruction. These were machined to standardized measurements before testing. All pieces shown are tensile test specimens except the pair on the right. This pair is a Charpy impact test specimen after it has been fractured.
(General Electric Co.)

22-13. TENSILE DUCTILITY METHOD OF TESTING WELDS

To tensile test a weld, a specimen of the weld is mounted in a machine and stretched. The machine determines three values for the weld:

1. Tensile strength of metal.
2. Yield point of metal.
3. Ductility of metal.

The tensile strength is recorded as the number of pounds per square inch required before the metal stretches be-

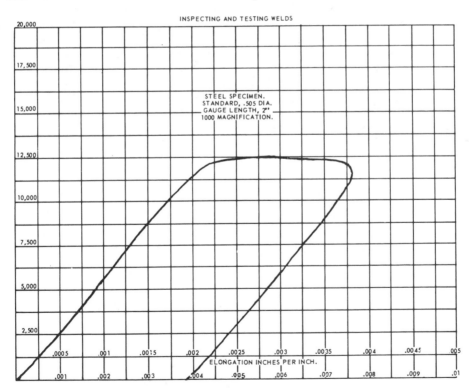

Fig. 22-23. Strength graph of a metal showing the yield point and the permanent set left in the steel when returned to a no-load condition.
(Tinius Olsen Testing Machine Co.)

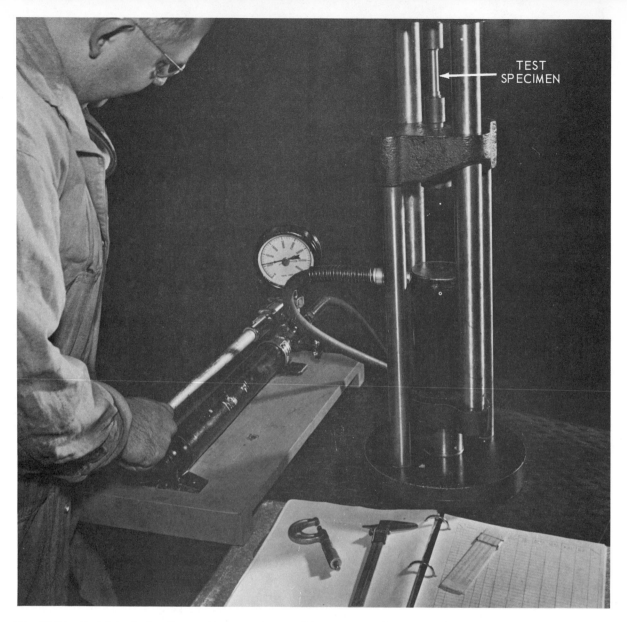

Fig. 22-24. *Portable, hydraulic tensile tester in use. This same machine may also be used to perform a guided bend test on a weld.* (Airco Welding Products, Div. of Airco, Inc.)

yond its elastic limit. The elastic limit of the metal means that it can be stretched just so far, and will return to its original length after the load is released. However, when more load is applied to the specimen after the elastic limit has been reached, the specimen loses its elasticity and the metal will not return to its original condition or size. When the specimen stretches instantaneously, or gives at a certain loading, but does not break, this is called the yield point. Fig. 22-23 is a

graph of the tensile load imposed on a metal sample. The yield point is important since it is not desirable to load metal to the point where it will stretch and not return to its former shape. Many machines have been developed to do this testing, and to record directly, even automatically, the tensile strength of the metal. Fig. 22-24 shows a hydraulic universal tester. Some machines are large, while others are portable and may be taken to field locations for testing metals on the

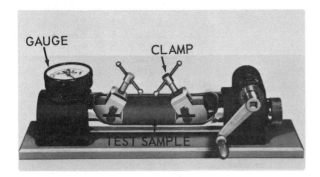

Fig. 22-25. *Portable tensile tester. The force applied to the test sample is obtained through the mechanical advantage of a threaded shaft which is moved by turning the operating crank.*
(Detroit Testing Machine Co.)

premises. Fig. 22-25 shows a portable screw-operated machine. Fig. 22-26 illustrates a portable universal tensile tester which is hydraulically powered.

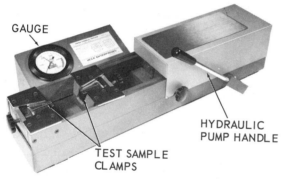

Fig. 22-26. *Portable universal tensile testing machine. Force applied to pull the test sample apart is obtained by operating a hydraulic jack built into the tester housing. The force applied is read on the gauge.*
(Vega Enterprises)

A hydraulically powered laboratory type universal tensile testing machine is shown in Fig. 22-27.

Fig. 22-28 shows the hydraulic circuits for a tester having a table moved by a hydraulic ram (piston). The valve spool, which controls the direction of hydraulic fluid flow, is moved by the manual control knob. When the control knob is in the load position, the valve spool is moved down opening passage B to connect the main pressure line to the radial pump. Pressure from the

pump acting on the bottom of the ram causes the ram and table to rise.

Another tensile tester model which

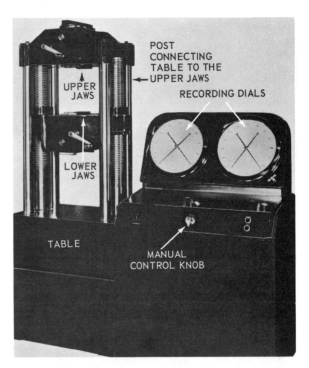

Fig. 22-27. *Laboratory-type hydraulic tensile test machine. Specimen is clamped in upper and lower jaws. A hydraulic ram moves table up to pull test specimen, which is mounted between clamping jaws.*
(Riehle Testing Machines Div., Ametek, Inc.)

may be either hand-powered or motor-powered is shown in Fig. 22-29. These machines can test the tensile strength of metals, fabrics, assembly strengths, etc. Some machines designed for tensile testing may also be used to test the compression strength of brittle substances by applying a crushing force on the material being tested.

It is desirable for these testers to be universal, so they can test metals of different cross sections and shapes; that is, the machine will hold and test round, oval, square, and rectangular specimens. At the same time the machine is testing the tensile strength of the metal, it also tests the ductility, meaning the stretchability (elongation)

Inspecting, Testing Welds 567

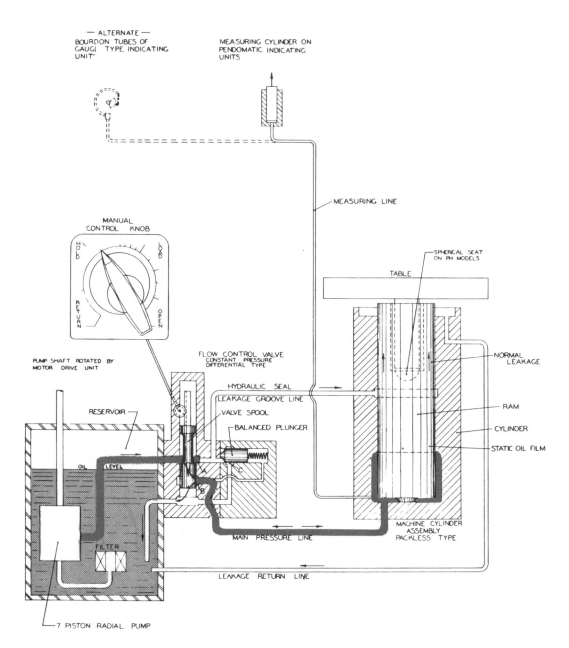

MEASURING CYLINDER ON
PENDOMATIC INDICATING
UNITS

MEASURING LINE

MANUAL
CONTROL KNOB

HOLD

LOAD

RETURN

OPEN

SPHERICAL SEAT
ON PH MODELS

TABLE

NORMAL
LEAKAGE

PUMP SHAFT ROTATED BY
MOTOR DRIVE UNIT

FLOW CONTROL VALVE
CONSTANT PRESSURE
DIFFERENTIAL TYPE

HYDRAULIC SEAL

RESERVOIR

LEAKAGE GROOVE LINE

VALVE SPOOL

RAM

BALANCED PLUNGER

CYLINDER

OIL LEVEL

STATIC OIL FILM

FILTER

MACHINE CYLINDER
ASSEMBLY
PACKLESS TYPE

MAIN PRESSURE LINE

LEAKAGE RETURN LINE

7 PISTON RADIAL PUMP

Fig. 22-28. *Hydraulic circuits of a moving table tensile tester.*

of the metal before it fails. This test is in relation to what is called the elastic limit of the metal. To measure the elongation of a weld, prick-punch or scribe two points on the weld specimen, 1/2 in. or 1 in. apart and then measure the change in this distance after the metal has been stretched to its elastic limit. The elongation is determined in percent by dividing the difference between the two readings by the original distance. For example: if 1 in. is the original distance between punch marks, and the distance between the punch marks is 1-1/4 in. after being stretched to its elastic limit, then 1-1/4 in. - 1 in. = 1/4 in. and 1/4 ÷ 1 = .25 or 25 percent. Fig. 22-30 is a table showing the tensile strength and percent of elongation for some common metals.

Fig. 22-29. Mechanical screw-type universal tester. The load is applied by hand, and the dial indicator registers the yield point and ultimate strength of the specimen. (W. C. Dillon & Co., Inc.)

22-14. TESTING THE HARDNESS OF WELDS

Another important factor which should be determined when testing a weld is the hardness of the weld metal. The hardness of welds is particularly important if the welds must be machined. There are many special metal alloys used for welding hard surfaces or for welding machine tools.

Methods used to determine metal hardness have been standardized. One of the most popular methods is to use what is called a Rockwell hardness testing machine, Fig. 22-31. This machine works somewhat like a press and is provided with a platform for holding the specimen. A point (which may be either a 1/16-in. diameter ball or a diamond cone ground at a 120 deg. angle) is pressed into the metal which is being tested by means of fixed weights operating through leverage, as shown in Fig. 22-32. When the ball-point is used, the distance the ball penetrates the metal between the first load (10 kilograms or 22 lbs. when the tester is calibrated and the dial set to zero) and the final or major load (100 kilograms or 220.5 lbs.) indicates the hardness on a gauge registering from 0-100.

The hardness is indicated on the

METAL	TENSILE STRENGTH WELDED PSI	ANNEALED	HEAT-TREATED ANNEALED	% OF ELONGATION ANNEALED
Low Carbon Steel	60,000	55,000	......	30
Medium Carbon Steel	40,000	55,000	55,000	..
Stainless Steel	75,000	95,000	180,000	55
Chrom-Moly Steel	40,000	....	90,000-180,000	..
Duralumin	20,000	....	55,000- 65,000	..
High Tensile Steel	75,000	50,000	50,000	25
Hard Surfacing	....	....	......	..
Soft 5% Phoylius	....	....	85,000	55
Bronze Monel	51,000	51,000	105,000	25
Duralumin 17ST	....	58,000	......	20

Fig. 22-30. Tensile strength and percent of elongation values for various metals.

Fig. 22-31. Laboratory-type Rockwell hardness testing machine. (Wilson Div., American Chain & Cable).

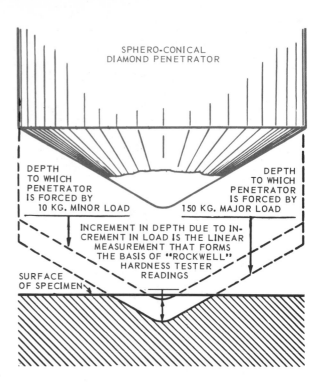

Fig. 22-33. Diamond penetrator used in Rockwell hardness tester. Note the depressions left in the metal by the diamond under the 10kg and 150kg in load.

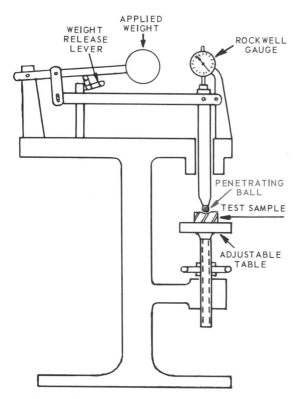

Fig. 22-32. Diagrammatic drawing showing how the weight is applied to the penetrator through leverage in a Rockwell hardness tester.

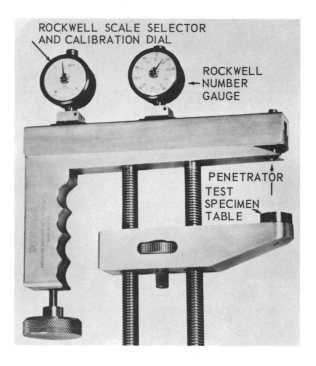

Fig. 22-34. Portable Rockwell hardness tester.

Weld being ground by a disk-type grinding wheel. The grinder motor is pneumatically powered. (Norton Co.)

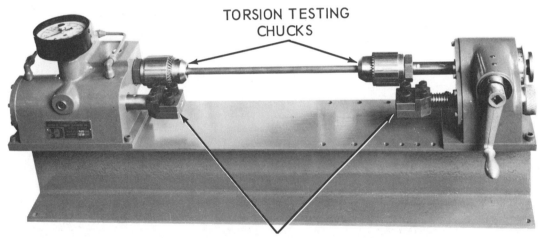

TORSION TESTING
CHUCKS

TENSILE/COMPRESSION
TEST CLAMPS

A testing machine which can be used for tensile tests, compression tests, and torsion tests. For tensile and compression tests, a load of 4000 lb. is possible with this tester. A capacity of 500 inch-pounds is possible for torsion tests. (Detroit Testing Machine Co.)

Rockwell B scale, when the 1/16 in. steel ball is used. When using the diamond cone penetrator, as shown in Fig. 22-33, the weights of 10 kilograms or 22 lb. and 150 kilograms or 330.7 lb. are used and the hardness is read on the Rockwell C scale.

Portable Rockwell hardness testers are available and may be used to check metals on the job. These units are easily calibrated and give accurate readings. A portable tester is shown, Fig. 22-34, while Fig. 22-35 shows this tester as it is used in production work.

Fig. 22-35. *Portable Rockwell hardness tester being used to check hardness of steel stock.*

Another method of testing hardness is to use a Shore, direct reading Scleroscope. This machine is based on the impact or rebound of a ball or hammer from a test specimen. The machine consists of a vertical glass tube channel of a certain height. At the top of this is mounted a steel hammer, having a diamond tip of a certain diameter and size. The specimen to be tested

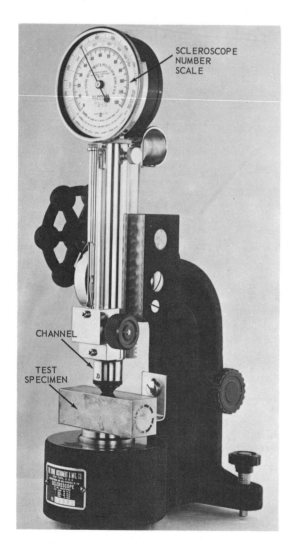

SCLEROSCOPE NUMBER SCALE

CHANNEL

TEST SPECIMEN

Fig. 22-36. *Scleroscope hardness testing machine. The distance the small hammer rebounds up the channel after it drops on the test specimen is read on the dial gauge.* (Shore Instrument & Mfg. Co.)

GAUGE MEASURES FORCE ON PENETRATOR BALL

PENETRATOR

SPECIMEN TABLE

Fig. 22-37. *A hydraulically-operated Brinell hardness tester.* (Detroit Testing Machine Co.)

is placed below the channel, and the hammer is released as shown in Fig. 22-36. The distance that the hammer rebounds after it contacts the metal may be read on the scale beside the channel. The hardness of the metal, as indicated by the scale number with this tester, will range from 0 to 140. The higher the number, the harder the metal. A high carbon steel will indicate approximately 95 points on the scale.

machine), as shown in Fig. 22-37. The ball-point is moved by hydraulic pressure (indicated on a dial scale). The dial indicates the number of pounds of

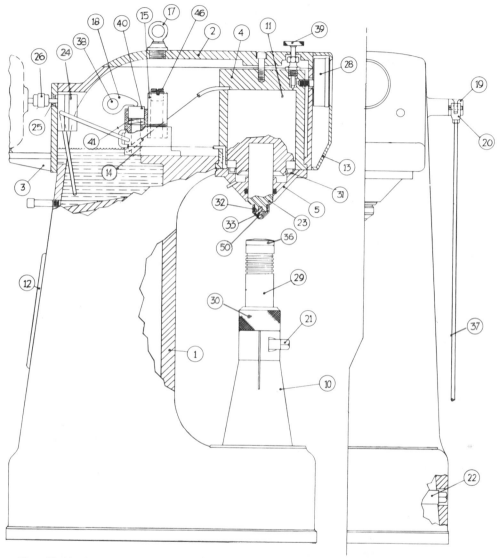

Fig. 22-38. Interior construction of Brinell hardness tester. The main parts are: 11 - Hydraulic cylinder. 14 - Hydraulic valve and pressure regulator, 24 - Hydraulic pump, 30 - Manual adjustment, 36 - Anvil, 37 - Hand-or-foot operated control lever, 50 - Ball (penetrator).

A rubber tube-and-bulb arrangement manipulates the hammer for testing purposes.

A third method of testing for hardness is to use a machine having a ball-point built into a press (Brinell

force exerted on the specimen. The specimen is mounted below the ball, and the ball (10 mm dia.) is pressed into the specimen under a load of 3,000 kilograms (6,614 lbs.) for 10 seconds. A microscope is then used to measure the diameter of the indentation in millimeters. The area of the depression divided by the load gives the Brinell-hardness number. A table is always supplied with the machine to permit

measurement indicates the hardness of the surface tested.

The length of an indentation made on hardened steel with the diamond penetrator under a load of 100 grams is

ROCKWELL C	ROCKWELL B	BRINELL	SCLEROSCOPE
69	- - -	755	98
60	- - -	631	84
50	- - -	497	68
40	- - -	380	53
30	- - -	288	41
24	100	245	34
20	97	224	31
10	89	179	25
0	79	143	21

Fig. 22-39. *Table showing comparison of hardness numbers in Rockwell, Brinell, and Scleroscope scales.*

one to determine the hardness number, once the diameter of the indentation is known. This method is often used for testing the softer metals. The construction of the machine is shown in Fig. 22-38. The table, Fig. 22-39 shows a comparison of hardness numbers in the Rockwell, Brinell, and Scleroscope scales. Microhardness testers are being developed which will injure the metal less because the indentations made by these machines will be very small. Because the metal surface is marked microscopically the tests can be performed more frequently over the surface of a metal object. Fig. 22-40 shows a microhardness testing machine.

Microhardness testers have been developed which make it possible to test the hardness of a part without appreciably injuring the part. A minute diamond penetrator is used to penetrate the surface to be tested. The loads on the diamond penetrator may be varied from 25 grams to 50,000 grams. After the surface has been penetrated, the size of the indentation is measured using a powerful microscope. This

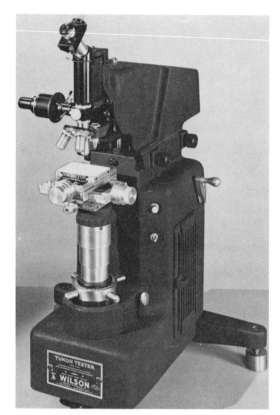

Fig. 22-40. *Microhardness testing machine. (Wilson Div., American Chain & Cable)*

about .0015 in. and the depth of the penetration is about .00005 in. Fig. 22-41 illustrates the indentation made with the Knoop diamond penetrator.

22-15. MICROSCOPIC METHOD OF TESTING WELDS

A test commonly used in the metallurgical laboratory for testing a weld is to procure a sample of the weld and polish it to a very smooth, mirror-like finish, which shows absolutely no scratches on the surface. The sample is then placed under a microscope which magnifies the surface of the metal from 50-5,000 diameters (usually 100 or 500 diameters). The appearance of the crystals, and the appearance of the metal in general under the microscope, reveals such things as the amount of impurities in the metal,

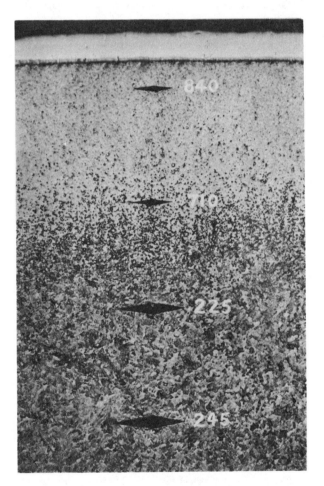

Fig. 22-41. The microhardness indentations made by a Knoop diamond penetration. The white layer at the top of the photograph is chromium (75X).

heat treatment, and grain size. In certain standard metals, a microscopic study of the surface can accurately determine the carbon content. Fig. 22-42 shows a microscope used

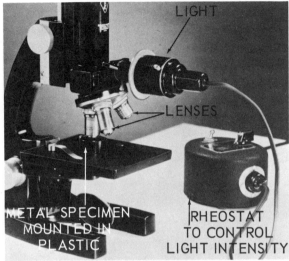

Fig. 22-42. Laboratory microscope used to examine polished metal specimens. Note the turret mount with three different magnifications, the plastic mounted specimen, and the rheostat-controlled light source. (Bausch & Lomb, Inc.)

to inspect metal specimens. Occasionally, the metal being worked on is treated with acid or etched. To etch a sample after it has been polished, immerse it in a weak acid solution (10 percent nitric acid, 90 percent alcohol, called nitrol, is used for steel) for a certain length of time. The metal is then studied under a microscope. Certain acids bring out certain features, such as the kind of impurities, slag spots, poor fusion, and shrinkage checks. A type of weld frequently given this test is the sample that is removed from boiler seams. The most important feature of the micrographic study of the metal is that photographs may be taken of the metal, by means of a specially adapted camera attached to the microscope. By taking photographs of each specimen, a very accurate comparison may be made between the samples studied. See CHAPTER 24 for microphotographs of the more common metals.

Sections for testing and examination may be cut from the base metal or the weld by a number of methods, one of which is shown in Fig. 22-43. The

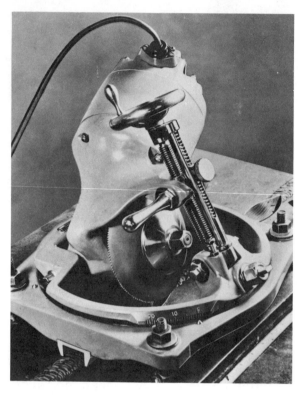

Fig. 22-43. *Machine for removing test samples of metal from welds. (Fibre Metal Products Co.)*

machine shown uses a circular blade to cut boat-shaped sections of any depth, as shown in Fig. 22-44.

The sections removed may be polished for micro and macroscopic examination; or they may be used for hardness tests, or chemical analysis. Defective areas of a weld may be removed by this method and the cavity filled with a sound weld.

Methods are being developed to perform micrometal analysis. By this method extremely small specimens may be removed from the metal to be

analyzed. This method enables an analysis to be made of metal parts that are to be placed in service.

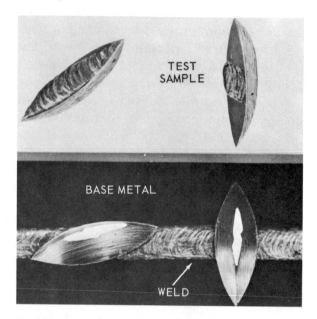

Fig. 22-44. *Typical samples removed from a weld for inspection and testing using machine shown in Fig. 22-42. The samples may be cut at any angle to the weld bead.*

22-16. MACROSCOPIC METHOD OF TESTING WELDS

A microscopic view of a weld does not cover enough area to obtain a picture of an entire weld for inspection purposes. Macroscopic pictures are only 10 to 30 magnifications (10X to 30X), and are better suited to this purpose. When the sample is deeply etched with hot nitric acid, the structure of the weld stands out more clearly. Fig. 22-45 shows a macroscope with 10X-30X magnifications. The crystalline structure of the metal is not so clearly revealed, but cracks, pits, and pin holes are more clearly seen, as shown in Fig. 22-46. Scale inclusions are easily detected by this method. This test also shows up grain (crystal) size. A large grain size indicates improper heat treatment after or during welding.

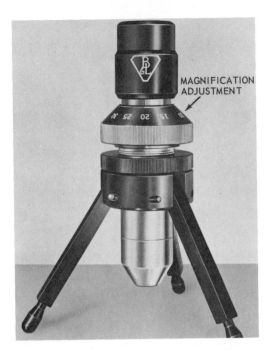

Fig. 22-45. A macroscope used to inspect welds at 10X to 30X magnifications. This instrument may be used for on-the-spot inspections. It weighs only 12 oz.

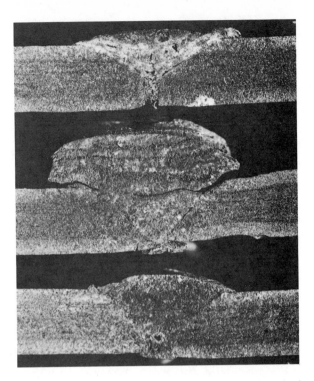

Fig. 22-46. A macroscopic photograph of etched welds in cross section. The quality of the weld is more easily judged than with the naked eye.

22-17. CHEMICAL ANALYSIS METHOD OF TESTING WELDS

The final complete investigation of the welded material consists of a thorough chemical analysis. The chemical analysis may be both qualitative and quantitative. Qualitative analysis determines the different kinds of chemicals in the metal while the quantitative analysis determines the kind and amount of each chemical in the metal. This type of investigation is necessarily tedious and expensive. The tests are not of direct value to a welder or to a welding company. It is only in cases of having trouble with large quantities of metal that these tests are resorted to. The weldability of a metal is dependent to a great extent on the impurities in the metal.

Usually the manufacturer can supply complete data on the physical and chemical properties of the metal.

22-18. THE PEEL TEST

Lap joints may be tested to destruction by means of the peel test. This test is most often used to check the strength of a resistance spot weld.

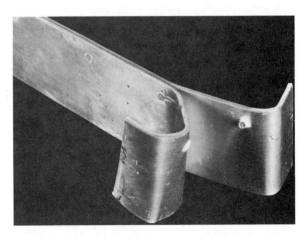

Fig. 22-47. Destructive peel test. Test welds are made to determine weld strength, size, and proper machine settings required for a given job.

All recommended machine settings are made on the spot welding machine before several spot welds are completed on test pieces.

The test pieces are then peeled apart. See Fig. 22-47. If the spot weld nugget is of the correct diameter and is torn out of one piece, the spot weld is considered to be properly made. All machine settings are considered to be correct for the parts to be welded when the peel test is made satisfactorily.

22-19. REVIEW OF SAFETY

Proper inspecting and testing of welds requires attention to certain safety factors. If there is a danger of parts flying when the weld is broken, a face shield should be worn to protect the operator. The operator should never stand in the probable trajectory of a flying test piece.

There are dangers from radiation. One should avoid exposing one's self to the types of radiation used to inspect metals. Only trained experts should handle isotopes (a radiation source). Thick concrete walls, lead and thick water sections are some of the shielding techniques used.

22-20. TEST YOUR KNOWLEDGE

1. Why is testing of welds considered important?

2. What two general methods are used for checking welds?

3. Why is the destructive test popular in shops?

4. What societies have standardized methods of testing welds?

5. Why are appearance tests considered the least accurate?

6. Why are shop destructive tests only comparable ones?

7. What is meant by the elastic limit of a metal?

8. What is meant by tensile strength of a metal?

9. Name three machines used for testing the hardness of metals.

10. Why is the hardness of machineable steel considered important?

11. Do the test of welds apply to both arc and acetylene welds?

12. Do tests apply to nonferrous metals as well as ferrous metals?

13. What is meant by the microstructure of a weld?

14. Are large crystals detrimental to the strength of a weld?

15. Name some special cases which involve special testing of welds.

16. How many times is pressure applied on the sample when using the Brinell Hardness Tester?

17. What advantage do X-ray tests have over magnetic flux-tests?

18. What is the usual magnification when viewing a weld through a microscope?

19. Is the ability to bend without breaking considered a good test for welds in mild steel? Why?

20. Is the template method of inspecting welds usable on pipe welds?

21. Is sound useful as a weld testing medium?

22. How small a leak can be detected with a mass spectrometer test?

23. What makes the particles adhere to the edges of a crack during a magnetic test?

24. Is the time element important when using the dye penetrant test?

25. When is blacklight used during testing?

Chapter 23

PRODUCTION OF METALS

It is important that a welder understand the origin, the production, and the refinement of metals in use today. One must have a thorough understanding of metals if one is to be a welding technician.

The processes used to change the ores to the commercial metals, and then to produce alloys, are constantly being improved.

Iron and steel are the most common metals in industrial use. Great quantities are produced each year. For many years copper and tin were the next most popular metals. At present, aluminum is rapidly increasing in use. Magnesium and titanium are also increasing in their applications, particularly in the aero-space industries.

Recently, many exotic metals have been successfully produced, mainly due to the needs of our nuclear and space age.

This chapter deals with metals that have industrial applications.

23-1. DEVELOPMENT OF METAL PRODUCTION

The first iron known to humans came to the earth in form of meteors. This explains why the Egyptians called it the "metal of heaven." The first iron was used as far back as 3500 B.C.

The forerunner of one modern blast furnace was the Catalan forge, developed in the 17th century. The Catalan forge was a shallow cavity made of brick and stone usually of an oval shape. A tube projected into the center of this cavity through which air entered. Air was supplied at first by a manually operated bellows and later by water power to force the air through the mixture of iron ore, wood, and charcoal. A man by the name of Dudley, substituted coal for the wood and charcoal. Abraham Darby introduced the use of coke as a fuel in the early part of the 17th century.

Aluminum is the most abundant metal on earth; iron is the fourth most abundant.

Except for minute amounts, aluminum was not used commercially until in the 1880's when the electrical process of refining the ore was developed. The growth of the use of this metal has been steadily increasing since.

Copper and its alloys have their use dating back to ancient times. The copper and tin alloy (bronze) was so common in ancient times that one segment of ancient history is called the Bronze Age. Copper is found as a metal in nature, and it was this metal that was used by the ancients for tools and weapons. Most copper (about 2,000,000 tons per year) is obtained by the electrolysis process.

23-2. METHODS OF MANUFACTURING STEEL

Steel is produced by adding carbon in controlled amounts to iron from which most, if not all, of the carbon and other impurities have been re-

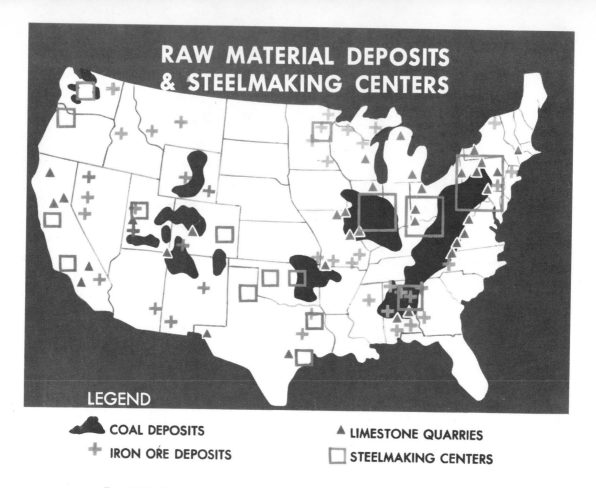

RAW MATERIAL DEPOSITS & STEELMAKING CENTERS

LEGEND

⬛ COAL DEPOSITS   ▲ LIMESTONE QUARRIES

✚ IRON ORE DEPOSITS   ▢ STEELMAKING CENTERS

*Fig. 23-1. Locations of raw material deposits, and steel making centers.
(U. S. Steel Corp.)*

moved. This alloy (mixture) of iron and carbon is called straight carbon steel.

Alloy steels are produced by adding elements, such as nickel, chromium, manganese and other elements to steel in controlled amounts.

The first step in the production of steel is to refine the iron ore containing many unwanted impurities, into cast or pig iron. The refinement of iron ore is usually done in the reducing atmosphere of a blast furnace. Pig iron contains some impurities and a relatively large amount of carbon. The high carbon content makes pig iron brittle. In order to produce steel, the carbon in the iron must be removed in the oxidizing atmosphere of a steel-making furnace or Bessemer con-verter. The oxidizing atmosphere "burns off" the carbon and impurities in the iron. Once the impurities and carbon have been eliminated or reduced to a minimum, controlled amounts of carbon and other alloying elements may be added to the iron to produce the type of steel desired.

The furnaces and processes used to produce steel and alloy steel are as follows:

1. Open hearth furnace.
2. Bessemer converter.
3. Basic oxygen furnace.
4. Electric furnace.
5. Vacuum furnace.

When it has been determined that the steel in the steel-making furnace contains the desired amounts of carbon and other elements, the molten steel is formed into ingots, slabs, blooms, or billets (all differ in cross sectional shape or area). From these rough stock shapes, the steel may later be formed into more exact shapes and products.

23-3. MATERIAL USED BY THE BLAST FURNACE

Iron seldom exists free in nature. It is mined from the earth in the form of iron ore or iron oxides mixed with impurities in the form of clay, sand, and rock. The most important types of iron ore are:

Hematite (red iron) _____ Fe_2O_3 _____ 70 % iron
Magnetite (black) _____ Fe_3O_4 _____ 72.4% iron
Limonite (brown) _____ $Fe_2O_3H_2O$ ___ 60 % iron
Siderite (iron carbonite) __ $FeCO_3$ _____ 48.3% iron

Hematite is the most important ore in the United States. It comes from the Lake Superior and Birmingham, Alabama districts.

A good flux that will melt and combine with impurities in the molten iron ore, such as limestone, must be used in the blast furnace. The limestone combines with the impurities and floats them in the combined state (slag) above the molten iron at the bottom of the furnace. The molten slag is drained from above the pig iron just before the pig iron is tapped or removed from the furnace.

One of the best fuels for the blast furnace is coke, as this furnishes enough heat for the reactions and is low in such impurities as sulphur and phosphorus. Some modern blast furnaces are using gas injection and solid fuel injection. These combustibles enter the furnace in the area of the tuyeres.

Coke is manufactured by heating soft coal (Bituminous) in a closed container until the gases and impurities are driven off. Coke which is practically pure carbon then remains. The gases which are driven off are condensed and furnish many useful by-products such as coal tar, gasoline, lubricating oil and fertilizer as well as coke oven gas that can be burned to furnish heat and power.

To operate a modern blast furnace for one day requires about 2,000 tons of iron ore, 1,000 tons of coke, 500 tons of limestone, and 4,000 tons of air.

Fig. 23-1 shows the location of the major coal, limestone, iron ore and steel-making centers in the United States.

23-4. BLAST FURNACE

The blast furnace has five major operations to perform:
1. Deoxidize iron ore.
2. Melt the slag.
3. Melt the iron.
4. Carbonize the iron.
5. Separate the iron from the slag.

The modern blast furnace is a huge tubular furnace made of steel and lined with firebrick. The average size is approximately 100-ft. high and 25-ft. in diameter. Some are much larger. Fig. 23-2 shows a typical blast furnace in cross section. Around the bottom of the furnace are openings (tuyeres) through which hot air may be forced. An opening near the top allows gases to escape. The ore, limestone, and coke are carried up to top of the furnace and dumped down into the furnace through a bell-shaped opening (hopper).

The coke burns and produces enough heat to melt the iron. The excess carbon from the coke unites with the iron and lowers its melting temperature.

The melted iron forms at the bottom of the furnace where it is drawn off when a sufficient quantity has collected. The flux melts and collects the impurities. It floats on top of the molten iron where it can be drawn off the furnace through an opening higher than the one hours a batch of blast furnace iron (pig iron) is drawn off from the bottom of the furnace.

The iron coming out of the blast furnace is called pig iron because it was formerly cast into bars called pigs.

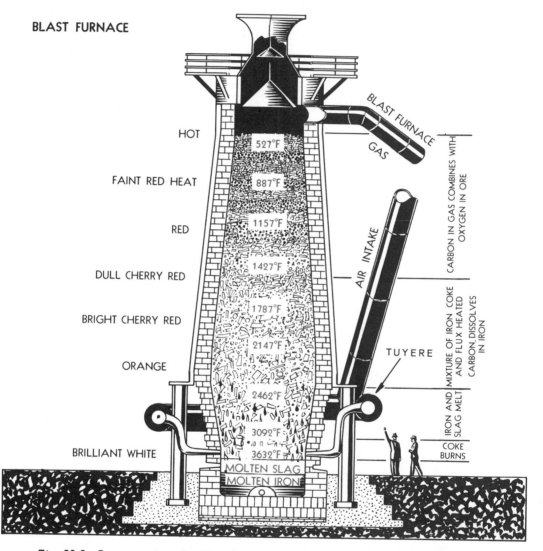

Fig. 23-2. Cross section of a blast furnace. Notice the temperatures and how the iron ore changes chemically as it moves down through the furnace. (General Motors Corp.)

through an opening higher than the one through which the iron is taken out. The operation of the furnace is continuous and the right proportions of ore, limestone, and coke are regularly dumped in at the top of the furnace. Every few

In modern practice, some of the iron from the blast furnace is not allowed to cool, but is taken directly to other furnaces such as the open hearth furnace where it is made into steel, or to the Cupola furnace where it may be further refined and made into castings.

(For information on blast furnace chemistry see CHAPTER 28.)

23-5. CAST IRON

Approximately 30 percent of all the pig iron produced is used in the manufacture of gray cast iron. Gray cast iron is simply a casting that has been cooled very slowly thus allowing some of the carbon to separate, forming free graphite. This graphite causes the gray appearance. White cast iron (very hard and brittle) is made by cooling the casting quickly. Cast iron is usually made by melting pig iron in a cupola furnace where the pig iron is refined to eliminate some of the excess carbon and other impurities.

23-6. MALLEABLE IRON

Cast iron is a desirable metal from which intricate metal parts may be cast as it is very fluid when molten and flows freely to all parts of a mold. It may be machined relatively easily. However, it has several undesirable characteristics such as brittleness and lack of malleability (the ability to be pounded into shape). In parts demanding malleability or resistance to shock, malleable iron may be used. Malleable iron is made by prolonged heating or annealing of white cast iron at a temperature of approximately 1400 deg. F. The period of time depends to a considerable extent on the thickness of the casting. The casting is allowed to cool slowly.

The heating period gradually reduces the carbon content in the skin of the casting leaving a low carbon surface. The depth of the low carbon metal depends on the time the casting is kept at the 1400 deg. F. temperature. When

cooled slowly the inner structure is gray cast iron and the skin is similar to mild steel, as shown in Fig. 23-3. This combination makes the structure more pliable and more resistant to shock.

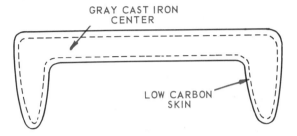

Fig. 23-3. *Structure of malleable iron shown in cross section.*

23-7. NODULAR CAST IRON

Gray cast iron has free graphite present in the metal. It obtains its gray appearance as a result of the combined inflection (bending) of light from the black graphite, plus the shiny metal. The graphite is in a flake form and the flakes create a weakness in the metal (brittleness).

A new process for making cast iron transforms the graphite flakes normally formed in cast iron into nodules of graphite (spheres). The resulting casting is more ductile and resistant to shock failure.

Nodular cast iron is obtained by adding a small amount of magnesium to the molten metal. Occasionally silicon and/or nickel is also added to provide other desirable properties.

23-8. WROUGHT IRON

Wrought iron contains the least amount of carbon of any of the ferrous metals used commercially. It is manufactured in a puddling furnace. To make wrought iron, pig iron is melted on the hearth of the reverberatory or puddling furnace which is lined with iron oxide. This process brings about the almost complete removal of the carbon, silicon, and manganese. As the carbon is removed, the fusion temperature of the iron rises, it becomes pasty and can be rolled up in balls and removed from the furnace. It is then squeezed through rollers to remove most of the excess slag. The wrought iron is rolled into muck bars and finally into commercial forms.

Wrought iron is soft, tough and malleable. It is ideal for ornamental work, as it is rust resisting, easily shaped, and may be easily welded.

23-9. STEEL

Steel may be defined as iron combined with .03 to 1.7 percent of carbon. See CHAPTER 24 for more technical specifications of steel.

It is produced by reducing cast iron and scrap steel in one of several types of furnaces. At the present time approximately 300,000,000 tons of steel are

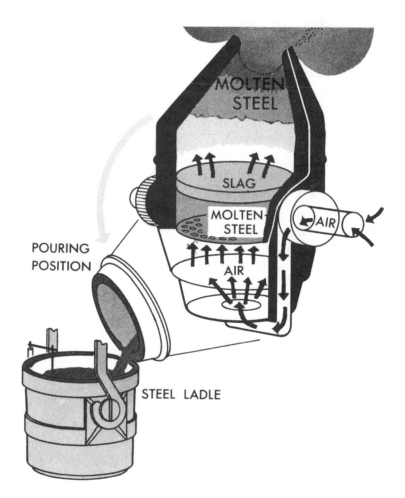

Fig. 23-4. Schematic drawing of a Bessemer furnace. Note how the furnace is tilted to pour the metal into a large ladle. (U. S. Steel Corp.)

produced in the world each year. The United States produces approximately one third of this total. Several types of steel producing furnaces are:

1. Bessemer.
2. Basic Oxygen.
3. Open Hearth.
4. Electric.
5. Crucible.

23-10. BESSEMER PROCESS

One of the earliest methods of manufacturing steel was invented in 1856 by Henry Bessemer. The Bessemer Converter, see Fig. 23-4, as the furnace is called, is a large pear-shaped container lined with firebricks which opens at the top.

Molten pig iron is poured into the furnace and a blast of air or oxygen is turned on through holes in the bottom. Each ton of pig iron contains about 75 lbs. of carbon, 25 lbs. of silicon, 1 lb. of sulphur, and 15 lbs. of manganese, most of which is burned out in the operation of the furnace. These elements in the pig iron become the fuel used in this process. Additional fuel is not necessary. The temperature of the molten pig iron is about 2060 deg. F. when it is poured in the converter. In a few minutes after the blast of air is turned on, the burning of the elements in the iron raises the temperature to 3500 deg. F. The converter holds 8 to 10 tons of molten pig iron. A "blow" or charge treatment requires 10 to 15 minutes. Most of the carbon, silicon, sulphur, and manganese are burned out, which causes the mouth of the furnace to belch forth bright flames. As the impurities are burned out, the flame changes color and an experienced Bessemer operator can tell by the color of the flames when the air blast should be turned off to terminate the blow. The

furnace is then tilted and the steel is poured into large ingots. Later these ingots are rolled to desired shapes, Fig. 23-5. If the ingots must be stored

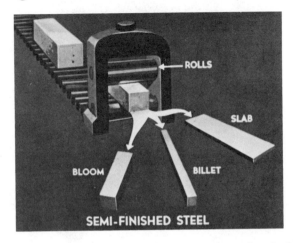

Fig. 23-5. An ingot being rolled into a semi-finished shape in a steel mill. The insert shows the semi-finished shapes generally produced from an ingot.

before they are rolled, they are placed in large ovens called soaking pits, as shown in Fig. 23-6, where they are kept at a temperature suitable for the rolling operation.

At present, only a small percentage of the steel produced in the United States is produced in Bessemer converters.

Portable Rockwell hardness tester. The dial indicates the Rockwell scale used and the hardness number is shown on the cylindrical scale.
(Ames Precision Machines)

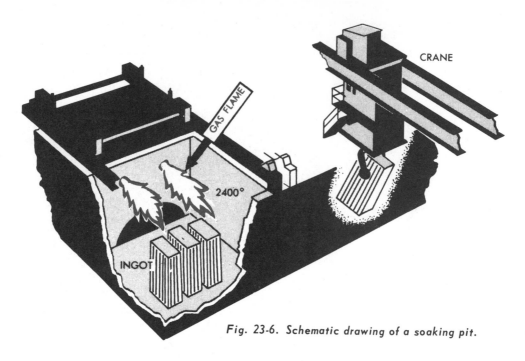

Fig. 23-6. Schematic drawing of a soaking pit.

Fig. 23-7. Schematic of a basic oxygen furnace. Oxygen is released above the molten metal in the furnace and the products of combustion are exhausted into a hood and are treated to prevent contamination of the atmosphere.

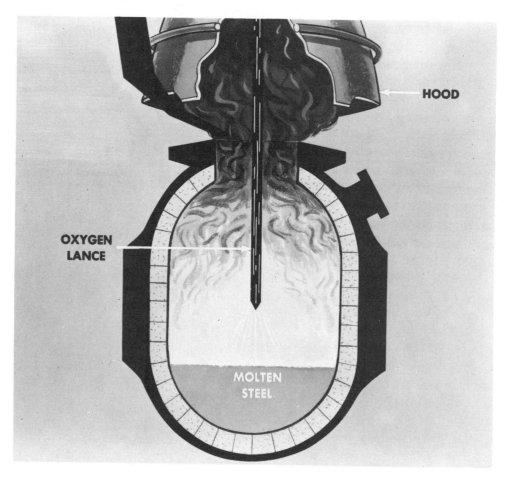

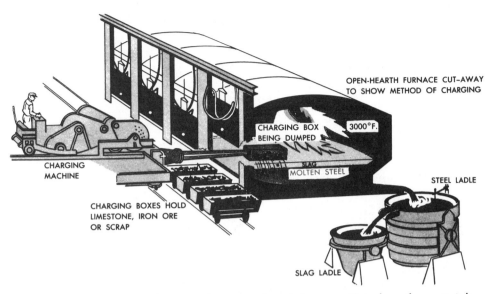

Fig. 23-8. Schematic drawing of an open hearth furnace. Note how the materials are charged into the furnace and also how they are removed from the furnace.

23-11. BASIC OXYGEN PROCESS

The basic oxygen furnace shown in Fig. 23-7 makes use of a furnace which resembles the Bessemer converter, but produces steel from iron in a different manner. The furnace is tipped on its side and molten iron and scrap steel are poured into the mouth of the furnace. The furnace is then rotated to a vertical position under an exhaust hood. A water-cooled oxygen lance is lowered to a position above the molten metal and oxygen is blown into the furnace. Oxygen from above burns off the impurities in the molten metal and produces steel. In 40 to 60 minutes, about 80 tons of quality steel can be produced. At the end of a prescribed time, the oxygen is turned off. The oxygen furnace is then tipped to pour off the steel produced and a new charge of molten iron and scrap steel is poured in to begin a new cycle.

23-12. OPEN HEARTH PROCESS

The open hearth furnace was invented in 1861 by Siemens. This method of making steel is sometimes called the Siemens-Martin Open Hearth Process, being named after the men who developed the method originally.

In such a furnace, Fig. 23-8, the metal is contained in a large shallow basin, holding from 150 to 300 tons of metal. At each end of the basin or metal container there is a preheating stove made of firebrick arranged in a checkerboard pattern. Air and a fuel gas enter the furnace at one end after passing through one of the preheating stoves. As the fuel gas and air burn above the metal in the furnace, they heat the metal and oxidize or burn off unwanted elements and impurities in the molten metal. As exhaust gases exit from the furnace, they travel through the unused preheating furnace on the exhaust side of the furnace. The hot gases heat the bricks in this stove as they travel out of the furnace. Fig. 23-9 shows the furnace in cross section to illustrate the method of preheating the incoming gas and air. Periodically the direction of the incoming fuel gas and air is reversed so that the cold input gases travel over the hot bricks in the preheat stove which were previously heated by the exhaust gases. This is called a regenerative process of heating. The preheating of the incoming fuel gas and air results in higher temperatures inside the open hearth furnace. The total time for processing each charge is 20 to 30 hours. Some open hearth furnaces are made so that the entire furnace may be tilted to make it easier to remove the molten steel at the end of the refining process.

Pure oxygen is now being used in steel. Some advantages of this process are:

1. A minimum quantity of steel is lost in the process.

2. Better control over the alloying elements is obtained.

3. Steel is cleaner because it contains fewer oxides.

4. Larger batches of steel may be made at one time.

5. Pig iron unsuitable for the Bessemer process may be made into steel by the open hearth method.

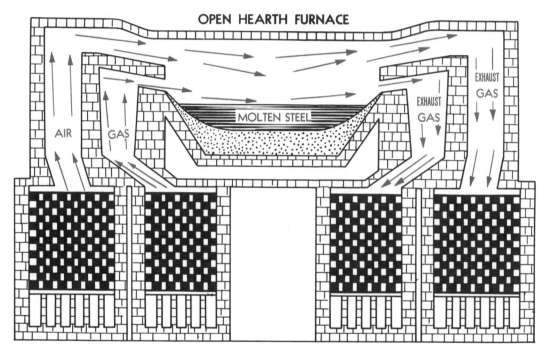

Fig. 23-9. Cross section of an open hearth furnace. Notice how fuel gas and air circulate through the furnace. (General Motors Corp.)

some open hearth furnaces to speed the production of steel. For example, an 82.5 ton heat in a furnace using oxygen may be completed in 61.5 minutes.

Since more steel can be made in a given time, the use of oxygen is economically feasible and is growing in popularity.

The high temperatures obtained by this process aid in burning out the carbon and impurities in the iron to make

6. Steel manufactured by the Bessemer process may be further refined in the open hearth furnace.

By taking periodical chemical analysis of the metal in the heat the exact composition may be determined. By adding alloying ingredients, the heat may be held to very close chemical tolerances and a much finer steel may be produced than is possible in the Bessemer furnace.

A copper crucible for a consumable electrode vacuum melting furnace being lowered into position. (Allegheny Ludlum Steel Corp.)

23-13. ELECTRIC FURNACE

A popular method used to produce special, high quality, steel alloys is the electric furnace, as shown in Fig. 23-10. In this type furnace the chemical constituents of the metal may be closely controlled. Various alloying elements may be added to create a steel with predetermined characteristics.

Generally steel from an open hearth furnace is used to charge the electric furnace. Chemical analysis of the steel placed in the electric furnace is made and all necessary alloying elements are placed in the furnace before the furnace is closed to the atmosphere.

Heat required to melt the metal in this furnace is produced by an electric arc.

Large diameter movable electrodes are installed in the top of the furnace above the metal. An arc is struck between the carbon electrode and the metal to furnish the heat required to melt the metal and the alloying elements. As the electrodes are consumed they are moved down to a given distance above the molten metal. The carbon electrodes may be changed as required from the top of the furnace.

Test samples of molten metal may be removed through small inspection ports. These samples are analyzed and

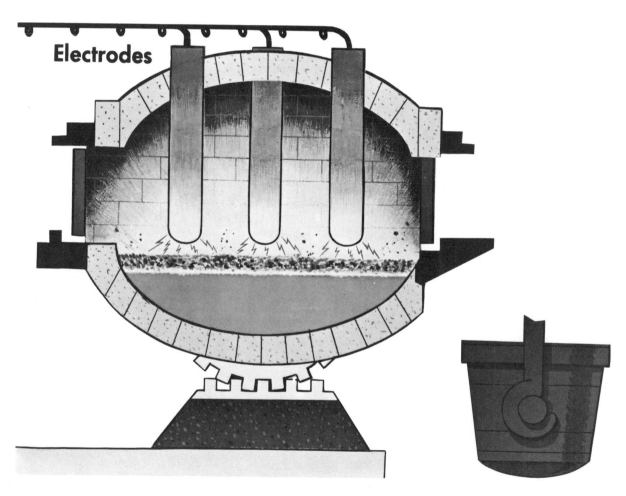

Electrodes

Fig. 23-10. An electric furnace. Note the carbon electrodes at the top. This furnace is tilted to pour molten metal into the ladle.

additional alloying elements are added to the furnace as required.

Electric furnaces vary in capacity from 5 to 50 tons. The entire furnace is generally built to tilt to make loading and unloading easy.

23-14. CRUCIBLE FURNACE

One of the oldest methods of refining steel is the crucible furnace. The heat for this furnace is produced by burning fuel gas. A crucible or covered pot made of ceramics and graphite is usually used to hold the charge. In this process, wrought iron or wrought iron and scrap steel are loaded into a crucible. The correct amounts of carbon and other alloying elements required to produce the desired finished steel are then charged into the crucible.

The use of scrap steel generally lowers the quality of the finished product. After the crucible is charged, it is covered and sealed. The crucible is then placed into the furnace where hot gases heat the crucible and its contents.

The quality of the steel produced in a crucible furnace is generally considered to be higher than that produced in an electric furnace, but the process is much slower and more expensive. The charge in a crucible may vary from a few pounds to several tons.

23-15. CONTINUOUS CASTING PROCESS

The continuous casting process for the manufacture of steel is a comparatively new method which is gaining in popularity. In this process, shown in Fig. 23-11, liquid steel is poured into a reservoir or tundish. From the tundish the metal flows vertically into a water-cooled mold. The molten metal in contact with the sides of the mold cools quickly and shrinks away from the side of the mold forming a shell around the molten metal in the center of the mold. This shell is supported by the withdrawing rolls as the column of steel is pulled from the mold. As the column of steel leaves the mold, jets of water are sprayed on the metal to cool and solidify the entire column. The metal as it comes from the mold may be in the form of a slab or a square bar. As the metal leaves the withdrawing rolls it is cut to desired lengths for further processing.

This process has the advantages of eliminating ingot pouring, removing of the ingots from their molds, the use

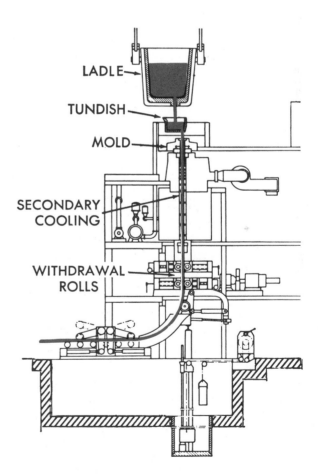

Fig. 23-11. Schematic drawing of the continuous casting process for making steel.

of soaking pits and reheating furnaces, and the rough rolling of ingots into semifinished forms.

23-16. INDUCTION HEATING PROCESS

Induction heating is defined as "raising of the temperature of a material by means of electrical generation of heat within the material and not by any other heating method such as convection, conduction, or radiation." The material being heated is not part of a closed electrical circuit as it would be in resistance welding or as it is in the electric furnace.

Induction heating is a phenomenon caused by an alternating magnetic field created in electrical conductors whenever alternating current flows through them.

Magnetic materials, such as iron and steel, when placed within the area of an alternating magnetic field, are heated by both hysteresis and eddy current losses, see Fig. 23-12.

Hysteresis loss is caused by friction among the molecules of the material as they move within the metal in the magnetic field.

The magnitude of this hysteresis loss and the heat created by it, is proportional to the frequency of the magnetic field. Eddy current losses are losses caused by resistance resulting from small circulating currents within a material placed in an alternating magnetic field. These resistance losses cause heat which is absorbed by the metal being heated. The amount of heat created by eddy currents is proportional to the square of the alternating frequency of the current, and the square of amperage flowing in the conductor, which produces the magnetic field.

In an induction furnace, the metal to be heated is contained in a vessel and electrical conductors are wound around the vessel to form a coil. Alternating current is passed through the coil and induction heating of the vessel and metal in the vessel takes place.

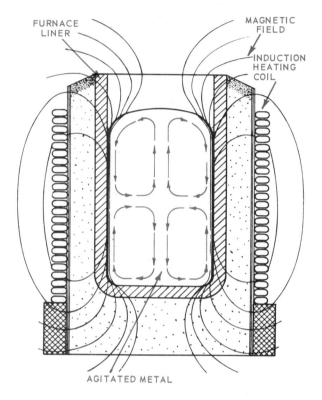

Fig. 23-12. Schematic cross section of a small induction furnace. Metal in the furnace is agitated and mixed by magnetic eddy currents, to produce a more homogenous metal.

By accurately controlling the frequency and amperage of the alternating current passing through the induction coil, it is possible to accurately control the temperature of the metal being heated. A small induction furnace unit is shown in Fig. 23-13.

A new type of low-frequency induction furnace with a 4-ton capacity is now in use. It is being used in conjunction with a cupola furnace to further refine cupola iron. Tests show that iron from the cupola-induction furnace combination shows superior tensile strength, fluidity, and other favorable casting metal characteristics. It also

Fig. 23-13. A complete induction melting furnace installation. (Ajax Magnethermic Corp.)

has desirable machining characteristics. Fig. 23-14 shows a schematic cross section of a large induction furnace.

The design of the furnace and its low frequency induction coil, in addition to heating the metal, sets up a controlled stirring action of the molten metal in the furnace. In addition, the stirring action keeps the slag on the surface agitated providing openings in the slag layer through which entrapped gases may escape.

This process produces a fine grained and homogeneous metal.

23-17. VACUUM FURNACES

The melting of steel, steel alloys, titanium and other pure metals in a vacuum, reduces the gases in the metal and prevents the absorption of gases by the molten metal. Steels, melted by more conventional methods, absorb gases which cause porosity and inclusions in the metal when it solidifies. Gases formed in a vacuum furnace are pulled away from the molten metal by the vacuum pumps. The absence of

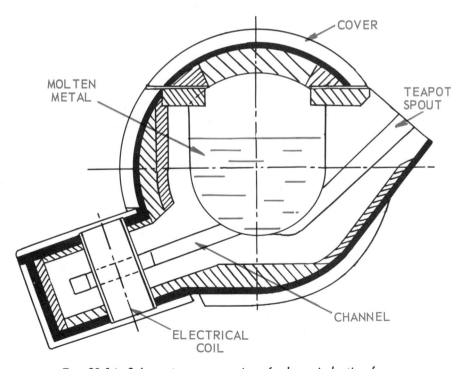

Fig. 23-14. Schematic cross section of a large induction furnace.

these gases improves many of the qualities of the steel including ductility, magnetic properties, impact strength, and fatigue strength.

There are two main types of vacuum furnaces used today. The two now in use are the vacuum arc, and vacuum induction-type furnaces. Other methods

Air and contaminating gases are constantly pumped out of the furnace by vacuum pumps. Fig. 23-15 shows such a furnace schematically.

This process is frequently used when close control of grain structure is required, or when the uniformity and purity of the metal is to be improved.

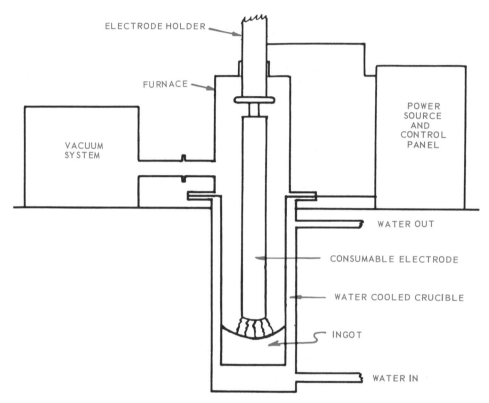

Fig. 23-15. Consumable electrode vacuum arc furnace shown schematically. (Allegheny Ludlum Steel Corp.)

are being developed.

In the vacuum arc furnace the metal, which is first produced by another method and formed into long round or square cylinders, is melted as a huge consumable electrode in the furnace. The metal is fed into the furnace at a controlled rate of speed to control the arc length.

As the metal drops from the end of the electrode, it falls into a water-cooled steel crucible which is the grounded part of the electrical circuit.

No provision is made in this furnace for adding alloying elements.

The vacuum induction furnace is used when close control of the chemistry of the metal is of prime importance. The metal is melted in an electric induction furnace which is airtight and attached to vacuum pumps so that contaminating gases are removed. Provisions are made to allow alloying materials to be added to the furnace without destroying the vacuum.

The heating of the metal, the pouring

Fig. 23-16. Removing hot ingots from vacuum chamber of induction vacuum melting furnace. The melting furnace is located in the tank on the left. Vacuum chambers are arranged to permit continuous operation of the unit.

of the metal into ingots, and the cooling of ingots are done under vacuum conditions to prevent contamination. Fig. 23-16 shows a large induction vacuum melting furnace installation.

23-18. MANUFACTURING STAINLESS STEELS

Stainless steel, the most common of the alloy steels, may be made in open hearth furnaces in the same manner as straight carbon steels except alloy metals are added during the time the metal is in the furnace. Stainless steels may also be made or refined in electric furnaces. Various types of stainless steels are described in CHAPTER 18.

23-19. MANUFACTURING COPPER

A great deal of copper ore contains a high percentage of pure copper. This ore is crushed and then washed in water to remove the lighter weight earth particles from the heavier copper ore. The ore is then mixed with coke and limestone and placed into a small blast furnace where the copper settles to the bottom of the furnace. Impurities in the copper ore are floated above the molten copper in the form of slag in the same manner as when refining iron ore in a blast furnace.

The copper removed from the blast furnace is further refined by the electrolysis process.

Electrolysis may be defined as a chemical change (decomposition) created in a material by passing electricity through a solution of the ma-terial or through the substance while it is in a molten state.

The electrolytic cell used to refine copper is shown in Fig. 23-17. Pure copper bars are used as the cathodes. Impure copper to be refined forms the anodes in the cell. Copper sulfate with some sulphuric acid is used as the electrolyte, or fluid, in the cell.

When the electric current is turned on, pure copper leaves the anode and deposits on or plates the cathode. The impurities fall to the bottom of the electrolytic cell. When the impure copper anode is consumed, it is re-placed and the heavily plated cathodes are removed and replaced with thinner bars. The sediment in the bottom of the cell often contains small quantities of silver and gold.

23-20. MANUFACTURING BRASS AND BRONZE

Brass is an alloy of copper and zinc. Bronze is an alloy of copper and tin. When a third or fourth element is added to brass or bronze to improve the physical properties, an alloy brass or an alloy bronze is created. Some of

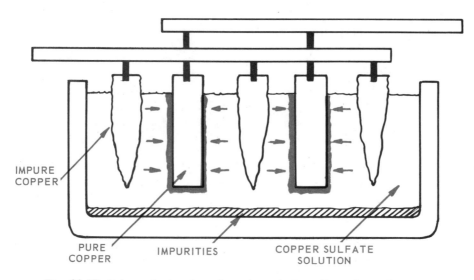

IMPURE COPPER

PURE COPPER   IMPURITIES   COPPER SULFATE SOLUTION

Fig. 23-17. Schematic drawing of an electrolytic cell used to refine copper.

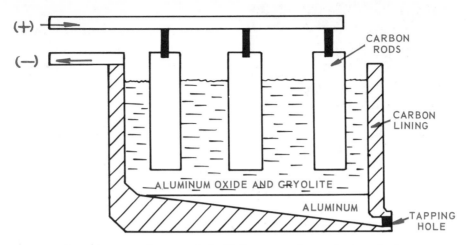

Fig. 23-18. Schematic drawing of the Hall process for producing aluminum.

the alloying elements added to brass are: tin, manganese, iron, silicon, nickel, lead, and aluminum. Alloying elements added to bronze are: nickel, lead, phosphorous, silicon, and aluminum.

To make brass or bronze, the alloying elements are heated in a crucible or an electric cupola.

The term bronze is often used with copper alloys which contain no tin or small quantities of tin in the alloy.

Hardware bronze, as an example, contains approximately 90 percent copper, 8 percent zinc, and 2 percent lead. Manganese bronze contains approximately 58.5 percent copper, 1.0 percent tin, about 39 percent zinc, .28 percent manganese, and 1.4 percent iron. See CHAPTER 28 for more information.

23-21. MANUFACTURING ALUMINUM

Aluminum is normally produced by separating it from the oxide (Al_2O_3) found in bauxite ore, although it may be found in many other forms. After the aluminum oxide is removed from the ore, it is dissolved in a molten bath of sodium-aluminum-fluoride (cryolite). An electric current is passed through this molten bath and pure aluminum is obtained by an electrolysis process (Hall Process).

The electrolytic cell used in this process is a carbon lined, open-top furnace with carbon electrodes suspended in a solution of aluminum oxide and cryolite. As the current passes through the solution the aluminum oxide is reduced. Pure aluminum is deposited at the cathode or lining of the furnace which falls to the bottom of the cell as shown in Fig. 23-18. Such a furnace is in continuous operation. Periodically the molten aluminum is poured from the cell into ingot molds which are stored for further processing.

23-22. MANUFACTURING ZINC

Zinc is principally produced by a distilling process. The zinc ore is heated with coke in a clay crucible and the zinc vapor is then condensed in a clay condenser. It may also be refined by an electrolytic process. It is used mainly as an alloying metal and for galvanizing.

23-23. PROCESSING METALS

Most metals as they come from the furnace are originally cast into ingots or molds for further processing. As

Fig. 23-19. A macrograph (4X) of etched cold rolled steel. The upper figure is across the grain of rolling, and the lower figure is along the grain of rolling.

of the metal, and to form them into more usable rough stock shapes, they are often rolled. The dense structure of steel after rolling is shown in Fig. 23-19.

In a steel-making plant, the large ingots cast from metal poured from a furnace, are rolled between large powerful rollers in a rolling mill and are reduced to blooms, billets, slabs, and even sheet stock as required. These blooms, billets, slabs, and sheet stock may be further processed in a rolling mill to produce rails, T beams, angle iron, bar stock, tubing, etc. Fig. 23-20 illustrates some typical shapes which are formed by rolling. Numerous operations are required to form some of the shapes.

Forging, either drop or press, is used to obtain shapes stronger than castings and which are not easily rolled into shape. Forge hammers and/or forming dies are used to pound the metal into the shape desired.

Extruding metal is a process by which metal normally in its plastic state is pushed with great force through dies which are cut in the shape of the

needed, the castings are reheated to a definite temperature depending on the metal and are then formed into a finished or semi-finished shape by one of the following methods:

A. Casting.
B. Rolling (Hot and Cold).
C. Forging.
D. Extruding.
E. Drawing.

Casting a metal in a sand or permanent mold is a popular method of producing objects with intricate shapes. Either a stationary or spinning (centrifugal) mold may be used.

To improve the physical properties

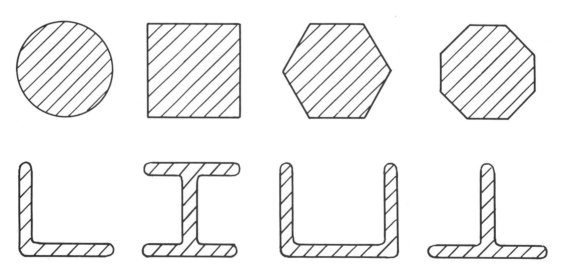

Fig. 23-20. Standard cross sectional shapes of metals. The welding process may be used to combine the principal forms into complex structures.

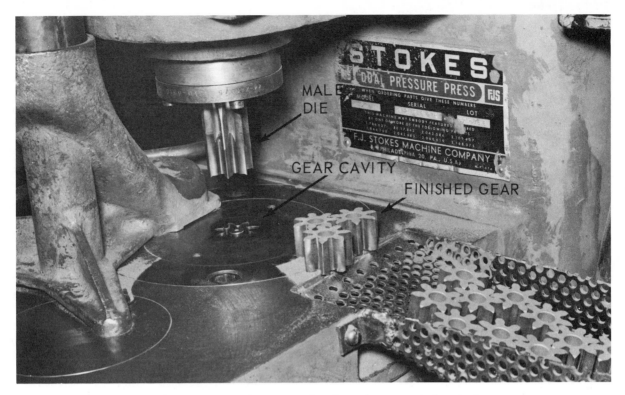

Fig. 23-21. Pump rotors made from powdered metals in a male-female die set.
(Ford Motor Co.)

desired cross section. This process produces long lengths of metal with uniform cross section.

Drawing metal through dies to form wires, tubing, and moldings is also a popular method of shaping metal to meet a certain requirement.

Many intricate shapes which would be difficult or impossible to make by any other method are now being produced by the powdered metals process. The metal to be used is reduced to a powder texture and is forced under pressure into a heated steel mold. The properties of the finished part approximate those of the original solid metal. Fig. 23-21 shows a die set being used to form gears from powdered metals.

These parts are made to very close tolerances. Most of them can be used without any additional machining or grinding.

Metal may also be reduced to actual fibers. The fibers are then laid down or woven to form mats. Mats of fibrous metal are finding increasing use in resistance welding where the mats are placed between parts to be welded.

23-24. REVIEW OF SAFETY

All furnaces which contain molten metal should be handled with great caution. Careless handling can cause great bodily harm and damage to equipment. Spilled molten metal will spread with great speed and will burn almost anything combustible. Special clothing, spats and special shoes must be worn. Special face guards should be worn. The eyes and face must be protected from flying particles and glare.

Gloves are needed at all times.

OSHA regulations for workers in metal refining plants and foundries must be obeyed.

23-25. TEST YOUR KNOWLEDGE

1. What materials are put into a blast furnace for producing pig iron?

2. How long does a charge remain on "blow" in a Bessemer furnace?

3. What fuel is used in a Bessemer furnace?

4. In which type of furnace is stainless steel most commonly made?

5. What types of steel does the crucible furnace produce?

6. What is the name of the most common aluminum ore?

7. Is molten pig iron part of the open hearth charge?

8. What is a tuyere?

9. What fuel is used for the open hearth furnace?

10. By what process is aluminum normally made?

11. How does malleable iron differ from gray cast iron?

12. Why does metal become heated in an induction furnace?

13. Why are vacuum furnaces used?

14. What type furnace is charged with some scrap steel?

15. Why is the open hearth furnace used more in steel making than the Bessemer furnace?

16. Why is oxygen used in place of air in some open hearth furnaces?

17. Describe the basic oxygen furnace.

18. How does the continuous casting process differ from other methods of making steel?

19. By what forming process may long pieces with intricate cross-sectional shapes be produced?

Chapter 24

METAL
PROPERTIES, IDENTIFICATION

It is necessary for a welder to have some accurate means of identifying metals. He must also have a good understanding of the constituents of metals in order that he may intelligently solve welding problems.

Metals are divided into two major fields. These are:

1. Ferrous metals.
2. Nonferrous metals.

Iron and its alloys are classified as ferrous metals.

The nonferrous metals include metals and alloys which contain either no iron or insignificant amounts of iron. Some of the more popular nonferrous metals a welder encounters are copper, brass, zinc, bronze, lead, and aluminum.

A great amount of welding is done on ferrous metals ranging from wrought iron to low, medium, and high carbon steel, tool steel, stainless steel and cast iron.

Recent developments in the various welding arts now makes it possible to satisfactorily weld practically all non-ferrous metals.

24-1. IRON AND STEEL

Iron is produced by reducing iron oxide, commonly called iron ore, to pig iron by means of the blast furnace.

Many types of furnaces are used to change pig iron into the various steels.

Some of these furnaces are: the open hearth furnace, the Bessemer furnace, the basic oxygen furnace, the electric furnace, and the cupola furnace. See CHAPTER 23 for more information on these furnaces.

Carbon steel is an alloy of iron and controlled amounts of carbon. Alloy steel is a combination of carbon steel and controlled amounts of other desirable metal elements. The percentage of carbon content determines the type of carbon steel. For example, wrought iron has .003 percent carbon, meaning three thousandths of one percent. Low carbon steel contains less than .30 percent carbon. Medium carbon steel varies between .30 and .45 percent carbon content. High carbon steel contains approximately .45 to .75 percent carbon, and very high carbon steel contains between .75 and 1.50 percent carbon. Cast iron contains 2-1/4 to 4 percent carbon of which some is in the noncombined form and is in the form of graphite in gray cast iron.

The carbon generally combines with the iron to form iron carbide, a very hard, brittle substance. This action means that as the carbon content of the steel increases, the hardness of the steel also tends to increase.

Various heat treatments are used to enable steel to retain the strength of its higher carbon content, and yet not have the extreme brittleness usually associated with high carbon steels. Also, certain other substances such as nickel, chromium, manganese, vanadium, and other alloying metals may be added to steel to improve certain physical properties.

A welder must also have an understanding of the impurities occasionally found in metals and their effect upon the weldability of the metal.

Two of the detrimental impurities sometimes found in steels are phosphorus and sulphur. Their presence in the steel is due to their presence in the ore, or to the method of manufacture. Both of these impurities are detrimental to the welding qualities of steel; therefore during the manufacturing process, extreme care is always taken to keep the impurities at a minimum (.05 percent or less). Sulphur improves the machining qualities of steel, but it is detrimental to its hot forming properties.

During a welding operation, sulphur or phosphorus tends to form gas in the molten metal, resulting in gas pockets in the welds, and in an increasing brittleness because of the sulphites and phosphates formed. Another impurity, usually resulting from the mechanical shaping of the metal to its final form, is dirt or slag (iron oxide) imbedded in the metal by the rollers, although some of the dirt may come from the by-products of the process of refining the metal. These impurities may also produce blow holes in the weld and reduce the physical properties of the metal in general. See PAR. 24-13.

24-2. PHYSICAL PROPERTIES OF IRON AND STEEL

A physical property is a characteristic of a metal which may be observed or measured.

As mentioned previously, the physical properties of steel are affected by the following:
1. Carbon content.
2. Impurities.
3. Addition of various alloying metals.
4. Heat treatment.

CHAPTER 22 describes various machines for determining these physical properties. Modifications of these testing machines are being introduced into the welding shop to enable operators to identify these metals and to check on the physical properties of weldments.

Some of the more important physical properties of steels are:
1. Tensile strength.
2. Ductility.
3. Hardness.
4. Brittleness.
5. Compression strength.
6. Elongation.
7. Malleability.
8. Toughness.
9. Grain size.

Fig. 24-1 illustrates the types of stresses imposed on structures.

Tensile strength is the ability of a metal to resist being pulled apart. This property may be measured on a tensile testing machine which puts a stretching load on the metal.

Ductility is the ability of a metal to be stretched. A very ductile metal such as copper or aluminum may be pulled through dies to form wire. The ductility of iron and steel may be measured during a tensile test.

Hardness is the quality which allows a metal to resist penetration. The Rockwell, Brinell or Scleroscope may be used to test hardness in iron or steel.

Brittleness in a metal means it will fracture easily if bent sharply or struck a sharp blow. Cast iron, especially a thin section, is very brittle.

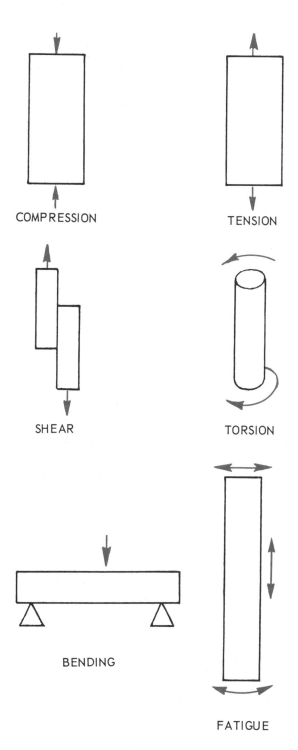

COMPRESSION

TENSION

SHEAR

TORSION

BENDING

FATIGUE

Fig. 24-1. Types of stresses (loads) imposed on structures: Compression, Tension, Shear, Torsion, Bending (lower surface is in tension, the upper surface is in compression), Fatigue (a vibration i. e. a cyclic reversing of load either compression, tension, torsion, or shear or any combination of these loads).

The compression strength of a metal is a measure of how much squeezing force it can withstand before it fails. The metal to be tested is mounted in a tensile tester, but instead of pulling on the metal, a squeezing (compression) force is applied.

Elongation is measured by how much a metal will stretch before it breaks. Elongation or the percent of elongation is measured during a tensile test. Two marks, exactly two inches apart are placed on the tensile sample and the distance between the marks is again measured after the sample breaks. The original distance between the marks is compared to the increase in length to determine the percent of elongation. The percent of elongation of cast iron is very low, since it is brittle.

Malleability is the property of a metal to be hammered into shape. Low carbon steel is malleable; cast iron is not.

A metal is said to be tough if it can withstand repeated application and release of force, such as in bending a metal back and forth.

The examination of the grain size of a known metal under a microscope will give a good indication of the brittleness of a metal, its tensile strength, the heat treatment, and its ductility.

Before these properties may be studied in detail, you must have an understanding of the effect of carbon on the properties of steel, and a thorough knowledge of alloys in general.

24-3. ALLOY METALS

As mentioned previously, an alloy metal may be defined as an intimate mixture of two or more metals.

Any ferrous or nonferrous metal may be alloyed to form an alloy metal with new and desirable characteristics.

Steel is an alloy of iron and controlled amounts of carbon. Alloy steels are created by adding other metals to plain carbon steel. Some metals which are alloyed with carbon steel and the qualities imparted to steel by each are:

Chromium - Increases resistance to corrosion; improves the responsiveness to heat treatment.

Manganese - Increases strength and responsiveness to heat treatment.

Molybdenum - Increases toughness and improves the strength of steel at higher temperatures.

Nickel - Increases strength and toughness; increases impact resistance qualities.

Tungsten - Produces dense, fine grain; helps steel to retain its hardness; helps steel to retain strength at high temperatures.

Vanadium - Retards grain growth and improves impact resistance strength.

The melting temperature of a metal is changed somewhat whenever an alloying metal is added. This may be illustrated by examining the cooling curve for a simple alloy.

A simple alloy consists of two metals in any proportion. An example of a simple alloy is the combination of lead and tin. The melting temperature of the lead is 635 deg. F. Tin has a melting temperature of 432 deg. F. However, as the two metals are mixed, any combination of the two results in a lower melting temperature than 635 deg. F. At a certain proportion of the metals, the lowest melting temperature is reached, which is lower in a simple alloy than the melting temperature of either of the two pure metals separately. This point is called the eutectic point, as shown in Fig. 24-2.

In this diagram at temperatures below line (1, 2) generally the lead is solid but the tin is liquid and then as the alloy cools, all of the tin becomes solid at line (4,2). Likewise under line (2, 3) the lead solidified and the tin remains a liquid until at line (2,5) it all becomes solid. Between points (4) and (5) the alloys have the same starting-to-melt (solidus) temperature. In the areas to the left of (1, 4) the alloy remains solid up to line (4, 5) but the crystal shape changes and is called alpha alloy. In the area to the right of (3, 5) the alloy remains solid up to the line (3, 5) but the crystal shape changes and is called beta.

24-4. COOLING CURVES

The temperature of a metal (not an alloy) under a certain pressure (atmospheric), remains constant during the change from the liquid to the solid state. As most metals are heated to their melting temperature, a thermometer will show a constant rise in temperature per unit of time, with a constant heat input, until the metal reaches its melting temperature. However, as the metal melts, the temperature will remain constant for a length of time. During this time, the molecular structure of the metal absorbs heat energy which is required to change the position of the metal atoms in the structure of the molecule as the metal changes from a solid to a liquid state. After the metal is melted, the temperature will again rise as the metal is heated.

As a metal cools from the liquid state, the temperature may be observed to drop until the point where the metal solidifies is reached. At this point, the temperature will remain constant while the atoms in the metal molecules change back to the solid state molecular structure. During this period of time while the atoms are changing position in the molecule the molecule releases heat.

After the metal has solidified, the temperature will again drop until it reaches room temperature.

The pause, even though there is absorption of energy on heating is called accalescence, while the release of energy on cooling even though there is no temperature change, is called decalescence. In these alloys the absorption of energy during heating, or energy release during cooling causes a delay in the temperature change while internal energy changes are taking place even though the alloy stays in the solid condition.

These points signify a crystalline structure change in the solid metal, and cause the metal to expand (absorbing heat), or contract while cooling as heat energy is suddenly released. These points are called critical points. The heat treatment of a metal through this crystalline structure change is of tremendous importance. These critical points, as they are sometimes called, are the important points one must observe when heat treating metals. The cooling curves for steels illustrate decalescence and accalescence very well. Some samples of steels show two of these points on cooling and heating while some show only one. If, for every

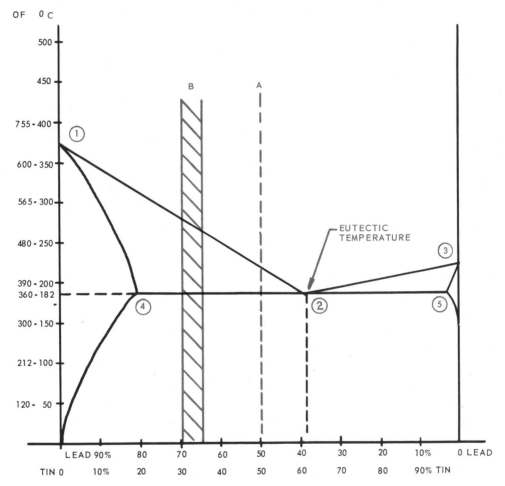

Fig. 24-2. Lead-tin diagram. Line A represents the popular 50-50 solder; Line B represents the range of body solders and lead pipe wiping solders.

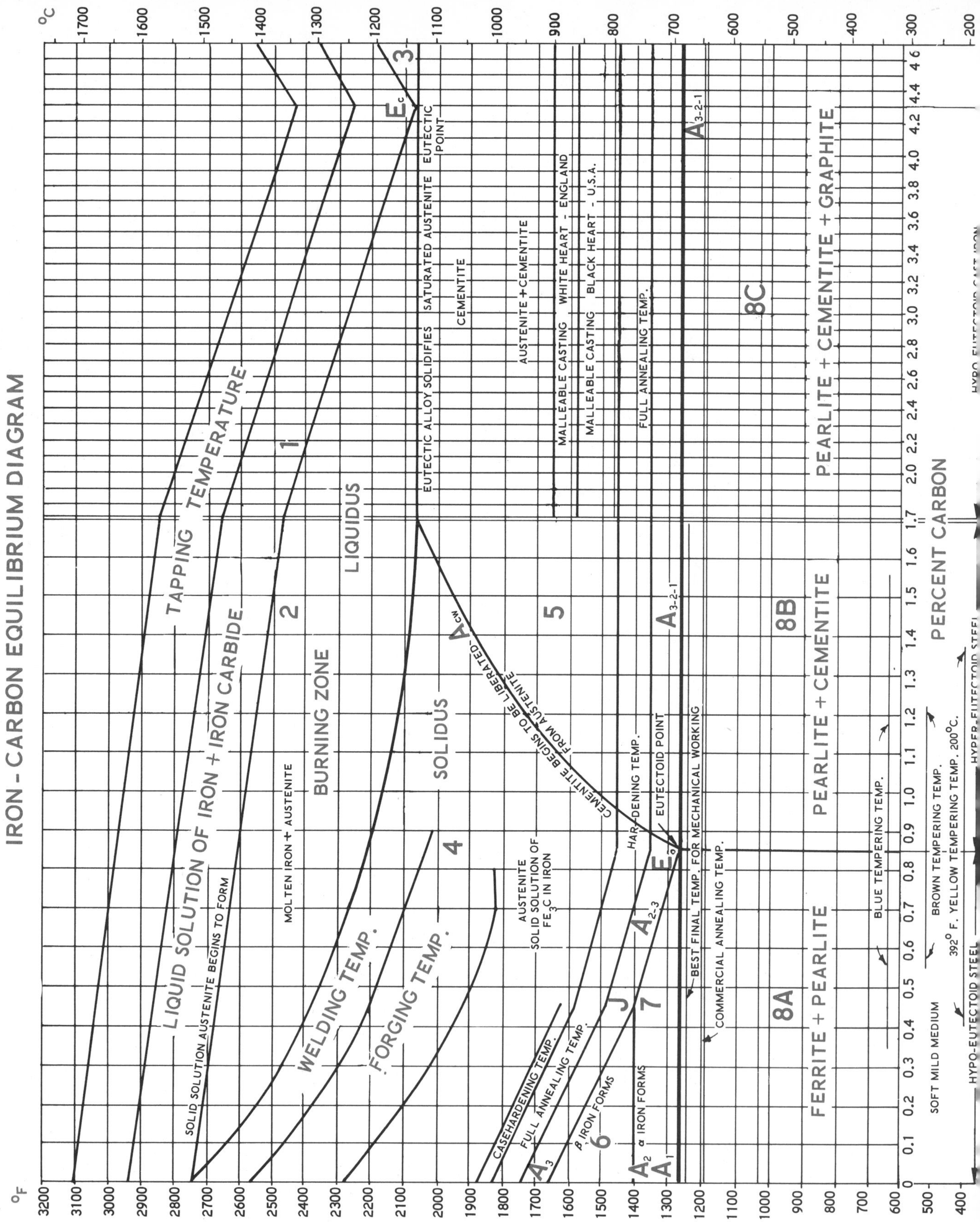

Fig. 24-3. Iron-carbon diagram.

carbon content, i. e., 5 point, 10 point, 15 point, etc., a cooling curve and heating curve were plotted, one would soon perceive that the points of accalescence and decalescence change with the carbon content. If a graph is made plotting the accalescence and decalescence points for the various carbon contents against temperature, the result is an iron-carbon diagram.

24-5. IRON-CARBON DIAGRAM

The iron-carbon diagram, Fig. 24-3, shows the melting temperature changes as the carbon content changes and, more important, it shows the critical points during the temperature changes. Note that some of the areas on the chart are labeled cold working, hot working, forging, welding, etc. Note that the amount of carbon in the steel is very small. In most steels the carbon content is less than 1 percent. Since a small change in carbon content changes the characteristics of steel so greatly, the carbon content must be specified in 1/100 of 1 percent. Steel men call each 1/100 of 1 percent carbon one point of carbon. One percent carbon steel (a high carbon steel) is called 100 point steel while 10/100 of 1 percent carbon steel is called 10 point carbon steel.

Some of the outstanding features of iron-carbon combinations are:

First, up to a percentage of 45 point carbon the steel goes through three molecular structure changes or three critical points. These points are at lines A_1, A_2, and A_3.

Second, in the range of 45 point carbon to 85 point carbon, there are two critical temperatures.

Third, at 85 point carbon content there is only one critical point and this is known as the eutectic point.

Fourth, from 85 point to 170 point carbon there are two critical points. These two points occur at line A_{3-2-1} and line A_{cw}.

Fifth, above 170 point carbon are included gray cast iron, white cast iron, and malleable iron.

All the critical points are of importance to the welder; but the lowest critical point is of the greatest importance, occurring at approximately 1250 deg. F. This is the temperature to which steels must be heated for hardening, annealing, etc. In addition, note that carbon steels are heated to 350-550 deg. F. and high speed steel is heated to 1000 deg. F. for tempering.

As a further example, follow the changes when 20 point carbon steel or .20 percent carbon steel, which is known as mild steel, or technically speaking as hypoeutectoid steel, is heated and cooled through the complete temperature range. As the temperature rises, (1) the combination is ferrite metal with very little cementite (iron carbide) up to 1250 deg. F., (2) if 20 point carbon steel is heated up to the temperature of 1350 deg. F. or slightly above the first critical point and then allowed to cool slowly, the steel will be partially annealed. If the steel is heated about 100 deg. above the A_3 line to the full annealing line and then cooled slowly, all crystals will be affected and the steel will be fully annealed. The temperature of 1250 deg. F. is the temperature where the crystalline structure changes. At this temperature the crystals are the smallest (finest). Some of the grains consist of pure iron which are called ferrite, while other grains consist of iron-carbide, called cementite. When the grains appear in alternate layers or rows of ferrite and iron-carbide, the metal has a pearly appearance under a microscope. This condition is called pearlite, a name developed from this appearance.

Just below the temperature of 1250 deg. F. a point is reached which is labeled the best heat for mechanical working. This means that when the steel is heated to this temperature, it may be worked by means of rollers and pounding, without fracture occurring. When a temperature slightly over 1400 deg. is reached, a new change occurs and the steel is labeled alpha iron. The steel, while at 1550 deg. F., is changed to the beta form. These terms (alpha and beta) are technical divisions, or iron-carbon combinations, and have no direct bearing on the practical treatment of steel. The terms alpha, beta, and gamma are the names given to the various shaped crystals (molecular structure) of iron and carbon steel.

Slightly above this temperature, however, at approximately 1600 to 1700 deg. F. the iron turns into gamma iron. If the steel is heated to this temperature and then allowed to cool slowly, the steel will be in its softest possible condition. The crystal size will be much larger than if the steel were heated only to 1250 deg. F.

When it is desired to increase the carbon content of mild steel, a temperature slightly above its full annealing temperature is used. If the steel is exposed to carbon at this temperature, the outer surface of the metal will combine with some of the carbon, thus producing an increased carbon content in the skin of the metal (case hardening process). This process is used when a hard surface is desired along with a strong tough interior.

If the metal is heated between 2100-2350 deg. F., it reaches the range in which the metal can best be forged (plastic range). Just above this temperature (from 2350 to 2550 deg. F.) is the welding temperature of the metal. Here the metal is molten, but at this temperature it will not burn. From 2550 deg.

F. and up, the 20 point carbon steel is in the burning zone, meaning the spontaneous oxidizing zones. This temperature is labeled the burning zone in Fig. 24-3.

From this description you can readily see that the temperature under which metal is worked, or to which the metal is heated, has a decided bearing upon its properties. Twenty point carbon steel is a very common steel, and you should try to remember the approximate temperatures that must be obtained to create the conditions listed in the preceding paragraphs.

It is interesting to note that as the carbon content is increased, the commercial annealing temperature and the mechanical working temperature remain approximately the same. However, the full annealing temperature decreases as the carbon content increases up to 85 point carbon, as does the forging temperature and the welding temperature.

Beyond the 45 point carbon steel point, the case hardening temperature is of no special significance as the metal may be hardened without having any carbon added.

The heating of steel during a fusion welding process will ruin the heat treatment, or any properties of the metal that are the result of mechanical working. To renew these properties, the metal must again be heat treated and/or mechanically worked.

You may also notice that the forging temperatures and welding temperatures are not listed above 70 to 75 point carbon. This is because forging metal that has more than 70 point carbon is not practical. Braze welding, rather than fusion welding, is recommended for steels from 70 point carbon up to the cast iron range.

Above 170 point carbon, welding may be used again. To attempt to weld steel between 85 and 170 point carbon usually subjects the steel to the burning point temperature which will affect the carbon content. High carbon steel, during welding, boils vigorously and becomes brittle and porous.

Above the horizontal line passing through eutectic point labeled E_0 at 85 point carbon steel, the temperature between 1350 and 1450 deg. F. is labeled the hardening temperature. If steel of 85 point carbon to 170 point carbon is heated to this temperature and then cooled very rapidly, the metal will be very hard.

To anneal this metal it should be reheated to a temperature between 1350 and 1450 deg. F. and held at that temperature until it is a uniform temperature throughout its thickness and then cooled as slowly as possible.

The size of the crystals in a metal determine to a great extent its strength and ductility. If 15 point carbon steel is heated, the crystal size will not change until a temperature of approximately 1350 deg. F. is reached. At this point the crystals change to their smallest and most refined state. If the metal is cooled rapidly from this temperature, the crystals will be very small. The metal is now at its maximum strength, but will not be extremely ductile. Above the critical point, grain growth is a result of two things--time and temperature. If the metal is heated to a temperature higher than the 1350 deg. F. point, the crystals gradually increase in size as a result of the temperature increase. If the metal is cooled slowly from any temperature higher than the 1350 deg. F., the crystals will continue to grow to a maximum size because of the time spent above the critical point until 1350 deg. F. is reached. But from 1350 deg. F. down, the crystal remains the same size.

24-6. IDENTIFICATION OF IRON AND STEEL

Because of the effect on the properties of steel caused by the three variables, carbon content, temperature, and time, a welder must be able to determine quite accurately the nature of the steel being handled. The manufacturers' specifications for the particular steel are most desirable. Whenever possible the welder should obtain these specifications and keep them on file. At the same time he should mark the metal to correspond with filed information. Where manufacturers' specifications are not available other methods may be used to determine the nature of the metal.

Many tests have been developed, but the following are the most common for shop use:

1. Spark test (with the power grinder).
2. Oxyacetylene torch test.
3. Fracture test.
4. Color test.
5. Density or specific gravity test.
6. Ring or sound of the metal upon impacting with some other metal.
7. Magnetic test.
8. Chip test.

Of these tests, numbers three through eight should take place almost subconsciously in the welder's mind, as he works on the metal. The spark test and the gas torch test must be done under carefully prepared conditions. These tests indicate to a remarkably accurate degree the properties and constituents of the metal.

24-7. SPARK TEST

A method of identifying metals, which was developed and carefully analyzed by John F. Keller, of Purdue University, is extensively used by welders to identify irons and steels. A power

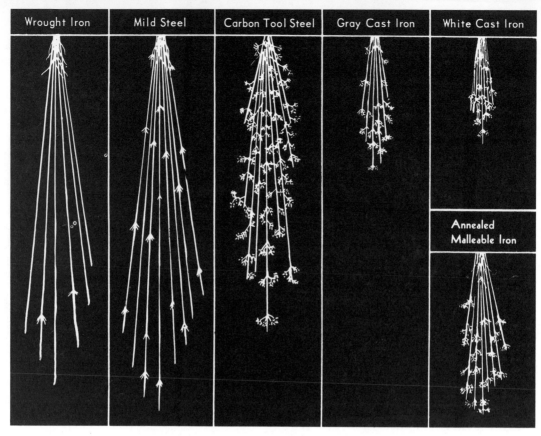

Fig. 24-4. Spark test for common cast irons and steels.
(Norton Co.)

grinder is used as the test equipment. When grinding you must always wear goggles. The grinder must be inspected to see that it is in good condition before proceeding with the test. When testing a sample, if you touch the rim of the revolving wheel lightly with the metal, the friction of the wheel surface as it contacts the metal heats the particles removed to the incandescent and burning temperature. The sparks resulting from the contact are found to differ in character for different steels. The lighter the contact, the better; and one should use a black background to better identify the spark. The theory of the spark test is when a metal is heated, the different parts of the metal oxidize at different rates and the oxidation colors are different. Relatively pure iron when heated by the grinding wheel does not oxidize quickly; therefore, the sparks are long and fade out

on cooling. As the carbon content of steel or cast iron increases, the compounds of carbon and iron have different ignition temperatures, and therefore the characteristics of the sparks differ. Four characteristics of the spark generally denote the nature and condition of the steel; they are:

1. Color of spark.
2. Length of spark.
3. Number of explosions (spurts) along the length of the individual sparks.
4. Shape of the explosions (forking or repeating).

For example, a mild steel containing 20 point carbon (.20 percent) will show a long white spark which will jump approximately 70 in. from the power grinder (using a 12 in. wheel). Some of these sparks will suddenly explode, shooting off smaller sparks at approximately 45 deg. angles to the direction of travel of the original spark. A 30 point carbon steel will have sparks almost identical to the 20 point carbon steel with the exception that more streamlined sparks will explode while the total length of the sparks will decrease slightly. This serves to show that as the carbon content of steel increases, the explosions of the sparks become more frequent. Also, as the carbon content increases the length of the spark is decreased because of the interruption during their flight of the individual explosions or extremely rapid oxidation of some of the materials in the incandescent particle. In Fig. 24-4, for example, a high carbon tool steel, containing 80 point carbon, has a very short spark with explosions occurring so rapidly that it is hard to identify the longer streamlined iron shapes. The sparks dissipate themselves very quickly. See Fig. 24-5.

carbon is white, is faster disappearing and the length of the spark is usually a little longer. Both of these metals show a small spark explosion at the extremity of the spark flight. The amount of the spark in these two cases is considerably less than one would expect from such a high carbon content.

24-8. OXYACETYLENE TORCH TEST

Even if you know the physical composition and the chemical composition of a metal, you must also know whether the metal has good welding properties. For example, some cold rolled sheet steels may show very good physical and chemical properties, but during some part of the manufacturing process, impurities have been added to it or certain work was done to the metal affecting its properties to the extent

METAL	VOLUME OF STREAM	RELATIVE LENGTH OF STREAM, INCHES*	COLOR OF STREAM CLOSE TO WHEEL	COLOR OF STREAKS NEAR END OF STREAM	QUANTITY OF SPURTS	NATURE OF SPURTS
Wrought iron	Large	65	Straw	White	Very few	Forked
Machine steel (AISI 1020)	Large	70	White	White	Few	Forked
Carbon tool steel	Moderately large	55	White	White	Very many	Fine, repeating
Gray cast iron	Small	25	Red	Straw	Many	Fine, repeating
White cast iron	Very small	20	Red	Straw	Few	Fine, repeating
Annealed mall. iron	Moderate	30	Red	Straw	Many	Fine, repeating

* Figures obtained with 12" wheel on bench stand are relative only. Actual length in each instance will vary with grinding wheel, pressure, etc.

Fig. 24-5. *Table of spark test characteristics of common cast irons and steels.*

When higher carbon content metals are tested, you may become confused between extremely high carbon steel and cast iron because the spark is somewhat the same. Heat treatments have some effect on the nature of the spark. The appearance of the cast iron and high carbon steel on the basis of the spark are almost identical. However, the cast iron spark, when leaving the point of contact, is a dull red and the spark jumps only 20 to 25 in. from the wheel; while the spark from a high carbon steel of approximately 130 point

the metal will not melt readily for welding, will not fuse readily, and the final weld may be unsatisfactory. The usual cause of this condition is that there are impurities imbedded in the metal in the form of slag and roller dirt or excessive sulphur and phosphorus. For these reasons a welder should subject steel to the torch test.

The actual test consists of melting a puddle in the steel. If the metal is thin, the puddle penetrates through the thickness of the steel until a hole is formed. This puddling should be done with a neutral flame, held at the proper distance from the metal. The puddle should not spark excessively; and it should be fluid. The puddle should not boil, and it should possess good surface tension. The appearance on the edge of this aperture is very indicative of the weldability of the steel. If the metal that was melted has an even, shiny appearance upon solidification, the metal is generally considered as

24-9. MISCELLANEOUS IDENTIFICATION TESTS

As an added precaution to the above tests, six other tests are occasionally used by the welding jobber. The fracture test is used extensively and consists of breaking a portion of the metal in two. If it is a repair job, the fractured surface is ready for inspection. The appearance of the surface where the

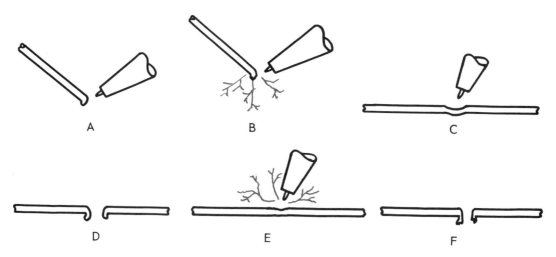

Fig. 24-6. Torch (blowpipe) flame test: A. Good quality filler rod. B and C. Good quality base metal. D. Poor quality filler rod. E and F. Poor quality base metal.

having good welding properties. However, if the molten metal surface is dull or has a colored surface, and if this surface is rough, perhaps even broken up into small pits or porous spots, the metal is generally unsatisfactory for welding.

This test is accurate enough for most welding. The test is very easily applied with the equipment on the premises and determines the one thing that is fundamentally necessary in any welding job, that is, the weldability of the metal. Fig. 24-6 shows how this test is conducted. While performing the weldability test of the metal, it is important to note the amount of sparking emitted from the molten metal.

metal is cracked shows the grain structure of the steel. If the grains are large, the metal is brittle and weak; if the grains are small, the metal is usually hard and brittle, or soft and ductile. The fracture shows the color of the metal which is a good means of identifying one metal from another, and the test also indicates the type of metal by the ease with which it may be fractured. Fig. 24-7 shows the appearance of the fractured surface of several different metals. The two main divisions of metals include the irons and steels which are indicated by their typical gray white color, and the nonferrous metals which come in two general color classifications of yellow and white. Copper may be rather easily identified by a welder and the same applies to brass and bronze. Aluminum, white metal, aluminum alloys, zinc, and the like, are all of somewhat the same silver-gray color although they may vary in shade.

The metals may also be differentiated by means of the weight or density test of the specimen. A good example of identification by density or specific gravity is identifying aluminum and lead. Roughly speaking, their colors are somewhat similar, but anyone may readily distinguish between the two metals because of their respective weights.

The ring test, or the sound of the metal test, is an easy means of identifying certain metals, after one has had some experience with this method. It is used extensively for identifying heat-treated steels from annealing steels. It is also used to detect alloys from the base metal. An example of the latter is aluminum and duralumin. The pure aluminum sheet has a duller

The magnetic test is an elementary test used to identify iron and steel metals from the nonferrous metals. Generally speaking, all steels are affected by magnetism while the nonferrous metals are not. However, some stainless steels are not magnetic.

Another test which must be accompanied by considerable experience is the chipping test of a metal. In this test, the cutting action of the chisel indicates the structure and heat treatment of the metal. Cast iron, for example, when being chip-tested, breaks off in small particles, whereas a mild steel chip tends to curl and cling to the original piece. The higher-carbon, heat-treated steels cannot be tested this way because of their hardness.

24-10. CHEMICAL TESTS TO IDENTIFY METALS

At the present time chemicals are not used extensively for identifying metals in the welding shop. However,

Fig. 24-7. Fractured surfaces of several different metals.
(Linde Div., Union Carbide Corp.)

sound, or ring, than the duralumin which is somewhat harder and has a more distinct ringing sound.

as different chemicals are developed this usage will grow. The tests in themselves are rather simple. The chemical is applied to the metal surface of a sample, or the metal is immersed in some chemical solution and the resulting reaction noted. This test is especially useful for alloy metals.

To identify aluminum from its alloys a caustic soda solution is used. The metal is exposed to a solution of 25 percent sodium hydroxide and 75 percent water by weight for 2 or 3 minutes. The pure aluminum will stay bright while the dural or any copper bearing aluminum will turn dark.

Nickel steel may be identified by dropping some 50 percent nitric acid solution on the cleaned metal for one half minute. Be certain that the metal being tested is clean and free from dirt, oil and/or scale. The nickel content is indicated by a bubbling or boiling action. The bubbling is caused by the evolution of a gas from the chemical reaction.

A popular aircraft steel, chrome-molybdenum steel, is identified by immersing some filings or drillings in dilute sulphuric acid. As the steel finishes dissolving, the solution will turn a deep green.

A rough test between mild carbon steel and chrome-moly steel may be indicated by the relative hardness of the metals while being hacksawed.

24-11. IDENTIFICATION OF IRON AND STEEL ALLOYS

During the last decade, extensive research has been carried on with iron and carbon steels in which other metals have been added to improve or bring out certain properties. The two main fields in which this development has taken place are in the stainless and low carbon alloy steels. These metals are difficult to identify from ordinary steels, and their composition greatly

affects their welding properties. The development of these new metals may result in the development of more accurate identifying equipment for the welding shop.

Alloys with iron-carbon metals are provided to develop or improve the various physical properties of the metal. A common illustration includes the stainless steels in which alloy metals are used in the steel to make the steel corrosion-proof, and to develop other desired physical properties such as toughness and strength. These metals consist of various combinations of chromium, nickel, and copper. Other metals may be added to the steel to increase its strength, hardness and/or toughness. A good example of this type of alloy is the use of tungsten. The addition of a small amount of tungsten to steel produces an extremely hard metal without sacrificing its other properties to any appreciable extent. Low-carbon alloy steels are sometimes improved by alloying these steels with various amounts of nickel or copper. These alloys serve the purpose of producing stronger steels at a minimum cost. Heat treatment is of great importance when one wants to obtain the best physical properties of alloy steels.

24-12. SPARK TEST FOR ALLOY STEELS

The appearance of the power grinder spark for alloy steels varies considerably, depending on the alloys included in the metal. The basic test, that is the iron-carbon spark, is identical with the straight, iron-carbon steels. The number of the spark explosions increases as the length of the spark flight decreases, and the color of the spark becomes brighter and brighter

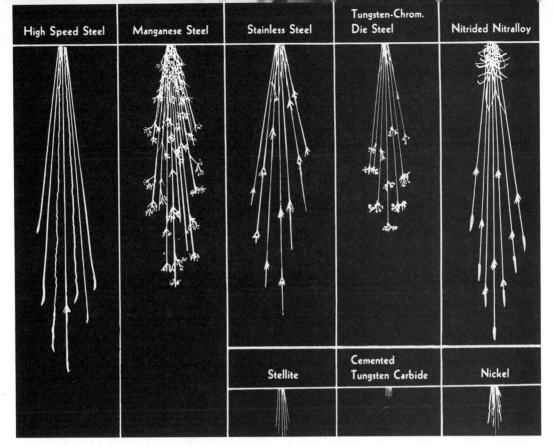

High Speed Steel	Manganese Steel	Stainless Steel	Tungsten-Chrom. Die Steel	Nitrided Nitralloy
		Stellite	Cemented Tungsten Carbide	Nickel

Fig. 24-8. Spark test for some alloy steels and alloying metals.

as the carbon content increases. The principal variations the alloy elements give to a steel are the multitude of right-angle, branch-off sparks which result from the different incandescent colors of the alloys in the metal (the change of color of the spark).

As an example, a manganese steel has only a slight indication of the spark explosion, but the manganese causes the sparks to shoot out at right angles to the flight of the original spark, and these sparks tend further to explode (repeating sparks) producing the appearance of a leafless tree branch, as shown in Fig. 24-8.

Tungsten-chrominum steels, which are used for high speed work, show the typical, high-speed steel spark with the exception that the spark turns to a chromium yellow (straw color) at the end of its flight. The chromium and the tungsten have a deteriorating effect on the carbon spark, causing it to be very fine or thin with repeating spurts. Fig. 24-9 shows the spark charac-

METAL	VOLUME OF STREAM	RELATIVE LENGTH OF STREAM, INCHES*	COLOR OF STREAM CLOSE TO WHEEL	COLOR OF STREAKS NEAR END OF STREAM	QUANTITY OF SPURTS	NATURE OF SPURTS
High speed steel (18-4-1)	Small	60	Red	Straw	Extremely few	Forked
Austenitic manganese steel	Moderately large	45	White	White	Many	Fine, repeating
Stainless steel (Type 410)	Moderate	50	Straw	White	Moderate	Forked
Tungsten-chromium die steel	Small	35	Red	Straw**	Many	Fine, repeating**
Nitrided Nitralloy	Large (curved)	55	White	White	Moderate	Forked
Stellite	Very small	10	Orange	Orange	None	
Cemented tungsten carbide	Extremely small	2	Light Orange	Light Orange	None	
Nickel	Very small***	10	Orange	Orange	None	
Copper, brass, aluminum	None				None	

*Figures obtained with 12" wheel on bench stand and are relative only. Actual length in each instance will vary with grinding wheel, pressure, etc. **Blue-white spurts. ***Some wavy streaks.

Fig. 24-9. Table of spark test characteristics of alloy steels.

teristics for alloy steels. The spark is also an interrupted one, that is, it disappears for a portion of its flight and then appears again.

24-13. TORCH TEST FOR ALLOY STEELS

It is difficult to give rigid specifications of the torch-melting test for all alloy steels. However, it is generally known that the higher the alloy constituents, the more difficult it is to weld the metal. Under the influence of the torch flame, the metals have a tendency to boil, because of the action of the alloys in them, and special fluxes are often used when welding them.

Stainless steels are difficult to flame weld, and a flux is needed to produce good welds. Without the flux the metal boils and a porous puddle is produced.

Manganese steel is often used for surfaces which are exposed to abrasive wear, such as power shovel buckets. The usual application of welding in this case is to build up the worn surfaces. The steel melts readily under the torch flame and presents few difficulties in welding.

Nickel steels can be identified under the torch flame by the boiling action of the metal.

Nickel-chrome steels behave in somewhat the same manner.

24-14. MISCELLANEOUS TESTS FOR ALLOY STEELS

The color test, ring test, magnetic test, fracture test, and the chip test are all applicable to alloy steels, but perhaps the easiest one is the color test. The different alloy constituents tend to change the color of the metal, which in certain compositions is very noticeable and characteristic.

Stainless steels, for example, have a distinct silvery color, which sets them apart from the other steel alloys. Some of the metals are magnetic, whereas others are not. Identification can be accomplished to a certain extent by using this test. The chip test is not used extensively for identifications because of the similarity in the appearance of practically all the stainless steel alloys. However, some of these alloys are considerably harder than others, and the chip test will bring this out quite clearly.

The ring or metallic sound of the metal is another test which can be applied to steel alloys with considerable accuracy. However, it is recommended that this test be used in conjunction with other tests.

24-15. CHEMICAL TESTS FOR ALLOY STEELS

There is no doubt that chemical tests will someday be commonly used to identify various steel alloys. The alloys can be identified by chemical reactions much more accurately and easily than by any other method. The application of different etching acids brings out the color of the alloy, or reacts with the alloys in the metal to the extent that they may be readily identified.

The method is: A chemical test rack, holding bottles which contain the test chemicals, is set up in the shop. These chemicals are listed as having the property of acting upon the different metals with different results. Perhaps one chemical will react with only one of the different alloy metals, or the chemical may react with more than one bringing out different results, such as color and speed of reaction.

24-16. NUMBERING SYSTEMS FOR METALS

The Society of Automotive Engineers (S.A.E.) has long been a leader in standardizing in the automotive field. The Society has developed a numbering system for identifying practically all steels. The code is based on a number of four digits, i.e., 2115. The first digit 2 classifies the steel, the number 2 representing nickel steels. The second digit represents the percent of the alloy metal in the steel, i.e., 2115 represents 1.25 to 1.75 percent nickel.

The last two digits represent the carbon content of the metal in points (hundreths of one percent), i.e., .15 to .25 percent carbon. Fig. 24-10 is a table of sample S.A.E. steels. S.A.E.

first number designations for steel and steel alloys are as follows:
1000 Carbon steels.
1100 Are special, sulphur-carbon steels that have free cutting properties.
1200 Phosphorus-carbon steels.
1300 Manganese steels.
2000 Nickel steels.
3000 Nickel-Chromium steels.
4000 Molybdenum steels.
5000 Chromium steels.
6000 Chromium-Vanadium steels.
7000 Tungsten steels.
9000 Silicon-Managanese steels.
Some special divisions are as follows:
30000 and 51000 are corrosion and heat-resisting steels.
30000 Nickel-Chromium steels.
51000 Chromium steels.
The American Society of Testing

SAE-AISI CARBON STEEL COMPOSITION NUMBERS

SAE	AISI	CARBON RANGE	MANGANESE RANGE	PHOSPHORUS MAX.	SULFUR MAX.
1010	C1010	0.08-0.18	0.30-0.60	0.040	0.050
1015	C1015	0.13-0.18	0.30-0.60	0.040	0.050
1020	C1020	0.18-0.23	0.30-0.60	0.040	0.050
1025	C1025	0.22-0.28	0.30-0.60	0.040	0.050
1030	C1030	0.28-0.34	0.60-0.90	0.040	0.050
1035	C1035	0.32-0.38	0.60-0.90	0.040	0.050
1040	C1040	0.37-0.44	0.60-0.90	0.040	0.050
1045	C1045	0.43-0.50	0.60-0.90	0.040	0.050
1050	C1050	0.48-0.55	0.60-0.90	0.040	0.050
1055	C1055	0.50-0.60	0.60-0.90	0.040	0.050
1060	C1060	0.55-0.65	0.60-0.90	0.040	0.050
1065	C1065	0.60-0.70	0.60-0.90	0.040	0.050
1070	C1070	0.65-0.75	0.60-0.90	0.040	0.050
1075	C1075	0.70-0.80	0.40-0.70	0.040	0.050
1080	C1080	0.75-0.88	0.60-0.90	0.040	0.050
1085	C1085	0.80-0.93	0.70-1.00	0.040	0.050
1090	C1090	0.85-0.98	0.60-0.90	0.040	0.050
1095	C1095	0.90-1.03	0.30-0.50	0.040	0.050

RESULPHURIZED CARBON STEELS

SAE	AISI	CARBON RANGE	MANGANESE RANGE	PHOSPHORUS MAX.	SULFUR MAX.
1115	C1115	0.13-0.18	0.60-0.90	0.040 max.	0.08-0.13
1120	C1120	0.18-0.23	0.70-1.00	0.040	0.08-0.13
1125	C1125	0.22-0.28	0.60-0.90	0.040	0.08-0.13
1140	C1140	0.37-0.44	0.70-1.00	0.040	0.08-0.13

(Continued on next page).

REPHOSPHURIZED AND RESULPHURIZED CARBON STEELS

SAE	AISI	CARBON RANGE	MANGANESE RANGE	PHOSPHORUS MAX.	SULFUR MAX.
1211	C1211	0.13 max.	0.60-0.90	0.07-0.12	0.08-0.15
1213	C1213	0.13 max.	0.70-1.00	0.07-0.12	0.24-0.33

MANGANESE STEELS

SAE	AISI	CARBON RANGE	MANGANESE RANGE	PHOSPHORUS MAX.	SULFUR MAX.	SILICON MAX.
1330	1330	0.28-0.33	1.60-1.90	0.040	0.040	0.20-0.35
1335	1335	0.33-0.38	1.60-1.90	0.040	0.040	0.20-0.35
1340	1340	0.38-0.43	1.60-1.90	0.040	0.040	0.20-0.35
1345	1345	0.43-0.48	1.60-1.90	0.040	0.040	0.20-0.35

NICKEL STEELS

SAE	AISI	CARBON RANGE	MANGANESE RANGE	PHOSPHORUS MAX.	SULFUR MAX.	NICKEL RANGE
2315	----	0.10-0.20	0.30-0.60	0.040	0.050	3.25-3.75
2330	----	0.25-0.35	0.50-0.80	0.040	0.050	3.25-3.75
2340	----	0.35-0.45	0.60-0.90	0.040	0.050	3.25-3.75
2345	----	0.40-0.50	0.60-0.90	0.040	0.050	3.25-3.75
----	2515	0.10-0.20	0.30-0.60	0.040	0.050	4.75-5.25

NICKEL CHROMIUM STEELS

SAE	AISI	CARBON RANGE	MANGANESE RANGE	PHOSPHORUS MAX.	SULFUR MAX.	NICKEL RANGE	CHROMIUM RANGE	SILICON
3140	3140	0.38-0.43	0.70-0.90	0.040	0.040	1.10-1.40	0.55-0.75	0.20-0.35
3310	E3310	0.08-0.13	0.45-0.60	0.025	0.025	3.25-3.75	1.40-1.75	0.20-0.35

MOLYBDENUM STEELS

SAE	AISI	CARBON RANGE	MANGANESE RANGE	PHOSPHORUS MAX.	SULFUR MAX.	CHROMIUM RANGE	NICKEL RANGE	MOLYBDENUM RANGE	SILICON
4130	4130	0.28-0.33	0.40-0.60	0.040	0.040	0.80-1.10	----	0.15-0.25	0.20-0.35
4140	4140	0.38-0.43	0.75-1.00	0.040	0.040	0.80-1.10	----	0.15-0.25	0.20-0.35
4150	4150	0.48-0.53	0.75-1.00	0.040	0.040	0.80-1.10	----	0.15-0.25	0.20-0.35
4320	4320	0.17-0.22	0.45-0.65	0.040	0.040	0.40-0.60	1.65-2.00	0.20-0.30	0.20-0.35
4340	4340	0.38-0.43	0.60-0.80	0.040	0.040	0.70-0.90	1.65-2.00	0.20-0.30	0.20-0.35
4615	4615	0.13-0.18	0.45-0.65	0.040	0.040	----	1.65-2.00	0.20-0.30	0.20-0.35
4620	4620	0.17-0.22	0.45-0.65	0.040	0.040	----	1.65-2.00	0.20-0.30	0.20-0.35
4815	4815	0.13-0.18	0.40-0.60	0.040	0.040	----	3.25-3.75	0.20-0.30	0.20-0.35
4820	4820	0.18-0.23	0.50-0.70	0.040	0.040	----	3.25-3.75	0.20-0.30	0.20-0.35

CHROMIUM STEELS

SAE	AISI	CARBON RANGE	MANGANESE RANGE	PHOSPHORUS MAX.	SULFUR MAX.	CHROMIUM RANGE	SILICON
5120	5120	0.17-0.22	0.70-0.90	0.040	0.040	0.70-0.90	0.20-0.35
5140	5140	0.38-0.43	0.60-0.90	0.040	0.040	0.70-0.90	0.20-0.35
5150	5150	0.48-0.53	0.60-0.90	0.040	0.040	0.70-0.90	0.20-0.35
52100	E52100	0.95-1.10	0.25-0.45	0.025	0.025	1.30-1.60	0.20-0.35

(Continued on next page).

CHROMIUM VANADIUM STEELS

SAE	AISI	CARBON RANGE	MANGANESE RANGE	PHOSPHORUS MAX.	SULFUR MAX.	CHROMIUM RANGE	VANADIUM	SILICON
6118	6118	0.16-0.21	0.50-0.70	0.040	0.040	0.50-0.70	0.10-0.15	0.20-0.35
6120	6120	0.17-0.22	0.70-0.90	0.040	0.040	0.70-0.90	0.10 min.	0.20-0.35
6150	6150	0.48-0.53	0.70-0.90	0.040	0.040	0.80-1.10	0.15 min.	0.20-0.35

NICKEL CHROMIUM MOLYBDENUM STEELS

SAE	AISI	CARBON RANGE	MANGANESE RANGE	PHOSPHORUS MAX.	SULFUR MAX.	CHROMIUM RANGE	MOLYBDENUM RANGE	NICKEL RANGE	SILICON
8115	8115	0.13-0.18	0.70-0.90	0.040	0.040	0.30-0.50	0.08-0.15	0.20-0.40	0.20-0.35
8615	8615	0.13-0.18	0.70-0.90	0.040	0.040	0.40-0.60	0.15-0.25	0.40-0.70	0.20-0.35
8720	8720	0.18-0.23	0.70-0.90	0.040	0.040	0.40-0.60	0.20-0.30	0.40-0.70	0.20-0.35
8822	8822	0.20-0.25	0.75-1.00	0.040	0.040	0.40-0.60	0.30-0.40	0.40-0.70	0.20-0.35

SILICON MANGANESE STEELS

SAE	AISI	CARBON RANGE	MANGANESE RANGE	PHOSPHORUS MAX.	SULFUR MAX.	SILICON RANGE	NICKEL RANGE	CHROMIUM RANGE	MOLYBDENUM RANGE
9260	9260	0.55-0.65	0.70-1.00	0.040	0.040	1.80-2.20	----	----	----
9840	9840	0.38-0.43	0.70-0.90	0.040	0.040	0.20-0.35	0.85-1.15	0.70-0.90	0.20-0.30

CHROMIUM NICKEL STEELS

SAE	AISI	CARBON	MANGANESE MAX.	SILICON MAX.	PHOSPHORUS	SULFUR	CHROMIUM RANGE	NICKEL RANGE	MOLYBDENUM
30304	304	0.08 max.	2.00	1.00	0.040	0.040	18.00 min.	8.00 min.	----
30317	317	0.10 max.	2.00	1.00	0.040 max.	0.040 max.	16.00-18.00	10.00 min.	2.00 max.
51410	410	0.15 max.	1.00	1.00	0.040 max.	0.040 max.	11.50-13.50	0.60 max.	0.60 max.

Fig. 24-10. Typical SAE-AISI carbon steel composition numbers.

Materials (A.S.T.M.) has standardized a code system for identifying and labeling steels.

The American Iron and Steel Institute (A.I.S.I.) has also prepared standards for steels and steel alloys. In general their code system is identical with that of the S.A.E. code system.

24-17. NONFERROUS METALS

Nonferrous metals are metals not containing iron (ferrous) or which contain such minute quantities of iron, as an alloying element, that they are not considered ferrous metals. This group includes copper, brass, bronze, aluminum, solder, stellite, lead, zinc, nickel, etc.

These metals can be identified by various means. They have distinctive colors. They are nonmagnetic (or have very weak magnetic attraction). They are usually soft metals. Very few of the nonferrous metals will spark when touched to the grinding wheel.

CAUTION - Grinding of nonferrous metals is not recommended except for the extremely short intervals of time required for spark testing. The oxides of the nonferrous metals are toxic and therefore the operator must wear an air filtering breathing apparatus and protective clothing. The grinder must be equipped with an adequate exhaust system.

24-18. COPPER

Copper is a chemical element of the metal family which has many uses because of its electrical conductivity and its ability to resist corrosion. Most copper has a reddish brown color. Copper melts at a temperature of approximately 1,980 deg. F. which is higher than the melting temperature of silver and somewhat lower than the melting temperature of cast iron, as shown in Fig. 24-11. Manufacturing processes

METAL	MELTING TEMPERATURES	
	°F.	°C.
Aluminum	1217	659
Armco iron	2795	1535
Bronze 90 Cu Bronze 10 Sn	1562-1832	850-1000
Brass 90 Cu 10 Zn	1868-1886	1020-1030
Brass 70 Cu 30 Zn	1652-1724	900-940
Copper	1981	1083
Iron	2786	1530
Lead	621	327
Mild Steel	2462-2786	1350-1530
Nickel	2646	1452
Silver	1761	960
Tin	450	232
Zinc	786	419

Fig. 24-11. Melting temperatures of some common metals.

are such that commercial copper usually contains sulphur, phosphorus, and silicon as its impurities. Each of these impurities has a tendency to make the copper more brittle and to reduce its weldability. However, a very small amount of phosphorus in copper is an aid to the welding of the metal, because of the dissolving property that the phosphorus has for copper oxides, permitting it to act as a flux.

The only copper recommended for fusion purposes is deoxidized copper. This copper has had a very small amount of silicon added to it, which has the property of dissolving whatever cupric oxides are present in the metal. Enough silicon is added to the copper during its manufacture so that an excess is left in the copper after the deoxidizing action takes place. If this amount is too much, as mentioned previously, the copper tends to become brittle.

Another feature of copper, which is typical of practically all nonferrous metals, is its behavior called "hot shortness." As copper is heated to its melting temperature, the copper becomes very weak at a certain temperature even though it is still a solid and the slightest shock or weight will tend to distort the metal unless it is very firmly supported and firmly clamped to prevent distortion while the metal is passing through this temperature. The approximate point at which this hot shortness will occur may be determined by the welder by the color of the metal as it is heated. When the color of the metal becomes a medium cherry red the metal is at the hot shortness temperature.

24-19. BRASS

Brass is an alloy of copper and zinc, although small amounts of other metals are frequently added. The amount of the zinc in the alloy may vary from 10 to 40 percent. A very common alloy of copper and zinc to form brass is 70 percent copper and 30 percent zinc. This metal is used principally because of its acid resistance qualities, its appearance, and because it is a good brazing alloy. There are two common types of brass: one type is called machine brass, which contain 32 to 40 percent zinc, and red brass which contains from 15 to 25 percent zinc. Some metals which are added to improve the physical properties of brass thus making it a triple alloy are tin, manganese, iron, and lead.

Brass may be identified by its color which is an opaque yellow.

24-20. BRONZE

Bronze is an alloy of copper and tin. A common ratio is 90 percent copper and 10 percent tin. Bronze is more coppery in color than brass. It behaves much like brass when being welded. Generally speaking, one uses the same' filler rod and the same flux for both.

Like brass, bronze is highly resistant to corrosion. Because of its attractive appearance, it is often used for decorative parts and objects.

24-21. ALUMINUM

Aluminum is an element of the metal family known for its electrical conductivity, heat conductivity, resistance to corrosion, and light weight. It is obtainable either in rolled (wrought) or cast form, and may be combined with many other metals to form alloys. In its pure form, the metal has a white color, and is very ductile in the rolled aluminum sheet form. Cast aluminum is very brittle. The strength of the pure metal is considerably less than that of steel. Its melting temperature is approximately 1,220 deg. F. This metal also has a critical point called "hot shortness" and for that reason, must be carefully supported when being welded. An element which may be added to aluminum to decrease its "hot shortness" is silicon.

Aluminum sheet metal may be obtained in several qualities and grades. The purest commercial aluminum contains 99-1/2 percent aluminum, while the more popular commercial grades contain 99 percent aluminum. Two metals which are added to aluminum to increase its desirable physical qualities are manganese and magnesium. Amounts of these are relatively small, varying from 1 to 5 percent. Special alloys of aluminum are becoming very popular for aircraft work. These alloys are noted for their high tensile strength, but they are usually very difficult to weld satisfactorily. Any mechanical working of the metals tends to increase their brittleness.

Due to the activity of the metal, aluminum oxidizes very readily upon being heated. Aluminum oxide melts at about 5,000 deg. F., and therefore chemicals must be used to dissolve the oxide. The oxide is more dense than the molten metal and settles into a puddle, causing a porous weld. Therefore, special fluxes must be used at all times during the welding process unless an inert gas process is used.

Another feature of aluminum which adds to the difficulty of welding is that it does not change in color before it reaches the melting temperature. In other words, the metal upon being heated maintains the same color, but when reaching the melting point, it suddenly collapses. The operator can determine the melting temperature of the metal, when welding aluminum, by using the filler rod and scratching the surface to reveal any softening. If one uses the ANSI lens shade recommendation, Shade No. 4, for gas welding aluminum, the weld is more visible and a slight change in color can be noted as the melting point is reached.

Aluminum may be identified by its

DUCTILE IRON TYPE	TENSILE STRENGTH PSI	YIELD STRENGTH PSI	ELONGA- TION %	HARDNESS BRINELL
*60-45-10	60,000 to 80,000	45,000 to 60,000	10 to 25	140 to 200
80-60-03	80,000 to 100,000	60,000 to 75,000	3 to 10	200 to 275
100-70-03	100,000 to 120,000	70,000 to 90,000	3 to 10	240 to 300
120-90-02	120,000 to 150,000	90,000 to 125,000	2 to 7	270 to 350
x Heat Resistant	60,000 to 100,000	45,000 to 75,000	0 to 20	140 to 300

* This number is decoded as follows:
    60 means 60,000 psi tensile strength, 45,000 psi, yield strength, and 10% elongation.

Fig. 24-12. Properties for five types of nodular cast iron.

silvery white color when fractured, and by a comparison of weights with other metals. Aluminum is about one third the weight of iron for a given volume.

Another test for aluminum is to burn some chips of the metal. Aluminum will burn to a black ash.

Magnesium may be mistaken for aluminum. However, magnesium chips when heated will actually ignite and will form a white ash. Caution must be taken when igniting magnesium chips so that only a small quantity of the chips are ignited, since magnesium chips, also powder, burn violently.

24-22. HARD SURFACING METALS

Research has succeeded in producing alloys of metals having the property of extreme hardness. There are several different types:

1. Ferrous metal with up to 20 percent alloying elements such as chromium, tungsten and manganese.

2. Ferrous metal with over 20 percent alloying elements such as chromium, tungsten and manganese. (Some cobalt and nickel is sometimes added.)

3. Nonferrous metal with alloying elements such as cobalt, chromium and tungsten.

4. Tungsten carbide (fused), with some other alloying elements.

5. Granules of tungsten carbide.

See CHAPTER 20 for information on metal surfacing.

These metals are difficult to identify. It is best to keep them in their identifying packages. Fig. 24-8 shows the characteristic spark of Stellite and for Cemented Tungsten Carbide, two hard surfacing materials.

These metals range from extremely hard to very tough. Their best application, therefore, is as a thin coating on metal of a more ductile nature. This combination produces a long wearing surface and also one of great strength.

24-23. SOFT SURFACING METALS

The soft surfacing metals are usually the nonferrous metals. These may be identified by the methods described earlier in this chapter.

The properties of the deposited surface metal depend upon the surfacing metal and the method used to apply the metal to the surface of the parent metal. Surfacing metals can be applied by:

1. Dipping.
2. Electroplating.
3. Spraying.
4. Brazing or soldering.

24-24. DUCTILE IRON

Ductile iron is a patented type of cast iron. This cast iron is not as brittle as gray or white cast iron due to the manner in which the carbon (graphite) forms itself within the iron. The graphite in gray cast iron normally forms in long lines or in flakes, causing the metal to be weaker, more brittle, and therefore not ductile. The graphite in ductile iron forms into nodules or spheroids (tiny balls). This makes the iron more ductile and stronger.

Ductile iron because of the shape of the graphite is also known as "nodular" or "spheroidal" iron. Ductile iron may be distinguished from gray cast iron by the nodular shape of the graphite under a microscope. Fig. 24-12 shows the physical properties of five types of nodular cast iron.

24-25. NEW METALS

Many formerly rare and/or little used metals have recently been used as alloying elements to improve the properties of the more common metals. A few of these newer metals are: (1) columbium, (2) titanium, (3) lithium, (4) barium, (5) zirconium, (6) tantalum, (7) beryllium, (8) nobelium, (9) plutonium, (10) aluminum.

The manufacture and use of these new metals is increasing at a very rapid rate. Aluminum and titanium have been gaining in their practical use as metals for manufactured products. Welding engineers are also developing and improving methods to weld and braze these metals. See CHAPTER 18 for instructions on the welding of the newer metals.

24-26. TITANIUM

Titanium is rapidly becoming an important structural metal. Its use has rapidly increased since 1949. It is the fourth most abundant metal exceeded only by aluminum, iron and magnesium.

The commercially pure titanium melts at about 3,140 deg. F. Many of the alloys are weldable under certain conditions. This metal has good weight to strength ratio, high temperature properties, and it is very corrosion resistant. The aircraft, chemical, and transportation industries are especially interested in its uses. It is about 67 percent heavier than aluminum, and about 40 percent lighter than stainless steels. It retains its strength very well up to 1000 deg. F.

Titanium-carbon alloys are finding wide acceptance. The carbon content varies from .015 to 1.1 percent. The titanium-carbon alloys become more brittle as the carbon content increases, but they also become more corrosion resistant. The maximum tensile strength is reached at about .4 percent carbon and is approximately 112,000 psi (pounds per square inch). However, .04 percent carbon alloy has a strength of 92,000 psi.

Titanium alloys using tungsten and carbon have a tensile strength of about 130,000 psi. In combination with aluminum it has a tensile strength of about 114,000 psi.

Chromium-nickel 195,000 psi
Chromium-molybdenum 175,000 psi
Chromium-tungsten 150,000 psi
Manganese-aluminum 160,000 psi

Titanium-manganese alloys such as 8 percent manganese are popular in aircraft structure. Some other alloys consist of 3 percent aluminum and 1/2 percent of manganese; 6 percent aluminum and 4 percent vanadium; and 5 percent aluminum and 2-1/2 percent tin.

Titanium is weldable by either the gas tungsten arc process (TIG), or resistance welding. The gas tungsten arc process usually uses argon gas with a direct current straight polarity machine. For example, 1/16 in. thickness metal needs about 125 amperes at 14 volts using a 3/32 diameter electrode and about 22 cubic feet per hour of argon. The gas metal arc process (MIG) has been used, with a 40V direct current, reverse polarity, and with helium, or argon gas or mixtures of the two.

Commercially pure titanium has been successfully resistance welded. The timing of the event must be carefully controlled such as a pulse time of 2 cycles, weld time of 1 cycle, squeeze time of 25 cycles, and hold time of 60 cycles. Each cycle means 1/60 of a second. The resistance electrodes are usually cooled by refrigeration.

For good results, titanium and its alloys should be postheated. This heat treatment improves the weld ductility.

24-27. LITHIUM

The welding industry has also become very much interested in lithium. When this metal is alloyed with some other brazing metal, the brazing can usually be performed without flux. For example, titanium has been successfully brazed in an inert atmosphere using an alloy of about 98 percent silver and 2 percent lithium. Lithium has a melting temperature of 367 deg. F.

24-28. ZIRCONIUM

Zirconium is a rare metal which is found in nature combined with silicon as zirconium silicate (Zr Si O4) commonly known as the semiprecious jewel zircon. Zirconium oxide melts at 2700 deg. C and is used as linings for high temperature furnaces.

This metal oxidizes easily. Its strength is affected by oxygen, nitrogen and hydrogen with which it may combine. Zirconium is used as an alloying element in alloy steels.

Zircalbys are a family of zirconium alloys containing about 1-1/2 percent tin, nickel, chromium and up to 1/2 percent iron. These alloys have improved corrosion resistance and higher strength than unalloyed zirconium.

Because zirconium readily unites with oxygen and nitrogen when heated it is welded best in an inert gas filled chamber. The shielding gases normally used are argon or helium.

24-29. BERYLLIUM

Beryllium is a rare metal which is expensive to produce, but is becoming increasingly available to industry. Beryllium is a light metal with a density (1.845 g/cc) slightly higher than the density of magnesium. It is often used as an alloying element with other metals.

It has been used as an alloying element in copper and nickel to increase elasticity and strength characteristics.

Beryllium has a high thermal and electrical conductivity and a high heat absorption rate. Because of these qualities it is difficult to weld.

Some beryllium copper alloys are very strong and are used to replace forged steel tools in places when an explosive atmosphere may be present since these alloys are non-sparking.

Beryllium has been welded by the inert-gas, resistance, ultrasonic, electron-beam, and diffusion welding processes. Brazing and soldering methods of joining have also been used.

Beryllium is used with magnesium to reduce its tendency to burn during melting and casting.

This metal and its compounds are reported to have dangerously poisonous properties and special precautions must be taken when working with the metal and alloys which use this metal.

CARBON CONTENT IN PERCENTAGES	TENSILE STRENGTH LB./SQ.IN.	ELONGATION IN PERCENTAGES	YIELD POINT LB./SQ.IN.
0	20,000	—	12,000
0.1	25,000	60	17,000
0.2	30,000	33	22,000
0.3	37,000	29	27,000
0.4	44,000	27	30,000
0.5	49,000	25	34,000
0.6	52,000	24	37,000
0.7	57,000	22	38,000
0.8	59,000	20	39,000
0.9	60,000	17	40,000
1.0	58,000	13	39,000
1.1	55,000	9	37,000
1.2	52,000	6	35,000
1.3	48,000	5	30,000
1.4	45,000	4	—
1.5	40,000	3	—
1.6	35,000	2	—

Fig. 24-13. Approximate physical properties changes of steel as carbon content changes.

24-30. EFFECT OF CARBON ON STEEL PROPERTIES

The carbon content of a steel has considerable effect on its physical properties. Three of these physical properties are shown in Fig. 24-13. Note how the tensile strength and yield point strength first increase and then decrease as the carbon content increases. Also note that the elongation decreases as the carbon content increases.

24-31. REVIEW OF SAFETY

Spark testing metals for identification requires the tester to wear goggles and/or face shield. The grinder must be in good condition, including balanced grinding wheels and wheel rest clearance not to be greater than 1/16 to 1/8 in.

When chemicals are used to etch metals for identification, the eyes and face should be protected from acidic or caustic fluid splashing.

When a torch flame or an arc is used to identify the metal by its melting and sparking behaviour, all the safety pre-cautions for gas welding and/or arc welding must be followed. See Chapters 1, 3 and 5.

24-32. TEST YOUR KNOWLEDGE

1. How do properties of steel change as carbon content increases?
2. What is the highest carbon content of steel?
3. May cast iron be heat treated?
4. What are the different forms of cast iron?
5. What is the difference between the various forms of cast iron?
6. What is meant by an alloy steel?
7. What properties do alloys give to steel?
8. What is meant by a low-carbon-alloy steel?
9. What characteristic does tungsten steel have?
10. What is stainless steel?
11. Describe the term "hot shortness."
12. Name some methods of identifying metals.
13. What is meant by the blowpipe test?
14. What is meant by the critical temperature of a metal?
15. What is meant by the fracture test?
16. Is there a difference in the appearance of copper, bronze, and brass?
17. What is the carbon range of steel that responds to heat treatment?
18. What is the tensile strength of 25 point carbon cold roll steel?
19. What is the spark test?
20. What is the explanation of the variation in sparks?
21. What method is used to secure accurate analysis of a metal?
22. What is the carbon content of a medium carbon steel?
23. Explain the eutectic condition.
24. Does the metal have to melt to have a critical point?
25. What do the initials A.S.T.M. represent?

A drum gate for lake level control. This structure is being welded using a stainless steel clad metal.
(R. C. Mahon Co.)

Chapter 25

HEAT TREATMENT OF METALS

The welder should be particularly interested in heat treatment of metals, because he must know what welding does to the heat treatment of the metal and the effect on the physical properties of the metal. It is necessary to know if a metal must be preheated for welding, and if the metal should be heated during the welding operation. It is necessary to know too, what heat treating procedure to use to bring back as nearly as possible, the original properties of the metal (postheat treatment). The most common applications of heat treating in a welding shop are annealing to relieve the metal of internal stresses and strains, caused by the expansion and contraction during welding; and to improve the properties of the metal in the weld. Most structural welding involves only the knowledge of how a metal may be annealed. In large shops and in manufacturing plants, the preheating required, the concurrent heating, and the postheat treating of a metal are a part of the procedure determined by the engineering office.

It is in job welding, involving the repair of broken parts, that an extensive and accurate knowledge of all phases of heat treating by the welding operator is most needed.

25-1. THE PURPOSES OF HEAT TREATMENT

All metals can be heat treated. Some metals are affected very little, but some, particularly most steels, are greatly affected. Heat treating may serve the following purposes:

1. Develop ductility.
2. Improve machining qualities.
3. Relieve stresses.
4. Change grain size.
5. Increase hardness or tensile strength.
6. Change chemical composition of metal surface as in casehardening.
7. Alter magnetic properties.
8. Modify electrical conduction properties.
9. Induce toughness.
10. Recrystallize metal which has been cold worked.

When heat treating there are three factors of great importance:

1. The temperature to which the metal is heated.
2. The length of time that the metal is held at that temperature and the speed of cooling (a time factor).
3. The material surrounding the metal when it is heated, as in casehardening.

When a weld is made, the metal in and around the weld joint is heated to a variety of temperatures as the distance from the weld joint increases. This drop in the metal temperature from the weld joint outward is called a temperature gradient. In steels some of the metal in the area of the weld is heated through only one critical point, some two, and some may be heated to only 500 deg. F. as shown in Fig. 25-1.

Because of the uneven heating, the strength, ductility, grain size and other metal properties may vary greatly and may affect the strength of the metal in the weld area.

The welding operator may use preheating, concurrent (continuous) heating and/or postheating to avoid temperature gradients in the weld area. Heating the metal before welding will help to prevent internal stresses and strains which may cause weld failure. The welder may use continuous heating for the same purpose, and may use postheating to relieve stresses and bring all the metal in the area of the weld to the same heat treat condition.

25-2. METHODS OF HEATING

Heating metals is a complex operation. All metals expand when heated and rather drastic changes in volume take place as metals are heated above their critical temperatures, or, are cooled below their critical temperatures where a crystalline structure change takes place.

Unequal heating or cooling may cause cracks, or at least, a warping of the structure. However, this action is not always a disadvantage, as warped articles can be straightened using this technique.

Local heating can be done by using a flame such as:

1. Air-fuel gas.
2. Oxy-fuel gas.

Or, it can be done by:

1. Electrical resistance heating.
2. Induction heating.
3. Localized furnaces (usually specially built around that part of an object to be heated).

Electrical resistance heating is done by either putting electrical resistance units against the parts to be heated as

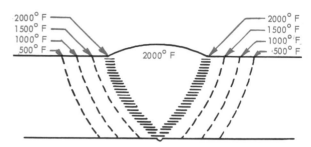

Fig. 25-1. Temperature zones in weld area.

HOW CHROMALOX TUBULAR PREHEAT UNITS ARE INSTALLED AND OPERATED

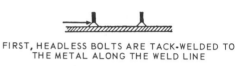

FIRST, HEADLESS BOLTS ARE TACK-WELDED TO THE METAL ALONG THE WELD LINE

SECOND, THE PREHEATERS ARE SLIPPED OVER THE BOLTS AND DRAWN UP TIGHT TO THE METAL

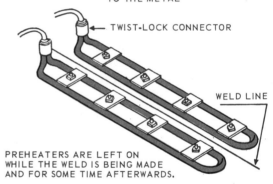

TWIST-LOCK CONNECTOR

WELD LINE

PREHEATERS ARE LEFT ON WHILE THE WELD IS BEING MADE AND FOR SOME TIME AFTERWARDS.

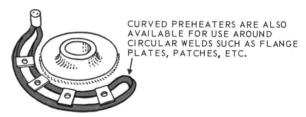

CURVED PREHEATERS ARE ALSO AVAILABLE FOR USE AROUND CIRCULAR WELDS SUCH AS FLANGE PLATES, PATCHES, ETC.

Fig. 25-2. Electrical resistance units for preheating. (Edwin L. Wiegand Div., Emerson Electric Co.)

shown in Fig. 25-2, or by using electric resistance units to heat a furnace.

Induction heating is a means of heating using high frequency alternating current to create eddy currents in the metal. This method heats a metal throughout its thickness.

General heating is usually done in a

furnace. The furnace can be heated with direct heat or indirect heat. The furnace may be heated by air-fuel gas, by electrical induction or electrical resistance units.

25-3. METHODS OF COOLING

The speed and evenness of cooling a metal object determines to a great extent the physical properties of the metal.

Welds performed on steels at room temperature cool quickly, and because the cooling takes place from the molten condition, the crystals are large.

Cooling may be accomplished in various ways. The heat loss is a combination of:

1. Convection.
2. Conduction.
3. Radiation.

Cooling in a gas is mainly convection and radiation. Cooling in a liquid is mainly convection and conduction. The cooling can only be done to approximately the temperature of the cooling medium. The colder the cooling medium the faster the cooling will take place. Cooling in a gas (by convection) is the slowest. Cooling in a liquid is fastest. The liquid can be any desired temperature. For example: water at room temperature (70 deg. F.) or liquid nitrogen (-230 deg. F.) (cryogenics).

The problem in heat treating is to cool the inside of the metal at the same speed as the surface of the metal.

Air is a common gas cooling medium.
Thermal convection cooling (still air) (static cooling).
Forced convection cooling (fans).
Water is a common liquid cooling medium.
Submerging in a water tank.
Spraying water on the metal to be cooled.

Oil, molten low temperature metals, liquid air, liquid nitrogen are other cooling liquids.

Either water cooled or refrigerant cooled fixtures are another means of cooling and controlling the cooling of metals.

25-4. CARBON CONTENT OF STEEL

Steel may be obtained with various carbon contents and alloying metals. Fig. 25-3 shows the carbon content of

ARTICLE	CARBON CONTENT
Axles	.40
Boiler Plate	.12
Boiler Tubes	.10
Castings, Low Carbon Steel	Less Than .20
Casehardening Steel	.12
Cold Chisels	.75
Files	1.25
Forgings	.30
Gears	.35
Hammers	.65
Lathe Tools	1.10
Machinery Steel	.35
Metal Tools	.95
Nails	.10
Pipe, Steel	.10
Piano Wire	.90
Rails	.60
Rivets	.05
Set Screws	.65
Saws for Wood	.80
Saws for Steel	1.55
Shaft	.50
Springs	1.00
Steel for Stamping	.90
Tubing	.08
Wire, Soft	.10
Wood Cutting Tools	1.10
Wood Screws	.10

Fig. 25-3. *Steels and how the carbon content varies depending on the use of the steel.*

some common items made of steel. The family of steels starts with low-carbon steel (almost wrought iron). As the carbon content increases, the steel becomes harder, stronger, and more brittle until approximately 1.75 percent

MATERIAL	CHARACTERISTICS
Austenite	Soft, ductile, tough, and wear resistant.
Martensite	First stage of austenite decomposition and a brittle constituent.
Troostite	Not quite as hard as martensite, But is considerably tougher. This is the constituent of steel used in files.
Sorbite	Not as hard as troostite, but it is tougher; because of its high yield point, it is the structure of spring steels.
Pearlite	The toughest structure with the highest breaking strength.

Fig. 25-4. A table of hardnesses for different grades of steel.

of combined carbon is reached. This is the maximum percent of carbon that will combine with iron to form steel.

PAR. 24-2 describes the physical properties of steel.

25-5. CRYSTALLINE STRUCTURE OF STEEL

Practically all heat treatments deal with the crystalline structure of steel, or the grain size. They are also concerned with combinations of iron and carbon, and with the distribution of iron carbide in the metal. In 10 point carbon steel, which contains 1/10 of one percent carbon, all of the carbon is in chemical combination with iron, forming iron carbide, Fe_3C. There is considerable free iron remaining, which in metallurgy is called ferrite. Steel containing a large amount of ferrite cannot be hardened. The only heat treatment possible is to alter the grain structure and size. Heating a steel to the upper transformation range temperature (critical temperature) does not greatly affect the steel because it is the status of the iron carbide chemical which determines the hardness.

Fig. P-32 shows a heat treatment chart. Iron carbide, mixed in various

proportions with ferrite, gives several combinations of hardness and brittleness, depending on various heat treatments. The hardness levels have been given names. The steels are listed in the order of their hardness, Fig. 25-4.

These various forms are obtained by the following steps:

1. Heating.
2. Quenching.
3. Drawing.
4. Tempering.

Eighty-five point crystalline steel has the correct amount of carbon to form steel that has the lowest critical temperature. This is called the "eutectic point" as shown in Fig. 24-3, CHAPTER 24 (Iron Carbon Diagram). When properly heat treated, the whole structure of this steel becomes pearlite. Pearlite is a steel which under the microscope has alternate layers or rows of iron carbide and ferrite. These beads look similar to a fingerprint printed with black ink on white paper, Fig. 25-5. This is the only steel that has just one critical temperature.

The hardness or physical structure of a steel which has been heated above the upper transformation range temperature (critical temperature), as in welding, may be affected by controlling the rate of cooling. If the steel sample is heated to 1,250 deg. F., and is then quenched in cold brine, austenite is formed. If it is heated to 1,250 deg. F. and is quenched in cold water, martensite is formed. By heating the steel to its critical temperature and quenching it in oil, troostite is formed. If it is heated to 1,250 deg. F., and quenched in a molten metal bath, sorbite and pearlite are formed. The above steps show that the rapidity of cooling affects the distribution and clustering of the iron carbide. Quenching, or the rate of cooling therefore, has much to do with the type of steel obtained.

To obtain the results described in the preceding paragraphs in another

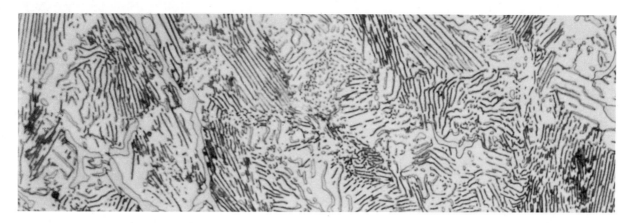

Fig. 25-5. *Photomicrograph of high carbon steel with pearlite grain structure. This is magnified 1000 diameters.*

manner, proceed as follows: Heat the eighty-five point carbon steel to 1,250 deg. F., then by quenching in cold brine, austenite is formed. If martensitic steel is desired reheat the sample to 400 deg. F. and cool again. If troostite is desired, reheat the sample to 600 deg. F. and cool. If sorbite is desired, reheat to 800 deg. F. and cool. When the metal is reheated to 1000 deg. F. and cooled, pearlite is formed. The temperatures to which these various forms of steel are heated overlap so that in many cases, one may have a steel which is part sorbite and part troostite, or part sorbite and part pearlite.

These temperatures must be held constant ($\pm$ 25 deg. F.) and the article must be left in the furnace at this temperature until the article is this temperature throughout its thickness.

25-6. ANNEALING STEEL

The term annealing is a common term in heat treating, and includes several types of heat treating operations. Generally speaking, annealing is considered to be that type of heat treatment which refines the grain and leaves the metal in a soft condition. An annealing operation may be performed for several reasons:

1. So the metal may be worked cold.

2. To make the steel machinable.

3. To relieve the internal stresses and strains produced while the metal was being shaped or welded.

When performing field welds, practical ways to preheat and anneal metal are with the oxyacetylene torch, or by induction heating. In induction heating, electrical coils are wrapped around the part and high frequency AC electricity passed through, as illustrated in Fig. 25-6. The changing magnetic fields,

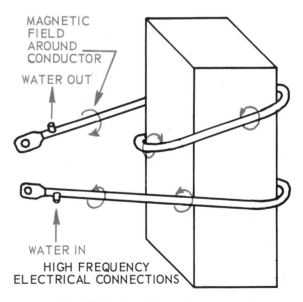

Fig. 25-6. *An induction heating coil.*

as these fields pass through the metal, heat the metal by the magnetic eddy currents formed. In the case of steels,

Fig. 25-7. Annealing and preheating weld joint on
a larger storage tank with resistance heating coils.
(J. B. Nottingham and Co.)

heat is also produced by magnetic hysteresis. Fig. 25-7 shows a large storage tank being annealed and preheated with resistance heating coils. The temperature is controlled by the amount of current flowing.

The general procedure for fully annealing most steels is as follows:

1. Heat the steel slightly above the critical temperature, Fig. 24-3 (Iron carbon diagram).

2. Cool the steel as slowly as possible.

If the grain size is of no importance, you do not have to be careful of the

limit and ductility of the metal decreases.

Steel, upon being heated to the critical temperature, changes its crystalline structure and reaches its most refined grain condition at its lowest critical temperature. However, if the steel is heated above this temperature, the grain size starts to grow and becomes larger as the temperature rises. The steel upon being cooled, will retain the size of the maximum crystalline structure reached. The crystals will continue to grow if the article is kept at the maximum temperature, or

NO. OF GRAINS PER LINEAR INCH	ELASTIC LIMIT TONS PSI	REDUCTION OF AREA, PERCENT
210	44.0	20.0
222	44.5	22.0
322	47.0	35.0

Fig. 25-8. Effects of grain size on elastic limit and ductility.

exact temperature to which the steel is being heated. As seen in Fig. 25-8, as the grain size increases, the elastic

if kept above its critical temperature until a maximum size is reached. Therefore, the speed of welding and the rapidity of cooling has some effect on crystal size. This is particularly important in welding because both the weld metal and the metal adjacent to the weld are heated considerably above the critical temperature.

Upon cooling a weld, regardless of the speed of cooling, the crystalline structure in the weld, and in the steel adjacent to the weld will be large and the physical properties of the metal will above the critical temperature of the metal and then allow it to cool slowly.

Peening affects the grain size only in the area worked, usually at the surface. Therefore, the latter method (2) is preferred. Industries with the facilities for doing so put the whole steel structure, which has been welded, in an annealing furnace; thus the whole structure is annealed instead of just the weld and the parts adjacent to it. Fig. 25-9

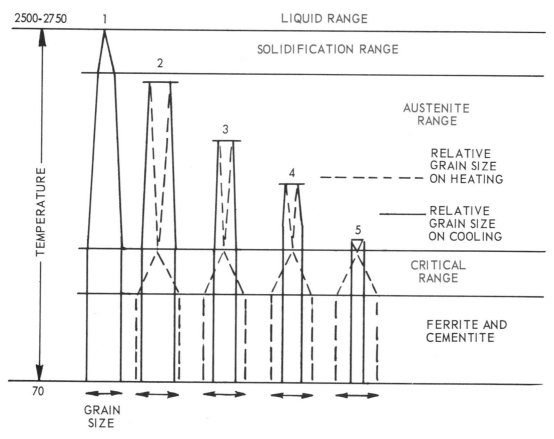

Fig. 25-9. Influence of heat on grain structure of steel. The dash lines indicate change in grain size as the metal is heated. The solid lines indicate the change in grain size as the metal is cooled to the room temperature.

be reduced. Two methods may be used to remedy this condition:

1. Pound the metal with a hammer as it is cooling (peening). The peening will refine the grain structure.

2. Allow the metal to cool to room temperature; reheat to just slightly shows how the grain size can be controlled by the temperature to which the metal is heated. The grain size is shown by the spacing between the solid lines on the figure.

Commercial annealing (partial annealing) may be done at a temperature between 1100 and 1200 deg. F. This process is used mainly to relieve internal stresses.

25-7. HARDENING STEEL

Steels of less than 50 point carbon content cannot be successfully hardened by heat treatment. Steels above 170 point carbon content are also difficult to harden by heat treatment. Within these extremes it is possible to obtain almost any degree of hardness by using the proper carbon content and the proper heat treatment. It is also com-

cooling determines to a great extent the hardness and brittleness of the metal. Cold brines, cold water, air, and molten metal baths, are used as mediums for cooling metal rapidly. The water cooling and the brine cooling result in extreme brittleness along with hardness. Air cooling and molten-metal-bath cooling tends to relieve much of the brittleness, but at a sacrifice of some hardness. The hardness of the metal seems to depend on the distribution and structure of the iron carbide throughout the steel. When cooled rapidly, there is little oppor-

PERCENT CARBON	TEMPERATURE °F. CRITICAL TEMPERATURE FOR HARDENING AND FULL ANNEALING	COMMERCIAL ANNEALING
.10	1675–1760	
.20	1625–1700	
.30	1560–1650	
.40	1500–1600	
.50	1450–1560	1020° F.
.60	1440–1520	to
.70	1400–1490	1200° F.
.80	1370–1450	
.90	1350–1440	
1.00	1350–1440	
1.10	1350–1440	
1.30	1350–1440	
1.50	1350–1440	
1.70	1350–1440	
1.90	1350–1440	
2.00	1350–1440	
3.00	1350–1440	
4.00	1350–1440	

Fig. 25-10. Critical temperatures for various carbon steels.

mon practice to obtain a refinement of the grain structure along with the hardness.

Generally speaking, to harden a steel specimen in this range of carbon content, the steel is heated just above its critical temperature and then cooled rapidly. The critical temperatures for steel with various carbon contents are shown in Fig. 25-10. As mentioned in a previous paragraph, the rapidity of

tunity for the iron carbide to separate or segregate, but as the rate of cooling is slowed, the carbides have time to separate and the metal becomes progressively softer.

25-8. TEMPERING STEEL (DRAWING)

As mentioned previously, quenching a steel sample in a molten-metal bath

An induction furnace in use. (Ajax Magnethermic Corp.)

will produce a hard metal without too much brittleness. However, a more effective method is a method called "drawing the hardness." By reheating the steel sample to a temperature between 400 and 1,000 deg. F. after the hardening operation, much of the brittleness in the steel may be relieved: most of the internal stresses and strains are also relieved. Reheating the metal to 600 deg. F. (blue color on the steel surface) is probably the most common drawing or tempering operation.

25-9. TEMPERING A COLD CHISEL

A common tempering operation in a school shop is to heat treat a cold chisel. Most cold chisels are made of approximately 80 point carbon steel and must have the following properties:

1. Cutting edge must be extremely hard.
2. Metal just back of the cutting edge must be hard and tough, not brittle.
3. Body of the chisel and the adjacent, or hammering end must be soft and ductile. To obtain these various properties from the same carbon content steel, one must heat treat it as follows:
   1. Heat the cold chisel slowly to just above its critical temperature (1350 deg. F., cherry red).
   2. Allow enough time for the body of the chisel to assume this temperature throughout its thickness; then quench about one inch of the cutting edge end of the chisel in cold water.
   3. Wait until the end in the water turns dark; then withdraw the chisel quickly from the water while the other end is still cherry red.
   4. Heat from the cherry red end will travel to the chilled cutting

edge, reheating it slowly. With a small pad faced with emery cloth, polish the flat surfaces near the cutting end of the chisel. Be careful of burns.

Now watch the polished surface carefully. As the heat from the shank of the chisel travels into this part, the polished surface will gradually change color. It will first become yellow, then brown. When the surface starts to turn purple, because of the oxidation of the metal as the temperature rises, this indicates the steel has been reheated to approximately 600 deg. F. Now quench the cutting edge in the water slowly. If the shank of the chisel has cooled below a cherry red color as it should have done, the complete chisel may be immersed in a bucket of cold water.

If the timing is correct on the heat treatment, and if the original temperature is not too high, the chisel will have a hard cutting edge, with a fairly hard and tough body and finally a soft ductile shank or hammer end. In case the original temperature was too high, the chisel will be brittle and will crack easily because of the large crystalline structure. The cutting edge will be too hard and will chip off as it is being used. If the cutting edge was requenched before it should have been, the cutting edge will be hard and the edge will break. If the shank of the chisel is quenched too soon (before its color has become a dark red), the shank will be too hard and brittle and will crack when being used. A chisel with brittle qualities is an extremely dangerous tool because if it cracks or chips the flying particles of metal may inflict injuries.

25-10. FLAME-HARDENING

One of the outstanding modern contributions to industry has been the development of the oxyacetylene flame-hardening process, which is being widely used today in many manufacturing operations. In many cases such as gears, shafting, connection rods, and the like, it is advisable to produce as hard a wearing surface as possible, and yet maintain the internal portion of the structure in a ductile and tough condition. The hard surface provides long wear and maintains an exact contour while the tough interior insures that the part will withstand shocks without breaking. Previous to the de-

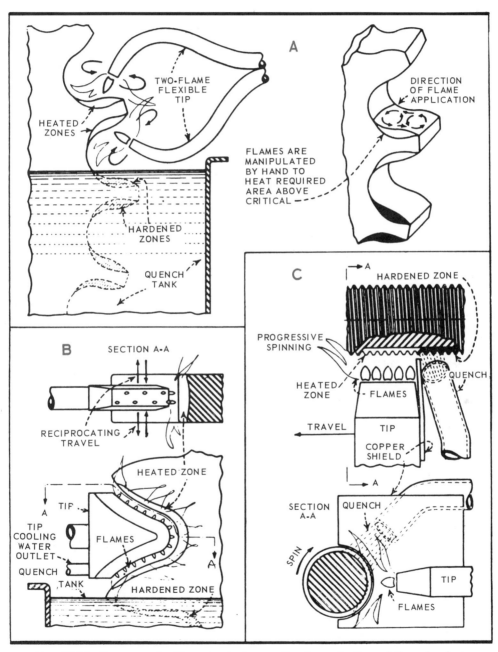

Fig. 25-11. Some popular flame-hardening applications. A. Manual flame-hardening allows concentration of heat at areas requiring maximum depth of hardness. B. Root and opposing faces of adjacent teeth are heated with contour tip. C. Progressive spinning is employed for fine threads, using water cooled nonquenching tip.

Heat Treatment of Metals 637

velopment of the flame-hardening process, complicated heat treatments had to be performed with temperatures under extremely accurate control, and with exact timing. Any change in the size or shape of a product requiring heat treatment necessitated a complete, new experimental setup to determine the proper heat treat conditions required.

Flame-hardening has eliminated much of this experimentation. The article is heat treated and then drawn to produce a tough, ductile structure throughout. The surfaces to be hardened are then put in a flame-hardening machine. A multiple-tipped oxyacetylene flame is passed over the surface to heat it quickly to a high temperature. Following the flame, a heavy stream of water is automatically fed to the surface, which quickly cools the entire surface, Fig. 25-11. The depth of the hardness can be accurately controlled by the temperature of the cooling agent. This method has proved to be a rapid and economical means of hardening that may be quickly adaptable to any type of structure. This application of the oxyacetylene flame is of real importance to the metal industry.

25-11. CASEHARDENING

Another solution to the problem of producing a metal article with a tough interior and a hard, accurate, hard-wearing surface is known as case-hardening. The fundamental plan of the process is to make the article out of low carbon steel, so any machining required may be done easily. The surfaces to be hardened are then exposed to carbon while they are at a high temperature. The high temperature iron will absorb some of the carbon, increasing the carbon content of the surface which

may then be hardened by the simple heat treating process. This method is called the spherodizing process. Typical examples of this method are piston pins and camshafts in all types of engines, and many other articles. In case where it is not desirable to case-harden certain portions of the article exposed to the carbon at high temperatures, these portions may first be copper plated. Generally speaking, the carbon will penetrate into the steel to a depth of 1/64 in. per hour. However, the depth will vary with the temperature, kind of carbon used, and the steel to be casehardened.

A cyanide bath may also be used to caseharden steel, but the danger from the toxic fumes makes its application limited. Thorough ventilation is imperative when using cyanide salts required in this process.

To heat treat a casehardened object, the procedure is: Heat the steel to its critical temperature and quench. This produces a glass-hard exterior on parts which have been exposed to the carbon bath. This also produces a tough interior with a refined grain which is of low carbon content and which is therefore strong and ductile. Then reheat to a temperature between 400 and 900 deg. F. to draw (temper) the surface. Fig. 25-12 shows a photomacrograph of casehardened gear teeth.

Fig. 25-12. A cross section of a casehardened steel gear tooth, showing casehardened surface (light). (General Motors Corp.)

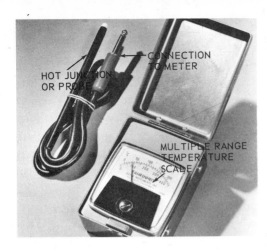

Fig. 25-13. Thermoelectric temperature measuring instrument. The electric probe draws minute quantities of heat energy from the surface to be measured. (Royco Instruments, Inc.)

25-12. TEMPERATURE MEASUREMENTS

Correct heat treatments can be obtained only if the temperature of the metal can be accurately measured.

One of the more difficult measurements is to determine if the surface temperature is also the inner temperature of the structure.

Several ways used to measure high temperatures are:

1. Optical pyrometers.
2. Gas thermometers.
3. Thermoelectric pyrometers, Fig. 25-13.
4. Resistance thermometers.
5. Radiation pyrometers.
6. Color change of the metal.
7. Temperature indicating crayons, cones, pellets, and liquids. The use of a temperature indicating pellet is shown in Fig. 25-14.

One of the most important factors in tool and die welding is the proper preheating and postheating of the metal. The temperature to which the metal is heated is of vital importance as a few degrees too little or too much will prevent getting the proper results. Automatic furnaces with heat source

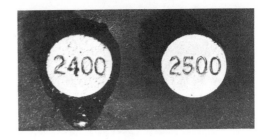

Fig. 25-14. Temperature indicating pellets. The fact that the 2400 deg. F. pellet is melting, and the 2500 deg. F. pellet is still solid, indicates a temperature between 2400 and 2500 deg. F. (Tempil° Div., Big Three Industries, Inc.)

controls and thermocouple temperature indicators are desirable. However, their expense makes them impractical for many small shops. An excellent solution to the problem for small shops is the use of temperature indicators such as crayons, pellets, or liquids. The operation of liquid temperature indicator is shown in Fig. 25-15. Fig.

Fig. 25-15. Liquid used to determine temperature is applied to surface of metal. After the liquid dries, it will melt at the temperature marked on the bottle. (Tempil ° Div., Big Three Industries, Inc.)

Fig. 25-16. Temperature-indicating crayon which will melt at 300 deg. F.

25-16 shows a 300 deg. F. crayon. Fig. 25-17 shows the crayon in use. These materials are obtainable to indicate a great variety of temperatures.

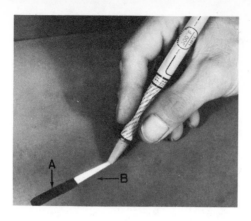

Fig. 25-17. *Temperature-indicating crayon in use. The 300 deg. F. crayon has just produced a mark (A) over a metal surface of 300 deg. or more. Note that the crayon mark melted. Over the part of the metal cooler than 300 deg. (B), the crayon mark was unaffected. (Tempil° Div., Big Three Industries, Inc.)*

They indicate temperatures in 12 1/2 deg. F. steps, in the 113 to 400 deg. F. range, and in 50 deg. F. steps in the 400 to 2000 deg. F. range. The pellets are also obtainable in 100 deg. F. steps from 2000 to 2500 deg. F. Fig. 25-18 shows one way a welder can use such a crayon to measure the preheat temperature of a pipe prior to welding.

25-13. HEAT TREATING TOOL STEELS

There are several classifications of tool steels. A cold chisel usually has a carbon content of 75 to 85 point car-

Fig. 25-18. *A welder preheating a pipe joint to 300 deg. F. prior to welding. Note at (A) the crayon mark and how it has melted in the area where the torch is pointing.*

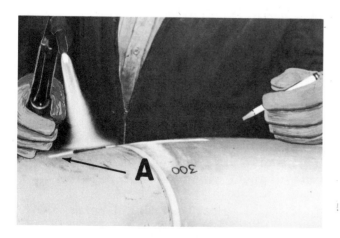

bon. It is considered a lower grade tool steel, and is similar to all tool steels having between 50 point carbon and 100 point carbon. To fully anneal this steel one should heat the 50 point carbon steel to 1500 deg. F. and cool slowly. To anneal 85 point carbon steel, heat to 1350 deg. F. and cool slowly. To anneal 100 point carbon steel, heat to 1350 deg. F. and cool slowly. These temperatures were obtained from the iron carbon diagram, CHAPTER 24, Fig. 24-3. To use this diagram, find the carbon content on the lower (horizontal) scale. Draw a line vertically until you reach the area of the diagram marked full annealing temperature, then draw a line horizontally to the vertical temperature scale, and note the required temperature for annealing the steel sample in question. To harden this range of steel, one must heat it to a temperature within the full annealing temperature zone and cool it rapidly in the proper cooling medium. Fig. 25-19 shows the grain structure of hardened high carbon steel.

Fig. 25-19. *Microphotograph of hardened high carbon steel magnified approximately 500x.*

To draw (temper) the tool steels mentioned in the preceding paragraph, one must heat them to the temperatures within the full annealing zone and then cool them rapidly in the proper medium. The metal must then be reheated to a temperature of 400 to 900 deg. F. and cooled. Fig. 25-20 shows the annealing effect of successive passes in a multiple-pass weld. The higher the tempera-

Fig. 25-20. Cross section of an arc weld that was welded in several passes. Note how successive passes have annealed the earlier passes. (General Electric Co.)

ture to which the metal is reheated, and the longer it is held at this temperature, the softer and more ductile the steel becomes. The yellow, brown, and blue metal color tempering lines are shown in Fig. P-32. Fig. 25-21 shows the grain structure of high carbon steel in the annealed condition. The word draw as applied to heat treating, means to draw from or remove from the steel, some of its hardness and brittleness.

Spherodizing is a heat treatment which is used to make high carbon steel even softer than is possible with the annealing process. The steel is heated to a temperature just below the transformation range for a number of hours, and is then slowly cooled through the upper cooling range. The cementite in the steel collects in small spherically-

Fig. 25-21. Microphotograph of grain structure of high carbon steel in annealed condition. (General Motors Corp.)

Fig. 25-22. Microphotograph of grain structure of spherodized high carbon steel.

shaped particles leaving a matrix of ferrite which is soft. Fig. 25-22 shows a photomicrograph of spherodized high carbon steel.

Tool steels between 100 and 170 point carbon must be heated from 1350 to 1450 deg. F. and cooled slowly to anneal.

To harden, heat to the same temperature and cool rapidly. To relieve the brittleness by drawing, heat to the annealing temperature, cool rapidly, and then reheat to a temperature between 400 and 900 deg. F. Most of these steels are not simple, iron-carbide steels, but are usually alloy steels. Fig. 25-23 shows an example of tempered high carbon steel.

Fig. 25-23. Microphotograph of grain structure of tempered high carbon steel.

25-14. HEAT TREATING ALLOY STEELS

The heat treatment of alloy steels is difficult because of the multiplicity of alloy metals (ingredients), and the hundreds of combinations of these alloy metals with the iron carbon steels. Alloy metals, generally speaking, lower the critical temperature of the iron carbide steel. Some alloy metals have a special influence on the characteristics of the heat treatment. For example, iron carbide steel, with manganese added to it as an alloying metal, will become a self-hardening steel. If a sample of manganese steel is heated to

its critical temperature (1350 deg. F.) and cooled in air, the sample is as hard as an ordinary iron carbide steel sample quenched in cold water. Further, this alloy has the characteristic of maintaining its hardness at high temperatures. This characteristic makes it adaptable as a metal-cutting tool. Chromium-tungsten high speed steel also has the property of maintaining hardness at a high temperature. It is difficult to anneal these steels because of the self-hardening properties. The iron carbon diagram should not be used to determine heat treat temperatures for alloy steels. In order to heat treat alloy steels properly, one should, if possible, obtain the exact percentage of the constituents in the steel, and use the recommended process as specified by the manufacturer of the steel.

25-15. HEAT TREATING TOOL AND DIE STEEL

It is very difficult for the welder to identify the alloy of a particular tool steel, so he will know the type of heat treatment required. Therefore careful records must be kept of the specifications for each tool from the time of manufacture or purchase. It is expected that the tool steel welder will be given this information before making a tool or die weld.

25-16. HEAT TREATING CAST IRON

As mentioned previously there are several forms of cast iron, namely, white cast iron, gray cast iron, nodular iron and malleable iron. They are similar cast irons in reference to their carbon content and method of manufacture, except the nodular irons which have other alloying metals in them.

They differ because of the heat treatments. White cast iron is formed by heating cast iron to the critical temperature (1650 to 2088 deg. F.) and cooling rapidly. This cast iron derives its name from the appearance of the fracture, which is white as a result of the carbon being in solution with the iron, forming cementite and pearlite.

If cast iron is heated to its critical temperature, and then cooled slowly (in asbestos, sand, or a furnace), the slow change through the temperature ranges permits enough time for some of the excessive carbon to separate from the iron to form microscopic flakes of graphite. When a sample of this type of cast iron is broken, the fracture has a grayish color because of the number of graphite flakes imbedded in the cast iron body.

White cast iron, because of the excessive amount of carbide, is too hard to machine whereas gray cast iron, because of the presence of the graphite flakes and the reduction of the amount of carbide is machinable.

Malleable iron is cast iron heat treated to make it more ductile, and more resistant to shock. This type of cast iron is formed by heating a casting to its critical temperature and cooling it rapidly, forming white cast iron; then reheating to a temperature between 1450 and 1650 deg. F., and keeping it at this temperature for 24 hours for each inch of casting thickness. Being maintained at this high temperature for this duration of time causes much of the carbon to separate slowly from the iron in the casting, forming globular carbon (spheroids of carbon or small microscopic pools of carbon surrounded by low carbon iron). After heat treatment at this high temperature for the proper length of time, the casting is cooled slowly.

By noting the conditions described in preceding paragraphs for the production of malleable iron, one will readily realize that if a malleable iron casting is welded, the casting will lose its malleable qualities in the weld and in the vicinity of the weld. For this reason malleable iron castings (recognized easily by a low-carbon spark on the surface and a cast iron spark in the center) should be brazed, or braze welded to keep the temperature below the point where white cast iron is again formed.

A special application of an effective heat treatment on cast iron is the heat treatment given to railroad car and street car wheels. These wheels are cast in a mold which provides a special means of rapidly cooling the rim or periphery of the wheel, whereas the remainder of the wheel is cooled slowly. This produces a wheel that has a white cast iron rim (very hard) and a gray cast iron body that is not so readily fractured if subjected to shock.

25-17. HEAT TREATING COPPER

Copper becomes hard and brittle while being mechanically worked. However the metal can be made soft again by annealing. To anneal copper, it should be heated to a temperature between 700 and 900 deg. F. It may then be cooled either rapidly or slowly since the rate of cooling has no effect on the heat treatment. This heat treatment eliminates the fractured crystalline structure caused by the slipping of the original crystals on their slip bands and makes the copper soft, pliable, and ductile. It brings back its original ductile qualities.

One must be careful when heating copper to its annealing temperature because, at approximately 900 deg. F., copper undergoes a physical phenomenon or change, called hot shortness. At this temperature the copper suddenly loses its tensile strength. If not adequately supported or subjected to strain, it will fracture easily.

25-18. HEAT TREATING ALUMINUM

Aluminum is heat treated much like copper. As it also has the characteristic of hot shortness one must be very careful when aluminum is passing through this temperature. Mechanical working also makes aluminum harder and more brittle. This may be relieved by heating and cooling. The critical temperature is about the same as that of copper.

Aluminum has a number of alloys. Each requires special heat treatment to bring out the best properties.

Some aluminum alloys age harden at room temperatures. They are refrigerated to keep them soft, ductile, and malleable while being shaped.

25-19. REVIEW OF SAFETY

Heat treatment of metals requires one to use safe working habits with the hot metal. Gloves are required and goggles are recommended, in most cases. Hot metals should be safeguarded to prevent other persons from injuring themselves.

Quenching metals requires a safety shield to protect the face from high-temperature flying fluids. Since the heat source used for heat treating varies, one should investigate the required safety steps depending on the heat source.

25-20. TEST YOUR KNOWLEDGE

1. What is the purpose of annealing?
2. How is a steel hardened?
3. What is tempering?
4. Why must aluminum be well supported when being welded?
5. What causes stresses and strains when metal is heated as in welding?
6. What carbon steel has the lowest critical temperature point?
7. What may be done to relieve the strains in a weld without reheating it?
8. How is a steel sample made both hard and tough?
9. Why is time critical when heat treating a sample?
10. What are the three primary factors to be be considered in heat treating?
11. How may gray iron be turned into white cast iron?
12. May a cold chisel be correctly heat treated by only heating it once? How?
13. Where is the best place to cool large articles annealed in a gas furnace?
14. How may a sample of brittle copper be annealed?
15. What is meant by the term "drawing" as applied to heat treating?
16. Explain preheating.
17. Will steel expand or contract when heated through a critical temperature?
18. What is the similarity between cast iron and the as-welded condition of the metal in the weld of a mild steel weld?
19. Is peening a hot working process, a cold working process, or both?
20. Does hysteresis take place in all ferrous metals when subjected to an alternating magnetic field?
21. In what medium does cooling take place when much of the heat loss is by radiation and convection?
22. Can molten metals be used to cool other metals?
23. When do the smallest crystals in steel exist?
24. Why is some steel heated to 400 deg. F. and then cooled?
25. List several high temperature measuring instruments.

Chapter 26

THE WELDING SHOP

Welding shops may be divided into three principal groups:

1. The independent shop which specializes in repairing and fabricating metal structures by the various methods of welding.

2. The welding shop, or department, attached to a factory or manufacturing establishment.

3. The welding equipment shop which overhauls and repairs welding equipment and sells equipment and supplies.

26-1. THE WELDING SHOP

The architectural design of the welding shop is important. A typical shop, would incorporate these features:

1. Heavy-duty load bearing floors, preferably of concrete.

2. Fire-resistant structure.

3. Building which is well-ventilated and has provision for localized exhaust ventilation.

4. Some means to move heavy equipment and material into and out of the shop.

5. Heavy-duty electrical service readily available.

The concrete floors should be at the ground level. The utilities should be arranged on the periphery (around the outside) of the room and overhead. A monorail or a double-rail crane system should be installed to provide easy movement of equipment and material to any spot in the room. If the business

is of sufficient size to justify accessory rooms such as paint booths, storerooms, toilets, showers, locker rooms, and offices, these should be in an annex to the shop for safety, cleanliness and to reduce the noise level.

26-2. WELDING SHOP EQUIPMENT

Equipment used in the welding shop depends to a considerable extent upon the kind of work handled. Some of the common equipment used is:

1. Oxyacetylene welding stations.

2. DC arc welding stations.

3. AC arc welding stations.

4. Inert gas arc welding stations.

5. Resistance welding machines.

6. Metal cutting equipment.

7. Preheating and postheating furnaces.

8. Overhead crane and/or heavy-duty hoists.

9. Forges.

10. Benches, vises, anvils.

11. Jigs and fixtures.

12. Heavy-duty power tools such as lathes, drill presses, grinders, nibblers, shears, brakes.

13. Sandblasting equipment.

14. Weld inspecting and testing equipment.

15. Paint booth.

Local and State Building and Safety Codes must be followed when designing, building and using all welding shops. The shop must be adequately equipped with personnel safety equipment and fire safety equipment such as fire extinguishers and blankets.

26-3. OXYACETYLENE STATION

In cases where more than one oxyacetylene station is used, manifolds may be used for the gases, and the cylinders can then be kept outside the shop area for greater safety. In some shops acetylene generators are used in place of cylinders. These generators may also be connected to a manifold system for distribution of the gas around the shop. Details of construction of the oxyacetylene welding equipment are given in CHAPTER 2. A station consists of gas cylinders, regulators, hose, torch, and a bench upon which the work to be welded is assembled. It is recommended that at least one of the oxyacetylene stations be portable so it may be moved around the shop. Fig. 26-1 shows a portable station with an acetylene generator.

The accessories of an oxyacetylene welding station should be attached to the station, or they may be kept in a portable kit to save time and effort in assembling the necessary apparatus to do a welding job. The accessories include leather gloves, pliers, wire brush, welding goggles, lighter, tips, and wrenches. Special wrenches needed for adjusting and operating the welding station are best handled by attaching them to the welding station rack, or to the welding bench near the station by means of chains. Portable units usually provide space for commonly used filler rods and fluxes.

26-4. ARC WELDING STATION

The arc welding station, either AC or DC, or AC-DC inert gas arc welding

stations, should have the following equipment:

1. Enclosed booth to protect other workmen in the shop from the harmful rays radiating from the arc.

2. Bench upon which the work is to be assembled and welded.

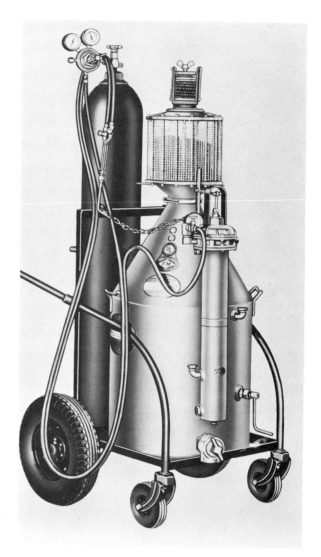

Fig. 26-1. Portable oxyacetylene welding station with an acetylene generator. (Rexarc, Inc.)

3. Ventilating fan to move fumes from the booth to the outside of the building before they can be dissipated into the shop.

4. Complete arc welding outfit. Some shops which use many arc welding sta-

Fig. 26-2. Four arc welding machines which may be stacked to save space and to minimize electrical primary power connections. (Miller Electric Mfg. Co.)

tions arrange the welding machines as shown in Fig. 26-2.

The booth should be painted on the inside with a special arc ray absorbing paint. This paint is of a gray white color and has the peculiar property of absorbing and not reflecting the infrared and ultraviolet rays.

For safety when working on jobs out in the shop, a portable canvas booth may be used as shown in Fig. 26-3. This screening eliminates much of the flashing, which otherwise might interfer with other workmen in the shop. It also eliminates much of the glare which creates reflections around the back of the welder's shield, or helmet, and may injure the operator's eyes.

26-5. SHIELDING GAS ARC WELDING STATION

An ever increasing part of the work done in a welding shop involves certain nonferrous metals and ferrous metals which require that they be welded in a shielding gas atmosphere to obtain consistently acceptable welds.

The welding shop operator, to bid on these special metal jobs, is finding that at least one shielding gas arc welding station is necessary to complete his shop facilities. The equipment is usually more expensive than the standard DC or AC welding station, but the results which can be obtained with shielding gas arc welding equipment far outweigh the added expense. Space requirements, ventilation equipment, and the booth are the same as when using a typical stick electrode arc welder. However, one must always remember that shielding gas arc welding emits more powerful infrared and ultraviolet rays.

In a small shop with only a few arc stations, the shielding gas arc welding machine may be installed so it can be used for either stick electrode or shielding gas arc welding. This dual purpose station is obtained by attaching the shielding gas arc welding cables and the metallic arc cables to the two terminals of the machine in parallel. The metallic arc electrode holder is hung on

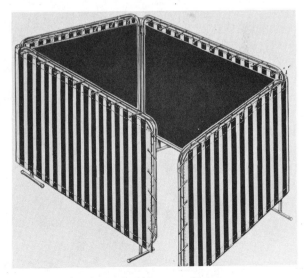

Fig. 26-3. Portable arc welding screen. (Singer Safety Products, Inc.)

an insulated hook when the shielding gas arc is used. By this method, the usefulness of the machine is increased when there is no need for TIG or MIG welding.

26-6. RESISTANCE WELDING MACHINES

More and more welding shops are using resistance welders. The most popular type of resistance welder is the spot welder. With this welder, a shop can become more versatile and capable of bidding on jobs involving the welding of thin materials.

The spot welder may be used to tack assemblies for alignment prior to other welding. The operator should have the following equipment for work on a resistance welder, gloves, clear tempered goggles or a plastic shield, pliers or tongs to hold the work, and a variety of electrode sizes and shapes plus tools for conditioning and changing the electrodes.

A small portable spot welder is often used in fabricating shops. There are many spot welding situations where the assembly is too awkward to move to a fixed spot welder and much time is saved by moving the portable unit to the job. More detailed information about resistance welding units will be found in CHAPTERS 13 and 14.

26-7. VENTILATION EQUIPMENT

Air contamination in the shop must be kept to a minimum for health and good housekeeping. The operators must have sufficient clean air, oxygen to eliminate respiratory problems, and

Fig. 26-4. Adjustable exhaust ventilator installed to exhaust fumes from arc welding station. Pickup for this exhaust system may be adjusted to keep fumes away from the operator. (Ruemelin Mfg. Co.)

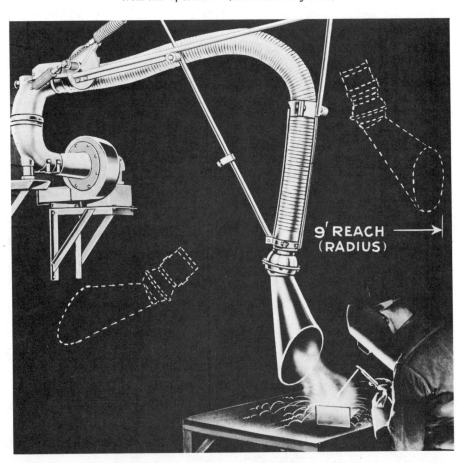

to be comfortable. The temperatures and humidity conditions should be comfortable. Toxic gases and toxic dust particles must be held to an absolute minimum.

Welding stations which produce dust particles and gases should have adequate exhaust ventilation. The amount produced varies with the size of the equipment, kind of metal being heated, fluxes used and the like. It is desirable to have an industrial hygiene technician take air samples under operating conditions and to adjust or modify the ventilating system until safe working conditions prevail.

Venting exhaust fumes from a welding area is shown in Fig. 26-4.

Air which is exhausted must be replaced. When the outdoor temperatures are comfortable, open doors and windows provide the air replacement. However, during cold weather conditions, the replacement air must be both heated and humidified. Adequate provisions must be included in the heating plant to provide the shop requirements.

26-8. METAL CUTTING EQUIPMENT

Metal may be cut to size and beveled to specifications prior to welding by several different processes:
1. Abrasive cutting wheel.
2. Power hacksaw.
3. Metal bandsaw.
4. Shears.
5. Nibbling.
6. Flame-cutting torch.
7. Arc cutting.
8. Plasma arc cutting.

Power hacksaws, shears and power cutoff saws are needed to cut standard size stock to needed lengths. Fig. 26-5 illustrates an abrasive wheel cutoff machine.

Where 90 deg. or other standard

Fig. 26-5. Dry abrasive cutting machine. This circular cutoff tool which is mounted on a pivot is held in the up position by spring force. The wheel is lowered to the metal being cut. In addition to rotating, the circular abrasive wheel also oscillates to produce a better cutting action.
(Allison Campbell Div., ACCO)

angle cuts are needed a reciprocating type power hacksaw or a band type power hacksaw are commonly used. Metal-cutting bandsaws are used for both straight and contour cuts.

Metal shears of all types, both manual and power-operated find extensive use in welding shops. Squaring shears are used for thin metal while alligator-jaw shears (some with special provisions for round stock cutting and angle iron cutting) are used for stock up to 8-in. in width. Fig. 26-6 shows a power-driven sheet metal and thin plate shears.

Fig. 26-6. Power shear for cutting sheet metal and thin plate. (Niagara Machine and Tool Works)

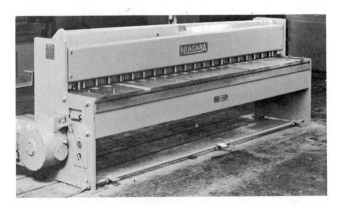

Fig. 26-7. Enlarged view of cut made by a nibbler machine.

Fig. 26-9. Portable nibbler being used to cut curves in sheet metal. This tool is capable of cutting along either straight or curved lines. (Black and Decker Mfg. Co.)

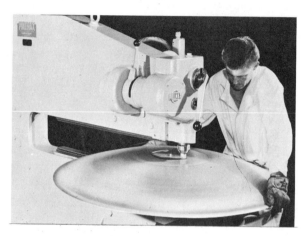

Fig. 26-8. Electrically-powered floor model nibbler machine making a circular shearing cut. (American Pullmax Co.)

Fig. 26-10. Steel plate being cut with a portable nibbler machine. (Fenway Machine Co.)

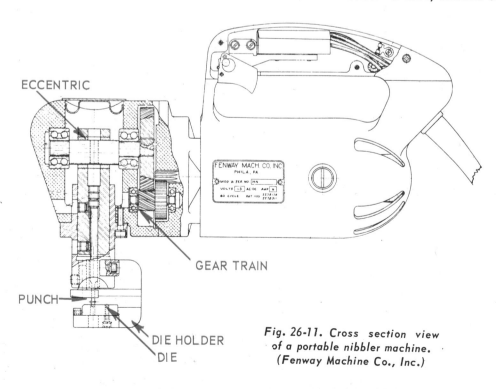

ECCENTRIC

GEAR TRAIN

PUNCH

DIE HOLDER
DIE

Fig. 26-11. Cross section view of a portable nibbler machine. (Fenway Machine Co., Inc.)

A nibbler machine complete with attachments provides a means of cutting irregular shapes for assemblies fabricated of sheet metal and plate. This machine uses two small, sharp steel blades, one stationary and one powerful moving blade. The cutting is done by shearing as shown in Fig. 26-7. It is possible to cut straight or curved lines with this type machine. Fig. 26-8 shows a floor model and Fig. 26-9 a portable model.

The portable nibblers will cut metal up to 1/4-in. thick. Fig. 26-10 shows a steel plate being cut in a straight line while the curved section shows the versatility of the tool. The internal construction of a nibbler showing the mechanism which produces the shearing action is illustrated in Fig. 26-11.

Tools have been developed for cutting tubing and pipe of all diameters and thicknesses to produce the special notched ends needed to weld or braze the joints. Fig. 26-12 shows a manually-operated notching machine. The different shaped ends of tubing which may be obtained are shown in Fig. 26-13.

When metal exceeds 1/8-in. in thickness, the metal to be gas or arc welded is usually beveled. The bevels may be

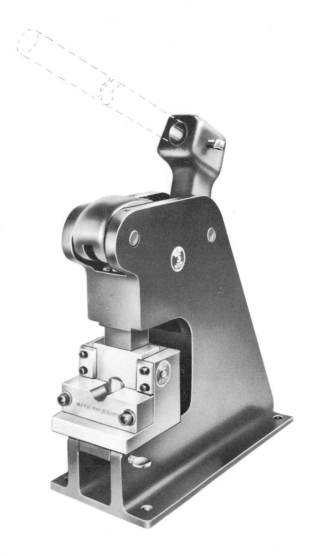

Fig. 26-12. A pipe and tube notching machine. (Vogel Tool and Die Corp.)

Fig. 26-13. Both rectangular and round tubing ends shaped for various type joints.

made by grinding, by flame or arc cutting, or may be made by a beveling machine as shown in Fig. 26-14.

Since a great deal of cutting takes place in a welding shop, either of various shapes prior to welding in a fabricated construction or in cutting risers from castings, an arc or flame cutting station is desirable. One or more of the oxyacetylene cutting stations should be portable for greater flexibility in the shop.

Fig. 26-14. *Power-operated beveling machine in action. This machine uses a cutter similar to a milling machine cutter. It can bevel metal up to 1-in. in thickness.*

26-9. FURNACES

A preheating furnace is a necessity for practically all welding shops. As previously explained, some metals and practically all complicated metal structures, when welded, are subject to excessive straining and perhaps breakage if heated or cooled unevenly. To minimize these conditions, the metal or the structure must be heated gradually to the correct temperature before the necessary welding is performed. The structure is then allowed to cool slowly and evenly after the welding is completed. This practice makes possible a minimum of warpage; it also prevents excessive stresses and cracking of the metal as it cools to room temperature. Three styles of preheaters are:

1. Portable torch type.
2. Flat-top open type.
3. Enclosed furnace.

The torch type may be a large portable blowtorch such as shown in Fig. 26-15. A preheat furnace may use city gas, oil, gasoline, or kerosene as the fuel. It may use air or oxygen as the

combustion supporting gas. These furnaces, or heaters, may be used for the smaller metal structures or for localized preheating.

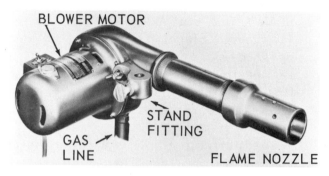

Fig. 26-15. *A preheating blowtorch. This unit with blower propelled air produces a temperature of 2200 deg. F. It uses natural gas as a fuel. A flexible gas line permits positioning. The torch may be used for either preheating, postheating, or metal heat treating.* (Clements Mfg. Co.)

The open flat-top preheater consists of a grate, or series of bars, with gas burners underneath. The article to be preheated is placed on top of the bars or grates. Occasionally a part too large for any existing shop furnace may be brought in for repair. In such cases, a preheat furnace may be built around the part using firebrick and asbestos sheets as needed and a fire built under the part. In some cases, large gas-fired industrial space heaters are used to provide heat for improvised preheat furnaces such as shown in Fig. 26-16.

Clay and firebricks are usually built up around the article to be preheated in order to enclose it. This shield may also protect the welder from the heat from the furnace which is dissipated into the room, making working conditions uncomfortable, especially in warm weather.

An advantage of this type of furnace is its flexibility in that the article being preheated, or welded, is readily accessible if certain bricks or asbestos sheets are removed. The article may be welded while it is still located in the furnace.

The enclosed type of preheating furnace is a typical furnace which may be used for heat treating, carburizing, or preheating. It consists of a firebrick-lined, steel structure and usually uses gas for fuel. A forced air blower is generally used for supporting the combustion and for raising the temperature. An advantage of this type furnace is its economy, its ability to cool the article slowly for annealing purposes, and the dissipation of the heat outdoors rather than into the shop. It has the disadvantage of usually requiring that the article to be welded must be removed from the furnace after preheating in order that it may be welded. In some cases this prevents the use of this type furnace.

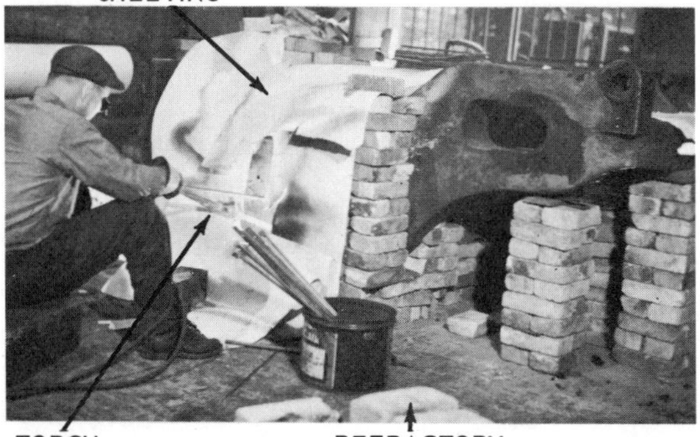

26-10. OVERHEAD CRANE

The size of the work to be handled by the welding shop may necessitate the use of power equipment for moving the article to be welded from place to place.

In the small shop, a chain fall, either hand driven or motor driven, may be suspended from rails or beams, which run the length of the shop. Or, a portable crane which runs on tracks mounted along the sides of the shop may be used. Fig. 26-17 illustrates a powered cable hoist.

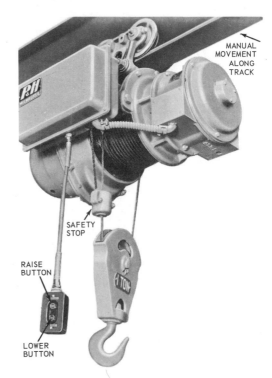

Fig. 26-17. Electrically-operated overhead crane hoist, of one ton capacity. Continuous pressure must be maintained on a button to operate the motor. (Harnischfeger Corp.)

26-11. FORGE

A welding shop is sometimes called on to shape and/or heat treat articles after they have been welded. The forge has some advantages over a welding torch for this purpose. Fig. 26-18 shows a typical blacksmith forge which uses forced air and has a vent to the outdoors through a hood located over the forge fire.

The forge temperature and amount of heat may be controlled by a valve on the blower duct, and the amount of coke used.

Fig. 26-16. Shop-built furnace made of firebrick and asbestos sheeting.

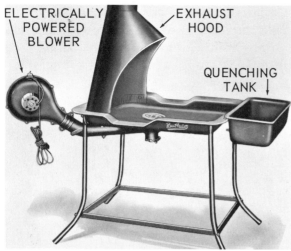

Fig. 26-18. Blacksmith type forge equipped with electrically powered blower, exhaust hood, and quenching tank. (Buffalo Forge Co.)

Fig. 26-20. Table-mounted fast-action clamp used for production work. (United Clamp Mfg. Co.)

26-12. JIGS AND FIXTURES

One of the greatest problems encountered in a welding shop is to prevent the metal from warping or buckling during the welding operation, or during the cooling after the welding operation. Many devices have been used to hold the metal during the welding so that the warpage and bending of the metal will be reduced to a minimum. Clamps, V-blocks, vises, and special

holding jigs are used extensively for this purpose. Fig. 26-19 shows a fast-action clamp. This clamp has a beryllium copper alloy spindle which resists injury from arc welding spatter. Fig. 26-20 shows a table-mounted type fast-action clamp. Fig. 26-21 shows how close one may arc weld to the clamp without injury to the clamp. When a large quantity of articles of the same type are to be made, it is economical to design and make jigs to hold the

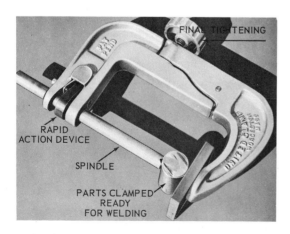

Fig. 26-19. Fast-acting clamp. The shaft or spindle is made of beryllium copper alloy. This permits arc welding within 1/4-in. of the spindle without injury from arc spatter. (United Clamp Mfg. Co.)

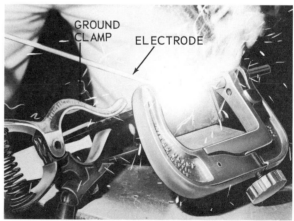

Fig. 26-21. An assembly being arc welded (note ground clamp) while the parts are held together with a fast-action clamp.

articles so the finished products will be identical in shape and size, Fig. 26-22.

Fig. 26-22. Assembly of toggle clamps mounted on steel plate to form clamping fixture. (De-Sta-Co Div., Dover Corp.)

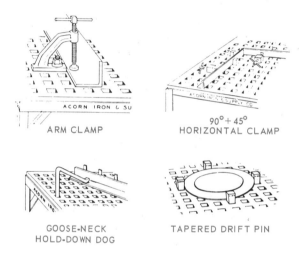

ARM CLAMP

90° + 45°
HORIZONTAL CLAMP

GOOSE-NECK
HOLD-DOWN DOG

TAPERED DRIFT PIN

Fig. 26-23. A faceplate (platen) used with various clamping and bending devices. (Acorn Iron and Supply Co.)

Angle iron, cast iron, or steel-grooved face plates may be used to advantage for fixtures of this kind. Fig. 26-23 shows a cast iron platen with various accessories used to clamp and/or shape the metal.

Turntables; horizontal, vertical, or tilting, are also popular in shops in both custom work and in production work as they help to provide down-hand flat welding. Fig. 26-24 shows a typical powered turntable. The turntable consists of a fixture and faceplate, mounted on rollers, or ball bearings, to which

any assembly to be welded may be fastened. The operator starts in one position while welding and the work is

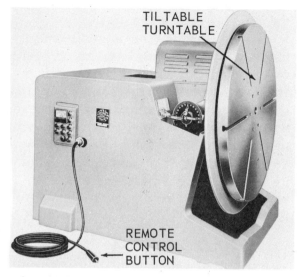

TILTABLE
TURNTABLE

REMOTE
CONTROL
BUTTON

Fig. 26-24. Power-operated welding turntable. The turntable may be power tilted through 90 deg. (flat to vertical). The turntable can be rotated at variable speeds. (Weldma Co.)

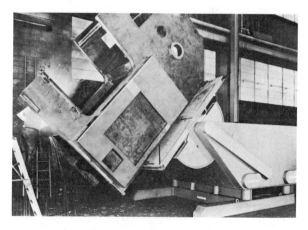

Fig. 26-25. Large positioner holding an assembly during a welding operation. (Aronson Machine Co.)

turned to position the joint where the operator can most conveniently do the welding. Such fixtures also minimize the physical exertion on the part of the welder. This type of jig may be purchased from equipment companies. Fig. 26-25 shows a large positioner holding an assembly while it is being welded.

26-13. POWER TOOLS

Power tools of various types are a necessity in high production welding, to cut overall production costs.

The drill press or portable drill is used to make holes in parts of weldments and for drilling holes for reinforcing rods in large cast iron welds. Electrically or pneumatically powered wrenches can be used when removing parts from large jobs. Power chisels save a great deal of labor when opening up a crack for a welding repair, or in removing scale from ferrous or nonferrous castings prior to welding. The

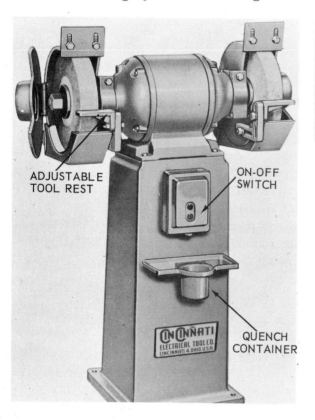

Fig. 26-26. A pedestal type grinder. (Cincinnati Elec. Tool Co.)

power chisel may also be used when removing excessive welding material from a surface to be ground.

The pedestal grinder may be used for snag grinding on finished parts, and for preparing small parts for welding, Fig. 26-26. The portable grinder finds wide usage in preparing and finishing the welded area. It is especially useful for grinding welds on large weldments that cannot be brought to a grinder. These machines are available in two types, the solid shaft type (similar to

Fig. 26-27. A portable electric grinder. (Miller Falls Co.)

Fig. 26-28. A portable electric sander-grinder combination of the disc type.

a portable electric drill) and the flexible shaft type. Figures 26-27 and 26-28 show two types of solid shaft or self-contained portable grinders.

An abrasive belt grinder is shown in Fig. 26-29.

In all grinding operations you should wear eye protection. The abrasive wheels must be in good condition. They should be inspected for cracks, looseness, and general good condition before being used. The operator should take precautions to stand in a safe and secure position. Small articles should be fastened before they are ground or sanded.

A flexible shaft grinder is shown in Fig. 26-30. The unit has several accessories such as wire brush, grinding disc, and chuck.

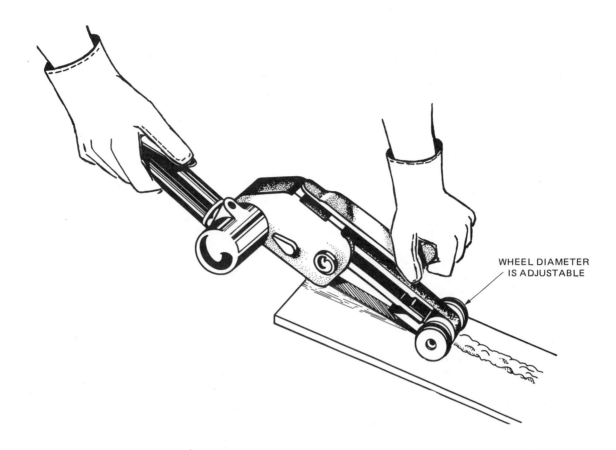

Fig. 26-29. *A unique belt machine designed to grind welds accurately and flat. The diameter of the wheel is adjustable to eliminate undercutting.* (Dynabrade, Inc.)

Fig. 26-30. *A 1 HP flexible shaft grinder, mounted on a portable base.* (Wyzenbeek and Staff)

26-14. BLAST CLEANING EQUIPMENT

A blast cleaning unit is handy when dirty or rusty metal stock must be cleaned before and/or after the welding has been done.

Air pressure is used to impinge (drive) steel grit, shot, sand or artificial abrasives against the surface to be cleaned. The operation is performed in a special cabinet with a suction unit and a recovery chamber for the abrasive. The operator usually stands outside the cabinet and uses specially built-in gloves to handle the blast nozzle and the parts as shown in Fig. 26-31.

The air pressure varies from 20 to

Fig. 26-31. Blast cleaning cabinet which has a nozzle using compressed air and grit which cleans the metal before and after welding. (Ruemelin Mfg. Co.)

100 psig. An exhaust system is used to remove the abrasive from the booth. Filters are mounted in the exhaust system to trap and remove the blasting material before the material can reach the exhaust fan.

26-15. WELD TEST EQUIPMENT

Welding shops may employ one or more of the following methods to check the quality of the welds and weldments produced:

1. Dye penetrates for surface cracks.
2. Magnaflux for surface cracks.
3. X-ray to detect internal flaws.
4. Supersonic to detail internal flaws.
5. Brinell or Rockwell tester to test metal hardness.
6. Tensile testers to check yield strengths in production welds.
7. Bend testers to determine qualifications of the operators.

These tests are described in CHAPTER 22 on Inspection and Testing.

26-16. THE WELDING JOB SHOP

There are two different types of shops which specialize in the fabrication of structures by welding, or repairing structures by welding. These two types do similar work and will be treated as one in the remainder of this chapter. These shops will be called "welding job shops." This type of shop will need all the necessary equipment explained in the previous paragraphs. In addition to these, it will need a certain number of other tools and supplies.

26-17. MOBILE WELDING EQUIPMENT

Many welding shops have portable welding stations mounted on trucks. These mobile units usually carry:

1. Arc welder - engine driven.
2. Combination gas welding and cutting unit.
3. Welding accessories.
4. Tools
5. Welding supplies.

These are excellent for making emergency repairs on stationary structures and various types of broken equipment. Fig. 26-32 shows a mobile welding station suitable for arc welding and cutting.

Fig. 26-32. Mobile welding truck being used to repair a farm disc. (Hobart Bros. Co.)

26-18. WELDING SHOP EQUIPMENT

The kind and number of tools the welding shop needs depends on the kind of work to be handled by the shop. The shop may prepare the metal for welding and weld it, or, it may specialize only in actual welding operations. In the former case, a complete metalworking department is a necessary part of the setup, including a complete machine shop and a complete sheet metal department. Some of the necessary metalworking tools for the department are a sheet metal brake and power shears. Other power tools often needed are a metal lathe, drill press, etc.

A necessary part of the welding shop is the cutting equipment. Every welding shop, regardless of its size, should have adequate cutting equipment to provide accurate shaping of the metal prior to welding.

26-19. WELDING SHOP TOOLS

The accessory tools, needed in a welding shop, are:
1. Wrenches
   A. Welding equipment wrenches.
   B. Open end wrenches for dismantling and assembling various articles.
   C. Box wrenches.
   D. Cylinder wrenches for acetylene cylinder.
   E. Pipe wrenches for preparing pipe material for welding.
2. Hammers of several sizes such as one, two, and four pound ball-peen for general work, sledge hammers for straightening and bending heavy stock, and chipping hammers.
   3. Chisels (manual and power).
   4. Files of all types and sizes.
   5. Screwdrivers.
   6. Wire brushes.
   7. Power grinder dresser.
   8. Grinder safety goggles.
   9. Squares.
   10. Levels.
   11. Clamps of all types.
   12. Mallets.
   13. Soldering coppers.
   14. Hacksaws and blades.

26-20. WELDING SHOP SUPPLIES

It is recommended that the reader refer to CHAPTERS 2, 4 and 6 for information on welding supplies. A list of the more common supplies is as follows:
1. GASES
   A. Oxygen.
   B. Acetylene.
   C. Argon, Helium, or CO_2.
   D. Occasionally some preheating gas such as natural, propane, etc.
2. WELDING ROD (OXYACETYLENE WELDING)
   A. Steel
      1/16, 3/32, 1/8, 3/16, and 1/4 in., dia.
   B. Cast iron
      1/8 and 1/4 in. round.
   C. Aluminum
      1/8-in. round drawn, and 1/4-in. square cast rod.
   D. Miscellaneous welding rods for special welding tasks.
3. ELECTRODES
   A. The shop operator must keep a stock of electrodes as required by the types of metals being used. Today a myriad of electrodes are available for all types of metal and various thicknesses. The welding shop should follow the detailed instructions of the electrode manufacturer in order to secure the best results.

4. METALS

A. Sheet Metal. A welding shop should have on hand all times a quantity of the various standard sizes of sheet steel, which are supplied in standard plate sizes, and may be stored in the warehouse of the shop until used. Fig. 26-33 is a table of U. S. Standard Gauge Steel Sheet. The various alloy metals and analysis for the various colors. This is becoming increasingly important since the availability of a multitude of special steel alloys.

B. Pipe. A quantity of pipe stock should be kept on hand for standard fabricating use.

C. Angle Iron. Angle iron of various sizes is also an important item.

GAUGE NO.	DECIMALS	FRACTION	WEIGHT PER SQ. FT. (LBS.)
0000000	.5	1/2	20
000000	.469	15/32	18.75
00000	.438	7/16	17.5
0000	.406	13/32	16.25
000	.375	3/8	15.0
00	.344	11/32	13.75
0	.313	5/16	12.5
1	.281	9/32	11.25
2	.266	17/64	10.625
3	.25	1/4	10.0
4	.234	15/64	9.375
5	.219	7/32	8.75
6	.203	13/64	8.125
7	.188	3/16	7.5
8	.172	11/64	6.875
9	.156	5/32	6.25
10	.141	9/64	5.625
11	.125	1/8	5.
12	.109	7/64	4.375
13	.094	3/32	3.75
14	.078	5/64	3.125
15	.070	9/128	2.812
16	.0625	1/16	2.5
18	.05	1/20	2.0
20	.0375	3/80	1.5
22	.0312	1/32	1.2
24	.025	1/40	1.0
26	.0187	3/160	.75
28	.0156	1/64	.625

Fig. 26-33. Table of U. S. Standard gauge size for steel sheet.

variations in the carbon content of the iron-carbon steels should be identified by careful marking on the metals. It is preferable to paint each piece a different color and to keep a code on hand indicating the steel

D. Round Solid Rod. Round solid rod is often used in a welding shop, and a limited quantity should be kept on hand for special jobs.

5. FLUXES

Fluxes are used principally when

oxyacetylene welding is done. The common fluxes which should be kept on hand are:

    A. Brazing flux.

    B. Cast iron welding flux.

    C. Cast iron brazing flux.

    D. Aluminum flux.

It is important that these fluxes be kept in sealed containers and in a cool, dry place when not in use. It is further recommended that when using the flux, enough should be transferred to a smaller container to handle the job at hand. This will keep the larger supply clean and fresh.

6. CARBON BACKING MATERIAL AND CARBON ELECTRODES

Carbon is a suitable material for backing welds and for forming small forms or molds, particularly for places that are hard to reach. The carbon helps to shape or form the metal as it is being puddled. This may be either flat, square, or round stock. The paste is sold in sealed containers and may be formed to any shape. Upon contact with the air the paste slowly solidifies.

Carbon electrodes are used for carbon-arc welding and may be obtained in different diameters.

7. VITRIFIED FIREBRICK

Firebrick is used for bench tops and for enclosing articles that are to be preheated. They are also used to slow up the cooling rate of heated articles.

8. MISCELLANEOUS

Glycerine for lubricating oxyacetylene moving parts, and wiping cloths for cleaning purposes.

26-21. WELDING SHOP POLICY

It is difficult to list in this text the policies that a welding shop should pursue in order to maintain correct business relationships. The work should all be done under a legal contract basis with clearly and definitely defined provisions for all emergencies. The welding shop should be extremely careful as to the metals being worked on if the metal is being supplied by the customer. This metal stock should be carefully investigated in order that the use of any poor stock will not be detrimental to the welding shop. The welding shop should list the specifications of the metal when purchasing and should hold the supplier of the material to these limits.

Frequently the welding shop must estimate the final cost of fabricating an article prior to starting work. This takes considerable training and experience. To estimate it properly, these items must be considered:

1. Length, depth, and shape of weld.

2. Type of weld--straight-in-line, flat, curved, vertical, or overhead weld.

3. Type of metal to be used.

4. Type of electrode or welding rod required.

5. Labor costs involved.

Such other variable operating cost items as fluxes, power costs, gas costs, must also be calculated. Depreciation of equipment and other overhead costs also affect the final cost of the jobs. By calculation, the estimator can determine the operating costs. For example, accurate information may be obtained from welding supply houses as follows: how long, how much welding rod, or electrodes, and how much gas or electricity may be involved in making welds of certain lengths and in various thicknesses of metal.

With this basic information, the estimator can determine the labor cost, the cost of material, and the power cost. Another item that should not be neglected in estimating the cost is the matter of handling the material to be welded. When dealing with small articles, this may be neglected; but with large cumbersome forms which have to be moved from one part of the shop to another and turned in various positions, this item is important.

There should be a clear understanding about moving costs. Either party may take this responsibility, which involves moving the articles to the shop to be welded, and returning them to the person contracting for the work.

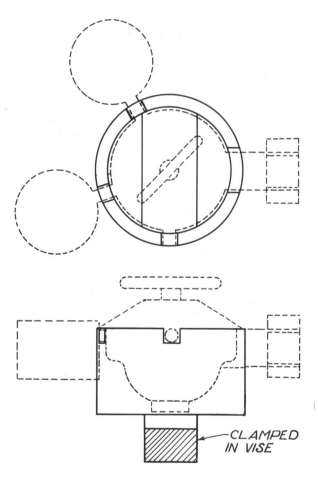

Fig. 26-34. Jig for holding regulator bodies while they are being repaired.

As with many other kinds of work, estimators, after gaining considerable experience, can estimate by eye and by mental calculations the approximate cost of a weld job. However, this method is not infallible and should always be backed by a complete, detailed calculation. Inspection and quality testing costs should be included in any agreement. A welding shop should use carefully worded contract forms. A lawyer who specializes in drawing up industrial contracts should be employed to draw up a standard form that may be applied to various types of welding jobs handled by the shop.

Many skilled welders have failed in their own welding businesses because they neglected the business aspects of the business. Accurate records must be kept at all times. The total cost of operating a shop includes not only the actual operating cost but the overhead cost as well. Consultation with an experienced accountant is highly desirable.

It is important to keep an accurate record of each job. These records can then be used as a base for estimating future jobs similar in nature.

26-22. THE WELDING EQUIPMENT REPAIR SHOP

Most manufacturers of welding equipment maintain shops to overhaul welding equipment and to repair it for various welding equipment jobbers. Some large manufacturers, who do an extensive amount of welding, have included a welding equipment repair shop as a part of their plant. Welding equipment repair may be divided into four principal divisions:

1. Oxyacetylene equipment repair.
2. Arc welding equipment repair.
3. Resistance welding equipment repair.
4. Electronic and electrical controls repair.

26-23. REPAIRING GAS WELDING EQUIPMENT

The repair of gas welding equipment should be attempted only by an experienced person, or the repair should be supervised by an experienced person.

Regulators, fittings, gauges, hose, torches and tips must be expertly dismantled, checked, repaired, assembled and tested to assure safe and efficient operation.

26-24. EQUIPMENT FOR REPAIRING GAS WELDING PARTS

Equipment needed to repair various oxyacetylene equipment includes:

1. Lathe with accessories such as universal chuck, drill chuck, tool bit holders and centers.
2. Drill press.
3. Grinder and buffer.
4. Soldering apparatus.
5. Welding station for testing.
6. Repairing equipment (jigs and fixtures).
7. Bench and vises.
8. Storage bins for parts.

26-25. TOOLS FOR REPAIRING GAS WELDING PARTS

Hand tools needed include:

1. Set of open-end wrenches.
2. Set of numbered and lettered drills.
3. Large open-end wrenches and box wrenches.
4. Assembling and dismantling jigs.
5. Calipers, both inside and out.
6. Micrometers.
7. Set of center punches.
8. Set of socket wrenches and handles ranging from 7/16 to 1-3/8 in.
9. Small ball-peen hammer.
10. Small torch or soldering copper.
11. Set of dies and a die stock.
12. Set of taps with a tap wrench.

It will be necessary to use special jigs and fixtures for holding different parts of the apparatus while being repaired. See Fig. 26-34.

26-26. SUPPLIES FOR REPAIRING GAS WELDING PARTS

The supplies needed in a welding repair shop include:

1. Sandpaper.
2. Emery paper.
3. Solder and flux.
4. Silver brazing wire and flux.
5. Cleaning solutions made especially for brass and copper cleaning.
6. Glycerine.
7. Litharge.
8. Clean cloth for use with clean polishing abrasives.

Equipment replacement parts necessary to maintain a welding repair shop include:

1. Nozzles and seats for nozzle-type regulators.
2. Diaphragms for all type regulators.
3. Gaskets for all type regulators.
4. Needles and seats for stem-type regulators.
5. Torch valves for all type torches.
6. Hose, both oxygen and acetylene.
7. Hose nipples and nuts for both oxygen and acetylene.
8. Torch valve packing for both oxygen and acetylene torch valves.
9. Hose clamps for both oxygen and acetylene hose.
10. Several main parts of different types of torches, such as handles, mixing chamber, barrels, and tips.
11. Several main parts of regulators, such as bonnets, regulator adjusting screw springs, seat retaining cage and the like.

12. Adaptor fittings for connecting hose to regulators.

13. Gauge parts such as needles, dials, crystals, bezels, Bourdon tubes, and the like.

26-27. REPAIRING GAS WELDING TORCHES

As explained previously, a torch (blowpipe) consists of shutoff valves, handle, mixing chambers, barrel, and tip. Three different fundamental torch designs are commonly used. These are based on the location of the mixing chamber.

1. A chamber formed between the tip and the tip socket is used as the mixing chamber. This allows for a different size mixing chamber for each size tip.

2. The mixing chamber is located inside the handle of the torch and is usually placed between the handle and the barrel junction.

3. The barrel and tip are made in one piece with the mixing chamber located between the barrel and the handle, thus enabling the mixing chamber size to be changed for each sized tip.

The latter type incorporates features of the first two because it is ideal to change the size of the mixing chamber with each tip size; it is also a good practice to keep the mixing chamber remote from the source of heat in order that pre-ignition may not take place.

26-28. REPAIRING TORCH VALVES

Troubles usually encountered with torch valves are leaks past the needle, or leaks around the valve packing nuts. The typical torch valve is of needle design, using a drop forged brass body, drilled and threaded, and a brass threaded needle with either a V tip or a ball bearing tip. The body of the valve is

threaded to receive the threads of the needle, Fig. 26-35. Being made of brass, the threads strip easily in case of abuse. The only repair in this case is to replace the complete valve.

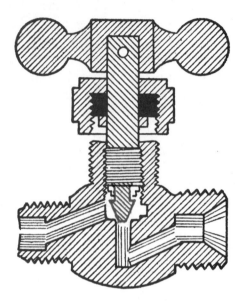

Fig. 26-35. Cross section of a torch needle valve. (Weldit/Winona)

To replace a torch valve:

1. Remove the packing and any gaskets from the torch.

2. The torch should be firmly mounted in a vise (do not mar or scar the torch body).

3. Clean the surface of the silver brazed joint.

4. Flux the joint.

5. Heat until the brazing material is molten.

6. Unthread the valve from the torch and replace it with a new one. Position the valve on the torch in a convenient position relative to the torch body before silver brazing it to keep the operator's hand from interferring with the adjustment while using the torch.

It is often necessary to repack around the valve needle to prevent leaks at this point. Leaks around the valve needle may be dangerous because they

Fabrication of heavy assemblies is facilitated by the use of suitable worktables, fixtures and holding devices. Note the rapid action clamping devices. (De-Sta-Co Div., Dover Corp.)

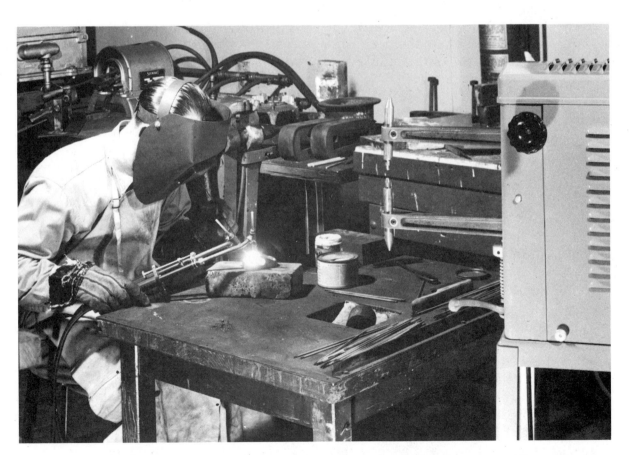

Small compact welding shop. This shop has facilities for doing jobs which require spot welding, arc welding, gas welding, brazing and braze welding.

mean a continuous loss of gas. When repacking a valve needle, it is recommended that the old packing be completely removed and that only new material, as recommended by manufacturer, be used. Asbestos rope, impregnated with graphite, is often used as packing material. It is important to use a correct size wrench on the valve packing nut to prevent stripping the threads, and to prevent rounding the hexagonal corners of the packing nut. NEVER USE PLIERS ON A WELDING TORCH.

The handle, being made of a thin brass tube, sometimes becomes loose and turns freely on the torch. This makes the torch difficult to hold. To remedy this looseness the handle may be silver brazed to the stationary parts of the torch, or it may be tightened if the torch barrel is provided with retaining screws. The handle may also become kinked or otherwise abused. It should then be replaced.

26-29. REPAIRING MIXING CHAMBERS

It is difficult to recommend repairs for the mixing chamber because of the importance of the surfaces of this part of the torch. The mixing chambers are accurately designed and are carefully machined to promote the proper mixing of the two gases. If the parts forming the mixing chambers are carelessly handled, the welding ability of the torch may be completely ruined. If the parts are scratched or bent out of shape, a replacement of these parts is essential. Fig. 26-36 shows a mixing chamber cross section. The most common difficulty encountered with mixing chambers is the carbon formation which sometimes takes place and which coats the walls, thus changing the size of the

chambers and preventing good mixing of the two gases. It is necessary that the carbon be removed and the surfaces given their original bright finish. To do this, carbon dissolving agents should be used in conjunction with a soft cloth. NEVER USE SANDPAPER OR EMERY CLOTH TO CLEAN MIXING CHAMBER SURFACES.

Polishing paper such as 000 or 0000, French Hubert paper is permissible, but should be used very carefully. When repairing parts of a mixing chamber, it is important that parts obtained from the manufacturer be used, if possible. If the shop must make a part for replacement, it cannot be emphasized too strongly that the tolerance during machining should be kept to less than one thousandth of an inch. Also, if a repair shop is to make its own parts, it is emphasized that careful consideration be given the material from which the part is being made. If the material is of poor quality a rapid deterioration may take place in the presence of acetylene or oxygen. Sometimes the chemical reaction between the gases and certain metals becomes dangerous. A welding shop doing parts manufacturing should secure and use an alloy which is inert to the action of the gases.

26-30. REPAIRING TORCH BARRELS

The design of the torch barrel depends on the type of torch. One type of torch does the mixing of the two gases in the handle and another type mixes the gases in the head. In one type torch where the mixing takes place in the handle, the barrel is usually made of drop-forged, or extruded brass, rifle bored, and bent to shape. Three parts of the barrel may need service. First, where the barrel is attached to the handle, a clamp nut is used to provide a leakproof joint. If abused, the torch will leak at this point and relapping, remachining, or replacement is necessary to stop leaks.

The barrel itself sometimes crystallizes and breaks because of excessive twisting or pounding. The barrels have sometimes been repaired by bronze welding, or silver brazing. However, this is only a makeshift repair; replacement of the complete barrel is recommended.

The socket where the tip is fastened to the barrel is another source of trouble. This is due to stripping the threads or abuse of the sealing surfaces, which may permit the gas to leak out around the threads. If the gases burn at this leaky joint, they will preheat the tip, causing the torch to backfire. If the threads of the head are stripped, replacement of the complete barrel is necessary. If the seat used as a sealing surface is abused, an end reamer may sometimes repair the trouble. If this is not successful, replacement will be necessary.

Occasionally small leaks, which occur at the point where the tip threads into the barrel, may be stopped by applying some bar soap to the threads. The soap will act as a sealing compound when the two are threaded together. It should be emphasized here that most of the abuses to which a torch is subjected are the results of the operator's forgetting that brass and copper parts are soft. It should also be emphasized that brass parts when threaded together, need not be tightened as tight as steel in order to secure a leakproof joint.

The type of torch which uses an integral torch and tip barrel for each size of welding tip has about the same trouble as the torch with the mixing chamber in the handle. This design, however, does eliminate a joint between the tip and the barrel.

Holders, made of wood or soft metal such as lead should be used for holding the parts while they are being repaired.

26-31. REPAIRING TORCH TIPS

The part of the torch which is subjected to the most severe abuse is the torch tip. The tip is immediately

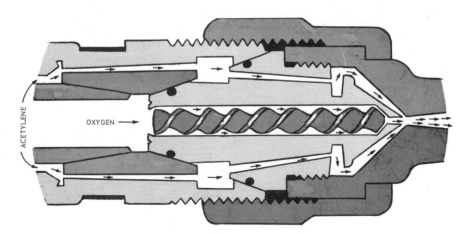

Fig. 26-36. Torch mixing chamber showing oxygen and acetylene passageways. Flow is from left to right. Spiral oxygen flow produces a whirling, turbulent gas that is said to provide better mixing of acetylene and oxygen in the mixing chamber (extreme right). (Victor Equip. Co.)

adjacent to a high temperature flame and in many cases much of the heat is radiated against the tip. To withstand this service, the tip is usually made of hardened copper. Steel cannot be used to advantage for this purpose because

be carefully treated. This orifice is usually drilled to size and is accurate to a thousandth of an inch. The inside orifice wall must also be smooth. The exact size will vary with each manufacturer. Fig. 26-37 shows a table

	WELDING TIPS					CUTTING TIPS							
						OXYGEN JETS				PREHEAT JETS			
TIP NUMBER	0	1	3	5	7	0	1	3	5	0	1	3	5
DRILL SIZE	65	60	53	43	30	58	54	50	40	73	70	64	55

Fig. 26-37. Average torch tip sizes and drill sizes for some welding torch tips and cutting torch tips.

it does not cool rapidly enough, and the tip soon becomes hot enough to preignite the gases.

Even though hard copper is used for the tip, this material is still soft enough to be easily injured. The three parts of the tip which most frequently give trouble are the orifice, threads, and seat.

The orifice, the opening through which the gas is fed to the flame, must

listing drill sizes for different tip sizes. Sizes given are approximate.

As the tip is being used, metal particles adhere to the outside of the tip. Some may find their way to the interior of the orifice where they will interfere with the gas flow. Foreign particles in the orifice must be removed. A special reamer is obtainable for this purpose. Because of their small size, these reamers may break if care-

Fig. 26-38. Torch tip cross sections showing undesirable conditions which prevent a good flame.

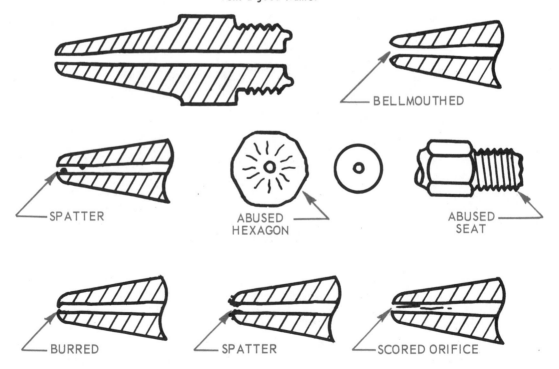

lessly used. This makes it imperative that they be used carefully. Start with an undersize reamer and finish with the correct size. Avoid forcing the reamer into the orifice. These reamers may also enlarge or score the orifice wall to such an extent that the tip may become worthless.

If an operator reams the tip from the outside (while the tip is fastened into the barrel), the oxygen should be turned on while the reaming is in process. This will help to blow particles out of the torch as soon as they are loosened.

The same repairs, recommended for the seat and threads of a barrel are also applicable to threads and seat of the torch tip.

The shape of the tip at the end where the gases burn is very important when one is attempting to secure a proper flame. Because of the constant high temperature the edges of the orifice tip wear away, thus widening or enlarging the orifice. Fig. 26-38 shows various defects which may affect proper tip operation. This bell-mouth prevents accurate and complete combustion, and enables the torch to blow out too easily. To remedy this condition the repairman may remachine the tip in a lathe, or he may file the tip until the bell-mouth is removed. The end of the tip must be 90 deg. to the orifice. If the end of the tip is not square to the orifice, an askew flame will result. A special tip end reamer is now available that enables the operator to recondition the end of the tip without removing it from the torch.

26-32. REPAIRING REGULATORS

As mentioned in CHAPTER 2, there are various types of regulators. The two fundamental designs in reference to

repair work are the stem type and the nozzle type. Practically all regulators, regardless of the trade name, are repared by following the same general procedure. The two most common regulator repairs are to the valve proper, which should not leak, and the diaphragm which should not be buckled or injured in any way if the regulator is to maintain a constant pressure.

26-33. REPAIRING NOZZLE-TYPE REGULATORS

This type of regulator embodies the use of a diaphragm which controls the movement of the seat against a hard copper nozzle. The construction is sometimes called reverse valve and seat construction. The seat carriage is fastened to the diaphragm, and when the adjusting spring tension is released, the compensating spring forces the seat tightly against the nozzle. It is important that the diaphragm be able to move freely in all positions to eliminate any jumps or catches caused by the warping of the diaphragm. A buckled diaphragm can be detected by inspection. Old diaphragms have a tendency to dish, or have a snap action, meaning that as the diaphragm moves in one direction, it will snap a definite amount after passing the neutral point, preventing accurate adjustment of the regulator over any pressure range (oil can action). This action is usually the result of worn diaphragms that have been in use long enough to cause weakening of the diaphragm metal, producing the dishing or snapping action.

A buckled diaphragm, having a surface wrinkle, is usually the result of unskilled installation. A dished diaphragm, or a warped diaphragm, may force the seat carriage from its true line and cause serious troubles by making the seat carriage bind in its guides. It may also produce a leak at the nozzle and seat. It is important upon assembly of these regulators that all the parts work freely in order to secure the desired accuracy of seating.

After a certain period of use, the seat should be either replaced or reversed to provide a smooth surface to contact the nozzle. It is important that the part of the carriage holding the seat should be clean before the seat is installed, as a small dirt particle will throw it out of alignment.

The nozzle, being made of hard copper and finely machined with an accurately drilled opening or orifice, does not often require repair. Fig. 26-39

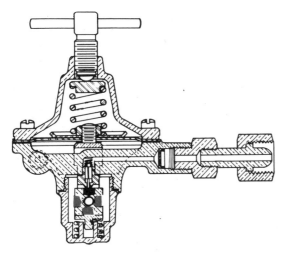

Fig. 26-39. Nozzle-type regulator equipped with three reserve seats (shown in red).
(Modern Engineering Co., Inc.)

shows a regulator equipped with three spare seats installed in the regulator. However, under excessive abuse the end of the nozzle may become scored. In such cases the recommended procedure is to replace the nozzle, inasmuch as filing or sanding will not produce a true surface.

The inlet fitting of the high-pressure gases on all makes has been standardized as to thread size. The standards are explained in CHAPTER 2. The fittings are different for different gases; but all oxygen cylinders, for example, use the same type and size of fitting. It is standard practice to fasten all of the fittings to the regulator body by means of a 1/4-in. Briggs Standard pipe thread.

The gauges or pressure indicators are also fastened into the regulators by means of 1/4-in. pipe fittings. The joint is often sealed by the use of litharge and glycerine paste. The inlet opening to the regulators is usually provided with a fine mesh screen (approximately 100 mesh) or a sintered metal filter. This screen should be cleaned as a part of the regulator repair operation. It is recommended that during the overhauling of regulators the various parts be buffed with a cloth buffing wheel, to remove any surface dirt clinging to them. Also the parts may be acid dipped either in a cold solution of cold nitric acid, or in a warm solution of sulphuric acid. Other cleaning solutions are also available. Great care must be exercised to prevent injury to the person when doing acid cleaning. Safety goggles are strongly recommended. The regulator adjusting screw usually threads into the bonnet and frequently requires repair because of sticking or stripped threads. To lubricate these threads, glycerine or soap may be used (OIL MUST NEVER BE USED ON OXYACETYLENE EQUIPMENT). Diaphragms are usually fastened to the body of the regulator by being clamped between the body and the bonnet. The bonnet is either threaded to the body regulator, or is fastened by small cap or machine screws. The latter construction is more easily repaired. However, the assembly in either case must be exceptionally well done or buckling of the diaphragm

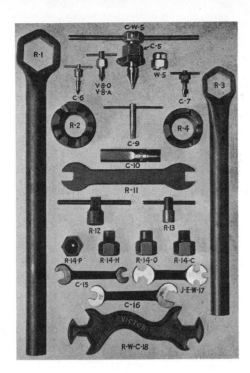

Fig. 26-40. Set of regulator repair and assembly tools.
(Victor Equip. Co.)

results, spoiling the correct operation.

When assembling nozzle and seat type regulator, the regulator adjusting screw is turned all the way in. This applies particularly to installation of valve seat. If not, soft valve seat, sliding across nozzle opening, may become permanently injured and prevent it from working correctly. Special tools are available for dismantling, repairing and assembling regulators, Fig. 26-40. See Par. 26-35 for instructions on how to test a regulator.

26-34. REPAIRING STEM TYPE PRESSURE REGULATORS

Care required in handling fittings, adjusting screws, diaphragms, body, and bonnet apply to stem type regulator as well as to nozzle type. Construction of valve proper is different. Stem type of regulator, as explained in Chapter 2, uses soft valve seat, usually rubber, against which a poppet valve is pressed

by means of a spring and the gas pressure. The poppet valve stem passes through the valve seat opening, and the end of it is manipulated by the regulator diaphragm. This stem is very seldom fastened to the diaphragm. Fig. 26-41 illustrates a stem type regulator. It is forced against the seat by

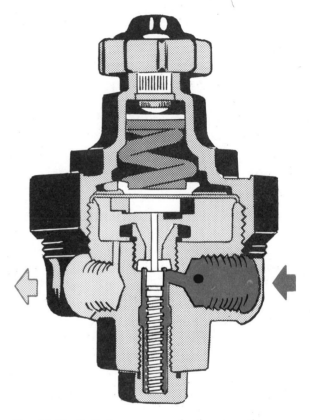

Fig. 26-41. Cylinder pressure (red) enters the regulator at the right. The valve mechanism is controlled by a stem which touches, but is not connected to diaphram. The low-pressure gas (light red) is gas supplied to torch. The adjustment screw (top) is used to vary pressure supplied to torch.

means of a spring which presses against the head of the valve. Also the high gas pressure tends to force the valve against its seat. The valve and head are usually assembled in a housing or cage to assure their proper alignment. It also minimizes the amount of labor in assembling and insures that the valve and seat are gas tight.

26-35. TESTING REGULATORS

To test a regulator after assembly, oxygen should be used for oxygen regulators, acetylene for acetylene regulators. Or, an inert gas may be used for either or both. A test for a regulator consists of:

1. Mounting the regulator on the cylinder with the adjusting screw turned all the way out.

2. Open the cylinder slowly. Watch the low-pressure gauge; if it starts to build up, close the cylinder valve quickly as the regulator control valve is leaking.

3. Then turn the regulator adjusting screw in to permit a certain pressure flow of gas.

4. Stop the flow of gas by capping the outlet of the regulator (hose connection) and note the amount of pressure increase on the low-pressure gauge. If this pressure increases more than 2 or 3 lbs. above the flowing gas pressure, either the nozzle fitting, or the stem is leaking slightly, or the parts are warped out of line. If the gas pressure creeps up steadily, it indicates that the moving parts are frozen within the guides, or that the valve is leaking badly. Occasionally after assembly, the regulator indicates a leaking valve. By tapping on the body of the regulator, it should immediately start to function correctly. This action usually indicates the presence of dirt between the valve and the valve seat, which lodged there during the assembly of the regulator.

To test for smoothness of operation, or the accuracy of the setting of the regulator, proceed as follows: A small cylinder, with leakproof connections, should be attached to the outlet opening of the regulator (hose connection). This cylinder is equipped with an exhaust valve. After the regulator has been mounted on the high-pressure cylinder, and after the small cylinder is attached to the regulator, the regulator adjusting screw may be turned in slowly. For each turn of the regulator adjusting screw there should be a corresponding increase in the small cylinder pressure. Any fluctuations in the small cylinder pressure will indicate that the diaphragm is not moving smoothly, or that the valve or valve fitting is sticking or catching in its guide. With the low-pressure part or the regulator under a pressure of 10 or more psig, the body of the regulator should be tested for leaks, using a glycerine or soap solution, preferably a soap solution. (NEVER USE OIL.)

It is important to remember that acetylene regulators and gauges are tested on acetylene cylinders, with master gauges also connected to determine the operation; while oxygen regulators and gauges are tested on oxygen cylinders to determine their operation. NEVER TEST ACETYLENE PARTS WITH OXYGEN, OR OXYGEN PARTS WITH ACETYLENE.

26-36. REPAIRING GAUGES

Pressure gauges used as part of the oxyacetylene welding equipment are very important to the welder. He must rely on their accuracy implicity, both as a safeguard to his own safety and as a means of informing him that his torch is adjusted accurately. It is, therefore, needless to say that these gauges must at all times give accurate readings. A welding equipment repair shop should have master gauges. These master gauges in turn should be periodically calibrated by the use of a dead-weight tester. By using these master gauges and by connecting the master gauge and the gauge being tested to a common cylinder, or a common regulator, a repairman may easily detect discrepancies in gauge readings.

Simple inaccuracies of gauge readings may be easily corrected. However, such items as a buckled Bourdon tube or a ruptured Bourdon tube cannot be satisfactorily repaired. Also broken hair springs usually necessitate the replacement of the complete gauge. The minor adjustment of a gauge for slight inaccuracies is usually taken care of by changing the length of the linkage which connects the end of the Bourdon tube to the gear sector. A jeweler's eyepiece and a set of jeweler's tools are handy for this type of work. Slight errors in gauge readings may also be repaired by removing the needle from its stem and setting it back on the stem in such a position as to correct the previous error. Use only a special puller to remove the needle. The use of pliers may bend the needle stem.

To test for the accuracy of a gauge before returning it to service, the repairman usually proceeds as follows: The oxygen high-pressure gauge (3,000 pound scale) should be tested with an inert. gas. The other gauges having lower scale reading, may be tested by using the cylinder pressures of the gases. The repairman should keep a quantity of gauges and replacement parts on hand to replace gauges which have been injured in service. A complete range of sizes varies from the 2-1/2-in. diameter to the 4-1/2-in. diameter gauge, and from the 0 to 30 lb. gauge, up to the 0 to 3,000 lb. gauge.

Some regulators are now equipped with pressure indicator pin devices. Such units are less bulky than the gauges.

26-37. REPAIRING WELDING HOSE AND FITTINGS

Worn and/or leaky hose and fittings should always be replaced. Any attempts to repair cracked, cut, or otherwise leaky hose is dangerous. Much can be done to keep hose in good condition. Keep the hose protected from hot, sharp or abrasive situations. Protect the hose from being squashed by tires or heavy objects. Always roll up the hose and hang it up when it is not in use. Keep the hose away from oil and grease.

Fasten the hose to the regulators and torch using ONLY approved ferrules, nipples, and nuts. These fittings must be in good condition. Clamp the hose to the nipples using only the correct size approved hose clamps. If a hose is removed from a nipple, cut away that part of the hose and insert the nipple in an undamaged end of the hose. Use only the proper size nipples, as the use of an improper size may damage the hose or cause a dangerous leak.

26-38. REPAIRING ARC WELDING EQUIPMENT

The maintenance and repair of arc welding equipment is an important part of a welding shop activity. This equipment too must be in good operating order if good welding results are to be obtained. Manufacturers maintain repair departments, large welding shops also have repair departments. Smaller shops usually have their repair work done by an equipment repair company.

The work includes:
1. Electrode holder repair.
2. Cable repair.
3. Motor-generator repair.
   A. Mechanical.

Fig. 26-42. Testing welding generator brush spring force. (Hobart Bros. Co.)

B. Electrical.
4. Transformer repair.
5. Controls repair.
6. Instrument repair.

26-39. EQUIPMENT FOR REPAIRING ARC WELDING PARTS

Repair shop equipment includes:
1. Monorail or portable crane.
2. Teardown bench.
3. Cleaning station.
4. Lathe large enough to hold motor-generator armatures.
5. Arbor press.
6. Electrical test equipment.
   A. Voltmeter.
   B. Ammeter.
   C. Ohmmeter.
   D. Continuity testers.
7. Repair benches equipped with vises.
8. Grinder, buffer.
9. Compressed air equipment.

26-40. TOOLS FOR REPAIRING ARC WELDER PARTS

Hand tools needed for repairing arc welding parts are the same as used for gas welding, with a few additions, such as:
1. Cable connector tools.
2. Set of Allen setscrew wrenches.
3. Soldering guns.

26-41. SUPPLIES FOR REPAIRING ARC WELDING EQUIPMENT

Supplies for this type repair work include:
1. Sandpaper.
2. Solder and solder flux.
3. Silver brazing wire and silver brazing flux.
4. Cleaning solvent.
5. Cleaning brushes (fiber and wire).
6. Cleaning cloths.
7. Electrodes.
8. Grease.
9. Paint.

Replacement parts for welding equipment should be kept on hand or should be readily available, including:
1. Generator brushes.
2. Exciter brushes.
3. Cable lugs.
4. Cable connectors.
5. Electrode holders.
6. Cable ground clamps.
7. Meters.

26-42. MAINTENANCE OF ARC WELDING GENERATORS

Arc welding motor-generators should be serviced regularly. The armature is a moving part and its bearings should be inspected at least twice a year. Covers should be removed, grease checked, and if oxidized or dirty, it should be replaced. Bearing grease space should only be one half filled to allow for expansion and movement of the grease. Oxidized grease sometimes has a "sour" odor. This can be used as a means of identifying old grease. Excessive bearing wear can be determined by using a dial indicator, and a lever to attempt to move the shaft (.001 in. is the usual maximum allowable clearance).

Brushes and commutator need regular care. The commutator must run true, brushes must fit commutator, contact surface of brushes should not be glazed, brushes should move freely but snugly in their holders, and the spring force pushing the brush against the commutator must be correct (manufacturer's specifications).

Only brushes recommended by the manufacturer of the machine should be used. Brushes vary considerably and a brush either too soft or too hard will result in poor service.

If the commutator is sparking noticeably while the generator is in use, the excessive sparking may be due to:

1. An overloaded generator - check current output.

2. Dirty, rough, or high mica commutator - clean the commutator. If this does not stop the sparking, commutator reconditioning is necessary.

3. The spring force may be too low - check with a scale.

4. The brushes may be sticking - check for free movement.

5. There may be loose or dirty connections - check with voltmeter or ohmmeter.

6. There may be shorts, opens or grounds in the armature - check with a continuity light.

7. There may be worn bearings - check with dial indicator.

First it is important that the brushes and the commutator of an arc welding generator be kept in good condition. It is also important that the pressure of the brushes against the commutator be accurately measured and adjusted for correct tension. The welding equipment repairman should, therefore, be capable of turning the commutator on a lathe and he should also be able to accurately fit brushes to the commutator. The spring pressure of the brushes against the commutator should be carefully measured, using a small size, spring scale. This force should be tested at the point where the spring pressure presses on the brush. As the size of the brush and the brush material determines the spring force, secure the recommended force from the manufacturer, and test as shown in Fig. 26-42. The brushes should fit on the commutator so that full surface contact is obtained. This is done by placing fine sandpaper (2/0) on the commutator with the sanding surface out and then putting the brush in its retainer. Slowly revolving the commutator will sand the end of the brush into a true fit on the commutator. The commutator should be turned only in the direction it runs when the generator is in use.

The machine and its parts should not be allowed to overheat. Air is usually used as the cooling medium. The cooling air must flow through the unit. The unit is equipped with a fan or fans and air flow spaces are built into the unit. The maintenance man must keep the fans and air passages clean. Do not install the unit in a restricted space or in a hot place. It is recommended that the unit be cleaned periodically--every two to three months, using clean dry compressed air. If air filters are used they should also be cleaned regularly.

26-43. REPAIRING ARC WELDING MOTOR-GENERATORS

Repairing of arc welding equipment is best performed by shops specializing in this work, or by shipping the parts needing repair back to the manufacturer.

Motor-generator are of two types:

1. Electric motor driven.
2. Gasoline engine driven.

Repairing the gasoline engine is a specialty, and should be done by auto engine mechanics. Such engines are usually heavy-duty industrial engines but their design and service is similar to the auto engine. The one specialty the engine does have is a governor which maintains a near constant engine speed when the welding load (current draw) changes. These engines should be given a complete overhaul after approximately each 2000 hours of operation (follow the manufacturer's recommendation).

The electric motor and the generator usually have eye bolts or provisions for eye bolts to enable a portable hoist or an overhead crane to handle the parts.

The brushes should be lifted from their brush holders while dismantling takes place. Be careful not to injure or burr the machined surface where the end bells fit on the frame. The unit should be completely dismantled and cleaned.

Special pullers should be used to remove end bells, and the parts should be firmly supported while the end bells are being removed.

Use cleaning solvents that are not injurious to the motor or generator electrical windings. The cleaning should be done in a well-ventilated area. Special safety precautions must be taken when using cleaning solvents that are flammable or toxic. All grease should be removed from the bearings. The bearings should then be checked for wear and greased as recommended by the manufacturer.

The armature should be checked for trueness. Mount the bearing journals in V blocks, and use a dial indicator to check the shaft alignment. Some shafts can be straightened by using an arbor press. The trueness of the commutator can also be checked at the same time. The commutator can be turned on a lathe if necessary. Cut away a minimum amount of metal, using a fine feed to obtain a true clean surface. An undercutting tool mounted on the tool rest of the lathe is used to groove the commutator, if this action is needed.

The armature windings should be checked for continuity (a test light), shorts (an ohmmeter), and grounds (a test light).

The field windings must also be checked electrically i. e. for continuity, shorts, and grounds.

The brush holders must be checked electrically. The brush springs should be tested for possible loss of strength.

All brushes should be replaced if worn. New brushes should be shaped to fit the commutator.

The bearing grease seals should be replaced if the bearings are not the self-sealing type. Bearings should be replaced, if worn.

During assembly of the motor-generator, all electrical connections should be cleaned and tightened.

Great care should be taken when assembling the unit:

1. Protect the windings from injury.
2. Protect the mating machined surfaces from nicks and burrs.

The accessories of the unit require as much attention as the main parts:

1. Rheostats.
2. Multiple winding switches (current adjustments).
3. Start and stop switches.
4. Relays.
5. Overload protection devices.

The contact points of the switches should be carefully inspected and cleaned. If badly worn or pitted, they should be replaced. The moving parts of the controls should be checked for wear. All the electrical circuitry should be checked for continuity and grounds. A voltmeter provides a quick way to determine if a resistance exists between contact points while the power is turned on. An ohmmeter can also be used for this purpose. It has the advantage of detecting poor connections with the power off.

26-44. MAINTENANCE OF TRANSFORMER TYPE ARC WELDERS

Transformer type arc welders produce the following currents:
1. AC.
2. DC (Rectifier).
3. AC or DC.

Maintenance of these units is usually less than on the motor-generator units because there are fewer moving parts.

These units need to be cooled. Air circulation is the medium used. Because the basic unit has no moving parts to which a fan could be coupled, a separate fan motor is used to circulate the air into and out of the unit. This fan should be checked periodically, cleaned and oiled if necessary. The unit should be cleaned with dry compressed air each two to three months. It is important that the unit be located in a place where sufficient cool air is available for cooling. An abrasive dust condition should be avoided.

DC and the AC-DC units use rectifiers to change the AC to DC. These rectifiers should also be kept clean.

26-45. REPAIRING TRANSFORMER TYPE ARC WELDERS

Repairing the transformer type welder is best done by a complete tear down operation, a thorough cleaning, a visual inspection, a mechanical inspection, and an electrical inspection of all the components.

These units quite often have hoisting connections to enable a chain fall, a portable hoist, or a crane to handle the heavy parts of the unit.

26-46. REPAIRING ARC WELDING INSTRUMENTS

Some arc welding machines are provided with ammeters and voltmeters, either built as one meter or as two separate meters. These meters are delicate instruments and should be handled with care. In these instances where the ammeter and voltmeter are the same meter (use the same pointer or needle), the operator uses a switch button to change the meter reading from voltage to amperage. The dial is calibrated with two scales, a volt scale and an ampere scale.

Occasionally the meters read from 0 to maximum both ways, but usually the meter swings only in one direction. For example, some voltmeters read 100 volts to 0 volts to 100 volts, and the ammeter reads from 400 amperes to 0 amperes to 400 amperes, but the more popular types usually read from 0 volts to 100 volts, and the same meter, 0 amperes to 400 amperes. As previously explained, the voltmeter registers both when arc welding is in progress and when it is turned on but not being used. The welding arc must be operating, however, before an ammeter reading may be taken.

The voltmeter, when indicating the open circuit voltage gives only a rough approximation of the capacity of the machine. The closed circuit voltage, as shown in Fig. 26-43, is more important. It is a good policy for the operator to know at all times what the

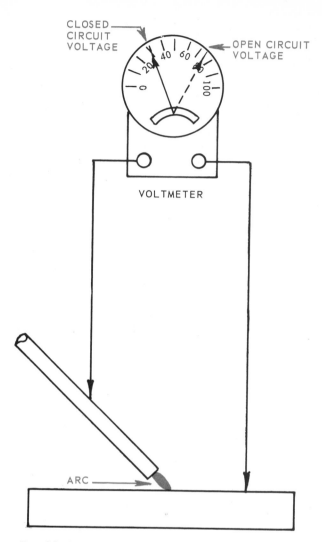

Fig. 26-43. *Testing arc potential (voltage drop across the arc) with a voltmeter.*

ampere reading of the machine is, when doing a particular welding job. He may obtain this reading by having the arc maintained in a protected booth, while he is standing in front of the machine, away from the arc, to observe the ammeter reading while the arc is being maintained. It is essential that the arc be of the correct length so the reading will be correct and not misleading. The longer the arc the less the ampere consumption; while the shorter the arc, the more the amperage. By having the electrode grounded and then by drawing an arc to the proper length, the

variation of ampere flow will be as high as 50 amperes which shows it is essential that the arc be of the correct length. Fig. 26-44 shows the change in ammeter and voltmeter readings during open circuit, and after the arc is obtained.

Before one may rely entirely on meter readings, the accuracy of the meters must be checked. This may be done by means of connecting master meters to the same machine, and by checking the reading of both meters while the machine is in operation. This may best be done by securing the services of a local electrical power engineer. An ammeter which may be applied to the circuit to obtain current flow readings is shown in Fig. 26-45. Electrical connections to the meters must be clean and tight. Meter repairing should be done only by a company that specializes in this work.

26-47. REPAIRING ARC WELDING CABLES AND CONNECTORS

Cables and the cable connections for arc welding are important. The diameter of the cables determines to a considerable extent, the efficiency of the electrical conductor. The larger the cable, the more current the cable can carry and the less the resistance. Resistance decreases as the cable increases in diameter, as the cable gets shorter, and as the cable temperature lowers. On this same basis the longer the cable the larger the cable should be. If too small a cable is used the resistance to current flow becomes so large that the efficiency of the machine is destroyed. Poor cable installation results in poor arc welding. CHAPTER 6 lists cable diameters for various current loads and for different lengths of run. In addition to the current carrying capacity of the cables, tight connections where the cables are fastened to the arc welding machine, electrode holder, and ground are important.

To test the efficiency of cables and which the ground cable is fastened. Record the reading of the ammeter and the voltmeter. Next, use a large size cable of short length, connect the positive and negative leads of the weld-

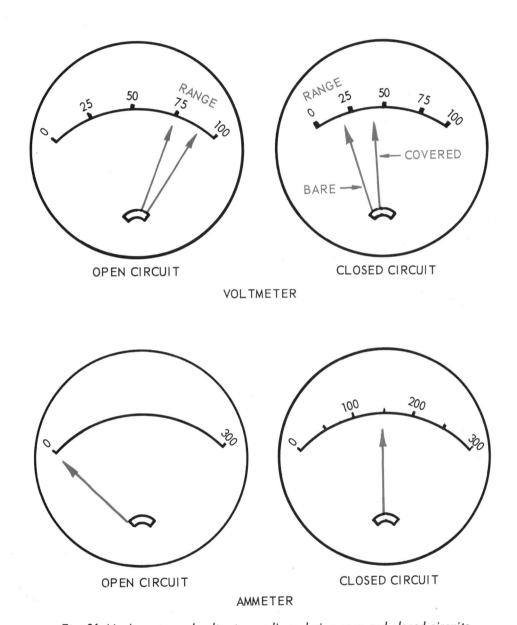

Fig. 26-44. Ammeter and voltmeter readings during open and closed circuits.

connections, the operator may proceed as follows: Start the machine and adjust it to the medium capacity of the unit. Ground the electrode holder securely to the table or to the article to

ing machine to each other (disconnect the regular cables). This action eliminates the cable runs. Start the machine and record the new voltage reading and ammeter reading. The difference in the readings indicates the efficiency of the cables and connections. If the discrepancy is large, the cable should be overhauled and new connections made. However, small differences in the reading should be disregarded as

Fig. 26-46. Cable lug used for connecting welding cable leads to ground and to arc welding machine.

Fig. 26-45. An ammeter used to quickly check the current flow in arc welding cables. (Columbia Elec. Mfg. Co.)

cables are not 100 percent efficient. Cable efficiency can also be checked by checking the potential (volts) drop along each cable and at each connection. An accurate voltmeter with needle point probes may be used to test these parts.

A quick indication of a poor connection is overheating of the connection. The connection should be the same temperature as the cable. For example, if an electrode holder handle becomes too warm, it is an indication that the lead to the electrode holder connection is faulty.

The connections where the cables are fastened to the machines and to the ground are usually made with large copper lugs, as shown in Fig. 26-46. The cable is either mechanically fastened, silver brazed, copper welded or

soldered to the lug. This connection must be carefully made or an inefficient joint will result. Many types of connectors are on the market. Some provide for a quick disconnect joint between various cable lengths as shown in Fig. 26-47. A semipermanent cable-

Fig. 26-47. Connections used to add cable lengths or remove cable lengths from an arc welding lead. Note setscrew method to fasten cable to fitting and quick disconnect cam lock. (Lenco, Inc.)

to-cable connector using a mechanical screw to clamp the wire to the fitting is shown in Fig. 26-48.

To solder a lug to a cable properly proceed as follows:

1. Carefully clean the surfaces to which the solder is to adhere, using cleaning materials.

2. Heat the lug to a temperature sufficient to melt and flow the solder.

3. Put clean solder paste in the lug cavity (to clean it chemically).

4. Fill the cavity in the lug (into which the cable end is to be inserted) with solder.

5. After the solder has been worked until it wets the inner surfaces of the cavity, excess solder is removed.

6. Clean the cable with sandpaper and cover the wires with soldering paste.

7. Insert the cable end into the cavity. The two are then reheated to make the solder flow. The lead must be firmly supported as any movement during the time the solder is solidifying will result in a poor joint.

8. The soldered connection is then cooled. The joint can be cooled quickly by using wet cloths or spraying water on the joint.

A good mechanical joint may be made by forming the copper lug in a

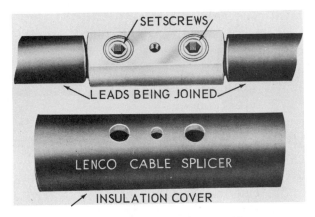

Fig. 26-48. *Semipermanent cable to cable connector fitting. Note setscrews for clamping cable to fitting. The middle hole is for holding insulation cover in place.*

fixture to create a clamping action on the copper wires as shown in Fig. 26-49. The method can be used for either fastening cables together or for fastening a cable to a lug. The procedure for making the connection is

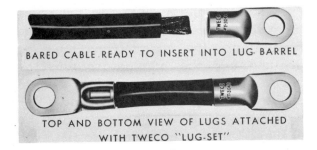

Fig. 26-49. *A metal forming method for making cable connections. (TWECO Products, Inc.)*

shown in Fig. 26-50. It is important to clean the surfaces to be clamped together before assembly. Because this is a cold-forming process, there is no danger of overheating the cable wires or the insulation.

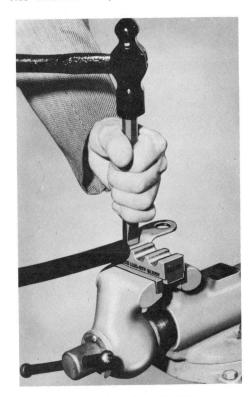

Fig. 26-50. *Forming barrel of a cable lug to obtain a connection between the cable and lug.*

26-48. REPAIRING ARC WELDING SWITCHES AND STARTERS

Most arc welding machines are provided with a device which will stop the power supply if the motor starts to consume more current that it should. These devices, called overload relays, consist of a heating coil through which part or all of the electrical current is passed. This control is of sufficient size to carry any normal welding load without becoming overheated. However, if too much current attempts to pass through the wire, the temperature rises to the point where it will either melt some of the solder and release a trigger device which will open the electrical circuit, or the heat will bend a bimetal strip which, upon bending, actuates a trigger device to interrupt the electrical circuit. These devices are provided with means by which they may be reset by the operator. Such devices offer very little trouble unless the heating element is pressed too close to the solder or to the bimetal strip which it actuates. Also the contact points, which the trigger device manipulates, may become corroded, resulting in overheating and an inefficient current supply.

Inasmuch as the average welding motor size varies between 10 and 15 HP, it is impractical to start the motor by using an ordinary knife switch or button switch. This type of switch would arc excessively. Such machines are equipped with a magnetic starting apparatus which is controlled by a simple push button control. This magnetic starter consists of a large point switch usually of three point contact construction. The switch must be operated with considerable snap action to minimize arcing. To operate with snap action an electromagnet is used to pull the contact points into place. When the operator presses the starting button, a small quantity of current is passed through an electromagnet. This electromagnet immediately moves a soft steel laminated core which is connected to the moving part of the large trigger contact switch. When the core and contact move, this completes the electrical circuit to the motor. It may be of interest to know that practically all arc welding motors use three-phase current and, therefore, the motor is not provided with brushes or centrifugal switches. In these, motors need only occasional cleaning and oiling.

26-49. MAINTENANCE OF RESISTANCE WELDING EQUIPMENT

Resistance welding is used extensively in production work. A factory that has sufficient units will generally have its own maintenance, repair and setup department.

Manufacturers maintain a force of technicians to help the users of their equipment. Most manufacturers also have overhaul facilities.

A good maintenance policy is essential to assure high quality work, and to minimize repair work.

Resistance welding machines consist of:
1. Frame.
2. Transformer.
3. Electrodes.
4. Mechanism for moving electrodes.
5. The mechanism for holding work.
6. Electrical circuits.
7. Pneumatic circuit.
8. Hydraulic circuit.
9. Oil circuit.
10. Cooling circuit.

The proper installation of the right capacity unit will, as in other mechanisms, reduce maintenance and repair.

The electrodes must be kept clean and in proper shape. The operator usually does this job, or he may simply change electrodes.

The maintenance man checks the pneumatic system. The air pressure must be correct, the filters must be kept clean, the water traps must be drained and the lubricators must be kept in good condition. The system should be periodically checked. Hydraulic system pressure must be checked, filters cleaned, and the system checked for leaks. The maintenance man should have a maintenance manual available for the particular machine and make frequent use of the manual for reference. He should understand about the circuits--water, oil, and air. He should know how to adjust each control valve and how to check the controls. Instruments are a necessity when checking resistance welding equipment. CHAPTERS 14 and 21 describe instruments for measuring electrode force, current flow, welding cycles and the like.

Some of the most common resistance welding "DOs and DON'Ts" as identified by P. R. Mallory and Co. are as follows:

DOs

1. Use the proper electrode material for the job you are doing.

2. Use standard electrodes wherever possible.

3. Use the most suitable tip diameter for the thickness of the stock being welded.

4. Connect the water inlet hose to the proper holder inlet so that water flows through the center cooling tube first.

5. Cool spot-welding tips internally with cool water flowing at a rate of at least 1-1/2 gal. per minute through each tip.

6. Be sure that the internal water-cooling tube of the holder projects into the tip water hole to within 1/2 in. of the tip-hole bottom.

7. Be sure that the top of the adjustable water-cooling tube in the holder is cut at an angle to avoid jamming the tip down and shutting off the water.

8. Place a thin film of cup grease on the tip taper prior to inserting it in the holder. This will make it easier to remove.

9. Use ejector-type holders for easy

removal of tips and to avoid damage to tip walls.

10. Keep the tip taper and holder taper clean, smooth and free from foreign deposits.

11. Dress spot-welding electrodes often enough to maintain weld quality.

12. Dress electrodes to their original contour in a lathe whenever possible.

13. Clean and tighten regularly all electrical connections in the secondary circuit.

14. Keep the welder throat area to the minimum permissible dimensions.

15. Lubricate the resistance welder's motor, gears and slides at periodic intervals.

16. Calibrate all pressure-applying mechanisms so that the exact welding pressure can always be known.

17. Clean--either chemically or mechanically--all material to be welded that has dirty, scaly, oxidized or otherwise contaminated surfaces.

18. Use suitable automatic current-timing devices.

19. Have all timing devices suitably calibrated.

20. Use synchronous electronic timing controls for short timing periods or in the welding of materials having critical weldability.

21. Mount timing devices rigidly to protect them from vibration or abuse and mount them for ready servicing.

22. Use synchronous controls, providing interrupted timing periods, in all seam-welding applications.

23. Use electronic or magnetic welding contactors developed specifically for welding applications.

24. When using an ignitron control with a welder drawing less than 40 amperes from the line, provide a load resistor across the primary so that the current does not fall below 40 amperes.

25. Keep all electrical contacts in re-

lays, switches and contactors in good condition by frequent inspection and dressing or replacement if necessary.

DON'Ts

1. Never use unidentified electrodes or electrode materials.

2. Avoid special, offset or irregular tips when the job can be done with a standard straight tip.

3. Don't use small tips on heavy-gauge welding jobs, or large tips on small work.

4. Never use water hose that will not fit the holder water-connection nipples snugly.

5. Do not allow water connections to become leaky, clogged or broken.

6. Avoid using holders with leaking or deformed tapers.

7. Do not permit the adjustable water tube to be "frozen" by an accumulation of deposits. A few drops of oil periodically will keep the tube free.

8. Do not allow electrodes to remain idle for extended periods in tapered holder seats.

9. Don't use pipe wrenches or similar tools to remove electrodes.

10. Avoid the use of white lead or similar compounds to seal a leaking taper.

11. Never permit a spot-welding tip to mushroom enough to make dressing difficult.

12. Never dress electrodes with a coarse file.

13. Don't permit flash particles to accumulate on the welding transformer.

14. Do not use a long welder arm extension or a wide arm separation unless the shape of the parts to be welded necessitates it.

15. Shun lubricants which are not recommended by the welder manufacturer.

16. Never permit sliding pressure members to become too tight or jammed since this will increase the frictional forces and inertia of the moving members.

17. Never weld unclean metals, scaly stock or metals with poor surface condition.

18. Avoid the use of slow-operating magnetic contactors if you are doing precision work.

19. Do not employ nonsynchronous timing devices for welding materials with critical welding characteristics--i.e., aluminum, magnesium and brass.

20. Avoid expensive synchronous timing devices for long timing, noncritical applications.

21. Don't use welding contactors or ignitron controls of capacity not within the rating of the welding machine and duty cycle of the application.

22. Never operate a synchronous ignitron control with the secondary of the welder open circuited.

23. Never use nonsynchronous timing devices for short timing (less than 4 or 5 cycles).

24. Don't expect the timer to operate satisfactorily if the limit relay or switch in the timer-welder circuit chatters, or the contacts are oxidized or pitted.

25. Do not expect the timer to provide uniform timing pulses if the primary voltage regulation is poor. Ascertain timer manufacturer's recommended limits of voltage variation.

From Resistance Welding Data Book published by P. R. Mallory and Co., Inc.

26-50. REPAIRING RESISTANCE WELDING EQUIPMENT

Resistance welding machines should be reconditioned after extensive use. However, before repairs or overhauls are attempted, it is important to accurately justify this procedure. Sometimes a machine will fail to perform without the actual machine being at fault.

The maintenance staff should check:

1. The electrical supply. There should be sufficient voltage and current, and the supply should be constant.

2. The water supply. The pressure and volume should be sufficient and should be steady.

3. The hydraulic, pneumatic, and oil circuits should be checked in a similar manner.

Only after local resources have checked the equipment and diagnosis proves that the machine needs reconditioning, should the factory representative or field service man be asked to help.

The electronic controls on the resistance welders are in most cases too complex for the ordinary shop to repair, and it is recommended that a manufacturer's representative be consulted and his recommendations followed in reference to making the necessary repairs and adjustments. Ordinarily, in a radio or television, one can determine if a tube is bad if it does not light or if some characteristic of the set changes when the tube is tapped. Welding control tubes light only intermittently and for very short periods, so it is hardly possible to check by sight, or by tapping.

Electrical controls can be checked in the shop with a minimum of equipment, but best results are obtained if a manufacturer's representative does the repairing. The solenoids used for controlling the flow of the gases or coolant are normally the trouble area when no flow takes place or when flow occurs with the machine turned off. The solenoid can be removed and tested to determine if the valve is moving freely and if not, it should be replaced. With the gas solenoid closed and the electrode holder removed from the economizer valve, if the flowmeter indicates a reading, the solenoid valve must be leaking slightly.

With the water solenoid closed there should be no water flowing in the water line. This can be checked by disconnecting the water line at the electrode holder or at the outlet of the solenoid valve.

The cables, wires, and electrical components in the various circuits may be checked for excessive resistance by testing the voltage drop in a line or across an electrical component with a voltmeter and checking the readings with the manufacturer's allowable loss. If the voltage drop is found to be excessive, the connections should be cleaned and tightened or the wire, cable, or part replaced and retested. Complete overhauls which involve complete teardowns, and rebuilding are best done by the manufacturer.

Parts, such as condensers, transformers, and rheostats, may be checked by a qualified electrician.

26-51. REVIEW OF SAFETY

All the safety procedures described throughout the text also apply to the various types of welding shops explained in this chapter.

Adequate ventilation, eye protection, protective clothing are all necessary. Safety shoes should be worn at all times when handling weldments and when repairing equipment.

Some of the precautions one must take when working in a welding shop are as follows:

1. Cranes and hoists must be peri-

odically inspected for condition of cables, hooks, clamps, rails, and the like.

2. Ventilation ducts and air-moving equipment must be periodically inspected, cleaned, and moving parts lubricated.

3. Aisles in the shop should be clearly marked and kept clear of material and equipment.

4. When heavy equipment is being moved all personnel in the vicinity should be alerted and should be moved out of potentially dangerous positions.

5. Periodic inspections should be made of all safety equipment and supplies in the shop.

26-52. TEST YOUR KNOWLEDGE

1. Describe the different types of welding shops.

2. Name some fundamental functions of a preheating furnace.

3. How are jigs and fixtures used in welding?

4. Describe two uses of a power grinder in the welding shop.

5. What parts of a pressure regulator need most frequent repair?

6. What materials are regulators and torches mostly made of?

7. How is a leaky regulator valve detected?

8. What is a common method of fastening the diaphragm to the regulator body?

9. How may a buckled diaphragm be detected?

10. When assembling a regulator, should the adjusting screw be turned all the way into the bonnet or all the way out?

11. Of what use is a portable spot welder in a welding shop?

12. Why is it important not to mar the surface of the mixing chambers?

13. What is the result of particles partly clogging the tip orifice?

14. What lubricant may be used on oxyacetylene equipment?

15. What may be the result of a bell-mouth tip orifice?

16. What material may be used to seal pipe threads?

17. What is the result of a loose connection in an arc welder?

18. How may one determine if the commutator brushes do not fit the commutator correctly?

19. What is the harmful result of a poor connection in an arc welding generator circuit?

20. What source of energy is used to energize a power wrench?

21. What method is used to shape the generator brush to correct fit quickly?

22. What is used to drive the generator in mobile arc welding equipment?

23. What is the difference between voltage and amperage?

24. Do all arc welding machines have both an ammeter and a voltmeter?

25. What is the approximate voltage drop across the arc when using covered electrodes?

26. What are some of the causes of excessive voltage drop?

27. What is the result of excessive voltage drop?

28. Does electrical resistance increase as the cable temperature increases?

Chapter 27

PROCEDURE AND WELDING PERFORMANCE QUALIFICATIONS

Whenever a structure such as a building, bridge, ship, or pressure vessel is welded, it is necessary that the manufacturer and buyer reach an agreement on how each weld will be made.

The agreement includes the welding method, base metal specifications and thickness, filler metal composition, preheat and postheat treatment, and other variables which will affect a welding procedure. This agreement also includes the performance qualification tests for the welders who are to make the welds agreed upon in the welding procedure. All these agreements are written in the form of a code.

To eliminate the necessity of writing a new code for each new job, several government agencies, societies and associations have developed codes which may be used. Federal, state, and local governmental agencies, for example, have written building and safety codes. Insurance companies which must insure welded structures have also prepared procedures and welder performance qualification codes. Codes are also written for brazed joints.

27-1. GOVERNMENT AGENCIES

Some of the governmental codes and standards are established by the following agencies:

Interstate Commerce Commission-- cylinders, design, construction, content, etc.

Federal Aeronautics Administration--aircraft welding.

Military Specifications (MIL)
Bureau of Ships (BUSHIPS).
Bureau of Ordnance (BUORD).
U. S. Army Ordnance.
U. S. Navy.
Bureau of Aeronautics.
Air Force.
U. S. Army.
Bureau of Standards.
Bureau of Mines.
The American Bureau of Shipping.
U. S. Coast Guard.

27-2. ASSOCIATIONS AND SOCIETIES

Some associations and societies that have established codes, standards and/ or tentative standards are:

1. American Standards Association - Code for Pressure Piping.

2. American Welding Society (AWS) - Structural Welding Code (D1.1-79).

3. American Petroleum Institute - Standard for Field Welding of Pipe Lines (1104).

4. American Society of Mechanical Engineers (ASME) - Boiler and Pressure Vessel Code: Section I - Power Boilers, Section VIII - Pressure Vessels, Section IX - Welding and Brazing Qualifications.

5. American Society of Testing Materials (ASTM).

6. Society of Automotive Engineers (SAE).

7. American Institute of Steel Construction.

8. Association of American Railroads, Operation and Maintenance Division.

9. Heating and Piping Contractors National Association.

10. American Society of Metals.

11. Resistance Welder Manufacturers' Association (RWMA).

12. American Institute of Electrical Engineers.

13. Tubular Exchanger Manufacturers' Association, Inc.

27-3. INSURANCE COMPANIES AND ASSOCIATIONS

Some of the insurance companies or associations that have established standards for eligibility for insurance where welding is done or used are:

National Board of Fire Underwriters.

Underwriters Laboratories.

National Fire Protection Association.

Factory Assurance Corporation.

The Hartford Steam Boiler Inspection & Insurance Co.

Lloyd's of London.

Lloyd's Register of Shipping Rules and Regulations.

27-4. IMPORTANCE OF PROCEDURES AND WELDING PERFORMANCE QUALIFICATION CODES

After a structure such as a building, bridge, or pressure vessel is designed, contracts must be written for construction. Agreements must be reached on a number of welding variables in order to insure weld quality, structural strength, and safety. Companies generally use the established codes rather than write new codes of their own.

If a bridge were constructed which also carried oil and pressure pipes over a span of water, several codes may be involved. The American Welding Society (AWS) "Structural Welding Code" may be used for the bridge structure. The oil pipe lines would be welded using the American Petroleum Institute (API) "Standard for Field Welding of Pipe Lines." Pressure pipes may be welded using the American Society of Mechanical Engineers (ASME) "Boiler and Pressure Vessel Code."

All codes applying to a welding job must be available to the welders and assigned inspector on the job. Copies of any welding code are available, for a fee, from the agency, society, or association which publishes them.

27-5. PROCEDURE QUALIFICATIONS

Procedure qualifications are limiting instructions written by a contractor or manufacturer which explain how welding will be done on a weldment in accordance with a code. These limiting instructions are listed in a document known as a "Welding Procedure Specification" (WPS).

The welding procedure specification must list in detail:

1. The various base metals to be joined by welding.

2. The filler metal to be used.

3. The range of preheat and postweld heat treatment.

4. Thickness, and other variables described for each welding process.

The variables are listed as essential or nonessential.

Each manufacturer or contractor must qualify his WPS by welding test coupons and by testing the coupons in accordance with the code. The results of these tests are recorded on a document known as a "Procedure Qualification Record" (PQR).

More than one welding process or procedure may be used in a single production joint. Each procedure must be qualified separately or in combination with other processes or procedures.

27-6. WELDING PROCEDURE VARIABLES

Most welding codes list the welding procedure variables which are essential and nonessential.

Essential variables are those which, when changed, will affect the mechanical properties of the weldment. Changes of essential variables require that the welding procedure specifications (WPS) be requalified.

Some of the essential variables for welds made with the shielded metal-arc welding (SMAW) process are changes in:

1. Base metal thickness (beyond accepted limits).
2. Base metal strength or composition.
3. Filler metal used (strength or composition).
4. Preheat temperature.
5. Postheat temperature.
6. Thickness of postheat treatment sample.

Nonessential variables are those which, when changed from the approved welding procedure specifications, do not require requalification of the procedure.

For SMAW, some of the nonessential variables are changes in:

1. Type of groove.
2. Deletion of backing in single welded butt joints.
3. Size of the electrode.
4. Addition of other welding positions to those welding positions which are already qualified.
5. Maintenance or reduction of preheat prior to postheat treatment.

6. Current or polarity or the range of amperage or voltage.

27-7. P NUMBERS

To reduce the number of welding procedure qualifications required, the ASME has developed a system of grouping metals with similar characteristics. The grouping is done by P numbers. The P number grouping is based on comparable base metals characteristics, such as composition, weldability, and mechanical properties. P numbers may be further subdivided into group numbers. There are 61 P groups in the ASME Code. Fig. 27-1 lists metal classified as P1, with subgroups 1, 2, and 3.

27-8. PROCEDURE QUALIFICATIONS SPECIMENS

To qualify for a welding procedure specification, a contractor or manufacturer must prepare several procedure test specimens (samples) for testing and approval.

The number and thickness of the required tension test and transverse bend test specimens are shown in Fig. 27-2. Also shown are the requirements for tension and longitudinal bend test specimens.

When a test plate is qualified, it will serve to qualify a range of thicknesses. If a test plate 1/8 in. thick is qualified, it will serve to qualify all thicknesses from 1/16 to 1/4 in., as shown in Fig. 27-2. The maximum thickness in the qualified range is "2t" or twice the test plate thickness.

27-9. WELDING PERFORMANCE QUALIFICATIONS

Each manufacturer or contractor must qualify each welder or welding

QW	P No.	Group No.	Base Metal Specification	Minimum Specified Tensile ksi	Type of Base Metal (Nominal Composition)
			Steel and Steel Alloys		
422.1	1	1	SA-31 Grade A	45	Carbon Steel Rivets (C)
			Grade B	58	Carbon Steel Rivets (C)
			SA-36	58	Carbon Steel Plate (C-Mn-Si)
			SA-53 Acid Bessemer	50	Carbon Steel Furnace Welded Pipe
			Open Hearth	45	Carbon Steel Furnace Welded Pipe
			Grade A	48	Carbon Steel Seamless or Welded Pipe (C)
			Grade B	60	Carbon Steel Seamless or Welded Pipe (C-Mn)
			SA-106 Grade A	48	Carbon Steel Pipe (C)
			Grade B	60	Carbon Steel Pipe (C-Mn)
			SA-135 Grade A	48	Carbon Steel Electric-Resistance-Welded Pipe (C)
			Grade B	60	Carbon Steel Electric-Resistance-Welded Pipe (C-Mn)
			SA-178 Grade A		Carbon Steel Electric Welded Boiler Tube (C)
			Grade C	60	Carbon Steel Electric Welded Boiler Tube (C)
			SA-179 . . .		Carbon Steel Seamless Low-Carbon Steel Tubes (C)'
			SA-181 Grade I	60	Carbon Steel Pipe Flanges (C-Si)
			SA-192	47	Carbon Steel Boiler Tubes, Seamless (C-Si)
			SA-210 Grade A-1	60	Carbon Steel Tubes (C)
			SA-214		Carbon Steel Electric-Resistance-Welded Steel Tubes (C)
			SA-216 Grade WCA	60	Carbon Steel Castings (C-Si)
			SA-226	47	Carbon Steel Electric-Welded Tubes (C-Si)
			SA-266 Class 1	60	Carbon Steel Seamless Drum Forgings (C-Si)
			SA-283 Grade A	45	Carbon Steel Plates (C)
			Grade B	50	Carbon Steel Plates (C)
			Grade C	55	Carbon Steel Plates (C)
			Grade D	60	Carbon Steel Plates (C)
			SA-285 Grade A	45	Carbon Steel Plates (C)
			Grade B	50	Carbon Steel Plates (C)
			Grade C	55	Carbon Steel Plates (C)
			SA-306 Grade 45	45	Carbon Steel Bars (C)
			Grade 50	50	Carbon Steel Bars (C)
			Grade 55	55	Carbon Steel Bars (C)
			Grade 60	60	Carbon Steel Bars (C)
			Grade 65	65	Carbon Steel Bars (C)
			SA-333 Grade 1	55	Carbon Steel Pipe for Low Temp. Service (C-Mn)
			Grade 6	60	Carbon Steel Pipe for Low Temp. Service (C-Mn-Si)
			SA-334 Grade 1	55	Carbon Steel Tubes for Low Temp. Service (C-Mn)
			Grade 6	60	Carbon Steel Tubes for Low Temp. Service with 0.10% min. Silicon (C-Mn-Si)
			SA-350 Grade LF1	60	Carbon Steel Forgings (C-Mn)
			SA-352 Grade LCB	65	Carbon Steel Castings (C-Si)
			SA-414 Grade A	45	Carbon Steel Sheet (C)
			Grade B	50	Carbon Steel Sheet (C)
			Grade C	55	Carbon Steel Sheet (C)
			Grade D	60	Carbon Steel Sheet (C-Mn)
			Grade E	65	Carbon Steel Sheet (C-Mn)
			SA-442 Grade 55	55	Carbon Steel Plates (C-Mn-Si)
			Grade 60	60	Carbon Steel Plates (C-Mn-Si)
			SA-515 Grade 55	55	C-Si Steel Plates (C-Si)
			Grade 60	60	C-Si Steel Plates (C-Si)
			Grade 65	65	C-Si Steel Plates (C-Si)
			SA-516 Grade 55	55	C-Si Steel Plates (C-Si)
			Grade 60	60	C-Si Steel Plates (C-Si)
			Grade 65	65	C-Si Steel Plates (C-Mn-Si)
			SA-524 Grade I	60	Carbon Steel Pipe (C-Mn-Si)
			Grade II	55	Carbon Steel Pipe (C-Mn-Si)
			SA-556 Grade A2	47	Carbon Steel Tubes—Seamless (C)
			Grade B2	60	Carbon Steel Tubes—Seamless (C-Si)
			SA-557 Grade A	47	Carbon Steel Tubes—Resistance Welded (C)
			Grade B	60	Carbon Steel Tubes—Resistance Welded (C)

QW	P No.	Group No.	Base Metal Specification	Minimum Specified Tensile ksi	Type of Base Metal (Nominal Composition)
			Steel and Steel Alloys (Cont'd)		
			SA-587	48	Low Carbon Steel Pipe (C)
			SA-662 Grade A	58	C-Mn Steel Plate (C-Mn-Si)
			Grade B	65	C-Mn Steel Plate (C-Mn-Si)
			SA-695 Type B Grade 35	60	C-Si Steel Bars (C-Si)
422.1	1	2	SA-105	70	Carbon Steel Pipe Flanges (C-Mn-Si)
			SA-106 Grade C	70	Carbon Steel Pipe (C-Mn)
			SA-181 Grade II	70	Carbon Steel Pipe Flanges (C-Si)
			SA-210 Grade C	70	Carbon Steel Electric-Resistance-Welded Steel Tubes (C-Mn)
			SA-216 Grade WCB	70	Carbon Steel Castings (C-Si)
			Grade WCC	70	Carbon Steel Castings (C-Mn-Si)
			SA-266 Class 2	70	Carbon Steel Seamless Drum Forgings (C-Si)
			Class 3	75	Carbon Steel Seamless Drum Forgings (C-Si)
			SA-299	75	C-Mn-Si Steel Plates (C-Mn-Si)
			SA-306 Grade 70	70	Carbon Steel Bars (C)
			SA-350 Grade LF2	70	Carbon Steel Forgings (C-Mn-Si)
			SA-414 Grade F	70	Carbon Steel Sheet (C-Mn)
			Grade G	75	Carbon Steel Sheet (C-Mn)
			SA-455 Type I	75	Carbon Manganese Steel Plates (C-Mn)
			Type II	73	Carbon Manganese Steel Plates (C-Mn)
			SA-508 Class 1	70	Forgings (C-Si Steel 0.35 max. C) (C-Si)
			SA-515 Grade 70	70	C-Si Steel Plates (C-Si)
			SA-516 Grade 70	70	C-Mn-Si Steel Plates (C-Mn-Si)
			SA-537 Class 1	70	C-Mn-Si Steel Plates (C-Mn-Si)
			SA-541 Class 1	70	Forgings (C-Si)
			SA-556 Grade C2	70	Carbon Steel Tubes—Seamless (C-Mn-Si)
			SA-557 Grade C	70	Carbon Steel Tubes—Resistance Welded (C-Mn)
			SA-695 Type B Grade 40	70	C-Si Steel Bars (C-Si)
		3	SA-537 Class 2	80	C-Mn-Si Steel Plates (C-Mn-Si)

Fig. 27-1. P1 grouping of metals for welding procedure and welder qualification. (ASME-- IX)

operator for each welding process to be used in production welding. The welding performance qualification test requires the welder or welding operator to complete a weld made in accordance with the welding procedure specification (WPS). Some departures from the exact WPS may be allowed by the code used.

The type and number of test specimens required for mechanical testing is specified by the code used. Also specified is the manner by which the specimens are removed from the weld sample. Specimens may also be tested by radiographic examination (X-ray).

The welder or welding operator is qualified by welding position. Welders may be qualified to perform a welding procedure in only one or possibly all positions. A welder qualified only in the flat position is not qualified to make welds in any other position.

Each welder or welding operator must be qualified for each process used. A welder qualified to weld in accordance with one qualified WPS is also qualified to weld in accordance with another WPS which uses the same process, within the limits of the essential variables of the code used.

Welders may be required to re-

Thickness, t of Test Plate or Pipe as Welded, in.	Range of Thickness of Materials Qualified by Test Plate, in. (Note 2)		Type and Number of Test Required (Guided Bend Tests)			
	Min	Max	Tension QW-462.1	Side Bend QW-462.1	Face Bend QW-462.3(a)	Root Bend QW-462.3(a)
Less than 1/16	t	2t	2		2	2
1/16 to 3/8, incl.	1/16	2t	2		2	2
Over 3/8, but less than 3/4	3/16	2t	2	(Note 1)	2	2
3/4 and over	3/16	2t	2	4		

NOTES:
(1) Four side bend tests may be substituted for the required face and root bend tests.
(2) See QW-403 for further limits on range of thickness qualified.

Tension Tests and Longitudinal Bend Tests

Thickness, t, of Test Plate or Pipe as Welded in.	Range of Thickness of Materials Qualified by Test Plate, in.		Type and Number of Tests Required (Guided Bend Tests)		
	Min	Max	Tension QW-462.1	Face Bend QW-462.3(b)	Root Bend QW-462.3(b)
Less than 1/16	t	2t	2	2	2
1/16 to 3/8, incl.	1/16	2t	2	2	2
Over 3/8	3/16	2t	2	2	2

Fig. 27-2. ASME procedure qualification specimens. The QW numbers are article numbers in the ASME Code.

qualify if they have not used the specific process for three months or if there is a specific reason to question their ability to make welds which meet the welding procedure specification.

A welder shall be requalified whenever a change is made in one or more of the essential variables for each welding process in the code used. The essential variables for shielded metal-arc welding (SMAW) listed in the ASME Boiler and Pressure Vessel Code are:

1. The deletion of backing in single-welded butt joints.

2. A change in the ASME electrode specification number or AWS electrode classification number.

3. The addition of other welding positions than those in which the welder has already qualified.

4. A change from upward to downward, or downward to upward, in the progress of the weld.

27-10. WELDING QUALIFICATION SPECIMENS

Welding positions and test specimen specifications illustrated in the following paragraphs are taken from the American Society of Mechanical Engineers (ASME) Boiler and Pressure Vessel Code, Section IX.

This information is provided for those who may wish to practice making specimens similar to those which may be required for welder qualification. This information is only a small part of the complete code and should not be used as a code. The complete code should be obtained from the agency, society, or association which writes the code.

27-11. WELDING TEST POSITIONS

The weld positions shown in Figs. 27-3, 27-4, 27-5 and 27-6 are typical

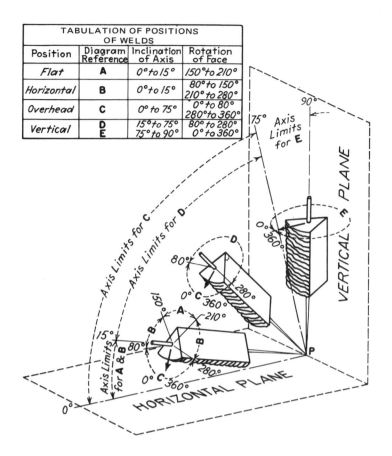

TABULATION OF POSITIONS OF WELDS			
Position	Diagram Reference	Inclination of Axis	Rotation of Face
Flat	A	0° to 15°	150° to 210°
Horizontal	B	0° to 15°	80° to 150° 210° to 280°
Overhead	C	0° to 75°	0° to 80° 280° to 360°
Vertical	D E	15° to 75° 75° to 90°	80° to 280° 0° to 360°

Fig. 27-3. Welding test positions. The horizontal reference plane is taken to lie always below the weld under consideration. Inclination of axis is measured from the horizontal reference plane toward the vertical. Angle of rotation of face is measured from a line perpendicular to the axis of the weld and lying in a vertical plane containing this axis. The reference position (0 deg.) of rotation of the face invariably points in the direction opposite to that in which the axis angle increases. The angle of rotation of the face of the weld is measured in a clockwise direction from this reference position (0 deg.) when looking at point P. (ASME--Section IX)

of those used for many welding codes. A welder or welding operator must be tested to become qualified for each of the positions required in a welding procedure specification (WPS). A qualification in one position does not automatically qualify the welder to weld in any other position. When the welding position changes, the welder must be requalified for that position in order to receive a qualified WPS.

27-12. METHODS OF TESTING SPECIMENS

There are six different weld specimen tests:
1. Tension-reduced section.
2. Side bend.
3. Face and root bend - transverse.
4. Face and root bend - longitudinal.
5. Fillet weld - procedure.
6. Fillet weld - performance.

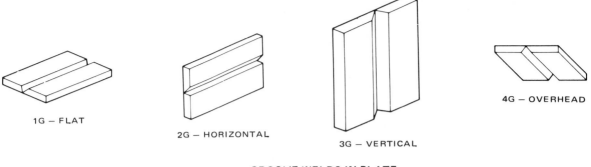

GROOVE WELDS IN PLATE

1G — FLAT

2G — HORIZONTAL

3G — VERTICAL

4G — OVERHEAD

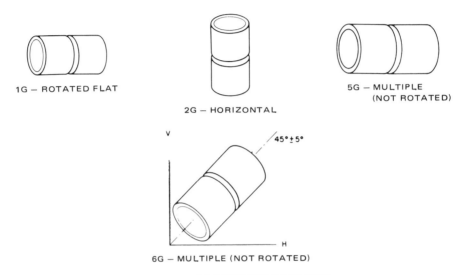

1G — ROTATED FLAT

2G — HORIZONTAL

5G — MULTIPLE
(NOT ROTATED)

45°±5°

6G — MULTIPLE (NOT ROTATED)

GROOVE WELDS IN PIPE

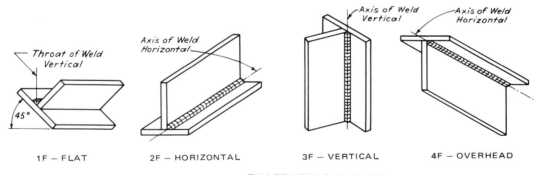

Throat of Weld
Vertical

45°

1F — FLAT

Axis of Weld
Horizontal

2F — HORIZONTAL

Axis of Weld
Vertical

3F — VERTICAL

Axis of Weld
Horizontal

4F — OVERHEAD

FILLET WELDS IN PLATE

Fig. 27-4. Test positions for various joints. Note that test position 5G and 6G are not rotated and the weld position changes as the weld is made. (ASME-- Section IX)

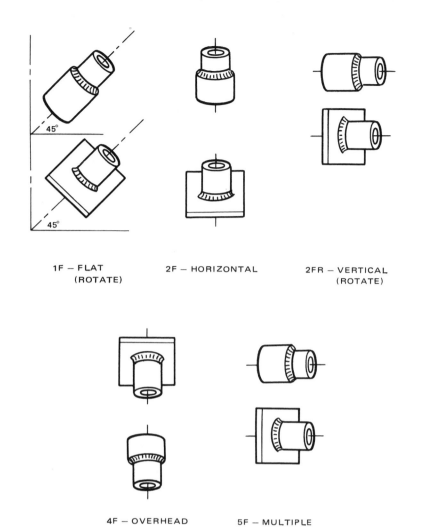

1F — FLAT
(ROTATE)

2F — HORIZONTAL

2FR — VERTICAL
(ROTATE)

4F — OVERHEAD

5F — MULTIPLE

Fig. 27-5. Test positions for fillet welds in pipe joints.
(ASME-- Section IX)

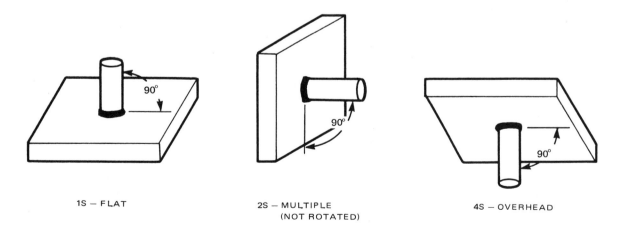

1S — FLAT

2S — MULTIPLE
(NOT ROTATED)

4S — OVERHEAD

Fig. 27-6. Test positions for stud welds.   (ASME-- Section IX)

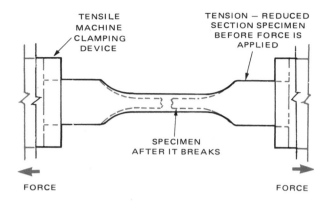

Fig. 27-7. Tension-reduced section test. When force is applied, specimen is stretched until it breaks. Maximum force is noted and tensile strength calculated.

Fig. 27-7 shows how the tension-reduced section specimen is tested and Fig. 27-8 demonstrates the side bend test.

Fig. 27-9 shows the transverse face and root bend samples before and after bending. The longitudinal face and root bend samples are shown before and after testing in Fig. 27-10.

27-13. TENSION-REDUCED SECTION TESTING

The tension-reduced section test sample should be cut from the test

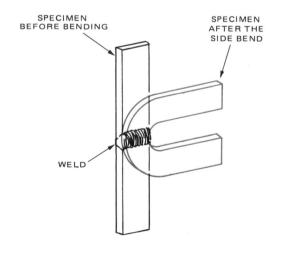

Fig. 27-8. The side bend. Note direction of bend.

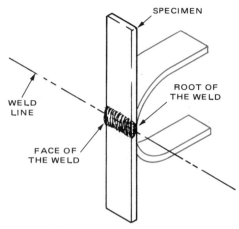

A — TRANSVERSE FACE BEND
The weld face is on the outside of the bend

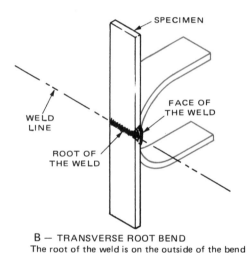

B — TRANSVERSE ROOT BEND
The root of the weld is on the outside of the bend

Fig. 27-9. Transverse face and root bend. Transverse sample is cut across weld line.

weld from the area shown in Fig. 27-11, A, B, and C.

The sample must then be prepared according to the dimensions shown in Fig. 27-12, A or B.

Before subjecting the weld sample to a tension load, the thickness and width of the weld at the narrowest point must be measured.

The cross-sectional area of the weld is obtained by multiplying the width times the thickness. Tensile strength

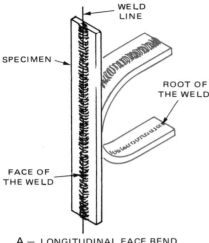

A — LONGITUDINAL FACE BEND
The weld face is on the outside of the bend

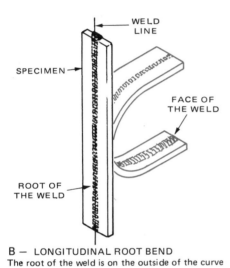

B — LONGITUDINAL ROOT BEND
The root of the weld is on the outside of the curve

Fig. 27-10. Longitudinal face and root bend. The longitudinal sample is cut in same direction as weld line.

is found by dividing the maximum tensile load by the cross-sectional area of the weld. Example:

1/4 in. thickness x 3/4 in. width = 3/16 in.^{2} area.

If the tensile load is 11250 lbs., then:

$$\text{Tensile strength} = \frac{\text{tensile load}}{\text{area}}$$

$$\text{Tensile strength} = \frac{11250}{3/16} = \frac{11250}{.1875}$$

Tensile strength = 60000 lbs. per sq. in.

27-14. SIDE BEND TEST

The side bend test sample is removed from the test weld as indicated in Fig. 27-11, part B. Each sample must be bent in a jig with the dimensions as shown in Fig. 27-13.

To be acceptable, the side bend test and all other bend tests must not break. According to the ASME, the bend test may not show defects larger than those indicated in the code article QW-163:

The weld and heat-affected zone of a transverse-weld bend specimen shall be completely within the bent portion of the specimen after testing.

The guided-bend specimens shall have no open defects exceeding 1/8 in. (3.2 mm), measured in any direction on the convex surface of the specimen after bending, except that cracks occurring on the corners of the specimen during testing shall not be considered, unless there is definite evidence that they result from slag inclusions or other internal defects. For corrosion resistant weld overlay cladding, no open defect exceeding 1/16 in. (1.6 mm) measured in any direction shall be permitted in the cladding, and no open defects exceeding 1/8 in. (3.2 mm) shall be permitted in the bond line.

27-15. FACE AND ROOT BENDS

Weld samples are welded from one side only generally. The face of the

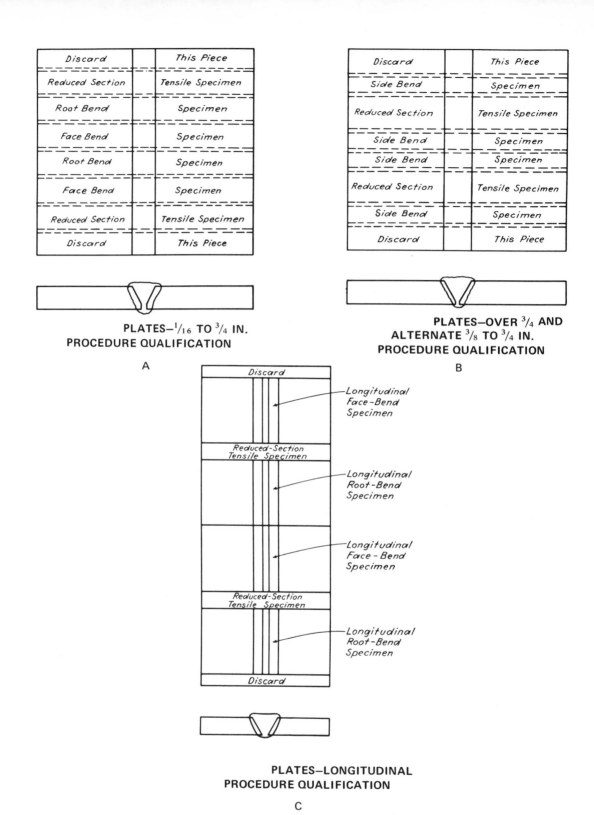

Discard		This Piece
Reduced Section		Tensile Specimen
Root Bend		Specimen
Face Bend		Specimen
Root Bend		Specimen
Face Bend		Specimen
Reduced Section		Tensile Specimen
Discard		This Piece

**PLATES—$1/16$ TO $3/4$ IN.
PROCEDURE QUALIFICATION**

A

Discard		This Piece
Side Bend		Specimen
Reduced Section		Tensile Specimen
Side Bend		Specimen
Side Bend		Specimen
Reduced Section		Tensile Specimen
Side Bend		Specimen
Discard		This Piece

**PLATES—OVER $3/4$ AND
ALTERNATE $3/8$ TO $3/4$ IN.
PROCEDURE QUALIFICATION**

B

Discard

Longitudinal
Face-Bend
Specimen

Reduced-Section
Tensile Specimen

Longitudinal
Root-Bend
Specimen

Longitudinal
Face-Bend
Specimen

Reduced-Section
Tensile Specimen

Longitudinal
Root-Bend
Specimen

Discard

**PLATES—LONGITUDINAL
PROCEDURE QUALIFICATION**

C

Fig. 27-11. Order of removal of test specimens from the test weld. (ASME- Section IX)

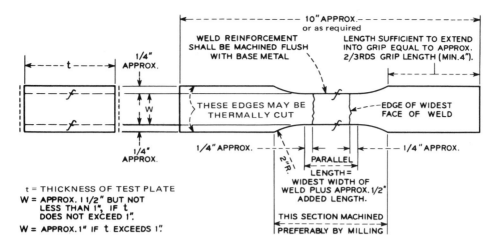

t = THICKNESS OF TEST PLATE
W = APPROX. 1 1/2" BUT NOT
    LESS THAN 1", IF t
    DOES NOT EXCEED 1".
W = APPROX. 1" IF t EXCEEDS 1".

TENSION—REDUCED SECTION—PLATE

A

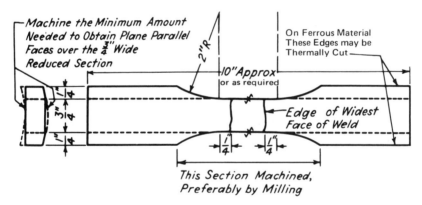

TENSION—REDUCED SECTION—PIPE

B

Fig. 27-12. Specifications for preparing tension-reduced section specimens from plate and pipe samples.
(ASME-- Section IX)

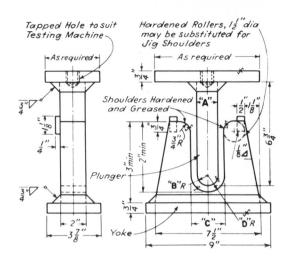

Thickness of Specimens, in.	A, in.	B, in.	C, in.	D, in.	Material	Refer To
3/8	1-1/2	3/4	2-3/8	1-3/16	All	
t	4t	2t	6t + 1/8	3t + 1/16	Others	
1/8	2-1/16	1-1/32	2-3/8	1-3/16	P-23 and SB-171, Alloy 628	QW-422.23 QW-422.35
3/8	2-1/2	1-1/4	3-3/8	1-11/16	P-11,	QW-422.11
t	6-2/3t	3-1/3t	8-2/3t + 1/8	4-1/2t + 1/16	P-25,	QW-422.25
					SB-148	QW-422.35
					Alloys CDA-952 & 954 and	
					SB-271 Alloy CDA-952	
1/16 – 3/8 in. incl	8t	4t	10t + 1/8	5t + 1/16	P-51	QW-422.51
1/16 – 3/8 in. incl	10t	5t	12t + 1/8	6t + 1/16	P-52	QW-422.52
1/16 – 3/8 in. incl	10t	5t	12t + 1/8	6t + 1/16	P-61	QW-422.61

Fig. 27-13. Guided-bend jig specifications. Notice that a different jig is required for various thicknesses and metals. The QW-422 numbers referred to are ASME P number metal specifications. (ASME-- Section IX)

weld is the surface of the joint where the weld bead is applied. The root or bottom of the weld is the surface opposite the weld bead.

After a test weld is completed, test specimens (samples) are cut from the test weld as shown in Fig. 27-11.

When the test specimen is cut across the weld, the specimen is called a transverse specimen. The test specimen may be cut parallel (in the same direction) to the weld. In this case it is a longitudinal specimen.

There are four types of face and root bends. They are:
1. Transverse face bend.
2. Transverse root bend.
3. Longitudinal face bend.
4. Longitudinal root bend.

Each face and root bend specimen must be bent in a jig constructed as shown in Fig. 27-13.

All bends, to be acceptable in the ASME Code, must not have flaws (defects) larger than stated in article QW-163. See Para. 27-14.

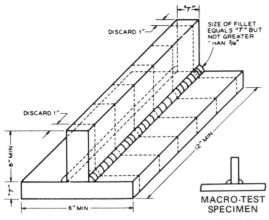

"T" MAXIMUM THICKNESS OF BASE METAL IN THE VESSEL AT POINT OF WELDING OR 1", WHICHEVER IS SMALLER.

MACRO TEST: THE FILLET SHALL SHOW FUSION AT THE ROOT OF THE WELD BUT NOT NECESSARILY BEYOND THE ROOT. THE WELD METAL AND HEAT AFFECTED ZONE SHALL BE FREE OF CRACKS. BOTH LEGS OF THE FILLET SHALL BE EQUAL TO WITHIN $1/8$ INCH.

Fig. 27-14. Specifications for preparing a fillet weld procedure qualification weld. This figure also shows how the test specimen is removed from the weld sample. (ASME-- Section IX)

27-16. FILLET WELD PROCEDURE TEST

A weld sample must be prepared according to the dimensions shown in Fig. 27-14. This illustration also shows how specimens for the procedure test are cut from the test weld.

In order to pass the ASME test, a specimen must satisfy article QW-183:

QW-183 Macro-Examination - Procedure Specimens

One face of each cross section shall be smoothed and etched with a suitable etchant (See QW-470) to give a clear definition of the weld metal and heat-affected zone. In order to pass the test:

Visual examination of the cross sections of the weld metal and heat-affected zone shall show complete fusion and freedom from cracks; and

There shall not be more than 1/8 in. (3.2 mm) difference in the length of the legs of the fillet.

27-17. FILLET WELD PERFORMANCE TEST

The performance test sample for the fillet weld must be prepared and made as shown in Fig. 27-15, A or B. Test specimens are removed from the weld sample as shown in the illustration.

The test specimens are placed under a load to bend the two parts (stems) flat on each other. To pass this performance test, the specimen must agree with the ASME article QW-182:

QW-182 Fracture Tests

The stem of the 4 in. (102 mm) performance specimen center section in QW-462.4 (b) or the stem of the quarter section in QW-462.4 (c), as applicable, shall be loaded laterally in such a way that the root of the weld is in tension. The load shall be steadily increased until the specimen fractures or bends flat upon itself.

If the specimen fractures, the fractured surface shall show no evidence of cracks or incomplete root fusion, and the sum of the lengths of inclusions and gas pockets visible on the fractured surface shall not exceed 3/4 in. (19 mm) to pass the test.

A macro-examination is then made as shown in article QW-184:

QW-184 Macro-Examination - Performance Specimens

The cut end of one of the end sections from the plate or the cut end of the quarter section from the pipe, as applicable, shall be smoothed and etched with a suitable etchant (See QW-470) to give a clear definition of the weld metal and heat-affected zone. In order to pass the test:

Visual examination of the cross section of the weld metal and heat-affected zone shall show complete fusion and free-dom from cracks, except that linear indications at the root, not exceeding 1/32 in. (0.8 mm) shall be acceptable; and

The weld shall not have a concavity or convexity greater than 1/16 in. (1.6 mm); and

There shall be not more than 1/8 in. (3.2 mm) difference in the lengths of the legs of the fillet.

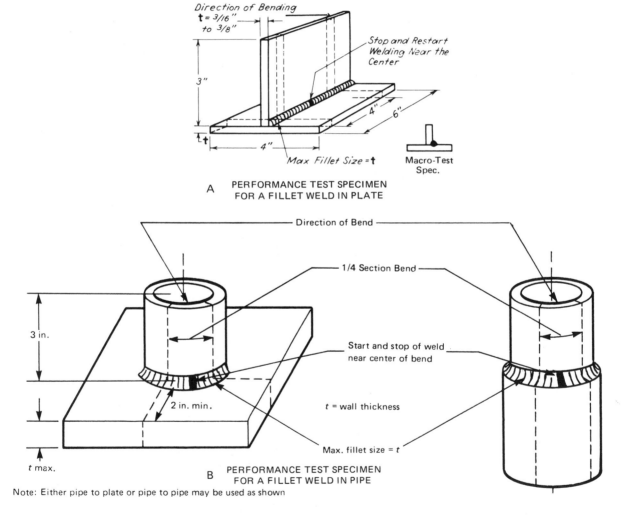

Fig. 27-15. Specifications for preparing a fillet weld performance qualification weld. This figure also shows how the test specimen is removed from the weld sample. (ASME-- Section IX)

27-18. TEST YOUR KNOWLEDGE

1. Name four organizations that have prepared welding codes.

2. What organization has prepared a code for structural welding, such as bridges and the like?

3. What organization has prepared a welding code for pressure vessels?

4. What is the meaning of P numbers as used in connection with the ASME code?

5. If a test plate of 1/8 in. thickness is qualified, what is the range of thicknesses that may be brought in under this qualification?

6. Under what conditions is it necessary to requalify a welder qualified in a given procedure?

7. Is it necessary that a welder be qualified for different welding positions?

8. Name four specimen tests for qualifying welders.

9. What do the letters WPS mean?

10. What are three of the essential welding procedure variables for shielded metal-arc welding?

Trade name	Series	71	70	69	68	67	66	56	55	54	53	52	51	50	49	48	45	44	43	42	41	40	36	35	34	1/8	30	29	28	27	26
Airco	All				1			3		4				5		6		7				8		9			10[5]				
Canadian Liquid Air	All				1			3		4				5						7					9		10				
Craftsman	AA		1							3										5								8[6]			
Dockson	4EC, 4SC, 7SC)																														
	3EC, 5EC, 6EC, 7EC)						3			5		6				7															
	All																		4				5						7	8	
Gasweld	G25, G35				1					4	5	6				7		8		9		10	12		13		16				
	G55				2					4	5	6						8													
	AVG				2							6	7		8																
Harris	2890-F	00						1					2																		
	6290			00				1						2																	
	7490-A							1			2					3															
	23, 13-F, 23A swedged)																		9			10					15[7]			19	
	17F swedged)																														
K-G	AP, APM, APL				1			3			4		5[1]			6		7				8[8]		9			10[9]				
Liquidweld	90, 70, 72		00					2			3				4				5				6				7	8	9	10	11
	80, 82		00					2			3				4				5				6				7	8			
Marquette	A		0			1				4				5					7			8	9				11			12	
	B				2		1	4		5		6				7						8	9				11			12	
	F	0			1			4		4	5		5			6		7				8	9				11				
	G	0					1	3		4		5	5		6							8	9							12	
	H & J				2					4	5		6																		
Meco	All		1							4		5								7		8						9	10		
National	B	2		3				7		8		9		10		11		12				13	14								
National	G, P							2		3	3			4					5		6		8				10				
	R		00																5								7	8	9	10	11
Oxweld	W-15				2					4		5	6		7																
	W-29		2							6	9	12			15		30					40					55	70			100
	W-17, W-22, W-26									6	9	12			15	20	30					40					70		85		
	W-45, W-47										9	12		15	20		30					40					70		85		100
Powr-Craft (Montgomery Ward)	84-5881						1			5					7				9												
Prest-O-Lite	420										15				20			30													
Prest-O-Weld	W-109	2			2			4			5					6		8													
	W-110, W-111				3				6[2]		7					8		9				10	11					29			
	W-120											9	12	15		20	30					40					70		85		
	W-121, W-122											9	12	15		20	30					40					70		85		
Purox	33				2			4			5					6		8													
	34						2			4						6							10					13			
	35						2				5																	13			
	W-200											9	12	15		20	30					40					70		85		
	W-201, W-202											9	12	15		20	30					40					70		85		
	00-D				2			4		5																					
Rego	GX, GXU, SX				68				55		53			50[10]						42			36[11]								
Smith	Pipeliner MW[3]	101		102		103		106	107		108				109			110				111		112							
	Airline AW[3]		101		102	103		106		107		108			109			110				111			112						
	Silver Star LW[3]		101		102	103		106		107		108				109		110				111			112		113				114
	LW[4]		700								703			704		705		706				707	708						710	711	
Torchweld	GP 570, 870				68				55		53			50						42			36[12]								
	71, 370, 170				68				55		53			50						42			36								
Victor	All		00			00½		2	2½		3		3½		4				5				6				7	8	9	10	11
Weldit	All	1								4	5		6			7		8	9			10	12				16				

[1] APL to No. 5 only. [2] W-110 to No. 6 Tip only. [3] Soft-flame tip. [4] Heavy-duty Tip. [5] Airco Tip. No. 11-Drill Size No. 25; No. 12-Drill Size No. 20; No. 13-Drill Size No. 10; No. 14-Drill No. 2; No. 15-1/4 in. Drill. [6] Tips No. 7 and 8 require special gooseneck. [7] 13-F and 17-F Swedged to No. 15 Tip only. [8] APM to No. 8 Tip only. [9] AP to No. 10 only. [10] SX to No. 46 Tip only. [11] GXU to No. 31 only. [12] GP 570 to No. 31 only. (Welding Engineer)

Fig. 28-1. Table of gas welding tip numbers and their orifice drill sizes.

Chapter 28

TECHNICAL DATA

28-1. TECHNICAL DATA ARRANGEMENT

The technical data on welding given in this Chapter will be arranged in main divisions as follows:

Physics of temperatures and heat.
Chemistry.
Gases.
Metals.
Welding Symbols.
Health.
Cable.

28-2. PHYSICS OF TEMPERATURE AND HEAT

The Molecular Theory of heat is generally accepted by chemists and engineers as the best explanation of heat energy. The three most important forms of energy are Heat Energy, Mechanical Energy, and Electrical Energy. It is well known that one form of energy can be easily converted into another form of energy. For example: an electric motor may be used to turn electrical energy into mechanical energy. The bearings of this motor warm up a little while it is running, showing that some of the mechanical energy is being turned into heat. Thorough research has shown that:

One-horse power (mechanical energy) equals 2545.6 Btu per hour (heat energy).

One-horse power (electrical energy) equals 746 watts.

Therefore, 746 watts equals 2545.6 Btu per hour

1 watt equals 3.412 Btu per hour
1 kW equals 3412 Btu per hour

The Molecular Theory of heat, briefly explained, is as follows: All matter consists of molecules and atoms; atoms further consist of protons, electrons, neutrons and other short-life particles. It is believed that these molecules, atoms, etc., are always in motion. That is, the molecule in a sheet of paper is continually in motion and that the rapidity with which it moves determines the amount of heat energy in the sheet of paper.

The rapidity with which the molecule moves determines the heat level or intensity of heat and is known as temperature. Whether one atom is moving at a certain speed, or whether a thousand atoms are moving at the same speed, the temperature will be the same. It is, therefore, necessary to know the number of molecules, or atoms, in a substance to determine the total amount of heat energy stored in that substance. This in brief is the difference between temperature and the amount of heat. It may be seen that to know the temperature you need to know only the speed of molecule motion. But to know the amount of heat in the substance one must know both the temperature and the weight of the substance.

As everyone knows all substances may exist in three forms, solid, liquid, and gas. It is usually conceded that there is no change in the chemical composition of the same substance in any of these three forms. It is, therefore, explained that in a particular substance in the solid form, the molecules have a vibrating motion. They stay in the same position relative to each other, but they are vibrating. When energy is applied to the substance, the molecules vibrate faster than before and this is indicated by an increase in temperature. However, only a certain definite amount of heat energy may be put into the substance, and only a certain temperature rise can be obtained in a solid. After this amount has been absorbed by the substance, if any additional energy is added the molecules will travel at such a rate that the molecules cannot stay within their vibrating bonds. At this heat level an internal change of structure occurs within the molecule, accompanied by considerable absorption of energy, and the substance changes slowly from the solid to the liquid state. The substance will absorb heat during the transformation, but there is no temperature rise. All the heat being applied, regardless of how fast or slow, produces the internal structural change, rather than an increase of the motion of the molecule.

The theory of the energy in a liquid substance is that the molecule now travels in a straight line until it comes into contact with another molecule, instead of vibrating. This necessarily means that the substance will now have no definite shape and will have no rigidity. It must, therefore, be kept in a container. However, the structure of the molecule is such that the individual molecules still have considerable attraction one for the other; one molecule will attract another enough to divert its path and also to prevent it from traveling too far apart.

As energy is applied to the liquid molecule, the rate of travel increases and this is indicated by a temperature rise (increase in the heat level). After a certain amount of energy has been absorbed by the liquid, it will reach a certain heat level of temperature, where energy applied results not in a temperature change, but in an internal structure change of the molecule with the result that the liquid now turns into a gas. While heat is being turned into the gas, the temperature cannot rise inasmuch as the heat being applied results only in a structural change. Upon becoming a gas, the molecules lose their attraction one for the other and travel in straight paths until contacting another molecule or some other substance. This means that gas must be confined to sealed containers.

Ice, water, and steam are the best examples of the above theory. All three conditions are easily obtained, and changing from one to the other does not reveal any change in chemical composition.

The welding industry is interested in the energy in heating gases; it is interested in turning solids to liquids and, therefore, the welders should know some of the theory of molecular energy.

The heat which turns solids to liquids, liquids to solids, liquids to gases, or gases to liquids is called latent heat (hidden heat) because the thermometer gives no indication of the amount of this heat. For example, it requires 970 Btu to change one pound of water at 212 deg. F. to steam at 212 deg. F. While welding, one may sometimes note that a certain sized tip on a particular welding job heats the metal to the melting point, but it has difficulty in actually melting the metal. This shows that the torch tips must be large enough to

PROPERTY	PREFERRED UNIT	SYMBOL	TO CONVERT FROM	TO	MULTIPLY BY
area	millimetre squared	mm^2	$in.^2$ mm^2	mm^2 $in.^2$	$6.451\ 600 \times 10^2$ $1.550\ 003 \times 10^{-3}$
current density	ampere per millimetre squared	A/mm^2	$A/in.^2$ A/mm^2	A/mm^2 $A/in.^2$	$1.550\ 003 \times 10^{-3}$ $6.451\ 600 \times 10^{-2}$
deposition rate	kilogram per hour	kg/h	lb./h kg/h	kg/h lb./h	0.45[a] 2.2[a]
electrode force	newton	N	lb. N	N lb.	4.448 222 $2.248\ 089 \times 10^{-1}$
flow rate	litre per minute	l/min.	cfh l/min.	l/min. cfh	$4.719\ 475 \times 10^{-1}$ 2.118 880
fracture toughness	meganewton$^{-3/2}$ metre$^{-3/2}$	$MN{\cdot}m^{-3/2}$	$ksi{\cdot}in.^{1/2}$ $MN{\cdot}m^{-3/2}$	$MN{\cdot}m^{-3/2}$ $ksi{\cdot}in.^{1/2}$	1.098 855 0.910 038
heat input	joule/metre	J/m	J/in. J/m	J/m J/in.	$3.937\ 008 \times 10^1$ $2.540\ 000 \times 10^{-2}$
linear measurements	millimetre	mm	in. mm	mm in.	$2.540\ 000 \times 10^1$ $3.937\ 008 \times 10^{-2}$
pressure	pascal	Pa	psi Pa	Pa psi	$6.894\ 757 \times 10^3$ $1.450\ 377 \times 10^{-4}$
tensile strength	same as pressure	Pa	—	—	—
travel speed, wire speed	millimetre per second	mm/s	ipm mm/s	mm/s ipm	$4.233\ 333 \times 10^{-1}$ 2.362 205

(a) Approximate values

Fig. 28-1A. Terms and conversions from U.S. conventional system to SI metric. These units will be useful in welding. (American Welding Society)

heat the metal, and must also furnish enough heat to supply the latent heat (heat of fusion).

28-3. TEMPERATURE SCALES

Many means have been used to measure the temperature level. The Fahrenheit scale is based on a 0 setting at the lowest temperature obtained with sodium salt, ice, and water mixture and on the premise that the temperature level between the melting of ice at sea level pressure and boiling of water at sea level pressure shall be divided into 180 equal spaces or increments. This results in a 0 deg. F., a 32 deg. F. for ice melting, and 212 deg. F. for boiling water.

The Celsius scale (metric scale) is based on the melting of ice at 0 deg. C and the boiling of water of 100 deg. C.

It has been calculated that molecular motion (that is, thermal motion) stops at −273.16 deg. C, and at −459.69 deg. F., 0 K, and 0 R.

28-4. WELDING METRIC TERMS AND VALUES

The metric system of measurements is now used in most of the world. The United States, too, is slowly moving toward adoption of this system. Since 1960, the metric system is standardized by the International System of units (SI). It is now usually called the SI Metric System. Fig. 28-1A lists some of the more common metric units used and their conversion factors.

The SI is a decimal system. All numbers are either whole numbers or decimals. Fractions are not used. Another desirable feature of the SI is its use of one unit of measure for each physical quality. The same SI unit is used for force, energy, and power, regardless of whether the process is mechanical, electrical, or thermal (heat). Power is measured in watts, whether it is from an engine or air conditioner.

Common SI base units include:

QUANTITY	UNIT	SYMBOL	APPROXIMATE COMPARISON
length	metre	m	39.37 in.
mass (weight)	kilogram	kg	2.2 pounds
time	second	s	same

Prefixes, such as "kilo," "deci," and "milli," are used in the SI system to indicate quantity or size. The common prefixes and their quantities are:

PREFIX	EXPONENTIAL EXPRESSION	QUANTITY
kilo	10^3	1000
hecto*	10^2	100
deka*	10	10
base unit	1	1
deci*	10^{-1}	0.1
centi*	10^{-2}	0.01
milli	10^{-3}	0.001

*Use of these prefixes is not recommended.

An example of the use of the table of prefixes is:

1 meter (base unit) x 10 = 1 dekametre
1 meter x 10^3 (1000) = 1 kilometre
1 meter x 10^{-3} (0.001) = 1 millimetre
International welding drawings may

ELECTRODE SIZES		FILLET SIZES	
in.	mm	in.	mm
0.030	0.76	1/8	3
0.035	0.89	5/32	4
0.040	1.02	3/16	5
0.045	1.14	1/4	6
1/16	1.59	5/16	8
5/64	1.98	3/8	10
3/32	2.38	7/16	11
1/8	3.18	1/2	13
5/32	3.97	5/8	16
3/16	4.76	3/4	19
1/4	6.35	7/8	22
		1	25

Fig. 28-2. Conversion tables from U.S. conventional to SI metric, give equivalents for electrodes and fillet sizes.

refer to electrodes or fillet welds sizes in SI units of length. Equivalent conversion tables are shown in Fig. 28-2.

28-5. TEMPERING TABLE

As Chapter 25 explains, the weld, as welded, is a casting. The correct heat treatment and/or mechanical working is necessary to maximize the desired physical properties of the metal. The table below shows the relationship between steel's color and its temperature. Likewise shown is importance of treatment for good service in the item. Welding is not recommended for all; some items are best brazed or braze welded.

TEMPERATURE FOR 1 HOUR		COLOR	TEMPERATURE 8 MIN.		USE FOR
F.	C.		F.	C.	
370	188	Faint Yellow	460	238	1
390	199	Light Straw	510	265	2
410	210	Dark Straw	560	293	3
430	221	Brown	610	321	4
450	232	Purple	640	337	5
490	254	Dark Blue	660	349	6
510	265	Light Blue	710	376	7

SUGGESTED USES:

1. Scrapers, brass turning tools, reamers, taps, milling cutters, saw teeth.
2. Twist drills, lathe tools, planer tools, finishing tools.
3. Stone tools, hammer faces, chisels for hard work, boring cutters.
4. Trephining tools, stamps.
5. Cold chisels for ordinary work, carpenters' tools, picks, cold punches, shear blades, slicing tools, slotter tools.
6. Hot chisels, tools for hot work, springs.
7. Springs, screwdrivers.

28-6. PROPERTIES OF ELEMENTS AND METAL COMPOSITIONS

ELEMENTS	SYMBOL	MELTING TEMPERATURE		SPECIFIC GRAVITY	WEIGHT PER CU. FOOT	GRAMS PER CU. Cm	SPECIFIC HEAT	
		°F	°C				BTU/lb/°F	Cal/g/°C
Aluminum	Al	1,218	659	2.7	166.7	2.67	0.212	0.226
Antimony	Sb	1,166	630	6.69	418.3	6.6	0.049	0.049
Armco Iron	..	2,795	1,535	7.9	490.0	7.85	0.115	0.108
Barium	Ba	1,600	850	3.6	219.0	....	...	0.068
Beryllium	Be	2,348	1,285	1.84	....	1.845	...	0.46
Bismuth	Bi	520	271	9.75	612.0	....	...	0.029
Boron	B	3,990	2,200	2.29	143.0	....	...	0.309
Brass (70Cu 30Zn)	..	1652-1724	900-940	8.44	527.0	....	0.092	...
Brass (90Cu 10Zn)	..	1868-1886	1020-1030	8.60	540.0	....	0.092	...
Bronze (90Cu 10sn)	..	1562-1832	850-1000	8.78	548.0	....	0.092	...
Cadmium	Cd	610	321	8.64	550.0	....	...	0.055
Carbon	C	6,510	3,600	2.34	219.1	3.51	0.113	0.165
Cast Pig Iron	..	2012-2282	1100-1250	7.1	443.2	....	0.13	...
Cerium	Ce	1,184	640	6.8	432.0	....	...	0.05
Chromium	Cr	2,770	1,520	6.92	431.9	6.92	0.104	0.12
Cobalt	Co	2,700	1,480	8.71	555.0	....	...	0.099
Columbium	Cb	3,124	1,700	7.06	452.54	7.25	...	...
Copper	Cu	1,980	1,100	8.89	555.6	8.9	0.092	...
Gold	Au	1,900	1,060	19.33	1205.0	19.2	0.032	0.031
Hydrogen	H	−434.2	−259	0.070	0.00533	....	...	3.415
Iridium	Ir	4,260	2,350	22.42	1400.0	22.4	0.032	0.032
Iron	Fe	2,790	1,530	7.865	490.0	7.85	0.115	0.108
Lead	Pb	621	327	11.37	708.5	11.32	0.030	0.030
Lithium	Li	367	186	.534	....	32.8	...	0.79
Magnesium	Mg	1,204	651	1.74	108.5	....	...	0.249
Manganese	Mn	2,300	1,260	7.4	463.2	7.40	0.111	0.107
Mercury	Hg	−38	−39	13.55	848.84	13.6	0.033	0.033
Molybdenum	Me	4,530	2,500	10.3	638.0	....	...	0.065
Nickel	Ni	2,650	1,450	8.80	555.6	8.9	0.109	0.112
Open Hearth Steel	..	2462-2786	1350-1530	7.8	486.9	....	0.115	...
Osmium	Os	4,890	2,700	22.48	1405.0	....	...	0.031
Palladium	Pd	2,820	1,550	12.16	750.0	....	...	0.059
Platinum	Pt	3,190	1,750	21.45	1336.0	21.4	0.032	0.032
Rhodium	Rh	3,540	1,950	12.4	776.0	....	...	0.060
Ruthenium	Ru	4,440	2,450	12.2	762.0	....	...	0.061
Silenium	Se	424	218	4.8	300.0	....	...	0.084
Silicon	Si	2,590	1,420	2.49	131.1	2.10	0.175	0.176
Silver	Ag	1,800	960	10.5	655.5	10.5	0.055	0.056
Tantalum	Ta	5,160	2,800	16.6	1037.0	....	...	0.036
Tellurium	Te	846	452	6.23	389.0	....	...	0.047
Thallium	Tl	576	302	11.85	740.0	....	...	0.031
Thorium	Th	3,090	1,700	11.5	717.0	....	...	0.028
Tin	Sn	450	232	7.30	455.7	7.30	0.054	0.054
Titanium	Ti	3,270	1,800	5.3	218.5	3.50	0.110	0.142
Tungsten	W	5,430	3,000	17.5	1186.0	19.0	0.034	0.034
Uranium	U	....	....	18.7	1167.0	18.7	0.028	0.028
Vanadium	V	3,130	1,720	6.0	343.3	....	0.115	...
Wrought Iron Bars	..	2,786	1,530	7.8	486.9	....	0.11	...
Zinc	Zn	787	419	7.19	443.2	....	0.093	...
Zirconium	Zr	3,090	1,700	6.38	398.0	....	...	0.066

28-7. STRESSES CAUSED BY WELDING

Stress is a force which causes or attempts to cause a movement or change in shape of parts being welded (strain).

During welding the heat created by the welding process causes the metal to expand. If the parts are not clamped into a fixture while welding, the parts will move due to expansion caused by the heat.

When the welded metal cools, it contracts (shrinks). Usually, it does not return to the original shape or position. Distortion has occurred.

When unclamped parts cool down, they are often distorted (changed in shape) but have no stresses remaining in them. Stresses which remain in parts after welding is completed are known as residual stresses. See Fig. 28-3.

When parts are clamped into a jig or fixture for welding, expansion, contraction, and resulting distortion is held to a minimum. Even though the distortion has been reduced, residual stresses remain in the metal after cooling. See Fig. 28-4. These residual stresses may cause the metal to distort at any time if not removed. To remove the residual stresses after welding, parts should be given a heat treatment.

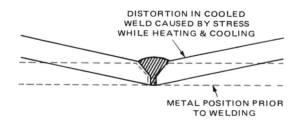

Fig. 28-3. *Distortion caused by welding stresses in an unclamped part. Residual stress after distortion may be zero.*

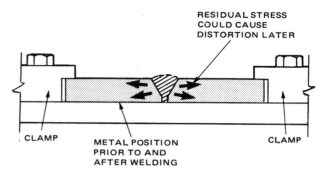

Fig. 28-4. *Residual stress remaining in a clamped part after welding. After weld has cooled, distortion may be near zero, but internal residual stress may be considerable. Residual stress may be removed by stress relieving.*

28-8. WEIGHT AND EXPANSION PROPERTIES OF VARIOUS METALS

METAL	WEIGHT PER CU. FT. IN POUNDS	EXPANSION FOR EACH 1 DEG. F. RISE IN TEMPERATURE IN .00001 IN.
Aluminum	165	1.360
Brass	520	1.052
Bronze	555	.986
Copper	555	.887
Gold	1200	.786
Iron (Cast)	460	.556
Lead	710	1.571
Nickel	550	.695
Platinum	1350	.479
Silver	655	1.079
Steel	490	.689

28-9. THE CHEMISTRY OF GAS WELDING

Chemical formula of combustion of Acetylene and Oxygen.

Acetylene = $C_2 H_2$

Oxygen = O_2

Carbon Monoxide = CO

Carbon Dioxide = CO_2

Water (Vapor) = H_2O

$$2 C_2 H_2 + 3 O_2 = 4 CO + 2 H_2 O + Heat$$

In the above equation, a chemical change is taking place as the gas welding torch does its work. This is what is happening: two molecules of acetylene (C_2H_2) combine with three molecules of oxygen (O_2). The result of this new combination of molecules is four molecules of carbon monoxide (CO), plus two molecules of water vapor (H_2O) plus heat.

Carbon monoxide is a very unstable gas. It will unite readily with oxygen to form carbon dioxide. The carbon dioxide accounts for the fact that surrounding the cone of the welding flame is an area where the flame is of lesser intensity.

It is in this area that the carbon monoxide is mixing with the atmosphere oxygen to form carbon dioxide. The layer of carbon monoxide tends to keep the molten weld metal from oxidizing. The carbon monoxide absorbs any free oxygen present.

The chemical action in the outer flame becomes:

$$2\ CO + O_2 = 2\ CO_2 + heat$$

The oxygen, in this case, comes from the atmosphere surrounding the welding flame.

This principle must be remembered when welding in a confined space in which a free movement of air cannot exist above the torch tip; under these conditions more oxygen will need to be fed to the torch tip in order that a neutral flame may be maintained.

28-11. BLAST FURNACE OPERATIONS

The blast furnace is used to convert iron ores to pig iron.

The following raw materials are fed to the furnace:

ORES	% IRON
Hematite (red iron) Fe_2O_3	70
Magnetite (black) Fe_3O_4	72.4
Limonite (brown) $Fe_2O_3\ H_2O$	60
Siderite Iron Carbonate $FeCo_3$	48.3

Hematite, an important ore, is obtained mostly from the Birmingham and the Lake Superior districts including the Marquette, Menominee, Gogebic, Vermillion, Mesabi, and Cayuna ranges.

FLUX

A good flux must melt and unite with impurities, called gangue, and carry them away in form of slag. Sand and Alumina are chief impurities. (Acid)
Flux is Limestone. (Basic)

FUEL

The fuel used must melt the charge and furnish heat for the reactions in the furnace. It must be low in phosphorus and sulphur.
Coke is ideal for this purpose.

AIR

Preheated compressed air is forced into the lower part of the furnace.

28-10. FLAME CHARACTERISTICS: HEAT UNITS AND FLAME TEMPERATURE OF VARIOUS FUEL GASES

GAS	CHEMICAL FORMULA	BTU PER CU. FT.	FLAME TEMPERATURE DEG. F. WITH AIR	FLAME TEMPERATURE DEG. F. WITH PURE OXYGEN
Acetylene	C_2H_2	1475	4800	5600
Hydrogen	H_2	275	4000	4390–5200
City Gas	- - -	600	3750	4400
Propane	C_3H_8	2520	3800	5300–5610
Butane	C_4H_{10}	3250	3900	5400
Natural Gas	CH_4 & H_2	1000	3800	5025
Mapp	- - -	2400	2680	5300

The main chemical reactions in the blast furnace are:

1. $C + O_2 \rightarrow CO_2 + heat$ In lower zone of furnace
2. $CO_2 + C \rightarrow 2CO -- heat$ the CO acts as a reduc-
3. $Fe_2O_3 + 3CO \rightarrow 3CO_2 + 2Fe + heat$ ing agent.
4. $MnO + CO \rightarrow CO_2 + Mn + heat$ In upper zone of furnace.
5. $SO_2 + CO \rightarrow CO_2 + Si + heat$

The Al_2O_3, CaO, and MnO go through both zones unchanged as not enough is furnished to cause the (4) & (5) reactions to go to completion.

6. $Fe_2O_3 + 3C \rightarrow 3CO + 2Fe -- heat$ In lower zone as much heat
7. $CaCO_3 + heat \rightarrow CaO + CO_2$ is required in upper part
     and goes practically to com-
     pletion (at about 1500 deg.F.).

8. $MgCO_3 + heat \rightarrow MgO + CO_2$
9. $CaSO_4 + 2C \rightarrow CaS + 2CO_2$ In lower zone and only part-
     ly complete.

10. $CaO + Al_2O_3 \rightarrow CaO . Al_2O_3$
11. $CaO + SiO_2 \rightarrow CaO . SiO_2$

MgO will form similar products. These products plus CaS form the slag.

SUMMARY

$CaCO_3 + heat \rightarrow CaO + CO_2$
$MgCO_3 + heat \rightarrow MgO + CO_2$
$Fe_2O_3 + 3CO \rightarrow 2Fe + 3CO_2$ In reduction or upper zone.
$SiO_2 + 2CO \rightarrow Si + 2CO_2$
$MnO + CO \rightarrow Mn + CO_2$

TECHNICAL DATA

$C + O_2 \rightarrow CO_2$
$CO_2 + C \rightarrow 2CO$
$Fe_2O_3 + 3C \rightarrow 3CO + 2Fe$
$CaO + Al_2O_3 \rightarrow CaO + Al_2O_3$ In lower or melting zone.
$CaO + SiO_2 \rightarrow CaO + Si O_2$
$CaSO_4 + 2C \rightarrow CaS + 2CO_2$

Pure Iron melts at 2700 deg. F. and, with impurities at lower temperatures.

Iron and slag separate in bottom of blast furnace because slag being lighter floats on top of the molten iron. Impurities such as Silicon Manganese and Carbon are soluble in iron and remain in the iron. Iron Sulphide and Iron Phosphide are also soluble in iron so must be kept as low as possible.

Blast furnace, in production of pig iron, must: (1) deoxidize the iron ore; (2) melt the iron; (3) melt the slag; (4) carburize the iron; (5) separate the iron from slag.

BLAST FURNACE

The furnace charge and the materials resulting are shown below:

MATERIALS IN		MATERIALS OUT	
ORE		PIG IRON	
Fe_2O_3	$CaSO_4$	Fe	FeS
SiO_2	$Ca_3(PO_4)_2$	C	Fe_3P
Al_2O_3	$MgCO_3$	Si	
MnO			
FLUX		SLAG	
$CaCO_3$		Al_2O_3	SiO_2
		CaO	CaS
FUEL			
C			
AIR		GAS	
O_2 plus N_2		CO_2	CO
		N_2	

28-12. CHEMICAL REACTIONS IN THE MANUFACTURE OF IRON AND STEEL

Pig iron, now graded by chemical analysis, is the starting product for wrought iron, cast iron, malleable iron and steel.

CAST IRON

Cast iron is pig iron remelted and somewhat refined in Cupola, and cast in its final form.

WROUGHT IRON

Wrought iron is produced in a puddling furnace.

Wrought iron is refined at a temperature below the melting temperature.

Wrought iron is quite pure iron but contains slag.

PUDDLING FURNACE

Pig iron is charged into hearth and melted.

Iron ore is charged into hearth and heating continues.

$$3Si + 2Fe_2O_3 \rightarrow 4Fe + 3S\ O_2 + A$$
$$3Mn + Fe_2O_3 \rightarrow 2Fe + 3MnO + A$$

after enough heat is produced then

$$3C + Fe_2O_3 \rightarrow 2Fe + 3CO + A$$
$$3CO + 3/2O_2 \rightarrow 3CO_2 + A$$

Phosphorus and Sulphur cannot be removed in this process.

The melting point is lowered as impurities are burned out and metal becomes pasty and is puddled into balls 100-200 lbs. and passed through rollers to squeeze out the slag.

The composition of pig iron, as compared to wrought iron or steel, follows:

	PIG IRON (in %)	W. IRON or STEEL (in %)
Total Carbon. (C)	3.5 — 4.25	.02 — 1.6
Silicon (Si)	1 — 3	.01 — .30
Manganese (Mn)	.5 — 1	.01 — 1
Sulphur (S)	.06	.04 — .06
Phosphorus (P)	.08 — 1	.04 — .10
Iron (Fe)	91 — 94	99

W. Iron contains 1 to 3 percent slag

28-13. OXYGEN CUTTING CHEMISTRY

The chemical reactions of the preheating flames are the same as those in the oxyacetylene torch, and the products of combustion are the same. However, the cutting torch uses a jet of pure oxygen. When the metal to be cut is heated to a cherry red by the preheating flames, the oxygen jet is started (usually a lever valve). The action is

$$O_2 + Fe_2 = FeO,\ FeO_2,\ Fe_2O_3,\ \text{etc.},$$

indicating all types of iron oxides.

When the oxidation starts the chemical combination gives off heat (releases heat). This release of heat is called an exothermal action. However, the heat provided is usually not enough to permit the cutting to be self-supporting. The preheat flames are usually kept burning to provide the needed additional heat.

28-14. THERMIT REACTION CHEMISTRY

The welding of parts or making of castings using the exothermic thermit process has been used for many years. The chemical reaction of the steel welding or casting process is:

$$8\ Al + 3\ Fe_3O_4 \rightarrow 9\ Fe + 4\ Al_2O_3 + \text{heat}$$

The aluminum and iron oxide mixture must be heated to approximately 2200 deg. F. to start the reaction described above.

Copper, nickel and manganese have also been welded or cast using this process. The word Thermit is a registered trade name commonly used to identify this process.

28-15. GASES

Gases are an important part of many welding processes.

Gas welding (or flame welding) uses many different gases including:

Fuel Gases:
  Acetylene
  Propane
  Hydrogen
  Natural Gas
Combustion Supporting Gases:
  Air
  Oxygen
Inert Gas Arc Welding Gases:
  Helium
  Argon
  Nitrogen
  Carbon Dioxide

The properties of these gases and their actions and reactions must be accurately understood if good welding results are to be obtained.

28-16. MANUFACTURE OF GASES

The manufacture or production of gases used in various welding systems must meet two standards:
  1. High purity.
  2. Low moisture content.
Three of the gases are generally produced by the liquefication and distillation of air process. These gases are:
  1. Oxygen
  2. Nitrogen
  3. Argon
The other gases and liquids are produced by various processes.

28-17. OXYGEN

Oxygen is the most abundant element. It is essential to animal and plant life. It is a vital part of all gas welding and cutting processes which involve combustion.

Symbol   O
Atomic Number   8
Atomic Weight   16.0

Specific Gravity 1.105 (based on air having a specific gravity of 1 therefore the gas is heavier than air).

Liquefies at   -182.9 deg. C
                    -297.2 deg. F
Solidifies at   -219 deg. C
                    -362 deg. F

LOX is a trademark for a liquid oxygen explosive. This symbol should NEVER be used to denote liquid oxygen.

Fig. 28-5 illustrates the effect of temperature on pressure in an oxygen cylinder.

TEMPERATURE OF OXYGEN, DEG. F.	GAUGE READING PSIG
100	2147
90	2098
80	2049
70	2000
60	1951
50	1902
40	1853
30	1804
20	1755
10	1706
0	1657

Fig. 28-5. Effect of temperature on the oxygen pressure within a fully charged oxygen cylinder (2000 psig at 70 deg. F.)

28-18. ACETYLENE

Acetylene is a hydrocarbon, that is, it is a substance formed of hydrogen and carbon. Its chemical formula is C_2H_2. It is made by adding calcium carbide to water which creates acetylene gas and calcium hydroxide. The chemical equation is:

$$CaC_2 + 2 H_2O = C_2H_2 + Ca (OH)_2$$

Acetylene C_2H_2 must never be stored as a free gas (in the gas state) at pressures above 15 psig because it is unstable at or above this pressure and may disintegrate violently. If it is dissolved in acetone, the pressure may then be safely increased to approximately 250 psig. However, there must be an absolute minimum of free gas space. It is for this reason that acetylene cylinders are filled with a porous material.

Acetylene has a great range of flammability. That is, it will burn with the oxygen in air in a great range of mixtures. This ability makes the storing and handling of acetylene dangerous if any leaks develop.

Symbol C_2H_2
Molecular Weight 26.038
Specific Gravity 0.907as
  compared to air standard of 1;
  therefore acetylene is lighter
  than air.
Liquefies at -115.6 deg.F.
Flame Temperature,
  Oxyacetylene 5900 deg. F.
Btu/cu. ft. 1483

Oxyacetylene flame burns at the rate of approximately 25 feet per second (about 20 miles/hour) and this speed is called flame propagation. To have the torch operate successfully, the gas speed coming out the torch tip orifice must be equal to this burning velocity.

28-19. PROPANE

Propane is a hydrocarbon with the chemical arrangement of C_3H_8. It is found in petroleum and natural gas.

The chemical equation when combined with oxygen is:

$$C_3H_8 + 5\ O_2 \rightarrow 3\ CO_2 + 4\ H_2O$$

Symbol C_3H_8
Molecular Weight 44.09
Specific Gravity 1.53 (based
  on air = 1). Therefore, it is heavier
  than air
Liquefies at -48 deg. F.
Flame Temperature,
  Oxypropane 5650 deg. F.
Btu/cu. ft. 2600

Flame Propagation
  with Oxygen 12 ft/sec.

Even though propane has almost twice as much heating value as acetylene per cu. ft. its speed of burning is approximately one half of acetylene. Propane gas produces about one half the heating value per unit of time as acetylene. Most propane tips are of the multiple orifice type, to provide better flame conditions due to the slow flame propagation. The lower temperature of the oxypropane flame means that for welding, oxyacetylene is p r e f e r r e d while oxypropane is useful for preheat flames in oxycutting operations.

28-20. HELIUM

Helium is an inert element and therefore an inert gas. It does not combine with other elements. This property makes helium a good shielding gas for all kinds of arc welding.

It has the chemical symbol He; atomic number 2; atomic weight 4.003, boiling temperature -432 deg. F. It is a light gas (low density) and quickly leaves the vicinity of the arc when being used as a shielding gas. Argon gas is approximately nine times as heavy, and therefore less argon gas is needed to shield the arc. Helium formerly was used extensively as a shielding gas for aluminum and magnesium welds. However, a r g o n and argon-helium mixtures are in more general use at the present time.

28-21. ARGON

Argon is a colorless, odorless, and tasteless gas.

A cubic foot of argon at 70 deg. F. and at atmospheric pressure weighs .1034 pounds (or 9.671 cubic feet weighs one pound). Argon gas is therefore heavier than air. This gas will condense into a liquid at -302.55 deg. F. (-185.86 deg. C) and will become a solid at -308.67 deg. F. (-189.26 deg. C) assuming atmospheric pressure.

It will dissolve in water at the ratio of four volumes to 100 volumes of water. Argon gas is chemically inert. It will not burn or explode. It will not react (unite) chemically with any other elements or compounds.

Commercial argon is obtainable in two grades:

1. Incandescent lamp grade.
2. Welding grade.

The incandescent lamp grade contains higher amounts of nitrogen and is the most common gas used in electric light bulbs.

The welding grade has the following composition:

.001% maximum oxygen
.001% maximum hydrogen
.008% maximum nitrogen
99.990% minimum argon

Argon is ideal for both gas tungsten-arc welding and gas metal-arc welding shielding purposes.

Argon is manufactured by:

1. Liquefying air.
2. Distilling air (oxygen, nitrogen, argon, etc., have different boiling temperatures and can be separated).
3. Rectification.
4. Reliquefying.
5. Distillation (to reduce the oxygen and nitrogen proportions to the correct limits).
6. Drying.

Argon is passed through a desiccant (moisture absorber - usually alumina) under high pressure and enough moisture is removed to produce a dew point temperature of -76 deg. F. (-60 deg. C). This means less than .02 milligrams per liter and the moisture content is less than .00084 percent by weight.

28-22. NITROGEN

Nitrogen is an odorless, colorless, and tasteless gas which will not burn or explode.

At normal temperatures it will not combine with other elements or compounds. At high temperatures it will combine with oxygen, hydrogen and certain metals (magnesium and chromium).

Nitrogen gas weighs .07247 pounds per cubic foot at 70 deg. F., and at atmospheric pressure (13.8 cu. ft. per lb.). It therefore is approximately the same weight as air.

The gas will liquefy at -320.46 deg. F. (-195.808 deg. C) and will become a solid at -346.04 deg. F. (-210.02 deg. C).

It is manufactured by:
1. Liquefication of air.
2. Distillation.
3. Rectification.
4. Drying.

28-23. CARBON DIOXIDE

Carbon dioxide is used in welding as a shielding gas. It is not an inert gas. It is composed of 1 part carbon and two parts oxygen (CO_2). This molecule is an oxidizing agent equal in oxidation action to a 91 percent argon - 9 percent oxygen mixture. Therefore the welding wire must have oxidation elements (sometimes called deoxidizers). Two common ones are manganese and silicon. It is known by test that approximately half of these two elements are turned into oxides as the welding wire passes through the arc.

In the arc, the carbon dioxide tends to break down. About 7.7 to 12 percent of the gas turns into Carbon Monoxide.

a toxic gas. Fortunately, the carbon monoxide can exist only in the high temperature zones and it recombines into carbon dioxide rapidly when good ventilation practices are followed. At 7 in. from the arc, only .01 percent of the gas ($\frac{1}{10,000}$) is carbon monoxide. This is 1/10 of the safe limit for carbon monoxide.

Carbon dioxide is usually stored in the liquid form in cylinders. The cylinders are filled approximately two thirds full. The pressure in the cylinder is a function of the vapor pressure characteristic of carbon dioxide.

For example:

Temp. deg. F.	Pressure in cylinder, psig
0	290
30	476
70	835
100	1450

Carbon dioxide gas must be kept dry as moisture in the gas creates porous welds. Only welding grade carbon dioxide should be used. This gas is so dry that the gas must be cooled to -40 deg. F. before moisture will start to condense out of the gas. This condition means that the gas has only .0065 percent moisture (1 percent = .01 therefore .0065 percent = .01 x .0065 = .000065 or 65 parts per million). Gas with a -80 deg. F. is much better in this respect, because it has only 30 parts per million of moisture .003 percent.

28-24. DRILL SETS AND SIZES

Drill bits come in sets of four different sizes. These are fractional sets, number sets, letter sets, and metric sets. The shank of each drill, when large enough, carries a stamped identification of the drill size. The size is given as a number, fraction, letter, or decimal.

Depending on the quality and use, drill bits are available in either high carbon steel (least expensive) or alloy steel, marked HSS (high speed steel), for high speed use.

Fractional drills usually begin with size 1/16 in. and go to 1/2 in. in steps of 1/64 in. Larger fractional sizes are available.

Number drill sets begin with No. 1 (0.228) and go to No. 80 (0.0135). Most commonly used number drill sets include No. 1 through No. 60. In number drills, the higher the number the smaller the drill.

Letter drill sets come in sizes "A" (0.234) to "Z" (0.413). Metric drill sets include sizes from 0.100 mm (0.0039 in.) to 25.50 mm (1.003 in.).

28-25. TABLE OF DECIMAL SIZES FOR NUMBER DRILLS

DRILL NUMBER	DECIMAL SIZE	DRILL NUMBER	DECIMAL SIZE	DRILL NUMBER	DECIMAL SIZE
1	.2280	28	.1405	55	.0520
2	.2210	29	.1360	56	.0465
3	.2130	30	.1285	57	.0430
4	.2090	31	.1200	58	.0420
5	.2055	32	.1160	59	.0410
6	.2040	33	.1130	60	.0400
7	.2010	34	.1110	61	.0390
8	.1990	35	.1100	62	.0380
9	.1960	36	.1065	63	.0370
10	.1935	37	.1040	64	.0360
11	.1910	38	.1015	65	.0350
12	.1890	39	.0995	66	.0330
13	.1850	40	.0980	67	.0320
14	.1820	41	.0960	68	.0310
15	.1800	42	.0935	69	.02925
16	.1770	43	.0890	70	.0280
17	.1730	44	.0860	71	.0260
18	.1695	45	.0820	72	.0250
19	.1660	46	.0810	73	.0240
20	.1610	47	.0785	74	.0225
21	.1590	48	.0760	75	.0210
22	.1570	49	.0730	76	.0200
23	.1540	50	.0700	77	.0180
24	.1520	51	.0670	78	.0160
25	.1495	52	.0635	79	.0145
26	.1470	53	.0595	80	.0135
27	.1440	54	.0550		

28-26. TABLE OF TAP DRILL SIZES

TAP		
Dia. of Screw	Threads per Inch	TAP DRILL
6	32	No. 36
6	40	No. 33
8	32	No. 29
8	36	No. 29
10	24	No. 25
10	32	No. 21
12	24	No. 16
12	28	No. 14
1/4	20	No. 7
1/4	28	No. 3
5/16	18	F
5/16	24	I
3/8	16	5/16
3/8	24	Q
7/16	14	U
7/16	20	25/64
1/2	13	27/64
1/2	20	29/64

28-27. TABLE OF WIRE AND SHEET METAL GAUGES

Gauge Number	American Wire Gauge (AWG) and Brown & Sharpe (B & S) for Non-Ferrous Metals Decimal Size	Manufacturers' Standard Gauge for Sheet Steel Decimal Size
0000	.460	
000	.410	
00	.365	
0	.325	
2	.258	
4	.204	.2242
6	.162	.1943
8	.128	.1644
10	.102	.1345
12	.081	.1046
14	.064	.0747
16	.051	.0598
18	.040	.0478
20	.032	.0359
22	.0253	.0299
24	.0201	.0239
26	.0159	.0179
28	.0126	.0149
30	.0100	.0120
32	.0080	.0097

28-28. COLOR CODES FOR MARKING STEELS

From Bureau Of Standards' Simplified Practice Recommendation R166-37)

S.A.E. Number	Code Color	S.A.E. Number	Code Color	S.A.E. Number	Code Color	S.A.E. Number	Code Color
	CARBON STEELS	2115	Red and bronze	T1340	Orange and green	3450	Black and bronze
1010	White	2315	Red and blue	T1345	Orange and red	4820	Green and purple
1015	White	2320	Red and blue	T1350	Orange and red		CHROMIUM STEELS
X1015	White	2330	Red and white		NICKEL-CHROMIUM STEELS	5120	Black
1020	Brown	2335	Red and white	3115	Blue and black	5140	Black and white
X1020	Brown	2340	Red and green	3120	Blue and black	5150	Black and white
1025	Red	2345	Red and green	3125	Pink	52100	Black and brown
X1025	Red	2350	Red and aluminum	3130	Blue and green		CHROMIUM-VANADIUM STEELS
1030	Blue	2515	Red and black	3135	Blue and green	6115	White and brown
1035	Blue		MOLYBDENUM STEELS	3140	Blue and white	6120	White and brown
1040	Green	4130	Green and white	X3140	Blue and white	6125	White and aluminum
X1040	Green	X4130	Green and bronze	3145	Blue and white	6130	White and yellow
1045	Orange	4135	Green and yellow	3150	Blue and brown	6135	White and yellow
X1045	Orange	4140	Green and brown	3215	Blue and purple	6140	White and bronze
1050	Bronze	4150	Green and brown	3220	Blue and purple	6145	White and orange
1095	Aluminum	4340	Green and aluminum	3230	Blue and purple	6150	White and orange
	FREE CUTTING STEELS	4345	Green and aluminum	3240	Blue and aluminum	6195	White and purple
1112	Yellow	4615	Green and black	3245	Blue and aluminum		TUNGSTEN STEELS
X1112	Yellow	4620	Green and black	3250	Blue and bronze	71360	Brown and orange
1120	Yellow and brown	4640	Green and pink	3312	Orange and black	71660	Brown and bronze
X1314	Yellow and blue	4815	Green and purple	3325	Orange and black	7260	Brown and aluminum
X1315	Yellow and red	X1340	Yellow and black	3335	Blue and orange		SILICON-MANGANESE STEELS
X1335	Yellow and black		MANGANESE STEELS	3340	Blue and orange	9255	Bronze and aluminum
	NICKEL STEELS	T1330	Orange and green	3415	Blue and pink	9260	Bronze and aluminum
2015	Red and brown	T1335	Orange and green	3435	Orange and aluminum		

28-29. COLOR CODE FOR MARKING NONFERROUS METALS

Metal	Grade	Specifications			Markings	
		Federal	Navy	A.S.T.M.	Background	Stripe
ALCLAD					Lead & tan	Black
ALUMINUM	B	QQ-A-451	46-A-2		Lead & tan	Brown
soft	A		46-A-3		Lead & tan	White
half hard	B		46-A-3		Lead & tan	Green
three-quarters	C		46-A-3		Lead & tan	Yellow
hard	D		46-A-3		Lead & tan	Red
ALUMINUM ALLOY, soft			46-A-4		Lead & red	None
half hard	B		47-A-4		Lead & red	Blue
hard	D		47-A-4		Lead & red	Brown
heat treated			47-A-3		Lead & red	Black
heat treated and rolled			47-A-3		Lead & red	White
Navy alloy A-2					Lead & white	None
Navy alloy A-2, heat treated					Lead & white	Black
Navy alloy No. 2, annealed					Lead & white	Brown
Navy alloy No. 2, annealed, heat treated					Lead & white	Green
Navy alloy No. 4, soft					Lead & yellow	Blue
BRASS			46-B-26		Brown & green	Black
Commercial, type 1	A	QQ-B-611		B15-18	Brown & green	Blue
Commercial, type 1	B	QQ-B-611		B16-18	Brown & green	Lead
Commercial, type 1	C	QQ-B-611			Brown & green	Tan
Commercial, type 2	B	QQ-B-611	47-B-2		Brown & green	Red
Commercial, type 2	C	QQ-B-611			Brown & green	White
Naval, rolled			46-B-6		Brown & green	Yellow
BRONZE	1	QQ-B-701	46-B-25		Brown & tan	Black
	2	QQ-B-701	46-B-25		Brown & tan	Blue
	3	QQ-B-701	46-B-25		Brown & tan	Green
	4	QQ-B-701	46-B-25		Brown & tan	Lead
	5	QQ-B-701	46-B-25		Brown & tan	Red
	6	QQ-B-701	46-B-25		Brown & tan	White
Aluminum			46-B-19		Brown & white	Black
Journal			46-B-9		Brown & white	Blue
Manganese		QQ-B-721	46-B-16	B7-27	Brown & white	Green
Muntz Metal					Brown & white	Lead
Phosphor			46-B-14		Brown & white	Tan
Rivet					Brown & white	Red
Special					Brown & white	Yellow
COPPER, hard drawn		QQ-C-501	47-C-2		Red & white	Black
Soft drawn		QQ-C-501	47-C-2		Red & white	Blue
Phosphor	A	QQ-C-571	46-C-3		Red & yellow	Black
Silicon		QQ-C-581	46-C-2	B53-27	Red & yellow	Brown
Nickel		QQ-C-541	46-M-7		Red & yellow	Green
NICKEL	A	QQ-N-301	46-N-2	B39-22	Green & lead	Black
	B	QQ-N-301	46-N-2	B39-22	Green & lead	Blue
MANGANESE-NICKEL			46-N-3		Green & tan	Black
NICKEL-SILVER	A	QQ-N-321	46-S-3		Green & tan	Lead
TIN, phosphor		QQ-T-351	46-T-2	B51-27	Green & yellow	Black
ZINC	A	QQ-Z-351	46-Z-1		Green & white	Black
	B	QQ-Z-351	46-Z-1		Green & white	Blue
	C	QQ-Z-351	46-Z-1		Green & white	Brown
	D	QQ-Z-351	46-Z-1		Green & white	Lead
	E	QQ-Z-351	46-Z-1		Green & white	Tan

28-30. HEALTH HAZARDS

As a part of your instruction in welding, it is important for you to be familiar with, and to be able to recognize conditions that may be hazardous to health. It is important too, to remember that the best way to attack the hazards is to eliminate or control the conditions that are responsible.

The oxygen-acetylene flame is generally safe, but there is a carbon monoxide (CO) problem in a poorly ventilated place. The reaction tends to stop at CO formation and needs heat to change the gas to carbon dioxide (CO_2). One must watch for cold metal (chilling the flame).

Coatings on metals are a problem when the metal is heated. Nitrogen dioxide, coated electrode fumes, and iron oxide fumes are examples.

Coated Surface Problems:

Red lead paint is often used outdoors for metal finishing and protection. Lead oxide fumes caused by burning lead paint coatings can produce acute lead poisoning.

Cadmium plate is frequently used on small parts. Cadmium oxide fumes at low levels produce a chronic condition, while at high levels the fumes are harmful to lungs and liver.

Silver brazing alloy fumes can be quite dangerous.

Termeplate is a metallic lead coating and is dangerous when heated.

Flux Problems:

Fluorides in fluxes are common. Fluoride fumes are harmful.

Manganese dioxide is not too toxic. It may cause trouble if ventilation is poor.

When inert gas arc welding aluminum, there is a possibility of real trouble as the ultraviolet frequency is right to form ozone. This gas is most toxic and may cause severe lung and body damage. It is irritating and causes coughing. It is advisable to keep the arc from traveling too far from overhead ventilation.

Ultraviolet rays are harmful to the welder's vision, but safety goggles with proper lenses will provide protection.

Ionizing radiations markedly increase the possibility of eye cataracts. Even small amounts of ultraviolet may generate ionization and accelerate cataracts. Those operating welding equipment, as well as helpers and other people in the area must take precautions at all times to prevent eye injury.

The use of carbon dioxide creates a carbon monoxide (CO) problem as it breaks down in the arc.

Some people are sensitive to beryllium - even very small amounts. Therefore, any operation involving beryllium must be contained.

Cobalt should be handled about the same as beryllium.

Thorium is toxic. Therefore, when thoriated electrodes are used, an alpha emission is produced and causes an ionization effect. Ventilate well.

Reducing furnaces may have carbon monoxide (CO) emissions. Such furnaces should be well vented at both the charging end and discharge end.

When vacuum furnaces or welding chambers are used, the pump exhaust must be vented away from people.

Oil smoke is a problem. The aromatics produced can be dangerous.

Fig. 28-6 shows a table of safe limits for some welding fumes.

NEVER FAIL TO PROTECT THE EYES, SKIN, AND RESPIRATORY SYSTEM. PROVIDE ADEQUATE VENTILATION DURING ALL WELDING, CUTTING, BRAZING, AND SOLDERING OPERATIONS.

| MATERIAL | GASES | | MILLION PARTS |
	ppm	mg/m^3	per Cu. Ft.
Acetylene	1000		
Beryllium		.002	
Cadmium Oxide fumes		.1	
Carbon Dioxide	5000		
Copper fumes		.1	
Iron Oxide fumes		10.0	
Lead		.2	
Manganese		5.0	
Nitrogen dioxide	5.0		
Oil Mist		5.0	
Ozone	.1		
Titanium Oxide		15.0	
Zinc Oxide fumes		5.0	
Silica, crystalline			2.5
Silica, amorphous			20.0
Silicates:			
Asbestos			5.0
Portland Cement			50.0
Graphite			15.0
Nuisance Dust			50.0

Fig. 28-6. A table of safe limits for some welding fumes. All gases tend to reduce oxygen by replacement. Such gases as argon, helium, carbon dioxide, etc., present this danger.

28-31. OCCUPATIONAL SAFETY AND HEALTH ACT (OSHA)

The Occupational Safety and Health Act (OSHA) provides many regulations and controls for the safe operation, handling and storage of welding supplies and equipment.

Instructional personnel should be familiar with OSHA requirements. Information concerning safety for any particular trade or industry can be gotten from the Department of Labor in the state where the program is located.

Publications are available also from the U.S. Department of Health, Education, and Welfare, Public Health Service, Center for Disease Control, National Institute for Occupational Safety and Health, Robt. A. Taft Laboratories, 4676 Columbia Parkway, Cincinnati, OH 45226.

Chapter 29
READING WELDING SYMBOLS

29-1. THE WELDING SYMBOL

The welding symbol developed by the American Welding Society (AWS) is used internationally. Currently, all structural, bridge, government, and nuclear fabricators use the symbols.

The symbol shown in Fig. 29-1 is used on drawings of parts to be welded. Whenever two or more pieces of a welded part (weldment) are placed to-

gether, their surfaces and edges form a joint. The drawing of the part to be welded indicates how the parts will be assembled and what type of welded joint will be made. See Fig. 29-2 for the types of welded joints and the types of welds used on the various joints.

The complete welding symbol will tell the welder how to prepare the base metal, the welding process to use, the

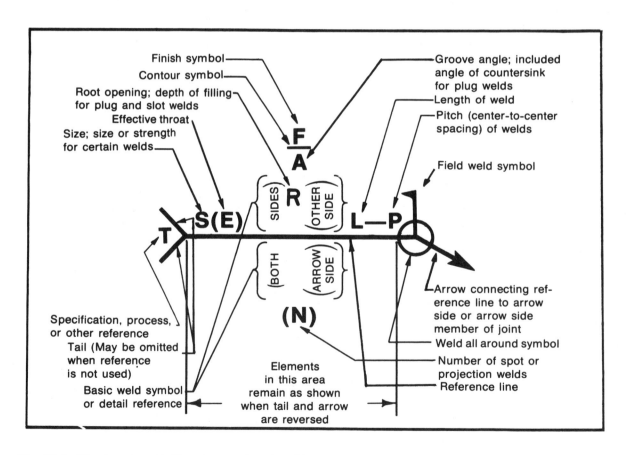

Fig. 29-1. The American Welding Society (AWS) welding symbol. Standard locations of information on a welding symbol are marked. (© 1979 by American Welding Society. Reprinted with permission.)

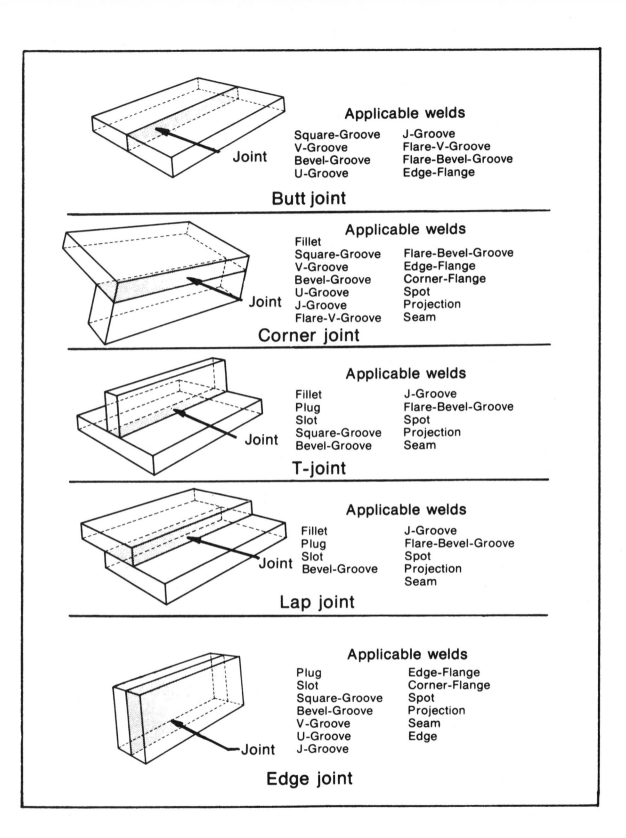

Applicable welds

Butt joint
- Square-Groove
- V-Groove
- Bevel-Groove
- U-Groove
- J-Groove
- Flare-V-Groove
- Flare-Bevel-Groove
- Edge-Flange

Joint

Butt joint

Applicable welds

- Fillet
- Square-Groove
- V-Groove
- Bevel-Groove
- U-Groove
- J-Groove
- Flare-V-Groove
- Flare-Bevel-Groove
- Edge-Flange
- Corner-Flange
- Spot
- Projection
- Seam

Joint

Corner joint

Applicable welds

- Fillet
- Plug
- Slot
- Square-Groove
- Bevel-Groove
- J-Groove
- Flare-Bevel-Groove
- Spot
- Projection
- Seam

Joint

T-joint

Applicable welds

- Fillet
- Plug
- Slot
- Bevel-Groove
- J-Groove
- Flare-Bevel-Groove
- Spot
- Projection
- Seam

Joint

Lap joint

Applicable welds

- Plug
- Slot
- Square-Groove
- Bevel-Groove
- V-Groove
- U-Groove
- J-Groove
- Edge-Flange
- Corner-Flange
- Spot
- Projection
- Seam
- Edge

Joint

Edge joint

Fig. 29-2. Basic types of joints. (© 1979 by American Welding Society. Reprinted with permission.)

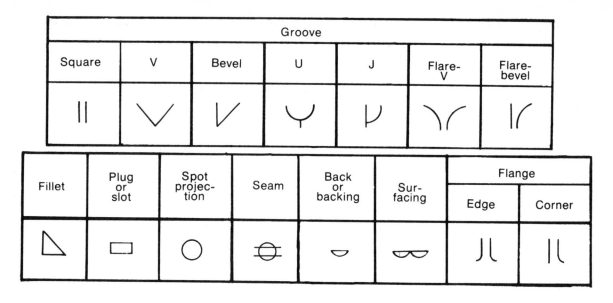

Groove						
Square	V	Bevel	U	J	Flare-V	Flare-bevel
‖	⋁	⋁	⋃	⋃	⋎	⋓

Fillet	Plug or slot	Spot projection	Seam	Back or backing	Sur-facing	Flange	
						Edge	Corner
◺	▭	◯	⊖	⌣	⌣	⏝	⏝

Fig. 29-3. Basic weld symbols. These are a part of the complete welding symbol.
(© 1979 by American Welding Society. Reprinted with permission.)

method of finishing, and much more information regarding each weld. Dimensions on a welding symbol may be in SI metric units or conventional U. S. units.

A complete welding symbol will have the "basic weld symbol" to show what type of weld to use. Fig. 29-3 shows the basic weld symbol used by AWS.

Information given in each part or area of the welding symbol will be explained in later paragraphs. A number of weld drawings, with their corresponding welding symbols, will be shown to illustrate the information given in the various areas of the complete welding symbol. The edges of the weld joint will be shown in red as it would be prepared and fitted up prior to welding. A completed weld for the welding symbol will also be shown to illustrate the use of the welding symbol.

29-2. THE REFERENCE LINE, ARROWHEAD, AND TAIL

The reference line shown in Fig. 29-4 is always drawn as a horizontal line. It is placed on the drawing near each welded joint. All other information to be given on the welding symbol is shown above or below this horizontal reference line. All information shown on a complete welding symbol is always shown in the same position as indicated in Fig. 29-1 and reads from left to right.

The arrow may be drawn from either end of the reference line. The welding symbol may appear in any view of the welding drawing. The arrow always touches the line which represents the welded joint.

The tail is used only when necessary. If used, it may give information on specifications, process, or other details required but not shown on the welding symbol. A number, such as 1, 2, 3, etc., may be used in the tail to refer the user to a note elsewhere on the drawing.

Companies may use their own number or letter codes in the tail to indicate the welding process, procedure, finishing method, or company specification.

If no tail is used, somewhere on the

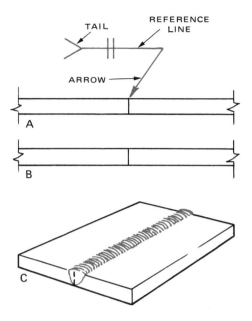

Fig. 29-4. Reference line, arrow, and tail of welding symbol. A—Welding drawing and welding symbol. B—Preparation of metal. C—Completed weld.

drawing there is a note such as, "Unless otherwise specified, all welds will be made in accordance with Specification No._____."

29-3. BASIC WELD SYMBOLS

The basic weld symbol shown on the complete welding symbol indicates the type of weld made on a weld joint. It is also a miniature drawing of the metal edge preparation, if any, required prior to welding.

Fig. 29-5 illustrates how some of the various types of weld symbols shown in Fig. 29-3 are used on a welding symbol. The vertical line used with a fillet, bevel, or J-groove weld is always drawn to the left.

29-4. THE ARROW SIDE AND OTHER SIDE

On the drawing of a welded part, the arrow of the welding symbol touches a line to be welded. The metal has two sides. The side of the metal which the arrow touches is always the arrow side. The opposite surface from the arrow is called the other side.

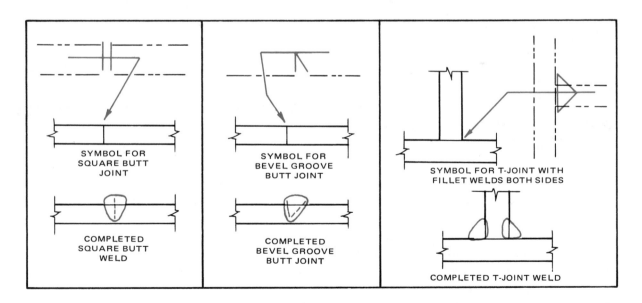

Fig. 29-5. Comparing some welding symbols and actual welds. Phantom lines are not shown on a basic weld symbol. They are used here, however, to illustrate that the basic weld symbol is a miniature drawing of the edge preparation and/or type of weld used.

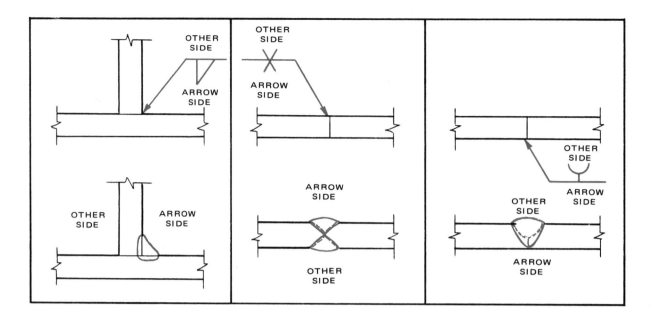

Fig. 29-6. Placement of weld symbol for welding on arrow side and on other side. Note: Side of metal that arrow touches is always the arrow side.

On many weldments there is no inside or outside, top or bottom, left or right. To simplify the location of the weld, the terms arrow side and other side are used.

It is not always possible to place the welding symbol on the side to be welded. The drawing is sometimes crowded and complicated. See Fig. 29-6 for examples of the use of the "arrow side" and "other side" on the welding symbol.

29-5. ROOT OPENING AND GROOVE ANGLE

The space between the metals being welded at the bottom of the joint (root) is the root opening. This root opening may be specified on the drawing in metric units, in fractions of an inch or as a single-place decimal of an inch. The root opening size appears inside the basic weld symbol on the complete welding symbol.

The included angle or total angle of

the groove weld is shown above the basic weld symbol. See Fig. 29-7. When preparing the edges for welding, half the groove angle is cut on each piece so that when placed together they will total the angle shown.

When a bevel or J-groove weld is used, only one piece of metal is cut or ground. The arrow of the welding symbol is bent to point to the piece to be prepared. See Fig. 29-7, view D.

29-6. CONTOUR AND FINISH SYMBOLS

On the welding symbol, the shape or contour of the completed weld is shown as a straight or curved line between the basic weld symbol and the finish symbol. The straight contour line indicates that the weld bead is to be made as flat as possible. The curved line indicates a normal convex or concave weld bead. See Fig. 29-8.

If the weld is not to remain in an "as welded" condition, a finish symbol is used on the welding symbol. See

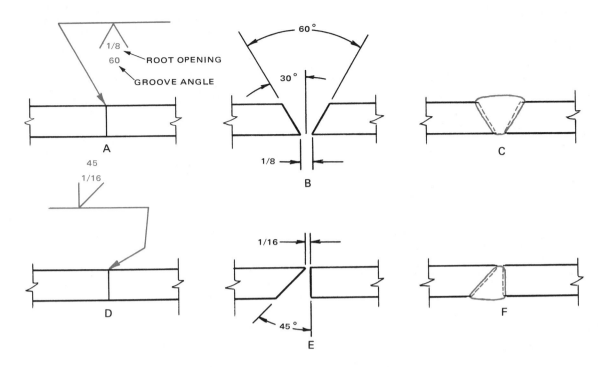

Fig. 29-7. *Root opening and groove angle. A and D show the weld symbol for a groove weld. B and E show the pieces cut and set up for welding. C and F show the completed weld. Note: Bend in arrow at D points to left piece, which is the part to be cut or machined.*

Fig. 29-8. The finish symbol indicates the method of finishing. A surface texture or degree of finishing may also be added if required. If all welds are to be finished in the same manner, a note on the drawing may indicate the

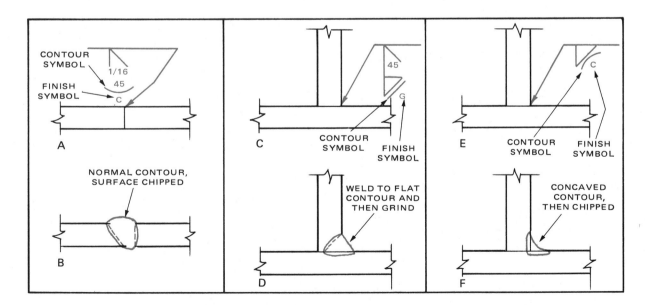

Fig. 29-8. *Weld contour and finish symbols. In A, C, and E, contour and finish symbols are shown on welding symbol; B, D, and F illustrate shape and finish of completed weld.*

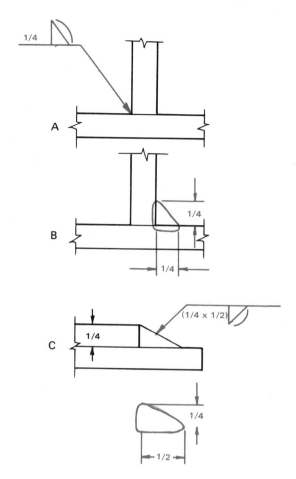

Fig. 29-9. Fillet weld size. A—A single dimension of 1/4 in. is shown. This indicates that both sides of the triangular fillet are 1/4 in. B—How weld at A will look. C—Two dimensions are shown. Size of each side of fillet is indicated by its relative size on welding drawing, as shown in color. D—How the weld will look.

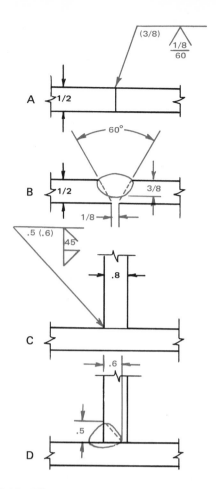

Fig. 29-10. Effective throat size. Effective throat or depth of weld penetration is shown in parenthesis at A and C. Note at B that effective throat size is less than metal thickness. At D, it is greater than groove depth.

finish used. Users of the finish symbol may create their own finish symbols.

The American Welding Society lists the following finish symbols: C - Chipping; G - Grinding; M - Machining; R - Rolling; H - Hammering.

29-7. SIZE OR STRENGTH OF THE WELD AND THE EFFECTIVE THROAT SIZE

The dimensions of the triangular shape of a fillet weld are shown to the left of the basic weld symbol. (See Fig. 29-1 for the placement of the size

dimension.) The size may be given as a fraction or a single-place decimal. The sides of a typical triangular fillet weld are equal in length. Thus, on the weld symbol for the typical fillet weld, only one dimension is given.

If the dimensions of the two sides of the triangular fillet are not equal, then two dimensions will be given, as shown in Fig. 29-9, views C and D. In the case of an unequal fillet weld, the shape of the fillet is shown on the welding drawing. The shape on the drawing will indicate to the welder which side of the fillet is the long dimension and which is the short dimension.

The effective throat or depth of the deposited weld metal is also shown to the left of the basic weld symbol.

When no size is shown for the single groove and symmetrical double grooved welds, complete penetration is required. See Fig. 29-10 for examples of effective throat size.

29-8. LENGTH AND PITCH OF THE WELD

In many welded parts it is not necessary to weld continuously from one end of the joint to the other. To save time and expense, where strength is not affected, short sections of weld may be spaced across the joint. This is called intermittent welding.

On intermittent welds, the length and pitch dimensions are used to indicate the length of each weld and how far apart to make each section of an intermittent weld. See Fig. 29-11 for examples of such welds.

On an intermittent weld, the pitch indicates the distance from the center of one weld to the center of the next.

Continuous and intermittent welds may be made on the same joint. In such a case, the drawing will use dimensions to show where each weld symbol's effectiveness begins and ends. See Fig. 29-12.

When intermittent welding is used between continuous welding, as in Fig. 29-12, a spacing different than the pitch size is used. This spacing is equal to the intermittent pitch minus the length of one intermittent weld. The spacing between the continuous and intermittent welds in Fig. 29-12, bottom drawing, equals the pitch minus the length, or 6" - 2" = 4", as shown.

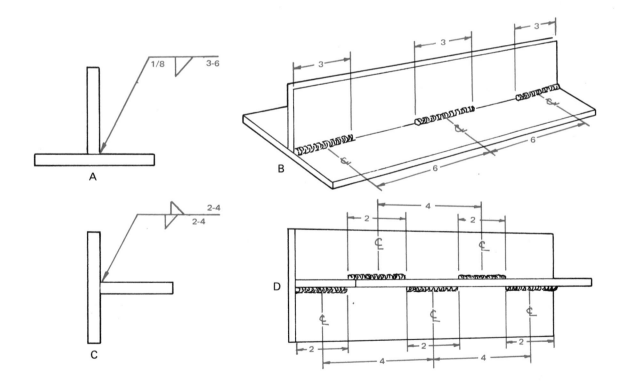

Fig. 29-11. Length and pitch dimensions of weld. A—Note placement of length (3) and pitch (6) on the welding symbol. B—Weld shows a series of 3 inch long welds which are 6 inches apart from center to center of the welds. C and D—Staggered weld. Notice staggered fillet symbols in C.

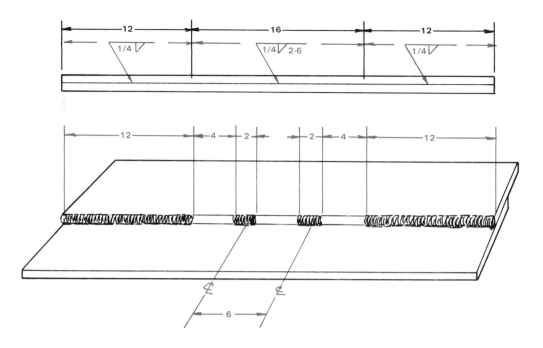

Fig. 29-12. Continuous and intermittent welds. Note that the dimensions on the top welding drawing limits use of welding symbol to distance shown. Note also that the spacing between continuous and intermittent weld is equal to pitch minus length of one intermittent weld (4 inches in this application).

29-9. BACKING WELDS AND MELT-THRU SYMBOLS

Weld joints that require complete penetration may be welded from both sides. A stringer bead (single pass weld without a weaving motion) may be all that is required on the side opposite a groove weld to insure complete penetration. In such cases, a back-

ing weld symbol may be used, Fig. 29-13.

The burn-through symbol is used when 100 percent penetration is required on one-side welds, Fig. 29-13.

29-10. WELD ALL-AROUND AND FIELD WELD SYMBOLS

Directions given on a welding symbol are no longer of any value when the weld

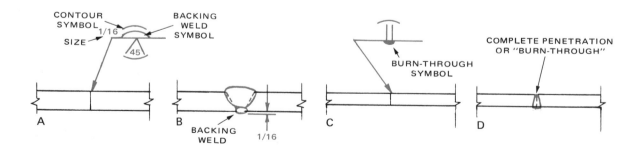

Fig. 29-13. Backing welds and burn-through symbols. The burn-through symbol is used on welds which are welded from one side only and which require 100 percent penetration, as shown in C and D. A backing weld may be used to obtain 100 percent penetration when welding is possible on both sides, as shown in A and B. Note: Contour symbols and dimensions may be used with these symbols, as in A and C.

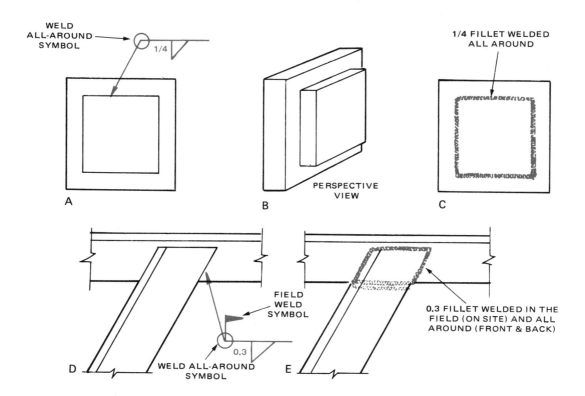

Fig. 29-14. Weld all-around and field weld symbols. Note the 0.3 fillet in D and E is welded in the field. It is welded all around the angle iron both front and back.

joint makes a sharp change in direction such as going around a corner. When the joint changes direction sharply, a new welding symbol must be used or a weld-all-around symbol may be used.

The weld-all-around symbol is used when the same type weld joint is used in all edges of a box or cylindrical part. See Fig. 29-14.

Some parts are assembled and welded in the shop. It is often necessary to take parts into the field to make final assembly and welds.

When welds are to be made in the field away from the shop, a field weld symbol is used, as in Fig. 29-14. If a weld is to be made in the shop, the field weld symbol is not used.

29-11. TEST YOUR KNOWLEDGE

1. List five things the welding symbol will tell the welder about the weld that is to be made?

2. Why is the tail used on the welding symbol?

3. What does the basic weld symbol tell the welder when used on the welding symbol?

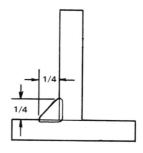

4. On a sheet of paper (do not write in the textbook) sketch the complete welding symbol for the welding sketched in the space above. The metal part shown is not to be ground or cut prior to welding, and the weld is to be continuous rather than intermittent.

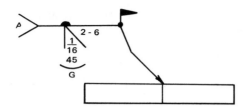

Refer to the welding symbol above when answering Questions 5 through 10:

5. Which piece of metal is to be ground, machined, or cut prior to welding?

6. What does the small, black, half-round symbol mean?

7. What shape is the weld face to be and what method is used to finish it?

8. Is the weld made in the shop or on site (in the field)?

9. At what angle is the one piece ground, and how far apart are the pieces at the root of the weld?

10. Is the weld made continuously, or is it intermittent? If intermittent, how long is each weld? How far apart?

Chapter 30
GLOSSARY OF WELDING TERMS

This chapter explains the meaning of terms most used by welders. Technical engineering terms have been simplified. For additional definitions, refer to the AWS WELDING TERMS AND DEFINITIONS (publication No. AWS A3.0-76). It is published by the American Welding Society.

ABRASION: Worn condition produced by rubbing.

AC or ALTERNATING CURRENT: That kind of electricity which reverses its direction of electron flow regularly.

ACETYLENE: Gas composed of two parts of carbon and two parts of hydrogen. When burned in an atmosphere of oxygen, it produces one of the highest flame temperatures obtainable.

ACETYLENE CYLINDER: Specially built container manufactured according to ICC standards. Used to store and ship acetylene. (Occasionally called "tank" or "bottle.")

ACETYLENE HOSE: See Hose.

ACETYLENE REGULATOR: An automatic valve used to reduce acetylene cylinder pressures to torch pressures and to keep the pressures constant.

ACTUAL THROAT: Distance from face of the weld to the root of the weld.

ADHESION: Act of sticking or clinging.

AIR GAP: In a magnetic circuit, a small gap which the magnetic flux must bridge. Cracks thus become small air gaps on a part's surface.

ALLOY: An intimate mixture of two or more metals.

AMPERE: Unit of electrical current. One ampere is required to flow through a conductor having a resistance of one ohm at a potential (pressure) of one volt.

AMPERE TURNS: The number of turns of wire in a coil plus the number of amperes.

ANNEALING: Softening metals by heat treatment.

ANODE: Positive terminal of an electrical circuit.

ARC: Flow of electricity through a gaseous space or air gap.

ARC BLOW: Tendency for an arc to wander or whip from its normal course during arc welding. It is caused by magnetic changes.

ARC CUTTING: Making a kerf in a metal using energy of an electric arc.

ARC VOLTAGE: Electrical potential (pressure or voltage) across the arc.

ARC WELDING: Fusing metals using an electric arc as source of heat.

ATOM: Smallest whole part of an element. Its nucleus is made up of protons and neutrons.

ATOMIC ARC WELDING: Welding with heat generated by hydrogen atoms created by an electric arc recombining to form hydrogen molecules. The heat welds delicate joints while the surrounding hydrogen atoms shield the weld.

AUTOMATIC WELDING: Welding in arc is mechanically moved while controls govern the speed and/or the direction of travel.

AXIS OF A WELD: An imaginary line along the center of gravity of the weld metal and perpendicular to a cross section of the weld metal.

BACKFIRE: A short "pop" of the torch flame followed by extinguishing of the flame or continued burning of the gases.

BACKHAND WELDING: Welding in the opposite direction the gas flame is pointing.

BACKING: Some material placed on the root side of a weld to aid control of penetration.

BACKING RING: Metal ring placed inside of a pipe before butt welding; insures complete weld penetration and a smooth inside surface.

BACK-STEP WELDING: Welding small sections of a joint in a direction opposite the progression of the weld as a whole.

BACKWARD WELDING: See backhand welding.

BASE METAL: Metal to be welded, cut, or brazed.

BAUXITE: Ore from which aluminum is obtained. Consists mostly of hydrated alumina ($Al_2O_3 3H_2O$).

BEAD: Appearance of the finished weld; the metal added in welding.

BEVEL: Angling the metal edge where welding is to take place.

BLACK LIGHT: Light waves below the visible range of violet light. The wave length reacts with certain dyes causing the dyes to fluoresce in a color range visible to the eye.

BLIND JOINT: A joint with no parts visible.

BLOWHOLE: A hole in a casting or weld which is caused by gas entrapment during cooling.

BLOWPIPE: Another name for an oxyacetylene torch.

BODY: Main structural part of a regulator.

BOND: Junction of the weld metal and the base metal.

BRAZE WELDING: Making an adhesion groove, fillet or plug connection with a brazing alloy.

BRAZEMENT: An assembly joined by brazing.

BRAZING: Making an adhesion connection with a minimum of alloy which melts above 800 deg. F. and which flows by capillary action between close-fitting parts.

BRINELL HARDNESS: An accurate measure of hardness of metal made with an instrument. Measurement is made as a hard steel ball is pressed into the smooth surface at standard conditions.

BRITTLENESS: Quality of a material which causes it to develop cracks with little bending (deformation) of the material.

BRONZE WELDING: See braze welding.

BUILDUP: Amount of a weld face extended above surface of joined metals.

BURNED METAL: Occasional term for metal which has been combined with oxygen so that some carbon becomes carbon dioxide and some of the iron becomes iron oxide.

BURNING: Violent combination of oxygen with any substance which produces heat. See Flame Cutting.

BUTT JOINT: An assembly in which the two pieces joined are in the same plane with the edge of one piece touching the edge of the other.

BUTTON: Part of a weld which is torn out in destructive testing of a spot, seam or projection welding.

CABLE: See lead.

CAPILLARY ACTION: Property of a liquid to move into small spaces if it has the ability to "wet" these surfaces.

CARBON: An element which, when combined with iron, forms various kinds of steel. In solid form, it is used as an electrode for arc welding. As a mold, it will hold weld metal. Motor brushes are made from carbon.

CARBONIZING: See carburizing or reducing.

CASEHARDENING: Adding carbon to the surface of a mild steel object and heat treating it to produce a hard surface.

CASTINGS: Metallic forms which are produced by pouring molten metal into a shaped container or cavity called a mold.

CAST-WELD ASSEMBLY: Cast parts fixed in an assembly by welding.

CATHODE: Electrical term for negative terminal.

CELSIUS: Temperature scale in SI metric.

CHALK TEST: Method of locating cracks by first applying a penetrating liquid, wiping it off and then applying chalk or whiting. Penetrant seeping out of cracks into the whiting or chalk causes a noticeable color difference.

CHAMPFER: Angled edge made to eliminate a sharp corner. Also: angled cutting edge at a tooth corner.

CHAMPFERING: See bevel.

CHARPY: Impact testing machine which strikes the specimen with a swinging hammer. The specimen is placed against anvil supports 40 mm apart.

CHILL: Cool rapidly.

CHLORINATION: Passing of dry chlorine gas through molten aluminum alloys to remove trapped oxides and dissolved gases.

CIRCUIT: The various connections and conductors of a specific device; the path of electron flow from the source through components and connections back to its source.

CLADDING: Somewhat thick layer of weld applied on a surface to improve resistance to corrosion or other agents which tend to wear away the metal.

CLEARANCE: Gap or space between adjoining or mating surfaces. Also: space provided between the trailing surface (relief) of a cutting tool and the surface being cut.

COATED ELECTRODE: See covered electrode.

COHESION: A sticking together through attraction of molecules.

COLD WELDING: Use of high pressure and no outside heat to force metal parts to fuse.

COLD WORK: Metal part on which a permanent strain has been placed by an outside force while the metal is below its recrystallization temperature.

COLD WORKING: Bending (deforming) metal at a temperature lower than its recrystallization temperature.

COLOR CONTRAST DYE: Dye added to a penetrant to give it higher color intensity and contrast when used in surface testing of metals.

COLOR CONTRAST PENETRANT: A penetrant having a dye with enough intensity to show flaws on a metal surface under white light.

COMBINED STRESSES: A state of stress more complicated than simple tension, compression or shear.

COMBUSTIBLE: Flammable, easily ignited.

COMPLETE FUSION: Fusion which has occurred over the entire surface of a base metal being welded.

COMPLETE JOINT PENETRATION: When weld metal completely fills the groove and fuses with the base metal through its entire thickness.

COMPRESSIVE STRENGTH: The greatest stress developed in material under compression.

CONCAVE WELD FACE: A weld having the center of its face below the weld edges.

CONDUCTIVITY: Ability of a conductor to carry current.

CONDUCTOR: A substance capable of readily transmitting electricity or heat.

CONE: Inner visible flame shape of a neutral or nearly neutral flame.

CONTINUOUS CASTING: Method of casting metal in an open ended mold so that metal is fed into and cools in the mold in a continuous form.

CONTINUOUS WELD: Making the complete weld in one operation.

CONVEX WELD: A weld with the face above the weld edges.

COOLING STRESSES: Stresses resulting from uneven distribution of heat during cooling.

CORNER JOINT: Junction formed by edges of two pieces of metal touching each other at angle of about 90 degrees (right angles).

CORROSION: The interaction of a metal – chemically and electro-chemically – with its surroundings causing it to deteriorate.

CORROSION EMBRITTLEMENT: Loss of ductility or workability of a metal due to corrosion.

CORROSION FATIGUE: Effect of repeated stress in a corrosive atmosphere characterized by shortened life of the part.

COULOMB: A unit of electrical charge equal to 3×10^9 electrostatic units of charge.

COUPON: Piece of metal used as a test specimen. Often an extra piece as in a casting or forging.

COVERED ELECTRODE: Metal rod used in arc welding which has a covering of materials to aid arc welding process.

CRACK: A break or separation in rigid material running more or less in one direction.

CRACKING: Term applied to action of opening a valve slightly and then closing the valve immediately.

CRATER: A depression in the face of a weld, usually at the termination of an arc weld.

CREEP: Permanent deformation caused by stress or heat or both.

CREVICE CORROSION: Deterioration of a metal caused by concentration of dissolved salts, metal ions, oxygen or other gases in pockets not disturbed by the fluid stream. Buildup eventually causes deep pitting.

CROWN: Curve or convex surface of finished weld proper.

CRYOGENICS: Study of physical phenomena at temperatures below −50 deg. F.

CUPOLA: Blast furnace in the shape of a vertical cylinder used in making gray iron.

CUPPING: Fracture of severely worked rods or wire. One end appears shaped like a cup and the other is shaped like a cone.

CUTTING FLAME: Cutting by a rapid oxidation process at a high temperature produced by a gas flame accompanied by a jet action which blows the oxides away from the cut.

CUTTING HEAD: The part of a cutting machine or cutting equipment to which a cutting torch or tip is attached.

CUTTING NOZZLE: See cutting tip.

CUTTING PROCESS: Action which causes separating or removal of metal.

CUTTING TIP: The part of an oxygen cutting torch from which the gases are released.

CUTTING TORCH: A nozzle or device which controls and directs the gases and oxygen needed for cutting and removing the metal in oxygen cutting.

CYLINDER: A container holding the supply of gas used in welding. (See oxygen, acetylene).

CYLINDER MANIFOLD: See manifold.

DC or DIRECT CURRENT: Electric current which flows only in one direction.

DECALESCENCE: A transformation which takes place during superheating of iron or steel. The metal surface darkens due to the sudden decrease of temperature during the rapid absorption of the latent heat of transformation.

DECARBURIZATION: Carbon loss from the surface of a ferrous alloy when the alloy is heated in a medium that reacts with the carbon on the surface.

DEEP ETCHING: Eating away (as with corrosive action of acid) of a metal surface for purpose of examining the surface (under a magnifier) to detect features such as grain flow, cracks or porosity.

DEFECT: An imperfection which, by its size, shape, location or makeup, reduces the useful service of a part.

DEGASIFIER: A substance which can be added to molten metal to draw off soluble gases before they are trapped in the metal.

DEGASSING: Process of removing gases from liquids or solids.

DEGREE: One unit of a temperature scale.

DEMAGNETIZATION: Removal of existing magnetism from a part.

DEMURRAGE: A charge made by a gas supplier to the gas user as rent on the gas cylinder. The user is allowed free use for a number of days. Then a daily charge is made for additional usage.

DENDRITE: A crystal development often found in cast metals as they are slowly cooled through the solidification range. The crystal shows a tree-like branching pattern.

DEOXIDIZER: Substance which, when added to molten metal, removes either free or combined oxygen.

DEOXIDIZING: Process of removing oxygen from molten metals with a deoxidizer. Also: the removal of other desirable elements using elements or compounds that readily react with them. Or: in metal finishing, removing oxide films with chemicals or electrochemical processes.

DEPTH OF FUSION: Depth to which base metal is melted during welding.

DIE CASTING: Production of parts by forcing molten metal into a metal die or mold. Also: the part made by this process.

DIE FORGING: A part shaped by use of dies in the forging operation.

DIFFUSION: Spreading of an element throughout a gas, liquid or solid so that all of it has the same composition. Also: spontaneous migration of atoms and molecules within a material.

DIRECT POLARITY: Direct current flowing from anode (base metal) to cathode (electrode). The electrode is negative and the base metal is positive.

DISCONTINUITY: Any abrupt change or break in the shape or structure of a part (cracks, seams, laps, bumps, or changes in density). Usefulness of part may or may not be affected.

DISTORTION: Warping of a part of structure.

DOWNHAND WELDING: See flat position.

DRAWING: Forming recesses in parts by causing metal to

flow in dies. Also: reducing the diameter of wire by pulling it through a die. Term sometimes mistakenly used for tempering metals.

DROSS: A scum consisting usually of oxidized metal or impurities which forms on top of molten metals.

DUCTILE CRACK PROPAGATION: Slow development of cracks and noticeable warping or deformation caused by an outside pressure.

DUCTILITY: Ability of a material to be changed in shape without cracking or breaking.

EDGE JOINT: Joint formed when two pieces of metal are lapped with at least one edge of each at an edge of the other.

EDGE WELD: A weld produced on an edge joint.

EFFECTIVE PENETRATION: Greatest depth at which ultrasonic transmission can properly detect discontinuities.

EFFECTIVE THROAT: The least distance from the root of a weld not counting any reinforcements. See joint penetration.

ELASTIC DEFORMATION: Temporary change of dimensions caused by stress; part returns to original dimension when the stress is removed.

ELASTIC LIMIT: Greatest stress to which a structure may be subjected without causing permanent deformation.

ELASTICITY: Ability of a material to regain its original size and shape after deformation.

ELECTROCHEMICAL CORROSION: Corrosion caused by current flow between areas on a metallic surface.

ELECTRODE: Terminal point to which electricity is brought in the welding operation and from which the arc is produced to do the welding. In electric arc welding, the electrode is usually melted and becomes a part of the weld.

ELECTRODE SKID: Sliding of an electrode along the work surface during spot, seam or projection welding.

ELECTROMAGNET: Magnet produced by a coil carrying an electric current. Coil surrounds a mass of ferrous material which also becomes magnetized.

ELECTROMOTIVE FORCE: Energy, measured in volts, which causes the flow of electric current.

ELECTRON: A fundamental part of an atom which has a small negative charge.

ELECTRON BEAM WELDING: Focused stream of electrons which heats and fuses metals.

ELECTROSLAG WELDING: Process using one or more arcs between continuously fed metal electrodes and base metal.

ELEMENT: A chemical substance which cannot be divided into simpler substances by chemical action.

ELONGATION: Percentage increase in the length of a specimen when stressed to its yield strength.

EMBRITTLEMENT: Reducing the normal ductility of a metal by a physical or chemical change.

EROSION: Reduction in size of an object because of a liquid or gas impacting on the object.

EUTECTIC ALLOY: A mixture of metals which has a melting point lower than that of any of the metals in the mixture, or of any other mixture of these metals.

EXPLOSIVE WELDING: Joins metals as powerful shock waves create pressure to cause metal flow and resultant fusion.

FACE OF A WELD: The exposed surface of the weld.

FAHRENHEIT: A temperature scale once used in most English-speaking countries. Symbol is F.

FATIGUE: Condition of metal leading to cracks under repeated stresses below the tensile strength of the material.

FATIGUE LIMIT: The stress limit below which a material can be expected to withstand any number of stress cycles.

FATIGUE STRENGTH: Most stress that a metal will withstand cracking for a certain number of stress cycles.

FERRITE BANDING: Bands of free ferrite which line up in the direction the metal was worked. Bands are parallel.

FILLER METAL: Metal added in making welded, brazed or soldered joints.

FILLER ROD: Metal wire that is melted and added to the welding puddle to produce the necessary increase in bead thickness. See welding rod.

FILLET: Weld metal in the internal vertex, or corner, of the angle formed by two pieces of metal, giving the joint additional strength to withstand unusual stresses.

FILLET WELD: Metal fused into a corner formed by two pieces of metal whose welded surfaces are approximately 90 deg. to each other.

FLAME CUTTING: Cutting performed by an oxygen-fuel gas torch flame which has an oxygen jet.

FLASH: Impact of electric arc rays against the human eye. Also the surplus metal formed at the seam of a resistance weld.

FLASH WELDING: Process using electric arc in combination with resistance and pressure welding.

FLASHBACK ARRESTORS: The check valves usually installed between torch and welding hose to prevent flow of burning fuel gas and the oxygen mixture back into hoses and regulators.

FLAT POSITION WELD: A horizontal weld on the upper side of a horizontal surface.

FLUORESCENT DYE: A dye which gives off or reflects light when exposed to short wave radiation.

FLUORESCENT PENETRANT: A penetrating fluid with a fluorescent dye added to improve visibility of flaws.

FLUX: Chemical used to promote fusion of metals during welding.

FLUX CORED ARC WELDING: Welding method in which heat is supplied by an arc between a flux cored electrode and the base metal.

FLUX CORED ARC WELDING ELECTROGAS (FCAW—EG): A type of flux cored arc welding process.

FLUX OXYGEN CUTTING: Oxygen fuel gas torch heats metal while powdered flux is fed into the cutting flame.

FOCAL SPOT (EBW and LBW): Spot where an energy beam's energy level is most concentrated and where it has the smallest cross-sectional area.

FORGING: Metallic shapes being made by either hammering or squeezing the original piece of metal.

FORMING: Changing the shape of a metal part without changing its thickness.

FORWARD WELDING: Fusing metal in the same direction as the torch flame is directed.

FREE BEND TEST: Bending the specimen without using a fixture or guide.

FREE CARBON: In steel or cast iron, that part of the total carbon content which is present in the form of graphite or temper carbon.

FRICTION WELDING: Weld in which welding heat is generated by revolving one part against another part under very heavy pressure.

FULL-WAVE RECTIFIED SINGLE-PHASE AC: Alternating current in which the reverse half of the cycle is made to travel the same direction as the other half of the wave. It produces pulsating direct current with no interval between pulses.

FULL-WAVE RECTIFIED THREE-PHASE AC: Conversion of alternating current by rectifier in which there is little pulsation in the resulting direct current.

FUSION: Intimate mixing or combining of molten metals.

FUSION FACE: The surface of the base metal which is melted during welding.

FUSION WELDING: Any type of welding using fusion as part of the process.

GAMMA RAYS: Electromagnetic radiation given off by a nucleus. Gamma rays always accompany fission.

GAS HOLES: Created by gas escaping from molten metal, these holes are round or elongated, smooth-edged dark spots. They appear in clusters or individually.

GAS METAL ARC WELDING (GMAW): Welding using a continuously fed consumable electrode and a shielding gas.

GAS POCKETS: Cavities in weld metal caused by entrapped gas.

GAS TUNGSTEN ART WELDING (GTAW): Welding using a tungsten electrode and a shielding gas.

GAUSS: Unit of magnetic flux density or induction. One gauss is one line of flux per square centimetre.

GENERATOR: A mechanism which generates or produces some substance, i.e., electric generator, acetylene generator.

GOUGING: Cutting a groove in the surface of a metal using a gas cutting torch of an arc-air cutting outfit.

GRAIN REFINER: Material which produces a finer grain when added to a liquid metal or alloy.

GRAPHITIZATION: Forming of graphite in iron or steel either during solidification or later during during heat treatment.

GROOVE WELD: A welding rod fused into a joint which has the base metal removed to form a V, U, or J trough at the edge of the metals to be joined.

GROSS POROSITY: Pores, gas holes or globular voids in a weld or casting. Called gross because they are bigger and more numerous than would be found in good practice.

GUIDED BEND TEST: Bending a specimen in a definite way by using a fixture.

HALF-WAVE RECTIFIED AC: Simplest manner of rectification in which the reverse half of the cycle is blocked out completely. Result is a pulsating direct current with intervals when no current is flowing.

HAMMER FORGING: Deforming of workpiece by repeated blows.

HAND SHIELD: See shield.

HARD FACING: Filler material placed on a surface by welding, spraying or braze welding. Purpose is to toughen the surface to resist abrasion, erosion, wear, galling, and impact wear.

HARDENING: Making metal harder by a process of heating and cooling.

HARDNESS: Ability of metal to resist plastic deformation; same term may refer to stiffness or temper, resistance to scratching or abrading.

HEAT: Molecular energy of motion.

HEAT AFFECTED ZONE: That part of the base metal altered by heat from welding, brazing or cutting operations.

HEAT CHECK: Pattern of surface cracks formed on surfaces that are alternately subjected to rapid heating and cooling.

HEAT-CHECKING: Crazing of a die surface, especially one subjected to alternate heating and cooling.

HEAT CONDUCTIVITY: Speed and efficiency of heat energy movement through a substance.

HEAT TINTING: Using heat to color a metal surface through oxidation for purpose of showing details of the microstructure.

HEAT TREATABLE ALLOYS: Aluminum alloys that reach maximum strength by solution heating and quenching.

HELIUM: An inert, colorless, gaseous element used as a shielding gas in welding.

HELMET: A protecting hood which fits over the arc welder's head, provided with a lens of safety glass through which the operator may safely observe the electric arc.

HORIZONTAL POSITION: A weld performed on a horizontal seam at least partially on a vertical surface.

HOT FORMING: Operations performed on metal while it is above the recrystallization temperature of the metal. Operations may include bending, drawing, forging, heading, piercing and pressing.

HOT SHORTNESS: Brittleness of metal which occurs in the hot-forming range.

HOT WORKING: Shaping of metal at a temperature and rate which does not cause strain hardening.

HYDROGEN: A gas formed of the single element, hydrogen. When combined with oxygen, it forms a very clean flame which, however, does not produce a very high temperature.

HYDROGEN EMBRITTLEMENT: Low ductility condition in metals due to absorption of hydrogen.

IMPACT ENERGY: Amount of energy that must be exerted to fracture a part; measurement is usually made in an Izod or Charpy test.

IMPACT STRENGTH: A material's ability to resist shock.

IMPACT TEST: A carefully measured test of how materials behave under heavy loading such as bending, tension or torsion. Charpy or Izod tests, for example, measure energy absorbed in breaking a specimen.

IMPEDANCE: Total resistance to flow of alternating current as a result of resistance and reactance.

IMPURITIES: Undesirable elements or compounds in a material.

INCLUSION: Foreign matter introduced into welds or castings.

INCOMPLETE FUSION: Less than complete fusion of weld material with base metal or with preceding bead.

INCOMPLETE JOINT PENETRATION: A lack of fusion between metals appearing as elongated darkened lines. May occur in any part of a weld groove.

INCOMPLETE PENETRATION: Incomplete root penetration or failure of two weld beads to fuse.

INDENTATION: The depression left on the surface of the base metal after spot, seam or projection welding.

INDENTATION HARDNESS: Degree of resistance of a material to indentation. This is the standard test of a material's hardness.

INDICATION, MAGNETIC: A magnetic particle pattern held magnetically on surface of material being tested.

INDICATION, PENETRANT: Visual evidence of a discontinuity, that is, the penetrant can be seen in the crack.

INDICATION, ULTRASONICS: A signal on the ultrasonic equipment indicating a crack in a material being tested.

INDUCED CURRENT: A secondary current which is set in motion as a second conductor in the shape of a closed loop is placed in the magnetic field around another current carrying conductor.

INDUCTANCE: In the presence of a varied current in a circuit, the magnetic field surrounding the conductor generates an electromotive force in the circuit itself. If another circuit is next to the first, the changing magnetic field of the first circuit will cause voltage in the second.

INDUCTION: The magnetism induced in a ferromagnetic material by some outside magnetic force.

INDUCTION HARDENING: Process of quench hardening using electrical induction to produce the heat.

INDUCTIVE REACTANCE: A force which opposes flow of alternating current through a coil. This force is independent of the resistance of a conductor to a flow of current.

INFRARED RAYS: Heat rays coming from both the arc and the welding flame.

INSIDE CORNER WELD: Two metals fused together.

One metal is held 90 degrees to the other. The fusion is performed inside the vertex of the angle.

INTERFACE: A surface which forms a common boundary between two bodies.

INTERGRANULAR CORROSION: Corrosion occurring, for the most part, between grains or on the edges of the grain in a ferrous material.

INTERMITTENT WELD: Joining two pieces and leaving sections unwelded.

ION: An atom or a group of atoms positively or negatively charged as a result of having gained or lost one or more electrons. Also: sometimes a free electron.

IONIZATION: Adding or removing electrons from atoms or molecules to create ions.

IZOD TEST: A test for impact strength made by striking the test piece with a measured downstroke of a pendulum. The specimen, usually notched, is held by one end in a vise. Energy absorbed, measured by the upward swing of the pendulum, indicates the impact strength of the specimen.

JIG: A fixture or template to accurately position and hold a part during welding or machining operations.

JOINT: Where two pieces are joined in an assembly.

JOINT EFFICIENCY: Strength of a welded joint given as a percentage of the strength of the base metal.

JOINT PENETRATION: The depth of weld metal and fusion in a welded joint.

KERF: Width of cut produced by a sawing operation.

KEYHOLE: A welding technique in which concentrated heat penetrates the workpiece leaving a hole at the leading edge of the weld. As the heat source moves on, molten metal fills the hole forming the weld bead.

KILLED STEEL: Steel which has been deoxidized with a strong agent to reduce the oxygen content. Deoxidizing is carried to the point where no reaction takes place between carbon and oxygen as the metal solidifies.

LACK OF FUSION: Defect in a weld caused by lack of union between weld metal and base metal.

LAMINATE: Sheets or bars made up of two or more metal layers built up to form a structural member. Also: forming a metallic product with two or more bonded layers.

LAP: A folded-over section of metal which is then rolled or forged into the surface. Considered a surface defect.

LAP JOINT: A joint in which the edges of the two metals to be joined overlap.

LASER BEAM WELDING: Process in which single frequency light beam concentrates minute spot of heat to fuse small, light metal materials.

LAYER: A certain weld metal thickness made of one or more passes.

LEAD WIRE: Electricity-carrying wire from the power source to the electrode holder or to the ground clamps.

LEG of FILLET WELD: Distance from point where the base metals touch to toe of the fillet.

LENS: A specially treated glass through which a welder may look at an intense flame without being injured by the harmful rays, or glare.

LIGHT METAL: Low-density metal such as aluminum, magnesium, titanium, beryllium or their alloys.

MACRO-ETCH: Eating away of the metal surface to make gross structural details stand out so that they can be observed with the naked eye or with magnification up to 10 times.

MACROGRAPH: Photographic or graphic reproduction of the surface of a prepared specimen which has been magnified up to 10 times normal size.

MACROSTRUCTURE: Physical makeup or structure of metals revealed under magnification of not more than 10 diameters.

MALLEABILITY: Ability of a metal to be deformed without breaking.

MALLEABLE CAST IRON: A cast iron made by annealing white cast iron while the metal undergoes decarburization, graphitization or both thus eliminating all or most of the cementite.

MALLEABLE CASTINGS: Cast forms of metal which have been heat treated to reduce their brittleness.

MANIFOLD: A pipe or cylinder with several inlet and outlet fittings. It is designed so that several cylinders can be connected together while the gas or oxygen in the cylinders can be piped to several locations or stations.

MAPP: A stabilized methyl acetylene-propadiene fuel gas.

MECHANICAL PROPERTIES: Description of a material's behavior when force is applied for purpose of determining the material's suitability for mechanical usage. Properties described, for example are, modulus of elasticity, elongation, fatigue limit, hardness and tensile strength.

METAL ARC CUTTING: Method of cutting metal with an electric arc. Molten metal flows away from the base metal.

METAL CARBON ARC CUTTING: Cutting method which uses an electric arc and a jet of oxygen.

METALLIC DISCONTINUITY: A break in the surface of a metal part or a void such as a gas pocket.

METALLOGRAPHY: The scientific study of the constitution and structure of metals and alloys as observed by the naked eye or with the aid of magnification and X ray.

METALLURGY: Science and technology of metal.

MIG: Term used for gas metal arc welding (metal shielding gas).

MIXING CHAMBER: Part of the welding torch where the welding gases are mixed prior to combustion.

MODULUS OF RUPTURE: In a bend or torsion test, the stress at which fracture occurs expressed as a constant.

NEGATIVE CONNECTIONS: Connections in an electric circuit through which the current flows back to its source.

NEUTRAL FLAME: Flame resulting from combusion of perfect proportions of oxygen and the welding gas.

NEWTON: SI metric unit of force. A force of 9.8 newtons is required to lift a mass of 1 kilogram.

NITRIDING: A casehardening process; adding nitrogen to a solid ferrous alloy by keeping the alloy at a suitable temperature while in touch with a material rich in nitrogen.

NODULAR CAST IRON: Cast iron containing primary graphite which is in a ball-like or globular form rather than in flakes as in gray cast iron. Also known as spheroidal graphite iron, it is more ductile and has greater strength than ordinary iron.

NONDESTRUCTIVE TESTING (NDT): Testing for defects using techniques which do not destroy or damage the part.

NORMALIZING: Heating steel above the temperature used for annealing and then cooling it in still air at room temperature; used as a preparation for further heat treatment. Normalized steel has a uniform unstressed condition with a grain size and refinement that makes the metal more suitable for heat treating.

NOTCH BRITTLENESS: Tendency of a material to break at points where stress is concentrated.

NOTCH SENSITIVITY: Measure of the extent of reduction of strength in a metal after introducing stress concentration (by notching).

NOZZLE: See tip.

OPTICAL PYROMETER: Temperature measuring device which compares the incandescence (white, glowing hot) of a heated object with an electrically heated filament whose incandescence can be regulated.

ORIFICE: Opening through which gases flow.

OUTSIDE CORNER WELD: Fusing two pieces of metal together with the fusion taking place on the underpart of the seam.

OVERHEAD POSITION: A weld on the underside of the joint with the face of the weld in a horizontal position.

OVERHEATING: Damaging the properties of a metal by too much heat. When original properties cannot be restored, the overheating is known as "burning."

OVERLAP: Extension of the weld face metal beyond the toe of the weld.

OXIDIZING: Combining oxygen with any other substance.

OXIDIZING FLAME: Flame produced by an excess of oxygen in the torch mixture, leaving some free oxygen which tends to burn the molten metal.

OXYACETYLENE WELDING: See oxygen-acetylene welding.

OXYGEN: Gas formed of the element oxygen. Called "burning" when actively supporting combustion; "oxidation" when it slowly combines with a substance.

OXYGEN CYLINDER: A specially built container manufactured according to ICC standards and used to store and ship certain quantities of oxygen.

OXYGEN HOSE: Reinforced, multilayered, flexible tube usually of rubber. Used to carry high pressure gases.

OXYGEN LANCE CUTTING: Oxygen fuel process which heats base metal and then blows away molten metal with jet of oxygen from an iron pipe.

OXYGEN REGULATOR: An automatic valve used to reduce cylinder pressures to torch pressures and to keep the pressures constant.

OXYGEN-ACETYLENE CUTTING: Cutting metal using the oxygen jet which is incorporated with an oxygen-acetylene preheating flame or flames.

OXYGEN-ACETYLENE WELDING: A method of welding using as a fuel a combination of the two gases, oxygen and acetylene.

OXYGEN-FUEL GAS CUTTING: Cutting method using an oxygen fuel gas to heat the metal. Any oxygen jet removes molten metal.

OXYGEN-HYDROGEN FLAME: The chemical combining of oxygen with the fuel gas hydrogen.

OXYGEN-LP GAS FLAME: Chemical combining of oxygen with the fuel gas LP (liquified petroleum).

PASS: Weld metal created by one progression along the weld.

PENETRANT: Either a liquid or a gas which, when applied to the surface of a metal, enters cracks (discontinuities) to make them visible.

PENETRATION: Extent to which the weld metal combines with the base metal as measured from the surface of the base metal.

PERCUSSION WELDING: The type of resistance welding in which the heat comes from an arc produced by an electrical discharge and instantaneous pressure applied during or immediately following the heating.

PHYSICAL PROPERTIES: Properties or qualities other than mechanical properties, that have to do with the physics of a material. Examples are density, ability to conduct electricity, ability to conduct heat, and thermal expansion.

PHYSICAL TESTING: Examination of a material to find out its physical properties.

PICKLING: Removing surface oxides from metals by chemical or electro-chemical reaction.

PLASMA: Temporary physical condition of a gas after it has been exposed to and has reacted to an electric arc.

PLASMA ARC CUTTING: Process of metal cutting using an electric arc and fast-flowing ionized gases.

PLASTIC WELDING: Process in which heated air softens and fuses plastic materials.

PLASTICITY: Ability of a metal to bend without breaking (rupturing).

PLUG WELD: Weld made in a hole of one piece of metal as it is lapped over another piece of metal.

POROSITY: Gas pockets or voids in a metal.

POSITIVE CONNECTIONS: Connections in an electric circuit into which current constantly flows.

POSTHEATING: Temperature to which a metal is heated after an operation has been performed on the metal.

POWDER METALLURGY: The art and technology of producing powdered metal and of utilizing it in the production of parts.

PREHEATING: Temperature to which a metal is heated before an operation is performed on the metal.

PRODS: Two hand-held electrodes designed to be pressed against the surface of a part for purpose of passing a magnetizing electric current through the part; used to find defects with magnetic particles.

PROJECTION WELDING: Type of resistance welding in which current flow is concentrated at points where metal pieces are in contact.

PUDDLE: Portion of weld that is molten due to the heat of welding.

PULSE ARC WELDING: A welding arc in which the current is interrupted or pulsed as the welding arc progresses.

PYROMETER: A device for determining temperatures over a wide range.

QUENCH AGING: A change in metal produced by rapid cooling after heat treating.

QUENCH ANNEALING: Process used to soften austenitic ferrous alloys by solution heat treatment.

QUENCH HARDENING: Hardening an iron alloy by austenitizing followed by rapid cooling so that some or all of the austenite becomes martensite.

QUENCHING: Rapid cooling of metal in a heat treating process.

RADIOGRAPH: A photograph made by passing X rays or gamma rays through the object to be photographed and recording the variations in density on a photographic film.

RADIOGRAPHER: One who performs radiographic operations.

RADIOGRAPHER'S EXPOSURE DEVICE: An instrument containing the X-ray source for making radiographic records on sensitized film.

RADIOGRAPHIC INTERPRETATION: "Reading" of the films to determine cause and significance of discontinuities below the surface of the material radiographed; one of the determinations of the reading is the suitability of the material.

RADIOGRAPHIC SCREENS: Sheets, either metallic or fluorescent, used to intensify the radiation effect on film.

RADIOGRAPHY: Use of radiant energy found in X rays or gamma rays to examine opaque objects and make a record of the examination.

RAYS: See infrared and ultraviolet.

RECARBURIZE: Add carbon to molten cast iron or steel. Also: process of adding more carbon to a surface which has lost some carbon in processing.

RECTIFIED ALTERNATING CURRENT: Alternating current made to flow in one direction only by use of a device which stops normal reversing of the current.

RECTIFIER: Device such as a tube or a circuit which acts like a one way valve. It converts one half of a waveform of alternating current to useful current flowing in the same direction as the other half of the waveform.

REDUCING FLAME: An oxygen-fuel gas flame with a slight excess of fuel gas.

REDUCTION OF AREA: Difference in cross sectional

area of a specimen after fracture as compared to original cross sectional area.

REGULATORS: See acetylene, oxygen.

REINFORCEMENT OF WELD: Excess metal on the face of a weld.

RESIDUAL STRESS: Stress still present in a body freed of external forces or thermal gradients.

RESISTANCE WELDING: A process using the resistance of the metals to the flow of electricity as the source of heat.

REVERSED POLARITY: Referring to direct current, a positive anode in which electrons flow from the base metal to the electrode.

ROCKWELL HARDNESS TESTER: A test which measures the hardness of materials based on depth of penetration of a standardized force.

ROOT CRACK: A crack in either the weld or the heat-affected metal at the root of a weld.

ROOT OF JOINT: Point at which metals to be joined by a weld are closest together.

ROOT PENETRATION: Depth to which weld metal extends into the root of a welded joint.

ROOT OF WELD: That part of a weld farthest from the application of weld heat and/or filler metal side.

ROSETTE WELD: See plug weld.

SCLEROSCOPE TEST: Hardness test which uses the height of rebound of a falling piece of metal to determine how much energy the material being tested absorbed.

SCRATCH HARDNESS: Hardness of a metal determined by measuring the width of a scratch made by a cutting point under a known pressure.

SECONDARY HARDENING: Tempering of some alloy steels at a higher temperature than normally used for hardening. Result is a hardness greater than is achieved by tempering at the lower temperature for the same period of time.

SEMI-KILLED STEEL: Incompletely deoxidized steel which contains sufficient dissolved oxygen to react with the carbon so that carbon monoxide is formed. This offsets solidification shrinkage.

SEQUENCE: The order in which operations take place.

SHEAR: Force which causes two parts of the same body touching each other to slide parallel to their contacting surfaces.

SHEAR FRACTURE: A break in which crystalline material separates by sliding under action of shear stress.

SHEAR STRENGTH: Stress required to fracture a part in a cross sectional plane when the forces of direction and of resistance are parallel and opposite.

SHIELD: An eye and face protector held in the hand. It enables a person to look directly at the electric arc through a special lens without being harmed.

SHIELDED ARC: A form of electric welding in which a heavy flux-coated electrode is used.

SHORT ARC: A gas metal arc process which uses a low arc voltage. The arc is continuously interrupted as the molten electrode metal bridges the arc gap.

SHOT PEENING: Working the surface of a metal by bombarding it with metal shot.

SIGMA WELDING: See gas metal arc welding.

SIZE OF WELD: The joint penetration in a groove weld. Also: the lengths of the nominal legs of a fillet weld.

SLAG: A nonmetallic byproduct of smelting and refining made up of flux and nonmetallic impurities.

SLAG INCLUSIONS: Nonfused, nonmetallic substances in the weld metal.

SOLDERING: Means of fastening metals together by adhering to another metal. Joining metal only is melted.

SOLID PELLET OXYGEN-FUEL GAS WELDING: Portable welding system using fuel gas with pellets. Pellets produce oxygen and eliminate need for bulky oxygen cylinders.

SOLID STATE CONTROLLER: An electronic controller which uses transistors, diodes, and other semiconductor devices.

SPHEROIDIZING: Heating and cooling to produce a spheroidal or globular form of carbide in steel.

SPRAY ARC: Gas metal arc process which has an arc voltage high enough to continuously transfer the electrode metal across the arc in small globules.

STRAIGHT POLARITY: In a DC circuit, a negative cathode electrode connecting DC current so that electrons flow from the electrode to the base metal.

STRAIN: Reaction of an object to a stress.

STRESS: Load imposed on an object.

STRESS RELIEVING: Even heating of a structure to a temperature below the critical temperature followed by a slow, even cooling.

SUBMERGED ARC WELDING: Process in which electric arc is submerged in granular flux.

SUPERFICIAL ROCKWELL HARDNESS TEST: A test for determining surface hardness of thin sections or small parts, or where a large hardness impression might be harmful.

TACK WELD: Small weld used to temporarily hold components together.

TANK: See cylinder.

TEMPER: Part of heat treating in which hardened steel or hardened cast iron is heated to a temperature below its melting point for purpose of decreasing the hardness and increasing the toughness.

TEMPERING: Reheating hardened or normalized ferrous alloys and then cooling at any rate desired.

TENSILE STRENGTH: Maximum pull stress in pounds per square inch that a specimen will withstand.

THROAT OF A FILLET WELD (actual throat): Distance from the weld root to the weld face.

TIG: Gas tungsten arc welding.

TINNING: In soldering, a coating of the soldering metal given to the metals to be soldered.

TIP: End of the torch where the gas burns, producing the high temperature flame. In resistance welding the electrode ends.

T-JOINT: Joint formed by placing one metal against another at an angle of **90 deg.**

TOE CRACK: Base metal crack at the toe of a weld.

TORCH: The mechanism which the operator holds during gas welding and cutting and from which issue the gases that are burned to produce heat.

TORSION: Twisting motion resulting in shear stresses and strains.

TOUGHNESS: A metal's ability to absorb energy and deform before breaking.

ULTIMATE COMPRESSIVE STRENGTH: The most compressive stress that a material can stand under a gradual and evenly applied load.

ULTIMATE STRENGTH: The greatest conventional stress, tensile, compressive or shear that a material can stand.

ULTRASONIC: Mechanical vibrations in a frequency above the range of humanly audible sound.

ULTRASONIC TESTING: Nondestructive testing method which transmits high frequency sound waves through them.

ULTRASONIC WELDING: Process using high sound frequencies to produce metal fusion.

ULTRAVIOLET RAYS: Energy waves that emanate from the electrodes and the welding flames of such a frequency that these rays are in the short light wave length known as the ultraviolet ray light spectrum.

UNDERBEAD CRACK: A crack in the base metal near the weld and beneath the surface.

UNDERCUT: A depression at the toe of the weld which is below the surface of the base metal.

VERTICAL POSITION: Type of weld where the welding is done in a vertical seam and on a vertical surface.

WELD BEAD: Deposit or row of filler metal from a single welding pass.

WELD CRACK: Crack in weld metal.

WELD METAL: Fused portion of base metal or fused portion of the base metal and the filler metal.

WELD NUGGET: Weld metal deposited in spot, seam or projection welding.

WELDING: Art of fastening metals together by means of interfusing the metals.

WELDING ROD: Wire which is melted into the weld metal.

WELDING SEQUENCE: Order in which the component parts of a structure are welded.

WELDMENT: Assembly of component parts joined together by welding.

WROUGHT IRON: A commercial iron made up of slag (also known as iron silicate) fibers entrained in a ferrite matrix.

YIELD STRENGTH: Stress in psi at which a specimen assumes a specified limiting permanent set.

INDEX